# CEPHALOPODS

## of Australia and Sub-Antarctic Territories

AMANDA REID

CSIRO PUBLISHING

National Library of Australia Cataloguing-in-Publication entry

Reid, Amanda Louise, author.

Cephalopods of Australia and sub-Antarctic territories / Amanda Reid.

9781486303939 (hardback)
9781486303946 (epdf)
9781486303953 (epub)

Includes bibliographical references and index.

Cephalopoda – Australia – Identification.
Cephalopoda – Australia – Maps.
Cephalopoda – Antarctica – Identification.
Cephalopoda – Antarctica – Maps.

594.50994

Published by

CSIRO Publishing
36 Gardiner Road, Clayton VIC 3168
Private Bag 10, Clayton South VIC 3169
Australia

Telephone: [+613] 9545 8555
Email: csiropublishing@csiro.au
Website: www.publishing.csiro.au

Front cover: *Sepia plangon* at Swansea, NSW. Photograph by Matt Dowse.
Back cover (left to right): *Sepia apama*, *Sepia plangon* and *Hapalochlaena fasciata*. Photographs by Matt Dowse.

Edited by Joy Window (Living Language)
Cover design by James Kelly
Typeset by Thomson Digital
Printed by Ingram Lightning Source



Feb26_RP_ILS

# Contents

# Acknowledgements

Many people have provided fabulous images for this book. For this I am extremely grateful to Louise Allcock, Greg Barord, Kat Bolstad, Clay Bryce, Roy Caldwell, Samantha Cheng, Darren Coker, Brandon Cole, David Cowdery, Nick Dawkins, Matt Dowse, Kim Fulton-Bennett, Martin Garwood, John Gaskell, Ken Graham, Roger Hanlon, John Keane, Tsunemi Kubodera, Steve Leporati, Sue Lindsay, Bob Macpherson, Edward McSweeny, Phil Mercurio, David Paul, Linda Mott, Mark Norman, Uwe Piatkowski, Martin Rauschert, Clyde Roper, Greg Rouse, Chris Rowley, Becca Saunders, Sascha Schulz, David Shale, Roger Springthorpe, Lynton Stephens, Darren Stevens, Ron and Valerie Taylor, Lyle Vail, Mike Vecchione, Peter Ward and Richard Young. Thanks to Alistair Graham and Mark Neighbour for providing images from the CSIRO database ScienceImage, the Australian Biological Resources Study (ABRS), and Patrizia Jereb, Jordi Lleonart and Michael Lambouef from the Food and Agriculture Organisation of the United Nations (FAO). Thanks also to illustrators Lyn Albertson, Emanuela D'Antoni, Kathie Hollis and Mike Shea.

Images captured by Darren Stevens were made available by NIWA (National Institute of Water and Atmospheric Research Ltd) as part of the NZ IPY-CAML project (voyage TAN0802). This research was funded by the New Zealand Government under the New Zealand International Polar Year Census of Antarctic Marine Life Project (Phase 1: So001IPY; Phase 2: IPY2007-01). Project governance was provided by the Ministry for Primary Industries Science Team and the Ocean Survey 20/20 CAML Advisory Group (Land Information New Zealand, Ministry of Fisheries, Antarctica New Zealand, Ministry of Foreign Affairs and Trade, and National Institute of Water and Atmospheric Research) and is gratefully acknowledged.

Special thanks to Chung Cheng Lu for introducing me to cephalopod taxonomy and cuisine, and Eric Hochberg, Takashi Okutani, Barry Reid, Cynthia Reid, Bert Roberts, Clyde Roper, Mike Sweeney and Mike Vecchione for their constant encouragement and support over many years. Thank you to many other wonderful friends and colleagues, particularly at the Australian Museum, who have steered me with patience through this endeavour, and my son, Hallam Roberts, who has endured this big project with me. Thank you also to Briana Melideo, Tracey Millen, Deepa Travers and Joy Window from CSIRO Publishing for their help and patience, Dr Kat Bolstad from Auckland University of Technology for her excellent suggestions and detailed review of the manuscript, and Karina Hall from the NSW Dept. of Primary Industries for her comments on *Uroteuthis*.

# Introduction

The impetus for this book developed not only from my fascination with cephalopods and interest in cephalopod taxonomy and systematics, but also from the identification of a need for a comprehensive guide to the fauna occurring in Australian waters, which host some of the highest diversity of cephalopods in the world.

The area covered roughly corresponds to the Australian marine jurisdictional area from the shore to the boundary of the Exclusive Economic Zone (EEZ). This comprises mainland Australia, Tasmania and offshore islands, including sub-Antarctic islands. The Australian Antarctic Territory has also been included (Fig. 1).

The Australian EEZ covers ~9.0 million km$^2$, with a further 2.04 million km$^2$ in Australia's Antarctic Territory, and an Extended Continental Shelf of 2.56 million km$^2$. The mainland EEZ has a coastline of ~36 000 km and spans > 5000 km from the tropics (9°S) to temperate latitudes (47°S). This extensive coastline, together with ~12 000 islands from the tropics to the polar region, contains a wide range of water conditions, habitats and ocean features which combine to result in our incredibly rich marine fauna (Butler *et al.* 2010). Two hundred and twenty-six cephalopod species are treated in this book, and this represents about one-third of the world's, cephalopod fauna.

Due to human overfishing of finfish, cephalopods are the new top marine predators. They function as a keystone group in understanding complex ecosystems. It seems timely to try to pull together as much information as possible specifically regarding Australian cephalopods into one volume to provide a solid launching pad for future studies and fisheries management. I hope this book will not only cover what we currently know about Australian cephalopods, but also indicate what we don't know in order to identify potential future research projects.

Of course, this book has limitations. An obvious one is treating taxa based on continental and political boundaries – a limit not observed by marine fauna in nature. Another is its attempt to restrict its scope to those studies that have been conducted on taxa occurring in Australian waters, though many often have much broader distributions. The reasons for this are threefold. Increasingly, particularly for inshore fauna, we are discovering that some (although clearly not all) widespread taxa comprise species 'complexes' and the 'true' species may have more restricted distributions. The widespread and commercially important *Sepia pharaonis* Ehrenberg, 1831 is an example. By focusing on literature pertaining primarily to taxa occurring within Australian waters, the likelihood of confusing information that may apply to more than one species is decreased (bearing in mind the caveat that cryptic species also occur *within* Australian waters). Also, for several reasons, it may be that aspects of the biology of a species that is found in Australian waters may, or may not, be significantly different from a population of that same species that also occurs elsewhere. On a more practical level, to try to include reference to studies that have been conducted on a particular taxon over its entire range would mean that the size of this book would blow totally out of proportion in its length and scope.

However, it is often necessary, or useful, to refer to publications pertaining to taxa that occur in Australian waters as well as other parts of the world. A clear example is in the case of original species descriptions (i.e. a species described from the Philippines may later be found to occur in Australian waters). Another is in the case of phylogenetic analyses, many of which have been conducted by northern hemisphere researchers using specimens obtained from a variety of geographic locations. Readers should also bear this in mind when consulting the original publications.

While crucially important, only scant mention is made of predator/prey relationships, and very little about host/parasite associations. The literature on the former subject is vast, so readers will need to consult other works for more detail. In addition, information on cephalopod beaks, often the only parts remaining in the stomach contents of predators, is not included here. Readers are referred to Xavier and Cherel's *Guide to Southern Ocean Cephalopod Beaks* available at www.researchgate.net/publication/260987002_Cephalopod_beak_guide_for_the_Southern_OCean and the invaluable electronic resource the Tree of Life Web Project at www.tolweb.org/Cephalopoda for more information on cephalopod beaks and other information about particular species. Nor are juvenile or paralarval forms fully described. A cursory summary of the biology and behaviour of cephalopods is

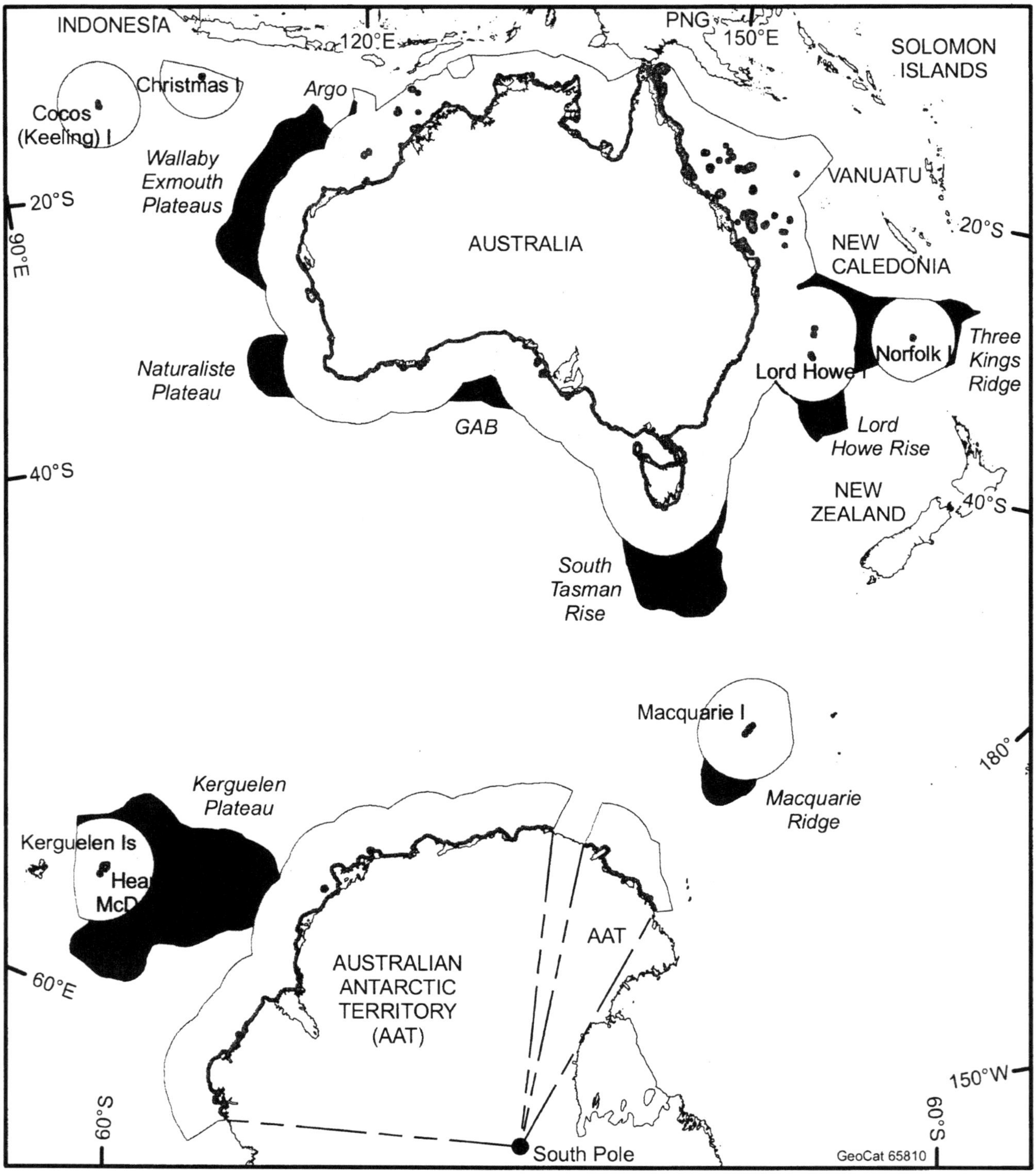


**Fig. 1.** Australia's, continental shelf confirmed by the Commission on the Limits of the Continental Shelf. Geosciences of Australia GeoCat 65810. Copyright © Commonwealth of Australia, Geoscience Australia (2008). Available at www.ga.gov.au/webtemp/image_cache/GA11214.pdf.

given; again the literature is vast and this book is intended primarily as an identification guide rather than a complete synopsis of cephalopod biology.

Despite these limitations, I hope this book will be a useful reference and inspire future research relating to Australian cephalopods.

## Using this book

Literature searches have been used to compile the list of species found in Australian waters (including the Antarctic Territories), using the Lu and Phillips (1985) checklist and the Australian Faunal Directory (ABRS 2009) as primary references. This information has been supplemented by checking species lists and information on taxa held in museum collections via the Atlas of Living Australia (www.ala.org.au), and identification guides published by the Food and Agriculture Organisation of the United Nations (Jereb and Roper 2005, 2010; Jereb *et al.* 2014). Species are only included if a definitive record of their occurrence in Australian waters exists.

This may result in some species being excluded from the list. For example Strugnell *et al.* (2008a) mention that *Thaumeledone rotunda* (Hoyle, 1885) is believed to have a circumpolar distribution; however, because a specimen (as far as can be ascertained) has not been collected from Australian waters, it has not been included here. Taxa have been grouped alphabetically by taxonomic order and family. Within families, species are ordered alphabetically by genus and species. The classification approximately follows the Classification of the Currently Accepted Taxa of Recent Cephalopoda (Sweeney and Roper 1998). Modifications published since that date have been incorporated (e.g. following Norman and Hochberg 2005, Jereb and Roper 2010, Strugnell *et al.* 2014 and Allcock *et al.* 2014). Where a taxon is suspected to be part of a species-complex, the species name is enclosed in quotation marks to alert readers to the fact that information pertaining to the particular taxon may in fact be referring to one or more species. The complete reference for each original species description is provided, with the page number corresponding to its first citation shown in square brackets after the reference. This information has been obtained primarily from the Type Finder and searching tool developed by Mike Sweeney (see www.jiscmail.ac.uk/cgi-bin/filearea.cgi?LMGT1=FASTMOLL&a=Zget&f=/Type_Finder_Fastmoll-1.pdf). Type repositories and type localities are also given for each species.

Individual taxon treatments focus on those characters that can be used to distinguish families, genera and species. Emphasis is placed, where possible, on external traits and features that can be recognised by non-experts. Details of internal anatomical features are included only if they appear to be diagnostic. Distinguishing characters are included to indicate particular traits that might assist in identifying a particular taxon as distinct from another, perhaps similar, taxon. Additional literature pertaining to each species, where not directly cited, is included under 'References' to indicate sources used for each species treatment and to facilitate a more detailed analysis of each species if required.

The distributions for each species are primarily based on online collection records obtained from Australian museums via OZCAM. Maps were downloaded from the Atlas of Living Australia (http://spatial.ala.org.au) during August to October 2013. There may be gaps in these data because not all museum collections are fully databased and many regions have not been well surveyed, so, in addition, recent FAO (Food and Agriculture Organisation of the United Nations) publications (Jereb and Roper 2005, 2010; Jereb *et al.* 2014) have been used to supplement this information. For some species, information on distributions is based on 'dot point' occurences while for others, particularily where obtained from FAO guides, these 'dots' have been 'joined' to give a broad distribution range, which may, or may not, be accurate.

While these distributions reflect the current state of our knowledge, species' distributions are changing due to the current unprecedented changes in water temperatures and movement due to climate change. Contraction and expansion of the distributions of some species as a result of climate change is already being documented in the literature. The southward range expansion of *Octopus tetricus* Gould, 1852, once found only on mainland Australia and now occurring in Tasmania, is an example (Pecl 2011). The data presented here may provide a baseline upon which to monitor such changes.

Some fisheries information is included. While relatively few cephalopods found in Australian waters are directly targeted at this time, this may change rapidly with the further decline in finfish fisheries worldwide due to overfishing on a massive scale and unsustainable human population increases. Those cephalopods (squids in particular) that have potential for fisheries exploitation in Australian waters should be a focus for study to learn more about their taxonomy, biology, trophic relationships and population structure if future fisheries are to be managed successfully.

### External characteristics and measurements used

Orientation, arm numbering and morphological terms are outlined in the glossary. Orientation is relative to the resting animal, with the arms and arm/tentacle crown being anterior. The body or mantle is considered posterior. The upper surface of the resting animal is considered dorsal and the underside ventral. Arms are numbered as left or right, commencing from the dorsal arms (i.e. the dorsal arms are pair 1). Arabic numerals have been used to number the arms; however, many authors use Roman numerals as will be discovered when consulting some other literature.

The mantle length (ML) is the standard length measurement used for all cephalopods (except *Nautilus* where shell diameter is used). In squids and cuttlefishes, the measurement is made along the dorsal surface from the anterior-most point to the posterior-most end of the mantle. In octopuses, mantle length is measured from the midpoint between the eyes to the posterior end of the mantle along the dorsal surface (Fig. 2).

### Determining sex

In many cephalopods, the sex of specimens can be determined externally by examining the arms to find modifications in maturing and mature males (hectocotylisation). Modifications include differences in the normal sucker arrangement, loss of suckers, increased sucker size, or thickening or lengthening of sucker stalks, protective membranes, and their supports. For many squids and cuttlefishes, one or both of the ventral arms are modified, although in some species no external modification is apparent. (This is commonly the case in those species, generally squids, that use their terminal organ to inseminate a female.) In some groups, such as the bobtail squids, arm modifications may be present on the dorsal rather than (or as well as) the ventral arms. In benthic octopuses, the third arm on the right-hand side (left arm in some genera) is usually modified. The modified arms are used to gather spermatophores from the males' genital opening inside the mantle and transfer them to the female. In poorly preserved or damaged material, for immature specimens, or for species which lack hectocotylised arms, dissection of the mantle cavity is necessary to determine sex.

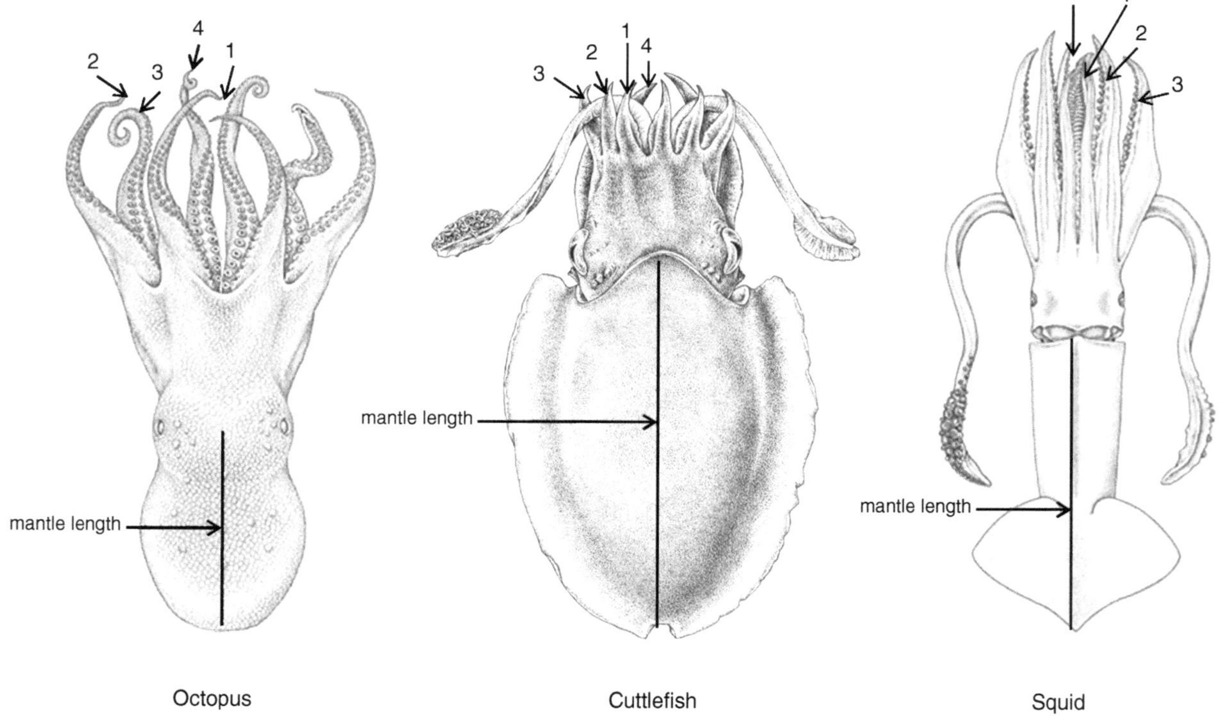


**Fig. 2.** Dorsal views of an octopus, a cuttlefish and a squid, showing standard mantle length measurements and numbering of arms from dorsal to ventral (K. Hollis, ABRS).

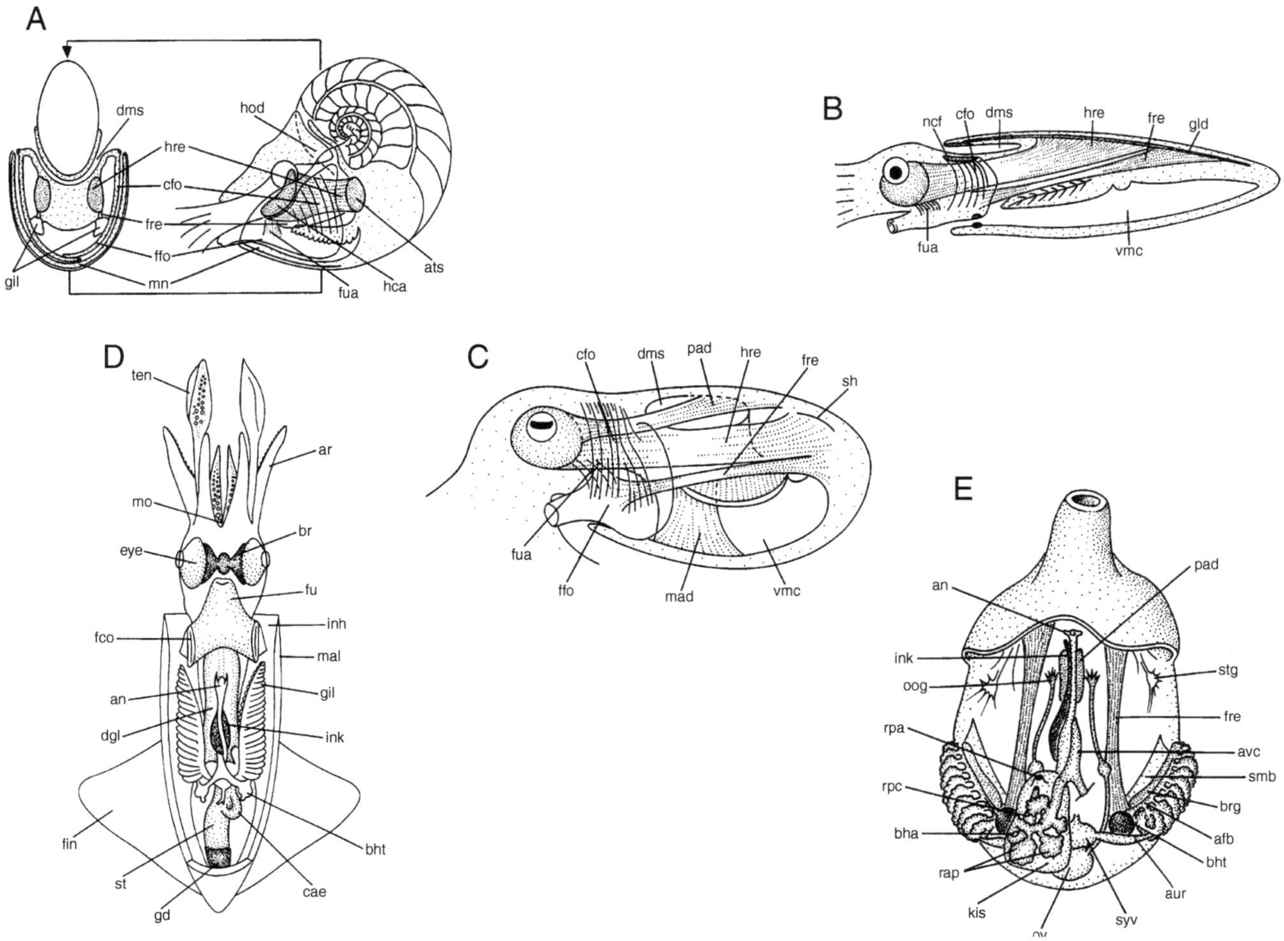


**Fig. 3.** Cephalopod mantle cavities, associated muscles, and organs: A, nautilus; B, squid; C, octopus; D, squid; E, octopus. *afb*, afferent branchial vessel; *an*, anus; *ar*, arm; *ats*, attachment to shell; *aur*, auricle; *avc*, anterior vena cava; *bha*, branchial heart appendage; *bht*, branchial heart; *br*, brain; *brg*, branchial gland; *cae*, caecum; *cfo*, collar fold; *dgl*, digestive gland; *dms*, dorsal mantle space; *eye*, eye; *fco*, funnel-locking cartilage; *ffo*, funnel fold; *fin*, fin; *fre*, funnel retractor; *fu*, funnel; *fua*, funnel adductors; *gd*, gonad; *gil*, gill; *gld*, gladius; *hca*, head cartilage; *hod*, hood; *hre*, head retractor; *ink*, ink sac; *inh*, inhalant opening; *kis*, kidney sac; *mad*, median pallial adductor; *mal*, cut mantle; *mn*, mantle; *mo*, mouth; *ncf*, nuchal cartilage; *oog*, opening of oviduct; *ov*, ovary; *pad*, pallial adductor; *rap*, renal appendages; *rpa*, renal papilla; *rpc*, renopericardial canal; *sh*, vestiges of shell (stylets); *smb*, suspensory membrane; *st*, stomach; *stg*, stellate ganglion; *syv*, systemic ventricle; *ten*, tentacle; *vmc*, ventral mantle cavity. (Fig. 11.11, Beesley *et al.* 1998.)

## Internal structures in the mantle cavity

In identifying certain species, or in sex determination, it is sometimes necessary to dissect the mantle cavity, exposing the funnel–mantle locking apparatus, gills and reproductive structures. This is done by making a mid-ventral longitudinal cut along the length of the mantle. The funnel and mantle elements of the locking apparatus are visible just inside the mantle immediately posterior to the funnel opening itself. Males are recognised by the spermatophoric complex (including the spermatophoric sac that stores fully formed spermatophores in mature squids and the coiled spermatophoric organ) and a penis-like structure, the 'terminal organ', on the left side (right in ventral view) of the midline. Females have paired nidamental glands and paired oviducal glands and oviducts in the anterior part of the mantle. In myopsid squids (such as *Uroteuthis*), only a single oviducal gland and oviduct is present on the left side of the animal and paired accessory nidamental glands are present anterior to the nidamental glands.

In cuttlefishes, mature females can readily be distinguished from males by the presence of a pair of leaf-shaped creamy yellow nidamental glands (Fig. 4). Eggs may also be seen in the ovary, below and posterior to the nidamental glands. In immature females, the nidamental glands

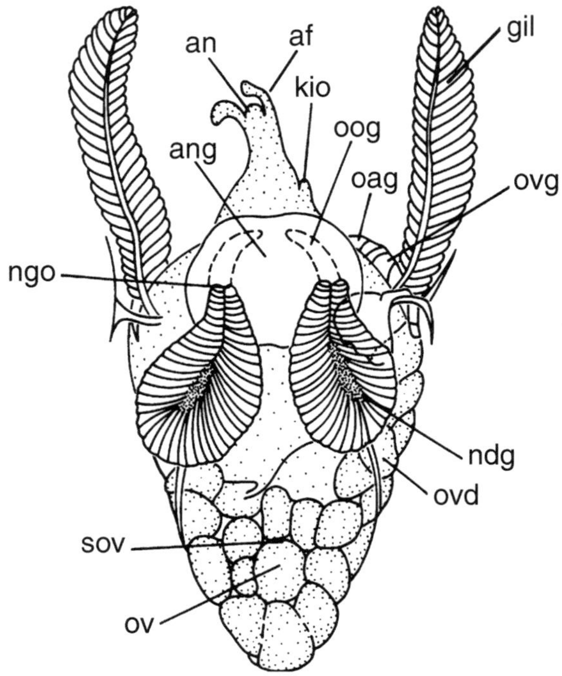


**Fig. 4.** Sepioid female reproductive organs: *af*, anal flap, *an*, anus; *ang*, accessory nidamental gland; *gil*, gill; *kio*, kidney opening (renal papilla); *ndg*, nidamental gland; *ngo*, nidamental gland opening; *oag*, opening of accessory nidamental gland; *oog*, opening of oviducal gland; *ov*, ovum, *ovd*, oviduct; *ovg*, oviducal gland; *sov*, beginning of oviduct.

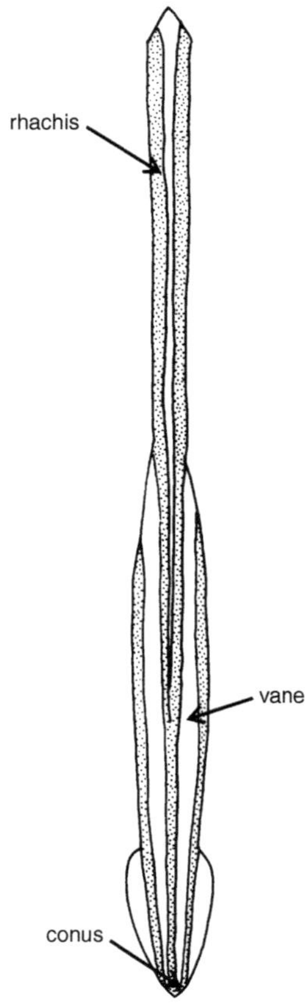


**Fig. 5.** Squid gladius.

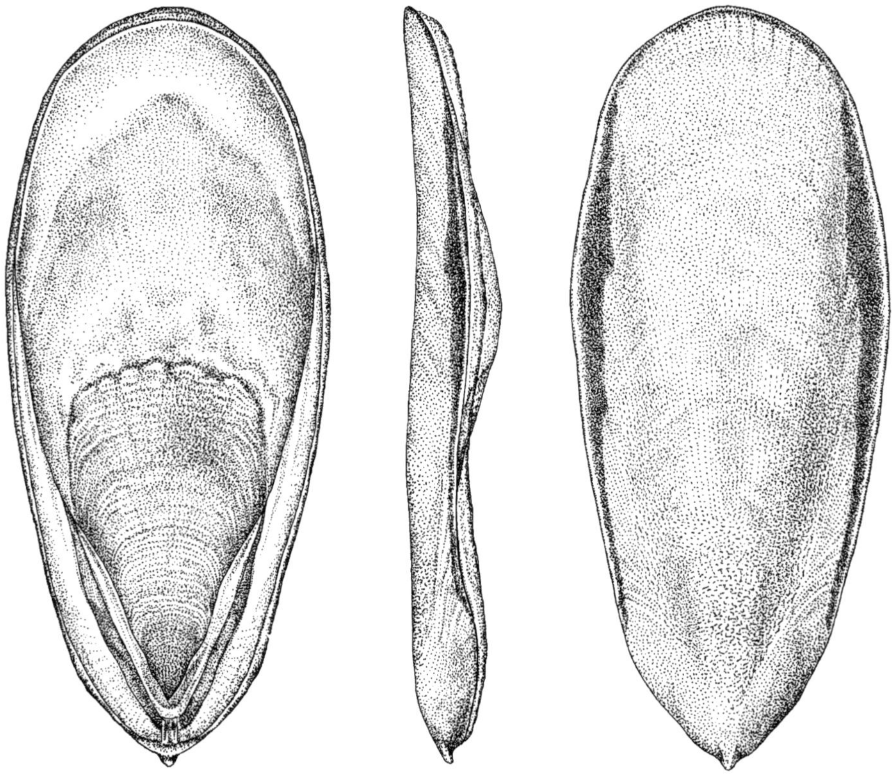

**Fig. 6.** Cuttlebone, ventral view (left), lateral view (middle), dorsal view (right).

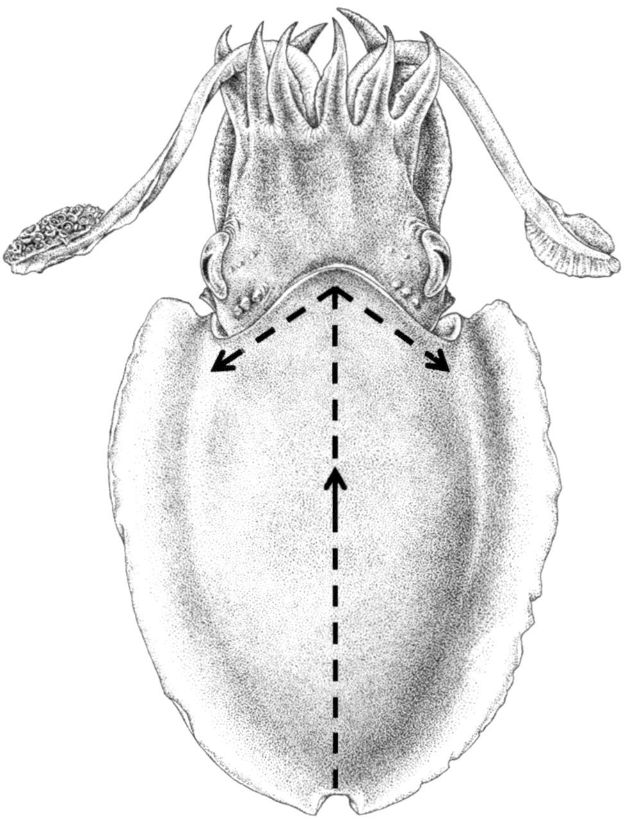

Fig. 7. Cuttlefish, indicating incision lines and direction for removing cuttlebone.

may be greatly reduced in size or visible only as two short slits. The shape of the male and female genital openings on the left side of the mantle cavity also differs slightly between the sexes.

## Removing internal shells

*Gladii.* In fresh and preserved specimens, the gladius (supporting structure in squids and some sepiolids) (Fig. 5) is removed by making a mid-longitudinal incision along the length of the dorsal mantle and peeling away the skin laterally. Care must be taken at the posterior end to ensure that all of the skin and muscle tissue is removed from around the cone.

*Cuttlebone.* The cuttlebone (Fig. 6) can easily be removed from a fresh animal by making a median longitudinal incision along the length of the dorsal mantle, and two shorter incisions at the anterior end of the mantle (Fig. 7), using scissors or a scalpel blade, taking care not to cut though the bone itself. The skin can then be peeled open to reveal the cuttlebone. This is more difficult to achieve without breaking the bone in preserved museum specimens. Again, the bone should be removed with considerable care from the posterior end of the animal.

*Stylets.* Octopus stylets (when present) are located towards the dorso-posterior end of the mantle (Fig. 3C).

# Glossary

*Aboral.* Away from or on the opposite side to the mouth.

*Accessory nidamental glands.* Glands of unknown function consisting of tubules containing symbiotic bacteria. Found in cuttlefishes and loliginid squids. Occur in both sexes, anterior to the nidamental glands in females (rudimentary in males) (Fig. 4).

*Acetabulum.* The central, open, water-filled cavity or cup of cephalopod suckers responsible for suction.

*Adult.* A female whose ovary is filled with mature eggs or a male that has spermatophores in the spermatophore storage sac.

*Ammoniacal.* The replacement of sodium chloride with lighter ammonium salts in the vacuoles of muscular tissue (or, in the Cranchiidae, in a fluid-filled coelomic cavity) in some squids. This is used as a buoyancy mechanism.

*Anal flaps.* The pair of fleshy papillae on each side of the anus involved in directing the release of ink (Fig. 4).

*Anal pads.* Ovoid pads of unknown function; appear glandular and lie either side of the rectum, just behind the anal opening in some squids.

*Antarctic Convergence.* See *Antarctic Polar Front.*

*Antarctic Polar Front.* The northern limit of Antarctic waters, forming a significant biological boundary; many species are typical of Antarctic water and rare on the other side of the Antarctic Polar Front. The surface temperature changes by about 2°C to 3°C from one side of the front to the other. (Formerly known as the *Antarctic Convergence.*)

*Anterior.* Towards the head end or towards the arm tips.

*Antitragus.* The knob that projects inward from the posterior surface of the central depression in the funnel-locking apparatus of some squids (Fig. 8).

*Anus.* The opening of the alimentary canal, or gut, through which undigested remains of food are expelled (Figs 3, 4).

*Areola spots.* Circular spots, generally light coloured but often with complex pigmentation and of unknown function, that occur in lines along the aboral surfaces of some arms and portions of the head in some opisthoteuthid cirrates.

*Arm.* One of the eight fixed circumoral appendages surrounding the mouth of coleoid cephalopods.

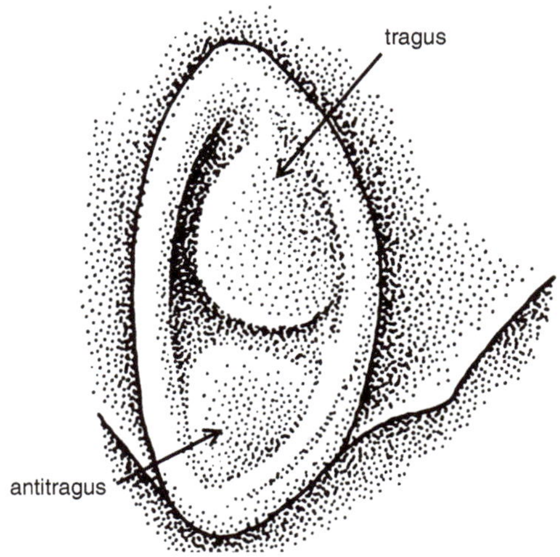


**Fig. 8.** Funnel-locking cartilage.

*Armature.* Refers to the presence and arrangement of suckers and/or hooks on the arms and tentacular clubs of cephalopods.

*Arm crown.* In octopods the ring of eight arms that surrounds the mouth, including the webs.

*Arm formula.* The relative order of arm lengths from longest to shortest; for example, '1 > 2 > 3 > 4' refers to arms decreasing in length from the dorsal-most pair (arm pair 1) to the ventral-most pair (pair 4), '4 = 3 = 2 > 1' refers to the dorsal pair being shorter than other arms that are of equal length (Fig. 2). (Arms are numbered in Roman numerals by some authors, i.e. I, II, III, IV.)

*Arm length index.* The length of the arm (measured from basal-most sucker) as a percentage of mantle length.

*Arm sucker index (ASIn).* Diameter of largest sucker on the designated arm as a percentage of mantle length.

*ASI.* Arm Sucker Index; the diameter of the largest arm sucker on each non-hectocotylised arm as a percentage of mantle length.

*Bathyl.* Pelagic zone from 1000 to 4000 m below the surface.

*Bathypelagic.* Midwater within the deep sea (~1000–3000 m depth).

*Beaks.* The chitinous jaws of cephalopods (Fig. 9).

*Benthic.* Bottom dwelling, living on or near the bottom of the sea (= *demersal*).

*Benthopelagic.* Describes a free-swimming organism that lives just above the ocean floor.

*Bioluminescence.* The production of light by living organisms.

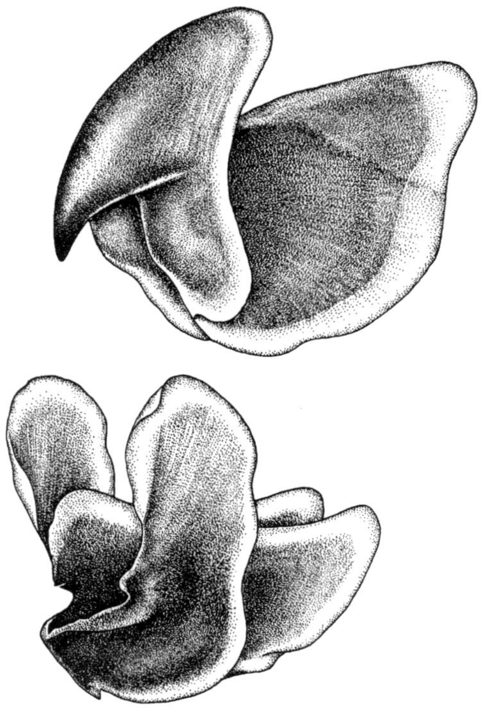

**Fig. 9.** Upper beak (top), lower beak (bottom).

*Brachial.* Pertaining to the arms.

*Brachial pillar.* The narrow, elongate anterior region on the paralarval or juvenile head of some families, between the eyes and the base of the brachial crown, well-developed in young cranchiid and brachioteuthid squids.

*Branchial.* Pertaining to the gills.

*Brooding.* Incubation of eggs by the female.

*Buccal.* Pertaining to the mouth.

*Buccal connective.* The thin, muscular band that attaches the buccal support of the buccal membrane to the base of the adjacent arm (Fig. 10).

*Buccal crown.* The umbrella-like structure that surrounds the mouth and, in turn, is enveloped by the brachial crown. It consists of buccal supports and the buccal membrane (Fig. 10).

*Buccal lappet.* The small, subtriangular flap at the tip of each buccal support of the buccal membrane.

*Buccal mass.* The muscular bulb at the anterior-most part of the digestive system consisting of the mouth, beaks, radula, muscles and anterior salivary glands.

*Buccal membrane.* The thin web of tissue that encircles the mouth, reinforced by six to eight buccal supports.

*Buccal membrane connectives.* Muscular bands that connect the supports of the buccal membrane to the bases of the arms.

*Buccal suckers.* Small suckers on the buccal lappets/membrane of some species.

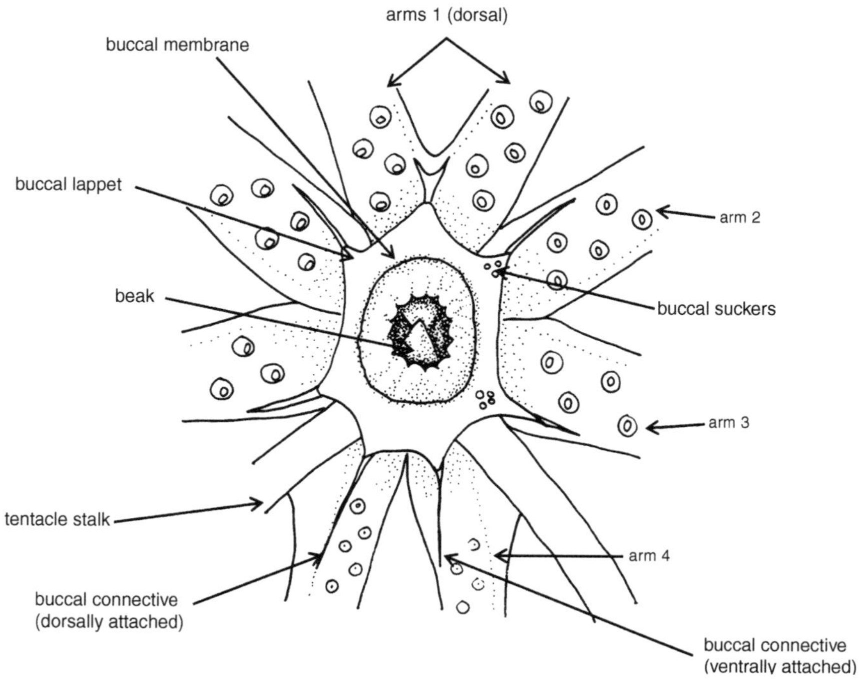


**Fig. 10.** Decapod buccal anatomy.

*Buccal support.* The muscular rod fused to the buccal membrane as a supporting rib; six to eight in number.

*Calamus.* The tongue-like projection at the base of the ligula on the tip of the hectocotylised arm of male octopuses. Found at the distal end of the spermatophore groove, distal to the last sucker (Fig. 11).

*Calcareous/calcified.* Chalky, calcified by deposition of calcium salts (calcium carbonate).

*Callus (umbilical callus).* The thickened layer that covers the umbilicus in some nautiluses.

*Carpal cluster.* An unusually distinct group of suckers and knobs on the carpus of the tentacular club (Fig. 12).

*Carpal knobs.* Small, rounded, muscular protuberances on the carpus to which the carpal suckers from the opposite club adhere during the locking of the clubs.

*Carpal-locking apparatus.* The arrangement of suckers and matching knobs on the carpal region of the tentacular club that permits the two clubs to be locked together (Fig. 12).

*Carpal suckers.* Small suckers on the carpus of the club that adhere to the carpal knobs on the opposite carpus during the locking of the clubs.

*Carpus.* The proximal zone of suckers and knobs on the tentacular club in some families (Fig. 12).

*Cartilaginous structures/scales.* Cartilage-like structures in the skin of some squids that may be overlapping and scale-like, or multifaceted platelets, knobs or papillae.

*Cement body.* The structure in the spermatophore that allows adhesion of the discharged spermatophore to a female (Fig. 13).

*Chitin.* The horny polysaccharide substance (fingernail-like) that forms the sucker rings, hooks and beaks of cephalopods.

*Chromatophores.* Pigment-filled, generally flat, muscular sacs in the skin under individual nervous control that collectively provide the background colour, colour patterns and colour play of cephalopods.

*Circumoral appendages.* Limbs that arise from the head and encircle the mouth.

*Cirri.* Elongate, fleshy, finger-like projections.

*Collar.* The muscular, flange-like structure that extends from the nuchal cartilage to the funnel. It closes and seals the mantle as it contracts, thus forcing water out through the funnel.

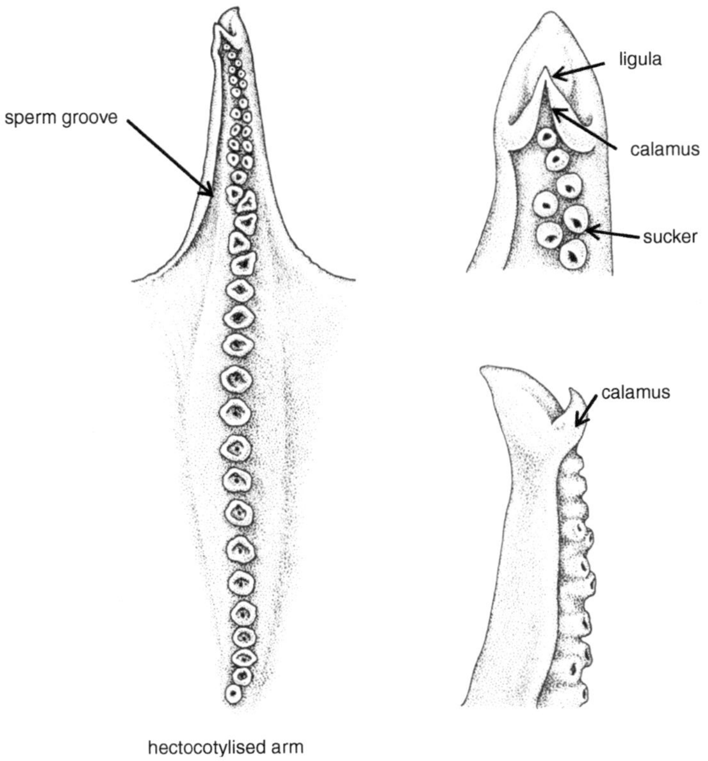


**Fig. 11.** Hectocotylised arm of an octopus and enlargement of arm tip, oral view (top), lateral view (bottom).

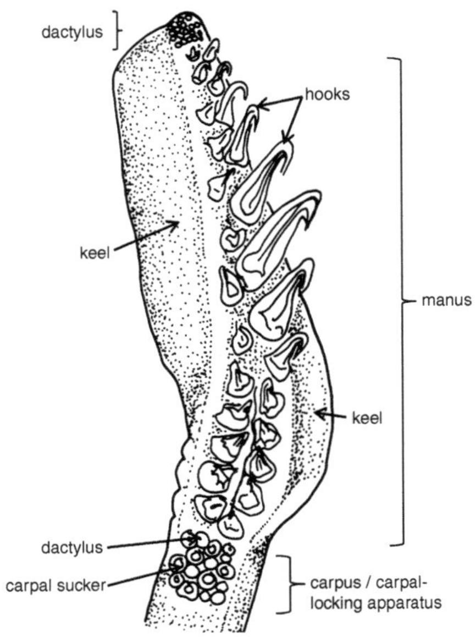


**Fig. 12.** Squid tentacular club.

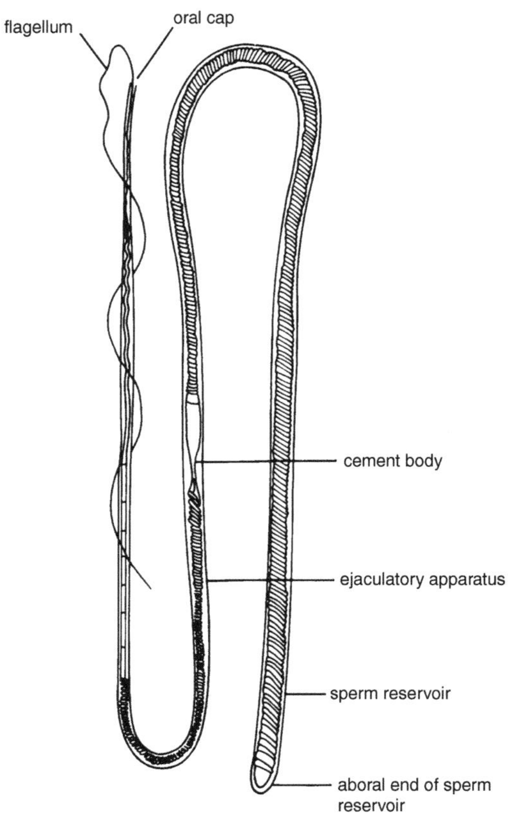


**Fig. 13.** Spermatophore.

*Complex photophores.* Photophores with various colour filters, reflectors, light guides, lenses and chromatophores surrounding the site of light production, the photogenic region.

*Cone, conus.* The spoon-like or cup-like conical posterior terminus of the gladius or cuttlebone (Fig. 5). (May or may not be present in some squids.)

*Corneal membrane.* The very thin, transparent skin that covers the eyes of myopsid squids and cuttlefishes.

*Counter illumination.* The production of bioluminescent light by an animal to conceal its silhouette against a lighted background, making an animal invisible from below under dim directional light.

*Cuttlebone.* The calcareous supporting plate in the dorsal part of the mantle of cuttlefishes. Organ used to maintain buoyancy. Consists of many thin plates, or septae, arranged in a thick bundle. The system of plates is called the phragmocone (= *sepion*). The septae are interconnected by supporting poles and pillars that are visible on the ventral side of the cuttlebone as striae (Fig. 14).

*Dactylus.* The distal, terminal section of the tentacular club in squids and cuttlefishes, often characterised by suckers of reduced size (Fig. 12).

*Decapods/decapodiformes.* Cephalopods with 10 limbs (eight arms and two feeding tentacles).

*Demersal.* Living near the ocean floor.

*DDVV.* Refers to the attachment arrangement of the buccal connective from dorsal to ventral. In this case the attachment is to the dorsal (D) side of arms 1 and 2 and the ventral (V) side of arms 3 and 4. See *buccal connective.*

*Diel vertical migration.* The daily vertical migration; usually refers to animals moving from deep to shallower water at twilight to feed.

*Digestive gland.* The liver of cephalopods, which secretes digestive enzymes and is also important in absorption and excretion (Fig. 3D).

*Distal.* Away from the body or point of origin; towards the peripheral parts (opposite of proximal).

*Doratopsis.* The paralarval stage that is characteristic of all members of the oegopsid squid family, Chiroteuthidae.

*Dorsal.* The uppermost surface of a cephalopod, opposite the ventral surface where the funnel is located.

*Ejaculatory apparatus.* Part of the spermatophore that performs the eversion of the spermatophore and extrusion of the sperm mass (Fig. 13).

*Epipelagic.* Living in the surface waters of the ocean, typically considered the upper 200 m.

*Epirenal bodies.* Glandular structures of unknown function that lie on either side of the renal papillae.

*Epithelial pigmentation.* Pigmentation contained in epithelial cells that are unable to change shape in the absence of muscles and nerves.

*Fin length.* The length from anterior lobe of the fin to the posterior-most attachment of the fin to the mantle or tail.

*Fin length index.* The length of the fin as a percentage of mantle length.

*Fins.* The pair of muscular flaps that arise along the lateral or dorsolateral surface of the mantle of sepioids, teuthoids and cirrate octopods (Fig. 3D). They are used for locomotion, steering and stabilisation. The fin angle is the angle between the longitudinal axis of the mantle and the posterior border of one fin (Fig. 15).

*Fin width index.* The greatest width of the fins (usually measured dorsally across both fins) as a percentage of mantle length.

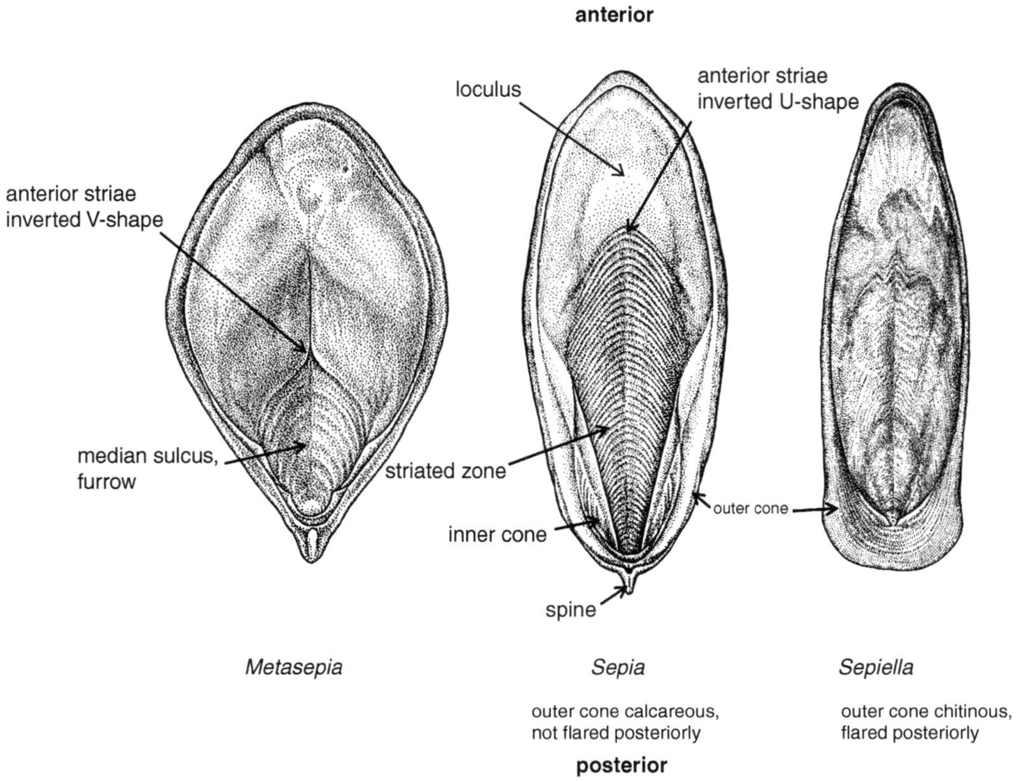


**Fig. 14.** Sepiid cuttlebones.

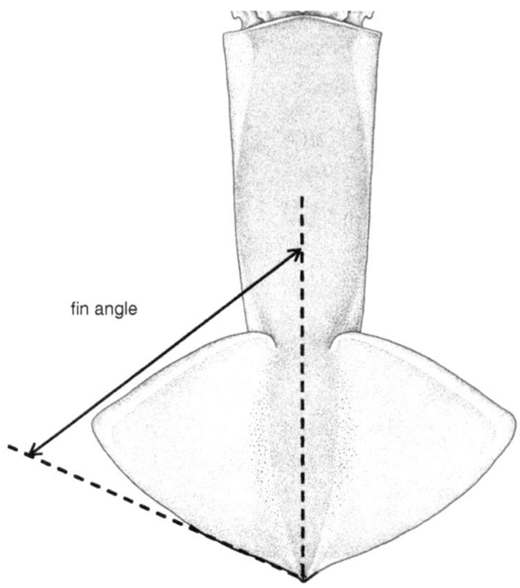


**Fig. 15.** Fin angle.

*Fixing apparatus.* The mechanism of suckers and knobs on the carpal region of the tentacular club that permits the two clubs to be locked together during prey capture.

*Foveola.* The transverse membranous fold of skin that forms a pocket in the funnel groove of some oegopsid squids (e.g. some ommastrephids) (Fig. 16).

*Frontal region.* An area, characterised by sharp horizontal gradients in hydrographic properties (such as temperature, salinity, density, oxygen and nutrients), that marks the boundaries of different water masses.

*Funnel.* The ventral, subconical tube through which water is expelled from the mantle cavity during locomotion and respiration (reproductive and waste products and the ink also pass through the funnel) (Fig. 17).

*Funnel-locking cartilage.* The cartilaginous groove, pit, pocket or depression on each ventrolateral side of the posterior part of the funnel that joins with the mantle-locking cartilage to lock the funnel and mantle together during locomotion, so water is expelled only through the funnel and not around the mantle opening (Fig. 17).

*Funnel–mantle locking apparatus.* The combination of the funnel-locking cartilage and the mantle-locking cartilage; also called the funnel–mantle locking apparatus. It is found laterally at the ventral mantle opening joining the posterior extension of the funnel to the mantle in squids and cuttlefishes (Fig. 17). The cartilages may be very simple in structure such as in loliginid squids or highly complex such as in the

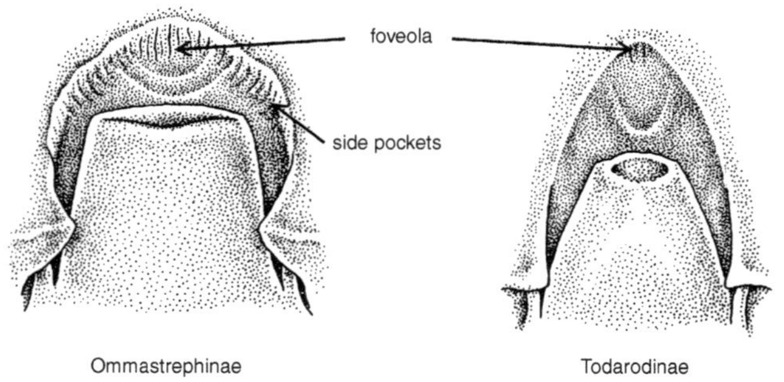


**Fig. 16.** Funnel groove.

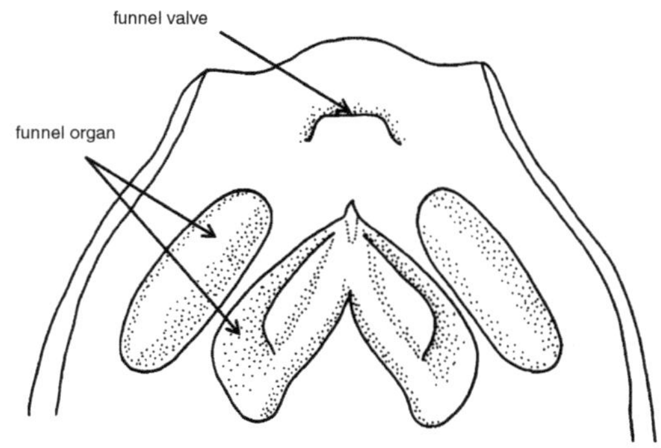


**Fig. 18.** Funnel organ and funnel valve exposed by a median longitudinal funnel incision.

ommastrephids. The apparatus may also involve a partial or complete muscular fusion between the funnel and mantle elements such as in the ommastrephid *Sthenoteuthis oualaniensis* (Lesson, 1830), the bobtail squid *Sepiadarium kochii* Steenstrup, 1881 and all cranchiid squids.

*Funnel organ.* Glandular pads on the internal surface of the funnel. Able to be seen following a ventral incision of the funnel (Fig. 18). An important diagnostic trait in octopods. Indistinct in poorly preserved or frozen material. A soluble dye, such as methylene blue, may make the outline more obvious. The funnel organ releases mucus that mixes with ink to alter its consistency.

*Funnel valve.* The muscular flap in the dorsal inner surface near the distal opening of the funnel in some species (Fig. 18).

*Genital opening.* The exit duct for eggs or spermatophores formed in the reproductive tract.

*Gills.* Paired structures, each consisting of many leaf-like lamellae, through which gas exchange occurs (Figs 3D, 4). To identify some cephalopods, it is necessary to count the number of gill lamellae on each side of each gill (= per demibranch). The count per demibranch excludes the central terminal (anterior-most) lamella; for example, a gill count of 10 refers to 10 lamellae on each side of each gill (Figs 3, 4).

*Gladius (pl. = gladii).* The feather or rod-shaped chitinous supporting structure (or shell) in the dorsal midline of squids and sepioids other than cuttlefishes (= *pen*) (Fig. 5).

*Gladius length index.* The length of the gladius as a percentage of mantle length.

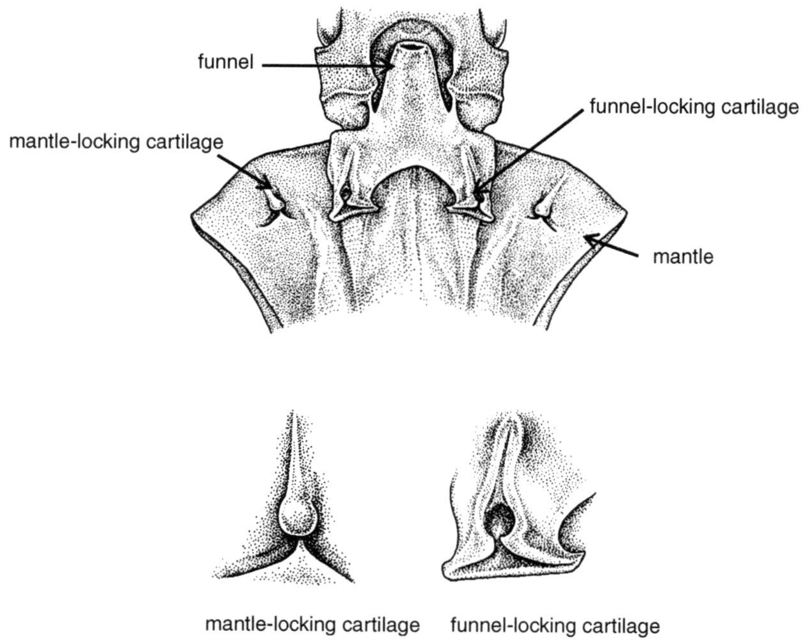


**Fig. 17.** Funnel, mantle-locking cartilage and funnel-locking cartilage.

*Gladius width index.* The greatest width of the gladius expressed as a percentage of gladius length.

*Hatchling.* A young cephalopod newly hatched from the egg.

*Hectocotylus.* The part of one (or more) arm(s) of male cephalopods modified for transferring spermatophores to the female. Modifications may involve suckers, sucker stalks, protective membranes, trabeculae in squids and cuttlefishes or a distinct ligula/calamus on the tip of the modified arm in octopods (Fig. 11).

*Hectocotylus length index.* The length of the modified portion of the arm expressed as a percentage of the total length of the hectocotylised arm.

*Hooks.* Chitinous, claw-like structures derived from the suckers on the arms and/or clubs of some oegopsids (Fig. 12).

*Infundibulum.* The disc or plate-like rim of cephalopod sucker rings that surrounds the opening to the acetabulum.

*Ink.* The secretion containing melanin and other substances, including proteins, produced in the ink sac of cephalopods. Mixes with mucus from the funnel organ to produce clouds of varying consistency. *Hapalochlaena* has tetrodotoxin in its ink. Inking is used as a visual defence (e.g. in the production of a pseudomorph behind which a cephalopod may escape a predator) or a chemical defence (containing attractant or deterrent chemicals that disrupt a predators' chemical senses).

*Ink sac.* The structure that stores the ink of cephalopods; it lies ventrally along the intestine or is embedded in the digestive gland (hepatopancreas) and empties anteriorly via a duct into the rectum (Fig. 3).

*Inner cone.* The structure composed of forked limbs on the ventral side of the cuttlebone, between the inner side of the outer cone and the phragmocone. It usually extends to the anterior end of the striated zone (Fig. 14).

*Iridophores.* Platelet-like components of the skin that form an iridescent or reflective sheen in cephalopods.

*Iteroparous.* Describes females having multiple reproductive episodes and spawn repeatedly over time.

*Jet propulsion.* The contraction of the circular muscles in the mantle to produce high hydrostatic pressure in the mantle cavity, resulting in a jet of water from the funnel and movement of the animal in the opposite direction to which the funnel is pointing.

*Juvenile.* The immature stage between the hatchling and the nearly mature subadult stage.

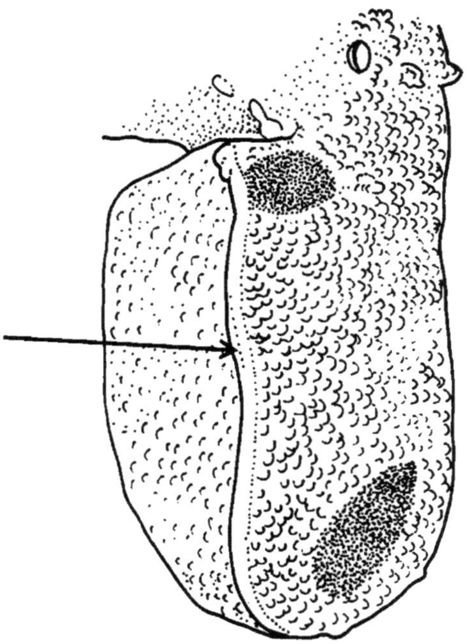

**Fig. 19.** Octopod lateral ridge (arrow).

*Keel.* The flattened, muscular extension along the aboral surface of some arms and tentacular clubs to render them more hydrodynamic (= *swimming membrane*); also a narrow longitudinal ridge on the keel of some cuttlebones, or a raised ridge around the lateral margin in octopods (also called a *lateral ridge*).

*Lateral ridge.* The muscular keel along the lateral mantle of some octopuses (Fig. 19).

*Leucophores.* White, reflecting components in the skin.

*Light organ.* An organ that produces and distributes light. See *photophore*.

*Ligula.* The spoon- or club-like tip of the modified (hectocotylised) arm of octopuses. Used to hold spermatophores (Fig. 11).

*Loculus.* The anterior part of the cuttlebone that is smooth, not striated (Fig. 14).

*Mantle.* The fleshy (muscular) tubular or sac-like body of cephalopods; provides propulsion through jet-like expulsion of water; contains the viscera (Fig. 3).

*Mantle cavity.* The space enclosed by the mantle.

*Mantle length (ML).* The standard measure of length in coleoid cephalopods. In octopods and other groups in which the head is fused to the mantle it is measured from the midpoint of the eyes to the posterior mantle margin. Where the mantle is not joined dorsally to the head, the measurement is made from the anterior mantle margin to the posterior tip of the mantle along the dorsal midline (Fig. 2).

*Mantle-locking cartilage.* The cartilaginous ridge, knob or swelling on each side of the ventrolateral, internal surface of mantle that locks into the funnel component of the funnel–mantle locking apparatus (see also *funnel-locking cartilage*) (Fig. 17).

*Manus.* The central portion of the club between the dactylus distally and the carpus proximally (Fig. 12).

*Mature.* Refers to sexual maturity, which is determined in females by the presence of eggs in the ovaries, and in males by the presence of spermatophores in the spermatophore storage sac.

*Medial (median).* Pertaining to a structure located towards, on or along the midline.

*Mesopelagic.* Living in the middle layers of the water column in oceanic waters, generally considered to be from 200 m to 1000 m depth.

*Neritic.* Inhabiting the sea over the continental shelf; arbitrarily taken to be the sea where it is shallower than 200 m.

*Nidamental glands.* Glands inside the mantle of females that secrete a fourth envelope for the eggs. Present in squids (except Enoploteuthidae), sepiolids, sepiadariids and cuttlefishes. The glands are tongue-like and bifurcated in squids, and oval in cuttlefishes (Fig. 4).

*Normal arms.* Non-modified arms. In males non-hectocotylised arms.

*Notalian.* Southern temperate marine realm.

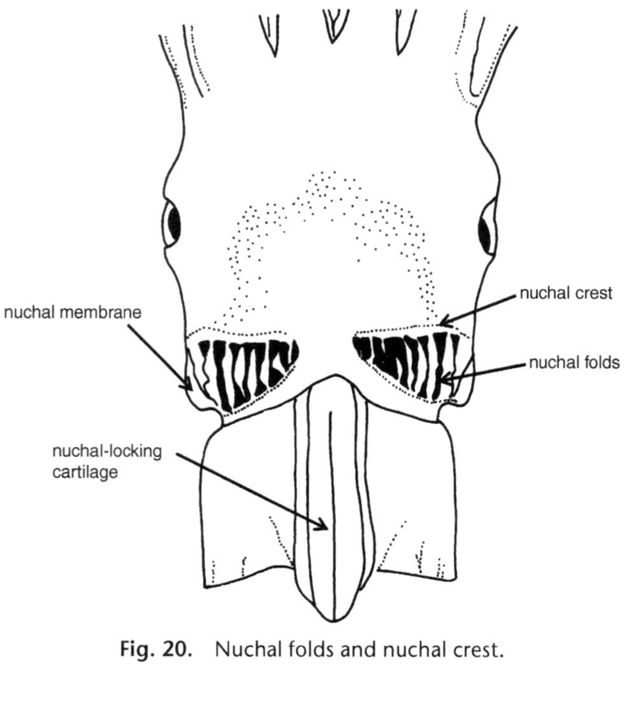


**Fig. 20.** Nuchal folds and nuchal crest.

*Nuchal crest.* The prominent transverse ridge that extends across the dorsal head and down the lateral head surfaces at its posterior end (Fig. 20).

*Nuchal folds.* Fixed folds or pleats on the head integument that are perpendicular to the nuchal crest (Fig. 20).

*Nuchal-locking apparatus/nuchal cartilages.* The oblong, cartilaginous locking structure located mid-dorsally just posterior to the head. It is composed of the nuchal cartilage, which forms an attachment site for the collar and head retractor muscles and an interlocking complementary cartilage on the mantle (Fig. 20).

*Nuchal membrane.* A thin membrane that connects the main nuchal folds at their posterior ends (= *occipital membrane*) (Fig. 20).

*Occipital band.* The band of tissue that joins the anterior mantle margin with the head.

*Occipital folds.* = *nuchal folds.*

*Ocellus.* The dark false-eye spot, found in 'ocellate' octopuses as a pair of ocelli, one spot below each eye on the lateral arm crown between the bases of arms 2 and 3 (as in *Octopus cyanea* Gray, 1849). In some taxa, ocelli are present on the dorsal mantle.

*Ocular photophore.* A photophore that lies on the eyeball.

*Olfactory organ.* The organ of the detection of smell in squids, cuttlefishes and vampyromorphs, represented by olfactory papillae, while in octopuses there are olfactory pits. Positioned on the sides of the head near the neck.

*Ontogenetic descent/ascent.* Refers to the movement at different life stages to inhabit a different depth in the water column.

*Oral.* Pertaining to the mouth.

*Orbit.* The cavity or depression housing the eyeball.

*Outer cone.* The rim surrounding the phragmocone in cuttlebones (Fig. 14).

*Oviducal gland.* The swollen glandular structure that secretes a coating onto spawned eggs and produces the egg or egg-string attachment (Fig. 4).

*Oviducts.* Tubular structures in females that conduct the eggs from the ovary to the mantle cavity (Fig. 4).

*Papilla.* A conical or flattened skin projection. Dorsoposterior to the eye, these are called supraocular papillae.

*Paralarvae.* The first free-living life history stage, usually planktonic.

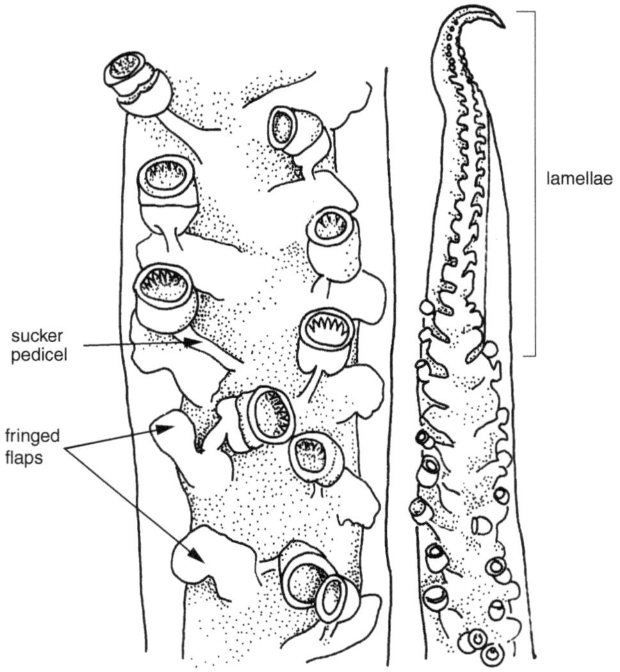


**Fig. 21.** Squid hectocotylised arm.

*Pedicel.* The short, tubular stalk that supports a sucker in sepioids and squids (Fig. 21).

*Pelagic.* Free swimming in the open ocean.

*Penis.* The male organ of insertion for the transfer of spermatophores. See also the *terminal organ.*

*Periostracum.* The epidermal covering of some shells.

*Photophore.* A simple or complex structure that produces bioluminescence by intrinsic (self-generated) or extrinsic (bacterial) means. Several species in two cephalopod families, the Loliginidae and the Sepiolidae, harbour bacteriogenic light organs in the mantle cavity. May be present in the skin, on the eyes, or embedded in muscle in the mantle, head, arms or tentacles, or on the viscera (= *light organ*).

*Phragmocone.* The system of plates comprising the cuttlebone.

*Pocket.* An open depression in the anterioventral surface of the head between the bases of arms 3 and 4 of cuttlefishes into which the tentacles are retracted when not in use.

*Posterior.* Away from the arms and tentacles, towards the tail or rear end of mantle.

*Protective membrane.* The thin, web-like integument along the lateral angles of the oral surface of the arms and clubs lateral to the suckers, supported by muscular rods called trabeculae.

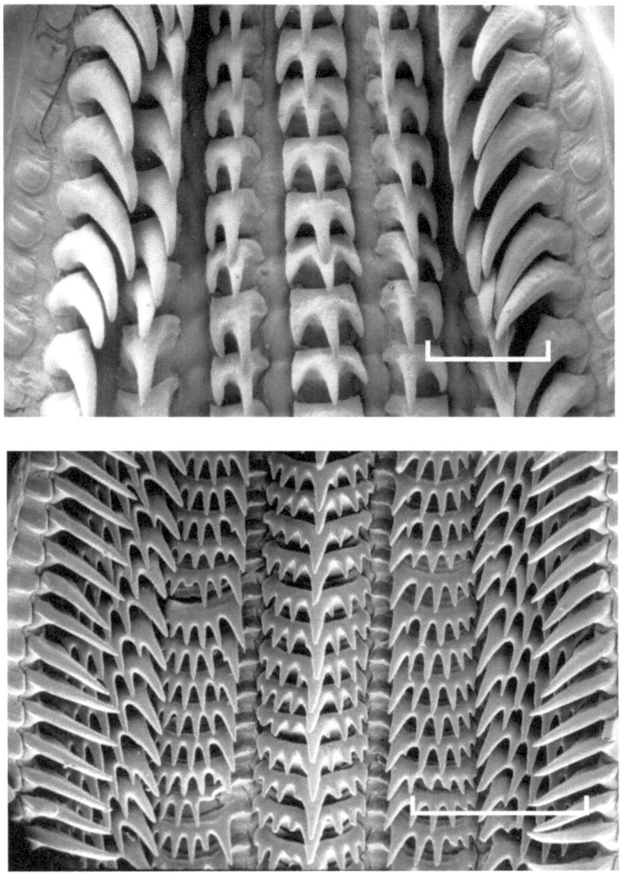

**Fig. 22.** *Architeuthis dux* Steenstrup, 1860 radula (top), radula of *Japetella diaphana* Hoyle, 1885 (bottom).

*Proximal.* Towards the body, or nearest or next to the point of origin or attachment (the opposite of *distal*).

*Pseudomorph.* An ejected mass of ink and mucus that appears like a body to a predator to fix its attention while the cephalopod escapes.

*Pygmy species.* A term used to define taxa, primarily octopods, that mature at small to tiny sizes (mantle lengths < 30 mm).

*Rachis.* The thickened central axis of the gladius that generally extends its entire length. The free rachis is the portion that does not support vanes (Fig. 5).

*Radula.* The ribbon-like band in the mouth containing up to seven transverse rows of teeth. Used to tear food and transport it to the oesophagus (Fig. 22).

*Renal papilla.* Kidney opening (Figs 3, 4).

*Renal sac.* Kidney.

*Rhynchoteuthion.* The paralarval stage of the Ommastrephidae characterised by the fusion of the tentacles into a trunk-like proboscis.

*Rows.* A transverse arrangement of structures (e.g. seven teeth per row on a radula). Counts are made at approximately the longitudinal midpoint of a structure unless stated otherwise.

*Salivary proboscis.* A muscular papilla equipped with teeth that lies below the radula in incirrate octopods. Used to drill holes in mollusc shells and crustacean exoskeltons in order to administer paralysing toxins.

*Secondary fold.* A fold on the eyelid, conspicuous in cuttlefishes.

*Semelparous.* Describing females that spawn once and then die.

*Seminal receptacle.* A cavity or invagination for deposition or storage of spermatangia (= *spermatheca*).

*Sepion.* See *cuttlebone*.

*Series.* A longitudinal arrangement of structures (e.g. two series of suckers on an arm). While this term is commonly used in the literature (often as 'biserial', 'tetraserial' etc.), the term 'rows' has been used in this book. While the number of transverse rows of suckers can be readily determined and in many cases is equivalent to the longitudinal series, in some cases the number of longitudinal series varies along the length of a structure. For example, there may be eight series of suckers halfway along the tentacular club of a cuttlefish, but this number may be different at the distal or proximal ends of the club, for example due to the structure tapering to a point at its distal tip.

*Side pockets.* Small membranous folds in the integument that form shallow pockets lateral to the foveola in the funnel groove (Fig. 16).

*Siphuncle.* A tube-like posterior extension of the body wall, coelom and covering sheath that penetrates the phragmocone chambers in cephalopods with a chambered shell (*Spirula*, nautiluses and cuttlefishes).

*Spadex.* Male reproductive structure used to transfer spermatophores to the female in nautiluses, formed by the fusion of four circumoral tentacles.

*Sperm cord.* The coiled string of sperm that lies within the sperm reservoir (Fig. 13) of the spermatophore.

*Spermatangium.* The encapsulated mass of sperm formed when spermatophores evert, often in the form of a round bulb.

*Spermatheca.* Seminal receptacle. In females, a bulbous structure in the buccal region or at the openings of the oviducts for deposition of spermatangia.

*Spermatophore groove.* The channel-like groove along the edge of the hectocotylised arm of mature male octopuses along which spermatophores are shunted to the tip for transfer to the female (Fig. 11).

*Spermatophore pad.* A fleshy patch of tissue, usually in the mantle cavity of some female squids, to which spermatangia adhere after mating.

*Spermatophores.* Encapsulated packets of sperm. Tubular structures manufactured by male cephalopods, capable of holding millions of spermatozoa, and transferred intact by a modified arm of the male and attached to the female, or inserted into the oviducts (as occurs in octopuses) where they remain until fertilisation occurs (Fig. 13).

*Spermatophore storage sac.* The organ of males where mature, functional spermatophores are stored. Opens to the exterior via the terminal organ. Sometimes called Needham's sac.

*Spine.* The sharp extension on the posterior tip of the gladius or cuttlebone, dorsal to the outer cone (Fig. 14).

*Statocyst.* The sense organ that detects gravity and sound. Contains statoliths.

*Statolith.* The calcareous concretion in the statocyst that detects acceleration and orientation.

*Stylet.* The pair of rod-like structures considered remnants of the molluscan shell in incirrate octopods. Found dorsolaterally in the mantle muscle.

*Subadult.* Describes an organism where defining features are present but the reproductive system is not mature and functional.

*Sub-Antarctic Front.* The northern limit of the Southern Ocean.

*Subequal.* Nearly equal.

*Subtropical Convergence.* The region characterised by a convergence of the current in the surface mixed layer where water is subducted into the permanent or oceanic thermocline.

*Subtropical Front.* A zone of enhanced meridional gradients of sea-surface temperature and salinity in the poleward part of the Subtropical Convergence.

*Sucker ring.* Chitinous, often serrated or toothed ring that encircles the opening of suckers of squids and cuttlefishes.

*Suckers.* Muscular, suction-cup structures on the arms and tentacles (rarely on the buccal membrane) of cephalopods. They are stalked and positioned on muscular rods that contract in squids and cuttlefishes, and are sessile and embedded without stalks on the oral surface of the arms of benthic octopuses. The number of sucker rows refers to counts made in oblique, transverse rows at around mid-length of the structure, such as the arm or club. The number approximately corresponds to the number of longitudinal series of suckers (see *series*).

*Swimming membrane.* The elongate, muscular vane along the aboral surface of the arms of cephalopods that streamlines and supports the arms during swimming (= *keel*).

*Tail.* The posterior extension of the mantle in some squids, frequently elongate. Fins or tapered terminations of fins may extend posteriorly along the tail.

*Tentacles.* The two elongate, stalked appendages used for prey capture. Their distal ends contain clubs with suckers (or hooks); the stalks are usually devoid of suckers. Tentacles in squids can only contract rather than retract into pockets as in cuttlefishes (Fig. 3D).

*Tentacular club.* The terminal portion of a tentacle armed with suckers (or suckers and/or hooks); used for capturing prey (Fig. 23).

*Tentacular organ.* In the Heteroteuthinae, a keel present at the base of the tentacular club as narrow elongate fold (except in *Sepiolina*) (Fig. 24).

*Tentacular sheath.* Formed when the keels of arms 4 are offset laterally and often enlarged to fully or partially conceal, protect or encase the adjacent tentacles.

*Terminal organ.* The muscular process of the male reproductive tract in octopuses which passes spermatophores into the base and spermatophore groove of the hectocotylised arm. (Alternative name for penis, as the true definition of a penis is 'organ of insertion'. Where the nature of spermatophores transfer is unknown it is difficult to determine whether the correct terminology should be 'terminal organ' or 'penis' as is the case in many squids. In this book, the term 'penis' is used if a hectocotylus is absent.)

*Terminal pad.* A small distinct pad or circle of small suckers at the tip of the club.

*Total length.* The length from the posterior tip of the mantle to tip of the longest arm in octopods or tentacles in squids and cuttlefishes. (Because of the various degrees of contraction of the highly extensible tentacles during capture or fixation/preservation, this is not generally a useful comparative measurement in squids and cuttlefishes.)

*Trabeculae.* Muscular rods that support the protective membranes on the arms and clubs of cephalopods.

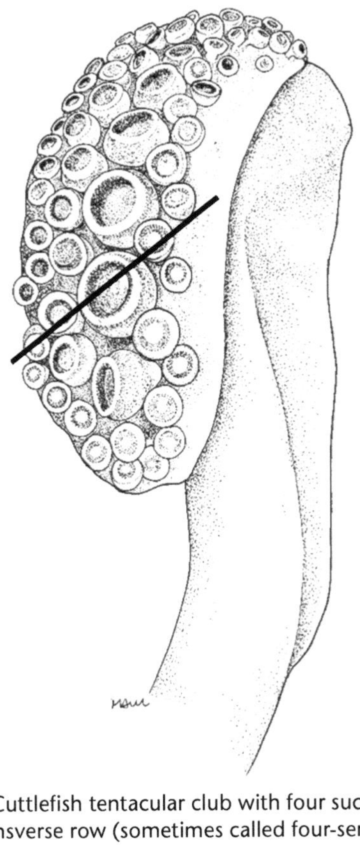

**Fig. 23.** Cuttlefish tentacular club with four suckers in each transverse row (sometimes called four-series).

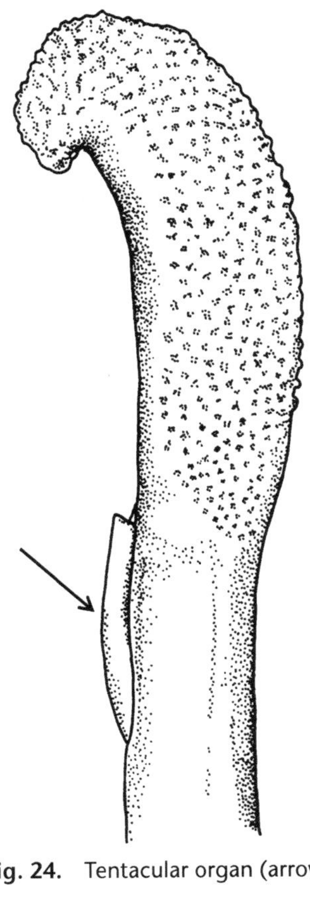

**Fig. 24.** Tentacular organ (arrow).

Occasionally the membranes are reduced and/or trabeculae are elongated, so they extend beyond the edge of the membrane, papilla-like.

*Tragus.* The inward projecting knob in the funnel-locking apparatus of some squids, found on the medial surface of the central depression (Fig. 8).

*Umbilical callus.* See *callus.*

*Umbilicus.* The central core of the chambered *Nautilus* shell, representing the juvenile shell with its initial coils.

*Vane.* Thin lateral expansion of the gladius that arises from the rachis (also see *rachis*) (Fig. 5).

*Ventral.* The lowermost or belly surface of a cephalopod; the surface on which the funnel is located; opposite the dorsal surface.

*Water pores/pouches.* Large cephalic openings at the base of the arms of some pelagic octopods, or eight small openings to the web pouches at the base of the arms in the octopod *Cistopus.*

*Web diameter index* (*WDI*). The depth of the deepest sector of the web measured from the mouth to the midpoint of the sector between the arms as a percentage of the longest arm.

*Web.* The membranous sheet that extends between the arms of many octopuses and some squids and cuttlefishes, giving an umbrella-like appearance when the arms are spread out.

*Web depth.* The distance from the mouth to the shallowest point of the web between adjacent arms.

*Web formula.* The comparative depth of each web sector measured from the mouth to the midpoint of the sector between the arms, expressed alphabetically in decreasing order (e.g. B/C/D > E > A).

# General biology

## Life history

### Reproduction

Cephalopods possess a variety of reproductive organs and show diverse sexual behaviours. When they mate, males transfer to a female sperm enclosed in complex structures called spermatophores which are accumulated in the spermatophoric storage sac (a storage organ inside the mantle cavity) before ejaculation through the terminal organ or penis. In cephalopods that inhabit shelf and epipelagic waters of the open ocean, the terminal organ is short and hidden inside the mantle. Males pick up the ejaculated spermatophores with a specially modified arm, the hectocotylus, to transfer them to the female. Females have special places for attachment on the body. These may be external, such as the back of the head, or the skin ring around the mouth, or internal (the oviducal gland openings near the gills).

With the exception of Ancistrocheiridae and Cranchiinae, most squids that spawn in deep water lack hectocotyli and have a longer penis (*see* Glossary) than shallow-water species. The Giant Squid, *Architeuthis dux* Steenstrup, 1860 is an exception: the species has a long penis and modified ventral arm tips, although it is not known whether these are for the transfer of spermatophores. The attachment positions of the spermatophores appear to be much more variable in deep-water squids and spermatophores can be located virtually everywhere outside the body and inside the mantle. Arkhipkin and Laptikhovsky (2010) observed the erection of the penis of *Onykia ingens* (Smith, 1881) to a length almost equal to that of the entire body (mantle, head and arms). This ability may explain why spermatophores are more variably placed in deep-sea squids than those that are more precisely placed using a hectocotylus. After attachment, the spermatophores dissolve the outer tissues of the female and penetrate the muscles where they remain until spawning. In some species (e.g. *Onychoteuthis banksii* (Leach, 1817)), cuts are made in the mantle tissue of the female using male arm hooks and spermatophores are placed on or near these.

Field and laboratory studies of some large *Sepia* species have shown that both sexes mate promiscuously, with males depositing spermatophores on the ventral portion of the female's buccal membrane in the paired seminal receptacles, but invariably some spermatophores are retained around the female's buccal area. All ovulated eggs are most likely to be fertilised around the female's buccal area after having been transferred into the female's arms through her funnel (Naud *et al.* 2005). Increased mating success via the removal of rival sperm by males has been documented. This has been well studied in *Sepia apama* (Hall and Hanlon 2002, Naud *et al.* 2004), in which males repeatedly flush a previously mated female's buccal area with jets of water from their funnel to dislodge retained spermatophores. In addition, Wada *et al.* (2010) have reported an increase in ejaculate volume as a method employed to increase the relative number of male sperm in *S. lycidas* Gray, 1849 from Japan.

The interplay between the duration of removal of rival sperm, variation in the volume of ejaculate and the number of ejaculations is complex. Wada *et al.*'s (2009) study indicates that these factors are affected by recognition of the female's mating history by subsequent rival males, male size (and therefore competitive ability) and spermatophore depletion – all of which influence strategic investment allocation and constraints.

Cephalopods are often described as having a live-fast-die-young lifestyle, with most species living for about a year, and a repeated spawning in a short period at the end of the individual's life. This knowledge of life history strategies is, however, biased toward commercially exploited epipelagic and inshore species, especially from the family of flying squids (Ommastrephidae). While this strategy may apply to many inshore species, scientific evidence is accumulating to suggest that members of most oceanic squid species do not, in fact, have exceptionally high growth rates, a short lifespan of 1 year or less and repeated spawning at the end of the individual's life (Hoving 2012). The same is true for species that inhabit colder and/or deep waters with consequent slower growth rates (Collins and Rodhouse 2006, Hoving *et al.* 2014).

An astonishing study that superbly illustrates our lack of knowledge and mistaken gross generalisations pertaining to cephalopod lifespan and reproductive strategies was recently illustrated by Robison *et al.* (2014). While extrapolations from data on shallow-water octopus species and laboratory studies indicated lower temperatures would prolong embryonic development and a longer brooding period, the known boundaries of physiological adaptations to life in the deep sea were dramatically extended through direct measurement of brooding in *Graneledone boreopacfica* Nesis, 1982. A single female was observed using a remotely operated vehicle (ROV) in the Monterey Submarine Canyon to brood eggs, seemingly without feeding, for a total of 53 months, or ~4.4 years. This is the longest known egg brooding period of any animal – a consequence of low temperature and the selective advantage of producing highly developed hatchlings. This example from the deep sea confers an extension of adult life that greatly exceeds most projections of cephalopod longevity (Robison, Seibel and Drazen 2014). If, as in other octopodoids, brooding takes up about one-quarter of the lifespan, then *G. boreopacifica* and possibly other deep-sea species may be among the longest lived of all cephalopods.

Most life history studies have, however, focused on inshore taxa that do generally have short life histories (with some small warm-water species completing their lifespans in a few months), but even here lifespans may differ between the sexes. Cross-generational mating and subsequent spawning may occur as a function of the disparity in longevity between males and females. This has been noted in many species. Following a study of the ecology of decapod reproduction (Teuthida and Sepiida) Laptikhovskii (2002) concluded that the duration of physiological maturation appears to depend neither on habitat (shelf or open ocean) nor taxonomic position of the species, but maturation is determined mainly by egg size and environmental temperature. Most species investigated were characterised by year-round spawning with peaks in spring to coincide with zooplankton outbreaks and autumn to coincide with an increase in body mass. This extremely expressed r-strategy (i.e. parental investment spread over many offspring) in reproduction is characteristic of surface ocean waters in tropical regions. The features of the K-strategy (i.e. parental focus on few offspring) become more pronounced when passing polewards, inshore, and into the ocean depths.

Jackson and O'Dor (2001) attribute fast growth rates and short lifespans to (1) a combination of efficient digestion with a protein-based metabolism; (2) the ability to sustain continued growth by a combination of both an increase in muscle fibre size and continual recruitment of new muscle fibres; (3) efficient use of oxygen; and (4) low levels of anti-oxidative defence. Short lifespans mean that their spawning strategy must ensure that these single-generation populations are maintained each year exclusively through young of the current year recruits. Recruitment failure in any year will result in a population crash as the current generation dies following completion of their spawning activities.

How this is achieved differs among the main groups, with spawning strategies (synchronous or repeated spawning) and the depth at which reproduction takes place influencing their life history strategies. Barratt *et al.* (2007), for example, have shown that among deep-sea incirrate octopuses, *Bathypolypus sponsalis* (Fischer and Fischer, 1892) has a synchronous ovulation pattern and therefore a simultaneous terminal spawning strategy, while greater variation in egg lengths suggest that spawning may occur over an extended period in *Benthoctopus levis* (Hoyle, 1885) and *B. normani* (Massy, 1907).

Rocha *et al.* (2001) reviewed cephalopod reproductive strategies. They defined five flexible reproductive strategies:

(1) simultaneous terminal spawning: spawning once, with synchronous ovulation, monocyclic spawning and the absence of growth between egg batches (semelparity) (most cephalopods are thought to be semelparous; examples include *Octopus cyanea*, *Todarodes pacificus* (Steenstrup, 1880));
(2) spawning more than once (iteroparity), with group-synchronous ovulation including:
  (a) polycyclic spawning with egg laying occurring in separate batches during different spawning seasons, and growth, including ovary regeneration, occurring between production of egg batches and spawning seasons (e.g. *Nautilus* spp.);
  (b) multiple monocyclic spawning, with egg laying in separate batches during the same spawning and growth between egg batches, but no ovary regeneration (e.g. *Sthenoteuthis oualaniensis*, *Ommastrephes bartramii* (Lesueur, 1821));
  (c) intermittent terminal monocyclic spawning, with egg laying in separate batches and no growth between egg batches (e.g. *Sepioteuthis lessoniana* Lesson, 1831, *Todaropsis eblanae* (Ball, 1841)); and
  (d) continuous monocyclic spawning, with asynchronous ovulation, with egg laying occurring over an extended and continuous spawning period and growth between egg batches (e.g. *Idiosepius pygmaeus* Steenstrup, 1881; *Argonauta hians* [Lightfoot], 1786). Laptikhovsky *et al.* (2013) have suggested that *Kondakovia longimana* Filippova, 1972 may possibly be iteroparous (i.e. spawn yearly, survive, feed, grow and regenerate their gonads for a further reproductive event the following year) as in category 2(a) above. Laptikhovsky's observations may be the first recorded case among coleoid cephalopods.

It appears that, with the exception of octopods, multiple spawning may be quite widespread among cephalopods, with several squid and sepiolid species producing large numbers of eggs throughout the adult lifetime. Species that spawn more than once generally produce large yolky eggs requiring a high level of energy input. As discussed by Kuipers *et al.* (2008), multiple spawning confers a number of advantages: (1) it enables small cephalopods to physically produce enough large eggs to manage the lifetime production of oocytes; (2) as reproduction is fuelled directly by ingested food and not stored reserves it enables the production of more eggs with large yolk reserves; and (3) producing large numbers of eggs over an extended period spreads the risk of mortality over time.

The strategy adopted by females has the capacity to affect how females allocate energy to egg production, patterns of lifetime fecundity, and survival. For example, females that 'trickle' eggs slowly over an extended period might risk having lower lifetime fecundity if mortality occurs early in their reproductive life. Producing eggs in batches requires more energy over short periods, potentially producing more strain on the individual and perhaps increasing its susceptibility to predation. Oocyte dynamics in the ovaries of mature females differ depending on which strategy is used and examining the ovaries of mature females can reveal the strategy adopted (Kuipers *et al.* 2008).

Eggs are either released into the plankton during spawning or attached to particular substrates. Many species of decapod cephalopods form spawning aggregations. Boal *et al.* (2010) have suggested that bioactive factors in the eggs of cuttlefishes may facilitate the formation of spawning groups. They postulate that the eggs of cuttlefishes may be a source of pheromones that coordinate reproductive behaviour, as has been reported for some squids.

## Growth

In cephalopods, growth rates and the age or size at maturation are highly variable. This is attributable to their typically indeterminate growth patterns; animals exhibit continuous growth and are capable of responding to a variety of biotic and abiotic factors (Tracey *et al.* 2003), such as temperature, food quality and quantity, age, size, gender, stage of maturity and level of activity, but the relationships between these factors are complex, particularly during early life history stages (André *et al.* 2008). For example, temperature, particularly in the early growth stages, has a dramatic effect on the growth of cephalopods, and hence age and size at maturity (Forsythe 1993). Temperature affects feeding and metabolic and growth rates, which have implications for survivorship, adult size and fecundity at the individual level, and subsequently influence population structure. Growth rates are directly influenced by temperature, with faster growth occurring at higher temperatures. This has been demonstrated in *Amphioctopus fangsiao* (d'Orbigny, 1839–1841) (as *O. ocellatus* in André *et al.* 2008) and *O. pallidus* Hoyle, 1885 (André *et al.* 2009): increases in temperature of as little as 1°C in their model implied up to 15.5% change in the threshold body mass for *A. fangsiao* and considerably altered the growth trajectory of individuals. The influence of temperature on threshold body mass is mostly driven by metabolism. This may substantially alter the population dynamics of most cephalopod species in the coming decades (André *et al.* 2009). However, the possible responses may be complex and difficult to predict. For example, increasing seawater temperatures, combined with a reduction in pH due to increasing levels of $CO_2$ in the oceans, will decrease the ability of cephalopods to bind oxygen for transport (Seibel and Fabry 2003), thereby altering metabolic rates at the individual level, thus influencing body mass at the time of transition out of exponential growth and affecting size at maturity.

Life history parameters, such as growth rates, age and size at maturity and final body, size show a high degree of temporal and spatial variability. The plasticity in life history strategies of cephalopods is an integral mechanism in maintaining survivorship of populations by allowing continuous response to varying biotic and abiotic conditions.

In species that have a broad latitudinal distribution, different life cycles from north to south have been demonstrated. The general trend appears to be that in tropical waters there is little seasonal variation and subannual life cycles generally occur (Jackson 1990, Jackson and Choat 1992). In temperate waters, squids grow larger, older, reproduce later, and are individually more fecund (O'Dor and Coelho 1993). This pattern has been demonstrated in a number of Australian species (Jackson and Choat 1992). This trend seems also to be true at a finer scale within these broad regions (i.e. tropical *v.* temperate). For example, Green (2011) reported that both male and female *Nototodarus gouldi* (McCoy, 1888) grew faster and were larger in cooler waters off Victoria compared with warmer waters off the Great Australian Bight.

Counting statolith increments is the standard method for ageing squids, but the results must be treated with caution because a number of factors may affect the periodicity of increment deposition in the statolith. Daily increment formation has been validated in one *Idiosepius* species (*I. pygmaeus*; Jackson 1989). In another study, it was suggested that stress may have affected increment formation in some, but not all, specimens in a laboratory sample of *I. paradoxus* (Ortmann, 1888) (Sato *et al.* 2008); daily increment formation occurred in only six of 16 specimens, with the rest forming less than one increment per day. In this study, field-sampled specimens from a single cohort did not exhibit the expected increase in statolith increments over time, suggesting that, in this case, ring number underestimated true age. Studies that involve field positioning and telemetry and recapture of tagged animals of known age could help to calibrate statolith information as could the analyses of statoliths from laboratory animals with known food consumption and growth. However, such validation is not always possible. Despite this, a number of researchers continue to base studies of population structure on the assumption of statolith increments increasing by one each day. For example, despite the fact that no validation studies on ommastrephids have been done for over a decade and their 'attempts at validation of *N. gouldi* were disappointing' these researchers conclude that 'the assumption of one increment per day is not unreasonable' (Jackson *et al.* 2005: 981).

Hunt *et al.* (2011) similarly concluded that the less-than-daily increment formation that they observed may be due to the periodicity of increment formation changing through the lifespan of *Sepioteuthis australis* Quoy & Gaimard, 1832, deleterious effects associated with rearing squids in captivity, or misinterpretation of statolith microstructure.

Stylet increment analysis is currently the key method used to age octopuses; however, it is not currently reliable for all species. For example, Doubleday *et al.* (2011) in a study of *O. maorum* (Hutton, 1880) from Australia and New Zealand found that low increment counts produced non-feasibly high growth rates based on the assumption of daily growth increments. This may have been due either to non-daily increment formation or the first increment counted perhaps not representing the first post-hatch day because it is not known at what stage in the early life history that stylets develop. While determining the periodicity of increment formation is a vital step, it is insufficient alone to obtain an accurate estimate of age. It is also necessary to determine the age at first increment formation. At this stage, this can only be validated through laboratory studies using known-age hatchlings, and this isn't easy for octopuses with pelagic, rather than benthic, young.

Clearly, caution needs to be applied to any age data derived from stylet increment analysis without validation of absolute age.

Doubleday and Semmens (2011) found in *O. pallidus* that the age pigment lipofuscin, visible in brain tissue using standard histological techniques, is highly exponentially correlated with age in known-age laboratory-reared animals and the concentration of lipofuscin is not related to sex, temperature or body weight in same-age individuals. In four of five wild animals, the predicted age, based on the relationship between lipofuscin and age was close to that determined using stylet increment analysis. These results suggest that lipofuscin has great potential as an alternative ageing method for octopuses.

### Diet

Most cephalopods are generalist predators and have a macerative feeding style, selectively ingest prey body parts, and have rapid digestion rates. Cuttlefishes are active carnivores that feed on living prey and are exclusively visual hunters of crustaceans and fishes. Cannibalism is common in most species in which diet has been studied (Ibáñez and Keyl 2010).

Most diet studies have been limited to the use of morphological methods for prey identification. These analyses rely on the preservation of diagnostic hard parts among gut or faecal contents. Recently, DNA-based analyses have been used to assess the diets of marine species. A study of the generalist predator, *N. gouldi*, using a combined methodological approach that included both morphological and DNA-based techniques, highlighted the limitations associated with restricting diet studies to a single method to identify prey species (Braley *et al.* 2010). In this study, morphological techniques identified 50% of the total number of identifiable prey taxa, while DNA-based techniques found 84%. In addition, the taxonomic resolution of prey species using molecular methods was increased. The underestimation of prey species using only morphological methods indicates that crucial links may be missing in current trophic models. This could have major implications for our current understanding of ecosystem structure and function and definitely warrants further research.

## Physiology and general behaviour

The tissues of most cephalopods are denser than seawater, meaning they need to adopt some sort of strategy to stay afloat, unless they live on the bottom (such as octopuses and many sepiolids). *Nautilus* is able to regulate its buoyancy by osmotically adjusting the levels of cameral fluid in its chambered shell. By this method it is able to traverse considerable depths. Cuttlefishes are neutrally buoyant, or nearly so, achieving this by osmotically adjusting the levels of cameral fluid in their hollow calcium carbonate cuttlebones. Because they can hover with very little effort, the energetic cost of travelling is less than that of squids. Remaining motionless is part of their camouflage, so they have little energy expenditure at rest and their tough skin lets them rest on the seafloor. Because they don't need to jet to hover, they make greater use of their fins for locomotion than do squids.

Unlike most other swimming cephalopods, cuttlefishes have a complex approach to locomotion, perhaps because they live in more complex benthic environments. They are masters of camouflage, as a means to both avoid predators and sneak up on prey. For this to work effectively they need to be able to remain still and this is achieved by controlling their buoyancy and through fine regulation of fin beating. According to Aitken and O'Dor (2004) they may actually be functionally better adapted to holding still rather than swimming. Although they are capable of rapid escape swimming, their lack of ability for sustained swimming is consistent with their sedentary lifestyle. Aitken and O'Dor (2004) also suggested (although based on respirometry measurements of three cuttlefishes only) that their swimming behaviour is variable and not predictable, perhaps due to their occupation of erratic environments. This has also been reported for some squid species (Finke *et al.* 1996).

Aitken *et al.* (2005) studied the behaviour and energetics of a population of *S. apama* at House Bay Reef, Boston Bay, South Australia following tagging with acoustic transmitters. This study showed that their home range (total area occupied over the monitoring period) was between 5300 $m^2$ and 23 700 $m^2$, with maximum ranges (distances between extreme points of the area covered by the cuttlefishes) for individual animals varying between 90 m and 550 m. They appear to be diurnal with greater distances travelled during the day than at night. They spend 95% of the day resting, foraging only 3.7% of the day and 2.1% at night. This strategy, which is akin to that of octopuses (combining predator avoidance, efficient foraging and a largely inactive lifestyle), allows energy to be effectively channelled into growth.

In contrast, many squids, such as *Loligo* and *Illex*, are constantly active (O'Dor and Webber 1991).

O'Dor *et al.* (2002) examined how currents affect squids (*Sepioteuthis lessoniana* and *Sepioteuthis australis*) and cuttlefishes (*Sepia apama*) behaviour, energetics and distribution. Using radio-acoustic positioning telemetry

(RAPT), they found that distinctive activity patterns indicated that tidal currents were key environmental influences. Continuous position and mantle pressure data correlated with visual and video documented behaviours. Cuttlefishes were diurnal, relatively inactive and spent their time within benthic boundary layers, hovering near or under structures. Squids, in contrast, were continually active, seeking particular current regimes to conserve energy using slope soaring tactics as previously seen in *Loligo forbesi* Steenstrup, 1856. (Slope soaring involves travelling on rising currents created by reefs and their boundary layers.) These behaviours indicate an energetic trade-off between neutral and negative buoyancy for access to prey in currents.

As top-level predators with a complex central nervous system, fascinating cognitive abilities (including conditional discrimination, a form of complex learning (Hvorecny *et al.* 2007)) and a suite of unique sensory organs, cephalopods are emerging models in research concerned with mechanisms and agonists governing the development of complex brains (Wollesen *et al.* 2010). In recent years, effort has been directed towards understanding the ontogeny of the cephalopod central nervous systems in classical neuroanatomical as well as developmental genetic studies and the potential for future research in this area is vast (see Wollesen *et al.* 2010 and references therein).

## Chemoreception

Food and other odours can be detected by cephalopods using chemosensory cells around their mouths and suckers. Waterborne odours can be detected using receptors in their olfactory organs. Chemical communication has also been shown to occur among conspecifics and has been demonstrated in *Loligo*, *Nautilus*, *Sepia* and, more recently, *Octopus* (Walderon *et al.* 2011) (and references therein). The detection of the odours of conspecifics may be utilised to identify the sex and reproductive status of other individuals, assist with the formation of mass breeding aggregations (thus facilitating reproduction), and used to avoid confrontation.

## Cephalopod ink

Many cephalopods release ink when attacked by predators. This ink acts as an anti-predator visual stimuli functioning as a 'smoke screen' or visual decoy. Work by Derby *et al.* (2007) suggests that the ink acts not only as a physical decoy but also as a chemical defence. In the six cephalopods studied, the ink was found to contain free amino acids that specifically match specific receptor systems in their major predators: crustaceans and fishes. Thus, the ink acts to mimic the food of predators to stimulate the predators' sensory pathways involved in feeding, and is an example of phagomimicry. While a predator is occupied investigating the ink secretion in this way, the cephalopod buys time to escape. Sensory disruption, in which the free amino acids are in concentrations that strongly stimulate the sensory pathways of predators (although not necessarily mimicking the composition of the food), also plays a role in chemical defence in cephalopods (Derby *et al.* 2007).

The role of ink in defence is based primarily on observational studies; experimental studies are needed to identify molecules and mechanisms underlying inking behaviour. In addition, little is known about the chemical constituents produced by the funnel organ that are mixed with ink to modify its consistency. Derby (2014) summarises what is known currently about cephalopod ink and lists a range of potential human benefits that may be offered by cephalopod ink, such as identifying antimicrobials and identifying new drugs for a variety of end uses (antimicrobials, anti-cancer and antioxidants).

## Skin colouration and vision

Unique in the animal kingdom is the ability of cephalopods to display an extensive repertoire of body patterns for camouflage and signalling. These body patterns are mediated by the dual action of thousands of chromatophores, or small pigmented organs, and by light-reflecting cells (iridophores and leucophores). The way both these cells function has been detailed by Mäthger and Hanlon (2007) and this paper is a useful starting point to the vast literature pertaining to this fascinating aspect of cephalopod biology.

Despite our perception of these amazing colour displays, most cephalopods are thought to be colourblind. However, the structure of their photoreceptors suggests they are able to detect polarised light. This has also been demonstrated behaviourally in *Sepioloidea lineolata* (Quoy & Gaimard, 1832) (Talbot and Marshall 2010) and *Sepia plangon* Gray, 1849 (Temple *et al.* 2012). In the marine environment this sensitivity could be used to enhance the levels of contrast in the visual field, perhaps by decreasing the deleterious effects of scattered light that falls between an animal and an object, and between that object and its surroundings. It might also reduce the confounding effects of flicker in shallow water as well as making objects appear more conspicuous against a background. Cartron *et al.* (2013) have demonstrated that the use of polarisation vision in cuttlefishes enhances their ability to detect objects in turbid water. This undoubtedly suits their nectobenthic habit in which they probably often experience low visibility conditions. Cuttlefishes also have a distinctive W-shaped pupil in bright light. Recent studies have shown that the W-shaped pupil is effective in balancing a vertically uneven light field, thus improving image contrast (Mäthger *et al.* 2013).

In addition, the ability to discriminate angular differences in the electric field vector (e-vector) significantly enhances visual information, providing fine-scale contrast information. This can be used to accurately pattern-match and detect background information. This has been demonstrated to be acutely developed in *Sepia plangon* and may enable this cuttlefish (and possibly many other species) to detect predators that are using an inferior or lower-resolution forms of polarisation pattern matching (Temple *et al.* 2012).

Many visual predators, such as fishes, have keen colour perception, so camouflage patterns should provide some degree of colour matching in addition to other visual factors such as pattern, contrast and texture. Chiao *et al.* (2011) have managed to quantify camouflage effectiveness in the eyes of potential predators using hyperspectral imaging (HSI), which records full-spectrum light data to simultaneously visualise colour match and pattern match in the spectral and spatial domains, respectively. These researchers have shown through HIS imaging and spectral analysis that most reflectance spectra of individual cuttlefishes and substrates were similar, rendering colour matching possible. Thus, cuttlefishes can produce colour-coordinated camouflage on natural substrates despite lacking colour vision.

Some cephalopods use the polarised patterns generated on parts of their bodies (for example, the light-reflecting iridophores) for intraspecific communication. This could be used as a covert communication channel invisible to animals lacking polarisation vision. These polarised signals can pass unchanged through the overlying chromatophores, so the body pattern could remain unchanged (such as a camouflage pattern) while communicating with conspecifics (Mäthger and Hanlon 2006).

Not surprisingly, this aspect of cephalopod biology has become the subject of US military research with millions of dollars currently being poured into studies on cephalopod camouflage (Birch 2010).

### Bioluminescence

In well-illuminated shallow water environments, visual signals play an important role. In poorly lit or dark environments, many cephalopods use biologically generated light (bioluminescence) for signalling and camouflage. It is produced in special light organs, or photophores. The distribution and arrangement of photophores may differ between species, between sexes and during growth, and is an important diagnostic trait in cephalopods. Photophores are common in oceanic oegopsid squids and are found also in *Spirula*, in many sepiolids, *Loligo*, *Vampyroteuthis*, in the octopods *Eledonella* and *Japetella*, and in some cirrates, such as *Stauroteuthis syrtensis* (Hanlon and Messenger 1996). There are usually more photophores ventrally than dorsally and they are especially common below the eye, on the mantle and arms (often at the tips) and sometimes on the tentacles, and they may be found on the posterior tip of the body. Their primary functions are ventral counter-illumination and signalling.

Concealment from predators located above or at the side of an animal is achieved by making the upper surface darker and having reflective sides. Concealment from a predator located below is more difficult and is achieved in many cephalopods by emitting light from their photophores. This 'counter-illumination' is used to conceal an animal from below when the emitted light matches that of light coming from above in intensity and wavelength, thus making the animal effectively invisible. Many photophores are complex and contain photogenic tissue and filters that enable them to produce a weak light whose maximum intensity matches the peak value for downwelling light in the open ocean. Photosensitive vesicles detect and monitor the light received and produced.

Stronger light production by a range of photophore arrays is used for communication. These may be used to signal species identity and identify sex, with sexual dimorphism in the photophores seen in some species. Herring (2007) lists photophore sexual dimorphism as occurring in Enoploteuthidae, Lycoteuthidae, Chtenopterygidae, Cranchiidae, and among the Octopodoidea in the Bolitaeninae. Photophores may differ between sexes in their number, position and presence. In some cephalopods, particular photophores may be present only in mature females and may degenerate after spawning. While none of these female-specific photophores have been seen illuminated, the implication is that they are involved in sexual communication. However, whether sexual dimorphism in photophores invariably indicates a sexual signal is open to debate. It is also possible that different sexes are able to select particular patterns of illumination from available photophores and may be able to produce light of different wavelengths. As squids have visual systems that are capable of discriminating different wavelengths, a spectrally selective sexual signal is a possibility. As Herring (2007) points out, a much clearer case for a sexual signal is provided by the unique photophores developed at maturity and resorbed after spawning in some female cranchiids and midwater octopodoids. Sexual signals are examples of sexual selection, which could take the form of female selection or interactions with competing males. Where population densities are low in the open ocean, female selection, is far more likely than male competition.

In addition, light may be used to influence other animals. The bright arm tip 'flashes' seen in *Taningia danae* Joubin, 1931 may be used to elicit a startle response to distract or confuse a predator. Perhaps luminescent clouds as seen in *Heteroteuthis* may have a similar function. Some photophores may possibly function as 'lures'. These ideas are yet to be supported by behavioural observations. Some cephalopods have many distinct types of photophores but their functions are yet to be determined. In summary, this area is an exciting one for future research.

## Sound

Whether and how cephalopods detect sound has, in the past, been a subject of debate. While anecdotal evidence has indicated the use of an auditory sense, the fact that they can detect sound has recently been demonstrated experimentally. Using near-field acoustic and shaker-generated acceleration stimuli, Mooney *et al.* (2010) showed in *Doryteuthis pealii* that squids detect sound similarly to most fishes, with the statocyst acting as an accelerometer through which squids detect the particle motion component of a sound field (rather than pressure). These researchers determined that the modality and frequency range audible to squids suggests that they probably detect acoustic particle motion stimuli from both predators and prey, as well as low frequency environmental sound signatures that may aid navigation. *Nautilus pompilius* Linnaeus, 1758, also responds to underwater vibrations (Soucier and Basil 2008). As they have limited vision and inhabit deep water in dark conditions during the day, vibrations may be another suite of stimuli to which they can respond.

Low-frequency swimming-generated sound detection is probably used for predator avoidance. Navigation is another potential use for low-frequency hearing, with sound generated by breaking waves or reef fishes providing cues for orientation. Detection of prey may also be one of the auditory functions.

Anthropogenic noise from shipping, airguns and other offshore activities also generates low-frequency noise, and sensitivity to these sounds may mask biologically relevant stimuli or induce acoustic trauma in cephalopods. Evidence of acoustic trauma has been reported by André *et al.* (2011). Research shows that low-frequency sound causes severe lesions in the in the sensory organs of squids, octopuses and cuttlefishes. The death and beaching of Giant Squid, *Architeuthis dux*, off Asturias, Spain in 2001 and 2003 has been attributed to excessive sound exposure following the use of airguns by offshore vessels. Damage to cephalopod statocysts when animals are exposed to low intensity, low frequency sound ranging between 50 and 400 Hertz (Hz) (similar to that experienced by the squids off Asturias) caused immediate statocyst hair cell damage followed by lesion formation. Because the statocyst is responsible for balance and spatial orientation, damage to the structure would undoubtedly affect the cephalopods' ability to hunt, evade predators and reproduce.

It is highly likely that the increase in offshore drilling around Australia (and worldwide) and the anticipated massive increase in cargo ship transportation (due largely to the export of environmentally toxic coal) and other large-scale offshore activities will result in increased noise pollution and therefore acoustic trauma in a wide range of marine organisms, including cephalopods.

## Movement and migration

Cephalopod movement and migration may occur during different life history phases in response to such factors as food availability, predator avoidance, reproductive status (as occurs during breeding aggregations) and life history stage. It can range from the passive drifting of egg masses and paralarvae to daily vertical migration in many species, and large-scale migrations of adults from feeding to spawning grounds. Sea temperatures and oceanographic processes influence movement, particularly in oceanic squids.

An example off south-eastern Australia is the formation of warm-core eddies off the East Australian Current. This is an ocean current that moves warm water from the tropical Coral Sea, where it splits from the South Equatorial Current, down the east coast of Australia. It is the largest ocean current close to the shores of Australia. Eddies form when a meander of the East Australian Current breaks off into a closed ring structure. These can last for up to 18 months, be resorbed by the East Australian Current, coalesce or divide. These eddies differ in nutrient cycling, phytoplankton productivity and fish species composition from surrounding waters. Brandt (1983) showed a clear association in the overall composition and relative abundances of cephalopod species with water temperature and location with respect to one of these eddies. Among the most abundant species caught during summer night-trawls, *Pterygioteuthis giardi* Fischer, 1895 and *Brachioteuthis 'riisei'* (Steenstrup, 1882) were most numerous inside the eddy, and *Abraliopsis gilchristi* (Robson, 1924) and *Pterygioteuthis gemmata* Chun, 1908 were larger in size outside than inside the eddy. *Abraliopsis gilchristi* was concentrated in warmer water and *Pyroteuthis margaritifera* (Rüppell, 1844) was more common at 250 m than at shallower depths. These results showed that warm-core eddies affect the distributions of oceanic squids in the Tasman Sea and water temperature is an important factor defining boundaries of oceanic squid communities, with depths of particular isotherms closely linked to the dynamics of an eddy, and eddy location itself is defined by temperature.

Studies of movement and migration are areas worthy of further investigation, with significant fisheries implications. Semmens *et al.* (2007) have summarised a range of approaches that can be employed in movement studies, and have outlined the efficacy of particular methods. These include tag–recapture studies and the use of electronic tags to study both small and large-scale movements. Chemical tagging of paralarvae through maternal transfer is suggested as possibly a viable technique and a way to study this poorly understood life history stage, as large numbers of individuals could be tagged at once. Indirect methods, such as chemical analyses of elemental and isotopic signatures of cephalopod hard parts, could be more widely utilised. With more detailed information on parasite–host specificity and parasite geographic distribution, geographic differences in parasite load could be used to indirectly provide movement information. In addition, molecular genetics offers a powerful approach to estimating migration rates among populations. Tracking fishing fleets is another indirect method of monitoring the movement of squid populations. This can be done via the analysis of satellite data showing the location of fishing lights.

To date, such studies indicate that migration in squids is extensive and modulated by major oceanographic features, while cuttlefish and octopus migration (in adults at least) is more locally restricted.

## Ecology

Squids (and other cephalopods) are likely to play an increasingly important role in marine ecosystems (Jackson *et al.* 2007). It appears that they may have a greater abundance now compared to past decades as a result of the removal by overfishing of many of their teleost predators (Myers and Worm 2003).

As has been discussed above, squid populations can undergo dramatic fluctuations in abundance due to changes in oceanographic conditions and this can have consequences for squid predators and squid prey within ecosystems (O'Dor and Dawe 1998, Dawe *et al.* 2000, Jackson and Domeier 2003). For example, in the Southern Ocean food chain, squids are an important link between lower trophic levels and higher vertebrate predators (Rodhouse and White 1995).

Many aspects of cephalopod ecology and their role in marine ecosystems still need considerable research for most species. This is particularly true for offshore and oceanic species. The gathering of this baseline data is imperative for ecosystem modelling and management.

The role of young stages and paralarvae in marine trophodynamics is still poorly understood. While they are difficult to study due to sampling challenges, this is an area much in need of additional research. As Jackson *et al.* (2008) point out, their high metabolism, feeding rates and rapid growth, together with their association with specific oceanographic features, allow them to rapidly reflect changes in the marine environment. Recruitment success of younger stages can provide data on oceanographic conditions in real time and help to predict adult recruitment. Only a few studies along these lines to date have been conducted in Australian waters.

Jackson *et al.* (2008) sampled cephalopod paralarvae and juveniles using light traps over two consecutive summers (1997/1998 and 1998/1999) in the southern North West Shelf and on Ningaloo Reef, Western Australia. Two trapping methods were deployed, one at the surface and one 3 m off the bottom, or at 75 m in deep water. Loliginid squids (*Uroteuthis* sp.) and octopodoids dominated in the catches and *Sepioteuthis* and *Sthenoteuthis* were well represented, with both *Uroteuthis* and *Sepioteuthis* predominantly collected inshore. If representative of abundance patterns these would seem to be the most abundant cephalopods in the North West Shelf region. A similar pattern and taxon composition was found on the tropical Australian east coast in Great Barrier Reef waters (Moltschaniwskyj and Doherty 1994, 1995). Little difference was found in taxon distribution and abundance between the summers despite the first summer being characterised by an El Niño event (with cooler water conditions and enhanced productivity due to upwelling) and the second summer being characterised by La Niña conditions that caused well-developed water column stratification, warmer temperatures and less productivity. *Sthenoteuthis oualaniensis* was more abundant in the second summer, perhaps related to the greater intrusion of warmer oceanic water on to the shelf that may have been preferred by this predominantly tropical oceanic species.

## Fisheries

Part of the impetus for writing this book has come from the inadequate identification material currently available to Australian fisheries experts to enable them to correctly identify catches, and therefore manage stocks. In many guides cephalopods are identified only as 'squid', 'cuttlefish' and 'octopus'. In some cases, species are very similar morphologically, making them difficult to identify, even by experts (an example are the commercially important *Uroteuthis* spp.).

Rapid growth in cephalopod fisheries has occurred worldwide in recent years. Their economic importance is increasing as evidenced by the rapid rise in their global landings over recent decades. This has significant implications as cephalopods are a key component of marine food webs, providing sustenance for a myriad of marine species. Hunsicker *et al.* (2010) have attempted to deduce the first measure of the ecosystem services that cephalopods contribute to fisheries as both a commodity and an ecological support service. They determined that commodity and support services provided by cephalopods contributed 55% of fishery landings (tonnes) and 70% of landed values (US$). This study focused on the contribution of cephalopods to commercial fisheries and did not include the value of cephalopods as a forage item for conservation species, such as marine mammals and seabirds. Consideration of these species could greatly increase the value of cephalopods to marine ecosystems. The contribution of cephalopods as a commodity was greatest in coastal ecosystems, whereas their contribution as a support service was highest in open-ocean systems. Both of these services provided by cephalopods have increased markedly in recent years, indicating a rising demand for cephalopods. As the current demands have no historical precedent, ecosystems in which cephalopods are highly exploited need to be carefully evaluated to prevent unsustainable development of marine fisheries within them. When predators are forced to compete with fisheries for their forage base, management plans for one target stock can affect the productivity of other target or non-target species. Trade-offs need to be made between the value of cephalopods as a commodity and their value as an ecological support service if simultaneous demands are to be sustained.

Leporati *et al.* (2009) raise a number of important points regarding octopus fisheries that can be extrapolated to other groups. Because of difficulties, or lack of attention paid to species identification, reliance on catch per unit effort has been widely used to assess the stock status of many cephalopod fisheries. In the case of octopuses, the lack of biological information has led to holobenthic and merobenthic octopus fisheries treated as biological equivalents. The resilience of octopus fisheries as a whole to increased fishing pressure is generally unknown because most research has focused on a small number of species that constitute most of the world's catch. It happens that these species are merobenthic, producing hundreds to thousands of eggs and have planktonic larvae. Other commercially exploited octopuses are given low research priority and the species are simply listed as 'octopus' in fisheries statistics. A large proportion of these are holobenthic, producing eggs batches in the hundreds and have benthic hatchlings. Recruitment dynamics for holobenthic species appear to be much more localised than for merobenthic species and are more likely to be dependent on the local population. If a localised area is subjected to continuous fishing pressure, the impact on recruitment could be much greater for holobenthic than merobenthic octopuses.

Other confounding factors that need to be considered in the development of management strategies for increasingly targeted cephalopod stocks are the climate and environmental conditions that govern their populations. Climate change will (and is) driving more variability in the environment and it is the environment and oceanographic conditions that drives recruitment and productivity.

Cephalopods form only a minor component of the Australian fisheries industry at present. However, this is likely to change if the fin fisheries in Australia and the rest of the world continue their present collapse, and continue to expand into lower trophic level species as new protein sources for vast human populations are sought. Heavy fishing pressure on large, late-maturing fishes may have already tipped the balance of marine ecosystems in favour of fast-growing, short-lived squids (Jackson and O'Dor 2001). Pressure on Australian stocks will undoubtedly increase. Their management is compromised not only because of remaining gaps in our understanding of species boundaries, but also in our knowledge of cephalopod life histories and ecosystem functions. Greater consideration of ecosystem structure and function will foster better ecosystem-based management decisions (Hunsiker *et al.* 2010).

In order to manage sustainable cephalopod fisheries, there has to be a greater understanding of population dynamics and the role of cephalopods in ecosystems. This needs to become a high research priority as relatively few such detailed studies have been undertaken on Australian species. One such study is that by Lansdell and Young (2007), who examined the distribution and biomass of cephalopods from pelagic waters off eastern Australia between 1997 and 2004 from studying the stomach contents of fishes taken in the domestic longline fishery: swordfish, yellowfin tuna, and dolphinfish. They recorded 38 taxa from 19 cephalopod families and noticed pronounced differences in species consumed by the three predators. Ommastrephid squids dominated in all predators sampled, indicating the importance of this taxon as a food stock. Among swordfish sampled for cephalopods, significant relationships were found between biomass, season, sea-surface fluorescence and year. (The presence of fluorescence correlated with cephalopod biomass in swordfish, indicating that cephalopods were concentrating in frontal regions where there are enriched prey concentrations.) In yellowfin tuna, cephalopod biomass was significantly correlated with season, area and sea-surface temperature. Studies such as this can give insights into individual cephalopod species distributions (bearing in mind the limits of the predator species) and behaviour across a range of spatial, seasonal and environmental parameters. This study showed gaps in the understanding of the vertical ranges of cephalopod species in the region and expressed a need for more targeted research in this area.

Population recognition and size estimates are also essential elements in the analysis of the conservation status of species under the International Union for the Conservation of Nature (IUCN) Red List Criteria (www.iucnredlist.org) and under the International Trade of Endangered Species (CITES) listing criteria for protecting species that are in international trade (www.cities.org). There are little data available on population sizes and density of Australian cephalopods to enable a preliminary baseline for such assessments. Population size estimation provides a temporal reference point from which to assess change.

Genetic studies of population structuring and dispersal patterns are important areas for future research and, to date, few such studies have been conducted on commercially important marine invertebrates (Doubleday *et al.* 2009). It is essential that any population division at the genetic level is identified and incorporated into fisheries management plans in order to prevent localised extinctions and to conserve diversity within and between genetically distinct stocks. Genetic diversity is essential for maintaining the adaptive potential of populations and the ability to deal with changing environmental conditions. Due to the low levels of genetic diversity observed in cephalopods, Doubleday *et al.* (2009) recommend the use of highly polymorphic microsatellite DNA markers in such studies rather than other more conservative markers such as allozymes and mitochondrial DNA, combined with the analysis of phenotypic variation in morphological characters.

There can be marked variations in mating behaviour and hence reproductive strategy among cephalopods; molecular markers are invaluable tools for studying the evolutionary significance and management implications of various mating systems (Van Camp *et al.* 2003). Future research and fisheries management would benefit from focus in this area of molecular research.

Doubleday *et al.* (2008) also showed that the elemental composition of the stylets of *O. maorum* provides a useful tool for examining population structure. The distribution and quantity of trace elements within hard biomineralised structures such as stylets are influenced by their availability in seawater and reflect changes in the chemical, physical and biological environment. As these structures have a temporally related incremental composition, they can be used as 'tags' for an individual's movement history, thus providing information on population structure, recruitment processes and dispersal patterns. These elemental signatures can be used to complement genetic data. Genetic methods provide estimates of dispersal over long time scales, with an exchange of only a few individuals per generation required to maintain genetic homogeneity among populations, but elemental signatures can indicate distinct subpopulations that are otherwise genetically homogenous due to periodic low-level mixing, or mixing during the early stages of the life cycle (particularly common for species with planktonic young).

Not so problematic in Australia, but particularly in the case of Asian markets, processing methods following collection result in a vast range of morphological transformations, making identification virtually impossible. This has led to the development of a number of analytical methods using molecular techniques to identify particular species (Espiñeira *et al.* 2010) and has resulted in the discovery of products labelled incorrectly, sometimes as a result of genuine misidentifications, but more often following substitutions between species of different commercial value. Espiñeira *et al.* (2010) used the PCR amplified cytochrome b gene in a study that included a number of species of Octopodidae, Sepiidae and Sepiolidae found in Australian waters to genetically identify species. (It should be noted, however, that samples of only one species, *S. apama*, were actually collected in Australian waters.) Of 20 commercial samples, 30% were incorrectly labelled. An earlier study by Santaclara *et al.* (2007) included ommastrephids and loliginids. These researchers have included a large number of species and have, in addition to providing reliable identification tools, produced valuable phylogenetic information. The DNA-based methods can be applied to fresh, frozen, precooked, canned or any other foodstuffs, including those that undergo intensive thermal treatment. Because it is anticipated that Australian cephalopod fisheries will expand in the future, it would be useful to broaden these kinds of studies to include more species found in Australian waters. The results can then be applied to questions relating to correct labelling, authenticity, traceability, and import and export control.

In another study, Bráullio *et al.* (2010) found that 5S rDNA was a useful marker, providing a low-cost and rapid forensic technique. They found distinct banding patterns among six species sampled belonging to the families Loliginidae and Ommastrephidae. Particularly in Europe, the Loliginidae are of higher quality and more expensive than the ommastrephids and members of the latter family are sometimes marketed as loliginids. Bráullio *et al.* (2010) found that 5s rDNA permits the identification of species as well as differentiating members of these two families. This may prove to be a useful tool when eventually more species are compared.

Fishing activity has direct and indirect impacts on target and non-target species, marine communities and habitats (Jennings and Kaiser 1998). An example to illustrate this is the effect of an artisanal cuttlefish trap fishery on cuttlefish eggs in the north Adriatic Sea (Melli *et al.* 2014), where high fishing pressure on *S. officinalis* Linnaeus, 1758 during spawning aggregations results in loss of eggs laid on traps, ropes and other gear. In this study, experiments were conducted to devise mitigation measures to reduce the impact of fishing on cuttlefish eggs. The results of such studies undertaken where cephalopods are fisheries targets will need to be carefully investigated should additional taxa become fished in Australian waters.

## Bioaccumulation

As a result of their relatively short lifespans and increasing fisheries importance, cephalopods are of major interest for monitoring variations of pollutant concentrations in their tissues. Accumulated concentrations reflect the bioavailability and metal variations in their immediate environment over a short time scale. A number of studies in recent years, particularly in the northern hemisphere, have focused on the bioaccumulation of trace elements and heavy metals in cephalopod tissue. A study by Lacoue-Labarthe *et al.* (2011, and containing references to earlier works) examined the bioaccumulation of heavy metals in the embryonic stages of four species: *S. officinalis*, *L. vulgaris* Lamarck, 1798, *O. vulgaris* Cuvier, 1797 and *Rossia macrosoma* (Delle Chiaie, 1830). The premise for the study was the fact that most coastal cephalopod species of commercial importance migrate into shallow water during the breeding season to lay their eggs, and are consequently subjected to coastal contamination due to anthropogenic activities. They found differing vulnerabilities among the different taxa to metal accumulation and differences reflecting the metals to which eggs were exposed ($^{110m}$Ag, $^{109}$Cd, $^{60}$Co, $^{54}$MN and $^{65}$Zn) based on differences in the permeability of egg membranes to different metals. Cuttlefish eggs, in which hard shells enclose the embryos, had the lowest concentration factor. Octopus eggs, protected only by the chorionic membrane, accumulated the most metal, with $^{110m}$Ag accumulated directly from the dissolved phase and indirectly through the yolk.

While no similar studies have as yet been undertaken on species found in Australian waters, it should be a consideration, particularly for those taxa that are already, or are likely to be, affected by future coastal development (Spencer Gulf, the north-western coast of Western Australia, and regions such as the Great Barrier Reef where dumping of mining waste is set to occur are places that immediately come to mind).

Kojadinovic *et al.* (2011) investigated 14 elements in the tissue of the oceanic squid *T. filippovae*, Adam 1975 from waters surrounding Île Amsterdam in the southern Indian Ocean and Tasmania. They found that the elemental concentrations differed between the two sites and, with the exception of As (arsenic) and Hg (mercury), the muscles showed relatively low trace element concentrations. Given their Hg content, they concluded that squids from Île Amsterdam are not recommended for human consumption on a regular basis. Regardless of source, cadmium and mercury in the digestive glands exceeded the European authorised limits in these organs. While most elements are assumed to be incorporated via the diet, direct uptake of elements from the seawater via the gills is also a known phenomenon in cephalopods. Kojadinovics *et al.*'s (2011) study indicated that seawater may be a direct source of arsenic and manganese.

Heavy metal bioaccumulation can occur at great distances from the source of water contamination. Bustamante *et al.* (1998) found very high levels of cadmium in *Benthoctopus thielei* Robson, 1932 and *Graneledone* sp. (likely *G. gonzalezi* Guerra, González and Cherel, 2000) collected close to the Kerguelen Islands despite the area not directly receiving anthropogenic inputs of this metal. This indicates that these cephalopods probably play an important role in the process of cadmium transfer to top vertebrate predators. Their work showed that the consumption of one 10 g wet weight octopus per day would result in a predator reaching cadmium levels that have been designated by the World Health Organisation as a 'provisional tolerable weekly intake' for humans. It is likely that Kerguelen top-level predators have high levels of cadmium in their kidneys and liver, which in humans can cause renal disease and other metabolic disruptions, with cadmium possibly competing with, and substituting, the essential element copper in these octopuses and their predators.

It seems that the concentrations of essential elements are more variable among species in relation to the specific physiological needs of each animal rather than purely dependent on the environment. Studies such as these and the setting of health guidelines are complicated as the toxicity of particular metals may be countered by others; for example, selenium offers a protective effect against mercury.

The study of bioaccumulation in southern hemisphere cephalopods is an area in much need of further research.

## Evolution and systematics

The majority of extant cephalopods are contained within the subclass Coleoidea, which includes octopuses, squids and cuttlefishes. A second subclass, Nautiloidea, contains the nautiluses and is represented by only a few species. The coleoids comprise a number of lineages that are well recognised as representing monophyletic groups. While morphological studies have proved useful in classifying species within subfamilies and/or genera and, in many cases, families, it has proved difficult to understand the evolutionary relationships among major higher-level coleoid lineages (Young and Vecchione 1996). The reasons for this include the large number of morphological differences between these lineages, the lack of shared characters, problems involving the determination of character independence, uncertainty regarding homology, apomorphic loss of characters, and differing rates of evolution across groups, in conjunction with a poor fossil record.

Cephalopod taxonomy and systematics have advanced markedly in the last few decades as a result of the application of molecular techniques to provide additional suites of characters that can be used to define species and examine their phylogenetic relationships. Allcock *et al.* (2014) have reviewed the contribution of molecular data to our understanding of cephalopod systematics, tracing progress in this field from its beginnings since the first cephalopod DNA sequence was published in 1983. They first discuss the numerous studies that have attempted to determine the position of Cephalopoda within the Mollusca. The most recent comprehensive study to date is that of Smith *et al.* (2011) who included representatives from all classes, including 35 operational taxonomic units, and sequences from 1185 gene regions, totalling 216 402 amino acids. Support was found for the Aculifera (Polyplacophora, Neomeniomorpha and Chaetodermomorpha) and the Conchifera clade (Monoplacophora, Cephalopoda, Gastropoda, Scaphopoda and Bivalvia) with a sister-taxon relationship between the Gastropoda and Scaphopoda, with Bivalvia sister to this clade. Cephalopoda was a sister taxon to Monoplacophora.

Within the Cephalopoda, molecular studies have historically produced conflicting results, perhaps due to limited taxon or gene sampling. High rates of evolution in cephalopods, coupled with their radiation into numerous habitats, continue to confound our ability to recover consistent phylogenetic relationships using both morphological and molecular data (Lindgren *et al.* 2012), particularly at the family level and above.

Allcock *et al.*'s (2014) review provides details and a comprehensively referenced account of the work that has been done to examine the relationships among major lineages using molecular data, so only a brief historical summary is provided here. Further details follow in the accounts pertaining to particular taxa throughout this book.

Initial studies were based on small fragments of single mitochondrial genes, initially 16S rRNA, and produced little resolution of the deep relationships among coleoid families. It was concluded that a more slowly evolving gene was required for examining deep evolutionary relationships. Some resolution was obtained using COIII datasets, which included some important groupings – for example, the recovery of the Octopodiformes, with two octopus species as sister taxa and *Vampyroteuthis* as a successive sister taxon. Decapodiformes were also recovered in the work of Bonnaud *et al.* (1997). In this paper GC bias (base composition heterogeneity) was discussed; this is a problem when trying to resolve relationships among distantly related groups that appear to have been evolving independently for some time, in that the branches become susceptible to long branch attraction and may be attracted to other, distantly related GC-rich branches. This is a feature of cephalopod phylogenies generally.

Work on COI followed (e.g. Carlini and Graves 1999), raising the potential problem of homoplasy in a dataset based on this fast-evolving gene. In addition, prior to studies examining relationships within the Mollusca, outgroup choice proved problematic, and still causes issues depending on the genes being studied because any outgroup is so widely diverged. However, as the range of genes used in cephalopod phylogenetics has continued to expand, together with increasing numbers of taxa studied, resolution of groups at the family level and above continues to improve. As Allcock *et al.* (2014) surmise, COI alone is unlikely to provide systematic resolution due to saturation in the third codon position and insufficient signal in the first and second codon positions. In addition, the evolution of a single mitochondrial gene may not reflect the evolution of the taxa as a whole.

Further attempts to resolve coleoid relationships included the use of additional mitochondrial genes and greater numbers of taxa. These analyses confirmed the monophyly of Sepiida and Decapodiformes, but other groups were not consistently resolved, including groups that are well established on morphological grounds. Thus, there were problems with using a small selection of relatively rapidly evolving mitochondrial genes on a broad range of taxa. It appeared that the phylogenetic placement of taxa varied depending on the genes targeted. Bonnaud *et al.*'s (2005) focus on the Idiosepiida is a prime example. The phylogenetic position of this taxon remains unresolved.

A major step forward involved the use of more slowly evolving nuclear and mitochondrial genes and some of these earlier studies, particularly in the case of internal octopodiform topology, is fairly consistent with current thinking (Carlini *et al.* 2000 *v* Strugnell *et al.* 2014). Bonnaud *et al.* (2002) revealed a difficulty in the use of 18S rRNA, and possibly 28S rRNA: there are a number of highly variable copies of 18S rRNA in some cephalopod species. In some cases, the differences between copies of rRNA in a single species were greater than those between species, indicating the importance of ensuring that non-homologous forms of 18S rRNA are not included in analyses.

Lindgren *et al.* (2004), the pioneer of large-scale combined analyses, introduced the use of the nuclear histone gene, together with 18S rRNA, 28S rRNA and COI in a dataset that included 18 outgroup taxa from a range of molluscan classes. Significantly, the study contained a cladistic analysis of a morphological dataset and analysis of a combined morphological/molecular dataset. Strugnell *et al.* (2004) introduced three additional nuclear genes (pxa6, rhodopsin and octopodine dehydrogenase) and results from this study suggested that the Octopodidae, as then defined, was not monophyletic. These results support those of an earlier study by Carlini *et al.* (2001).

Strugnell *et al.* (2005) built on this work of illustrating the advantages of using a multigene approach. (This paper also gives a well referenced account of 'early' (mid-1990s) studies of coleoid relationships that used molecular characters.) Strugnell *et al.* (2005) examined 35 taxa using 3415 base pairs from three nuclear genes (octopine dehydrogenase, pax-6 and rhodopsin) and three mitochondrial genes (12S rDNA, 16s rDNA and cytochrome oxidase I). The results of this study led to the following conclusions: (1) the Order Octopoda and suborders Cirrata, Incirrata, and Oegopsida are monophyletic groups; (2) Spirulidae is the sister taxon to the Sepiidae; (3) the Octopodidae are paraphyletic; (4) the Argonautoidea are basal within the Incirrata; and (5) *Benthoctopus* and *Enteroctopus* are sister taxa.

Strugnell *et al.* (2005) also noted that, in some instances, topology varied according to the model used. This was investigated further by Strugnell and Nishiguchi (2007) who attribute the lack of clear resolution in higher-level coleoid relationships to the different alignment and analysis combinations that have been used. Consistent among these results, however, is the monophyly of the Octopoda, Argonautoidea, Oegopsida and Ommastrephidae based on two sequence homology methods and three analysis methods, with the Bathyteuthidae the sister taxon of the Oegopsida in each case. The paper by Strugnell and Nishiguchi (2007) gives a useful summary of the outcomes of various molecular analyses to 2007 and advocates the use of a variety of different alignment and analysis strategies in phylogenetic analysis. Lindgren and Daly (2007) also explored factors affecting tree topologies.

Allcock *et al.* (2011) examined mitochondrial gene order and amino acid coded sequence data of all mitochondrial protein coding genes to construct phylogenies. These analyses supported the resolution of the recent Cephalopoda into Octopodiformes and Decapodiformes (with Spirulida basal among Decapodiformes), indicated well-established support for Teuthoidea (Oegopsida and Myopsida) to the exclusion of Sepiidae, and supported a monophyletic Oegopsida. Consideration of paleontological evidence and mapping morphological characters onto their phylogeny suggested that their phylogeny reflects true evolutionary relationships.

Lindgren *et al.* (2012) have generated the most comprehensive multi-gene phylogeny to date, based on 10 genes for over 400 cephalopod taxa (representing 42 of 47 families). In addition to providing a well-supported phylogeny, these researchers created a character matrix of habitat type and morphological characters from which to infer ancestral character states and test for correlation between habitat and morphology to study the phylogenetic and environmental forces driving character evolution. Convergent evolution was shown for all six morphological characters analysed, three of which correlate with habitat and, therefore, environmental pressures. The study also serves to underline the importance of a combined approach in formulating robust phylogenetic hypotheses. Their results confirmed monophyly for almost all major orders and families studied (Sepiida, Myopsida, Sepiolidae, Bathyteuthida and Oegopsida) except the Octopodidae, which is a highly morphologically divergent group with numerous species-complexes (Norman and Hochberg 2005). Bathyteuthida was supported as the sister taxon of Oegopsida. Sepiadariidae was supported as sister taxon to Sepiolidae, providing molecular support for Sepiolida (= Sepiolidae + Sepiadariidae). The position of Loliginidae was ambiguous (perhaps due to substitution saturation and/or rate heterogeneity) and the sepioid orders at the base of the decapodiform clade did not form a monophyletic group. It remains to be seen whether these relationships can ever be resolved, but Allcock *et al.* (2014) suggest that genomic studies are likely to contribute significantly to our knowledge in future.

Other studies involving sequencing entire mitochondrial genomes are currently underway and the list of fully sequenced cephalopod taxa is growing. Allcock *et al.* (2011) analysed the available whole mitochondrial dataset. A monophyletic Octopodiformes was again supported but relationships among Decapodiformes taxa were not well resolved. There was evidence that *Spirula* might be basal and the topology was generally congruent with morphological hypotheses. Allcock *et al.* (2011) suggest that either the arrangements in the mitochondrial genomes might be too extensive or the taxon sampling is not yet sufficient to provide additional insights.

Allcock *et al.* (2014) also tabulate the published chromosome numbers for cephalopods and suggest that additional karyological data might improve our understanding of cephalopod evolution and the number of genome duplication events that have occurred in the past. Interestingly, this table (Allcock *et al.* 2014: 18, 19) does not contain chromosome numbers for any species found in Australian waters, so gathering these data in future would be a worthwhile study. A method for preparing chromosome squashes is provided in Appendix 2. Cephalopod genomic studies will undoubtedly contribute enormously to our understanding of cephalopod evolution in future. Albertin *et al.* (2012) give an overview of this possibility.

Molecular techniques have very usefully been applied to species and population level questions. Although species have traditionally been defined based on morphology, this approach can underestimate the 'true' number of species if cryptic species are present. Similarly, DNA-sequencing techniques can over- or underestimate the 'true' number of species when inappropriate markers are used, so caution should be taken in determining species boundaries. Ideally, as many suites of character traits as possible should be included in both morphological and molecular analyses, incorporating combined analyses of multiple gene sequences from each individual (covering both nuclear and mitochondrially encoded sequences). In addition, to genetically distinguish different species in sympatry it is essential to independently confirm any phylogenetic hypothesis based on molecular data by either morphological or unlinked molecular characters (Milinkovitch *et al.* 2002, Feulner *et al.* 2007).

Aside from molecular techniques another indirect tool that can, and has been, applied to the study of cephalopod taxonomy and systematics is the study of dicyemid mesozoans (Dicyemida Van Beneden, 1876), small microscopic, worm-like organisms, are the most common and characteristic parasites in the renal appendages (kidneys) of benthic cephalopods where they can occur in densities of thousands of animals per cubic centimetre (Catalano and Furuya 2013). They spend most of their life cycle attached to the host's renal appendages, deriving all their metabolic requirements from the dissolved nutrients in the host's urine (Hochberg 1982). In general, dicyemid parasites are highly host-species specific. Nonetheless there are records of one dicyemid species being found in two or more host species, but this occurrence is quite rare and often between related host genera. An example in Australia is the same parasite species occurring in the closely related cuttlefishes *S. apama* and *S. novaehollandiae* Hoyle, 1909 (see the species accounts for these two taxa). Most dicyemid descriptions are based on morphological characters, but a different perspective of dicyemid diversity and specificity may be gained by incorporating molecular techniques in the future. (A method for preparing kidney smears for histological examination of parasites is described in Appendix 2.)

## Cephalopod biodiversity

Australian waters are extremely rich in marine biodiversity as a result of the vast size of this region and diverse range of water conditions and habitats. A diverse geological history has also influenced the distribution of the Australian marine fauna. Butler *et al.* (2010) have summarised what is known about marine biodiversity in the Australian region.

Australia was shaped by its rifting from Gondwana, beginning in the Cretaceous (~83 to ~124 million years ago), the formation of the Tasman Sea in the late Cretaceous to the Eocene, and collision with the Pacific Plate 10–15 million years ago. It is bounded by three oceans (Pacific, Southern and Indian) and four marginal seas (Timor, Arafura, Coral and Tasman). The region includes the sub-Antarctic islands in the Southern Ocean (Macquarie Island, and Heard Island, MacDonald Island on the Kergulen Plateau) as well as offshore islands near the continent (Christmas and Cocos-Keeling Islands in the Indian Ocean and Lord Howe and Norfolk Islands in the Pacific) (Fig. 1). Thus, the Australian region is heterogeneous (Condie and Harris 2005) with a range of diverse habitats and history, so despite the seas being considered oligotrophic (the continent is arid, waters generally clear and low in planktonic abundance, and there are no major coastal upwellings as dominate ocean systems in other parts of the world) this heterogeneity is reflected in the diversity of our fauna. While there has been a historical emphasis on the study of inshore fauna, museums are now building significant collections of deep-water shelf and slope species, which will (taxonomic expertise permitting) undoubtedly result in an increase in the number of described cephalopod fauna. Our understanding of pelagic systems is weak and this would be a useful focus for future cephalopod research effort. In addition to determining species diversity and occurrence, a shift to understanding how populations are structured in relation to the physical environment should be a focus for future research endeavours.

Regional current systems are an important influence on water conditions and dispersal, primary among these the poleward-flowing Leewin Current on the west coast and the East Australian Current and its eddies on the east cost of the Australian continent, as well as the eastward-flowing Antarctic Circumpolar Current. Both major continental boundary currents (the Leewin and East Australian) are influenced by basin-scale ocean–atmosphere processes (e.g. the El Niño–Southern Oscillation and the Indian Ocean Dipole) and are strongly influenced by climate change, which is already showing significant consequences for the biota.

## Australian Antarctic Territories

A work on Australian cephalopods would not be complete without the inclusion of those taxa occurring in the Southern Ocean within the Australian Antarctic Territories, an area of nearly 5.9 million square kilometres – about 42% of Antarctica – and nearly 80% of the total area of Australia itself.

An understanding of the fauna occurring in this region is also important as a background for potential future fisheries exploitation, particularly of some ommastrephids. Collins and Rodhouse (2006) have given a comprehensive review of the cephalopod fauna of the region. The main points with respect to the Australian fauna are given here, though clearly much applies to Antarctic species in general.

The Southern Ocean contains fast-moving circumpolar currents that form large gyres at their interstices. It contains the eastward-flowing Antarctic Circumpolar Current, which provides a link between three ocean basins: the Atlantic, Indian and Pacific Oceans. It has an important role in the transfer of environmental signals around the globe and provides a mechanism for linking environmental variability throughout the southern hemisphere. As a result of this link, variability in the abundance of cephalopod populations exhibits a degree of synchronicity, driven by global patterns of environmental variability (Waluda *et al.* 2004).

Movement of some Southern Ocean organisms by these currents may act to homogenise their populations by promoting high rates of gene flow. For other species, these currents and gyres may act as boundaries that restrict migration and may lead to genetic differentiation among populations in different regions (Rodhouse and Boyle 2010).

Much of the fauna is endemic, particularly the squids (which are all oceanic pelagic species) and octopods. The oegopsid squids are exclusively pelagic or benthopelagic and most have a circum-Antarctic distribution. In contrast, the incirrate octopods are speciose and abundant and have more restricted distributions. There are two endemic squid families (Psychroteuthidae and Batoteuthidae) and six endemic genera: *Mesonychoteuthis*, *Psychroteuthis*, *Kondakovia*, *Alluroteuthis*, *Slosarczykovia* and *Batoteuthis*. Sepiids and sepiolids are absent from the Southern Ocean. Octopods dominate with high levels of diversity associated with niche separation.

The Southern Ocean contains 13 genera and over 25 species of octopods. At least five genera are endemic to this region: *Pareledone*, *Megaleledone*, *Adelieledone*, *Praealtus* and *Bathypurpurata*. The endemic incirrates are thought to have evolved after the separation of the Antarctic continent and the subsequent formation of the Antarctic circumpolar current (25–28 million years ago) and the Antarctic Polar Front, which produced a thermal barrier to shallow fauna. Strugnell *et al.* (2011) have recently shown that the Southern Ocean is both an evolutionary source and sink for different octopod clades, with some having their origins in the Southern Ocean and others originating in the deep sea. *Benthoctopus*, for example, probably originated in deep water in the northern hemisphere. The members of the unique Antarctic octopod clade that are not restricted to the deep sea can be found throughout the Antarctic shelf and continental slope where they have undergone extensive radiation, particularly in the case of *Pareledone*. Among these octopods there are large overlaps in species distributions, often with several species occurring in sympatry.

Sympatry is often associated with trophic niche partitioning, which has undoubtedly led to a range of adaptations; these are largely yet to be studied. As an example, to explore one possible adaptive trait, Undheim *et al.* (2010) compared the venom of four Antarctic octopus species (*Adelieledone polymorpha* (Robson, 1930), *Megaleledone setebos* (Robson, 1932), *Pareledone aequipapillae* Allcock, 2005 and *P. turqueti* (Joubin, 1905)). These researchers found a range of toxic activities in the venom of these octopods, including evidence for cold adaptation, suggesting that venom plays an important role in the ecology of Antarctic octopodids. While their results did not show an obvious link in terms of toxic activity, morphology and diet due to a lack of reliable prey-preference data, this area of molecular adaptive evolution is definitely worth further investigation.

(Not all the abovementioned taxa are reported to occur in Australian Antarctic Territory.)

## Climate change

The literature on climate change and its possible effects on marine organisms is very extensive and a summary is not attempted here. However, a few points relevant to cephalopods in Australian waters are discussed below.

It is well known that global warming is affecting ocean water temperatures, currents and weather patterns worldwide. Ocean temperatures around most parts of the Australian coast have warmed at over twice the global average rate. This is occurring at even greater rates in south-eastern and south-western Australia, and has resulted in pattern changes in the East Australian Current (EAC), which is now extending further south into Tasmanian waters. The distributions of some cephalopod species have already begun to change as a result. For example, *O. tetricus*, common around mainland Australia, is now being observed in northern Tasmanian waters (Pecl 2011). The same may be true for *O. maorum*. Species may either be shifting poleward to avoid the warming waters of their previous habitats (Pecl 2011), or changes in currents may result in shifts in larval dispersal patterns. Among other studies, monitoring is now being attempted as part of a project called Redmap (Range Extension Database and Mapping project). This project was first launched in Tasmania and now extends Australia-wide (www.redmap.org.au).

As previously discussed, temperature has a profound effect on cephalopod growth rates. For example, significant negative associations have been found between sea-surface temperature and body size (Rosa *et al.* 2012), so future changes will influence reproductive biology and population structures. Due to the relatively short life cycles and fast growth rates (at least for inshore species), cephalopods have the capacity to respond quickly to environmental and ecosystem change. They exhibit a high level of plasticity in their life history strategies and changes in environmental conditions significantly influence population size, abundance and distribution. This could prove to be advantageous; for example, elevated temperatures accelerate cephalopod life histories, increasing their growth rates and shortening lifespans. This might suggest that, if food supply remains adequate, growth rates may increase, animals may be larger and population turnover more rapid. However, this scenario may actually be more complex, as suggested by Pecl and Jackson (2014); the size of emerging hatchlings becomes smaller as temperatures increase, and hatchling size may have a critical influence on the size at age that may be achieved as adults. Thus, the influence of higher temperature on the egg and adult stages may be opposing forces on life history: cephalopods may hatch out earlier and smaller, undergo faster growth over shorter lifespans and mature younger and at a smaller size, with individuals having greater food requirements and greater susceptibility to fluctuations in prey availability. The same may or may not be true for their prey or predator species, so it is difficult to predict outcomes with certainty.

Not only is embryonic development highly temperature dependent, but environmental variability influences cephalopod movement and migration (Semmens *et al.* 2007). Spawning regions and the distribution of egg masses are linked to specific temperature zones and the distributions of many paralarval and juvenile squids are linked to local regional oceanography. Environmental conditions during spawning and hatching are linked to recruitment success or failure. For example, local oceanographic processes during early life history stages may be important in retaining paralarval squids in areas of high food availability. Oceanographic variability alters the patterns and timing of migrations that may be linked to specific water masses. Interest in the influence of environmental parameters on cephalopod movement and migration is increasing because rising ocean temperatures will disrupt or change migration patterns for some species. The full implications of this remain to be determined.

A study by Spady *et al.* (2014) of *I. pygmaeus* from Townsville, Queensland found that exposure to projected levels of $CO_2$ in seawater affected behaviour, both at the lower- and upper end-of-century projections. Movement, squid vigilance, defensive behaviours and escape responses increased with higher $CO_2$ levels. They suggest that increased activity could lead to adverse effects on energy budgets in addition to increasing visibility to predators. An increased tendency to respond to a stimulus with escape behaviours could increase survival but may be energetically costly. These results showed that future acidification will affect the behaviours of a tropical squid species.

On a greater temporal and spatial scale, the work by Strugnell *et al.* (2008b, 2011) has shown that the initiation of the global thermohaline circulation around Antarctica provided a mechanism for dispersal and radiation of deep-sea octopodoids. Understanding how environmental factors such as this generate and maintain large-scale patterns of biodiversity is critical to predicting the impacts of global climate change.

Anthropogenic carbon dioxide emissions are acidifying the world's oceans. The pH of seawater has decreased by 0.1 units since the mid-1900s and is predicted to decrease by another 0.5 units by 2100. Together with declining pH vales, ocean carbonate ion ($CO_3^{2-}$) concentrations will decrease, which in turn will lead to a reduction of calcium carbonate saturation (Ω) in seawater. Those cephalopods that form shells or bones from calcium carbonate minerals (primarily aragonite or calcite) will potentially be the most affected (nautiloids, Sepiidae and *Argonauta*). Most marine invertebrates respond negatively to elevated $CO_2$ concentrations. However, in a study of *Sepia officinalis*, Gutowska *et al.* (2008) found no adverse affects. During a 6 week period, juvenile *S. officinalis* maintained calcification under ~4000 and ~6000 ppm $CO_2$, and grew at the same rate and the same gross growth efficiency as did control animals (held at ~700 ppm $CO_2$). From this they conclude that active cephalopods may possess a certain level of pre-adaptation to long-term increases in $CO_2$ levels.

In contrast, female *Argonauta* secrete a fragile calcitic spiral egg case. The case is held externally, not protected by mucus or epithelium (as are other cephalopod carbonate structures, such as cuttlebones, beaks and statoliths) and has a high surface area and low volume, so is potentially much more vulnerable than the calcified structures found in other cephalopods to changes in pH and dissolution. This has been shown experimentally by Smith *et al.* (2012) in a study of argonaut shells. Shell weight loss increased with lowering pH. It is not known whether argonauts will be able to compensate for this loss by modifying their biomineralisation strategy.

Rodhouse (2013) reviewed the role of squids in the Southern Ocean pelagic ecosystem and the possible effects of climate change on squid biology based on predictions of change by 2100. Eighteen species of squids are found in this region. The Antarctic Peninsula is one of the fastest warming regions worldwide but elsewhere the Southern Ocean is warming more slowly. Model predictions suggest that although the Southern Ocean surface waters will warm less than those in other oceans, sea ice will reduce in area by ~33%, especially in winter and spring. It seems that the Antarctic circumpolar current may shift slightly southwards with intensification of westerly winds. It is likely that decreases in pH may have a greater impact on Southern Ocean squid than predicted temperature increases. In the cold Southern Ocean, the solubility of calcium carbonate is high, so decreases in pH and therefore calcium carbonate will have a high impact on calcifying organisms. Temperature increases are unlikely to have impacts beyond changes in squid distributions near the limits of their ranges. Rodhouse (2013) suggests that small changes in large-scale oceanic circulation will have low impacts on squids but mesoscale oceanographic changes, the results of which are not possible to accurately predict based on current models, may have high impact. While changes in sea ice may not have a direct effect, ecosystem changes could have a major impact. The potential for fast evolution shown by many cephalopod species may act in favour of these ecological opportunists.

# Class Cephalopoda

**Key to recent cephalopod orders and families found in Australian waters (after Lu and Dunning 1998 and Jereb and Roper 2010)**

1a. More than 10 (63–94) circumoral appendages. Suckers absent. Chambered, coiled external shell.......................................... **Order Nautilida (Nautilidae)**

1b. Eight or 10 circumoral appendages. Suckers (and/or hooks) present. No external shell, or, if present, shell very thin and fragile (*Argonauta*)..............................**2**

2a. Suckers stalked with chitinous rings and/or hooks. Ten circumoral appendages, eight arms and two ventrolateral tentacles (tentacles may be lost in some species). Mantle cavity open to exterior via three openings (**Decapodiformes: Bathyteuthida, Idiosepiida, Myopsida, Oegopsida, Sepiida, Sepiolida, Spirulida**) .............................................................**3**

2b. Suckers without stalks, bases sometimes constricted in finned (cirrate) octopods and vampires. Suckers without chitinous rings. Eight arms, no ventrolateral tentacles. Mantle cavity communicates with the exterior via a single opening, rarely two (**Octopodiformes: Octopoda, Vampyromorphida**) .......................................................................................**4**

3a. Internal shell straight, laminate, calcified (**Sepiidae**), coiled, chambered (**Spirulidae**), rudimentary, chitinous (**Sepiolidae**, **Idiosepiidae**) or absent (**Sepiadariidae**). Tentacles contractile and retractile into pockets between arms 3 and 4. Fins not joined posteriorly. Mantle edge near mantle cartilages straight .........................................................................**5**

3b. Internal shell straight, feather- or rod-shaped, chitinous. Tentacles contractile, not retractile, no pockets. Fins usually joined posteriorly. Mantle edge near mantle cartilages with small projections or 'angles' (**Bathyteuthida, Myopsida, Oegopsida**) ... **9**

4a. Internal shell a chitinous, thin, broad plate. Pair of small filamentous appendages in pouches between bases of arms 1 and 2. Light organ (photophore) present at base of each fin and medial to each eye dorsally. Colour black to dark red or brown ............. ...................................................... **Vampyroteuthidae**

4b. Internal shell vestige either small cartilaginous rods or a U-shaped support or absent. Secondary filamentous appendages absent. Suckers sessile, without stalks and without chitinous rings. Colour variable but never black............................................**Octopoda**

5a. Internal shell calcified ................................................**6**

5b. Shell chitinous or lacking...........................................**7**

6a. Shell a spiral coil, round in cross-section, in posterior end of body. Radula absent................**Spirulidae**

6b. Shell a thick oval or rhomboidal plate on dorsal side of mantle. Radula present ............................ **Sepiidae**

7a. Mantle elongate, slightly narrowed and bluntly pointed posteriorly. Fins on posterior mantle. Rough, oval attachment organ on posterior-dorsal mantle...................................................... **Idiosepiidae**

7b. Mantle oval, broadly rounded posteriorly. Fins located at or slightly posterior to mantle midlength. No attachment organ ................................................ 8

8a. Web between third and fourth arms encircles bases of tentacles on outside and inside, forming a cutaneous sac. Fins with broad bases and slightly developed 'earlets' ....................................... **Sepiadariidae**

8b. Web between third and fourth arms envelops tentacle bases on outside only, without forming sac. Fins with short bases and well-developed 'earlets' ..... **Sepiolidae**

9a. Eyes covered by transparent membrane (cornea) **Myopsida** ................................................................. 10

9b. Eyes without transparent membrane (cornea) **Oegopsida** ................................................................. 11

10a. Fins joined at posterior end of mantle .... **Loliginidae**

10b. Fins not joined at posterior end of mantle .............. ....................................................... **Australiteuthidae**

11a. Funnel free from mantle. Funnel–mantle locking cartilage present ...................................................... 12

11b. Funnel fused on either side of mantle. No funnel–mantle locking cartilage present ......... **Cranchiidae**

12a. Funnel-locking cartilage a simple, straight groove and ridge ................................................................... 13

12b. Funnel-locking cartilage not a simple, straight groove and ridge ..................................................... 29

13a. Arms with hooks or with suckers in four series on proximal half of ventral arms ................................ 14

13b. Arms without hooks and with suckers in two series on proximal half of ventral arms ........................... 18

14a. Armature (suckers, hooks) of arms in two series ... 15

14b. Armature (suckers, hooks) of arms in four series ... ..................................................................... **Gonatidae**

15a. Tentacles and clubs absent in adults (although present in larvae or juveniles (*Taningia*)), but when present, always rudimentary with few suckers ......................................... **Octopoteuthidae**

15b. Tentacles and fully developed clubs present ......... 16

16a. Photophores on viscera but not on mantle or surface of head and/or arms ............ **Pyroteuthidae**

16b. Photophores on mantle and surface of head and arms, but not viscera ............................................... 17

17a. Photophores on tentacles, but not eyeballs ............... ...................................................... **Ancistrocheiridae**

17b. Photophores on ventral eyeballs, but not on tentacles .......................................... **Enoploteuthidae**

18a. Buccal membrane connectives attach to ventral sides of arms 4 .......................................................... 19

18b. Buccal membrane connectives attach to dorsal sides of arms 4 ................................................................... 24

19a. Hooks present on tentacular clubs ............................ ...................................................... **Onychoteuthidae**

19b. Tentacular clubs without hooks ........................... 20

20a. Mantle with cartilaginous scales (may be minute). Tentacular clubs with four series of suckers ......... 21

20b. Mantle without cartilaginous scales. Tentacular clubs with more than four series of suckers ......... 22

21a. Tentacles with numerous laterally compressed suckers ........................................... **Pholidoteuthidae**

21b. Tentacles lost in adults; tentacles in juveniles weak, with a few (around six) poorly differentiated suckers ............................................ **Lepidoteuthidae**

22a. Fins nearly as long as mantle, supported by strong, transverse muscular ribs. Minute suckers present on oral surface of buccal lappets ...... **Chtenopterygidae**

22b. Fins less than half the body length and without supporting ribs. No suckers or buccal lappets .... 23

23a. Tentacular clubs with six series of suckers. Long spike-like tail present, greater than fin length ...... ......................................................... **Batoteuthidae**

23b. Tentacular clubs with four series of suckers. No long spike-like tail present ........... **Brachioteuthidae**

24a. Ventral surface of eye with a single row of photophores. Buccal membrane with eight separate lappets ................................................. **Lycoteuthidae**

24b. No photophores on eyes. Buccal membrane with seven lappets or less ............................................... 25

25a. Surface of mantle, head and arms covered with numerous photophores (usually large and distinct) ............................................ **Histioteuthidae**

25b. Surface of mantle and head without photophores (base of arms may have a few photophores) ........ 26

26a. Minute suckers present on oral surface of buccal membrane/lappets .......................... **Bathyteuthidae**

26b. No suckers on oral surface of buccal membrane ... 27

27a. Many small to minute suckers (or suckers and knobs) at proximal (carpal) end of manus ........... 28

27b. No cluster of small suckers at proximal (carpal) end of manus ................................ **Psychroteuthidae**

28a. Posterior borders of fins slightly convex. Carpal knobs in a single dorsal row or absent. Adults small .................................................... **Neoteuthidae**

28b. Posterior borders of fins concave. Carpal knobs in a cluster alternating with carpal suckers. Adults attain a gigantic size ......................... **Architeuthidae**

29a. Funnel-locking cartilage with a longitudinal and a transverse groove: ⊥-shaped or ⊣-shaped ......... **30**

29b. Funnel-locking cartilage subtriangular or oval. With or without inward-projecting knobs........... **31**

30a. Funnel-locking cartilage with a longitudinal groove crossed by a transverse groove at its posterior end, ⊥-shaped. Fins less than two-thirds ML ........................................................................... **Ommastrephidae**

30b. Funnel-locking cartilage with a longitudinal groove from which a shorter groove branches medially, ⊣-shaped. Fins > 80% ML............. **Thysanoteuthidae**

31a. Funnel-locking cartilage oval with one or two knobs directed towards centre of the concavity................ **32**

31b. Funnel-locking cartilage oval or subtriangular, without knobs .......................................................... **33**

32a. Club with four series of suckers .... **Chiroteuthidae**

32b. Club with many (> 15) series of minute suckers....... ......................................................... **Mastigoteuthidae**

33a. Suckers on arms in six series on arms 1–3, four series of suckers on arms 4. Tail extremely long (greater than ML) as a spike-like extension of the gladius. No fins on tail.................. **Joubiniteuthidae**

33b. Suckers on arms in two series. Tail absent............... ........................................................... **Cycloteuthidae**

## Subclass Nautiloidea Agassiz, 1848

### *Order Nautilida*

### Nautilidae Blainville, 1825

#### DISTINGUISHING CHARACTERS

External chambered shell. Eyes without lenses. Funnel formed from overlapping flaps. Numerous tentacles with adhesive ridges.

#### DIAGNOSTIC FEATURES

Large calcareous external shell divided into chambers (phragmocone) pierced by a siphuncle. (The animal lives in the largest and most recently formed chamber.) Umbilical callus may be present. Shell sculpture of sinuous growth lines and fine spiral ridges in some species. Eyes simple, without lenses. Funnel consisting of two lobes that fold together to form tube-like structure. Up to 47 pairs of arm-like appendages around mouth; suckers and hooks absent; four tentacles in males fused to form spadex used to transfer spermatophores to females during copulation. Two pairs of gills. Radula with 13 elements including two lateral teeth and two inner marginal support plates. Ink sac absent.

*Colour.* Flesh creamy white, hood and dorsal side of arms orange-brown; chromatophores absent. Shell colouration variable by species. Irregular red to yellow-brown stripes radiate from umbilicus; cover entire shell in juveniles, restricted to dorsal chambered portion of shell in adults.

#### HABITAT AND BIOLOGY

The shell acts as a protective and buoyancy device. The animal can withdraw into the shell and seal the entrance with a large, tough hood. It swims by jet propulsion provided by contraction of the funnel and the piston action resulting from the withdrawal of the head into the living chamber. Buoyancy is controlled by pumping fluid into and out of the chambers by an osmotic mechanism that operates against hydrostatic pressure. The depth limit for this mechanism is ~300 m. Thus, nautilids can only exceed this depth for short periods of time. At ~800 m the shell implodes.

Nautilids are deep-water predators and scavengers that tend to be associated with isolated seamounts. These habitats act as reproductive and biological barriers for these animals as they cannot traverse deep water (often in excess of 1500 m) to cross ocean expanses to other mounts and reefs. They are generally considered to inhabit reef slopes to ~500 m and, although capable of travelling up to 4.3 km per day, they are not generally perceived to be open-water travellers. Ocean drift of living animals does occur, but is not considered to be a significant method of dispersal (Sinclair *et al.* 2007). Until recently they were commonly thought to inhabit deeper water during the day and migrate to shallower water at night. Such diurnal movements have been observed in Palau and Papua New Guinea. However, Dunstan *et al.* (2011b) have shown that this is not necessarily the norm for all populations. Using ultrasonic telemetry and remotely operated vehicle (ROV) observations of *Nautilus pompilius* Linnaeus, 1758 at Osprey Reef in the Coral Sea, Australia, they recorded virtually continuous nightly movement between depths of 130 m to 700 m with daytime behaviour split between virtual stasis at 160–225 m to active foraging in depths of 489–700 m. Juveniles live within the same vertical regime as adults. These movement patterns are most likely dictated by optimal feeding substrate, avoidance of daytime visual predators, upper temperature limits of ~25°C, implosion depths of 800 m and the need for resting periods at ~200 m to regain neutral buoyancy. Dunstan *et al.* (2011b) hypothesise that the slope, terrain and biological community of various *Nautilus* populations may result in different vertical movement strategies.

Mark–recapture studies indicate that nautiluses may live for > 20 years, maturing between 5 and 15 years. Eggs are laid in relatively shallow water (80–100 m) and after hatching the young move into deeper, colder waters. Female *Nautilus* are attracted to male conspecific odour (Basil *et al.* 2002). Most species have been successfully bred in captivity. Aquarium observations indicate a period from egg laying to hatching of 12 months, with juveniles 26 mm in shell diameter consisting of seven chambers.

Barord *et al.* (2014) used baited remote underwater video systems (BRUVS) to estimate the population abundance of *Nautilus* sp. (see below) at sites in the Philippine Islands, American Samoa, Fiji, and along a ~125 km transect on the fore reef slope of the Great Barrier Reef from east of Cairns to east of Lizard Island, Australia. Because, as acknowledged by Barord *et al.* (2014), nautiluses are attracted to bait using chemoreception over significant distances, there are inherent biases related to the use of baited traps to measure actual population numbers. Despite this, comparative densities between sites can indicate relative abundance. Barord *et al.*'s (2014) study found the greatest number of nautiluses at the unfished Australian sites (93 total, 2.01 $km^{-1}$), followed by Beqa Passage, Fiji (20 total, 1.58 $km^{-1}$), and Taena Bank, American Samoa (22 total, 1.48 $km^{-1}$). In stark contrast, the fished Panglao locality in the Philippines had the lowest number (six total, 0.25 $km^{-1}$).

Nautilids have existed for > 500 million years, and now only two genera, *Nautilus* Linnaeus, 1758 and *Allonautilus* Ward & Saunders, 1997, represent extant nautilids. Harvey *et al.* (1999) reanalysed the data presented by Ward and Saunders (1997) and concluded that *Nautilus* and *Allonautilus* could not be recognised as distinct monophyletic taxa based on Ward and Saunders' (1997) data; however, a study by Bonacum *et al.* (2011) based on molecular data has verified the distinctiveness of *Allonautilus* on phylogenetic grounds. In addition, this genetic work indicates that nautiluses have been undergoing a rapid adaptive radiation since the Pliocene. Barord *et al.* (2014) deduce from this and the limitations to dispersal imposed by an 800 m depth implosion limit, combined with a nektobenthic lifestyle, that there may be many currently unrecognised sibling species unique to islands and land masses separated by water depths greater than this physiological limit (hence the preference to simply refer to '*Nautilus* sp.' to define the taxa studied in this publication).

*Interest to fisheries.* None in Australia. Nautilids are actively fished mainly in the Philippines, but fishers from Indonesia and New Caledonia are also involved in the trade. They are traded internationally as live animals for the aquarium and possibly the pet trade. Whole and part shells are sold as products for the curio, tourist, jewellery and clothing industries, and the soft parts are sold as meat for exotic food markets. The low fecundity (10–20 eggs per year), late maturity (12–15 years), long gestation (10–12 months) and long lifespan of *Nautilus* (20+ years) suggest that representatives of this group are vulnerable to over-exploitation. Demand for the ornamental shell trade has contributed to their rapid decline in localised populations. Research has shown an 80% decrease in catch per unit effort from 1980 to the present in the Philippines (Dunstan *et al.* 2010) that can be attributed to fishing pressure. This represents fewer than three *Nautilus* generations. Together with the comparative population estimates documented by Barord *et al.* (2014), it is clear that this fishery is unsustainable, suggesting that the population is endangered. It is currently being assessed for CITES listing.

Unfortunately, the population structure and genetics of most species are poorly understood, making management difficult. Detailed molecular research is needed to fully elucidate the population genetics of *Nautilus* to determine whether the fishing of some populations should cease immediately.

## REMARKS

Bonacum *et al.* (2011) examined the evolutionary relationships among species of *Nautilus* and *Allonautilus* based on molecular characters from two mitochondrial gene regions, 16S rDNA and cytochrome oxidase *c* subunit I. Their results indicated that *Nautilus* is currently undergoing a period of evolutionary radiation in the Indo-Pacific region. They suggest that the basal divergence between *Nautilus* and *Allonautilus* occurred in the waters surrounding New Guinea and the northern part of the Great Barrier Reef in north-east Australia, followed by a migration to the east by the common ancestor of the *N. macromphalus* Sowerby, 1848 and the *N. pompilius* populations in Vanuatu, Fiji, and

American Samoa. A subsequent migration to the west led to the founding of populations off the west coast of Australia, the Philippines, Palau and Indonesia. The results suggest that while *N. macromphalus* and *A. scrobiculatus* ([Lightfoot], 1786) are phylogenetic species, taxa currently identified as *N. pompilius* are a paraphyletic assemblage of populations and do not represent a true phylogenetic species. This supports an earlier conclusion drawn by Wray *et al.* (1995) who also found that *N. pompilius* represents a paraphyletic group of diverging lineages and that these fall into three geographically distinct clades, one of which is the Australian/Papua New Guinea clade. Divergences within *Nautilus* appear to be driven by geographic isolation – a result of constraints on dispersal imposed by the ecology and biology of these animals.

Nautilid species are the only remaining cephalopods with an external shell. Although often cited as a 'missing link', genetic studies indicate that the current species may only have originated between 1 and 5 million years ago.

Juveniles are large (there is no larval phase) and dispersal is limited by shell implosion at ~800 m, thus restricting their movement in regions exceeding 800 m in depth. This results in morphologically and genetically distinct populations. This precludes, or drastically slows, recolonisation of depleted populations and supports a strong argument for protection of all *Nautilus* stocks to maintain genetic diversity (Dunstan *et al.* 2010, 2011a).

Five species – *N. pompilius*; *N. belauensis* Saunders, 1981; *N. macromphalus*; *N. stenomphalus* Sowerby, 1848 and *A. scrobiculatus* – are currently recognised worldwide, but the taxonomy remains uncertain. Only those species known to occur in Australian waters are included here: *N. macromphalus*, *N. pompilius* and *N. stenomphalus*. Shells of *A. scrobiculatus* may drift to Australia from New Guinea; however, this species is not known to occur in Australian waters.

#### DISTRIBUTION

Nautilidae are restricted to the Indo-Pacific from 30°N to 30°S latitude and 90°E to 185°W longitude. They inhabit steep coral reef drop-offs in depths from ~0–700 m.

The shells can drift for thousands of kilometres and have been found in Madagascar, South Africa, Japan, Tasmania and New Zealand.

#### REFERENCES

Jereb and Roper (2005), Young (2010).

## *Nautilus* Linnaeus, 1758

#### DISTINGUISHING CHARACTERS

*Nautilus* differ from *Allonautilus* in having a small or moderate umbilicus (from ~5% to ~16% of the shell diameter) and the whorl cross-section is oval. In *Allonautilus* the umbilicus is larger, ~20% of the shell diameter, and the whorl cross-section is quadrate in shape.

See the family treatment above for diagnostic features, habitat and biology, etc.

### Key to *Nautilus* species (after Nesis 1982/1987)

1a. Shell smooth with closed umbilicus ............................ 2
1b. Shell smooth or wrinkled, with open umbilicus ............................ *N. macromphalus*
2a. Coloured stripes on shell reach umbilicus and pass more or less far onto outer lateral walls of last chamber. Callus present. Hood papillae not pronounced ............................ *N. pompilius*
2b. Coloured stripes on shell do not reach umbilicus. Callus absent. Hood papillae with pronounced nodules ............................ *N. stenomphalus*

## *Nautilus macromphalus* Sowerby, 1848

*Nautilus macromphalus* Sowerby, G. B. (1848). Monograph of the genus *Nautilus*. *Thesaurus Conchyliorum 2*, 463–465 [464, pl. 98 figs 4, 5].

Type data: unresolved.

Type locality: not designated.

#### COMMON NAME

Bellybutton Nautilus

#### DIAGNOSTIC FEATURES

Shell with open umbilicus in which inner coils of shell are visible. Opening ~15% of shell diameter, a deep, round shouldered concavity. No thickened layer (callus) present.

#### HABITAT AND BIOLOGY

Inhabits continental shelf and slope waters associated with coral reefs, from the surface to a depth of ~500 m. In the Coral Sea it is caught in traps at depths between 300 m and 400 m.

#### SIZE

Shell ~160 mm diameter.

Shell, lateral view (AM slide collection).

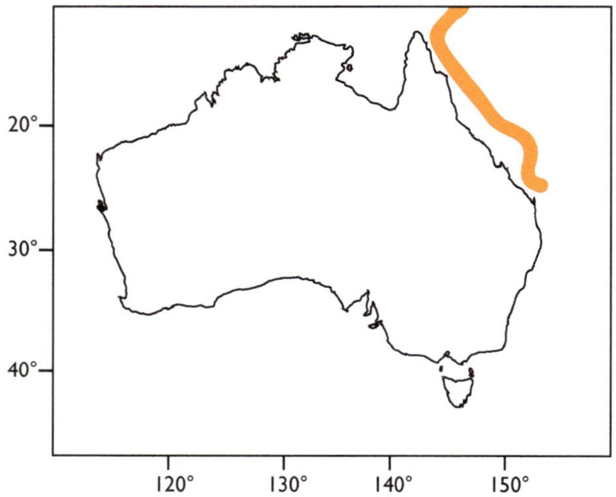


*Nautilus macromphalus* distribution. *Note:* The distribution is based on shell and whole animal data (shells may have drifted).

#### REMARKS

The complete mitochondrial genome of *Nautilus macromphalus* has been sequenced by Boore (2006). The mitochondrial genome is 16 258 nts in length and 59.5% A+T.

#### DISTRIBUTION

South-western Pacific Ocean. Known only from off north-eastern Australia, New Caledonia and the Loyalty Islands.

#### REFERENCES

Norman (2000), Jereb and Roper (2005).

## *Nautilus pompilius* Linnaeus, 1758

*Nautilus pompilius* Linne, C. (1758). Systema Naturae per Regna tria Naturae, Secundum Classes, Ordines, Genera, Species cum Characteribus, Differentiis, Synonymis, Locis. 824 pp. Holmiae [709].

Type data: unresolved.

Type locality: India.

#### COMMON NAME(S)

Emperor Nautilus, Chambered Nautilus, Pearly Nautilus

#### DIAGNOSTIC FEATURES

Umbilicus small, visible as shiny silver and black patch, closed; callus usually present. No inner coils visible. Hood covered with low, rounded, white elevations giving it a mottled, slightly bumpy appearance. Tentacle sheaths with smooth edges.

*Colour.* Shell colour patterns variable: irregular brown to reddish brown stripes radiating from umbilicus but striping can be reduced to various degrees. Hood reddish brown.

#### HABITAT AND BIOLOGY

*Nautilus pompilius* is the most common and widespread *Nautilus* species. Variations in shell colour patterns have suggested that different and possibly distinct populations are present. Sinclair *et al.* (2007) examined the genetic differences among populations of *N. pompilius* in the northern section of the Great Barrier Reef and the Coral Sea. They found that populations from these two regions are distinct: two discrete well-supported clades were detected when mitochondrial gene sequences (*coxI*) were compared. Within these clades there were no population-specific groupings, but Sinclair *et al.*'s (2007) results indicate that diversification is taking place in distinct populations across deep-water walls on the northern Great Barrier Reef and in the isolated seamounts of the Coral Sea.

In a later study, Sinclair *et al.* (2011) compared *N. pompilius* from the northern Great Barrier Reef, the Coral Sea and the Scott Reef off Western Australia. They found significant evolutionary separation among each of the three populations into strongly supported discrete clades matching geographic stratification. Genetic variation was evident within each of the three discrete populations. The east–west divergence is probably a result of the historical physical barrier, the Torres Strait land bridge.

Live animal (P. Ward).

Three *Nautilus pompilius* investigating bait (G. Barord).

Shell, lateral view (AM C.63205, L. Mott).

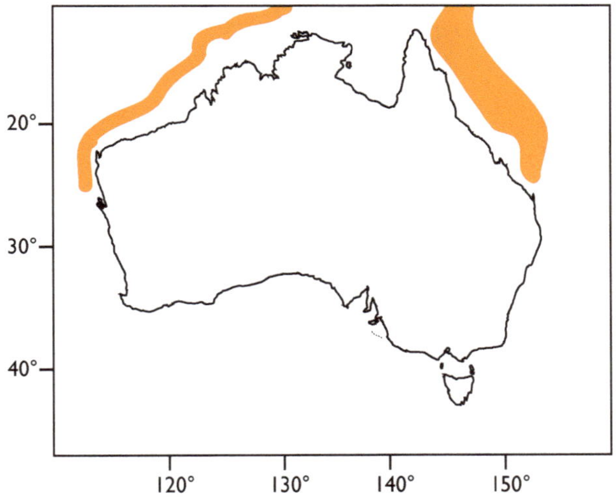


*Nautilus pompilius* distribution. *Note*: The distribution is based on shell and whole animal data (shells may have drifted).

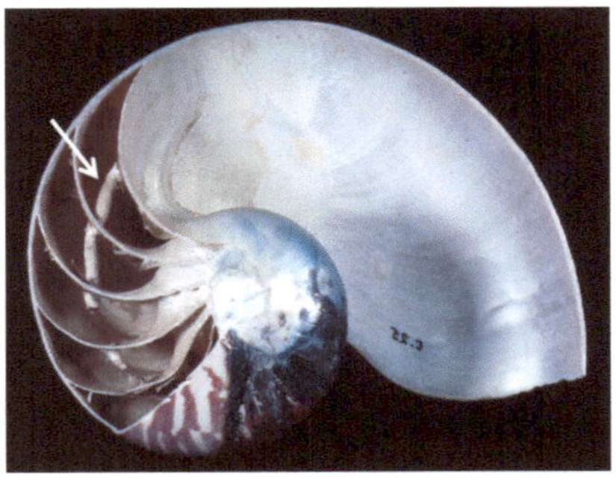

Shell cut lengthwise to show internal structure; siphuncle (partially intact) indicated by arrow (AM slide collection).

Sinclair *et al.* (2011) point to the lack of concordance under the current classification, where potentially two to five species are all grouped under the single species, *N. pompilius*. At the very least, the lack of contemporary gene flow between the two populations from north-eastern Australia and the Western Australian population, in addition to the large number of synapomorphies separating these stocks, supports subspecies delineation for these three populations.

Within populations, the genetic diversity suggests that sequence divergence occurs slowly in *Nautilus* and dispersal between populations is highly restricted. Sinclair *et al.*'s (2011) analysis indicates that the lineages are currently diverging among this polyphyletic group.

Dunstan *et al.* (2011a) used capture–mark–recapture, baited remote underwater video systems, ultrasonic telemetry and remotely operated vehicles to estimate population size, growth rates, distribution and demographic characteristics of an unexploited population of *N. pompilius* at Osprey Reef in the Coral Sea. They estimated a small and dispersed population of between 844 and 4467 individuals (14.6–77.4 $km^{-2}$) dominated by males (83:17 male:female) and few juveniles (< 10%). They suggest that baited remote underwater video systems could be used more widely to assess *Nautilus* populations elsewhere. Dunstan *et al.* (2011c) found sexual dimorphism in size at maturity of specimens from Osprey Reef, Queensland with males having a shell diameter of ~132 mm, and females ~119 mm.

Capture–mark–release studies conducted at Osprey Reef in the Coral Sea have shown that this unexploited population is stable from a catch-per-unit-effort basis over 12 years. These results contrast sharply with the dramatic decline in the Philippines reported by Dunstan *et al.* (2010, 2011a).

*Interest to fisheries.* Not fished in Australian waters at this time. (See family account above.) The recognition of distinct geographic patterns of genetic structuring as reported by Sinclair *et al.* (2007, 2011) indicates discrete Australian units that should be managed discretely as each has high conservation value containing unique genetic variation. As stocks of *Nautilus* elsewhere are being depleted, the possibility exists that exploitation of new fishing areas may be targeted in future. Should this occur, there is clearly a risk that unique genetic resources among the Australian populations, and possibly species and subspecies, may come under threat.

#### SIZE

Up to 170 mm diameter.

#### REMARKS

*Nautilus pompilius* is sympatric with *Allonautilus scrobiculatus* off New Guinea and *N. stenomphalus* off north-eastern Australia. Saunders and Ward (1987) suggested that, based on morphology, hybridisation of *N. pompilius* and *N. stenomphalus* may be occurring on the Great Barrier Reef, close to Lizard Island.

*Nautilus pompilius* is the most widely distributed and best known of all *Nautilus* species.

#### DISTRIBUTION

Andaman Sea to Fiji, southern Japan to Great Barrier Reef. North-eastern and north-western Australia.

#### REFERENCES

Saunders and Ward (1987), Norman (2000), Jereb and Roper (2005), Sinclair *et al.* (2007), Dunstan *et al.* (2011a, b, c).

### *Nautilus stenomphalus* Sowerby, 1848

*Nautilus stenomphalus* Sowerby, G. B. (1848). Monograph of the genus *Nautilus*. *Thesaurus Conchyliorum 2*, 463–465 [465, pl. 97 fig. 3].

Type data: repository unresolved.

Type locality: not designated.

#### COMMON NAME

White-patch Nautilus

Live animal (P. Ward).

Shell, lateral view (AM C.148211, L. Stephens).

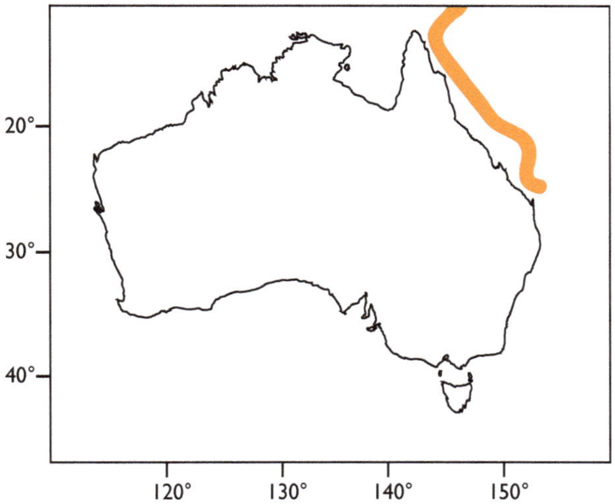


*Nautilus stenomphalus* distribution. *Note:* The distribution is based on shell and whole animal data (shells may have drifted).

#### DISTINGUISHING CHARACTERS

Very similar in size and weight to *Nautilus pompilius*, from which it can be distinguished by the absence of the callus and reduced banding colouration, lacking altogether in the umbilicus region. The nodose texture of the hood can be used to distinguish this species from *N. pompilius*.

#### DIAGNOSTIC FEATURES

Thickened callus absent. White patches in umbilical and shoulder regions of shell. Hood covered by elevated papillae in very characteristic rough, nodose texture that may extend to tentacular sheaths. Tentacle sheaths with scalloped edges.

#### SIZE

To ~170 mm in diameter.

#### REMARKS

Due to variations in shell patterning, it may sometimes be difficult to distinguish this species from *N. pompilius* based on the shell alone. The soft part morphology of the hood readily distinguishes the two species.

#### DISTRIBUTION

Only known from the Great Barrier Reef, Queensland. It may possibly hybridise with *N. pompilius* in this region.

#### REFERENCES

Saunders and Ward (1987), Norman (2000), Jereb and Roper (2005).

### *Allonautilus scrobiculatus* ([Lightfoot], 1786)

*Nautilus scrobiculatus* [Lightfoot, J.] (1786). A catalogue of the Portland Museum, lately the property of the Duchess Dowager of Portland, Deceased. London, 194 p. [169, 182].

Type data: repository unresolved.

Type locality: New Guinea.

#### DIAGNOSTIC FEATURES

Shell evolute and quadrate in cross-section. Shell ornamentation scrobiculate, or reticulate in all growth stages, largely non-bifurcating and concentrated on upper flanks of shell, coalescing and disappearing on shoulder. Umbilicus wide with perpendicular walls. Periostracum shaggy on freshly caught shells. Hood ornamented by multi-pronged tubercles (after Ward and Saunders 1997).

#### REMARKS

While not found in Australian waters, *Allonautilus scrobiculatus* is included here because drift shells are occasionally found in northern Australia. This is the rarest and most distinctive of the living nautiloids.

#### DISTRIBUTION

Papua New Guinea, Solomon Islands, Indonesia.

#### REFERENCES

Ward and Saunders (1997), Jereb and Roper (2005).

Shell, lateral view (AM C.113362, L. Mott). *Note:* Shaggy periostracum absent.

## Subclass Coleoidea Boettger, 1952

### Superorder Decapodiformes

This group includes all cephalopod taxa with 10 arms, suckers and/or hooks, and no external shell: the orders Bathyteuthida, Idiosepiida, Myopsida, Oegopsida, Sepiida, Sepiolida and Spirulida.

### *Order Bathyteuthida*

The Bathyteuthida comprises the Bathyteuthidae and the Chtenopterygidae. Some authors treat this taxon at the superfamily level Bathyteuthoidea. The head has a tentacle pocket and there are suckers on the buccal lappets (unlike oegopsid squids). This group comprises six species in two monogeneric families. In a molecular study that included five species, Lindgren (2010b) confirmed the monophyly of each genus and the order.

#### Key to families and genera of the Bathyteuthida

1a. Fins round, paddle-like; subterminal ..........................
..............Bathyteuthidae, *Bathyteuthis* (*B. abyssicola*)

1b. Fins extend full length of mantle in adults, margins fringe-like ........................................................................
................Chtenopterygidae, *Chtenopteryx* (*C. sicula*)

### Bathyteuthidae Pfeffer, 1900

#### DIAGNOSTIC FEATURES

Subterminal fins round or paddle-like. Funnel-locking cartilage simple, straight. Arm suckers in irregular series. Tentacular club small with numerous rows of minute suckers. Buccal connectives attach dorsally to ventral arms. A single photophore embedded at base of each of the dorsal, dorsolateral and lateral arms.

#### HABITAT AND BIOLOGY

Rarely caught except in Antarctic waters.

*Interest to fisheries.* No fisheries importance due to their small size and apparent rarity.

#### SIZE

Less than 100 mm ML.

### *Bathyteuthis* Hoyle, 1885

#### DIAGNOSTIC FEATURES

Mantle cylindrical, bluntly rounded posteriorly. Head wider than mantle. Eyes angled slightly anteriorly. Fins round or paddle-like; subterminal. Funnel-locking cartilage straight. Arm suckers arranged in irregular longitudinal series (two series proximally, increase to four series distally on arms 1–3; one or two series only on 4). Clubs small, unexpanded with many transverse rows (six to eight) of minute suckers, no carpal cluster. Buccal connectives attach to dorsal borders of arms 4; seven buccal lappets with minute suckers.

*Photophores.* A single, small, black-pigmented photophore embedded at base of each arm 1–3 (clear in young specimens but may be obscure in adults).

*Colour.* Uniform deep maroon over all surfaces.

#### HABITAT AND BIOLOGY

While most species of deep-sea squids produce ammonium to achieve neutral buoyancy, bathyteuthid squids concentrate high amounts of an unidentified low-density cation. This results in muscular tissue more comparable to shallow-living squids, which, in turn, enables bathyteuthids to be much stronger swimmers than other deep-sea squids.

Photosensitive vesicles on the dorsal surface of the head appear to function to detect bioluminescent light emitted from neighbouring organisms beyond the visual field.

#### SIZE

Small-sized squids; maximum ML 75 mm.

#### REMARKS

A molecular study by Carlini and Graves (1999) has indicated a close relationship between Bathyteuthidae and Chtenopterygidae. The single genus, *Bathyteuthis*, contains three species, with only *B. abyssicola* Hoyle, 1885 found in Australian waters.

#### REFERENCES

Roper and Jereb (2010a).

## *Bathyteuthis abyssicola* Hoyle, 1885

*Bathyteuthis abyssicola* Hoyle, W. E. (1885). Narrative of the Challenger Expedition. *Report on the Scientific results of the voyage of the H.M.S. Challenger during the years 1873–76, Narrative* 1(1), 269–274 [272, fig. 108].

Type data: holotype BMNH 1890.1.24.15.

Type locality: south-western Indian Ocean, 46°16′S, 48°27′E.

### COMMON NAME

Bathyal Deep-sea Squid

### DIAGNOSTIC FEATURES

Mantle robust, bluntly rounded posteriorly. Fins round, small, separate, short. Eyes orient slightly anteriorly. Arms short, blunt, with few (~100 on arms 1–3) small suckers; protective membranes present, with straight or gently scalloped borders; trabeculae not free, enlarged or elongate. Clubs unexpanded, short, with relatively few, small, similar-sized suckers. Gills short, blunt, narrow.

*Colour.* Deep maroon.

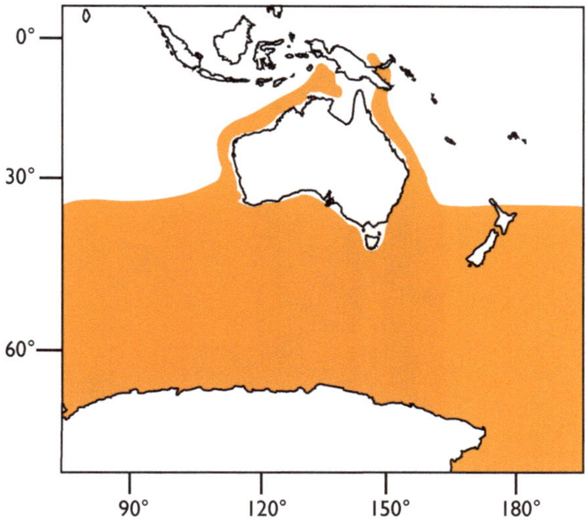


*Bathyteuthis abyssicola* distribution.

### HABITAT AND BIOLOGY

This is an oceanic species that occurs between ~100 and 4200 m depth, but normally it is encountered between 700 and 2000 m in the Southern Ocean where it carries

Preserved specimen, dorsal view (AM C.131861, L. Mott).

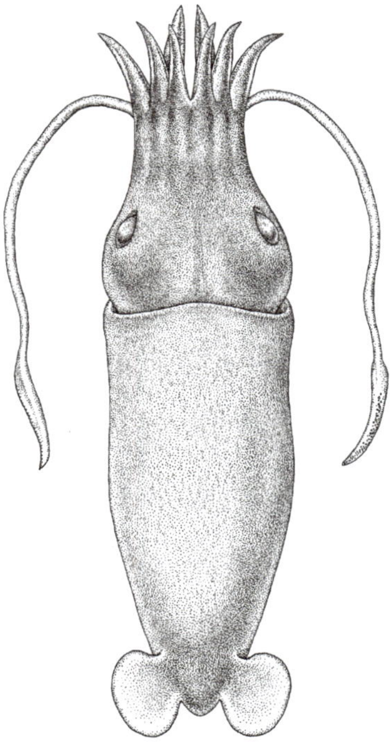

Dorsal view (ABRS, K. Hollis).

Tentacular club (ABRS, K. Hollis).

out a vertical diurnal migration. Paralarvae and juveniles tend to live at shallower depths than the adults. The ML at first maturity is ~40 mm to 50 mm in females and 35 mm in males.

*Interest to fisheries.* Although the species is abundant in the Southern Ocean, it presently has no commercial value.

#### SIZE
The maximum ML is 75 mm.

#### REMARKS
Kawashima *et al.* (2013) have determined the complete mitochondrial genome for *B. abyssicola*.

#### DISTRIBUTION
Circumglobal and cosmopolitan. It is most abundant in the Southern Ocean and in highly productive waters of the eastern Pacific, Atlantic and Indian Oceans and extends to south of the Antarctic Polar Front. Records from the Mediterranean Sea.

#### REFERENCE
Roper and Jereb (2010a).

## Chtenopterygidae Grimpe, 1922

#### COMMON NAME
Comb-finned Squids

#### DIAGNOSTIC FEATURES
Posterior end of mantle broadly rounded. Fins extend full length of mantle in adults, fused posteriorly; fins attach to lateral walls of muscular mantle, comprised of slender muscle bundles (fin ribs) connected by their membranes to produce a comb-like appearance. Fins very short in hatchlings and paralarvae, undergo strong ontogenetic growth in length. Funnel-locking apparatus straight, simple. Distal half of arms 1–3 with 6–14 longitudinal series of suckers. Arms 4 with very few small suckers in zigzag arrangement. Hectocotylus absent. Tentacular club suckers small to minute, in 8–20 irregular longitudinal series. Tentacles narrow, clubs not expanded, carpus absent. Mature females with accessory nidamental gland. Minute suckers on lappets of buccal membrane. Buccal connectives attach to ventral borders of arm 4.

*Photophores.* Mature males with large, dorsally directed photophore in posterior mantle cavity.

#### HABITAT AND BIOLOGY
Members of this monogeneric family are small, muscular, midwater squids that occupy tropical to subtropical waters at depths of 500–1000 m during the day and migrate into near-surface waters at night (upper 200 m).

During growth, the arms, tentacles and head all become longer relative to the body, which scarcely changes in shape. The fins become relatively longer, starting at the terminal end of the body and extending the whole length of the mantle when it reaches 30 mm. The structure of the fins suggests that chtenopterygids are not capable of sustained rapid movement.

Chtenopterygids are regularly encountered off the east coast in tropical oceanic surface waters. Larvae have been collected off New South Wales during summer and winter, suggesting that the spawning season is broad.

*Interest to fisheries.* They do not reach a suitable size or abundance to be of any commercial fisheries interest.

#### SIZE
Small squids to 100 mm ML.

#### REMARKS
The generic name formerly was spelled *Ctenopteryx*, but that name was preoccupied by a coleopteran. As a result, Appellöf (1890) spelled the genus *Chtenopteryx* and subsequently that spelling has been upheld by the International Code of Zoological Nomenclature through a petition by Bello and Giannuzzi-Savelli (1993).

Several undescribed species are present in this genus. Species-level systematic characters in *Chtenopteryx* currently seem limited to the following: (1) the maximum number of sucker series on the arms and tentacular clubs; (2) the presence/absence of visceral and eyeball photophores; (3) the relative size of the visceral photophores; and (4) the mantle width relative to ML. Two confirmed species are recognised: *C. sicula* (Verany, 1851) and *C. sepioloides* Rancurel, 1970. Two other species are currently of undetermined status: *C. chuni* Pfeffer, 1912 and *C. canariensis* Salcedo-Vargas & Guerrero-Kommritz, 2000. Both species are based on very few, small specimens. Little is known of the biology of any species.

#### REFERENCE
Roper and Jereb (2010b).

## *Chtenopteryx* Appellöf, 1890

#### DIAGNOSTIC FEATURES

Long fins containing muscular ribs. Buccal connectives attach to ventral border of arms 4.

### *Chtenopteryx sicula* (Verany, 1851)

*Sepioteuthis sicula* Verany, J. B. (1851). Cephalopodes de la Mediterranee. *Mollusques Mèditeranèens observé décrits figurés et chromolithographiés d'après le vivant ouvrage dedie à SM le Roi Charles Albert 1*, 1–132 [75, pl. 27].

Type data: holotype MHNN?

Type locality: Italy, Messina.

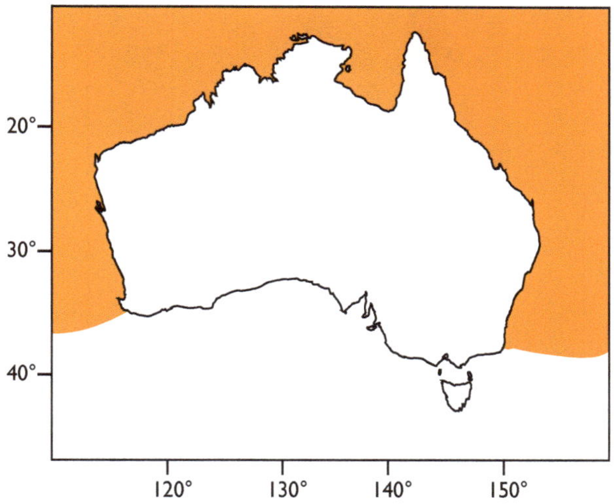

*Chtenopteryx sicula* distribution.

#### COMMON NAME

Comb-finned Squid

#### DIAGNOSTIC FEATURES

Three nuchal folds, with olfactory organ on first fold. Fins not extending to anterior margin of mantle. Arms 1–3 with biserial suckers proximally and tetraserial suckers distally, arms 4 with one or two series of suckers in zigzag arrangement; formula 4 > 3 > 2 > 1. Tentacular clubs with more than eight series of suckers. Each buccal membrane with two series of suckers. Flat, spoon-like conus present. Arm suckers in at least four series distally. Club suckers in more than eight series.

*Photophores.* Ocular photophore present, large elongate photophores on ventral surface of eyes; visceral photophores present (but see 'Remarks'). Nesis (1982/1987: 204) describes: 'A large photophore is situated on ventral side of the ink sac under the intestine; it looks like a small yellowish ball lying on the concave surface of the ink sac (like a fried egg on a black frying pan)'.

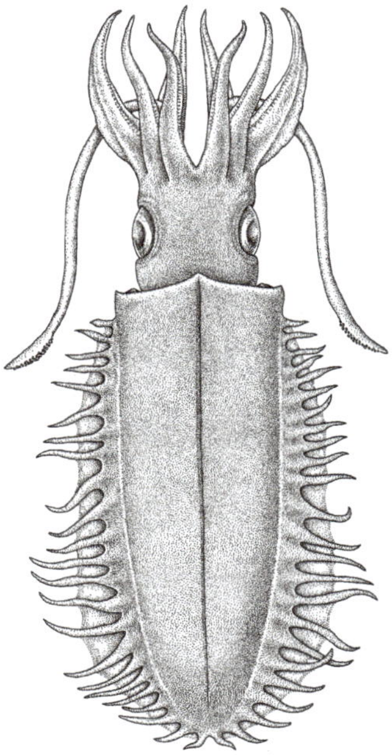
Dorsal view (ABRS, K. Hollis).

#### HABITAT AND BIOLOGY

Paralarvae epipelagic; adults descend to mesopelagic and bathypelagic depths and undergo strong diel vertical migrations.

#### SIZE

Maximum ML to 100 mm.

#### REMARKS

The original description of this species that reports the absence of the viceral photophore was based on squids of 13 mm ML and smaller. Subsequently, however, specimens from the Mediterranean Sea that have visceral photophores have been described. This suggests that the identity of *C. sicula* is uncertain because two species may occur in the Mediterranean Sea.

At present, two species in the genus are recognised, but according to Roper and Jereb (2010b) there may be more undescribed species in the genus. Specimens from Australian waters need detailed examination. The Australian representatives of the genus do have a visceral photophore.

There has been some confusion regarding the correct termination of the species name: *siculus*, or *sicula*. The correct termination is the feminine *sicula*.

### DISTRIBUTION

Cosmopolitan (circumglobal) in tropical and subtropical waters.

### REFERENCES

Nesis (1982/1987), Roper and Jereb (2010b).

## *Order Idiosepiida*

## Idiosepiidae Appellöf, 1898

While the recognition of this taxon as distinct on morphological and molecular grounds, the phylogenetic position of Idiosepiida remains unresolved (Allcock *et al.* 2014).

## *Idiosepius* Steenstrup, 1881

### COMMON NAME

Pygmy Squid

### DIAGNOSTIC FEATURES

Small animals (< 25 mm ML). Mantle elongate, obovate to cigar-shaped; posterior mantle margin bluntly pointed at distal tip. Glandular oval attachment organ (or 'adhesive gland') on dorsal posterior end of mantle, towards posterior end of fin region. Head prominent, eyes covered by a cornea. Dorsal mantle not fused to head. Nuchal cartilage absent. Fins small, kidney-shaped, attached laterally, and slightly obliquely to posterior end of mantle. Arms short; arm suckers biserial, enlarged in males. Hectocotylus present, both ventral arms modified: right arm flattened, wide, with protective membranes, sometimes with transverse ridges and grooves orally, distal tip of left ventral arm bilobed. Club with two to four suckers in each transverse row. In females, right oviduct non-functional. Eggs small, benthic.

*Gladius.* Reduced or a vestige only.

### HABITAT AND BIOLOGY

*Idiosepius* species generally live in shallow water among seagrass, algae, or among mangroves and are solitary in habit. Jackson (1986) noted niche separation in *I. pygmaeus v. I. paradoxus*,* with *I. pygmaeus* found in estuaries and having a nektonic life, while *I. paradoxus* forms part of the shelf plankton community. They are camouflaged during the day by sticking to the underside of seagrass or algae using an adhesive organ on the dorsal mantle.

Histochemical, histological and ultrastructural methods have been used to elucidate the nature of the secretory cells in the attachment organ of *I. paradoxus*, *I. pygmaeus* and *I. biserialis* (von Byern *et al.* 2008). Their work suggests that *Idiosepius* uses a transitory adhesion, perhaps induced by the secretion of a highly viscous carbohydrate-rich gel. Release may be caused by contraction of the mantle musculature and/or chemical release mechanisms. The precise mechanism has yet to be determined. This is in contrast to a 'duo-gland' adhesive system, as seen in *Euprymna*, in which adhesive secretions and releasing secretions are produced by different cells.

Females grow to larger sizes and live longer than males. Mating takes place head to head. Nabhitabhata and Suwanamala (2008) have noted that some species (for example, *I. paradoxus* and *I. pygmaeus*) use the hectocotylus for transferring spermatophores to the females, while others (*I. biserialis* Voss, 1962 and *I. thailandicus* Chotiyaputta *et al.*, 1991) use their tentacles to pass spermatophores to the females. Spermatophores are usually attached to the base of the arms. They are promiscuous with multiple matings occurring with different partners.

*Idiosepius* is unusual among decapods in having both ventral arms modified in males and these arms are distinctly different. (Both arms may be modified in some sepiolids, but they are usually mirror images of each other.) Sato *et al.* (2013) have shown, using high-speed camera observations, that *I. paradoxus* from Japan uses the right hectocotylus as a guide for spermatophore transfer by the left hectocotylus. They suggest that this may be because idiosepiids have unusually large spermatophores for their size (~20% ML) and both hectocotylised arms are needed to facilitate spermatophore transfer. Multiple spawning has been confirmed for *I. pygmaeus* and is likely true for other species. In *I. paradoxus* from Chita Peninsula, central Honshu Island, Japan, multiple spawning occurred with 17–64 eggs laid per event.

Development includes a pelagic stage but no metamorphosis occurs.

### SIZE

Idiosepiids are the smallest known cephalopods, with mature males ranging up to 17 mm ML and females up to 22 mm ML.

* Recent work has shown that *I. paradoxus* does not occur in Australian waters. Works referring to *I. paradoxus* from Australia possibly pertain to *I.* sp. nov. Reid and Strugnell (in prep.).

### REMARKS

The phylogenetic position of *Idiosepius* among the Decabrachia remains unclear. A relationship with the oegopsids has been postulated in the molecular studies of Bonnaud *et al.* (1997, 2005) and Takumiya *et al.* (2005). A study of the vascular anatomy of *I. paradoxus* by Yoshida *et al.* (2010a) suggests, as supported by the molecular studies of Carlini *et al.* (2000) and Strugnell *et al.* (2005), that they are more closely related to the sepiolids than the oegopsids. Lindgren *et al.* (2004) and Lindgren *et al.* (2012) suggest a relationship with the sepiids.

### DISTRIBUTION

Indo-west Pacific from Japan to Australia (all coasts) and South Africa. One species has been reported from Russia.

### REFERENCES

Jackson and Choat (1992), Lu and Dunning (1998).

### REMARKS

The Australian representatives of this species are currently under review by the author and colleagues. Three *Idiosepius* species occur in Australian waters: *I.* sp. nov. Reid & Strugnell (in prep.), *I. notoides* Berry, 1921 and *I. pygmaeus* Steenstrup, 1881. *Idiosepius* is currently thought to include eight species distributed throughout the Indo-West Pacific from Thailand to Australia and around South Africa. At this time it is difficult to distinguish female specimens belonging to this species.

## Key to species of the Idiosepiidae

1a. Seven to 11 pairs of suckers along entire length of ventral arms of male. Funnel–mantle locking apparatus a straight ridge corresponding to a straight furrow ........................................ *I. notoides*

1b. Ventral arms of males with suckers on basal part only. Funnel–mantle locking apparatus a curved ridge corresponding to a deep, rounded socket......2

2a. Ventral arms much shorter than remaining arms in both sexes. One to four suckers at base of ventral arms of male. Gills with 28–45 lamellae per demibranch........................................ *I. pygmaeus*

2b. Ventral arms slightly longer than remaining arms in males; similar in length to remaining arms in females. Six to 10 suckers at base of ventral arms in males. Gills with 14–20 lamellae per demibranch ........... *I.* sp. nov.

## *Idiosepius* sp. nov. Reid & Strugnell (in prep.)

*Idiosepius* sp. nov. Reid & Strugnell (in prep.).

### REMARKS

This species has been confused with *Idiosepius paradoxus* (Ortmann, 1888), but *I. paradoxus* does not appear to occur in Australian waters, despite earlier reports.

The complete mitochondrial genome of *I.* sp. nov. from Dunwich, Stradbroke Island, Queensland (~27°29′54″S,

Live animal (S. Schulz).

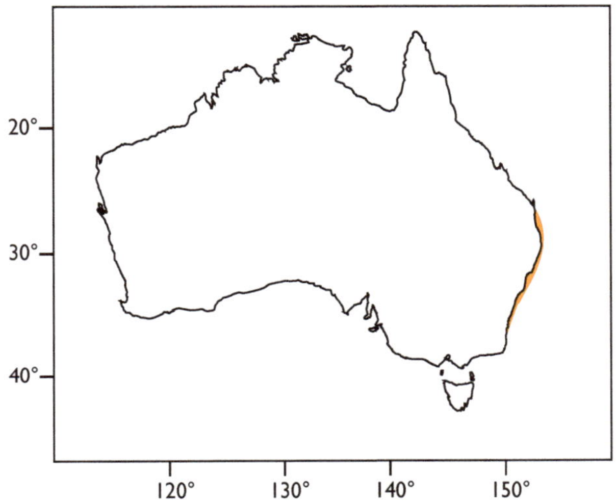


*Idiosepius* sp. distribution.

153°23′58″E) has been sequenced (Hall *et al.* 2014). This is the first complete mitochondrial genome of any pygmy squid that has been sequenced. The mtDNA genome is 16 183 bp long with an AT content of 75.4%. The gene order most closely resembles that of the bobtailed squid *Semirossia* (Sepiolidae) (a genus that does not occur in Australian waters).

Live animal, dorsal view (J. Gaskell).

**DISTRIBUTION**

Eastern Australia.

**REFERENCE**

Reid and Strugnell (unpublished data).

## *Idiosepius notoides* Berry, 1921

*Idiosepius notoides* Berry, S. S. (1921). A review of the cephalopod genera *Sepioloidea*, *Sepiadarium*, and *Idiosepius*. *Records of the South Australian Museum 1*(4), 347–364 [361, fig. 67].

Type data: holotype SAMA D.17495.

Type locality: Australia: South Australia, Goolwa.

Live animal swimming (J. Gaskell).

**COMMON NAME**

Southern Pygmy Squid

Live animal attached to seagrass (P. Mercurio).

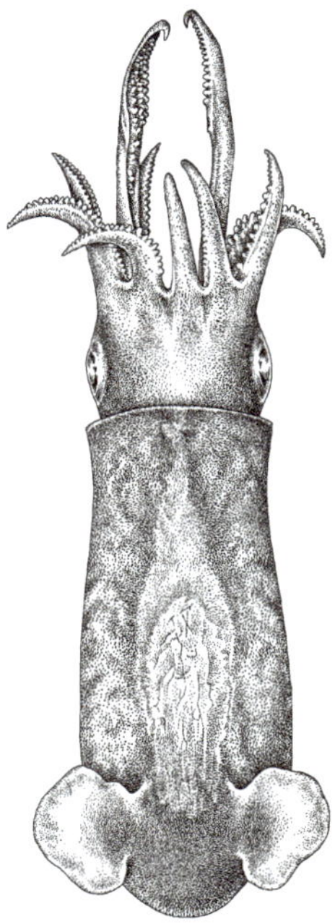

Dorsal view (ABRS, K. Hollis).

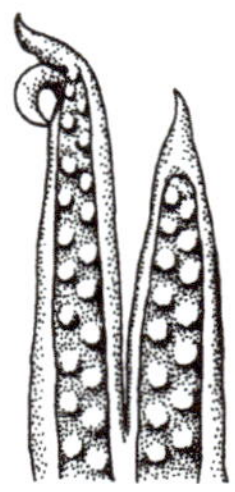

Hectocotylus (K. Hollis, redrawn from Nesis 1982/1987).

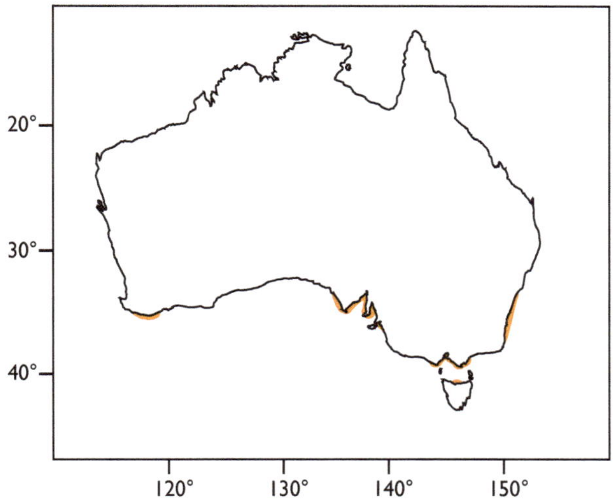


*Idiosepius nototoides* distribution.

## DIAGNOSTIC FEATURES

Mantle-locking cartilage a straight, narrow ridge corresponding to straight furrow in funnel. Males with > 10 suckers on both modified ventral arms.

*Colour.* Short white line around the eyes.

## SIZE

Males up to 15.8 mm ML; females up to 25 mm ML.

## REMARKS

Age, growth and maturation of *Idiosepius notoides* from Tasmania was compared with that of *I. pygmaeus* in a study by Tracey *et. al.* (2003). (*I. pygmaeus* occurs in more northerly tropical and subtropical latitudes.) They found, using statolith increment analysis, that the maximum age of *I. notoides* was 115 days, with males having a shorter lifespan than females despite both sexes having similar growth rates. The onset of maturity was ~68 days for males and 88 days for females. *I. notoides* grows to a larger size than *I. pygmaeus*. Both species had a similar size of onset of sexual maturity (6.5 mm ML for males and 14.0 mm ML for females), but these sizes were reached in a shorter time (45–60 days) in *I. pygmaeus*. Females appear to attain a greater size than males due to their increased longevity. Growth rates are directly influenced by hatching temperature, with faster growth occurring at higher temperatures. (It is important to note here, however, that this study was conducted without statolith age validation to determine whether increments are, indeed, added daily. A study by Sato *et al.* (2008), which included age validation, has suggested that the increment count in *I. paradoxus* occurring in temperate Japanese waters may actually underestimate the true age. As *I. notoides* is also a temperate water species, it may be that the ages reported by Tracey *et al.* (2003) are underestimates.)

The ovaries of mature *I. notoides* contain eggs at all stages of development with the exception of stage five oocytes. (In *I. pygmaeus* these are retained only for a few minutes immediately before spawning.)

## DISTRIBUTION

Southern and eastern Australia, including Tasmania. Shallow, inshore waters.

## REFERENCE

Reid (2005a).

## *Idiosepius pygmaeus* Steenstrup, 1881

*Idiosepius pygmaeus* Steenstrup, J. (1881). *Sepiadarium* og *Idiosepius* to nye Slaegter af Sepiernes Familie. Med Bemaerkninger om de to beslaegtede Former Sepiolidea D'Orb. og Spirula Lmk. *Danske Videnskabernes Selskabs Skrifter, 6 Raekke, Naturvidenskabelig og Mathematisk 1*(3), 211–242 [219, pl 1 figs 11–22].

Type data: syntypes ZMUC.

Type locality: South China Sea, 4°20′N, 107°20′E, and Zamboanga, Philippine Islands.

## COMMON NAME

Tropical Pygmy Squid

Live animal (M. Norman).

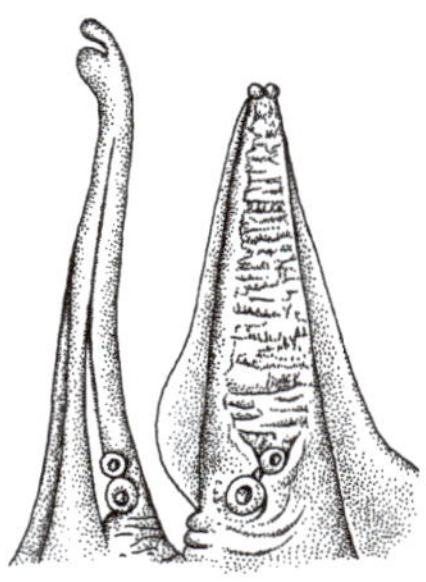

Hectocotylus (K. Hollis, redrawn from Nesis 1982/1987).

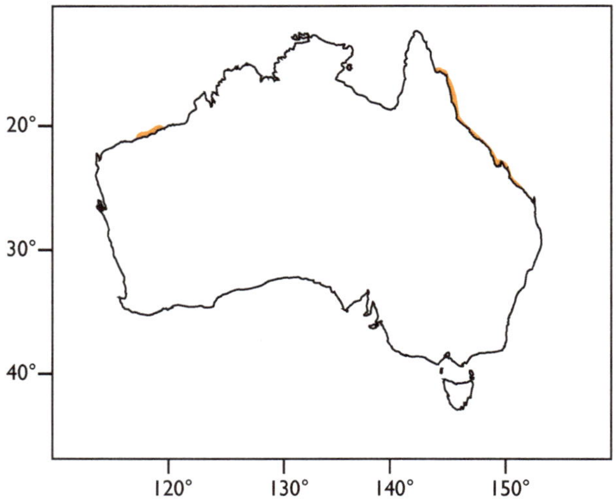


*Idiosepius pygmaeus* distribution.

**DIAGNOSTIC FEATURES**

Mantle–funnel locking cartilage a curved ridge corresponding to a deep, rounded socket. Male modified arms 4 with only one to four suckers at the base. Tentacles much thinner than arms.

*Colour.* Live animals often display reverse countershading (pale dorsally, dark ventrally).

**HABITAT AND BIOLOGY**

*Idiosepius pygmaeus* has a fast growth rate and short lifespan in tropical Australia with males maturing at 42 days and living for ~67 days, while females mature as young as 60 days and live for ~79 days (Jackson and Choat 1992).

Jackson *et al.* (2008) captured *I. pygmaeus* in light traps deployed in deep water (~75 m) off Ningaloo Reef, Western Australia. This suggests that the species may inhabit a wider range of habitats than previously understood.

**SIZE**

Up to 20 mm ML.

**REMARKS**

The life history of *I. pygmaeus* has been well studied. This species can produce egg clutches totalling up to 9 × its body weight (Lewis and Choat 1993), while still growing at a substantial rate (Jackson 1989). Females appear to attain larger sizes than males via a comparatively faster growth rate.

**DISTRIBUTION**

Indo-Pacific: Japan, South China Sea, Philippines, Palau Islands, Indonesia, northern and north-eastern Australia. Shallow, inshore waters.

**REFERENCE**

Reid (2005a).

## *Order Myopsida*

**DIAGNOSTIC FEATURES**

The head has a pocket for the tentacles. The eyes of myopsid squids are covered with a corneal membrane, with a minute pore in most species. There is no secondary ventral eyelid. The funnel lacks lateral adductor muscles. Suckers are usually present on the buccal lappets. The arms and the clubs have suckers, but never hooks. The club lacks a carpal-locking apparatus. The right oviduct is absent and accessory nidamental glands are present.

*Gladius.* The gladius extends along the full length of the mantle.

**Key to myopsid families**

1a. Aquiferous pores in the cornea present. Funnel–mantle locking apparatus a simple, straight groove and ridge ....................................................Loliginidae

1b. Aquiferous pores in the cornea absent. Funnel–mantle locking apparatus subcircular, bisected by a curved groove that extends anterioposteriorly.......... ..........................................................Australiteuthidae

## Australiteuthidae Lu, 2005

The family Australiteuthidae is monotypic, represented by *Australiteuthis aldrichi* Lu, 2005. This family has some typical sepiolid-like characters: no acquiferous pores in the cornea, deep tentacular pockets into which the tentacles are retracted, and a sepiolid-like photophore on the ink sac. It also has the following teuthid traits: eyelids absent, buccal membrane connectives attached to the arms, and tricuspid rhachidian teeth.

### *Australiteuthis* Lu, 2005

As for family.

## *Australiteuthis aldrichi* Lu, 2005

*Australiteuthis aldrichi* Lu, C. C. (2005). A new family of myopsid squid from Australasian waters (Cephalopoda: Teuthida). *Phuket Marine Biological Center Research Bulletin 66*, 71–82. [72, figs 1–9].

Type data: holotype NTM P.12300.

Type locality: Australia: Western Australia, Joseph Bonaparte Gulf.

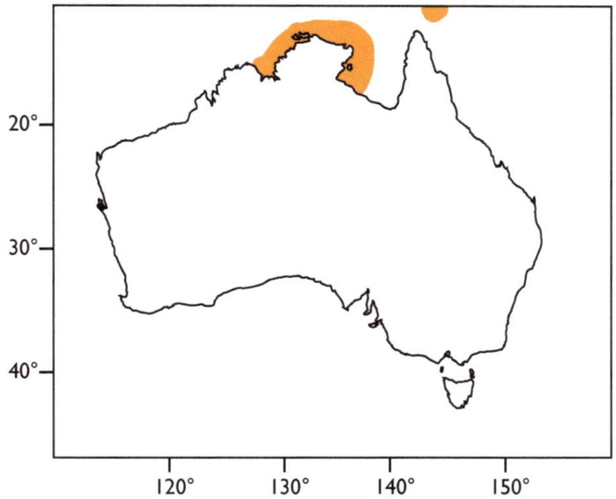


*Australiteuthis aldrichi* distribution.

### COMMON NAME(S)

Aldrich's Australasian Inshore Squid

### DISTINGUISHING CHARACTERS

Differs from the Loliginidae in shape of the funnel-mantle locking apparatus and the absence of aquiferous pores in the cornea.

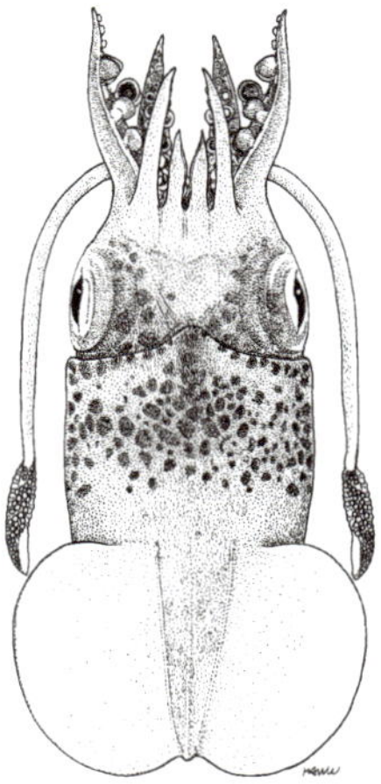

Dorsal view (FAO, E. D'Antoni).

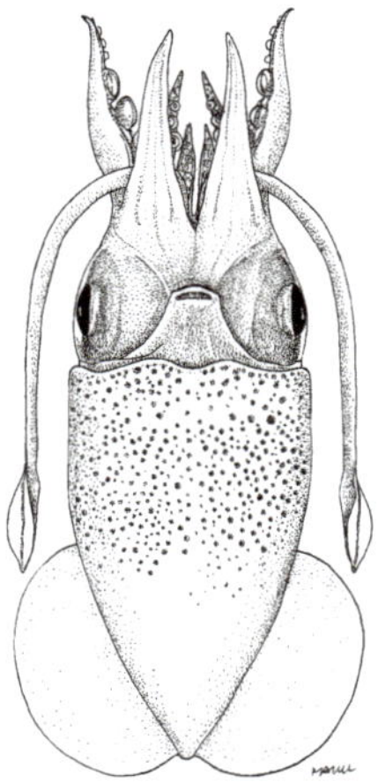

Ventral view (FAO, E. D'Antoni).

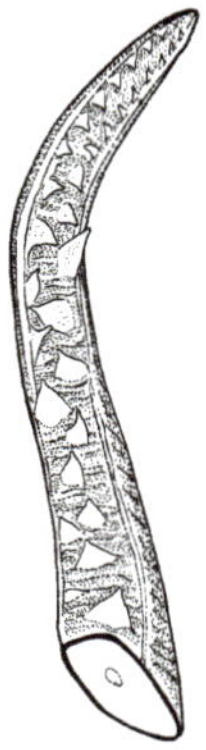

Hectocotylus (FAO, E. D'Antoni).

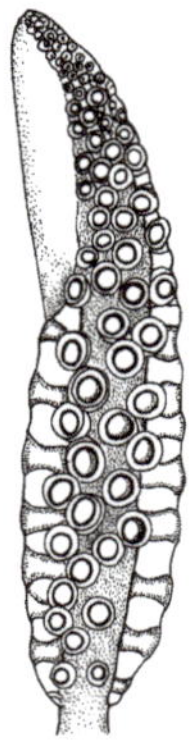

Tentacular club (FAO, E. D'Antoni).

### DIAGNOSTIC FEATURES

Eyes covered with transparent cornea, but without aquiferous pores. Fins separated, not united posteriorly. Funnel-locking cartilage large, subcircular, bisected by boomerang-shaped groove that runs anterioposteriorly; anterior wing of groove with wide anterior end; groove deepens and becomes slightly narrower posteriorly and curves medially to form posterior 'wing'. Left ventral arm hectocotylised by reduction in size of suckers of whole arm to tiny swellings on top of enlarged conical-shaped sucker stalks.

*Photophores.* Dumb-bell-shaped photophore of 'sepiolid type' on ventral surface of ink sac.

### HABITAT AND BIOLOGY

Probably benthic to epibenthic on soft substrates, *A. aldrichi* has been captured from 9 to 61 m using beam trawls in near-shore waters. Males mature ~17 mm ML, females larger, mature at 22 mm ML.

### SIZE

Very small: 15–28 mm ML.

### DISTRIBUTION

Western Australia, Joseph Bonaparte Gulf; inshore waters of the Northern Territory and New Guinea, Gulf of Papua.

### REFERENCE

Jereb *et al.* (2010b).

## Loliginidae Lesueur, 1821

### COMMON NAME(S)

Calamary, Inshore Squid, Pencil Squid

### DISTINGUISHING CHARACTERS

Other commercially exploited families (e.g. Ommastrephidae, Thysanoteuthidae, and Onychoteuthidae) have eye lenses that are open to the sea, not covered by a transparent corneal membrane. They all lack suckers on the buccal lappets that are present in some loliginids.

### DIAGNOSTIC FEATURES

Mantle elongate and tapered posteriorly. Fins (of variable shape and length) attach laterally to mantle, always reaching posterior end of body. Transparent skin (corneal membrane) covers eye lens. Aquiferous pore present. Funnel–mantle locking apparatus simple, straight groove and ridge. Arms with biserial suckers. Left ventral arm hectocotylised in males. Tentacular club usually with tetraserial suckers, arms partially retractile. Hooks absent. Buccal connectives attach to ventral margins of ventral arms 4; seven buccal lappets, with suckers in some species. Females with a single oviduct and accessory nidamental glands.

*Gladius.* As long as mantle, chitinous, pointed anteriorly, keeled on ventral side.

### HABITAT AND BIOLOGY

Jet propulsion is the most important method of locomotion in loliginids.

Eggs are spawned in finger like masses and attached to the substrate. Many (but not all) loliginids are multiple spawners that continue to feed while growing and reproducing though there may be deterioration in the condition of mature individuals. Paralarvae are like miniature adults. Growth is continuous and non-asymptotic.

Jackson (2004) summarised the results of extensive studies of growth and life histories in loliginid squids and concluded that validation studies on loliginids continue to support the assumption that statolith increments are laid down daily (although not all researchers concur with this view, see Hunt *et al.* (2011)) and that short lifespans are typical, with nine of 21 species studied having lifespans of < 200 days, eight of 21 species having lifespans between 200 days and 1 year, and only three species with lifespans > 1 year.

Loliginids eat a wide range of food, including crustaceans, polychaetes, fishes and cephalopods, with food preferences in the non-Australian species studied shown to vary with growth, habitat, sex and reproductive maturity. *Uroteuthis* (*Aestuariolus*) *noctiluca* (Lu, Roper & Tait, 1985), *U.* (*Photololigo*) *'chinensis'* (Gray, 1849) and *Sepioteuthis 'lessoniana'* have all been reared successfully in captivity on live sergestid shrimps, juvenile penaeid prawns and fishes (Jackson 1990). Prey detection and capture are visual. In turn, loliginids are preyed on by many vertebrates, including fishes, seabirds and marine mammals. As stated by Cotton (1940: 409), '[a] large variety of fishes prey upon the calamaries with eagerness'.

Extensive diel migration is thought to characterise members of this family.

*Interest to fisheries.* The loliginids are one of the major groups of commercially utilised cephalopods. In other parts of the world they are also used widely in neurophysiological research. Loliginids are important in fisheries, but are a largely underutilised Australian resource, with large standing stocks in the southern part of the Arafura Sea, the northern Timor Sea, the southern Timor Sea and the North West Shelf. The taxonomy of several purported loliginid species needs to be resolved if fisheries are to be sustainably managed.

### REMARKS

While the monophyly of the loliginids is well supported (e.g. Anderson 2000), systematics within Loliginidae has long been problematic. Conflicting generic-level classifications have been obtained through recent morphological and molecular studies. A workshop convened during the 2003 meeting of the Cephalopod International Advisory Council held in Phuket, Thailand addressed these concerns and reached a consensus regarding the classification of this family, which includes 10 genera and nine valid subgenera. That classification has been followed in the recently published FAO Guide (Jereb and Roper 2010) and is also followed here.

In a study primarily to fill knowledge gaps in the understanding of South Atlantic loliginids, Sales *et al.* (2013) examined a large molecular dataset that included several different loliginid taxa. Their results are generally consistent with Anderson's (2000) study using mtDNA, COI and 16S and Strugnell *et al.*'s (2005) combined data from mitochondrial and nuclear DNA sequences. As in the earlier studies, Sales *et al.* (2013) reported a paraphyletic arrangement for the genus *Sepioteuthis* using sequences of mitochondrial data, although the nuclear data supported the monophyly of this taxon. It also supported the hypothesis that *S. 'lessoniana'* is a species-complex. Although not including any specimens collected in Australian waters, this study found no evidence for cryptic speciation across the vast range of *U. (P.) 'chinensis'*, but suggested that *U. (P.) 'duvaucelii'* (Orbigny, 1835) likely represents a cryptic species-complex.

Two genera, *Uroteuthis* and *Sepioteuthis* are found in Australian waters. Determination of species boundaries within the loliginids, especially for the genus *Uroteuthis*, has proved problematic due to the considerable morphological variation that makes diagnosis difficult.

#### DISTRIBUTION

Shelf and upper slope worldwide. Not found in Antarctic regions.

### Key to genera of the Loliginidae

1a. Body cylindrical, fins very elongate and extend full length of the mantle. No photophores on ventral surface of ink sac........................................ *Sepioteuthis*

1b. Body elongate, tapering posteriorly, fins terminal and short. Pair of photophores on ventral surface of ink sac.......................................................... *Uroteuthis*

## *Sepioteuthis* Blainville, 1823

#### DIAGNOSTIC FEATURES

Mantle broad, relatively stout, tapered to a blunt posterior end. Posterior mantle not elongated into tail-like structure. Fins semi-oval extending nearly full length of mantle (except in very young squids). Arm sucker rings with pointed teeth around entire margin. Tentacular clubs expanded, with tetraserial suckers. Hectocotylus without crest; sucker stalks elongated along modified portion of arm to form papillae on both dorsal and ventral sides; proximal suckers unmodified. Spermatophore cement body short.

*Gladius.* Quill-shaped.

*Photophores.* Absent.

#### HABITAT AND BIOLOGY

*Sepioteuthis* occur primarily in coastal waters and bays, and are found on seagrass beds, coral reefs and over sandy substrates. Eggs are laid in a single, straight series in egg strands that are joined basally forming masses.

#### REMARKS

Two of the world's three nominal species are found in Australian waters: *S. 'australis'* and *S. 'lessoniana'*. (A fourth species has been described from the Red Sea but its status is currently uncertain (Jereb *et al.* 2010c)). Based on allozyme studies, cryptic species have been detected among both *S. 'australis'* (Triantafillos and Adams 2001), and *S. 'lessoniana'* in Australian waters (Triantafillos and Adams 2005); hence the species names for the two taxa are enclosed in quotation marks throughout this book.

#### DISTRIBUTION

Members of the genus are widely distributed in the tropical and subtropical waters of the Indo-west Pacific, all around Australia (including Tasmania), the western Atlantic and the Mediterranean Sea.

#### REFERENCES

Anderson (2000), Jereb *et al.* (2010c).

### Key to *Sepioteuthis* species

1a. Lateral margins of fins weakly angular; maximum fin width towards midpoint of mantle. Live animals without narrow transverse bars and spots.................. ...................................................................... *S. 'australis'*

1b. Fins more rounded than angular; maximum fin width towards posterior end of mantle. Live animals with pale translucence; iridescent pale transverse bars or spots present dorsally in all colour phases.... .................................................................. *S. 'lessoniana'*

## *Sepioteuthis 'australis'* Quoy & Gaimard, 1832

*Sepioteuthis australis* Quoy, J. R. and Gaimard, J. P. (1832). Mollusques. *Voyage de decouvertes de l'Astrolabe pendant les annees 1826–1827–1828–1829, Zoologie 2*(1), 1–320. Atlas of 93 plates [1833] [77, pl. 4 fig. 1].

Type data: syntype MNHN 2-5-430 (beaks only).

Type locality: Australia: [New Holland], probably Westernport [Port Western].

#### COMMON NAME

Southern Calamari Squid

#### DISTINGUISHING CHARACTERS

General appearance closely resembles *Sepioteuthis 'lessoniana'*, but lateral margins of fins are weakly angular. Maximum width of fins towards midpoint of mantle.

## DIAGNOSTIC FEATURES

Mantle robust, tapers to a blunt point posteriorly. Fins very long, exceeding 90% ML, fin width <50% of fin length; greatest width of fins at about midpoint. Arm 3 sucker rings with 25–30 sharp teeth. Tentacular clubs long, not greatly expanded; suckers moderately sized with 22–27 sharp teeth around rings of largest suckers.

*Colour.* Translucent in life; when captured change rapidly to rust red or brown. Narrow white to blue luminescent line along base of fins where they attach to body.

## HABITAT AND BIOLOGY

*Sepioteuthis 'australis'* is the most well-studied Australian cephalopod, with most of the research done, and described below, on southern Australian populations. It commonly inhabits coastal waters and bays, usually in depths of <70 m, and it is often found in depths from the surface to 10 m. In South Australia, offshore nursery and inshore spawning grounds have been identified, with small (<30 mm ML) and large (>150 mm ML) individuals found predominantly in shallow, inshore waters, while medium-sized animals are more abundant offshore. The distribution and abundance patterns of adult squids in the Gulf St Vincent (South Australia) were found to be highly variable, but they conform to a systematic, seasonal pattern consistent year to year; abundance typically increases for 6 months, peaks, and declines for the remainder of the year, with the peak timing varying among regions, following an anticlockwise direction around the Gulf. Males attain larger sizes than females and trawl data indicate a roughly 1:1 male to female sex ratio.

As with many other cephalopods, *S. 'australis'* shows extreme individual variation in reproductive investment. Ho *et al.* (2004) examined how the lifetime growth rates of adult *S. 'australis'* related to somatic and reproductive condition. They found that the allocation of energy given to growth and reproduction was gender-specific, with faster-growing males being in better somatic and reproductive condition. Females do not allocate the same level of energy to somatic growth, perhaps due to a greater requirement to allocate energy for reproductive growth. Both males and females in better somatic condition were in better reproductive condition. Fast growth rates by individuals was achieved primarily by muscle fibre growth rather than the production of new fibres.

Pecl and Moltschaniwskyj (2006), in a study of *S. 'australis'* found off the east coast of Tasmania, showed that males hatched in warmer months had greater reproductive investment, faster growth and better somatic and reproductive condition (as shown in Ho *et al.*'s 2004 study), while females hatched in spring and summer and had less reproductive investment relative to their autumn- and winter-hatched counterparts. There was also evidence of seasonal impacts on life history, perhaps due to abiotic and biotic conditions.

Jantzen and Havenhand (2002) observed mating behaviour and spawning in *S. 'australis'* close to the suburb of Marino, South Australia (~35°04′S, 138°50′E). Unpaired males regularly engaged in antagonistic behaviour with paired males that involved extending the arms and tentacles and colliding with each other. While the pair males were otherwise occupied, females were often approached by and copulated with other competitive males, as has been observed in the cuttlefish *Sepia apama* in a strategy called 'sneaker mating' (Hanlon and Messenger 1996). As in *S. apama*, these male fights can escalate into physical combat. As well as defending their female from other males, paired males guard females when they are egg laying. Field observations showed that sneaker mating comprised approximately one-third of all mating attempts. The documentation of multiple paternities in egg strands by Van Camp *et al.* (2004) indicates that multiple matings are successful. Promiscuity ensures a ready supply of spermatophores for females and also increases genetic diversity in a brood. The most common number of sires within a single egg strand was three.

Spawning involves complex body patterns. Body patterning in *S. 'australis'* as used for communication on this spawning ground has been studied in detail by Jantzen and Havenhand (2003). When the male approaches from below the receptive female, copulation involves the male depositing spermatophores into the mantle cavity; when the male approaches from above or beside the female they are deposited onto the arm region. Males observed by Jantzen and Havenhand (2002) presented an extended tentacle and club to the female and then moved above or slightly anterior to the female, and rotated 180° on the longitudinal axis to an inverted position before reaching down with an arm to deposit spermatophores in the buccal region of the female. This upturned mating behaviour has been observed in other *Sepioteuthis* species. Mating with 'sneaker males' was always observed to occur as the females added a new egg capsule to an egg cluster. (Reproductive behaviour, however, may differ among different populations.)

Van Camp *et al.* (2003) have developed seven polymorphic microsatellite loci for mating system analysis in *S. 'australis'*. They have found that observed heterozygosities range from 32 to 100% and with this amount of variation these loci will have sufficient power to be used in future studies to examine the relative success of alternate mating strategies in *S. 'australis'*.

Mature individuals are found year-round, indicating that spawning also extends throughout the year. Kuipers *et al.* (2008) have shown that, rather than producing oocytes in a continuous trickle throughout the year, eggs are produced in larger discrete batches. During the time of their study (August 2004 to January 2005 off the east coast of Tasmania) they found that the average batch fecundity declined over the 3-month spawning season but the number of oocytes in the ovary increased, suggesting that in December females might have deposited small batches more often.

Moltschaniwskyj and Pecl (2007) showed that a spawning aggregation of *S. 'australis'*, again off eastern Tasmania, studied over a 2-year period, comprised a series of 'microcohorts'; in each month the aggregation was composed of different individuals with different biological characteristics, thus ensuring phenotypic and genetic diversity is maintained in the population. Currently, management of many squid populations assumes that there is a single cohort in the aggregation. The measurement of stock biomass at the start of the spawning season cannot be used as a basis for fisheries management decisions. A more reliable measure may be obtained by quantifying the density of deposited eggs. Temporally structured protection from harvest will ensure maintenance of population diversity. In an earlier study Moltschaniwskyj *et al.* (2002) showed that short fishing closures are of value in protecting the spawning aggregation.

Eggs are laid in white, finger-like strands, ~60–70 mm long, each strand containing between three and nine eggs. Since females may store sperm from more than one male, individual egg strands may display considerable genetic diversity through multiple paternities. Clusters of egg strands are attached to macroalgae and seagrass in sheltered, shallow, inshore areas. The egg mass size varies considerably, ranging from solitary egg strands attached to the end of seagrass blades to large masses containing up to 1912 strands. Most egg masses contain <300 egg strands. Egg laying may be initiated by visual stimuli. Several *S. 'australis'* have been observed adding eggs to an already established egg mass.

Moltschaniwskyj and Pecl (2003) found that egg laying in inshore seagrass beds occurred predominantly during spring and early summer. Of two sites examined (Hazards Bay and Coles Bay, Tasmania), egg production was similar between the two years at Hazards Bay but differed between the two years at Coles Bay. There was considerable spatial variability in egg production among seagrass beds that could not be attributed to density or vegetation cover among beds. On one occasion there was considerable egg loss during storms.

Steer and Moltschaniwskyj (2007) examined the effects of egg mass size, egg position, substrate and biofouling on embryo mortality in *S. 'australis'*. They found that embryo mortality was related to egg mass size, and ranged between 2 and 4%. Small egg masses <300 eggs were generally attached to soft-sediment vegetation, whereas larger masses were attached to macroalgae holdfasts or were unattached. Mortality was highest in the largest unattached egg masses and in embryos within the interior of the mass. In attached egg masses, mortality was highest in egg strands near the point of attachment within the mass and in eggs closest to the substrate. They attribute this to the inhibition of water circulation and, therefore, the inability of the embryos to respire adequately and eliminate metabolic wastes. Biofouling did not have a marked effect on embryo mortality. Other studies have shown that while relatively unaffected by natural temperature fluctuations, embryos may be damaged by sudden salinity changes, like those caused by heavy rainfall and subsequent runoff of freshwater in coastal areas (Steer *et al.* 2003a).

Embryonic development takes 61 days at 13°C and 31 days at 20°C. Hatching rates decline significantly with increasing temperature. Embryos hatch at night, reducing the risk of predation by visual predators. Tasmanian *S. 'australis'* hatchlings range in size from 4.3 to 7.3 mm ML, which is slightly larger than those from South Australia (mean 4.75 mm ML). Once hatched, paralarvae are photopositive and actively swim to the surface. Observations of South Australian paralarvae indicate that they remain planktonic on the spawning grounds until ~8 mm ML, after which they become benthic. Small paralarvae (up to 7 mm ML) have been observed to feed on mysid shrimp and other zooplankton associated with low-relief seagrass beds. Although their dietary requirements are unknown, mysid shrimp and crab zoea have been successfully used as prey for hatchlings in captivity. Initial

Live animal (P. Mercurio).

Live animal (P. Mercurio).

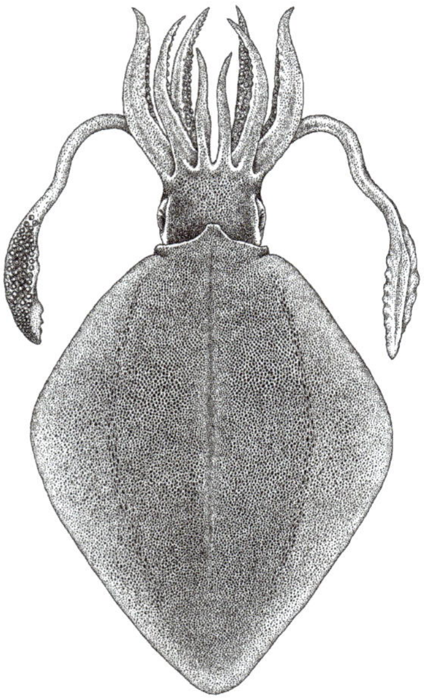

Dorsal view (ABRS, K. Hollis).

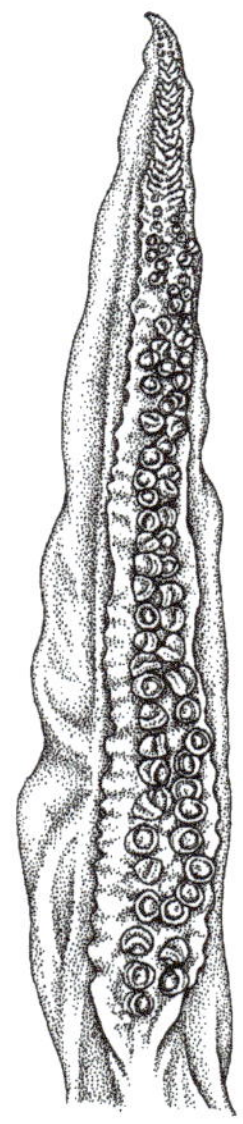

Hectocotylus (ABRS, K. Hollis).

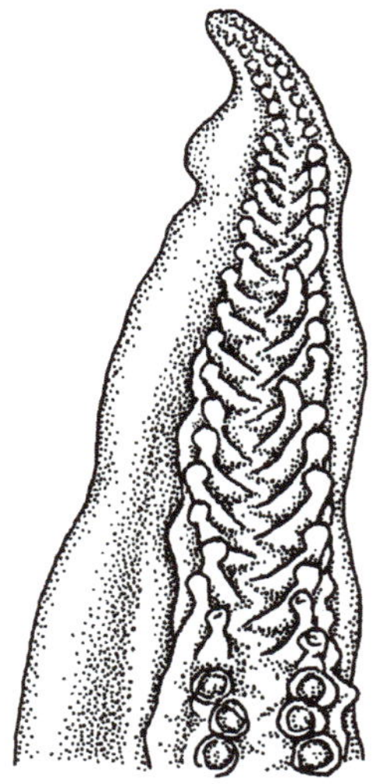

Enlargement of hectocotylus (ABRS, K. Hollis).

hatchling size may, by itself, explain the high variability in size in animals of the same age.

Growth is rapid, and the lifespan is ~1 year; males grow faster and attain larger sizes than females. Water temperature significantly affects growth rates, with individuals hatched in warmer seasons reaching a final size that could surpass that of earlier-hatched and, therefore, older individuals. The effect of temperature is not limited to initial growth stages, but continues to influence growth patterns throughout the adult life cycle. These seasonal differences in growth translate into substantial variations in size-at-age. The dynamic nature of oceanographic conditions in some areas of its distributional range (i.e. Tasmanian waters) may result in high variability in both the population structure and stock-recruitment relationships. In addition to environmental factors influencing growth, genetic variation plays a significant role. Triantafillos (2004) examined three sympatric genetic types from Western Australia, South Australia and New Zealand: two parental taxa and one hybrid. Significant differences in growth between genetic types were apparent, with the hybrids always growing faster (at least two-thirds larger at 150 days old) than the two parental taxa, a phenomenon commonly referred to as hybrid vigour. Spatial differences in growth also exist.

School of *Sepioteuthis. 'australis'* at Cabbage Tree Bay, New South Wales (N. Dawkins).

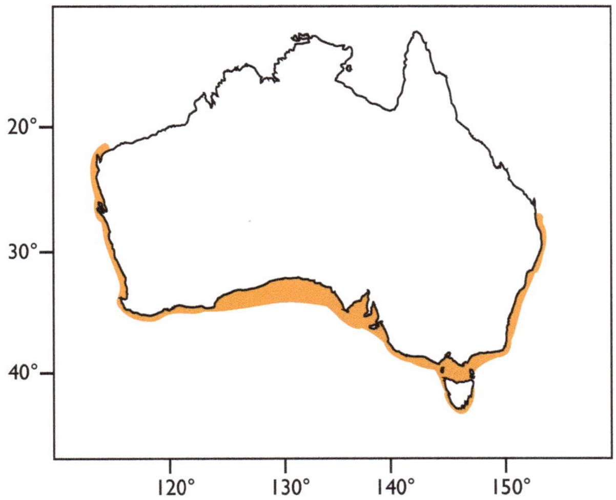


*Sepioteuthis 'australis'* distribution.

For example, individuals from Western Australia usually grow faster than those from South Australia and New Zealand.

Hunt *et al.* (2011) tested the 'one day one increment' relationship in the statoliths of hatchlings up to 40 days post-hatching and adult *S. 'australis'* held in captivity over summer and winter. The daily increment was verified for the hatchlings but not for adults, in which the number of days elapsed post chemical staining was an underestimate of the true age and was consistently greater for winter caught animals. These researchers concluded that the less than daily increment formation may be due to the periodicity of increment formation changing through the squid's lifespan, deleterious effects associated with rearing squids in captivity, or misinterpretation of statolith microstructure.

The statolith natal radius and its linear relationship to ML were utilised by Steer *et al.* (2003b) to determine whether any size-selective processes operated throughout the life history of *S. 'australis'*. Their work showed that smaller hatchlings are less likely to survive to adulthood, so there does appear to be an element of size-mediated mortality operating in populations of *S. 'australis'*.

*Sepioteuthis 'australis'* is a key component of the marine ecosystem as a primary consumer of crustaceans and fishes and as a food source for a variety of predatory species. It is the host taxon for the dicyemid parasite *Dicyema calamaroceum* Catalano and Furuya, 2013.

*Interest to fisheries.* *Sepioteuthis 'australis'* is the object of fisheries throughout its distributional range, where it is taken by commercial trawlers as well as by jigging and recreational fishing activities (Jereb *et al.* 2010c). Around southern Australia it is mainly fished in inshore habitats, particularly with trolled jigs over seagrass (Wadley and Dunning 1998). It is also a commercially important by-catch of the South Australian prawn fishery and is heavily fished by amateurs in South Australia. A study by Hibberd and Pecl (2007) examined the effects of commercial fishing on a population of *S. 'australis'* before, during and after a 3-month commercial fishing closure on the summer spawning grounds of Great Oyster Bay, Tasmania. Their work showed that the fishery is apparently selective for males, resulting in a 'natural' ratio biased towards males to one that led to a more even sex ratio. Their results highlight the need for closure during the spawning period to avoid modifying the process of sexual selection and the mating behaviours of the individuals within the spawning population.

Observations on spawning individuals in protected areas with acoustic telemetry (Pecl *et al.* 2006) demonstrate that animals spawn over several months, travelling hundreds of kilometres within the spawning areas during this period, without leaving the protected areas. Consequently, closures of critical spawning regions during periods of reproductive activity may protect spawners and enable undisturbed egg laying.

## SIZE

Maximum ML 550 mm in males, ~400 mm in females; maximum weight ~4 kg.

## REMARKS

Allozyme electophoresis identified three different 'genetic-types' from southern Australian and northern New Zealand waters (Triantafillos and Adams 2001) that were categorised as 'peripheral', 'central' or 'hybrid' types. The 'peripheral' types were almost exclusively found around the south-western and south-eastern limits of the distributional range (i.e. Perth and Albany on the western coast and

Newcastle, New South Wales on the eastern coast), whereas the 'central' type dominated in the Spencer Gulf and St Vincent's Gulf waters and the southern Great Australia Bight. Where these two forms overlap, a third, 'hybrid' form is found. Preliminary morphological and reproductive data supported the hypothesis of the existence of two taxa of *S. 'australis'* in South Australian waters, while mitochondrial DNA sequence data were inconclusive.

A recent study by Smith *et al.* (2015), however, using seven polymorphic microsatellite markers to assess connectivity and population structure of *S. 'australis'* across a large part of its geographic range in Australia, showed no significant genetic differentiation between any of the locations sampled, indicating high levels of gene flow and connectivity. Thus, *S. 'australis'* may be more resilient than previously thought against spatial and temporal mortality fluctuations.

### DISTRIBUTION

South-western Pacific Ocean: endemic to southern Australia and northern New Zealand, from Dampier in Western Australia (~20°S) to Moreton Bay in Queensland (~27°S), including Tasmania.

### REFERENCES

Semmens and Moltschaniwskyj (2000), Triantafillos and Adams (2001), Van Camp *et al.* (2004), Pecl and Moltschaniwskyj (2006), Pecl *et al.* (2006), Hibberd and Pecl (2007), Moltschaniwskyj and Pecl (2007), Jereb *et al.* (2010c).

## *Sepioteuthis 'lessoniana'* Lesson, 1831 (species-complex)

*Sepioteuthis lessoniana* Ferussac in Lesson, R. P. (1830–1831). Mollusques, Annelides et Vers. Voyage Autour du Monde, sur la Corvette de la Majeste, La Coquille, pendant les annes 1822, 1823, 1824 et 1825, sous le commandement du capitaine Duperrey. Paris, *Zoologie 2*(1), 471 pages (pp. 25–471 = 1831); Atlas, Mollusques, 16 plates (pls 11 = 1831) [241, pl. 11].

Type data: type, MNHN (not extant).

Type locality: not designated.

### COMMON NAME(S)

Bigfin Reef Squid, Northern Calamari, Oval Squid

### DISTINGUISHING CHARACTERS

Maximum width of fin towards posterior end of mantle; fins more rounded than angular. Pale translucent; iridescent pale transverse bars or spots present dorsally in all colour phases.

### DIAGNOSTIC FEATURES

Mantle long, robust, width ~40% of length. Fins large, length >90% ML, width up to 75% ML; greatest width posterior to midpoint of fins. Funnel very stout, large. Arm sucker rings with 18–29 long, sharp, regularly spaced, conical teeth around entire margin. Left arm 4 hectocotylised along distal quarter to a third; sucker pedicels elongate, blunt, conical and displaced laterally. Tentacular clubs long, expanded; median manus suckers enlarged; rings with 14–23 sharp, conical teeth.

*Colour.* Dark to light brown or green to pale translucent; iridescent pale transverse bars or spots present dorsally in all colour phases and are distinctive in this species.

### HABITAT AND BIOLOGY

*Sepioteuthis 'lessoniana'* is a neritic taxon occurring from the surface to a depth of at least 100 m. It occurs predominantly in subtropical coastal bays and inlets and offshore reefs (such as the outer Great Barrier Reef). The spawning season may be quite extended. Copulation (as observed in Australian representatives of this taxon) occurs by a male grasping a female from below and inserting his right ventral arm into her mantle. The right ventral arm is withdrawn and the left ventral arm carrying the spermatophores is then rapidly inserted. Wada *et al.* (2005) have observed more complex behaviours in a Japanese population, in which three mating behaviours were observed: male parallel to female, male upturned, and sneaking. Each method was postulated to result in different mating success as a result of differing spermatophore attachment sites. The method adopted was related to the size of the specimen, with male parallel mating the most successful due to the deposition of spermatophores at the opening of the oviduct. Eggs pass from the oviduct out through the funnel and are enveloped in secretions from the oviducal and nidamental glands. Finger-shaped egg capsules, each containing three to seven eggs, are attached in clusters to seaweeds, submerged mangrove roots, twigs, stones and corals in coastal waters.

This species grows to 200 mm ML in ~150 days in northern Australia. Based on the examination of statolith growth rings, Jackson (1990) concluded that *S. 'lessoniana'* reaches sexual maturity in <100 days in tropical north Queensland and reaches adult size in <200 days. (Age

Live animal (D. Cowdery).

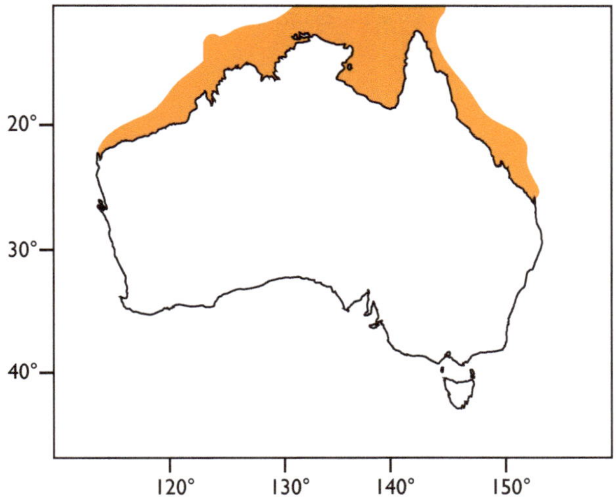


*Sepioteuthis 'lessoniana'* distribution.

Fresh-caught specimen, dorsal view (S. Cheng).

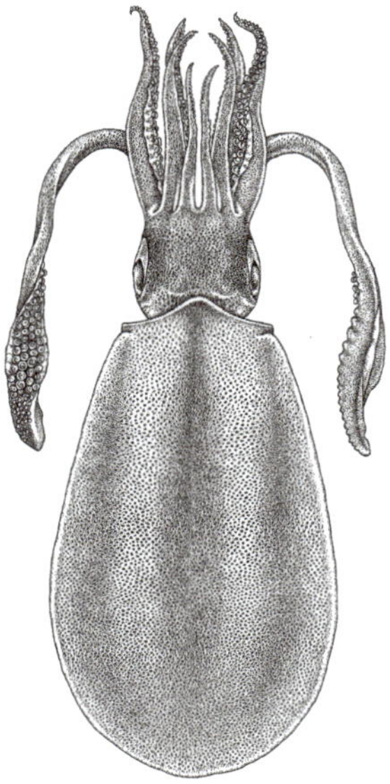

Dorsal view (ABRS, K. Hollis).

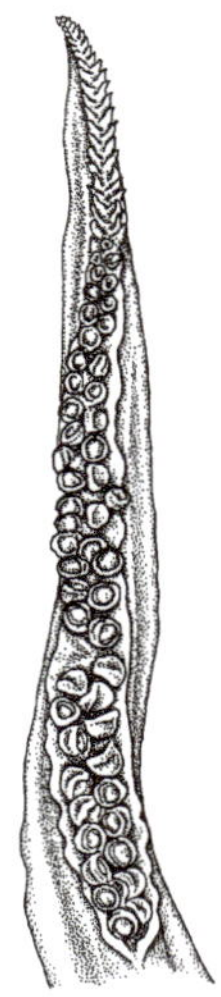

Hectocotylus (ABRS, K. Hollis).

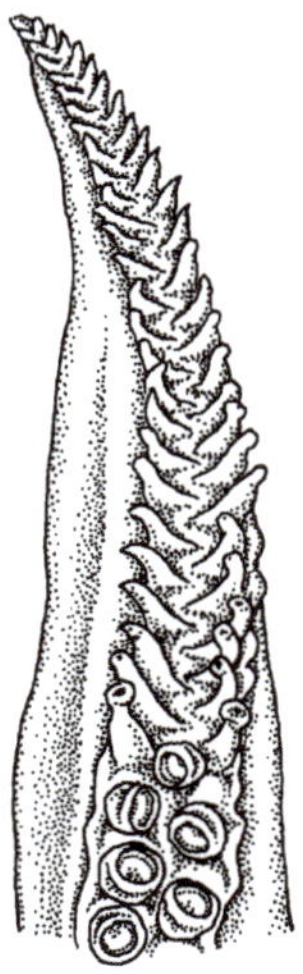

Enlargement of hectocotylus (ABRS, K. Hollis).

validation showing the addition of one statolith growth ring per day was determined under laboratory conditions.) Jackson and Moltschaniwskyj (2002) compared age, growth, and maturity parameters for *Sepioteuthis 'lessoniana'* both temporally and spatially (i.e. equatorial tropical and subtropical). The study revealed considerable plasticity in the size-at-age of this species. The tropical population had growth parameters that fluctuated between an equatorial strategy (fast growth, small body size, and small gonads) and a subtropical strategy (large body size, slower growth, and large gonads) depending on season.

Thomas and Moltschaniwskyj (1999) examined the relationship between statolith size and growth in this species. They found that statolith total length was correlated with age for squids < 60 days of age, but not beyond that age. The shape of the statolith also changed during ontogeny.

Schooling behaviour has been examined in a Japanese population of this species. Sugimoto and Ikeda (2012) found that they swim randomly for 20 days after hatching then form a school 30–60 days after hatching with an inter-individual distance of two mantle lengths.

*Interest to fisheries.* While of major commercial and artisanal value throughout south-east Asia, in Australia it is a minor by-catch of prawn trawling in Queensland, the northern prawn fishery, and Western Australian prawn fisheries. In southern Queensland it is taken by tunnel net fishers from March to mid-December. Recreational fishers take them on baited jigs or lures (Wadley and Dunning 1998).

The species has been successfully cultured through seven successive generations in a closed reticulated seawater system (Walsh *et al.* 2002), suggesting future aquaculture potential.

### SIZE

Males up to 420 mm ML; females smaller, up to 200–300 mm ML.

### REMARKS

This widespread taxon is likely to be a species-complex. Among other studies on the group, Triantafillos and Adams (2005) found that *S. 'lessoniana'* comprises two cryptic species from two sites in Shark Bay, Western Australia using allozyme electrophoresis. The taxonomy requires resolution given its high interest to fisheries. Fixed gene differences at four loci were found in four individuals at one of the sites (Acp, Got2, Idh2 and PepD). The genetic differences between these two groups of individuals (13% fixed differences and Nei's standard genetic distance, Nei D = 0.178) were about twofold greater than those between two cryptic taxa in the southern *S. 'australis'*, but smaller than those between *S. 'australis'* and *S. 'lessoniana'*.

Samantha Cheng and colleagues are examining this taxon over its entire range. These researchers have examined the phylogenetic relationships of representatives of this taxon from the Indian and Indo-Pacific Oceans using mitochondrial genes (COI 16s) and nuclear genes (rhodopsin, ODH) from ~400 individuals (Cheng *et al.* 2012). So far, their results suggest there are three distinct lineages in this region with no clear geographic delineations or morphological differences between them. They appear to occur in sympatry at a range of spatial scales, indicating that these lineages have pronounced differences in population structure. Levels of connectivity differ among the three lineages. Further work will investigate differences in their ecology and life history that may facilitate their co-occurrence. Future work will also investigate this taxon in the Australasian region to delineate the total number of cryptic species present in this complex.

### DISTRIBUTION

Northern Australia from approximately Geraldton in Western Australia to Moreton Bay on the east coast (27°S), widespread in the Indo-Pacific to central Japan and the Hawaiian Islands.

### REFERENCE

Jereb *et al.* (2010c).

## *Uroteuthis* Rehder, 1945

### DIAGNOSTIC FEATURES

A medium-sized squid; ML up to 490 mm. Proximal margin of arm sucker rings with semi-crescentric plate; distal margin with square teeth. Hectocotylus with two rows of papillae. Tentacular clubs expanded, with tetraserial suckers. Spermatophore cement body short. Eggs small.

*Photophores.* Pair of photophores on ventral surface of ink sac.

### REMARKS

All species in this genus have a pair of photophores on the ventral surface of the ink sac and similarities in the hectocotylus and arm-sucker dentition. These characters led Natsukari (1984) to establish the genus *Photololigo* for the Indo-West Pacific photophore-bearing species that were formerly assigned either to *Loligo* or *Doryteuthis*. However, *Uroteuthis bartschi* Rehder, 1945 also has similar photophores, a similar hectocotylus and arm suckers, and an Indo-West Pacific distribution. Nomenclatural priority belongs to *Uroteuthis* as the name of the genus, so the Australian species are assigned here to that genus. *Photololigo* and *Aestuariolus* are recognised as subgenera. The reason for this generic and subgeneric designation is explained further in Jereb *et al.* (2010c).

Yeatman and Benzie (1993, 1994) presented evidence that some purported species of *Uroteuthis* (*Photololigo*) from Australia are actually species-complexes. The complex contains at least two unnamed species that can occur sympatrically.

Allozyme and morphometric analysis of *Uroteuthis* (*Photololigo*) from two northern Australian locations (the North West Shelf and the Gulf of Carpentaria) revealed three taxa (Yeatman and Benzie 1993). Each was distinguished by several fixed gene differences. One species, *U.* (*P.*) *'chinensis'* (Gray, 1849) (occurring at both locations), was distinguished from the other two species by morphological and genetic data, with gladius width index, hectocotylus length index and sucker dentition the best criteria for distinguishing adult *U.* (*P.*). *'chinensis'*. The other two taxa (in Yeatmen and Benzie's (1993) paper, called *Loligo* species 1 (confined to the North West Shelf) and *Loligo* species 2 (from the Gulf of Carpentaria)), were best distinguished from each other by allozyme differences, although multivariate morphometric analyses differentiated most mature specimens. The gross morphology of the latter two taxa is most consistent with that of *U.* (*P.*) *'edulis'* (Hoyle, 1885).

In a later study, Yeatman and Benzie (1994) recognised four species based primarily on allozyme differences. This work included a greater number of populations and a greater distributional range than the previous study and also examined distributions with respect to depth. Two species (as above) fitted the gross morphology of *U.* (*P.*) *'edulis'* and two species shared *U.* (*P.*) *'chinensis'* features. They suggested that depth constraints, or factors associated with depth, act as effective barriers to gene flow and, therefore, provide mechanisms for allopatric speciation in the genus.

These data indicate that the current view of *Uroteuthis* species having broad distributions needs revision and several taxa are likely to comprise a series of unrecognised cryptic species, some of which occur sympatrically. Further studies are required to clarify the problematic systematics of *Uroteuthis* (*Photololigo*) species to devise some reliable diagnostic tools. Because the loliginid squids represent the most commercially important cephalopod resource from northern Australia it is important to gain a better understanding of the taxonomy and biology of the species fished so they can be managed appropriately. Determining the taxonomy of this group based on morphological criteria is difficult because, to date, it is often only mature males that can be confidently identified. The taxonomy is complicated by both sexual dimorphism and ontogenetic variation.

A study by Sin *et al.* (2009) of *U.* (*P.*) *'chinensis'* from Hong Kong and Xiamen (China) and *Uroteuthis* (*P.*) *'edulis'* from Yamaguchi, Japan and Shanghai, China did not reveal any new morphological characters to enable the identification of these two taxa based on a multivariate analysis of 27 morphometric indices. Genetic analyses, however, based on the mitochondrial COI and 16S rRNA revealed a high divergence of 15.5% and 7.5% respectively, supporting the recognition of these forms as distinct species.

Sin *et al.*'s (2009) study suggests that, in addition to molecular data, useful characters appear to be arm sucker ring tooth shape and number, fin angle, the shape of the posterior fin margin, hectocotylus length (i.e. length of modified portion compared to the rest of the arm) and gladius length and width.

#### REFERENCES

Yeatman and Benzie (1993, 1994), Dunning (1998), Vecchione *et al.* (1998), Vecchione *et al.* (2005), Sin *et al.* (2009), Jereb *et al.* (2010c).

### Key to *Uroteuthis* subgenera

1a. Mantle stout. Fins heart-shaped, rounded. Hectocotylus modified along entire arm. Mature at small size (< 50 mm ML) ............ *Uroteuthis* (*Aestuariolus*)

1b. Mantle elongate. Fins rhomboidal with posterior borders straight or slightly concave. Modified portion of hectocotylus less than entire arm, proximal portion not modified. Generally mature at > 50 mm ML ......................... *Uroteuthis* (*Photololigo*)

## *Uroteuthis* (*Aestuariolus*) *noctiluca* (Lu, Roper & Tait, 1985)

*Loliolus noctiluca* Lu, C. C., Roper C. and Tait, R. W. (1985). A revision of *Loliolus* (Cephalopoda; Loliginidae), including *L. noctiluca*, a new species of squid from Australian waters. *Proceedings of the Royal Society of Victoria* 97(2), 59–85 [59, figs 1–6].

Type data: holotype MV F.31119.

Type locality: Australia: Victoria, north Port Phillip Bay.

#### COMMON NAME

Luminous Bay Squid

#### DISTINGUISHING CHARACTERS

This species may be confused with juveniles of other, larger, more commercially important species from which it can be separated by fin shape and, in freshly caught specimens, by the presence of yellow and pink chromatophores on the fin margins and mantle.

#### DIAGNOSTIC FEATURES

Mantle cylindrical; mantle width ~28–30% ML. Fins roundish. Arms short; arm sucker rings with four to seven teeth. Left ventral arm of males hectocotylised along entire length. Suckers modified into conspicuous elongate interconnected papillae with minute apical suckers, sucker stalks on hectocotylus of males modified as papillae along entire arm length. Tentacles relatively short, robust, clubs large, broadly expanded; medial manal suckers three to six times the diameter of marginal manal suckers; large manal-sucker rings smooth.

*Photophores.* Pair of photophores on ventral surface of ink sac.

#### HABITAT AND BIOLOGY

This species occurs in shallow coastal habitats on sandy bottoms with seagrass beds, to depths of ~50 m. It can

Fresh-caught specimen, dorsal view (K. Graham).

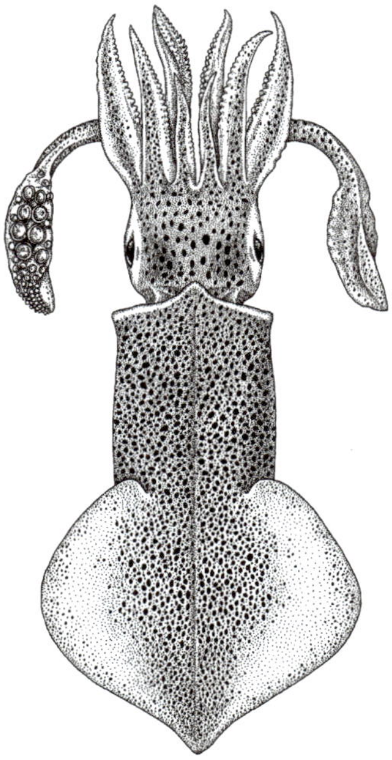

Dorsal view (ABRS, K. Hollis).

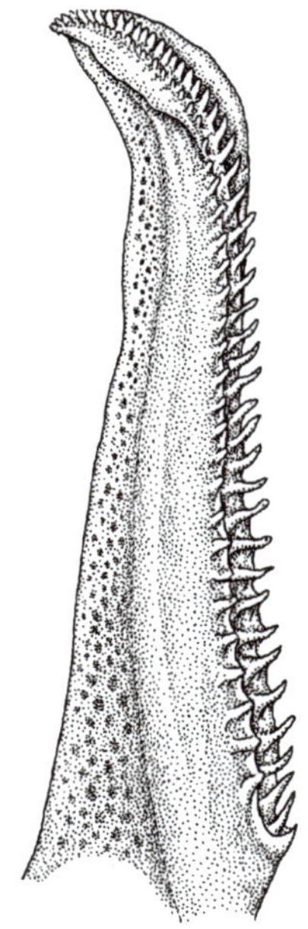

Hectocotylus (ABRS, K. Hollis).

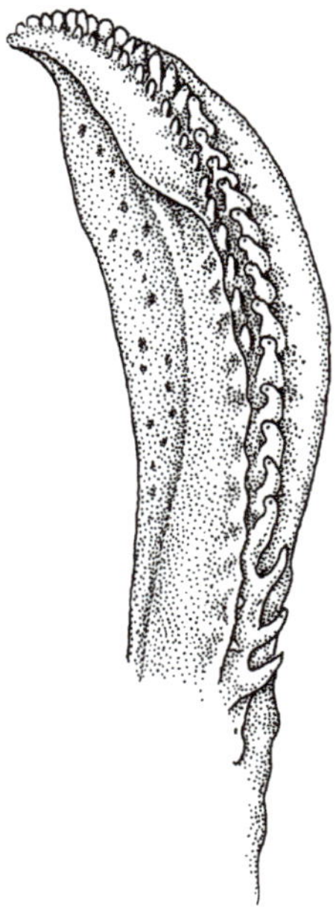

Enlargement of hectocotylus (ABRS, K. Hollis).

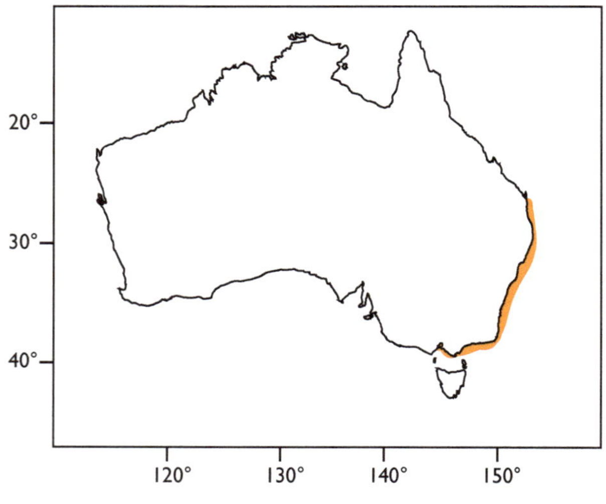


*Uroteuthis (Aestuariolus) noctiluca* distribution.

enter estuarine water with bottom-water salinities as low as 24 PSU, surface salinities as low as 17 PSU and a temperature of ~11°C (Lu *et al.* 1985).

It is short-lived, reaching 60 mm ML in tropical Australian waters in ~70 days; here, winter-caught individuals are the fastest growing and achieve the largest size compared with summer- or autumn-caught individuals. There is also a marked seasonal influence on the onset of sexual maturity. Females grow faster than males during the winter. A comparison between tropical north Queensland and temperate New South Wales populations indicated that temperate individuals live longer and have slower growth rates than tropical forms (Jackson and Moltschaniwskyj 2001). Off north Queensland the life cycle was just >4 months (based on daily statolith increments). In Tasmanian waters, near the southern limit of its distribution, *U. (A.) noctiluca* appears to have a life cycle of <6 months (Ceriola and Jackson 2010), with July (winter) hatched females growing faster than May (autumn) hatched individuals. Thus, over the extensive latitudinal distributional range of this species (New Guinea to Tasmania) there appears to be a trend towards increased lifespan and decreased growth rate with increasing latitude. Ceriola and Jackson (2010) suggested a multiple-spawning, rather than a single-spawning strategy was a characteristic of this species.

*Interest to fisheries.* This species is caught incidentally in inshore and estuary prawn trawls along the north-eastern Australian coast.

## SIZE

Maximum ML 90 mm (females larger than males).

## REMARKS

Two forms exist on the eastern Australian coast, with males differing in the morphology of sucker rings on the right ventral arms. No differences have been found so far in females. Dimmlich and Hoedt (1998) studied growth in this species as measured using statolith growth increments. Tetracycline staining experiments verified previous work on tropical forms that showed that growth increments are deposited daily. However, they noted major differences in the form of growth, longevity and life history pattern between tropical and temperate forms of this species.

#### DISTRIBUTION

Inshore Australian waters from Queensland, along the eastern Australian coast, south to Tasmania.

#### REFERENCES

Lu *et al.* (1985), Dunning (1998), Okutani (2005), Jereb *et al.* (2010c).

### Key to *Uroteuthis (Photololigo)* species

*Note:* These species, particularly females, are difficult to identify based on their morphology and cryptic species are present.

1a. Tentacular club large suckers with 20–30 sharp, conical teeth, 6–10 larger teeth interspersed with small teeth; median manal suckers 1.5 × diameter of lateral suckers.................................... *U. (P.) 'chinensis'*

1b. Tentacular club suckers truncate, blunt; median manal suckers < 1.5 × diameter of lateral suckers.....2

2a. Modified portion of hectocotylised arm in males < 50% of arm length. Tentacular clubs short, slightly expanded......................................... *U. (P.) singhalensis*

2b. Modified portion of hectocotylised arm in males > 50% of arm length. Tentacular clubs expanded....3

3a. Mature males with cutaneous ridge on ventral surface of mantle. Median suckers on manus of tentacular club with 30–40 sharp, conical teeth (10 large interspersed with small teeth)..... *U. (P.) 'edulis'*

3b. Mature males without cutaneous ridge on ventral surface of mantle. Median suckers on manus of tentacular club with 14–22 sharp, regularly spaced subequal conical teeth .................. *U. (P.) 'duvaucelii'*

## *Uroteuthis (Photololigo) 'chinensis'* (Gray, 1849) (species-complex)

*Loligo chinensis* Gray, J. E. (1849). Catalogue of the Mollusca in the British Museum. Part I. Cephalopoda Antepedia. 164 pages. London [74].

Type data: syntypes (2) BMNH 1993021.

Type locality: China: Canton market.

#### COMMON NAME(S)

Broad Squid, Mitre Squid, Pencil Squid, Slender Squid

#### DISTINGUISHING CHARACTERS

Arm sucker rings with 10–18, conical, pointed teeth (unlike *Uroteuthis (Photololigo) 'edulis'*, with which it co-occurs, with truncate teeth).

#### DIAGNOSTIC FEATURES

Mantle elongate, slender, bluntly pointed posteriorly; mantle width ~20–30% ML, widest in its anterior part. Cutaneous ridge present or absent on ventral surface of mantle in adult males. Fins rhombic, long, over two-thirds ML in adults, fan-shaped with straight or slightly concave borders towards tail in adults. Arm suckers biserial; largest sucker rings of arms 2 and 3 with 10–18 sharp conical teeth distally, and with degenerate, plate-like teeth or smooth proximally. Left arm 4 hectocotylised: distal 33–40% with modified suckers and stalks, > 30 sucker stalks slender, elongate, conical, largest in ventral row. Mature males with enlarged proximal suckers on arms 2 and 3. Tentacular club suckers tetraserial; ~12 medial manal suckers on tentacular clubs 1.5 × diameter of lateral suckers and 2 × diameter of largest arm sucker; large rings with 20-30 sharp, conical teeth, 6–12 larger teeth separated by 1–4 smaller teeth.

*Photophores.* Oval photophores on each side of rectum on ink sac.

*Colour.* Live specimens, at least in New South Wales, bright red or brick red (Karina Hall, personal communication). Recently dead or preserved animals flesh coloured, mantle pale tan to creamy pink.

#### HABITAT AND BIOLOGY

The biology of this species is poorly known. It is a neritic species, inhabiting shallow inshore waters, coastal bays and inlets ranging from intertidal areas to 170 m depth. It is positively phototaxic and periodically forms large aggregations. Spawning occurs throughout the year, with peaks in spring and autumn. Fecundity varies from 3000 to 11 000 demersal eggs. The ML at 50% maturity is 160 mm and 140 mm ML for males and females, respectively. Age analyses suggest that maturity is governed more by individual size than by age. This is one of the largest loliginid species of the Indo-Pacific region. It feeds on a variety of fishes and crustaceans.

*Interest to fisheries.* This taxon (a species-complex) is targeted or a welcome by-product of numerous commercial and small-scale fisheries throughout its range. It is taken using a variety of gear, often involving attraction to lights. In northern Australian waters (particularly off the

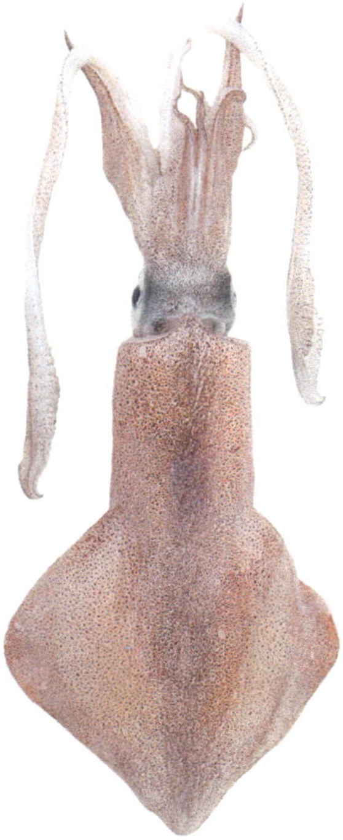

Fresh-caught specimen, dorsal view (K. Graham).

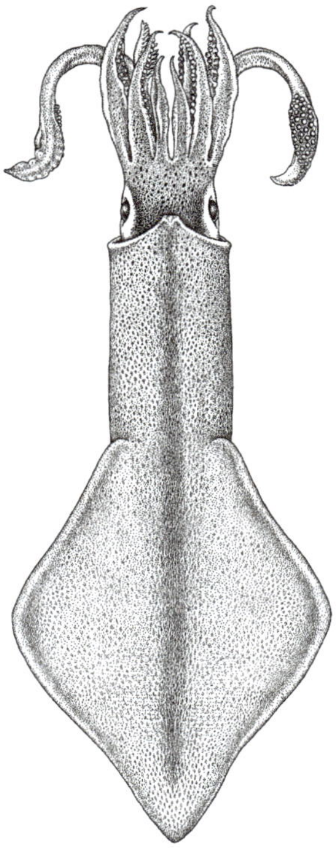

Dorsal view (ABRS, K. Hollis).

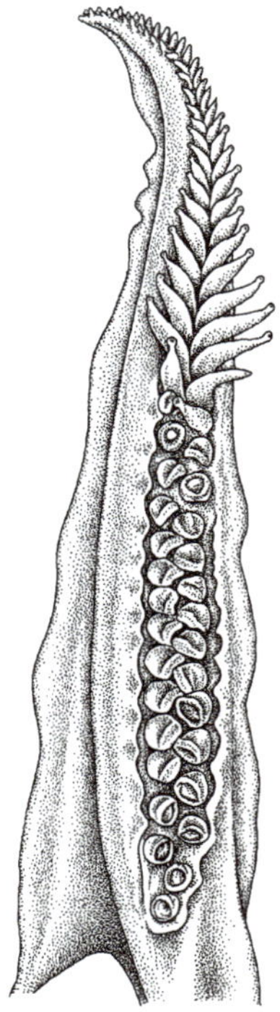

Hectocotylus (ABRS, K. Hollis).

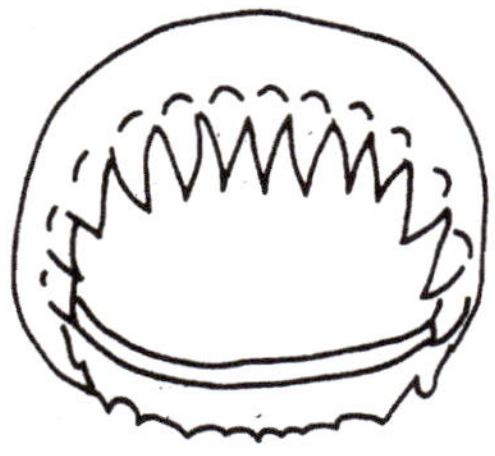

Arm 3 sucker ring (A. Reid, redrawn from Dunning 1998).

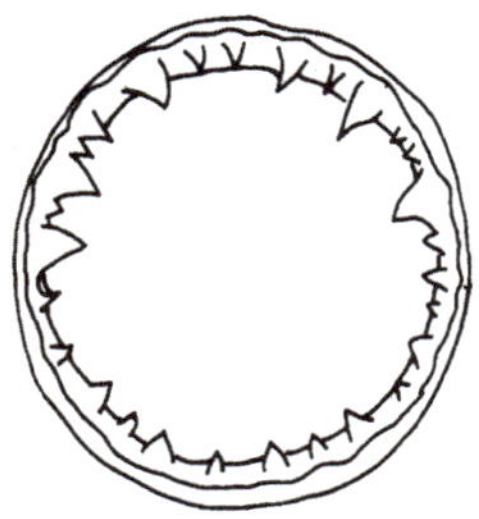

Largest tentacular club sucker ring (A. Reid, redrawn from Dunning 1998).

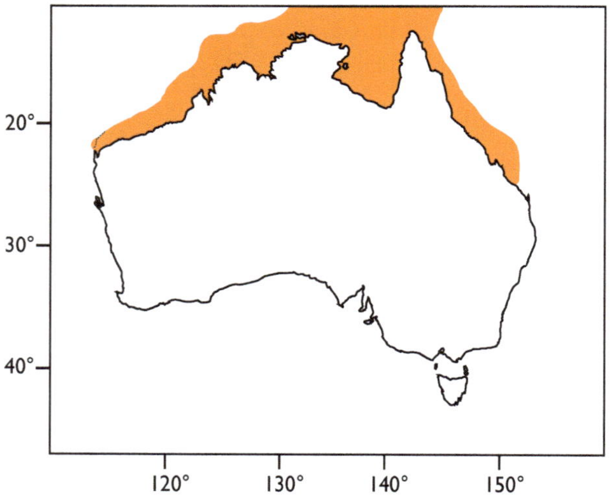


*Uroteuthis (Photololigo) 'chinensis'* distribution.

Kimberley's and western Gulf of Carpentaria) it is targeted with banana prawn nets, together with *U.* (*P.*) *'edulis'* and is one of the most important commercial cephalopod species in northern Australian waters. It is also commonly trawled in Moreton Bay, along the northern New South Wales coastline and in the Hawkesbury River estuary, and is sold as seafood or as bait. It is also taken by recreational fishers with baited jigs or lures.

## SIZE

Maximum reported size 490 mm ML for males, 310 mm for females; commonly to 200 mm ML.

## REMARKS

Following its original description, this species has been redescribed several times under different names, as summarised by Natsukari and Okutani (1975), who also give a redescription of the type-specimen, along with a taxonomic review. In Australia, the species-complex contains two unnamed species with overlapping distributions. It also occurs with *U.* (*P.*) *'edulis'*. (Of the two unnamed taxa, the one occurring more commonly in estuarine and inshore waters has been referred to as *U.* (*P.*) *etheridgei* (formerly *Loligo etheridgei*) which is now considered to be a synonym of *U.* (*P.*) *'chinensis'*.)

The long-term confusion inherent in the convoluted regional loliginid nomenclature and presence of cryptic species makes it very difficult to assess the true importance of *U.* (*P.*) *'chinensis' sensu stricto* to fisheries, and it seems probable that catches are under-reported in many places.

Recent studies indicate that extracts from the ink of *U.* (*P.*) *'chinensis'* show potent anti-tumour activity and low toxicity.

## DISTRIBUTION

An Indo-Pacific species, extending from the western Pacific (Japan, South China Sea, Hong Kong, Philippines, Indonesia, northern, western and eastern Australian waters), to the Indian Ocean (Andaman Sea, Thailand and the Bay of Bengal). Within Australia it is found from Shark Bay, Western Australia through the Arafura and Timor Seas to Queensland and possibly New South Wales.

## REFERENCES

Jackson (1993), Yeatman and Benzie (1994), Dunning (1998), Jereb *et al.* (2010c).

### *Uroteuthis* (*Photololigo*) *'duvaucelii'* (Orbigny, 1835)

*Loligo duvaucelii* D'Orbigny, A. (1835) In: Ferussac, A. E. de and A. D'Orbigny, A. (1834–1848). Histoire Naturelle Generale et Particuliere Cephalopodes Acetabuliferes Vivants et Fossiles. 96 pages + lvi pages + 361 pages, Atlas with 144 plates. Paris [318, pls 14, 20, dated from pl. 14].

Type data: syntypes MNHN (not extant).

Type locality: India, Sumatra.

## COMMON NAME

Pencil Squid

## DIAGNOSTIC FEATURES

Mantle relatively short, stout, cylindrical for about half its length; tapers gently into blunt tip. Anterior margin with small rounded lobe in dorsal midline. Fins gently rhombic, broad, ~50% ML (up to two-thirds ML). Arm suckers of females about equal size on arms 2 and 3; suckers with about seven broad, large, square teeth on distal margin, smooth proximally. In males, suckers of arm 2, and especially 3, greatly enlarged; up to 5–13 squared to rounded teeth in distal two-thirds of ring, proximal one-third smooth. Left ventral arm of male hectocotylised for more than half its length (up to 75%), with two series of large papillae, some with minute suckers on tips; ventral series of papillae larger, turned outward, comb-like. Tentacles long; tentacular clubs expanded, large, up to 45–50% ML; large median manal suckers (1.5 × diameter of marginal suckers), with 14–22 short, sharp teeth, subequal in size, regularly spaced around entire margin.

*Photophores.* Oval photophores on each side of rectum on ink sac.

## HABITAT AND BIOLOGY

The biology of this species has been well studied, due to its importance in Thai and Indian fisheries. (Details pertaining directly to the Australian representatives are currently unknown.) Off India and Thailand, *Uroteuthis*

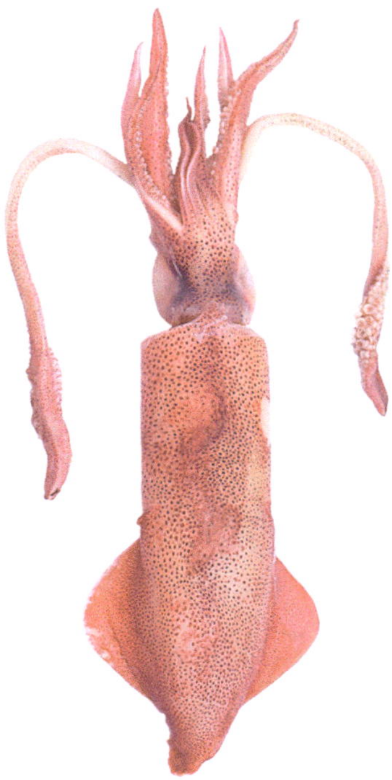

Fresh-caught specimen, dorsal view (K. Graham).

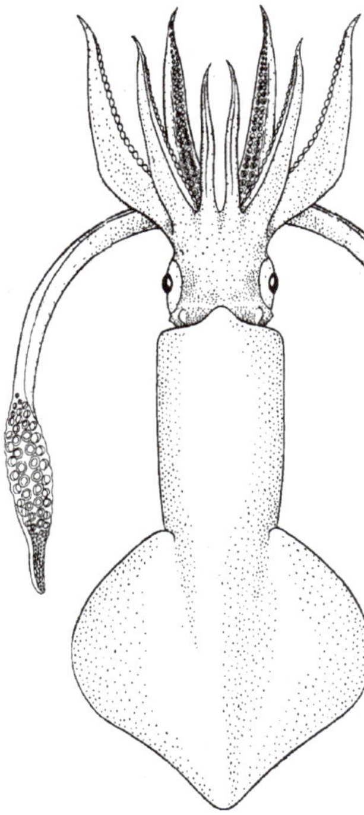

Dorsal view (FAO, E. D'Antoni).

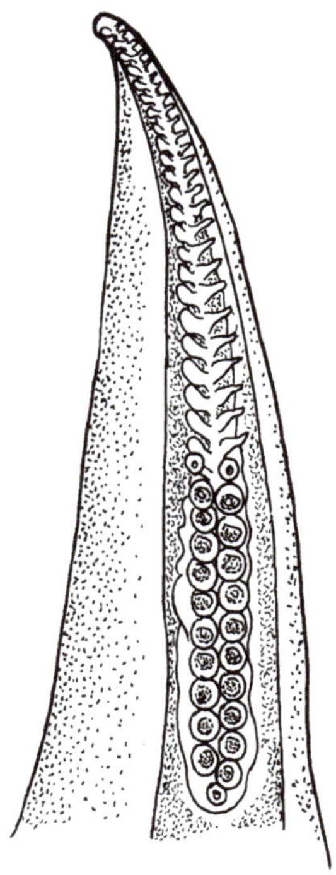

Hectocotylus (FAO, E. D'Antoni).

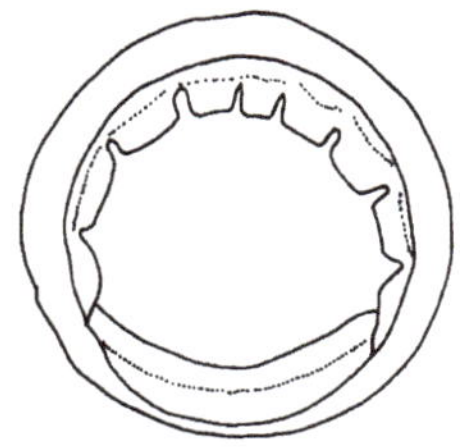

Arm 3 sucker ring (A. Reid, redrawn from Dunning 1998).

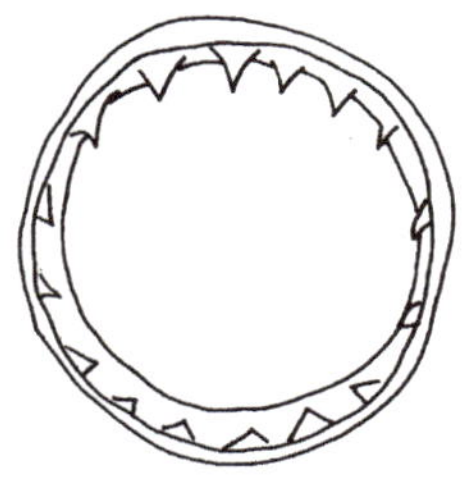

Largest tentacular club sucker ring (A. Reid, redrawn from Dunning 1998).

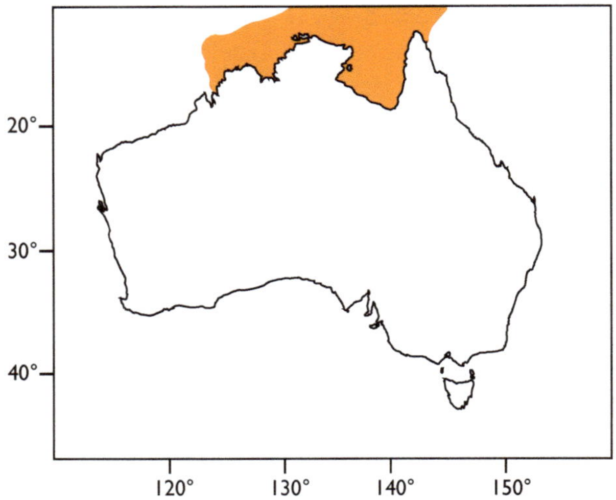


*Photololigo 'duvaucelii'* distribution (FAO, E. D'Antoni).

*(Photololigo) 'duvaucelii'* is found at depths between 30 and 170 m, and it forms large aggregations during the spawning season. The spawning period appears to be almost year-round, with peaks mainly in spring and autumn. Observations on growth after sexual maturity suggest an extended reproductive phase within the life cycle (i.e. not strictly semelparous reproduction). Egg masses consist of many egg capsules which contain 125–290 eggs. Eggs are ~2 mm long and 1.75 mm wide and very yolky. They take between 7 and 10 days to develop at temperatures ranging between 28 and 32°C. Hatchlings are ~1–1.8 mm ML and are planktonic. Growth is allometric. Females grow faster than males and are heavier at the same length; however, males ultimately attain a greater size and age. The maximum reported size is 330 mm ML from Indian waters. According to statolith-based age analysis, the life cycle of the species is 1 year, which is shorter than estimates obtained by length–frequency analysis. The Indian squid feed on crustaceans, fishes and squids, and are preyed upon by a variety of predators, including fishes and marine mammals. Cannibalism is common.

*Interest to fisheries.* *Uroteuthis (Photololigo) 'duvaucelii'* is one of the most common species among the Indo-Pacific loliginids and is fished throughout its range and is overexploited in some areas. No fishery exists in Australia. Research in Thailand aims to determine the feasibility of commercial culture (Nabhitabhata *et al.* 2005).

### SIZE

Medium-sized species; maximum reported size is 330 mm ML for males (west coast of India). They are most commonly up to 150 mm ML. Maximum weight 1.5 kg.

### REMARKS

*Uroteuthis (P.) 'duvaucelii'*, like other lologinids is polymorphic and is possibly a species-complex (Sales *et al.* 2013). Different forms are recognised in the commercial fishery in the eastern Indo-Pacific region (Okutani 2005) and two differing forms have been reported from the Gulf of Aden and the Arabian Sea (Nesis 1982/1987).

### DISTRIBUTION

An Indo-West Pacific species, *Uroteuthis (P.) 'duvaucelii'* occurs throughout the Indian Ocean, from the South African coasts, the Red Sea and the Arabian Sea, eastward to the Bay of Bengal, Sri Lanka and the Andaman Sea. It is common in Malaysia and the Philippines and has been recorded in the western Pacific north to Formosa Island and the Taiwan Strait and south to the Java and Arafura Seas. It occurs off northern Australia.

### REFERENCES

Dunning (1998), Okutani (2005), Jereb *et al.* (2010c).

## *Uroteuthis* (*Photololigo*) *'edulis'* (Hoyle, 1885) (species-complex)

*Loligo edulis* Hoyle, W. E. (1885). Diagnosis of new species of Cephalopoda collected during the cruise of H.M.S. Challenger. – II. The Decapoda. *Annals and Magazine of Natural History (series 5) 16*, 181–203 [186].

Type data: holotype BMNH 1889.4.24.83.

Type locality: Japan, Yokohama, purchased in market.

### COMMON NAME(S)

Mitre Squid, North-west Pink Squid, Pencil Squid, Swordtip Squid

### DISTINGUISHING CHARACTERS

Largest medial arm suckers with 6–12 squared, truncate teeth (in contrast *Uroteuthis* (*Photololigo*) *'chinensis'* have pointed teeth on the arm sucker rings).

### DIAGNOSTIC FEATURES

Mantle moderately stout to elongate, slender in mature males. Head small, cube-shaped, narrower than mantle. Mature males with cutaneous ridge on ventral mantle surface. Fins rhombic, 60–70% ML in adults, lateral angles rounded, anterior margin slightly convex, posterior margins slightly concave, width two-thirds ML (usually slightly larger in females). Arms moderately long, with biserial suckers. Arm formula variable:

Fresh-caught specimen, dorsal view (CSIRO ScienceImage AF9590).

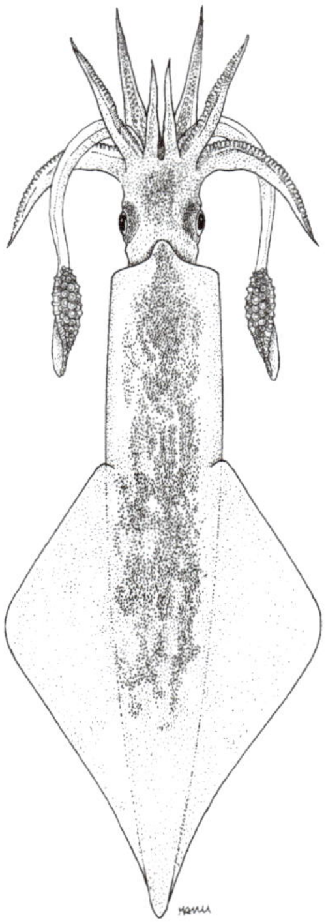

Dorsal view (FAO, E. D'Antoni).

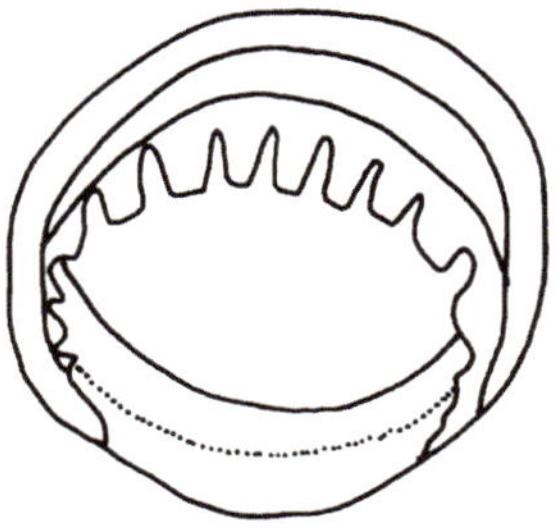

Arm 3 sucker ring (A. Reid, redrawn from Dunning 1998).

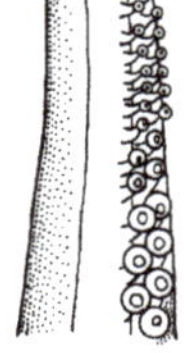

Hectocotylus (FAO, E. D'Antoni).

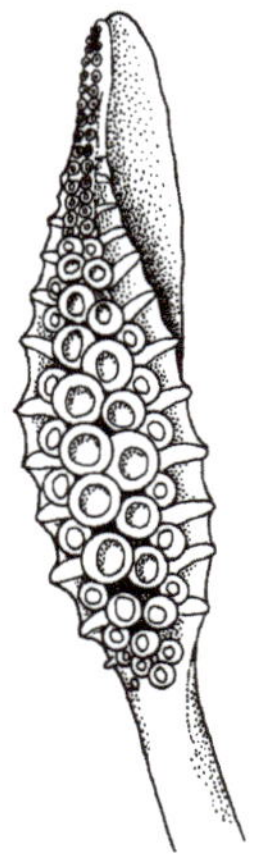

Tentacular club (FAO, E. D'Antoni).

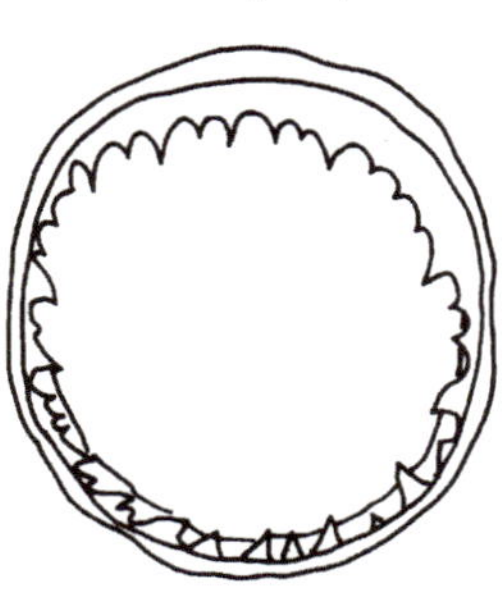

Largest tentacular club sucker ring (A. Reid, redrawn from Dunning 1998).

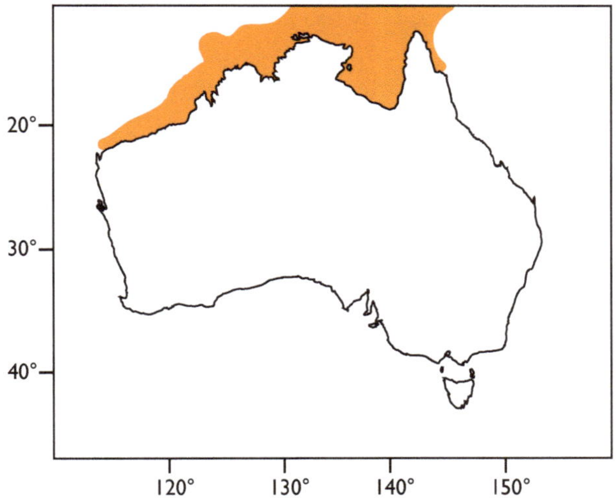


*Uroteuthis (Photololigo) 'edulis'* distribution.

3 > 4 > 2 > 1 or 4 > 3 > 2 > 1. Suckers on arms 2 and 3 of mature males enlarged. Arm sucker rings with up to 12 (more often six to eight) long, square, truncate teeth on distal two-thirds, proximal margin smooth or only irregularly denticulate with inconspicuous teeth. Left ventral arm in male modified for < 50% of its length. Tentacular clubs expanded, lanceolate; ~16 medial manal suckers 1.2 × larger than marginal suckers, approximately equal in diameter to largest arm suckers, with 30–40 sharp conical teeth, 10 larger teeth interspersed with 20–30 smaller teeth. Arm sucker rings with 6–12 distinct, squared, truncate teeth in distal two-thirds, smooth or with rudimentary denticles proximally. More than 50% of left arm 4 hectocotylised by enlargement of ~50 pairs of sucker pedicels into swollen papillae, each with a minute rudimentary sucker on tip; papillae slightly larger in ventral row.

*Photophores.* Fusiform photophore on each side of rectum, beneath ink sac.

*Colour.* Live animals flesh-coloured with small pink spots on mantle.

## HABITAT AND BIOLOGY

*Uroteuthis (P.) 'edulis'* is a neritic species occurring in depths from 30 m to 170 m. It may undergo seasonal inshore–offshore migrations, forming large aggregations. Data from northern Australia suggest an extended spawning and shows that in the Gulf of Carpentaria they reach sexual maturity at smaller sizes (70–80 mm ML) than in the northern waters of the western Pacific. Eggs are demersal.

*Interest to fisheries.* At least two taxa in this complex are taken as by-catch in prawn trawling in northern

Australian waters. They have excellent flesh quality and constitute one of the most important commercial cephalopod species in northern Australian waters. They are excellent for human consumption and fetch a high market value in Asia where they are processed into a dry product and also used raw for sashimi. They are attracted to light.

#### SIZE

Medium- to large-sized squid, maximum ML in Australian waters ~100–120 mm. (Up to 400 mm ML in the South China Sea.)

#### REMARKS

*Uroteuthis (Photololigo) 'edulis'* is characterised by marked polymorphism, both by locality and by season. The existence of such a variety of 'forms', the taxonomic relationships of which still are not clear, makes the confident identification of species and the assessment of populations difficult. In Australian waters, four different species of *Uroteuthis (Photololigo)* from the northern area were identified using allozymes by Yeatman and Benzie (1994). Two of these species fit the gross morphology of *U. (P.) 'edulis'* from Japan, while the other two share the distinguishing features of *U. (P.) 'chinensis'*. However, no major population genetic structuring was evident in any of the species over the geographical range sampled. It was suggested that depth constraints, or factors associated with depth, may act as effective barriers to gene flow and, therefore, provide mechanisms for differentiation in this genus.

#### DISTRIBUTION

Northern Australia from North West Cape to the North West Shelf, Arafura Sea, Gulf of Carpentaria and northeast Cape York. Philippine Islands, South China Sea to central Japan. It also extends throughout the Indian Ocean, from its south-eastern waters: the Andaman Sea, Thailand, and the south-eastern Bay of Bengal, to the Arabian Sea, including the Gulf of Aden, the Gulf of Oman, and the Persian Gulf, the Red Sea, and southward to Mozambique (east African coast, Saya de Malha Bank).

#### REFERENCES

Voss (1963), Yeatman and Benzie (1994), Jackson and Yeatman (1996), Okutani (2005), Jereb *et al.* (2010c).

### *Uroteuthis* (*Photololigo*) *singhalensis* (Ortmann, 1891)

*Uroteuthis singhalensis* Ortmann, 1891 Cephalopoden von Ceylon, gesammelt von den Herren Dres. Sarasin. *Zoologische Jahrbücher (Systematisches) 5*, 669–678. [676, pl. 46 fig. 3a–d].

Type data: not resolved Syntypes (25+): BMNH Syntypes (7) 1930.1.9.6–12 [only 5 syntypes present; *fide* Lipinski *et al.* (2000: 106); not traced by Lu *et al.* (1995)].

Type locality: Indo-China, Gulf of Siam (Thailand), southern coast of Annam.

#### COMMON NAME

Long Barrel Squid

#### DIAGNOSTIC FEATURES

Mantle long, slender, cylindrical; tapers to sharply pointed tip. Fins narrow, rhombic in outline; relatively short in young specimens, up to 50–60% ML in adults. Arm sucker rings with 5–11 long, wide, square, plate-like teeth distally, smooth proximally. Left ventral arm hectocotylised <40–50% of total arm length, with slender, cone-shaped fleshy papillae without suckers at distal tips. Papillae on ventral series only slightly longer than papillae in dorsal series. Tentacular clubs short, slightly expanded. Suckers in medial series of manus of tentacular club only ~25% larger than those on lateral series; manus suckers with 15–25 conical, sharply pointed teeth around margin interspersed with small, inconspicuous pointed teeth.

#### HABITAT AND BIOLOGY

The biology of this species is poorly understood.

*Interest to fisheries.* Possibly caught with *Uroteuthis (Photololigo) 'edulis'* in northern Australian waters.

#### SIZE

Females up to 310 mm ML, males up to 500 mm ML, weight 800 g to 1 kg respectively (in non-Australian parts of its range).

#### REMARKS

The subgeneric status of this species is unsettled and it perhaps should be placed in a new subgenus (Jereb *et al.* 2010c).

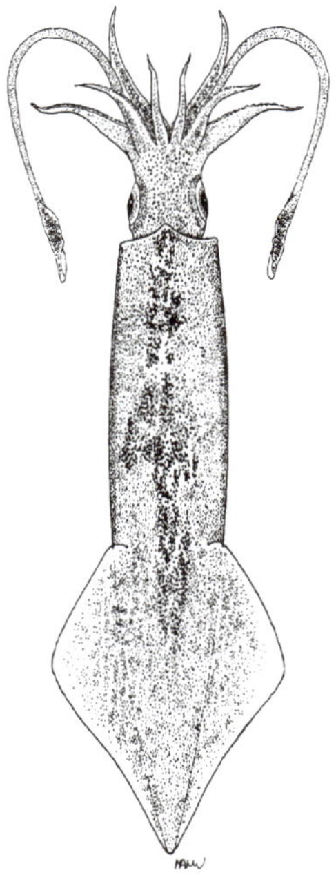

Dorsal view (FAO, E. D'Antoni).

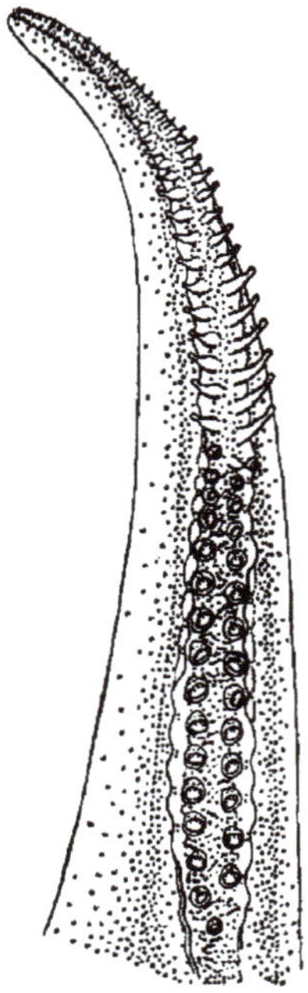

Hectocotylus (FAO, E. D'Antoni).

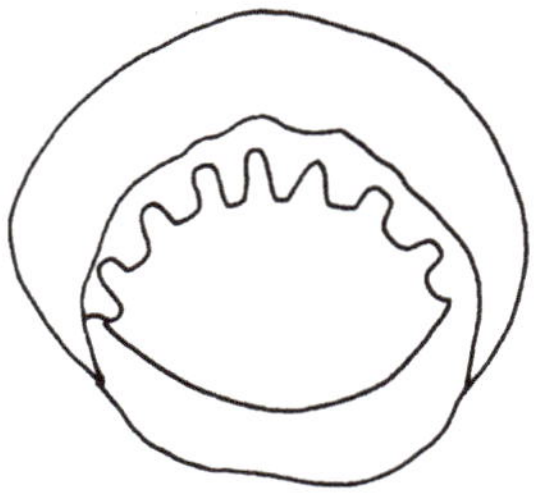

Arm 3 sucker ring (A. Reid, redrawn from Dunning 1998).

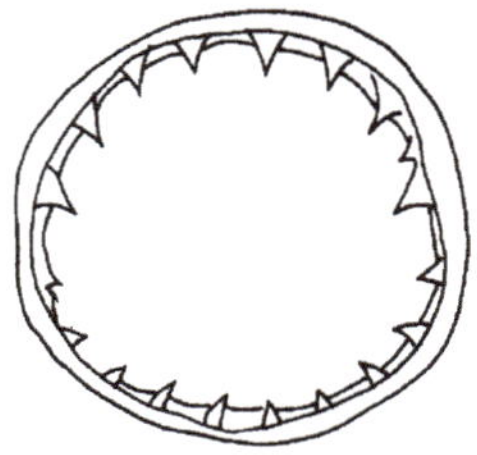

Largest tentacular club sucker ring (A. Reid, redrawn from Dunning 1998).

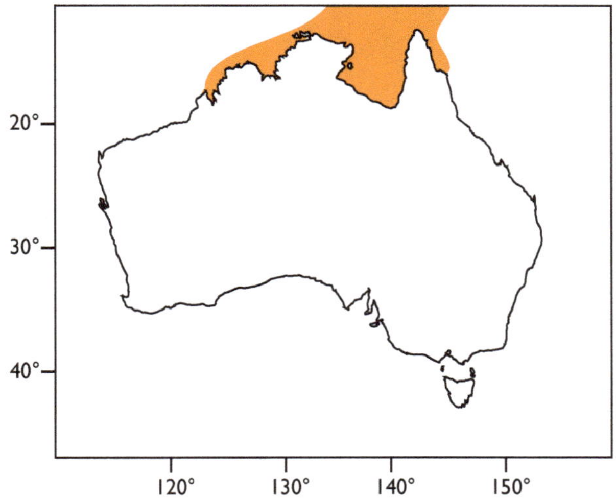


*Uroteuthis (Photololigo) singhalensis* distribution.

**DISTRIBUTION**

Indo-Pacific from the South China Sea through to Indonesia and northern Australia. Indian Ocean from Andaman Sea through to the east coast of Africa. Known depth range 30–120 m.

**REFERENCE**

Jereb *et al.* (2010c).

## *Order Oegopsida*

Oegopsid squids lack a corneal membrane covering the eyes. Unlike myopsids, the head does not have a tentacle pocket. The funnel does not have lateral adductor muscles. The arms and clubs have suckers and/or hooks in some species. The tentacular club usually has a carpal-locking apparatus. The gladius extends the full length of the mantle. The female gonoducts are paired and accessory nidamental glands are absent.

Lindgren (2010b) and Lindgren *et al.* (2012) have extensively studied many representatives of this diverse family using nuclear and mitochondrial genes. Of the families with multiple representatives, only Mastigoteuthidae was not recovered as monophyletic. Close relationships, also supported by morphological data, were found for the groups Pyroteuthidae and Enoploteuthidae; Histioteuthidae and Psychroteuthidae; Lepidoteuthidae and Octopoteuthidae; Chiroteuthidae and Batoteuthidae; and between Magnapinnidae, Joubiniteuthidae and Mastigoteuthidae. Members of many of these families occur in Australian waters.

### Key to oegopsid families (after Jereb *et al.* 2010a)

1a. Funnel free from mantle, funnel–mantle locking apparatus present..........2

1b. Funnel fused to mantle on each side, no funnel–mantle locking apparatus present ........ 22
2a. Funnel–mantle locking apparatus a simple, straight groove and ridge ........ 3
2b. Funnel–mantle locking apparatus not a simple, straight groove and ridge ........ 17
3a. Arms with hooks, or with suckers in four series on proximal half of ventral arms ........ 4
3b. Arms without hooks, or with suckers in two series on proximal half of ventral arms ........ 8
4a. Armature (suckers, hooks) biserial ........ 5
4b. Armature (suckers, hooks) tetraserial ........ Gonatidae (*Gonatus antarcticus*)
5a. Tentacles and clubs absent in adults; clubs in larvae or juveniles rudimentary ........ Octopoteuthidae
5b. Tentacles present; clubs fully developed ........ 6
6a. Photophores on viscera and eyes but not on mantle or surface of head and/or arms ........ Pyroteuthidae
6b. Photophores on mantle, surface of head and arms, and eyes, but not on viscera ........ 7
7a. Photophores on tentacles but not on eyeballs ........ Ancistrocheiridae (*Ancistrocheirus lesueurii*)
7b. Photophores on ventral eyeballs but not on tentacles ........ Enoploteuthidae
8a. Buccal membrane connectives attach to ventral side of ventral arms ........ 9
8b. Buccal membrane connectives attach to dorsal side of ventral arms ........ 13
9a. Tentacular clubs with hooks ........ Onychoteuthidae
9b. Tentacular clubs without hooks ........ 10
10a. Mantle with cartilaginous scales (may be minute). Tentacular clubs with tetraserial suckers ........ 11
10b. Mantle without cartilaginous scales. Tentacular clubs with more than four series of suckers over part of their length ........ 12
11a. Tentacles lost in adults. Tentacles in juveniles small, with about six poorly differentiated suckers ........ Lepidoteuthidae (*Lepidoteuthis grimaldii*)
11b. Tentacles present with numerous, laterally compressed suckers ........ Pholidoteuthidae (*Pholidoteuthis massyae*)
12a. Tentacular clubs with six series of suckers. Photophores on arm tips, but not on eyes. Tail present, long (greater than fin length), spike-like ........ Batoteuthidae (*Batoteuthis skolops*)
12b. Tentacular clubs with four series of suckers distally, many series proximally. Photophores on eyes, but not on arm tips. Tail absent ........ Brachioteuthidae
13a. Eye with row of photophores on ventral surface. Buccal membrane with eight lappets ........ Lycoteuthidae
13b. No photophores on eyes. Buccal membrane with seven or fewer lappets ........ 14
14a. Surface of mantle, head and arms covered with numerous photophores (usually large and distinct) ........ Histioteuthidae
14b. Surface of mantle and head without photophores, arms may have a few photophores ........ 15
15a. Many small to minute suckers (or suckers and knobs) at proximal end of manus ........ 16
15b. No cluster of small suckers at proximal end of manus ........ Psychroteuthidae (*Psychroteuthis glacialis*)
16a. Adult size small. Medioposterior borders of fins slightly convex. Carpal knobs in a single dorsal row or absent ........ Neoteuthidae (*Alluroteuthis antarcticus*)
16b. Adult size gigantic. Medioposterior borders of fins concave. Carpal knobs in cluster alternating with carpal suckers ........ Architeuthidae (*Architeuthis dux*)
17a. Funnel-locking cartilage with a longitudinal and transverse groove, ⊥-shaped, or ⊣-shaped ........ 18
17b. Funnel-locking cartilage triangular or oval (may have inward projecting knobs) ........ 19
18a. Funnel-locking cartilage ⊥-shaped. Fins < 60% ML ........ Ommastrephidae
18b. Funnel-locking cartilage ⊣-shaped. Fins > 80% ML ........ Thysanoteuthidae (*Thysanoteuthis rhombus*)
19a. Funnel-locking cartilage oval with one or two knobs directed towards the centre of the concavity ........ 20
19b. Funnel-locking cartilage oval or subtriangular, without knobs ........ 21
20a. Club with four series of suckers ........ Chiroteuthidae (except in *Grimalditeuthis*, which has none)
20b. Club with more than 15 series of minute suckers ........ Mastigoteuthidae (except in *M. cordiformis*)
21a. Tail longer than ML. Suckers on arms in four to six series ........ Joubiniteuthidae (*Joubiniteuthis portieri*)
21b. Tail under 50% ML, or absent. Suckers on all arms in two rows ........ Cycloteuthidae
22a. Mantle free dorsally, articulates with head by ridge and groove ........ Chiroteuthidae (*Grimalditeuthis*)
22b. Mantle fused dorsally with head ........ Cranchiidae

## Ancistrocheiridae Pfeffer, 1912

The Ancistrocheiridae are monotypic, represented by *Ancistrocheirus lesueurii* (d'Orbigny, 1842).

**REMARKS**

The Enoploteuthidae was formerly comprised of three subfamilies: Enoploteuthinae, Pyroteuthinae and Ancistrocheirinae (Roper and Jereb 2010c). Now these are recognised as closely related families of the enoploteuthid group. Young *et al.* (1998) present the history of these familial changes, diagnostic features of their genera and a list of currently accepted species and their distributions. A phylogenetic study has supported the validity of dividing the group into three families (Young and Harman 1998).

Enoploteuthids can be distinguished from the superficially similar Octopoteuthidae in having tentacles in adults, the buccal connectives attached to the dorsal borders of arms 4, and many small photophores on the ventral mantle and ventral surfaces of the head in adults. Hooks are strongly developed on all arms of all Australian enoploteuthids.

Enoploteuthids are among the most commonly caught squids in midwater trawls and are a significant food source for fishes and marine mammals. They feed predominantly on pelagic crustaceans, particularly euphausids.

Adults in Australian waters occur at depths of 300–600 m during the day and migrate into the upper 150 m at night (Lu and Dunning, 1998).

No enoploteuthids are fished commercially in Australian waters. They are readily caught in good condition and are ideal subjects for laboratory studies.

## *Ancistrocheirus* Gray, 1849

### *Ancistrocheirus lesueuri* (d'Orbigny, 1835)

*Enoploteuthis lesueurii* D'Orbigny (1842). In Ferussac and D'Orbigny, 1834-1848. *Histoire Naturelle Generale et Particuliere Cephalopodes Acetabuliferes Vivants et Fossiles.* 96 pages + lvi pages + 361 pages, Atlas with 144 plates. Paris. [339, Onychoteuthe pl 14].

Type data: holotype MNHN 2-14-614, 2-14-615 (gladius), 2-14-616 (beaks).

Type locality: not designated.

**DIAGNOSTIC FEATURES**

As for family, and as follows: mantle long, broad, conical, thick-walled. Head and tail with fleshy, gelatinous tissue; vesicles absent. Posterior end of mantle elongate, forms tail. Nuchal folds present, prominent. Fins thick, rhomboidal, long (70–80% ML); broad (80% ML), slightly subterminal, without posterior lobes. Arms robust, with biserial hooks; small, stalked suckers may be present on arm tips. Tentacles not modified near bases; stalk 'ligament' and vein leave tentacle at base and in membrane. Tentacular clubs narrow, with distinct carpal cluster; manus with two transverse series of sharp hooks; seven or eight in ventral series larger than dorsal counterparts; suckers absent. Buccal crown with eight supports; connectives attach to dorsal margins of arms 4. No secondary buccal connectives attach to ventral margins of arms 1 and 2. Nidamental glands present; oviducal glands normal. Oviducts unpaired. Ink sac not embedded in digestive gland.

*Photophores.* Regularly spaced photophores on ventral and lateral surfaces of mantle, head and tentacles. Ventral surface of mantle studded with 20–24 relatively large, separated photophores arranged in transverse rows; ventral surface of head with 8–10 similarly arranged photophores; large photophores on funnel, bases of dorsolateral arms and 12 photophores along aboral side of tentacular stalks. Numerous very small photophores on ventral surface of fins, mantle, funnel, head and ventral arms. No photophores on eyeballs or viscera.

**HABITAT AND BIOLOGY**

*Ancistrocheirus lesueurii* is an oceanic species, found in the lower epipelagic to mesopelagic and bathyal depths. It undergoes diel vertical migration and is able to regulate its bioluminescent counterillumination. As a nerito-oceanic species it can occur in association with seamounts and submarine ridges. Age and growth studies based on statolith analysis indicate a strong sexual dimorphism: males live for about 1 year, females begin to mature at 1 year and spawn at ~1.5 years. Following ontogenetic descent into bathyal waters (to at least 680 m) the growth rate decreases to lowest known values for squids. The relative fecundity is 200 000–800 000 ova, 1.9–2.6 mm diameter, of which 60–65% are spawned. Hatching occurs throughout the year.

*Interest to fisheries.* This species is believed to have some fishery potential because of its size, but development of a fishery beyond subsistence levels awaits a better understanding of its geographic and vertical distribution,

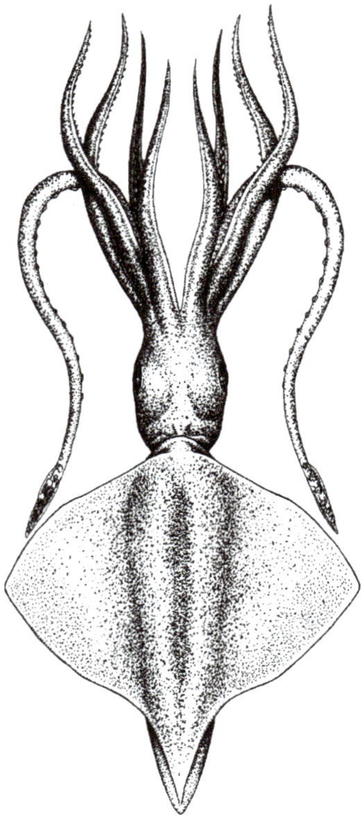
Dorsal view (FAO, E. D'Antoni).

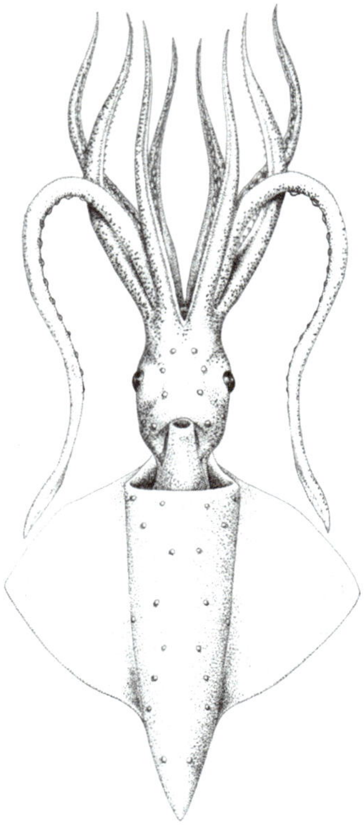
Ventral view (FAO, E. D'Antoni).

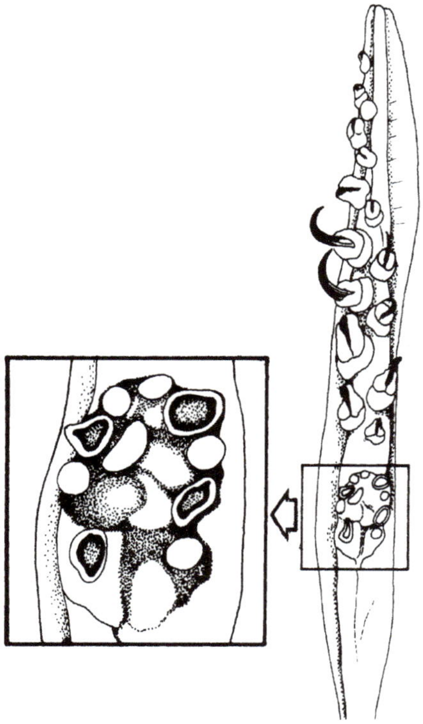
Tentacular club and enlargement of carpal region (FAO, E. D'Antoni).

*Ancistrocheirus lesueuri*

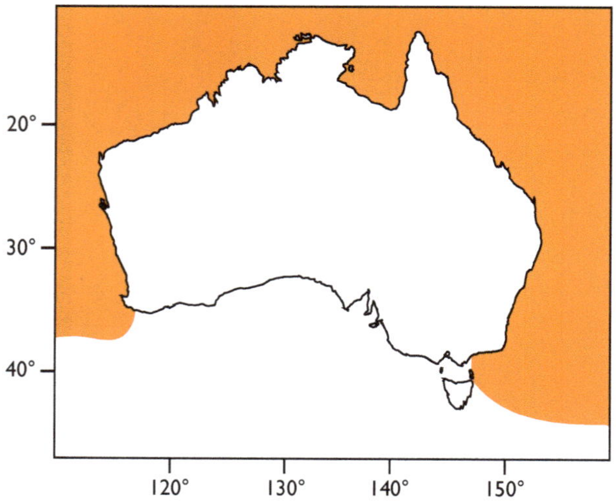

*Ancistrocheirus lesueuri* distribution.

as well as knowledge about its biology and population structure and density. The species, however, is unlikely to find local market acceptance at present because of the semi-gelatinous texture of its mantle.

## SIZE

The maximum ML reaches 410 mm. Gravid females attain 3 kg.

## REMARKS

Ripe eggs of mature females reach >3 mm in length, which is unusually large for oegopsids.

## DISTRIBUTION

Worldwide in tropical, subtropical and temperate open ocean waters (~40°N to 30°S); includes the Mediterranean Sea.

## REFERENCE

Roper and Jereb (2010c).

# Architeuthidae Pfeffer, 1900

## DIAGNOSTIC FEATURES

Long, narrow, tentacular clubs with tetraserial armature (pairs of alternating suckers and pads) along nearly their whole length. Manus of club with conspicuously large suckers in the two medial series; proximal end of carpus of club with cluster of numerous, small, alternating suckers and knobs in six or seven irregular transverse series.

The family is monotypic, represented by *Architeuthis dux* Steenstrup, 1860.

### *Architeuthis* Steenstrup, 1857

As for family.

### *Architeuthis dux* Steenstrup, 1857

*Architeuthis dux* Steenstrup In Harting, P. (1860). Description de quelques fragments de deux cephalopodes gigantesques. *Verhandelingen der Koninklijke Akademie van Wetenschappen, Amsterdam* 9(1)[1861], 16 pages, 3 plates. [Published as separate, 1860] [11, pl 1 fig. 1A].

Type data: holotype ZMUC.

Type locality: Atlantic Ocean, 31°N, 76°W.

#### COMMON NAME

Giant Squid

#### DIAGNOSTIC FEATURES

Mantle muscular. Eyes very large. Fins proportionally small, ovoid, without free anterior lobes. Straight, simple funnel-locking cartilage. Arm suckers biserial. Club, long, narrow, with tetraserial suckers; large suckers in two medial series of manus and small suckers in marginal series; triangular carpus with distinct cluster of numerous, small, alternating toothed suckers and knobs in six or seven irregular transverse series; fixing apparatus consisting of cluster of smooth-ringed suckers and knobs in approximately equal numbers. Hooks absent. Two longitudinal series of transverse pairs of alternating smooth-ringed suckers and knobs extend along almost entire length of extremely long tentacular stalks, increasingly widely spaced proximally. Proximal 14–28% of tentacular stalk is without suckers and knobs. Arms with very numerous suckers in two transverse series; hooks absent. Buccal connectives attach to dorsal border of arms 4. Ink sac present.

*Photophores.* Absent.

Live animal approaching bait (T. Kubodera).

#### HABITAT AND BIOLOGY

Giant and Colossal (*Mesonychoteuthis hamiltoni* Robson, 1925) deep-sea squid have extremely large eyes. Nilsson *et al.* (2012) postulate that, rather than evolving to detect

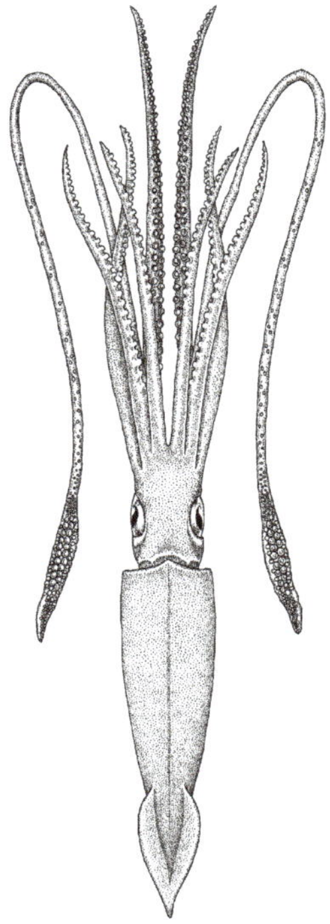

Dorsal view (ABRS, K. Hollis).

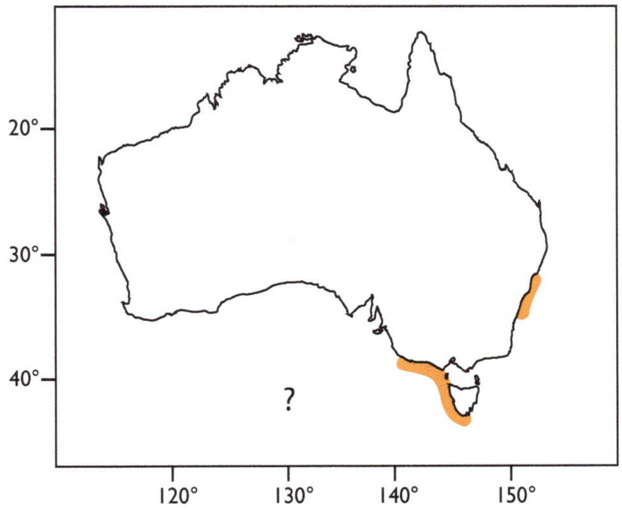


*Architeuthis dux* distribution.

prey or mates at long distance, they are uniquely suited for detecting large predators, such as sperm whales. At below 600 m depth, an eyeball of 27 cm diameter and a 9 cm pupil would allow detection of a sperm whale at distances > 120 m, enabling a well-prepared and powerful evasive response. This may have driven the evolution of huge bodies and eyes in *Architeuthis* and *Mesonychoteuthis*.

The extensive locking apparatus of the carpus and stalk suggests that the two tentacles are held together to capture targeted prey, as has been observed in other squids. Stomach contents include both demersal and pelagic prey, including fast swimming fishes and squids.

*Interest to fisheries.* Ammonium ions present in the muscle tissue of this species render it inedible.

#### SIZE

Giant squids have been reported to total lengths of 18 m and mantle lengths 3 m (to 5 m although this is a questionable record), but animals of this size are seldom recorded. Most specimens are within the range of 6–13 m total length. Weight to 500 kg, possibly greater.

#### REMARKS

A plethora of taxonomic names marks the history of Architeuthidae. Eight genera have been named, but all of them are synonyms of *Architeuthis*. Numerous species have been named, but most of the 21 nominal species are based on incomplete, damaged specimens and are so inadequately described and poorly understood that the systematics of the group has been thoroughly confused. A complete list of species associated with the Architeuthidae is presented by Sweeney and Young (2002). Three species were recently considered valid or 'probably valid' by some authors (e.g. Roper and Shea 2013) (namely *Architeuthis dux* Steenstrup, 1857; *A. martensi* (Hilgendorf, 1880) and *A. sanctipauli* (Verlain, 1877)), while others regarded these three purported taxa as subspecies of *A. dux*. The diagnostic morphological features currently are poorly understood and the entities are distinguised geographically. Molecular and morphological studies indicated that these entities possibly represent the single species worldwide: *A. dux* (Förch 1998, Carlini and Graves 1999, Roeleveld 2000). However, the resolution and breadth of these analyses were insufficient to provide conclusive evidence.

Winkelmann *et al.* (2013) provide a much more comprehensive study that examined mitochondrial genome diversity and population structure of 43 *Architeuthis* samples collected across the range of this taxon. They found no detectable phylogenetic structure at the mitochondrial level and showed that the level of nucleotide diversity is exceptionally low. These results are consistent with the hypothesis that there is only one global species of *Architeuthis*, and it is highly vagile, possibly dispersing through both a drifting paralarval stage and migration of large individuals. They also found evidence for a recent population expansion, which may explain the low level of genetic diversity.

Based on morphological characters, Roper and Young (1972) first proposed that the small deep-sea squid family Neoteuthidae (not known in Australian waters) was the sister taxon to the Architeuthidae. This has been supported by molecular studies (e.g. Lindgren, 2010a).

*Architeuthis* is the largest of all cephalopods in total length. Specimens are occasionally found moribund at the surface of the ocean, but most records are from strandings, sperm whale stomachs and, increasingly, from deep-sea trawling activity.

#### DISTRIBUTION

*Architeuthis* has a worldwide oceanic distribution, but the true extent of geographical and depth ranges has yet to be determined. Specimens are excluded from equatorial, tropical and high polar latitudes. The map here shows Australian records, but these are based on beached specimens.

#### REFERENCES

Roper and Young (1972), Roeleveld (2002), Roper and Jereb (2010d), Roper and Shea (2013).

## Batoteuthidae Young & Roper, 1968

#### DIAGNOSTIC FEATURES

Head disproportionately small and narrow. Buccal connectives attach dorsally to arm pairs 1 and 2 and ventrally to arm pairs 3 and 4. Club 80% or more of tentacle length, with six series of very small, numerous suckers. Club without distinct dactylus and carpus.

The family is monotypic, represented by *Batoteuthis skolops* Young & Roper, 1968.

### *Batoteuthis* Young & Roper, 1968

As for family.

### *Batoteuthis skolops* Young & Roper, 1968

*Batoteuthis skolops* Young, R. E. and Roper, C. F. E. (1968). The Batoteuthidae, a new family of squid (Cephalopoda; Oegopsida) from Antarctic waters. *Antarctic Research Series 2*, 185–202 [185, pls 1–6].

Type data: holotype NMNH 576029.

Type locality: South Pacific Ocean about 1000 km south-east of Wellington, New Zealand, in the region of the Antarctic Polar Front, 49°57′S, 169°01′W.

#### COMMON NAME

Bush-club Squid

#### DIAGNOSTIC FEATURES

Mantle long, slender, produced into long, pointed 'tail' posteriorly; tail ~50% ML in young specimens. Mantle wall thin, muscular (similar to most cranchiids). Head disproportionately small and narrow, eyes proportionally large. Fins short, ~20% ML. Funnel-mantle locking apparatus distinctly curved; funnel component deeply grooved, slightly broader and deeper posteriorly. Arms long, slender, very attenuate at tips. Inner rings of arm suckers with broadly spaced, sharp, pointed teeth on distal half. Tentacles, short, only 1.5 × longer than arms, robust. Club 80% or more of tentacle length, not expanded; six distinct transverse rows of very small, numerous suckers; no separate, distinct dactylus and carpus; numerous pointed trabeculae present, but protective membrane vestigal. Club suckers with small, conical, widely spaced teeth. Buccal connectives attach dorsally to arm pairs 1 and 2, ventrally to arm pairs 3 and 4. Viscera compact, located near anterior end of mantle cavity (except gonads). Gladius visible along dorsal midline and extends far posteriorly as supporting element of long, thin, spindle-like 'tail'.

*Photophores.* Large elongate photophores on aboral tips of arms 4 in large subadult males; small aboral photophores on tips of arms 4 of large subadult females.

#### HABITAT AND BIOLOGY

This species is a bathypelagic to abyssopelagic form. Its recorded depth ranges from 366 m for a 30 mm ML juvenile during the day to 1200–2500 m for larger juveniles and subadults at night (open net captures only). A few growth stages have been described, but paralarvae and mature adults are unknown. It is an ammoniacal species.

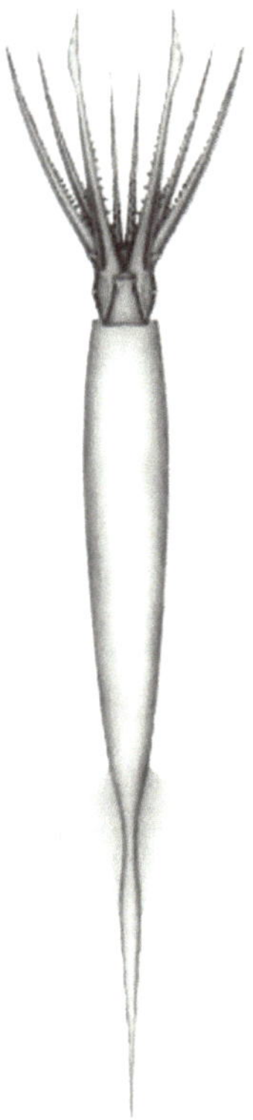

Ventral view (C. Roper).

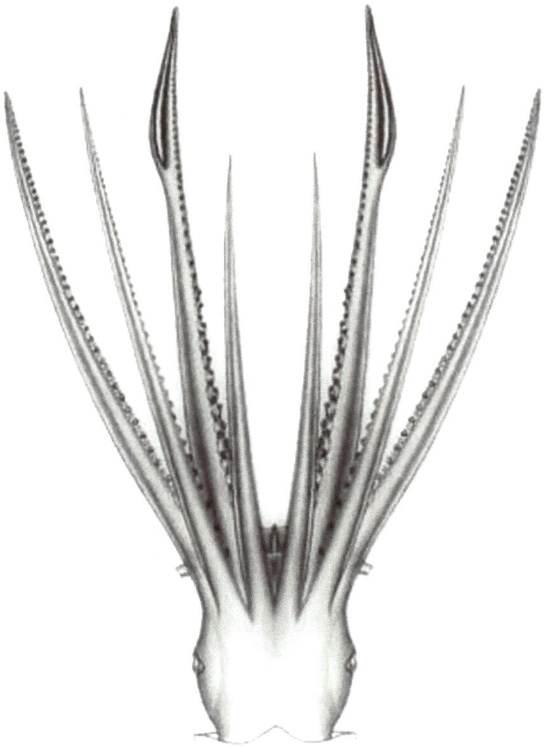

Head, dorsal view (C. Roper).

Tentacular club (Young and Roper 1968).

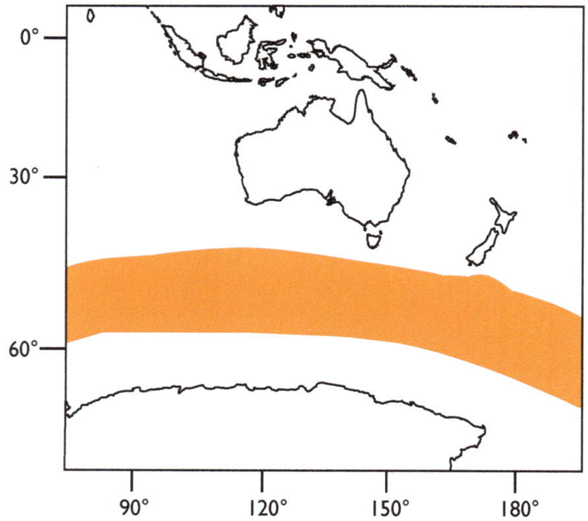


*Batoteuthis skolops* distribution.

*Interest to fisheries.* This species is rarely collected in deep-sea trawls, so has no current interest to fisheries.

### SIZE

The total mantle and 'tail' length extends to 350 mm ML up to 300–350 mm (excluding 'tail') posterior to the fins.

### REMARKS

This squid has been reported only infrequently since its original description.

### DISTRIBUTION

Bathypelagic to abyssopelagic found circumglobally in Antarctic and sub-Antarctic waters of the South Atlantic, South Pacific and southern Indian Oceans between ~45°S to 60°S. Antarctic endemic extending north to the Sub-Antarctic Front.

### REFERENCES

Young and Roper (2009), Roper and Jereb (2010e).

## Brachioteuthidae Pfeffer, 1908

### COMMON NAME

Arm Squids

### DISTINGUISHING CHARACTERS

The unusually long neck in paralarval brachioteuthids distinguishes them from all other families of squids except for chiroteuthid paralarvae. However, most chiroteuthids additionally have an elongate arm crown stalk (variable among species). Also, the structure of the elongated neck differs between the two families. Brachioteuthids have a neck that appears to be a single, muscular, fluid-filled tube continuous with a large, fluid-filled chamber in the mantle cavity. In contrast, the elongate neck of young chiroteuthids is supported by numerous separate chambers, so the paralarvae and juveniles of the two families are easily distinguished (Dunning and Lu 1998).

### DIAGNOSTIC FEATURES

*Adults.* Mantle long, slender, muscular, but thin-walled, sometimes produced as short 'tail'. Fins relatively short, terminal, heart-shaped or rhomboidal, generally < 50% ML, with free anterior lobes. Funnel-locking cartilage straight, simple groove. Arm suckers biserial. Lateral arms with swimming keels. Tentacular club with numerous transverse rows of very small, irregular, long-stalked suckers on greatly expanded proximal end of manus of tentacular clubs; dactylus normal with three or four (usually) series of suckers; carpal-locking apparatus extends along tentacular stalk; numerous small suckers extend proximally along club. Buccal connectives attached to ventral margins of arms 4. Digestive gland compact, located well posterior to cephalic cartilage in mantle cavity.

*Photophores.* Some species with a single photophore on ventral surface of each eyeball.

*Paralarvae and juveniles.* Mantle elongate to elongate/bulbous. Distinctive, long, slender neck (no arm-crown stalk); mantle opening wide relative to neck; eyes lateral but frequently situated towards ventral part of head; distinctive bump-like swelling on dorsal surface of head. Fins separate, terminal, paddle-shaped, transversely oval. Club suckers develop at hatching; adult-like pattern of numerous suckers on proximal manus well established by ~10 mm ML; tentacles large, present at hatching, robust relative to arms.

### HABITAT AND BIOLOGY

The occurrence of larvae in eastern Australian waters suggests that they spawn throughout the year. Paralarval *Brachioteuthis* species apparently orient in the water in a head-down posture to resemble a drifting medusoid bell with dangling tentacles to capture prey. Changes in morphology during development appear to be correlated with rapid ecological changes. Data obtained from other parts of the world collectively suggest an extensive diel vertical migration for larval *Brachioteuthis* species. The distribution of adults is not known, although they have been collected in midwater trawls off eastern Australia. They are prey items for numerous epipelagic and mesopelagic fishes – for example, yellowfin tuna and swordfish off eastern Australia.

*Interest to fisheries.* The relatively small size, thin-walled mantle and apparent rarity of *Brachioteuthis* species in catches make them unsuitable for commercial harvesting. However, research has shown the high value of these species as prey for fishes, marine mammals and other components of the marine fauna.

### SIZE

Small- to medium-sized squids up to 200 mm ML.

### REMARKS

Two genera are presently recognised in the family: *Brachioteuthis* and *Slosarczykovia*. Six species are currently recognised as valid in the genus *Brachioteuthis*, with three species found in Australian waters: *B. behnii* Steenstrup, 1882, *B. picta* Chun, 1910, and *B. 'riisei'* Steenstrup, 1882. The taxonomic status of *Slosarczykovia* Lipinski, 2001 has yet to be reviewed and a thorough revision of all taxa in the family is required. An increasing body of biological information on the species has developed since the early 1990s regarding the geographical distribution, ecology, predation and prey. By far the greatest research information has been attributed to *B. 'riisei'*, an apparently extremely wide-ranging species. However, because the identification of *Brachioteuthis* species is so challenging, it is difficult to know which species and growth stages are actually represented in published reports, both for distribution and biology.

### REFERENCE

Roper and Jereb (2010f).

### Key to genera of the Brachioteuthidae (after Lipinski 2001)

1a. Skin smooth; reticulate sculpture and/or warts in mature males. Tentacular club with sharply differentiated suckers; largest from proximal manus half the size of those on proximal dactylus; carpal-locking apparatus usually present ........................................ *Brachioteuthis*

1b. Skin in both sexes forms fibrous, delicate net that covers the whole body. Tentacular club with weakly differentiated suckers; largest from proximal manus one-third or larger than those from proximal dactylus; carpal-locking apparatus absent ................ ........................... *Slosarczykovia* (*S. circumantarctica*)

## *Brachioteuthis* Verrill, 1881

### DIAGNOSTIC FEATURES

Skin smooth; reticulate structure and/or warts may occur in mature males, especially in late subadult and adult stages. Tentacular club with distinctly differentiated suckers; largest suckers on proximal manus are at least 50% smaller than those on proximal dactylus. Carpal (tentacular)-locking apparatus normally present.

### REMARKS

See family.

### Key to *Brachioteuthis* species (after Nesis 1982/1987)

1a. Mantle very narrow, slightly widening anteriorly and then abruptly narrowing anterior to fin. Fin width to length ratio ~0.9–1.1........................................ *B. picta*

1b. Mantle cylindrical, without abrupt narrowing anterior to fin. Fin width to length ratio ~1.2–1.4...2

2a. Larvae with neck of moderate length. Eyes directed forwards and laterally at an acute angle to longitudinal axis of body............................................... *B. 'riisei'*

2b. Larvae with extremely long neck. Eyes directed laterally, perpendicular to longitudinal axis of body ......................................................................... *B. behnii*

## *Brachioteuthis behnii* (Steenstrup, 1882)

*Tracheloteuthis behni* Steenstrup, J. (1882). En ny Blaeksprutteslaegt: Tracheloteuthis. *Videnskabelige Meddelelser fra den Naturhistoriske Forening I Kjrbenhaven*, series (4)3, 293–294 [294].

Type data: syntypes ZMUC.

Type locality: Indian Ocean (purchased); west coast of New Guinea, western Pacific Ocean; no locality.

### DIAGNOSTIC FEATURES

Paralarvae have a very long neck and laterally directed eyes.

*Photophores.* Absent.

### SIZE

Maximum ML to 60 mm.

### REMARKS

Further study is needed to determine the validity of this species.

### DISTRIBUTION

Equatorial and central waters of the Indo-Pacific Ocean as well as the equatorial Atlantic.

*Brachioteuthis behnii* ventral view (D. Shale).

## *Brachioteuthis picta* Chun, 1910

Chun, C. (1910). Die Cephalopoden. Oegopsida. *Wissenschaftliche eregebnisse der deutschen tiefsee expedition auf dem dampfer 'Valdivia' 1898–1899, 18*(1), 1–401 [207, pls 29–32].

Type data: holotype ZMB Moll.-102155.

Type locality: South Atlantic Ocean, Benguela Current, 5°6′N, 9°58′E.

### DIAGNOSTIC FEATURES

Mantle long, very slender, slightly flared at opening, abruptly narrows anterior to fins, extends posteriorly into a pointed 'tail'. Fins saggitate, longitudinally rhomboidal; fin length and width ~50% ML; width: length ratio ~0.9: 1.1. Tentacular clubs expanded, covered with numerous minute suckers in carpal region that extend proximally along the club. Integument texture reticular, fibrous.

*Photophores.* A single sausage-shaped photophore on ventral surface of each eyeball.

*Colour.* Colour a light purple-brown or chocolate hue.

### HABITAT AND BIOLOGY

*Brachioteuthis picta* is an oceanic species occurring in equatorial, southern subtropical and southern peripheral water masses. Paralarvae are epipelagic and mesopelagic;

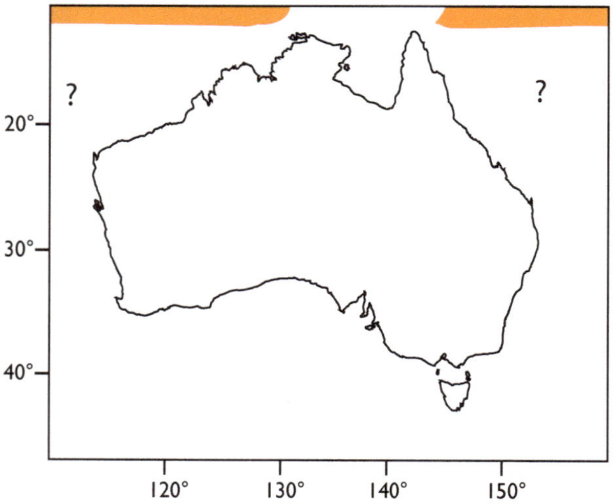


*Brachioteuthis behnii* distribution.

### REFERENCE

Roper and Jereb (2010f).

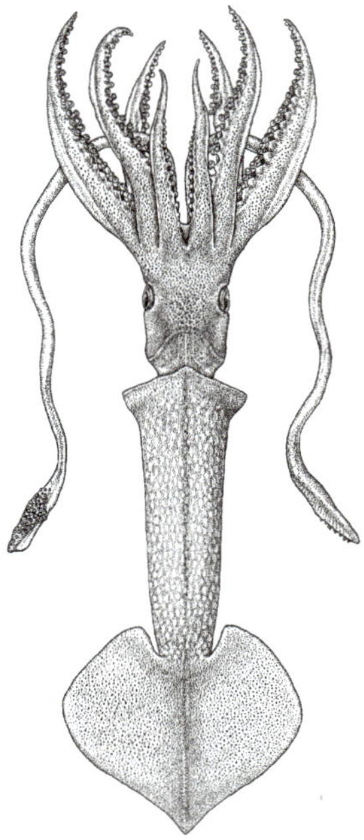

Dorsal view (Chun, 1910).

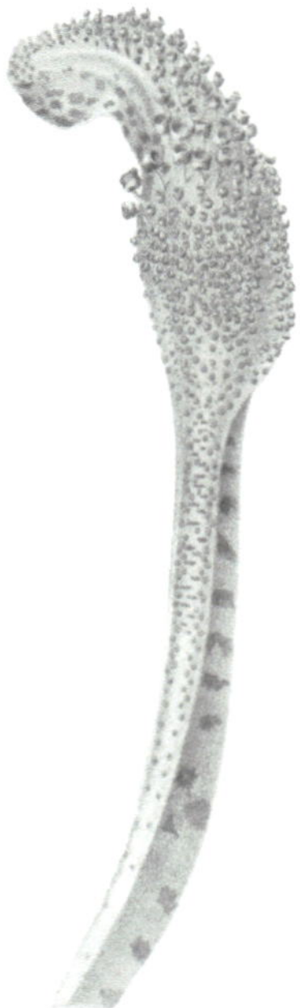

Tentacular club (Chun, 1910).

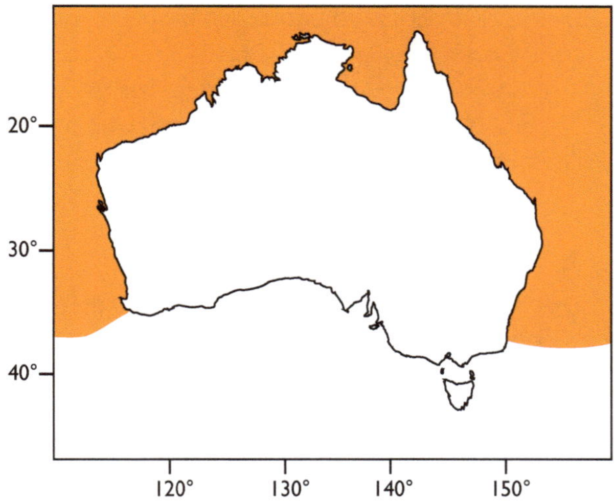


*Brachioteuthis picta* distribution.

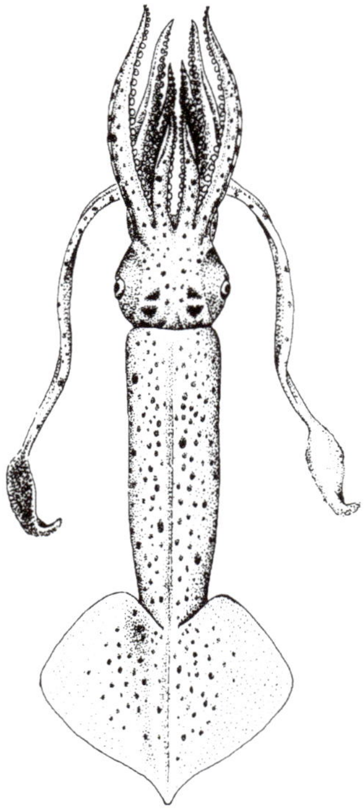

Dorsal view (FAO, E. D'Antoni).

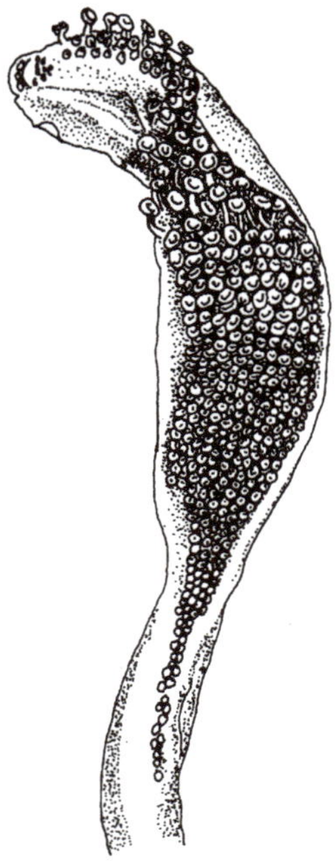

Tentacular club (FAO, E. D'Antoni).

adults are lower epipelagic, mesopelagic and bathypelagic. A principal prey item seems to be krill.

#### SIZE

Maximum ML to 90 mm.

#### DISTRIBUTION

Circumglobal, oceanic from ~30°N to at least 40°S to 45°S.

#### REFERENCE

Roper and Jereb (2010f).

### *Brachioteuthis 'riisei'* (Steenstrup, 1882) (species-complex)

*Tracheloteuthis riisei* Steenstrup, J. (1882). En ny Blaeksprutteslaegt: Tracheloteuthis. *Videnskabelige Meddelelser fra den Naturhistoriske Forening I Kjrbenhaven*, series 4(3), 293–294 [294].

Type data: holotype ZMUC.

Type locality: south-eastern Atlantic Ocean, 34°40'S, 7°W.

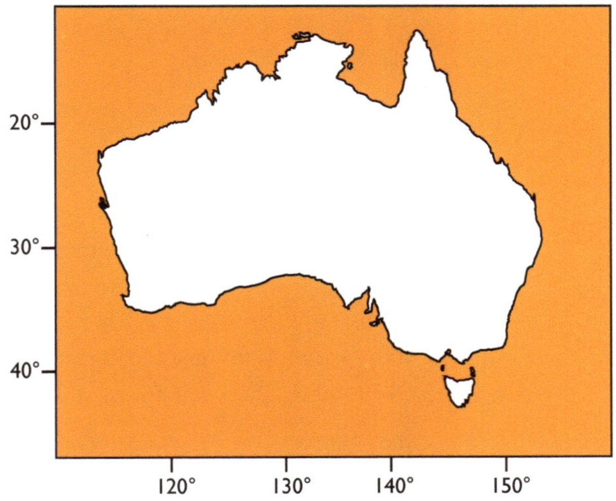


*Brachioteuthis 'riisei'* distribution.

#### COMMON NAME

Common Arm Squid

#### DIAGNOSTIC FEATURES

Mantle narrow, elongate, cylindrical, muscular; flares at mantle opening; conical posteriorly to tip. Head narrow, not retractable into mantle. Neck long to very long in paralarvae; eyes ventral to lateral midline of head. Fin length from anterior border of fins <50% ML and shorter than width; saggitate to slightly rounded posteriorly, angles rounded, lobes prominent. Tentacles long, tentacular clubs wide, with very numerous, minute suckers in carpal portion; four to six transverse series of larger suckers in distal portion; four series at dactylus. Tentacles in paralarvae and juveniles proportionally much longer than in adults; tentacular clubs broad with two to four series of larger suckers in carpal area, becoming minute in midportion, and then six to eight series of knobs in dactyl region. Carpal-locking apparatus with two irregular series of small suckers along oral surface of tentacular stalk. Buccal connectives attach to ventral edge of arms 4. Paralarvae of *B. 'riisei'* are characterised by moderately long neck and eyes that are directed anteriorly at acute angles to the longitudinal axis of the head.

*Photophores.* A single photophore on ventral surface of each eye developed in adults.

#### HABITAT AND BIOLOGY

*Brachioteuthis 'riisei'* is an oceanic taxon that occurs from near the surface to 3000 m depth. The spawning season appears to be considerably extended, reflecting the cosmopolitan distribution of the species-complex. Paralarvae and juveniles are epipelagic to mesopelagic, and then undergo ontogenetic descent to mesopelagic and bathypelagic zones as adults. It was among the most abundant species caught during summer night-trawls within a warm core eddy of the East Australian Current in the Tasman Sea (Brandt 1983). It is not known whether this species occurs commonly in the eddy source water of the Coral Sea.

#### SIZE

Maximum ML to 170 mm in Antarctic populations, to 80 mm in other areas.

#### REMARKS

The full extent of the distribution is undetermined due to uncertain identifications in the literature and the probable existence of a currently undifferentiated species-complex. This group is in need of taxonomic review; hence the species name has been put in quotation marks.

#### DISTRIBUTION

A cosmopolitan species. Circumglobal in the Southern Ocean, but rare in the tropics.

#### REFERENCE

Roper and Jereb (2010f).

### *Slosarczykovia* Lipinski, 2001

The genus is monotypic, represented by *Slosarczykovia circumantarctica* Lipinski, 2001.

### *Slosarczykovia circumantarctica* Lipinski, 2001

*Slosarczykovia circumantarctica* Lipinski, M. R. (2001). Preliminary description of two new species of cephalopods (Cephalopoda: Brachioteuthidae) from South Atlantic and Antarctic waters. *Bulletin of the Sea Fisheries Institute, Gdynia 1*(152), 3–14 [6, Figs 3E–G; 5B, C; 6B; 7A, B; 8, 9, 10C, 11B].

Type data: holotype SAMC S-3949.

Type locality: Southern Ocean, Antarctica, the waters adjacent to Wilkes Land.

#### DIAGNOSTIC FEATURES

Mantle long, slender. Surface integument of mantle, head, arms and fins of both sexes covered with reticulate fibrous, delicate tissue in specimens 60 mm or longer. Fins rhomboidal. Characteristic and strong V-folds (ridges) dorsally on neck and head, from nuchal cartilage to bases of arms 1. Funnel organ distinct, with relatively small ventral pads and very long papilla on dorsal component. Arms short, formula 2 > 3 > 4 > 1. Protective membranes low dorsally and well-developed ventrally, with long filiform trabeculae. Arm suckers globular and fleshy with short stalks; sucker rings with variable arrangement of teeth, from smooth to loosely packed, to densely packed. In females, sucker rings of arms 1–4 smooth; in males, one to four low, inconspicuous teeth may be present. Hectocotylus absent. Arms with median groove and no keel. Tentacles short, club long; with only weakly differentiated suckers; largest suckers on proximal manus ~33% (or slightly larger) of those on proximal dactylus. Carpal-locking apparatus absent (series of seven small carpal suckers, but no corresponding series of knobs has yet been recognised); 10 sparsely distributed rows of suckers on manus; diameter of largest sucker on proximal dactylus only about one-third larger than largest sucker on manus. Dorsal marginal row and suckers on stalk minute, rest approximately subequal. Club sucker rings with 11–13 sharp, long teeth. Buccal membrane fleshy with many papillae in females, or smooth in males; eight lappets with DDVV attachment.

#### HABITAT AND BIOLOGY

Unknown. It is a non-ammoniacal species.

#### SIZE

Mantle length reaching 170 mm.

Dorsal view (D. Stevens, NIWA, NZ IPY-CAML (voyage TAN0802)).

Ventral view (D. Stevens, NIWA, NZ IPY-CAML (voyage TAN0802)).

Enlargement of dorsal view (D. Stevens, NIWA, NZ IPY-CAML (voyage TAN0802)).

*Slosarczykovia circumantarctica*

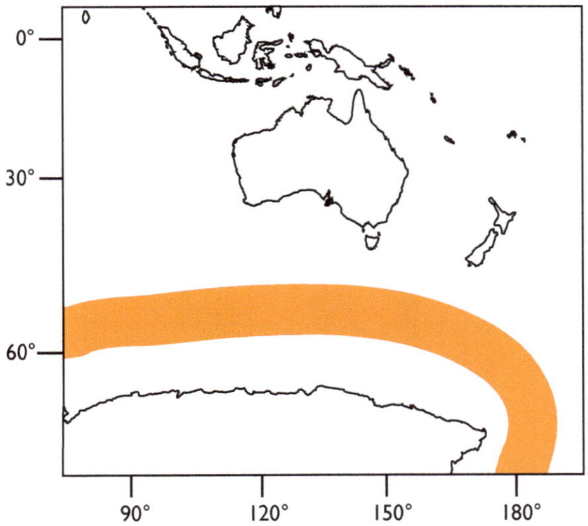


*Slosarczykovia circumantarctica* distribution.

### REMARKS

This is one of the most common squids in Antarctic waters.

### DISTRIBUTION

Antarctic endemic extending north to the Sub-Antarctic Front.

### REFERENCE

Lipinski (2001).

## Chiroteuthidae Gray, 1849

### DISTINGUISHING CHARACTERS

Ventral arms enlarged. Mastigoteuthids also have enlarged ventral arms, but differ in that the funnel-locking cartilage has a poorly developed posterior knob, and sometimes a poorly developed median knob (these are well developed in chiroteuthids). In addition, mastigoteuthids have hundreds of small suckers on the tentacular clubs and these are not arranged tetraserially as is the case in the Chiroteuthidae.

### DIAGNOSTIC FEATURES

Body gelatinous and slender. Indistinct eyelid sinus. Head with elongate neck; nuchal folds absent; olfactory organs on long, slender stalks. Mantle–funnel-locking apparatus oval, funnel cartilage with one or two knobs projecting towards centre of cavity (variable among genera, fused in *Grimalditeuthis*). Ventral arms enlarged. Hectocotylus absent. Arm suckers biserial, toothed; ventral pair greatly enlarged. Some species with very elongate tentacles (up to 5 × ML) and distinctive clubs with tetraserial suckers on long stalks; club divided into two or three portions by symmetrical protective membranes (except in *Planctoteuthis*). Buccal connectives attached to ventral border of ventral arms. Penis developed into an elongate, spearheaded structure. Tail supported by gladius posterior to fins with various embellishments.

*Photophores.* Abundant photophores present along tentacle stalks and ventral arms; large photophores present ventrally on the eyes, at tips of the tentacular clubs, and embedded in ink sac on both sides of intestine.

All species have a distinctive paralarva (the doratopsis stage). The gladius of paralarvae (and of the adults of some species) extends well posterior to fins and supports an unusual elongate 'tail' that may have oval, fin-like flaps or a series of small flaps and/or oval bulbs.

All doratopsis paralarvae possess all of these characters during some phase of their development: (1) elongate, chambered neck; (2) chambered brachial pillar; (3) gladius extending posteriorly beyond fins and supports flotation structures; (4) vesicular tissue in posterior region of mantle, just anterior to tip; (5) vesiculate arms in advanced stages; (6) ventral arms greatly elongate in advanced stages; and (7) tentacular clubs of advanced stages with keels parallel to dorsal protective membrane along most of club length.

### HABITAT AND BIOLOGY

Chiroteuthids are bathypelagic or meso-bathypelagic slow-moving squids. They occasionally occur on, or in association with, the bottom at bathyal depths. Chiroteuthids are considered to be primarily oceanic; however, *Chiroteuthis picteti* Joubin, 1894 has been taken in demersal trawls in continental slope waters at depths from 300–600 m off northern Australia. In deeper oceanic waters, adult *Chiroteuthis* apparently occur below 500 m during the day, but are distributed throughout the water column at night. Diel vertical migration is also evident in some species. Doratopsis paralarvae occur in oceanic waters in the upper 200–300 m.

Very little is known of their life history. The paralarvae live in the upper few hundred meters of the ocean. The duration of the doratopsis stage before transformation into the very different-looking subadult stage is unknown. Some changes involve the resorption of the paralarval clubs and formation of new clubs, extensive elongation of the tentacular stalks, development of photophores in some species, loss of the 'tail' in most species, and generally a marked change in the proportions of body components. Some of these ornately tailed doratopsis paralarvae are remarkably similar to some siphonophores (Vecchione *et al.* 1992). Behavioural studies have shown that juvenile and subadult *Chiroteuthis calyx* Young, 1972 in the Monterey Submarine Canyon appear to mimic the abundant siphonophore *Nanomia bijuga* (Chiaje, 1841) to deter predation (Burford *et al.* 2014).

The arms, head and mantle have numerous chambers filled with ammonium chloride, a light-weight fluid that provides near-neutral buoyancy.

*Interest to fisheries.* No fisheries potential due to their soft, gelatinous bodies, ammoniacal tissue, widespread deep-sea distribution and generally small size.

### SIZE

Small- to medium-sized squids; maximum ML to ~800 mm. Paralarvae large: up to 90 mm ML.

### REMARKS

Considerable morphological differences exist among the four nominal chiroteuthid genera, *Asperoteuthis*, *Grimalditeuthis*, *Chiroteuthis* and *Planctoteuthis*, all of which are represented in Australian waters. The group is very much in need of a comprehensive taxonomic review.

### DISTRIBUTION

Members of the family are represented in all geographic zones of all oceans of the world, from sub-Arctic to sub-Antarctic, circumglobally.

### REFERENCES

Roper and Jereb (2010g), Young and Roper (2011a).

## Key to genera of the Chiroteuthidae (after Roper and Jereb 2010g)

1a. Funnel–mantle locking cartilages fused; nuchal-mantle apparatus not fused. Tentacular clubs without suckers........ *Grimalditeuthis* (*G. bonplandi*)

1b. Funnel–mantle locking cartilages free, not fused. Tentacular clubs with suckers....................................2

2a. Mantle covered with numerous, minute 'cartilaginous' tubercles. Fins together oval to elongate/oval with long axis parallel to mantle. Arms 4 not longer or thicker than arms 2 and 3 ....................................... ................................*Asperoteuthis* (*A. acanthoderma*)

2b. Mantle without tubercles. Fins together circular or oval with long axis transverse to mantle. Arms 4 longer than arms 2 and 3............................................3

3a. Fins together roughly oval and lobed both posteriorly and anteriorly. Arms 4 not appreciably thicker than arms 2 and 3. Funnel valve absent; funnel-locking apparatus without tragus............................. ........................................... *Planctoteuthis* (*P. danae*)

3b. Fins together approximately circular, not lobed posteriorly. Arms 4 usually much thicker than arms 2 and 3. Funnel valve present; funnel-locking apparatus with distinct tragus present............. *Chiroteuthis*

## *Asperoteuthis* Nesis, 1980

### DIAGNOSTIC FEATURES

Funnel valve present. Arms 4 not enlarged; suckers present. Tentacle pads present. Club suckers absent proximally, tetraserial distally. Arms 4 photophores absent, visceral photophores absent. Olfactory papilla in doratopsis stage lies halfway between funnel and eye.

### REFERENCES

Roper and Jereb (2010g), Young and Roper (2010).

## *Asperoteuthis acanthoderma* (Lu, 1977)

*Chiroteuthis acanthoderma* Lu, C. C. (1977). A new species of squid, *Chiroteuthis acanthoderma*, from the Southwest Pacific (Cephalopoda, Chiroteuthidae). *Steenstrupia 4*, 179–188 [179, figs 1–14].

Type data: holotype ZMUC.

Type locality: western Pacific Ocean, Celebes Sea, 6°37′N, 122°02′E.

### COMMON NAME

Thorny Chiroteuthid Squid

### DIAGNOSTIC FEATURES

Small, pointed cartilaginous tubercles over surface of head, mantle and arms. Fins 50–55% ML; fin width 35–40% ML and secondary 'fin' comparable in length to primary fin. Funnel-locking apparatus with long, slender tragus and narrow antitragus that form inverted Y-shaped groove. Arms long; tentacles extremely long, may be 12 × ML (i.e. 5.5 m). Large arm suckers with three or four broadly rounded teeth on distal half of inner ring. Tentacles with nine blunt teeth over distal half of inner ring.

*Photophores.* Oval photophores on ventral surface of each eye. Numerous photogenic pads on tentacle stalk. Large club-tip photophores with short, broad, terminal papilla. Twelve pairs of small, embedded photophores on aboral surface of club in two series and of two sizes.

### HABITAT AND BIOLOGY

Mesopelagic to bathypelagic.

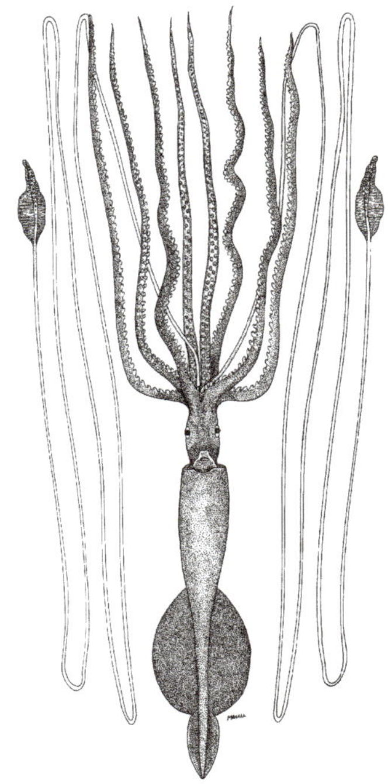

Dorsal view (ABRS, K. Hollis).

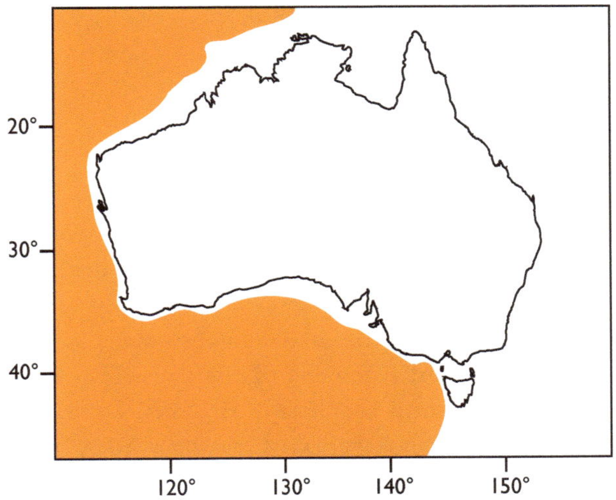


*Asperoteuthis acanthoderma* distribution.

### SIZE

Mantle length to 800 mm.

### DISTRIBUTION

Indo-West Pacific; Celebes Sea, Ryuku Islands, Molucca Sea, New Guinea, southern Africa; eastern Gulf of Mexico, Gulf Stream Florida, South Atlantic Ocean.

### REFERENCES

Roper and Jereb (2010g), Young and Roper (2010).

## *Chiroteuthis* d'Orbigny, 1839

### DIAGNOSTIC FEATURES

Olfactory organ on stalk posterior to each eye. Olfactory papilla in doratopsis stage is behind the eye. Fins semicircular, without lobes. Funnel–mantle locking cartilage present, with tragus and antitragus; funnel valve present. Arms 4 much longer, thicker than arms 1–3; suckers present. Hectocotylus absent. Club suckers tetraserial; suckers on proximal and distal regions of club. Tentacles very long, retractile into elongate tubes formed by broad lateral membranes of arms 4. Tentacles with pad-like photophores along tentacular stalk. Elongate penis present; can extend beyond mantle opening.

*Photophores.* Photophores present on ventral surface of arms 4 (alternate with lateral arm suckers), eyes and viscera (in Australian species). Club tip with enlarged photophore with muscular opening and closing lids. Large pad-like photophores along tentacular stalk (embedded distally), and numerous tiny photophores on oral surface. Ocular photophores present as stripes or series of round organs (or a combination of both).

*Colour.* Photophores silver or red in colour in live or fresh specimens.

### HABITAT AND BIOLOGY

The late doratopsis stages of the paralarvae have tentacles with a distal, expanded, paralarval club and proximal presumptive (developing) adult club. The developing adult club has four series of very small, regularly arranged suckers. The 'tail' of the late doratopsis stage is highly decorated, but lost at the end of this stage. They mimic siphonophores until they lose their 'tails'. Behavioural and associated ontogenetic changes have been studied and documented by Burford *et al.* (2014). The club tip photophores in adults may be used to lure prey.

### REMARKS

This genus contains at least six species, three of which are found in Australian waters.

### REFERENCES

Roper and Jereb (2010g), Young and Roper (2011a), Roper and Young (2013).

### Key to *Chiroteuthis* species

1a. Mantle covered with cartilaginous tubercles .................................................................... *C. veranyi*
1b. Mantle without cartlaginous tubercles .................... 2
2a. Tentacular club very long, slender. Buccal membrane pigmented. Funnel-locking cartilage with wide, low tragus that undercuts mantle component ..... *C. picteti*
2b. Tentacular club short. Buccal membrane unpigmented. Tragus of funnel-locking cartilage high and narrow .................................................. *C. joubini*

## *Chiroteuthis joubini* Voss, 1967

*Chiroteuthis joubini* Voss, G. L. (1967). Some bathypelagic cephalopods from South African waters. *Annals of the South African Museum 50*(5), 61–88 [79].

Type data: holotype ZMUC.

Type locality: western North Atlantic Ocean, 8°19′N, 44°35′W.

### DIAGNOSTIC FEATURES

Head length 34–37% ML. Tragus and antitragus of funnel-locking apparatus large; tragus high and narrow. Arms 1 50–55% ML; arms 3 85–90% ML. Largest arm suckers with 22–26 pointed, triangular teeth over two-thirds of rings; largest suckers not globular. Clubs short, 28–35% ML; club suckers with seven or eight pointed, triangular teeth over distal half of ring; teeth joined basally, no enlarged central tooth; suckers with stalks in two distinct parts; stalks of lateral and medial suckers similar in length. Club protective membrane in three distinctive regions: proximal region shorter than others with 7–12 trabeculae (14–18% of club length); middle region without distinctive tabeculae but opposite 9–12 marginal suckers (32–36% of club length); distal region with 20 or 21 narrow, broadly spaced trabeculae (46–54% of club length).

*Photophores.* Eyeball with oval photophores in two series. Lateral or outermost series with five or six organs, from posterior to anterior: one enlarged light organ followed anteriorly by three or four smaller, similar-sized light organs, and then one slightly larger anterior light organ. Medial, or innermost series with five light organs: anterior-most organ largest. Club tip photophores moderate in size with small papilla. Viscera with two large photophores.

*Colour.* Buccal membrane unpigmented. Olfactory organ with few chromatophores on sensory bulb. Most of club pigment in epithelial cells rather than chromatophores. Club sucker stalks with epithelial pigment, not pleated, pigmentation diminishes then disappears distally on club. Club tip photophores with chromatophores only.

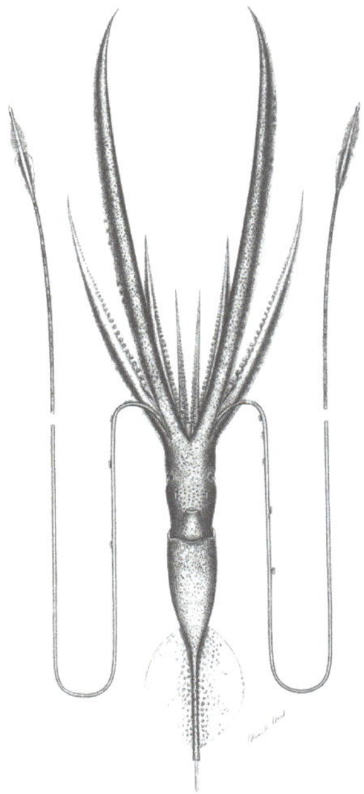

Ventral view (Roper and Young (1999), drawn by A. D. Hart).

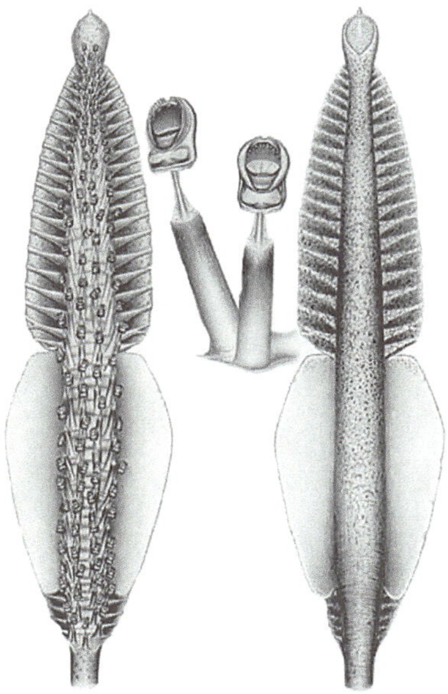

Tentacular club, oral view (left), aboral view (right) (Roper and Young (1999), drawn by A. D. Hart).

### HABITAT AND BIOLOGY

Mesopelagic to bathypelagic.

### SIZE

Mantle length to ~100 mm.

### REMARKS

The paralarval stages are unknown. The taxonomy of this genus is in need of review. The description here is based largely on the original description of Atlantic specimens and conspecificity with the Australian form has yet to be confirmed.

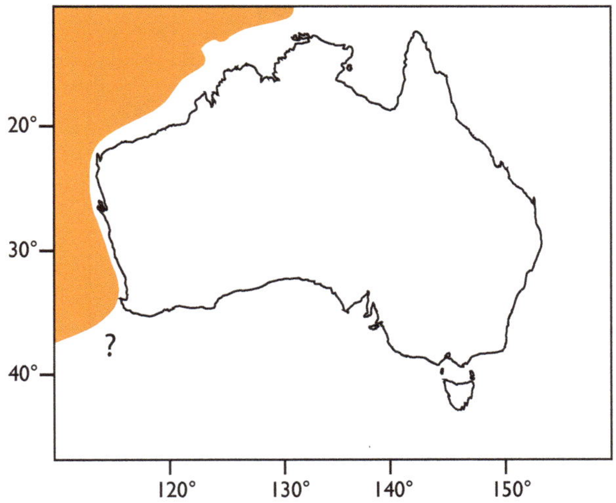


*Chiroteuthis joubini* distribution.

### DISTRIBUTION

Northern subtropical and tropical Atlantic Ocean, Sargasso Sea; western and southern Indian Ocean, 0°S to ~45°S, south-eastern Africa, south-western Australia; Banda Sea.

### REFERENCE

Roper and Young (1999).

## *Chiroteuthis picteti* Joubin, 1894

*Chiroteuthis picteti* Joubin, L. (1894). *Cephalopodes d'Amboine. Revue Suisse de Zoologie et Annales du Musee d'Historie Naturelle le Geneva 2*, 23–64 [40, pls 1, 2].

Type data: unresolved, syntypes MOM?.

Type locality: eastern Indian Ocean, Indonesia, Sumatra, Amboina, 0°15′N, 98°08′E.

### DISTINGUISHING CHARACTERS

*Chiroteuthis picteti* Joubin, 1894 can be separated from most other members of the genus by its very long, slender tentacular clubs; proximal region of club very short. Three rows of individual round photophores on eyeball.

### DIAGNOSTIC FEATURES

Head length 32–50% ML. Funnel-locking apparatus wide with wide, low tragus that slightly undercuts mantle component, antitragus deeply undercuts mantle component. Arms 1 54–65% ML; arms 3 73–103% ML, arms 4 120–200% ML. Large arm suckers with 10–20 usually separate, pointed to blunt teeth on distal two-thirds of sucker ring;

Dorsal view (AM C.476044, A. Reid).

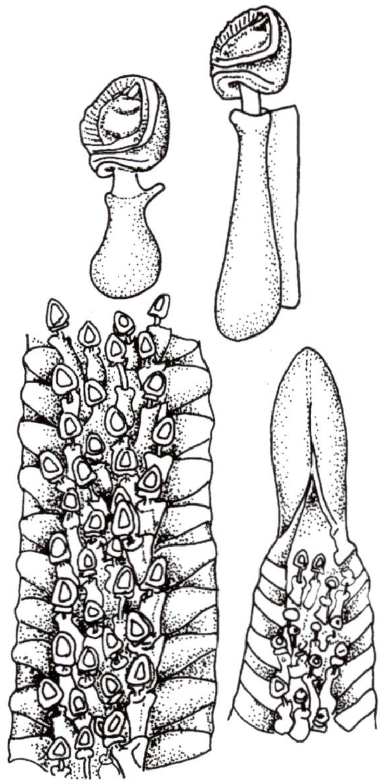

Oral views of portions of the tentacular club and club sucker: top left, medial club sucker; top right, lateral club sucker; bottom left, mid section of club showing shape of trabeculae; bottom right, distal tip of tentacular club showing closed club tip photophore (A. Reid, redrawn from Roper and Young 2011a).

largest suckers not globular. Hectocotylus absent. Clubs long (115% ML), with protective membranes in two sets: distal set with broad, triangular, closely spaced trabeculae, long (~93% of club length) with 83–96 broad but separate trabeculae; proximal set less than one-tenth length of distal set and only slightly broader than distal set, short (~7% club length) with 11–13 trabeculae. Club sucker rings with one large central tooth and 11–15 pointed teeth over distal three-quarters of ring, teeth joined basally; sucker stalks divided into two parts with flag-like lateral keel on proximal segment; stalks of suckers in lateral series about twice as long as those of medial series. Males with large penis that can extend well out of mantle cavity.

*Photophores.* Ocular photophores in three series/stripes on eyeball: lateral series with 6–9 photophores, medial series with 6–10 photophores, intermediate series with 6–10 photophores. Viscera with two photophores. Club tip photophore small with little or no terminal papilla. Aboral tentacular stalk with cushion-like photophores.

*Colour.* Pigment on clubs in chromatophore organs rather than epithelial cells; club sucker stalks without pigment; buccal membrane with dense epidermal pigment, particularly in immature animals. Dark epithelial pigment has been observed on oral surfaces of ventral arms and tentacular sheath in one female (Roper and Young 2011a).

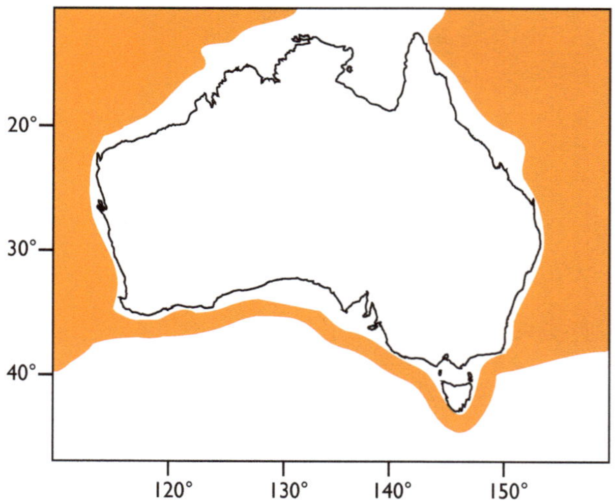


*Chiroteuthis picteti* distribution.

## HABITAT AND BIOLOGY

*Chiroteuthis picteti* is mesopelagic and bathypelagic to bathybenthic. Paralarvae and juveniles are found in the upper few hundred metres. They undergo ontogenetic descent to 2000 m or more as adults. Some adults are captured on the bottom in deep slopewaters. Eggs tiny (~1 mm in diameter).

**SIZE**

Mantle length to 250 mm, possibly to 400 mm. Advanced paralarvae are at least 44 mm ML.

**REMARKS**

Chun (1908) described *C. imperator* from off Sumatra, Indonesia, close to the type locality of *C. picteti*. Roper and Young (1999) consider *C. imperator* to be a synomym of *C. picteti* and that decision is followed here. Some northern Australian specimens may be referred to as *C. imperator* in older literature and museum collections. It is likely that *C. 'picteti'* is a species-complex.

The paralarval doratopsis is unusual in virtually lacking a brachial pillar, and at later stages, having the oesophagus centrally located in the now present brachial pillar.

The short, proximal segment of the club membrane can be difficult to recognise as distinct from the distal segment.

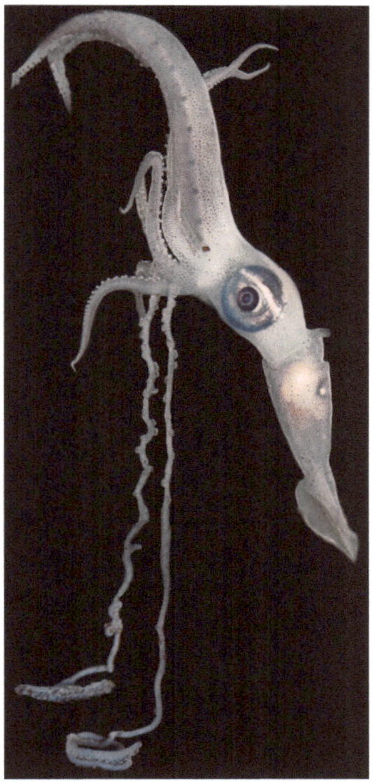

Live animal (E. McSweeny).

**DISTRIBUTION**

Indo-Pacific Ocean from Hawaii, Japan, Philippines, Indonesia to 60°E in the Indian Ocean; Australia, Tasman Sea, New Zealand.

**REFERENCES**

Young and Roper (1999), Roper and Jereb (2010g), Roper and Young (2011a).

### *Chiroteuthis veranyi* (Ferussac, 1834)

*Loligopsis veranii* Ferussac, A. E. de (1834). Notice sur deux nouvelles especes des Cephalopodes appartenant aux genres Calmaret et Cranchie. *L'Institut, Journal General des Societes et Travaux Scientifiques de la France et de l'Etranger* 2(77), 355.

Type data: syntype MHNN.

Type locality: Mediterranean Sea, from dolphin stomach.

**COMMON NAME**

Verany's Long-armed Squid

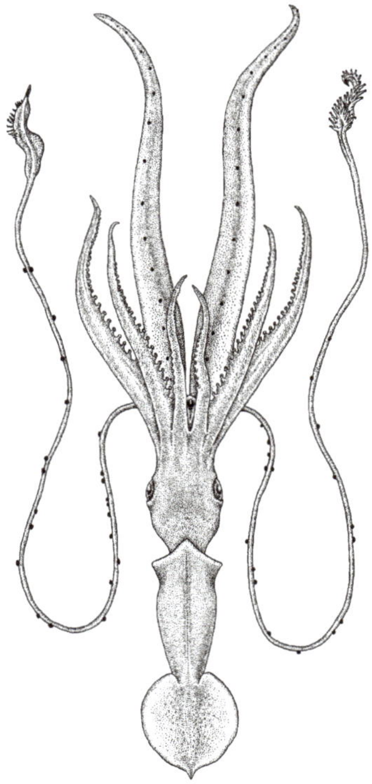

Dorsal view (ABRS, K. Hollis).

**DIAGNOSTIC FEATURES**

Mantle elongate, slender. Integument of mantle, head, arms covered with minute, conical, 'cartilaginous' tubercles. Head length ~55% ML. Fins elongate, lanceolate, secondary fins present, without lobes. Funnel valve present; funnel-locking apparatus with large, bulbous tragus and

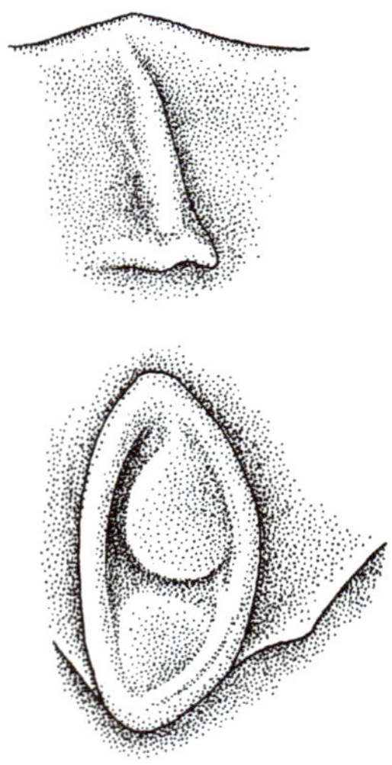

Funnel (bottom) and mantle-locking (top) cartilages (ABRS, K. Hollis).

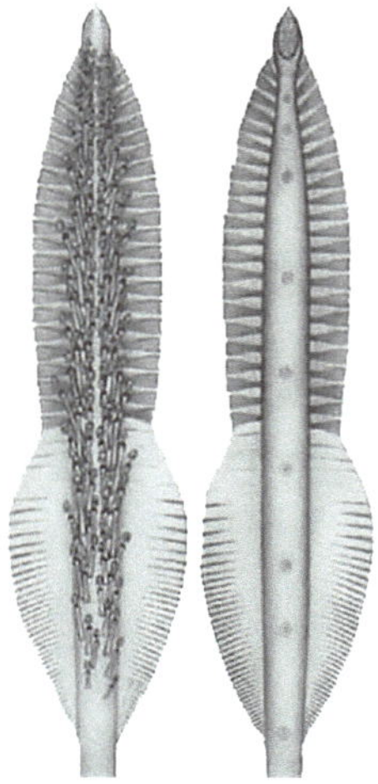

Tentacular club, oral view (left), aboral view (right) (after Young and Roper (1999), drawn by A. D. Hart).

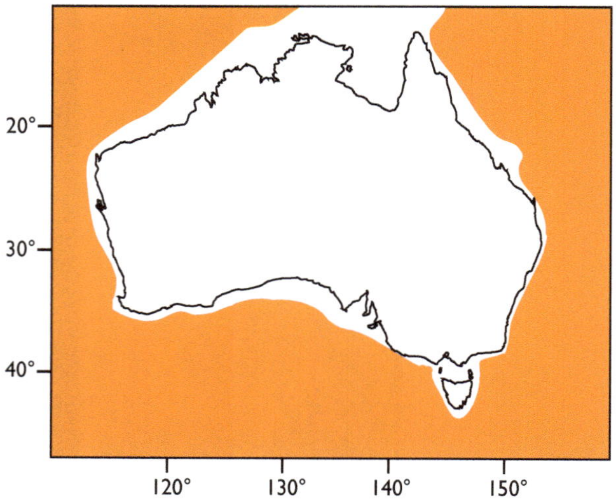


*Chiroteuthis veranyi* distribution.

smaller tubercle-like antitragus. Arms very long, thin; arms 4 not enlarged, subequal in length to arms 2 and 3: arms 1 ~85% ML, arms 3 ~145% ML, arms 4 ~220% ML. Largest arm suckers not globular or greatly enlarged, largest suckers with 16 narrow pointed teeth on distal two-thirds of ring. Tentacular stalks extremely long, with numerous knobs/pads and photophores along aboral surfaces. Clubs short, 50% ML, club in two nearly equal length parts, divided by protective membranes; trabeculae in proximal protective membrane of club fused at bases and undivided at tips; distal half with 19 or 20 well-separated trabeculae. Suckers tetraserial, stalks of lateral sucker series about twice as long as those of medial series. Club sucker rings with a single large central tooth and seven to nine pointed teeth over distal two-thirds of ring.

*Photophores.* Ocular photophores in two stripes and several round photophores: lateral and medial series, stripes; intermediate series with two posterior photophores (one large and one small), and one anterior photophore. Eleven photophores present on tentacular stalk, absent on arms 4 and ink sac. Club tip photophore small, without central midline and with small papilla. Pair of visceral photophores present.

*Colour.* Deep violet. Club pigment primarily in epithelial cells rather than chromatophore organs.

### HABITAT AND BIOLOGY

Adults are mesopelagic to bathypelagic. This species is frequently found in the stomach contents of several predator species, indicating it is common, even though it is rarely captured in commercial fishing gear. Its biology is poorly known; paralarvae and juveniles are epipelagic to mesopelagic and widespread in the water column. It may be gregarious, because frequently in deep net tows numerous specimens are captured together, which is generally an uncommon characteristic with deep-sea squids. It is an ammoniacal species.

*Interest to fisheries.* No direct interest for human consumption, but very important prey species.

### SIZE

Mantle length from 100–200 mm.

### DISTRIBUTION

Tropical, subtropical to sub-Antarctic; circumglobal extending to south of the Antarctic Polar Front. Occurs off southern Australia.

### REFERENCES

Roper and Jereb (2010g), Roper and Young (2011b).

## *Grimalditeuthis* Joubin, 1898

### DIAGNOSTIC FEATURES

Mantle elongate, gelatinous, conical anteriorly, very narrow, long, attenuate posteriorly from anterior margin of fins. Olfactory papilla in doratopsis stage lies near funnel. Funnel valve present; funnel-locking apparatus fused with mantle component. Dorsal nuchal mantle-apparatus free. Fins large, elliptical, 50% ML; secondary fins extend posterior to mantle on needle-like extension of gladius, elongate, heart-shaped. Arms of about equal length, slender. Arms 4 not enlarged; suckers present. Tentacular club reduced, entirely devoid of suckers. Tentacular stalks without knobs or pads along aboral surface.

*Photophores.* Single, elongate photophores on tips of arms 1–3. Photophores absent from arms 4 and from viscera.

**SIZE**

Mantle length to 250 mm.

**DISTRIBUTION**

Circumglobal in tropical, subtropical, temperate waters.

**REFERENCES**

Roper and Jereb (2010g), Young and Roper (2011b).

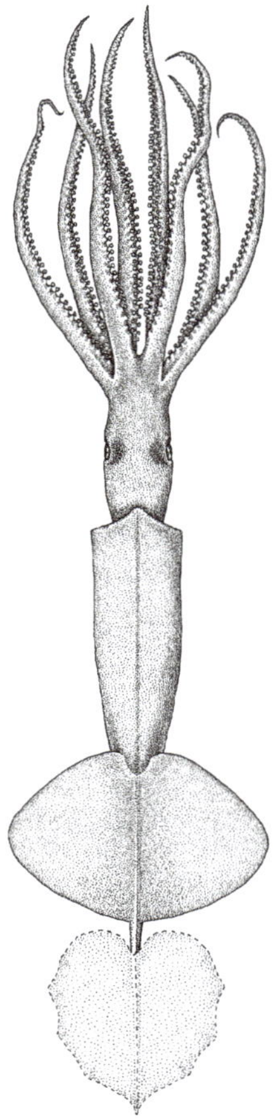

Dorsal view (ABRS, K. Hollis).

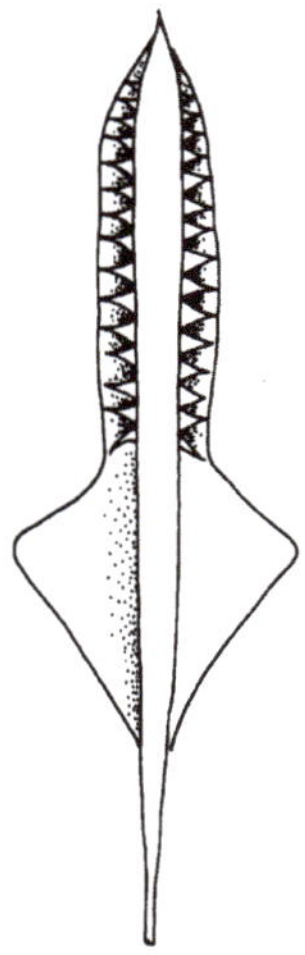

Tentacular club, aboral view (A. Reid, redrawn from Young and Roper 2011b).

## *Grimalditeuthis bonplandi* (Verany, 1839)

*Loligopsis bonplandi* Verany, J. B. (1839). Memoire sur six nouvelles especes de Cephalopodes trouves dans la Mediterranee a Nice. *Memorie della Reale Accademia delle Scienze di Torino, (series 2)1*, 99–101 [99, pl. 1].

Type data: unresolved MHNN? (uncertain).

Type locality: eastern Central Atlantic Ocean, 29°N, 39°W.

**COMMON NAME**

Grimaldi's Chiroteuthid Squid

**DISTINGUISHING CHARACTERS**

*Chiroteuthis*-like in appearance, distinguished by subequal arms and fusion of each funnel–mantle locking apparatus.

**DIAGNOSTIC FEATURES**

Gelatinous with vesiculate tissue in the head, arms and mantle. Head not fused to mantle in nuchal region. Funnel-locking apparatus fused to mantle component. Funnel valve present. Olfactory organ located lateral to base of funnel (anterior to collar at posterior end of neck). Eyes small. Arms approximately subequal in length, gelatinous. Suckers with three conical papillae basally. Protective membranes absent. Tentacular club without suckers; club divided into two portions by symmetrical protective membranes. Only scattered chromatophores present. Tail retained in adults with twin fin-like 'flotation devices' (secondary fins) arising from 'tail'.

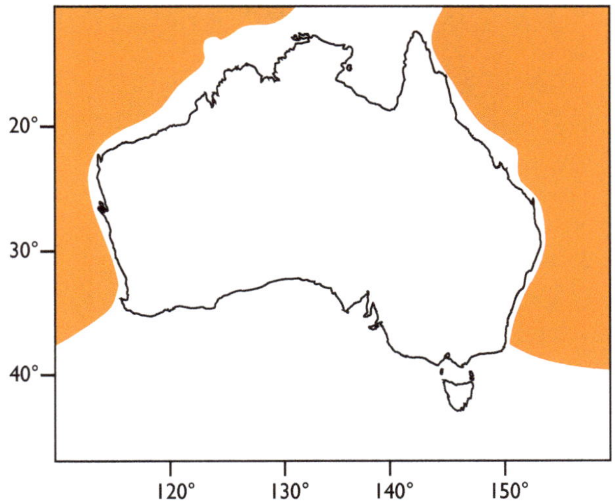


*Grimalditeuthis bonplandi* distribution.

*Photophores.* Photophores absent except on arm tips in females (possibly only in mature females).

*Colour.* Pattern of chromatophores on head distinctive. Line of chromatophores across ventral surface of head between anterior ends of eyes; another line runs along neck from each olfactory papilla posterior to each eye, and then anterior to each eye along brachial pillar, terminating at base of arms. Iridescent patches on ventral side of head.

### SIZE

Up to 250 mm ML.

### REMARKS

The tentacle stalks are very thin, fragile, almost always broken during capture. Live animal observations have shown that these unique tentacles that lack suckers, hooks and photophores are capable of rapid extension and retraction (like other squids) but they also manoeuvre the tentacles by undulating the club trabecular protective membranes (Hoving *et al.* 2013). These movements resemble the movements of small marine organisms and are postulated to function in mimicry to lure prey, perhaps by stimulating bioluminescence in the darkness of the surrounding water, create low-frequency vibrations, or produce a hydrodynamic wake. The growth series of paralarvae was described by Young (1991).

### DISTRIBUTION

Occupies the mesopelagic and bathypelagic zones of tropical and subtropical North Atlantic and tropical and temperate North Pacific. Off Hawaii adults have been collected at 700 m and smaller squid in the upper 350 m.

### REFERENCES

Young and Roper (2011b).

## *Planctoteuthis* Pfeffer, 1912

### DIAGNOSTIC FEATURES

Mantle elongate, goblet-shaped, narrows abruptly around anterior third of fins. Eyes project ventrally from head. Olfactory papilla in doratopsis stage is behind eye. Fins subterminal, transversely broad, proportionally large, muscular, width ~85% ML, length ~50% ML; with pronounced anterior and posterior lobes. Funnel valve absent; funnel component of locking apparatus with posterior distinct antitragus, oval, deeply concave (tragus absent); mantle component nose-shaped with tip directed posteriorly, curves laterally to conform to deep lateral depression of funnel component. Arms 1–3 short, subequal in length; arms 4 extremely long, slender; suckers absent distally, few suckers proximally in uneven single series. Tentacular clubs very short, simple, compact, without distinct keels, protective membranes or trabeculae. Club suckers present, few in number on short stalks, tetraserial. Tentacular stalks without knobs/pads or photophores. Posterior part of mantle joined to narrow conus of gladius; integument only continues posteriorly over gladius.

*Photophores.* Photophores entirely absent from arms 4, viscera, tentacular stalks, club tips.

### REMARKS

*Planctoteuthis* are usually small and very fragile deep-sea squids that are often badly damaged during capture. As a result, few species have been described. Unlike other chiroteuthids, the subadult retains the peculiar doratopsid paralarval tentacular club.

### REFERENCE

Roper and Jereb (2010g).

## *Planctoteuthis danae* (Joubin, 1931)

*Valbyteuthis danae* Joubin, L. (1931). Notes preliminaires sur les Cephalopodes des croisieres du DANA (1921–1922), 3e Partie. *Annales de l'Institut Océanographique, Monaco 10*(7), 169–211 [188, text figs 11–16].

Type data: holotype ZMUC.

Type locality: eastern central Pacific Ocean, 7°30′N, 79°19′W.

### DIAGNOSTIC FEATURES

Relatively large, muscular fins. Fin length more than 50% ML. Antitragus of funnel-locking apparatus with two distinct lobes. Arms 1 shortest, arms 2 and 3 similar length, arms 4 longest. Arms 2 with two enlarged suckers medially. Arms 4 with 12 or 13 suckers on proximal quarter only; remaining arms with suckers along their entire length. Clubs bilaterally symmetrical and lack keels; length 12–18% ML; holotype with ~60 suckers; carpal suckers with broader apertures than regular club suckers.

### HABITAT AND BIOLOGY

*Planctoteuthis danae* is a bathypelagic species, rarely mesopelagic. It is very rarely captured. In the eastern North Atlantic, paralarvae are found mostly between 200 m and 300 m. Between ~10 m and 15 mm ML they are found between 200 and 800 m and at larger sizes thay are found at depths >700 m during the day and sometimes >1000 m.

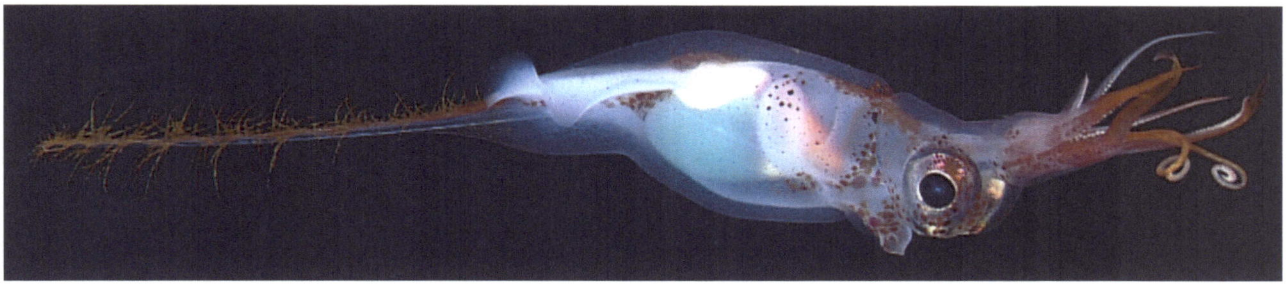

Live animal photographed at 1200 m in the Gulf of California (© 2012 MBARI).

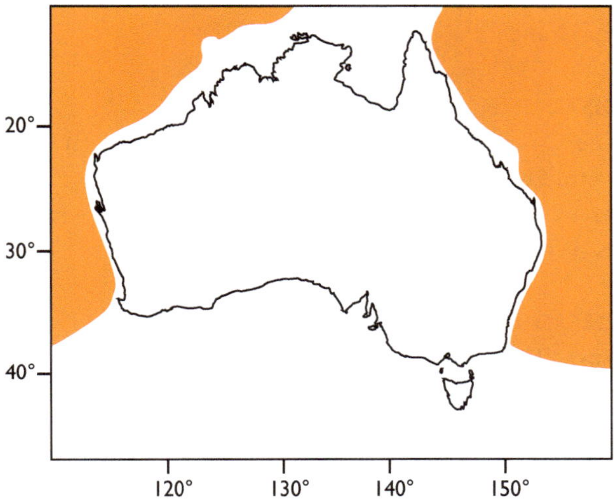


*Planctoteuthis danae* distribution.

## SIZE

Mantle length to 55 mm.

## DISTRIBUTION

Cosmopolitan in tropical and temperate waters.

## REFERENCE

Roper and Jereb (2010g).

# Cranchiidae Prosch, 1847

## COMMON NAME(S)

Cranch Squids, Glass Squids

## DISTINGUISHING CHARACTERS

One prominent character easily distinguishes all members of this group from all other species of squids: the mantle is fused to the head at the nuchal cartilage and to the funnel at its two posteriolateral corners. Paralarval/juvenile cranchiids are readily identified by their long eyestalks. The family contains a great diversity of species that exhibit a wide variety of basic characters.

## DIAGNOSTIC FEATURES

Mantle stout to elongate, cylindrical, conical or spindle-shaped; posterior end varies from blunt, rounded (*Bathothauma*) to sharply pointed, filiform (*Taonius*); thin walled, usually muscular, leathery; some species with cartilaginous-like tubercles. Mantle cavity divided into dorsal and ventral chamber by broad horizontal membrane formed by modification of funnel retractor muscles. Large buoyancy chamber extends around viscera for entire length of mantle cavity. Fins normally quite thin (except very thick, muscular, in *Mesonychoteuthis*), insert laterally along lanceola and conus of gladius; shape varies greatly among species. Head short, eyes large, sessile, prominent, or protruding, usually oriented anteriorly or anteriolaterally. Funnel moderate to very large; funnel valve present or absent; dorsal pad of funnel organ an inverted V- or U-shape, normally with two or three papillae or flaps or both (up to seven papillae in *Leachia*); two small to very large ventral pads (shape depending on species). Arms suckers biserial with wide apertures, often without dentition (hooded hooks midarm in *Mesonychoteuthis*); well-developed protective membranes occur along both margins of all arms. Right or left ventral arm (4) hectocotylised (or hectocotylus absent). Tentacular clubs with four transverse series of suckers (hooded hooks in median series in *Galiteuthis* and *Mesonychoteuthis*). Tentacular stalks, variable in length and diameter among species, with two to four series of alternating carpal suckers and pads for most of tentacle length. Buccal membrane with eight connectives (seven lappets) that attach to ventral borders of arms 4. Digestive gland usually spindle-shaped.

Paralarval and young cranchiids characterised by stalked eyes and stalked arm crown of variable length; very short arms and long tentacles; frequently ventrally directed rostrum on eyes; photophores absent.

*Gladius.* Midline of rachis and conus of gladius usually visible through skin.

*Photophores.* Photophores on ventral half to two-thirds (or more) of eyes, ink sac, and digestive gland of some species. Seven of the 13 genera have large photophores ('brachial end organs') in mature females (these are quite different to other photophores on the eye or mantle (Herring 2007)).

### HABITAT AND BIOLOGY

Cranchiids occur in the subtropics, tropics and Antarctic waters. Paralarvae and juveniles occur in the epipelagic zone and then, with growth, undergo gradual ontogenetic descent into mesopelagic and bathypelagic depths, with adults of some species known to inhabit depths in excess of 2000 m. Many species undergo diel vertical migration; some seem not to migrate vertically at all. Other species occur in transitional zones and a few are associated with near-coastal pelagic habitats.

Marked morphological changes occur during descent and maturation, including change of eye shape and position, change of fin shape, increased pigmentation, development of photophores on arm tips, modification of arm structure, and loss of the tentacles.

Fusion of the mantle, head and funnel in cranchiids and partitioning of the mantle cavity and coelom into two chambers results in very efficient respiration. The chambers contain a weak solution of low-density ammonium chloride, which provides near-neutral buoyancy – a further energy saving.

A few species of cranchiids have been observed alive in their deep-sea habitats from submersibles. Among other observations, several species have been observed in a stereotypic resting position, with their arms and tentacles held tightly together in a bundle and held dorsally over their heads. This position is known as the 'cockatoo' posture.

*Interest to fisheries.* Currently no directed fishery of any sort is conducted on any species of Cranchiidae. It is doubtful that a fishery would ever develop, because of their ammoniacal fluid content. Also, the mantle tissue, though muscular, is very thin and leathery, except in *Mesonychoteuthis hamiltoni*, which is thick but soft and gelatinous, making it unsuitable for human consumption.

### SIZE

Very diverse among species; adults range from 10 cm to well over 200 cm ML.

### REMARKS

Because the paralarvae do not closely resemble adults of their species, historically they were not connected with their conspecific adults. This resulted in considerable overnaming, creation of synonyms and taxonomic confusion, which has largely been resolved, but only in recent years. The Cranchiidae currently consists of 13 genera and in excess of 60 species. Among these, 12 genera and 14 species are known to occur in Australian waters. Many species in the family are very poorly known or undescribed. The genera are grouped under two subfamilies: the Cranchiinae and the Taoniinae.

### DISTRIBUTION

The Cranchiidae are represented in virtually all of the world's oceans and seas, with the notable exception of the Arctic Ocean, with the greatest species diversity in subtropical and tropical waters.

### REFERENCES

Voss (1980), Jereb and Roper (2010).

### Key to subfamilies and genera of the Cranchiidae (adults) (after Roper and Jereb 2010h)

1a. Ventral surface of mantle with one or two cartilaginous strips extending posteriorly from anterior apex of funnel–mantle-locking cartilages; funnel fused to head laterally; eyes with four or more small, round to oval photophores......... (Subfamily Cranchiinae) 2

1b. Ventral surface of mantle without cartilaginous strips extending posteriorly from anterior apex of funnel–mantle-locking cartilages; funnel free from head laterally; eyes with one, usually large, photophore, or two or three markedly dissimilar-sized photophores with the largest usually crescent-shaped.................................. (Subfamily Taoniinae) 4

2a. Ventral surface of mantle with two cartilaginous strips in inverted V-shaped pattern extending posteriorly from anterior apex of funnel–mantle-locking cartilages; funnel valve present; dorsal pad of funnel organ with three longitudinal, triangular flaps; gladius with short conus............................................3

2b. Ventral surface of mantle with one cartilaginous strip extending posteriorly from anterior apex of funnel–mantle-locking cartilages; funnel valve absent; dorsal pad of funnel organ with three to seven narrow papillae; gladius with long slender conus ........ *Leachia*

3a. Mantle covered with cartilaginous tubercles; eyes with 14 photophores; brachial photophore present on all arms in near-mature and mature females; suckers in four series on midportion of hectocotylised right arm 4 in males ........ *Cranchia* (*C. scabra*)

3b. Mantle without cartilaginous tubercles except on ventral cartilaginous strips and sometimes on dorsal median line; eyes with four or 14 photophores; brachial photophore present only on arms 3 in near-mature and mature females; suckers in two series on midportion of hectocotylised right or left arm 4 in males ........ *Liocranchia*

4a. Fins small, paddle-shaped, subterminal; eyes with one, usually large, photophore ........ 5

4b. Fins not paddle-shaped, terminal or laterotermінal; eyes with one large and one or two small photophores ........ 6

5a. Fins fused distally, insert on short rostrum of gladius that projects dorsally free of end of mantle; eyes small to medium ........ *Helicocranchia* (*H. pfefferi*)

5b. Fins widely separated, insert on lateral expanded ends of transverse extensions of posterior end of gladius; eyes proportionally large to huge ........ *Bathothauma* (*B. lyromma*)

6a. Gladial conus short, broad to narrow; fins short (<25% ML), oval to round; digestive gland long, narrow, spindle-shaped ........ 7

6b. Gladial conus medium to long, narrow, or needle-like to filiform; fins medium to long (30–60% ML), narrow, lanceolate to ovate; digestive gland stout, spindle-shaped or rounded ........ 8

7a. Posterior fin insertions do not extend to tip of gladius; no tubercles present on funnel–mantle locking cartilages; dorsal pad of funnel organ with large, triangular lobe on each lateral arm; eyes with small, round, anterior photophore indented into median anterior margin of large, round, posterior photophore ........ *Sandalops* (*S. melancholicus*)

7b. Posterior fin insertions extend to tip of gladius; two small tubercles present at anterior end of funnel–mantle locking cartilages; dorsal pad of funnel organ with large, spatulate papilla on each lateral arm; eyes with small, crescent-shaped, anterior photophore lying closely within concavity of large, crescent-shaped, posterior photophore ........ *Liguriella* (*L. podophthalma*)

8a. Anterior fin insertions on lateral margins of lanceola of gladius ........ 9

8b. Anterior fin insertions on lateral margins of mantle ........ 10

9a. Arms without hooded hooks; fins lanceolate becoming attenuate posteriorly ........ *Galiteuthis*

9b. Arms with hooded hooks on midportion; fins stout, ovate (nearly round in combined outline in juvenile), do not become attenuate posteriorly ........ *Mesonychoteuthis* (*M. hamiltoni*)

10a. Funnel valve present; dorsal pad of funnel organ with triangular flap on each lateral arm; eyes with two photophores (large, roughly crescent-shaped posterior photophore, and within its concavity, smaller, roughly elongate S-shaped anterior photophore); carpal suckers in two series on tentacular stalk ........ 11

10b. Funnel valve absent; dorsal pad of funnel organ with long, spatulate papilla on each lateral arm; eyes with three photophores (large, crescent-shaped posterior photophore, and within its concavity, a smaller, crescent-shaped anterior photophore and a third small, oval photophore); carpal suckers in four series set in zigzag pattern on tentacular stalk ........ *Teuthowenia* (*T. pellucida*)

11a. Photophores absent on ventral surface of stout, spindle-shaped, digestive gland; long, single united digestive duct ........ *Egea* (*E. inermis*)

11b. Large compound photophore present on ventral surface of rounded digestive gland; two short digestive ducts remain separate ........ *Megalocranchia*

## Subfamily Cranchiinae Pfeffer, 1912

### DIAGNOSTIC FEATURES

Ventral surface of mantle with one or two cartilaginous strips (with cartilaginous tubercles) that originate at either side of funnel where it joins mantle and extends posteriorly. Funnel fused to head laterally. Secondary sexual modification may be present on tips of arms 3. Right or left arm 4 hectocotylised in males. Caecum larger than stomach. One or more rows of four or more small, round to oval photophores on the eyes. Adult females may have brachial photophores on tips of third arms (or on all four arm pairs).

### REMARKS

This subfamily contains the genera *Cranchia*, *Leachia* and *Liocranchia*.

## *Cranchia* Leach, 1817

The genus is monotypic, represented by *Cranchia scabra* Leach, 1817.

## *Cranchia scabra* Leach, 1817

*Cranchia scabra* Leach, W. E. (1817). Synopsis of the orders, families, and genera of the class Cephalopoda., *Zoological Miscellany* 3(30), 137–141 [140].

Type data: holotype BMNH 1986265.

Type locality: West Africa, off Congo.

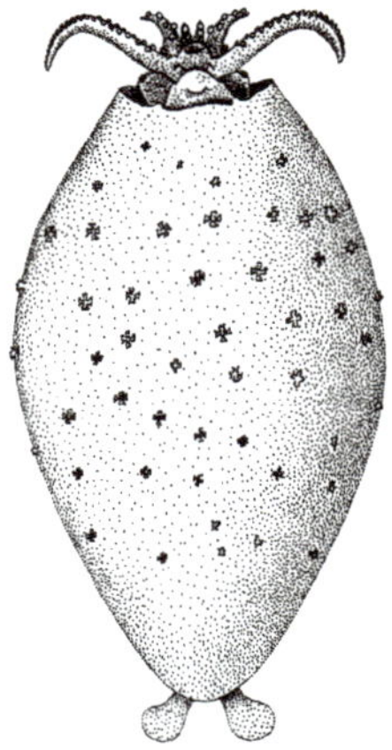

Larva, ventral view (Voss, 1980).

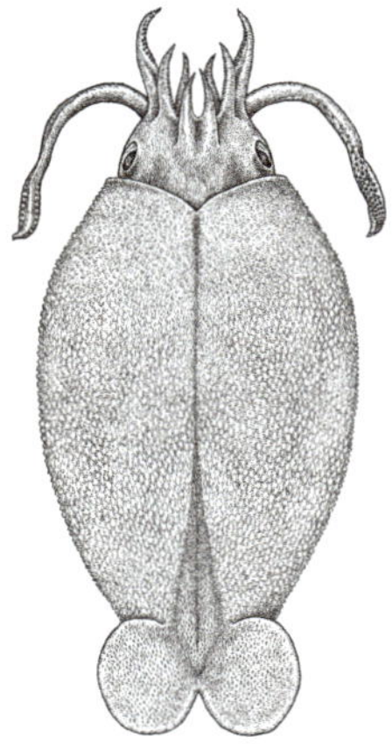

Dorsal view (ABRS, K. Hollis).

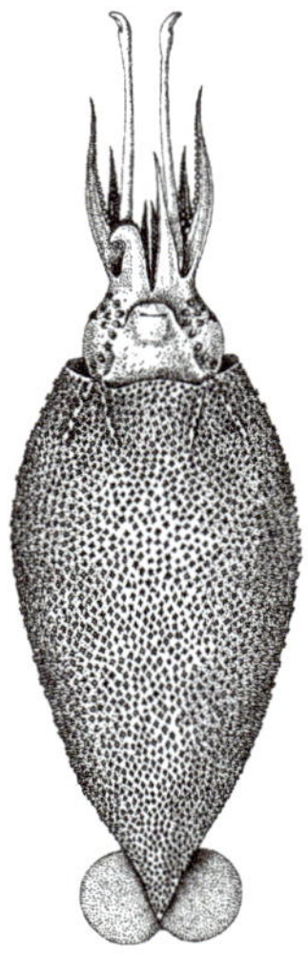

Subadult male, ventral view (Voss, 1980).

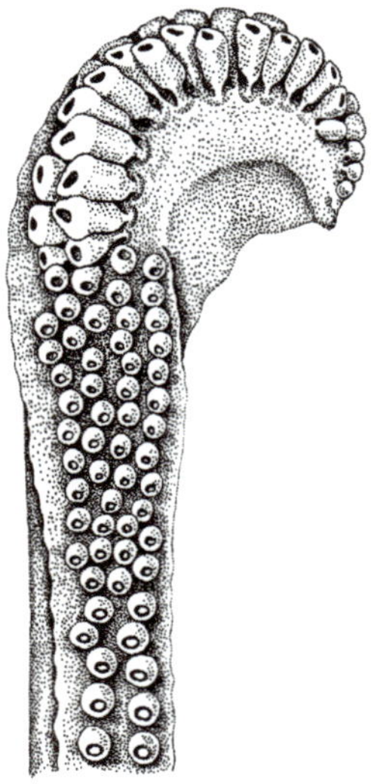

Hectocotylus, subadult male (Voss, 1980).

### COMMON NAME

Rough Cranch Squid

### DIAGNOSTIC FEATURES

Mantle stout, spindle-shaped, tapers to broad posterior point. Mantle covered with variable-sized, complex, cartilaginous tubercles that are square or triangular on their upper surface, each corner armed with triangular point, with tops bearing spines. Ventral surface of mantle with two subequal cartilaginous strips (with cartilaginous tubercles) that originate at either side of funnel where it joins mantle and extends posteriorly, strips V-shaped with complex multipoint tubercle at apex. Head with prominent eyes. Fins small and round, separate anteriorly, joined posteriorly with free posterior lobes. Funnel valve present. Funnel organ inverted V-shaped, ventral pads half-moon-shaped. Arms short with dorsal and ventral protective membranes. Suckers globular, biserial. Suckers tetraserial on midportion of hectocotylised right arm 4. Tentacle stalk with small, smooth-ringed widely spaced suckers. Club suckers tetraserial, small proximally, large medially, abruptly changing to small suckers distally, distal tip terminating in group of four large suckers on circular pad; club with protective membranes, extending on each side from carpus to tip of club. Club with swimming membrane originating at middle of club and extending to tip.

*Gladius.* Gladius visible along dorsal midline and bears rows of tubercles along most of its length.

*Photophores.* Each eye with 13 or 14 photophores on ventral surface of bulbus and around iris; brachial end organ (photophore) present on tips of all arms of females.

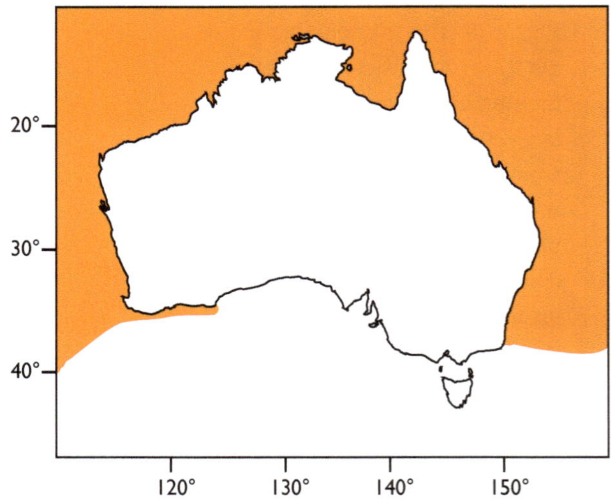


*Cranchia scabra* distribution.

**HABITAT AND BIOLOGY**

Paralarvae and juveniles are epipelagic to upper mesopelagic (upper 400 m), while adults descend to mesopelagic and bathypelagic zones, perhaps to 2000 m.

**SIZE**

Mantle length to 150 mm.

**DISTRIBUTION**

This species occurs circumglobally in epipelagic to meopelagic tropical and subtropical open oceanic waters.

**REFERENCES**

Voss (1963), Roper and Jereb (2010h).

## *Leachia* Lesueur, 1821

**DISTINGUISHING CHARACTERS**

*Leachia* is the only genus in the subfamily Cranchiinae in which the paralarvae have stalked eyes.

**DIAGNOSTIC FEATURES**

Mantle elongate, slender, spindle-shaped, tapers posteriorly to sharp point. Ventral surface of mantle with two cartilaginous strips (with cartilaginous tubercles) that originate on each side of the funnel where it joins mantle and extend posteriorly. Arms subequal, third pair longer and more robust than rest. Fins together transversely elliptical; barely unite posterior to gladius. Funnel fused to head laterally; funnel valve absent. Median suckers on manus of tentacular club greatly enlarged.

*Photophores.* Eyes with 5–21 oval photophores (species dependent); brachial photophores present on tips of arms 3 in mature or nearly mature females.

**HABITAT AND BIOLOGY**

Members of this genus are epipelagic, found from the surface to the bathypelagic zone, to ~2000 m.

**SIZE**

Small- to medium-sized squids, ML up to 200 mm.

**DISTRIBUTION**

*Leachia* is a circumglobal genus in tropical and subtropical waters.

**REFERENCE**

Roper and Jereb (2010h).

### Key to *Leachia* species

1a. Eight eye photophores, five in outer row, three near pupil ........ *L. cyclura*

1b. Eyes with six photophores, five outer, one inner near pupil ........ *L. pacificus*

## *Leachia cyclura* Lesueur, 1821

*Leachia cyclura* Lesueur, C. A. (1821). Description of several new species of Cuttle-fish. *Journal of the Academy of Natural Sciences of Philadelphia 2*(1), 86–107 [90, pl. 6].

Type data: holotype not extant ANSP.

Type locality: southern Indian Ocean, off South Africa, 33°S, 37°E.

**COMMON NAME**

Leach's Cranch Squid

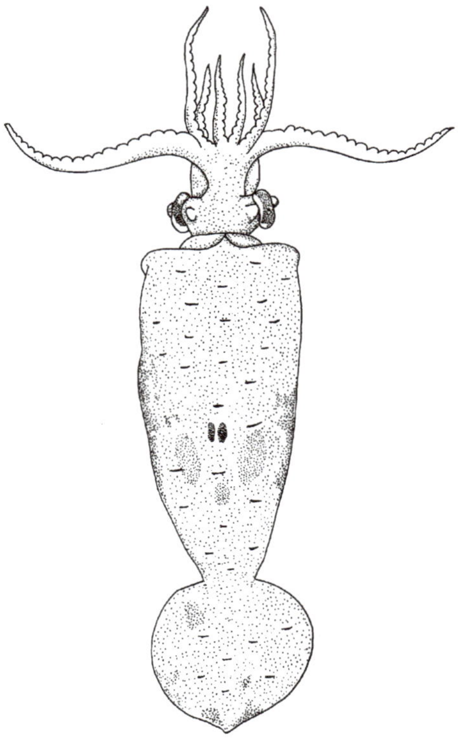

Dorsal view (A. Reid, redawn from original description).

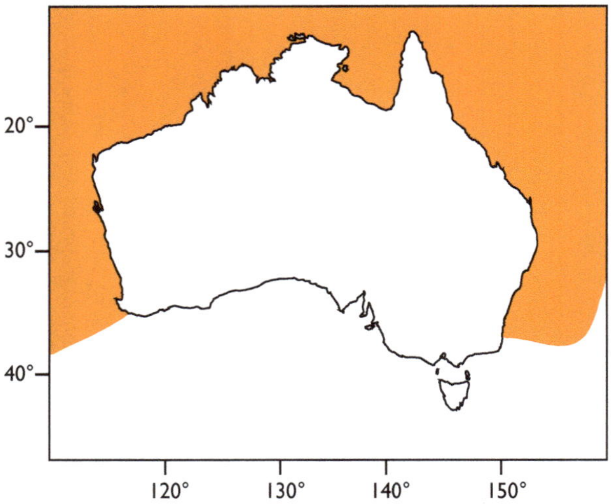


*Leachia cyclura* distribution.

## DIAGNOSTIC FEATURES

Body coniform. Length of cartilaginous tubercle strip ~20–30% ML. Head small, eyes large, prominent. Together, paired fins round in shape, extend beyond tip of mantle. Arm formula 3 > 2 > 1 = 4. Arms 3 very robust.

*Photophores.* Eight eye photophores, five in outer row, three near pupil (at 60–70 mm ML).

*Colour.* Fresh-caught specimen: dorsal side of head and arms light blue, body tinged bluish and red with small red chromatophores and several irregular deeper red spots and with scattered transverse black lines. Two large, light brown, oval spots on posterior half of dorsal mantle. Pair of black spots anterior to oval spots and one red spot posteriorly.

## HABITAT AND BIOLOGY

Bathymetric range from surface waters to bathypelagic, ~2000 m, following a gradual process of ontogenetic descent.

## SIZE

Maximum ML 150–200 mm.

## DISTRIBUTION

Tropical and southern subtropical Indo-West Pacific oceanic waters.

## REFERENCES

Nesis (1982/1987), Roper and Jereb (2010h).

# *Leachia pacificus* (Issel, 1908)

*Zygaenopsis pacifica* Issel, R. (1908). Diagnosi preliminari di un nuovo genre e di due nouve specie di Cefalopodi appartenenti alla fam. Cranchiidae raccolti dalla R. Nave 'Liguria'. *Monitore Zoologico Italiano 19*(4), 102–104 [103].

Type data: repository, unresolved.

Type locality: central western Pacific Ocean, between Tahiti and Pago Pago Islands, 14°32'S, 167°43'W.

## DIAGNOSTIC FEATURES

Large brachial photophores on arms 3 of mature females. Tubercular strips with second, large anterior tubercle noticeably offset medially from linear pattern.

*Photophores.* Eyes with six photophores, five outer, one inner.

## HABITAT AND BIOLOGY

Juveniles are found in shallow water in the epipelagic zone, from near the surface to ~150 m. As sexual maturity occurs, the subadult squid undergo a sudden ontogenetic descent. Animals of ~45–55 mm ML may be found throughout the water column between ~200–1300 m; mature animals occur at depths from ~1200–1800 m (possibly to 2300 m) where they become fully mature, mate and spawn.

## SIZE

Maximum ML to 60 mm.

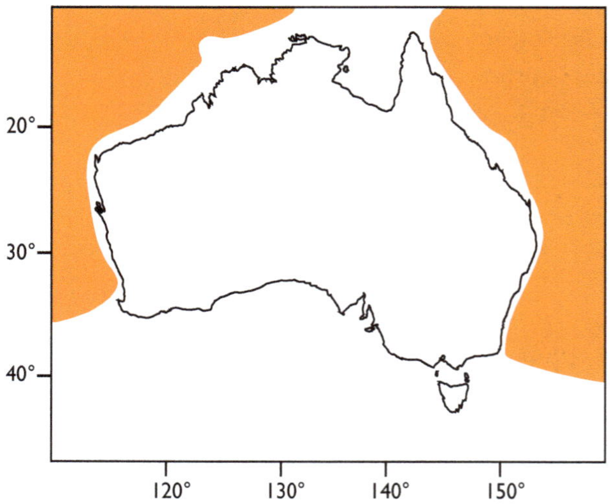


*Leachia pacificus* distribution.

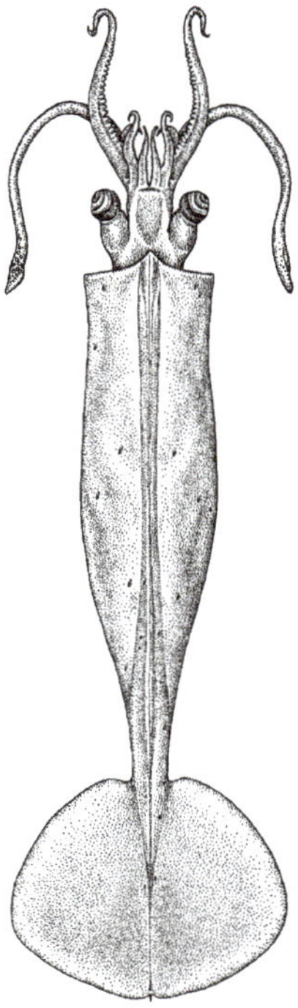

*Leachia pacificus* dorsal view (ABRS, K. Hollis).

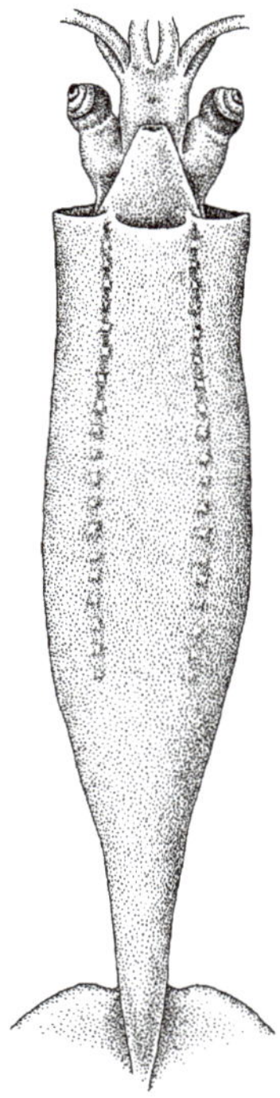

*Leachia pacificus* ventral view (ABRS, K. Hollis).

**DISTRIBUTION**

Tropical waters of the North Pacific and Indo-Pacific Oceans, and westward into equatorial waters of the Indian Ocean.

**REFERENCE**

Roper and Jereb (2010h).

## *Liocranchia* Pfeffer, 1884

**DIAGNOSTIC FEATURES**

Mantle elongate, spindle-shaped, tapers posteriorly to narrow point. Not covered with cartilaginous tubercles; two moderately long cartilaginous strips studded with cartilaginous tubercles extend from each funnel–mantle locking cartilage in inverted-V pattern; components equal in length or ventral component longer. Fins ovoid to nearly circular, join posterior to gladius. Funnel valve present. Hectocotylised right or left arm 4 in males with biserial suckers.

*Photophores.* Eyes with four or 14 oval photophores; brachial photophores present only on tip of arms 3 in submature and mature females.

**HABITAT AND BIOLOGY**

*Liocranchia* occupies epipelagic to bathypelagic depth zones.

**SIZE**

Medium-sized squids up to 250 mm ML.

**DISTRIBUTION**

Circumglobal in tropical to subtropical oceanic waters.

**REFERENCE**

Roper and Jereb (2010h).

### Key to *Liocranchia* species

1a. Cartilaginous tubercles along dorsal midline, overlaying gladius. Fourteen oval photophores on eye .... ..................... *L. reinhardti*

1b. No cartilaginous tubercles along dorsal midline. Four oval photophores on eyeball............ *L. valdiviae*

## *Liocranchia reinhardti* (Steenstrup, 1856)

*Leachia reinhardti* Steenstrup, J. (1856). Hectoctyldannelsen hos Octopodslaegterne Argonauta og Tremoctopus, oplyst ved Iagttagelse af lignende Dannelser hos Blacksprutterne i Almindelighed. *Kongelige Danske Videnskabernes Selskabs Skrifter, 5 Raekke, Naturvidenskabelig og Mathematisk, series 5, 4*, 185–216 [200].

Type data: syntypes (3) ZMUC.

Type locality: North Atlantic Ocean, 15°19′N, 24°54′W and 23°N, 32°W.

**COMMON NAME**

Reinhardt's Cranch Squid

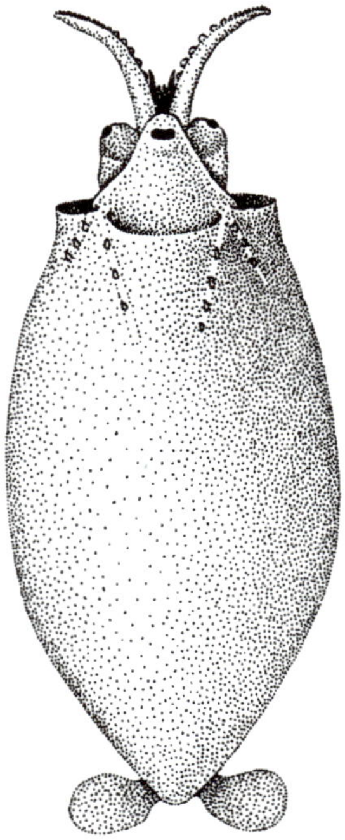

Larva, ventral view (Voss, 1980).

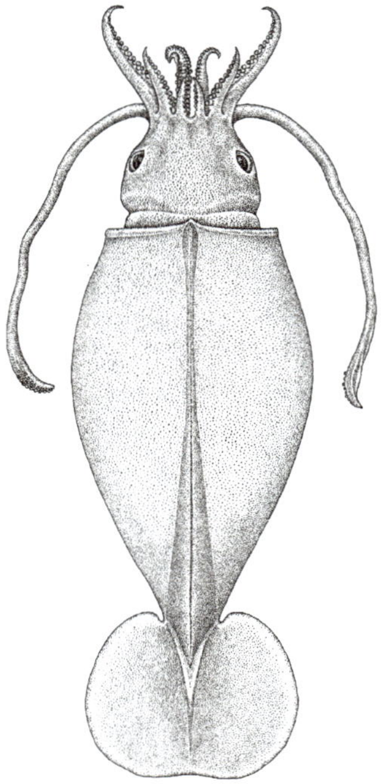

Adult, dorsal view (ABRS, K. Hollis).

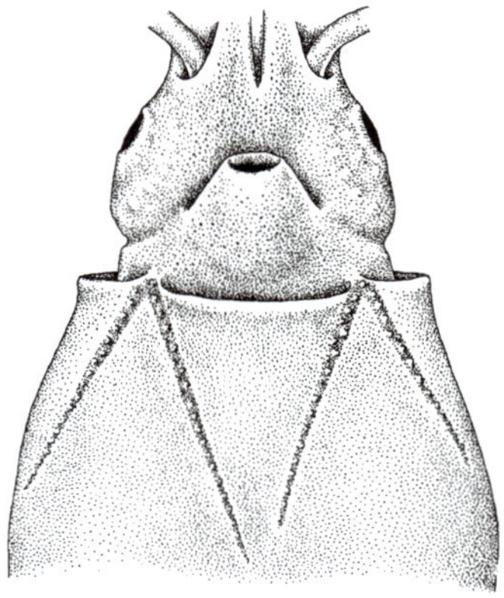

Ventral view (ABRS, K. Hollis).

Hectocotylus, mature male (Voss, 1980).

### DIAGNOSTIC FEATURES

Cartilaginous tubercles along dorsal midline, overlaying gladius.

*Photophores.* Fourteen oval photophores on eyeball: seven medial series, six in lateral series near lens and one in between. Large photophores on tips of arms 3 when mature.

### HABITAT AND BIOLOGY

Paralarvae and juveniles inhabit epipelagic to upper mesopelagic waters. Adults occur at lower epipelagic, mesopelagic and bathypelagic depths from the surface to 1200 m. They appear to occur primarily over bottom depth of less tham 700 m and appear to be land-associated.

### SIZE

Maximum ML to 250 mm.

### DISTRIBUTION

*Liocranchia reinhardti* is distributed circumglobally in tropical and subtropical waters, recorded from 50°N to ~35°S, generally bounded by the north and south subtropical convergences.

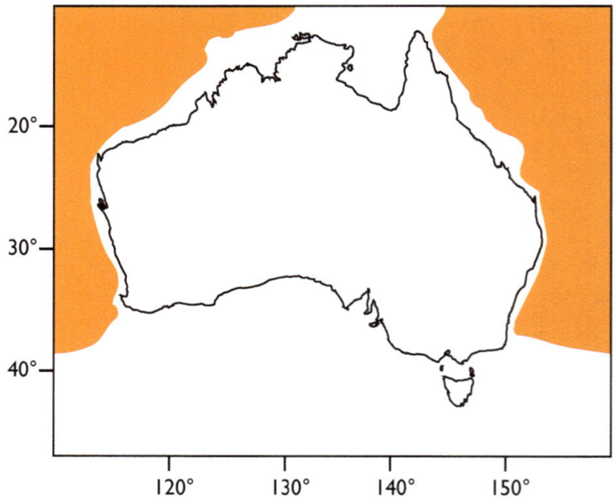


*Liocranchia reinhardti* distribution.

### REMARKS

The chromatophore pattern of the paralarvae differs from *L. valdiviae.* Young stages can be recognised by the single large chromatophore on the ventral tip of the mantle. Large paralarvae develop a row of large chromatophores near the anterior end of the ventral mantle.

### REFERENCE

Roper and Jereb (2010h).

## *Liocranchia valdiviae* Chun, 1906

*Liocranchia valdiviae* Chun, C. (1910). Die Cephalopoden. Oegopsida. *Wissenschaftliche Ergebnisse der Deutschen Tiefsee-Expedition dem Dampfer Valdivia, 1898–1899* *18*(1), 1–401 [337, pl. 48 figs 3–4, pl. 51 figs 1–4, 8–10, pl. 60 figs 7–11].

Type data: lectotype ZMB Moll.-110038.

Type locality: Indian Ocean, 4°45′S, 48°58′E.

### DISTINGUISHING CHARACTERS

This species is distinguished from *Liocranchia reinhardti* (Steenstrup, 1856) by the absence of cartilaginous tubercles along dorsal midline and the presence of only four oval photophores on the eyeball. Paralarvae can be recognised by the distinctive chromatophore pattern that includes three small chromatophores on the dorsoposterior tip of the mantle and a larger patch on the posterior half of the dorsal mantle.

### DIAGNOSTIC FEATURES

Body goblet-shaped, narrow posteriorly, partly transparent. Head short, broad with protruding eyes; ventral side of eyeball with short-stalked olfactory tubercle. Funnel organ with two large oval parts and three-pronged median dorsal part; medial funnel organ with short, protruding cone. Fins relatively large, circular in outline. Ventral mantle with inverted V-shaped cartilaginous ridges corresponding to funnel–mantle fusion. Ridges with conical tubercles; where ridges meet are three tubercles, and further ventrally a fourth tubercle. Arm formula 3 > 2 > 4 > 1. Suckers biserial. Left ventral arm hectocotylised: slightly shorter than right ventral arm; suckers much smaller than those on right ventral arm and form broad papillae at arm tip; after proximal four pairs of suckers, remaining suckers in a single row, decreasing in size distally, second uniserial sucker slightly larger than rest. Distal end of hectocotylised arm strongly recurved; some suckers biserial or irregularly arranged; with wide swimming keel that extends to base of arm. Tentacular stalk with some widely spaced suckers. Tentacular club carpus with two or three rows of suckers; manus suckers tetraserial.

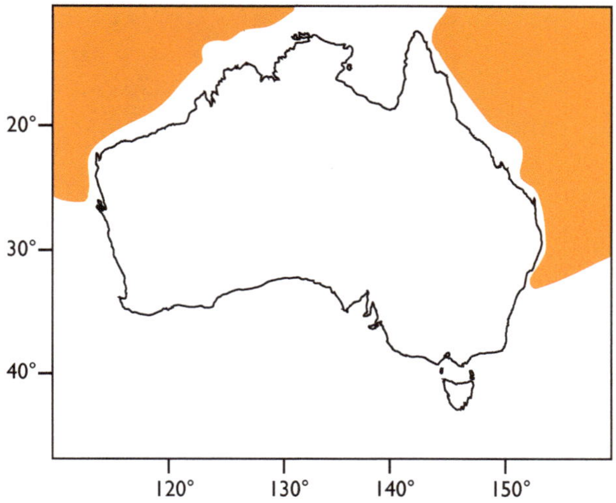

*Liocranchia valdiviae* distribution.

Adult, dorsal view (from Chun, C., 1910, plate XLVIII, Fig. 3).

Ventral view (from Chun, C., 1910, plate XLVIII, Fig. 4).

*Photophores.* Eyeball with four large photophores, appear as silvery spots in live animals and visible through the skin in preserved specimens.

*Colour.* Few chromatophores: longitudinal series of chromatophores along gladius, diverging into two rows posteriorly. Chromatophores scattered on dorsal and ventral mantle and aboral side of arms; fin chromatophores absent except at medioposterior margin. Tentacles vividly pigmented on aboral side; club with distinct chromatophores medially.

### HABITAT AND BIOLOGY

Five to 15 mm ML juveniles are epipelagic, concentrated in the upper few hundred metres; 15–25 mm ML animals descend to mesopelagic depths of 500–700 m and animals >25 mm ML are bathypelagic, moving progressively deeper to in excess of 1200 m. The species does not exhibit diel vertical migration.

#### SIZE

Maximum ML unknown, perhaps 200–250 mm.

#### DISTRIBUTION

*Liocranchia valdiviae* occurs mainly in the Indian and Pacific Oceans primarily in tropical and equatorial oceanic waters.

#### REFERENCE

Roper and Jereb (2010h).

## Subfamily Taoniinae Pfeffer, 1912

#### DIAGNOSTIC FEATURES

Taoniinae do not have cartilaginous strips that extend posteriorly along lines of funnel–mantle locking cartilages. Funnel free laterally (not fused to the head). Hectocotylus absent in males; secondary sexual modifications may occur on various arm pairs or may be absent entirely. Caecum smaller than stomach.

*Photophores.* Eyes with one to three markedly dissimilar-sized and shaped photophores (largest one usually crescent-shaped). Adult females may have brachial photophore on tips of various arms, or they may be absent altogether.

#### REMARKS

This subfamily contains numerous and diverse species that are arranged in several genera, many of which were described based on paralarval or juvenile forms (e.g. *Fusocranchia*, *Teuthowenia*, *Taonidium*). The currently recognised genera are: *Taonius*, *Bathothauma*, *Belonella*, *Egea*, *Galiteuthis*, *Helicocranchia*, *Liguriella*, *Megalocranchia*, *Mesonychoteuthis*, *Sandalops* and *Teuthowenia*. All but *Belonella* and *Taonius* are found in Australian waters.

#### REFERENCE

Roper and Jereb (2010h).

## *Bathothauma* Chun, 1906

*Bathothauma* is currently monotypic, represented by *Bathothauma lyromma* Chun, 1906.

## *Bathothauma lyromma* Chun, 1906

*Bathothauma lyromma* Chun, C. (1906). System der Cranchien. *Zoologischer Anzeiger 31*(2), 82–86 [86].

Type data: holotype ZMB Moll.-110019.

Type locality: southeastern Atlantic Ocean, west of Cape Verde Islands, 17°28′N, 29°42′W.

#### COMMON NAME

Lyre Cranch Squid

#### DISTINGUISHING CHARACTERS

Fins widely separated (unique to this genus within the Cranchiidae). Most of the posterioventral mantle surface of representatives of this genus is covered with a single, very large, half-round photophore that is indented in the medioanterior margin. There is also one photophore on each eye. These traits are shared only with *Helicocranchia*. Paralarva and juveniles are exceptionally distinct, with very long eye stalks, distinct rostrum on each eye, a long brachial pillar, widely separated fins and unusual chromatophore pattern.

#### DIAGNOSTIC FEATURES

Mantle elongate cylindrical, sac-shaped, rounded posteriorly. Eyes large to huge; protruding, anteriorly oriented, nearly spherical. Cartilaginous tubercles absent (or nearly absent) on funnel/mantle fusion. Fins small, widely separated, rounded, paddle-shaped, subterminal, inserted on lateral expanded ends of transverse extensions of posterior end of gladius (unique to this genus). Funnel valve absent, funnel organ with inverted V-shaped dorsal pad. Tentacular club with suckers only; tentacular stalk with biserial suckers and pads proximally, tetraserial suckers near club forming elongate carpal group.

*Photophores.* Most of posterioventral surface of mantle covered with a single, very large, half-round photophore (unusual in family, shared only with *Helicocranchia*), indented in midanterior margin. Brachial photophores on arm tips of adult females unknown. Paralarvae and juveniles with very long eye stalks, distinct rostrum on each eye, long brachial pillar, widely separated fins.

#### HABITAT AND BIOLOGY

Young *Bathothauma* occur from near-surface epipelagic waters in the open ocean, and the bathypelagic realm in excess of 2000 m as adults. The species shows clear ontogenetic descent.

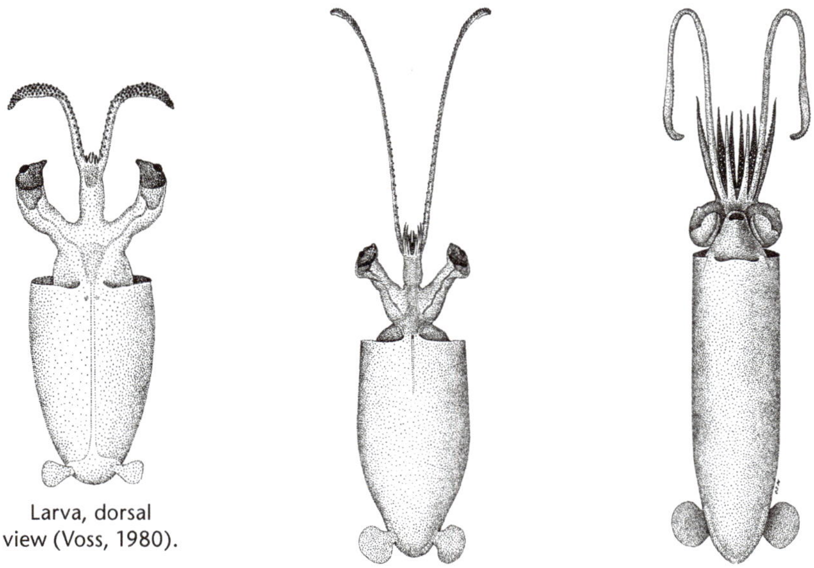

Larva, dorsal view (Voss, 1980).

Early juvenile, dorsal view (Voss, 1980).

Subadult female, ventral view (Voss, 1980).

*Bathothauma lyromma*

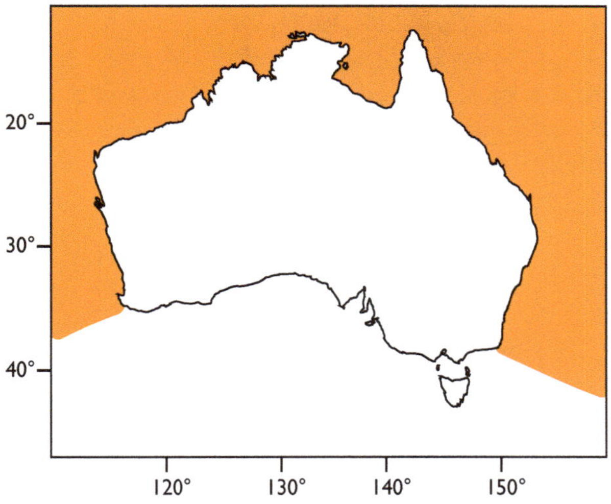


*Bathothauma lyromma* distribution.

**SIZE**

Maximum ML to 200 mm. Medium-sized species: paralarvae up to ~100 mm ML, adults with maximum ML to 200 mm.

**REMARKS**

*Bathothauma* currently is monotypic, with *B. lyromma* the only described species. However, ongoing research suggests that at least four very closely related species exist.

**DISTRIBUTION**

*Bathothauma lyromma* occurs circumglobally in tropical and subtropical zones, extending into temperate waters in the north-eastern Atlantic and South Pacific.

**REFERENCES**

Voss *et al.* (1992), Roper and Jereb (2010h).

## *Egea* Joubin, 1933

*Egea* is currently monotypic, represented by *Egea inermis* Joubin, 1933.

### *Egea inermis* Joubin, 1933

*Egea inermis* Joubin, L. (1933). Notes preliminaires sur les Cephalopodes des croisieres du DANA (1921–1922), 4e Partie. *Annales de l'Institut Océanographique 13*(1), 1–49 [48, figs 47, 48].

Type data: lectotype ZMUC.

Type locality: North Atlantic Ocean, 33°51′N, 66°43′W.

**COMMON NAME**

Unarmed Cranch Squid

**DISTINGUISHING CHARACTERS**

Fins lanceolate, postioned laterally on the posterior margin, taper posteriorly to tip of gladius, no anterior lobes; attachment one-third or more of ML. (This trait is unsual in the subfamily and is shared only with *Teuthowenia*.) Funnel valve well developed. (*Megalocranchia* is the only other genus in the subfamily with a funnel valve.) S-shaped lateral ocular photophores.

Larva, dorsal view (Voss 1980).

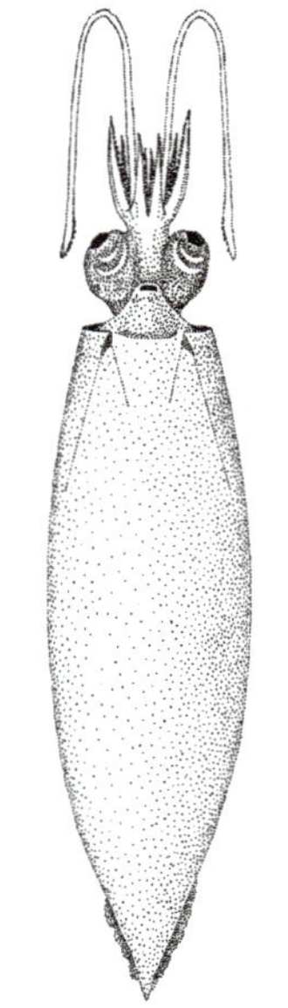

Juvenile, ventral view (Voss 1980).

Subadult, dorsal view (ABRS, K. Hollis).

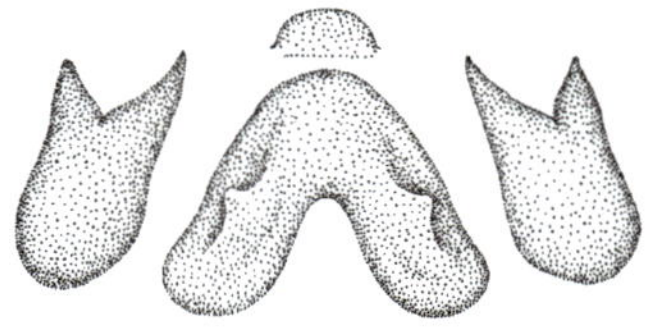

Funnel organ and funnel valve (Voss 1980).

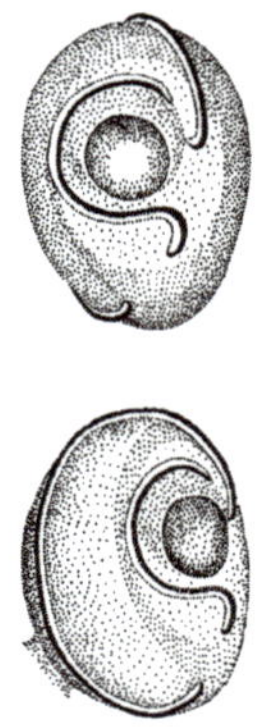

Left eye: upper, lateral view; lower ventrolateral view showing photophores (Voss 1980).

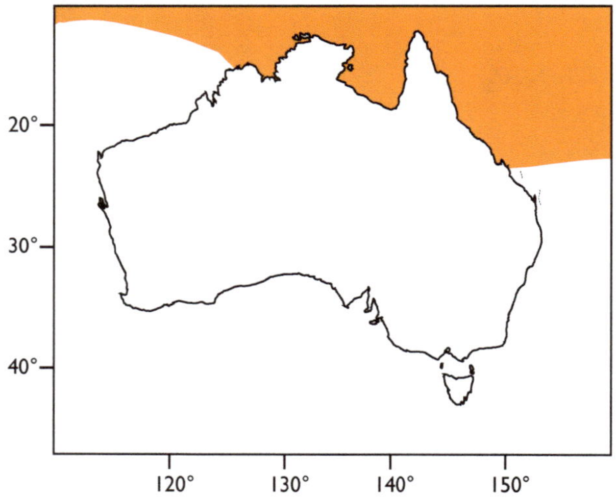


*Egea inermis* (distribution).

## DIAGNOSTIC FEATURES

Mantle long, narrow, spindle-shaped, thin-walled, leathery, tapers posteriorly to attenuated tip. Non-cartilaginous inverted V-shaped marks on each side of ventral mantle with tip of 'V' corresponding in position to mantle-locking cartilage. Eyes spherical, protruding, anteriolaterally oriented. Fins lanceolate, postioned laterally on posterior margin, taper posteriorly to tip of gladius, no anterior lobes; attachment one-third or more of ML. Funnel moderately large. Dorsal pad of funnel organ large, inverted V-shaped, broad anterior margin and broad, triangular flap on each lateral limb, no mid-anterior papilla, ventral pads with anterior V-shaped notch. Funnel valve present, well developed. Funnel–mantle locking cartilage elongate, triangular, without tubercles. Nuchal-locking cartilage conspicuous, spindle-shaped without tubercles. All arm tips with four transverse series of minute suckers. Hectocotylus absent. Tentacles of medium length, muscular, stalk with two series of suckers and pads on distal 75%; clubs slightly expanded, without carpal cluster, suckers on manus medium-sized, with long pedicels.

*Photophores.* Ventral surface of eyes with two long, narrow photophores: one long, crescent-shaped, positioned medioposteriorly; one anterior, positioned laterally, encircles two-thirds of lens, crescent-shaped without-curved terminal curve at posterior end. Photophores present on arms 3 in subadult females (adult females unknown).

## HABITAT AND BIOLOGY

Paralarvae up to ~35 mm ML are epipelagic, concentrated in subsurface water to 200 m both day and night. With growth and metamorphosis (~35–40 mm ML) juveniles broaden their range in the mesopelagic zone to 800 m or more. Subadults and adults descend into the bathypelagic zone where maturation occurs at ~2000 m. Captures of juveniles and large subadults at night in subsurface waters to ~300 m suggest they undergo a diel vertical migration.

#### SIZE
Maximum reported ML 420 mm.

#### REMARKS
The genus *Egea* is currently monotypic.

#### DISTRIBUTION
Circumglobal in equatorial, tropical to subtropical waters (35°N to 42°N, to 25°S), in the Atlantic, Indian and western Pacific Oceans.

#### REFERENCE
Roper and Jereb (2010h).

### *Galiteuthis* Joubin, 1898

#### DISTINGUISHING CHARACTERS
Species are recognised by the hooks on the tentacular clubs, the lack of hooks on the arms and the long, slender terminal fins. The large hooks on the tentacular club are unusual within the family, shared only with *Mesonychoteuthis*.

#### DIAGNOSTIC FEATURES
Mantle long, spindle-shaped, broadest in anterior half, tapers posteriorly to slender, pointed tip. Numerous rounded cartilaginous tubercles on mantle in animals from ~10 mm ML. Large multipointed tubercles present externally on mantle–funnel fusion and nuchal-locking cartilage sites. Anterio-laterally directed eyes dominate head. Fins terminal, medium to long, lanceolate, variously attenuate posteriorly, with stout gladial spine and low delicate fringe to tip of gladius. Anterior margin of fins insert on posteriolateral margins of lanceola. Funnel-locking cartilage comma-shaped, may have two elongate tubercles each with two or three cusps anteriorly. Funnel valve absent. Dorsal pad of funnel organ inverted U-shape with three large, slightly flattened papillae. Nuchal-locking cartilage broad, triangular, with two small turbercles on each side with two or three cusps. Arms short, without hooded hooks. Tentacles short to medium, muscular. Carpal suckers biserial along distal two-thirds to three-quarters of stalk, carpal cluster present. Club short, slightly expanded, with four to eight pairs of median suckers on manus that with growth (from ~35–60 mm ML) become conspicuous, large, hooded hooks. Club keels absent. Suckers on lateral rows greatly reduced or lost. Protective membranes equally developed dorsally and ventrally with weak trabeculae.

*Photophores.* Ventromedial surface of eye with one large, very elongate, crescent-shaped posterior photophore that envelops one small, slightly crescent-shaped to round anterior (lateral) photophore. Arm tips without brachial end-organ photophores.

#### HABITAT AND BIOLOGY
Paralarvae inhabit immediate subsurface depths. Adults occur at depths in excess of 1500 m.

#### REMARKS
The genus comprises five nominal species and at least one or two additional species are yet to be validated.

#### DISTRIBUTION
Representatives of this genus are pelagic, found circumglobally in all oceans in tropical, subtropical, temperate, subpolar and Antarctic waters (absent in Arctic Ocean).

#### REFERENCE
Roper and Jereb (2010h).

#### Key to *Galiteuthis* species (after Nesis 1982/1987)
1a. Mantle surface rough, with many rounded cartilaginous tubercles in subadults. Two linear, three-pointed tubercles on anterioventral mantle, corresponding to position of funnel–mantle fusion. Dorsally, two or three small, conical tubercles on each side corresponding to position of nuchal-locking cartilage. Arm sucker rings with low, rounded teeth ........................ *G. glacialis*

1b. Mantle surface smooth. One tubercle with several (five or six in adults) points on anterioventral mantle corresponding to position of mantle-locking cartilage. No tubercles on dorsal side of mantle at the mantle-head fusion site. Arm suckers with smooth rings ................................................................ *G. suhmi*

### *Galiteuthis glacialis* (Chun, 1906)

*Crystalloteuthis glacialis* Chun, C. (1906). System der Cranchien. *Zoologischer Anzeiger 31*(2), 82–86 [85].
Type data: holotype ZMB Moll.-110013.
Type locality: Antarctic Drift Current, 59°16′S, 40°13′E.

#### DIAGNOSTIC FEATURES
Mantle surface rough, with many rounded cartilaginous tubercles in subadults. Two linear, multicuspid tubercles on anterioventral mantle, corresponding to position of mantle-locking cartilage; tubercles with two or three

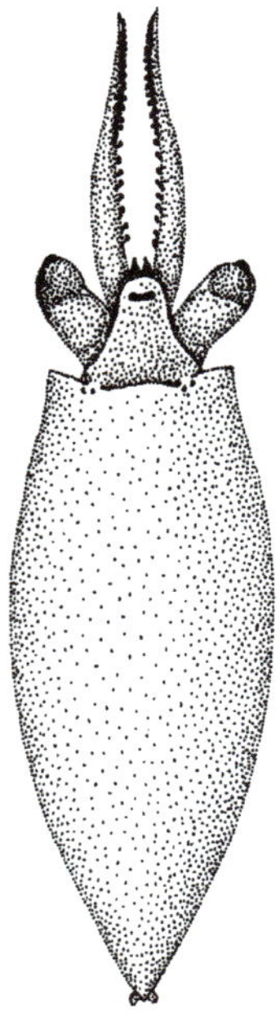

Larva (Voss 1980).

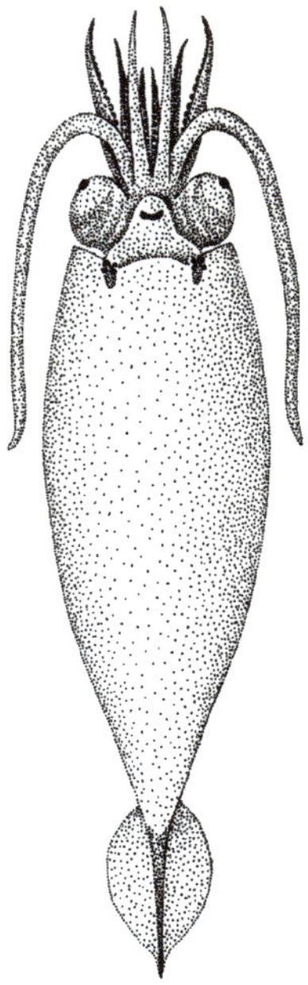

Juvenile (Voss 1980).

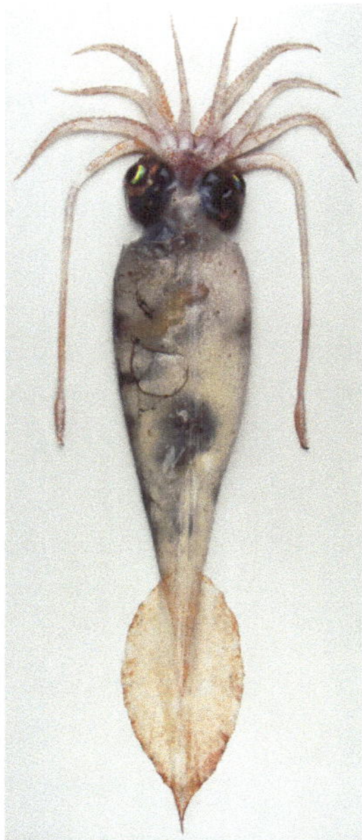

Fresh-caught specimen, dorsal view (Darren Stevens, NIWA, NZ IPY-CAML (voyage TAN0802)).

Fresh-caught specimen, ventral view (Darren Stevens, NIWA, NZ IPY-CAML (voyage TAN0802)).

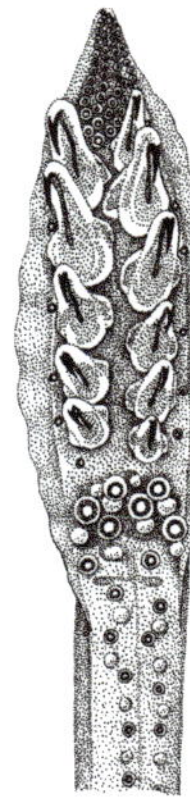

Left tentacular club (Voss 1980).

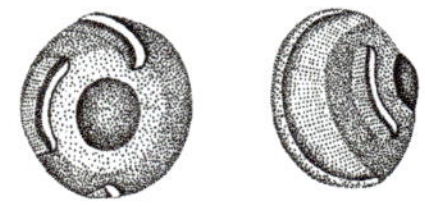

Left eye, lateral and ventral views showing photophores (Voss 1980).

cusps. Dorsally, two or three small, conical tubercles on each side of nuchal-locking cartilage. Tentacular club with 10–12 hooks in subadults; club manus with small lateral suckers.

Paralarvae (at ~11 mm) have eyes on short stalks and extremely small, widely separated fins. Tubercles at the nuchal fusion can be detected at ~5 or 6 mm. At 7 mm the tubercles at the funnel fusion can be detected. At ~30 mm ML, the club is well delineated and bears tetraserial suckers. At ~54 mm ML young have sessile eyes and fins are approaching adult shape. At 55–65 mm tentacular club hooks begin to develop and they are well formed in animals just over 100 mm ML. At just over 100 mm ML tubercles can be seen, and at 125 mm ML ocular photophores are fully developed.

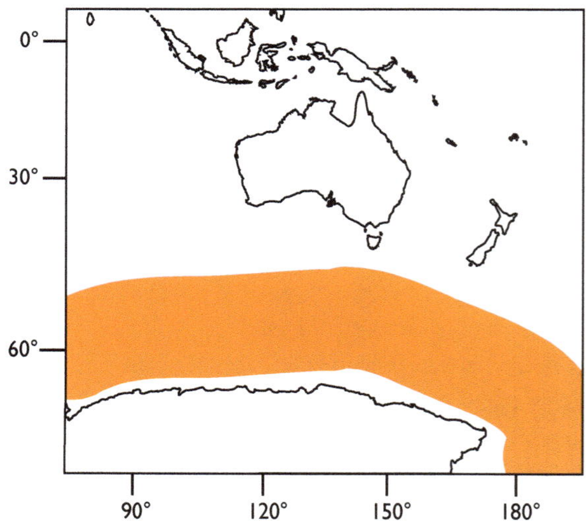


*Galiteuthis glacialis* distribution.

## HABITAT AND BIOLOGY

A mature female caught off the Falkland Islands had eight spermatangia inserted into the gelatinous mantle wall. The ovary contained ~3600 oocytes mostly 2.2–2.5 mm long and some ripe eggs, ~3.0 × 2.5 mm in the oviducts (Laptikhovsky and Arkhipkin 2003).

The vertical distribution of paralarvae and juveniles extends from the lower epipelagic zone, concentrated ~200 m, but as shallow as ~75 m. With continued growth and ontogenetic descent, larger juveniles are concentrated at 300–1000 m, subadults to 2000 m; then sexual maturation occurs in very deep water in excess of 2500 m. Significant diel vertical shifting appears to occur. It is an ammoniacal species.

*Galiteuthis glacialis* is one of the most abundant Antarctic squids.

## SIZE

Maximum ML to 500 mm.

### DISTRIBUTION

*Galiteuthis glacialis* occurs in oceanic Antarctic circumpolar water masses (45°S to 70°S) and northward only where great masses of Antarctic intermediate waters flow out northward of the Antarctic Polar Front. Antarctic endemic extending north to the Sub-Antarctic Front.

### REFERENCES

Young and Mangold (2006a), Roper and Jereb (2010h).

## *Galiteuthis suhmi* (Hoyle, 1886)

*Taonius suhmi* Hoyle, W. E. (1886). Report on the Cephalopoda. *Report on the scientific results of the voyage of H.M.S. Challenger during the years 1873–76, Zoology 16*(44), 1–245, 33 plates. [192, pl. 32, figs 5–11].

Type data: unresolved, syntype BMNH 90.1.24.19, NMGW syntype?.

Type locality: Southern Ocean, 47°25′S, 130°22′E.

### DIAGNOSTIC FEATURES

Body elongate, fusiform, acuminate posteriorly. Mantle surface smooth. Mantle margin tranverse. One tubercle with several (five or six in adults) points on mantle–funnel fusion sites. No tubercles on dorsal side of mantle at mantle–head fusion site. Head small, eyes large. Fins lanceolate. Arms subequal in length 4 > 3 > 2 > 1; suckers biserial with smooth, horny rings. Tentacular club with tetraserial suckers, median suckers largest, infundibulum with four bluntly pointed teeth.

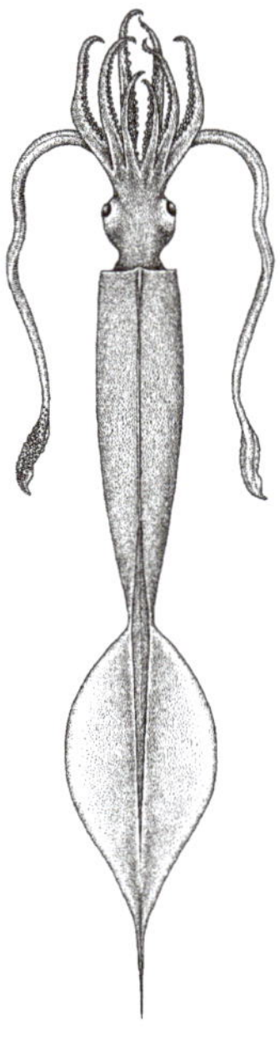

Dorsal view (ABRS, K. Hollis).

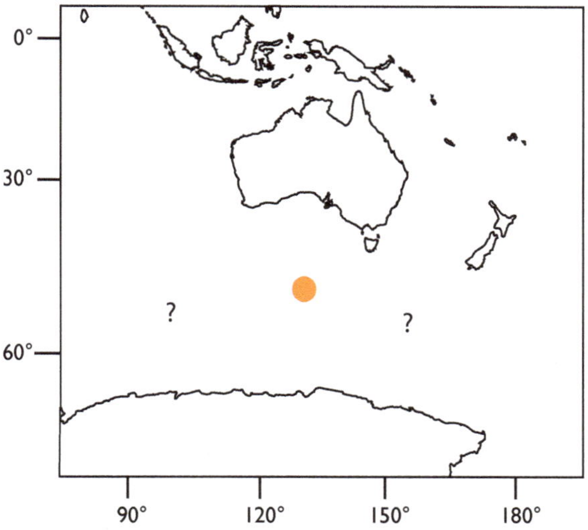



*Galiteuthis suhmii* distribution. (Australian records are based on few specimens, but the species is likely to occur more widely.)

*Colour.* Subadult: pale, almost white. Mantle semi-transparent even when preserved; numerous oblong chromatophores in eight transverse rows on ventral surface, fewer rows on dorsal surface (at least in subadults based on the original description).

### HABITAT AND BIOLOGY

Mesopelagic to bathypelagic species.

### SIZE

Up to 400 mm ML.

### REMARKS

The original illustrations and description are based on a juvenile specimen, and there is a possibility that the type series contains more than one species.

### DISTRIBUTION

Southern subtropical to sub-Antarctic circumglobal waters.

### REFERENCE

Nesis (1982/1987).

## *Helicocranchia* Massy, 1907

### DISTINGUISHING CHARACTERS

Members of this genus are readily identified by the exceptionally large funnel that extends anteriorly beyond the beak and the very small paddle-like fins that attach to the posterior gladius that projects dorsally above the mantle musculature.

#### DIAGNOSTIC FEATURES

Mantle long, slender, cylindrical, broadest medially; tapers posteriorly to blunt terminal point that is not supported by gladius. Fins very small, paddle shaped, subterminal, fused distally, inserted on short rostrum of gladius that projects dorsally free of end of gladius; posterior fin lobes free. Funnel very large, long, broad. Funnel–mantle locking cartilage broad, indistinct, without tubercles; funnel valve absent. Male arms more robust than female arms, all arms with enlarged suckers (especially arms 3); distal tips of male arms 1–4 with multiple series of suckers; variably expanded protective membranes, especially on arms 1 and 2. Females occasionally with few pairs of enlarged suckers on arms 3.

*Photophores.* Eyes with one, usually large, oval to round, protruding, anteriorly oriented photophore. Brachial photophores absent.

#### HABITAT AND BIOLOGY

Paralarvae and juveniles typically occur in upper epipelagic waters and then, with growth, gradually undergo ontogenetic descent through to the upper bathypelagic zone.

#### SIZE

Small-sized cranchiids; maximum ML to 100 mm.

#### REMARKS

*Helicocranchia* currently consists of three valid species and an additional 11 undescribed species worldwide.

#### DISTRIBUTION

*Helicocranchia* are widely distributed in tropical, subtropical and temperate (at least in the Atlantic Ocean) waters.

#### REFERENCE

Roper and Jereb (2010h).

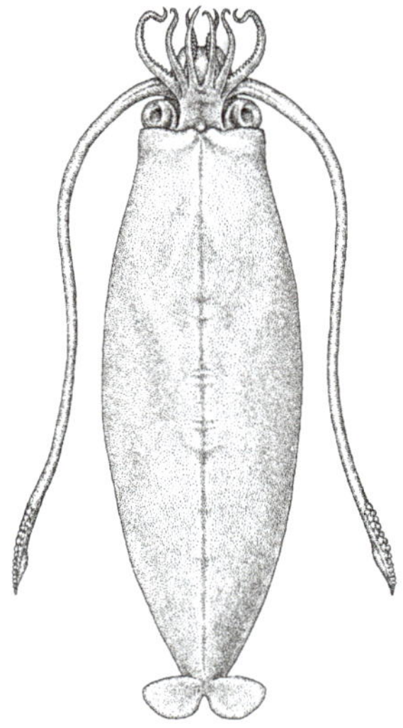

Dorsal view (ABRS, K. Hollis).

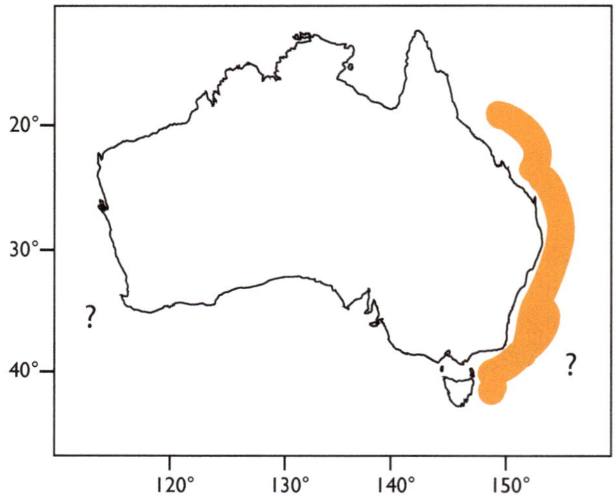


*Helicocranchia pfefferi* distribution.

### *Helicocranchia pfefferi* Massy, 1907

*Helicocranchia pfefferi* Massy, A. L. (1907). Preliminary notice of new and remarkable cephalopods from the South-west Coast of Ireland. *Annals and Magazine of Natural History, series 7 20*, 377–384 [382].

Type data: holotype BMNH 1890.1.24.10 (1909.11.27.10?).

Type locality: temperate western North Atlantic Ocean, off the southwestern Irish coast, 51°54′N, 11°57′W.

#### COMMON NAME

Pfeffer's Cranch Squid

#### DIAGNOSTIC FEATURES

No cartilaginous papillae on mantle. L-shaped ventral pads on funnel organ. No enlarged suckers on arms 3 of adult females (or on tentacular clubs). Moderately long, robust tentacles, shorter than ML.

*Colour.* Bands of orange-brown chromatophores on sides of mantle.

#### HABITAT AND BIOLOGY

Paralarvae and juveniles occur in the upper few hundred metres of the epipelagic zone, and then descend gradually to the upper mesopelagic; finally, adults descend into the lower mesopelagic and the bathypelagic zones to at least 2000 m. Some diel vertical migration might occur; paralarvae and mated/spent females captured at or near the surface suggest that adults return to the surface to mate and spawn.

**SIZE**

Maximum ML to 90 mm to 100 mm.

**DISTRIBUTION**

*Helicocranchia pfefferi* is probably a complex of species that occur circumglobally in open oceanic waters of the tropical, subtropical and North Atlantic temperate zones.

**REFERENCE**

Roper and Jereb (2010h).

## *Liguriella* Issel, 1908

*Liguriella* is currently monotypic, represented by *Liguriella podophthalma* Issel, 1908.

### *Liguriella podophthalma* Issel, 1908

*Liguriella podophthalma* Issel, R. (1908). Diagnosi preliminari di un nuovo genre e di due nouve specie di Cefalopodi appartenenti alla fam. Cranchiidae raccolti dalla R. Nave 'Liguria'. *Monitore Zoologico Italiano* *19*(4), 102–104 [103].

Type data: type repository unresolved.

Type locality: South Atlantic Ocean, 28°38′S, 47°31′W.

**DIAGNOSTIC FEATURES**

Mantle elongate, tapers to a blunt point posteriorly; anteriodorsal mantle margin free, except medially. Fins terminal, small; posterior fin insertions extend to tip of gladius and fuse along distal end of gladius vane. Funnel–mantle locking cartilage small, slightly curved, with two small cartilaginous tubercles at anterior end. Dorsal pad of funnel organ with three spatulate papillae; funnel valve absent. Tentacular club with suckers, no hooks, no carpal cluster, two series of suckers and knobs along tentacular stalk.

*Photophores.* Eyes with two contiguous photophores: anterior photophore small, crescent-shaped, lies closely

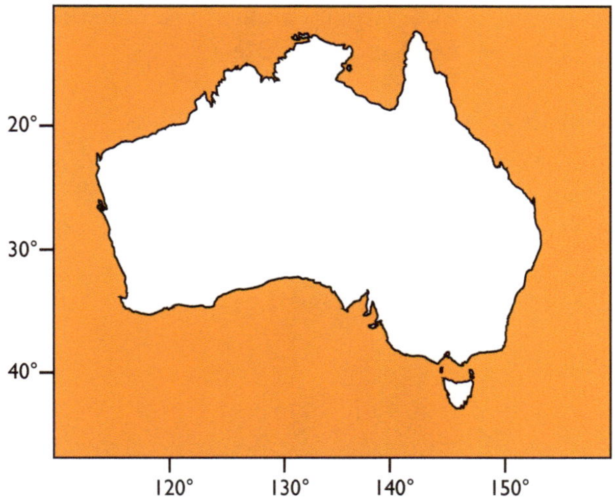


*Liguriella podophthalma* distribution.

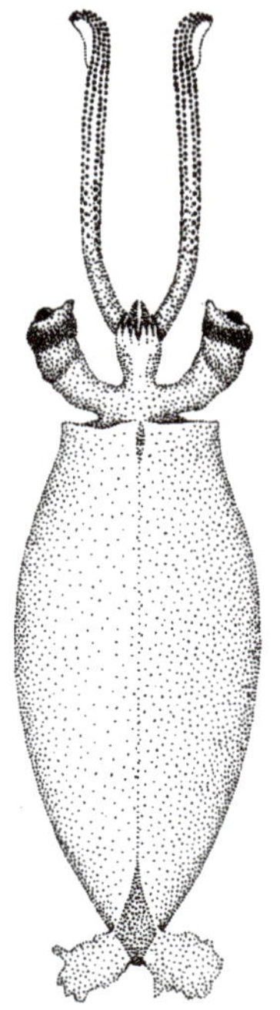

Larva, dorsal view (Voss 1980).

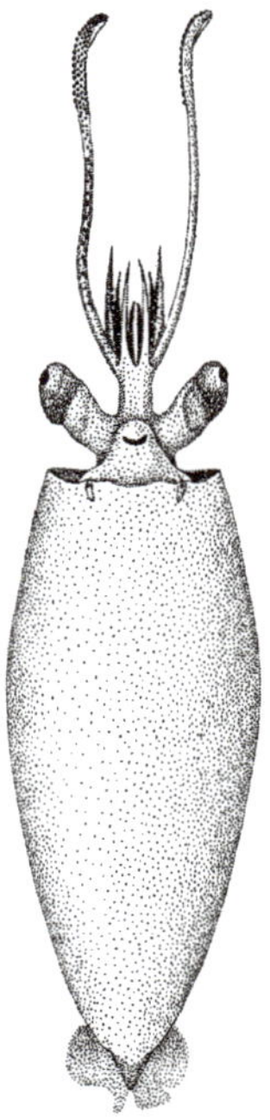

Juvenile, ventral view (Voss 1980).

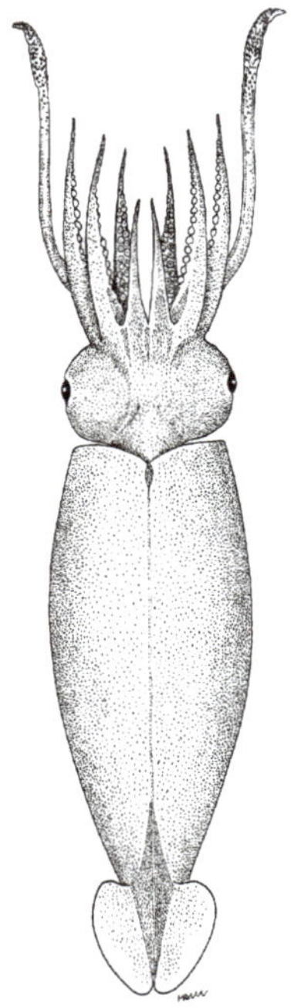

Subadult female, dorsal view (FAO, E. D'Antoni).

Right funnel–mantle locking cartilage (external) (Voss 1980).

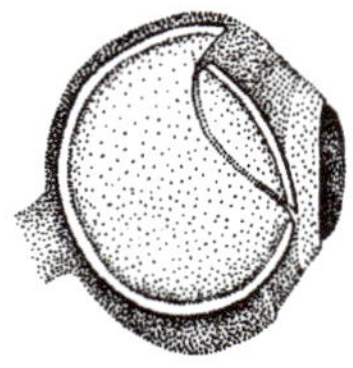

Left eye, ventral view showing photophores (Voss 1980).

within concavity of large crescent-shaped posterior photophore. Arm tip photophores absent.

### HABITAT AND BIOLOGY

The species occurs from subsurface waters to depths in excess of 1000–1500 m. Paralarvae and juveniles up to ~40 mm ML occur from subsurface depths to ~400 m, while older animals extend progressively deeper with growth. Little is known about their biology.

### SIZE

Maximum ML known 243 mm.

### DISTRIBUTION

*Liguriella podophthalma* occurs circumglobally in tropical, subtropical and northern sub-Antarctic waters.

### REFERENCE

Roper and Jereb (2010h).

## *Megalocranchia* Pfeffer, 1884

### DISTINGUISHING CHARACTERS

Unique within the family: paralarvae (to ~50 mm ML) with thick, gelatinous mantle dermis. Fusion of anterior 10–15% of fins to mantle is unique in the family; only in *Egea* and *Teuthowenia* fins are fused to mantle for 30% or more of their length; all other cranchiid fins attach to the gladius. The complex, bilobed, large, multilens (four), visceral photophore on the ventral surface of the digestive gland and ink sac is also unique within the family.

### DIAGNOSTIC FEATURES

Mantle slender, elongate, tapers posteriorly to thin tip; mantle thin, leathery, transparent. Paralarvae (to ~50 mm ML) with thick, gelatinous mantle dermis. Eyes very large, spherical, protruding, oriented anteriolaterally. Fins long (~50% ML), lanceolate, posteriolateral, anterior 10–15% insert along lateral margin of mantle for about half ML, no anterior lobes. Mantle-locking cartilage subtriangular without tubercles; nuchal-locking cartilage spindle-shaped. Funnel organ dorsal pad large, with triangular flap on each lateral limb, anterior margin broad; funnel valve well developed.

Arm tips 1–3 in large males abruptly attenuated with reduced, minute biserial suckers. Tentacles short, with two series of carpal suckers on mid-third of stalk and then four longitudinal series distally merge into strong carpal group; clubs moderate length, slightly expanded, with suckers in four series only moderately enlarged.

*Photophores.* Most of ventral hemisphere of eyes covered with two photophores: one large, crescent-shaped

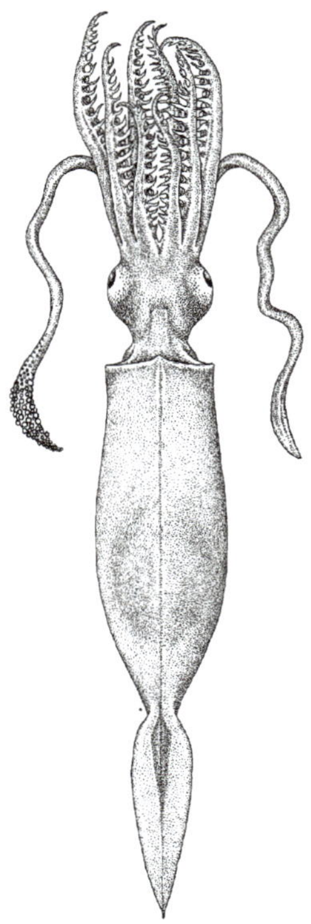

*Megalocranchia abyssicola?*, ventral view, juvenile (ABRS, K. Hollis).

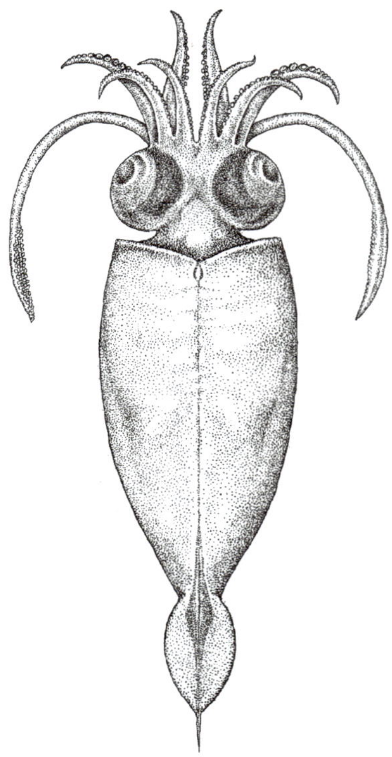

*Megalocranchia abyssicola?*, dorsal view, adult (ABRS, K. Hollis).

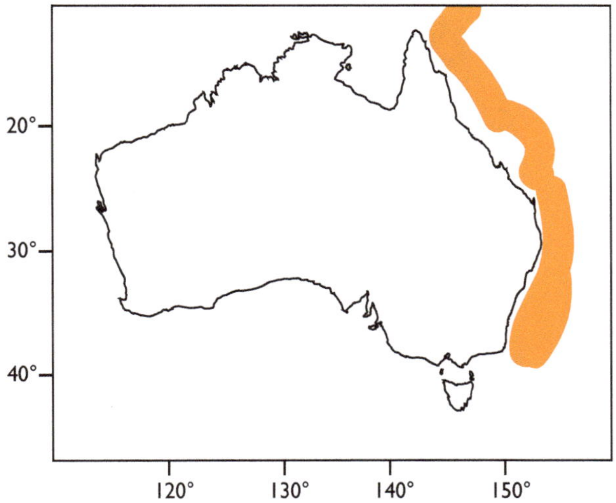

*Megalocranchia* sp. distribution.

posterior (medial) organ with irregularly shaped inner margin and central bulge, and within its concavity one smaller, narrow, roughly elongate, irregularly S-shaped, anterior (lateral) organ. Subadult and adult females with brachial end organs (photophores) on arms 1–3, or only on 3, or occasionally only on 2. Complex, bilobed, large, multilens (four), visceral photophore on ventral surface of digestive gland and ink sac.

#### HABITAT AND BIOLOGY

Paralarvae are epipelagic from very near-surface waters to ~200 m day and night. Juveniles metamorphose at ~40–50 mm ML, are concentrated in the upper 200 m at night and then descend to 600–800 m during the day. Subadults undergo further ontogenetic vertical descent through the mesopelagic and into the bathypelagic zone to >2000 m during daytime and some appear to migrate at night back towards lower epipelagic and mesopelagic depths (~100–700 m). While adults probably mature in the bathypelagic zone, they apparently reverse migrate back into epipelagic waters to spawn, as indicated by specimens caught at or near the surface at night. The species of *Megalocranchia* are neutrally buoyant.

#### SIZE

Moderately large to very large, up to 1.8 m ML.

#### REMARKS

The systematics of this complex genus is in need of review. Brandt (1983) reported three megalocranchids occurring in warm core eddies off eastern Australia: *M. abyssicola* (Goodrich, 1896), *M. megalops* (Prosch, 1849) and *M.* sp. Lu and Phillips (1985) state that Brant's (1983) record of *M. megalops* is referable to *Teuthowenia*. As *T. megalops* occurs only in the sub-Arctic and northern Atlantic Ocean is likely that Brandt's (1983) records refer to *T. pellucida* (Chun, 1910). Young and Mangold (2010) and Roper and Jereb (2010h) cite three species as currently valid: *M. maxima* Pfeffer, 1884, *M. oceanica* Voss, 1960 and *M. fisheri* (Berry, 1909). *Megalocranchia maxima* appears to have a disjunct distribution in the subtropical waters of the South Atlantic, sub-Antartctic, the south-west Indian Ocean and Japanese waters. *Megalocranchia oceanica* is found in the North and South Atlantic, and *M. fisheri* in the central North Pacific. *Megalocranchia abyssicola* and *M. pardus* Berry, 1916 are reported by Lu and Phillips (1985) as occurring in Australian waters, the former species in the Coral and Tasman Seas, and the latter off New South Wales and south-eastern Australia. A comprehensive revision of this group is required to determine the status of these two species.

#### DISTRIBUTION

*Megalocranchia* species occur circumglobally in tropical and subtropical waters.

#### REFERENCE

Roper and Jereb (2010h).

### *Mesonychoteuthis* Robson, 1925

*Mesonychoteuthis* is currently monotypic, represented by *M. hamiltoni* Robson, 1925.

### *Mesonychoteuthis hamiltoni* Robson, 1925

*Mesonychoteuthis hamiltoni* Robson, G. C. (1925). On *Mesonychoteuthis*, a new genus of oegopsid Cephalopoda. *Annals and Magazine of Natural History* (series 9) *16*, 272–277 [272, figs 1, 2].

Type data: holotype BMNH 1926.3.31.28.

Type locality: South Shetland Islands, from the stomach of a sperm whale 50 miles south of Smith Island.

#### COMMON NAME

Colossal Squid

#### DISTINGUISHING CHARACTERS

About 3–10 sucker pairs in midarm section develop into large, hooded hooks beginning at ~45 mm ML (increasing numbers of hooks on arms 1 through to 4). This is a unique character within the Cranchiidae.

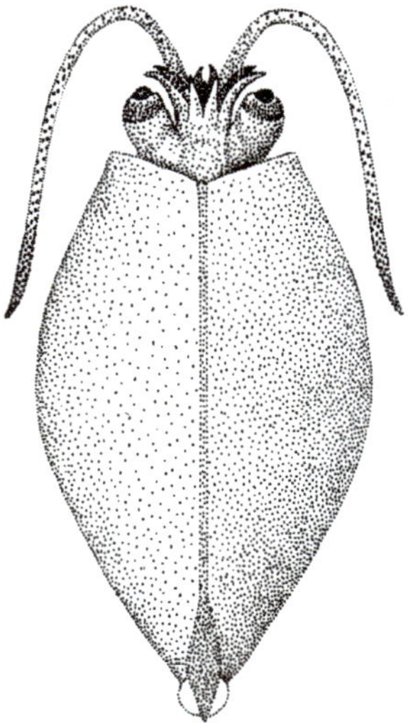

Larva, dorsal view (Voss 1980).

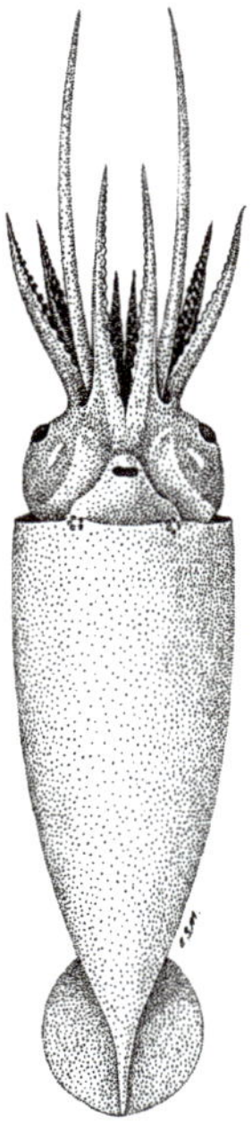

Juvenile, ventral view (Voss 1980).

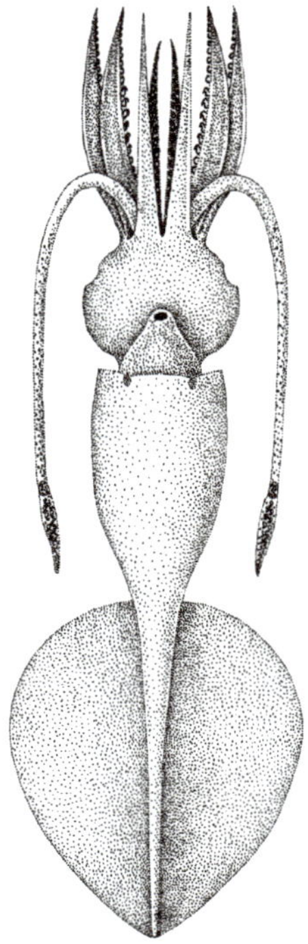

Subadult, ventral view (Voss 1980).

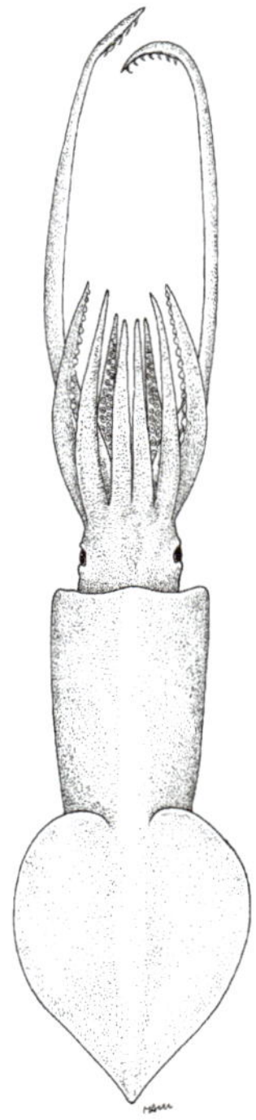

Adult, dorsal view (FAO, E. D'Antoni).

Left arms 1–4 (Voss 1980).

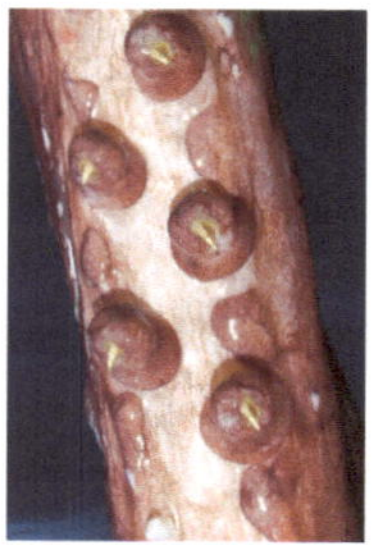

Portion of arm of freshly caught animal showing hooked suckers (B. Macpherson).

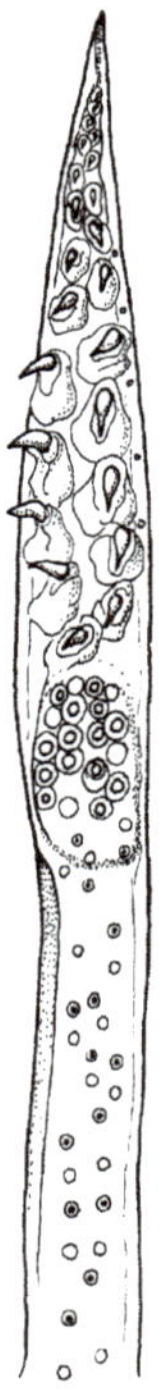

Tentacular club (FAO, E. D'Antoni).

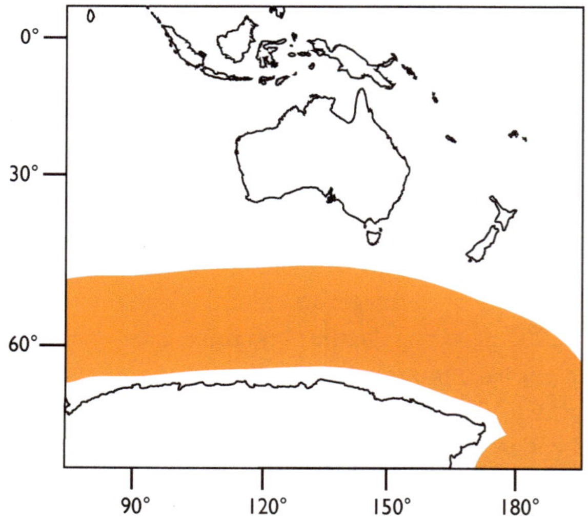


*Mesonychoteuthis hamiltoni* distribution.

## DIAGNOSTIC FEATURES

Mantle elongate, moderately stout in anterior quarter, tapers sharply anterior to fins and then tapers gradually to form long, very narrow posterior half. Mantle wall thin, muscular to leathery and tough in juveniles and adults, semigelatinous in mature females, up to 50 mm thick in adults, smooth. Head large, dominated by huge, nearly spherical, protruding anteriolaterally directed eyes. Fins terminal, large, stoutly ovate, without anterior lobes, taper posteriorly to a broad point at tip of gladius, anterior fin insertions extend along posteriolateral margins of lanceola, except at anterior-most margin; length up to two-thirds ML in adults, very muscular medially, taper to thin lateral margins.

Funnel–mantle locking cartilages short, stout, curved, four to six conical cartilaginous tubercles anteriorly in paralarvae and juveniles. Nuchal-locking cartilage triangular, without tubercles in adults, a single anterior tubercle in paralarvae and early juveniles. Funnel organ dorsal pad inverted V-shaped with short, stout papilla on each lobe and long, narrow conical papilla mid-anteriorly; funnel valve absent.

Arms robust, moderately long, ~50% ML, subequal; arm suckers biserial with dentate rings, ~3–10 pairs in midarm section develop into large, hooded hooks beginning at ~45 mm ML (increasing numbers of hooks in arms 1 through to 4). Tentacles moderately long, robust; stalks with diagonally oriented pairs of suckers and pads along distal two-thirds, carpal cluster present, club short, robust, slightly expanded, with all suckers in medial series of manus developed into ~12 pairs of large, hooded hooks; suckers of lateral series minute.

*Photophores.* Two photophores on ventral hemisphere of each eye, a small, oblong anterior one encircled by a large, wide, crescent-shaped posterior one. Brachial end organs (photophores) absent.

### HABITAT AND BIOLOGY

This is a fully oceanic squid that occurs as shallow as 20 m for paralarvae, down to mesopelagic depths ~500 m. Larger paralarvae, juveniles and subadults range into the bathypelagic realm of ~500 m to in excess of 2000 m. It is a relatively passive swimmer. Nearly mature female *Mesonychoteuthis hamiltoni* have been taken in Antarctic waters hooked on long lines set to fish the upper few hundred metres. This suggests that the females at least, may migrate vertically from great depths to near-surface waters to spawn; mating depths are unknown. It matures at mantle lengths > 1 m and 25–30 kg weight. Spermatophores are 170–270 mm long. It has a potential fecundity of 4–4.2 million oocytes. This species is believed to have some potential for a fishery. The flesh is rumored to be of 'excellent quality and very flavourful'.

*Mesonychoteuthis hamiltoni* is the major source of prey for sperm whales that migrate into Antarctic waters in summer.

See also biology of *Architeuthis* with respect to eye and body size above.

### SIZE

This is a gigantic squid with maximum mantle lengths of 2.25 to ~3 m; total length approaches 9–10 m; maximum recorded weight of 495 kg (Roper and Jereb 2010h).

### REMARKS

While shorter in total length than *Architeuthis dux* Steenstrup, 1857, this squid is very much more robust.

### DISTRIBUTION

This species is an Antarctic endemic, extending north to the Antarctic Polar Front: occurs circumglobally in Antarctic waters to 70°S only rarely crossing the Antarctic Polar Front to ~40°S, or very occasionally into southern waters off South Africa. A fresh specimen was recently found floating on the surface off Portland, Victoria, suggesting that it may also venture into southern Australian waters.

### REFERENCE

Roper and Jereb (2010h).

## *Sandalops* Chun, 1906

### DIAGNOSTIC FEATURES

Mantle stout, broadest at posterior third, cylindrical, tapers gradually to broad posterior point. Eyes large, protruding, nearly spherical, anteriorly oriented. Fins small, short, oval to round, subterminal, with anterior and posterior lobes, posterior insertion does not reach terminal end of gladius, anterior insertion at broadest point on vanes of gladius (attach to posteriolateral margins of lanceola). Funnel–mantle locking cartilage narrow, without tubercles; nuchal cartilage small, without tubercles. Dorsal pad of funnel organ with two large triangular flaps; funnel valve absent.

*Photophores.* Each eye with one very large, round posterior photophore and one very small, round to elliptical anterior photophore contiguous with, and indented into, anterior margin of posterior photophore. Brachial photophores absent.

### SIZE

Maximum ML 110 mm.

### REMARKS

Currently only one nominal species (*Sandalops melancholicus*), but Voss *et al.* (1992) indicate that the genus comprises three or four species.

### DISTRIBUTION

*Sandalops* occurs circumglobally in oceanic tropical and subtropical waters at depths from the upper epipelagic into the lower mesopelagic and bathypelagic, following ontogenetic descent.

### REFERENCE

Roper and Jereb (2010h).

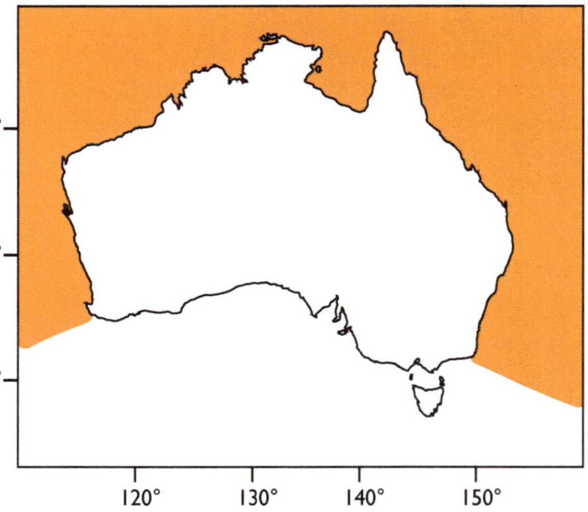


*Sandalops melancholicus* distribution.

## *Sandalops melancholicus* Chun, 1906

*Sandalops melancholicus* Chun, C. (1906). System der Cranchien. *Zoologischer Anzeiger 31*(2), 82–86. [86].

Type data: lectotype ZMB Moll.-110022.

Type locality: central South Atlantic Ocean, off NE Tristan da Cunha Island, 32°08′S, 08°28′W.

### COMMON NAME

Melancholy Cranch Squid

### DIAGNOSTIC FEATURES

Mantle fleshy, cylindrical. Eyes are long, tubular in paralarvae and particularly in juveniles when they occupy the upper mesopelagic zone; eyes oriented vertically upward. Eyes become nearly hemispherical in subadults. Fins small, wider than long, situated on lateral margins of gladius. Arms very small. Tentacles covered over most of their length with small, stalked suckers.

Fresh-caught specimen, dorsal view (K. Graham).

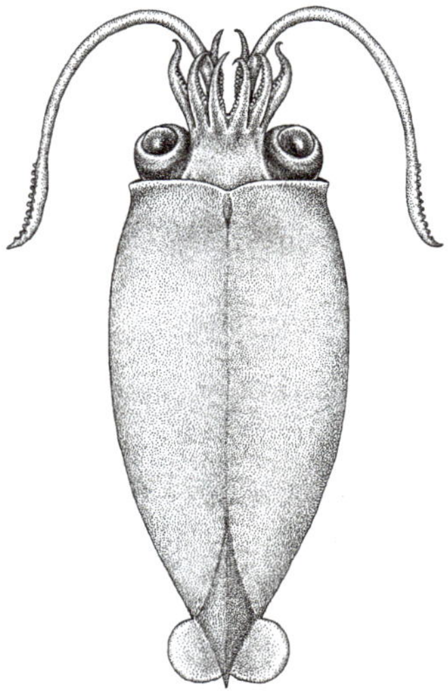

Dorsal view (ABRS, K. Hollis).

### HABITAT AND BIOLOGY

*Sandalops melancholicus* lives in the epipelagic, mesopelagic and bathypelagic zones, following the general cranchiid pattern of ontogenetic descent. By full growth, animals have descended into the bathypelagic zone beyond 2000 m depth, where maturation and mating occur. No evidence currently exists for diel vertical migration.

### SIZE

Maximum ML to 110 mm.

### DISTRIBUTION

The species is cosmopolitan in tropical and subtropical waters.

### REFERENCE

Roper and Jereb (2010h).

## *Teuthowenia* Chun, 1910

### DISTINGUISHING CHARACTERS

Anterior one-third or more of fins insert on lateral margin of mantle for 40–60% ML. This is unusual in the family and shared only with *Egea*.

## DIAGNOSTIC FEATURES

Mantle conical, elongate, tapers gradually posteriorly to narrow posterior tip, mantle wall thin, leathery. Glands (possibly pheromone secreting, or seminal vesicles) develop on dorsal midline of mantle of females at maturity. Head dominated by huge anteriorly directed eyes. Fins long, narrow, posteriolateral, taper gradually posteriorly, terminate in small lobes that extend beyond tip of gladius and fuse with median cleft, anterior lobes absent, anterior one-third or more of fins insert on lateral margin of mantle for 40–60% ML rather than on shell sac of gladius. Funnel–mantle locking cartilages small, oval to spindle-shaped, with one to four cartilaginous tubercles at mantle margin. Nuchal-mantle fusion cartilage small, elongate, diamond-shaped, without tubercles. Funnel large, long, broad; funnel organ dorsal pad with three long papillae, one lanceolate in middle section, two spatulate on narrow lateral arms; funnel valve absent. Tentacles short, muscular; four series of carpal suckers and pads in zigzag pattern on distal half to one-third of stalk; no carpal cluster; club slightly expanded, with suckers on long pedicels.

*Photophores.* Each eye with three photophores: one large crescent-shaped posterior (medial) organ with one smaller, narrow, crescent-shaped anterior (lateral) organ within concavity of larger, one small, oval organ on inner anteriodorsal margin near lens. Brachial organs (photophores) present on all arms of maturing and mature females.

## HABITAT AND BIOLOGY

*Teuthowenia* species generally show ontogenetic descent from near-surface, epipelagic to upper mesopelagic waters, to ~200–300 m for paralarvae (until they reach 50–100 mm ML); mesopelagic for juveniles (300–600 m) and increasingly deeper into the upper bathypelagic for early subadults to ~1000 m; late subadults and adults range to 1500 m and maturation occurs from there to ~2700 m where they mate and spawn. During mating, discharged spermatophores (spermatangia) attach to the dorsal surface of the female then penetrate the mantle wall and presumably discharge sperm into the mantle cavity. A degree of diel vertical movement is also evident in these near-neutrally buoyant species.

## SIZE

Medium-sized to moderately large cranchiids. Maximum reported ML ~380–400 mm.

## REMARKS

Typical of most genera in the family, *Teuthowenia* has undergone numerous nomenclatural changes. The chromatophore pattern and number and configuration of tubercles on the funnel–mantle locking cartilage are useful characters for distinguishing species.

## DISTRIBUTION

*Teuthowenia* includes three species that display allopatric (disjunct) distributions circumglobally in the southern Subtropical Front zone, and the sub-Arctic, north temperate and eastern tropical waters of the Atlantic Ocean in areas where primary productivity is reltively high.

## REFERENCE

Roper and Jereb (2010h).

### *Teuthowenia pellucida* (Chun, 1910)

*Desmoteuthis pellucida* Chun, C. (1910). Die Cephalopoden. Oegopsida. *Wissenschaftliche Ergebnisse der Deutschen Tiefsee-Expedition dem Dampfer Valdivia, 1898–1899 18*(1), 1–401 [357, pls 53, 54].

Type data: holotype ZMB Moll.-110025.

Type locality: South Atlantic Ocean, in Benguela Current, 33°20′S, 15°58′W.

## DIAGNOSTIC FEATURES

Funnel–mantle fusion with multipoint tubercle (two to four points, sometimes one to five points). Arms 1 of males with two or three series of suckers on modified tips; arms 2 of males with three or four series of suckers on modified tips. Arms 1 and 2 of males with 11–18 normal suckers in midportion of arm, proximal to modified ends but distal to modified basal suckers. Largest arm 3 suckers 2.5 × basal suckers. Tentacular club with poorly defined carpal cluster. Largest club suckers with 26-32 teeth. Mature males have a long penis that extends well beyond opening to mantle cavity.

## HABITAT AND BIOLOGY

The vertical distribution extends from the surface to 2400 m, with ontogenetic descent. Juveniles and subadults appear to undergo diel vertical movement from 700–800 m during the day into the upper 300 m at night. Mature adults seem to occur below ~1500 m and remain from there down to 2400 m or more. Females mature between 150–190 mm ML.

Fresh-caught specimen, dorsal view (K. Graham).

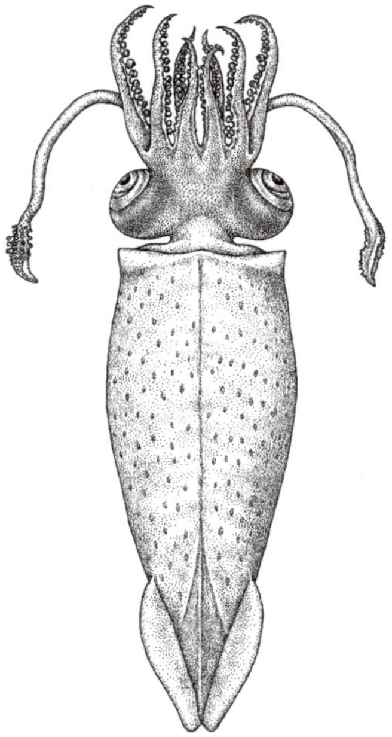

Dorsal view (ABRS, K. Hollis).

At maturity the female undergoes marked morphological changes. The musculature becomes flaccid, the body is distended with mature eggs, photophores develop on the arm tips and glandular organs develop in the dorsal midline. Mature females have nearly spherical eggs of 2.2 mm diameter. A single female carries ~6000–8000 eggs. Mature females commonly have suckers, possibly detached from males during mating, encysted within tissues in the mantle cavity. It is likely that the females are semelparous. Males mature at 140 mm ML. At maturity the modified tips of arms 1 and 2 elongate. The finely toothed suckers from the arm tips are commonly missing, presumably lost in the mantle cavity of a female during mating. The basal arm suckers become modified at maturity and the arms become more robust.

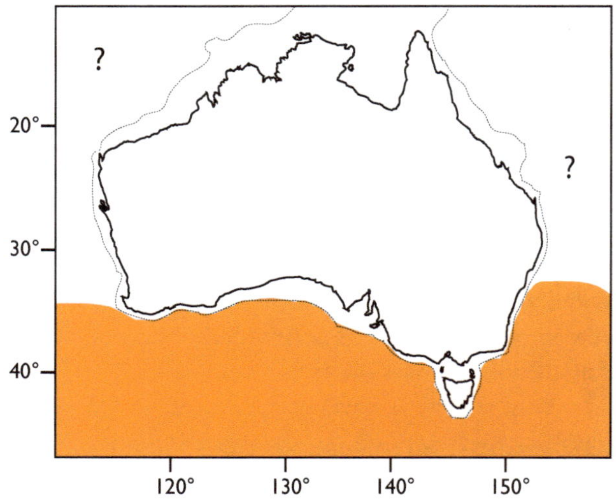


*Teuthowenia pellucida* distribution.

### SIZE

Maximum ML recorded is 210 mm.

### DISTRIBUTION

*Teuthowenia pellucida* occurs in a circumglobal belt centred on a band width of ~6–7° associated with the Southern Subtropical Front and to a certain extent into the boundary waters that fringe the convergence zone to the north and south. The total north–south extent of distribution is ~32°S to 48°S in waters with hydrologic characteristics of the convergence.

### REFERENCE

Young and Mangold (2006b).

## Cycloteuthidae Naef, 1923

### COMMON NAME

Disc-fin Squids

### DISTINGUISHING CHARACTERS

Cycloteuthids are superficially similar to octopoteuthids; however, the fins of *Discoteuthis* are separate and inserted directly onto the shell sac of the rigid gladius. In contrast, the fins of *Octopoteuthis* are fused and continuous across the dorsal midline and pass over the shell sac.

#### DIAGNOSTIC FEATURES

Mantle broadly conical, goblet-shaped; not strongly muscular; layer of gelatinous connective tissue lies between thin inner and outer layers of strong, circular muscle. Head also covered with moderately thick semi-gelatinous tissue. Fins long, broad, disc-like, > 70% ML in adults (including 'tail' when present). Funnel component of funnel–mantle locking apparatus subtriangular to irregularly oval, with deep pit to receive prominent, protruding, nose-shaped or tubercular mantle cartilage component; mantle component does not reach anterior mantle margin. Arm suckers biserial. Hectocotylus absent. Tentacular club suckers tetraserial. Club compact, broad with well-defined manus and dactylus, ill-defined carpus, with about seven small knobs and seven small suckers. Buccal connectives attach to ventral borders of arms 4.

*Photophores.* Present in adults, variously (depending on species) around eyes, head, ventral mantle, arm bases and ink sac.

#### HABITAT AND BIOLOGY

Very little is known about the biology of any species in the family. Members of the family occur in lower epipelagic to mesopelagic zones, occasionally to the bottom in bathyal waters. Little is known about their vertical distribution in Australian waters, but elsewhere cycloteuthid species appear to undergo diel vertical migration; specimens have been taken in the upper 200 m at night and at 500–1000 m during the day. The small size of most species, the soft consistency of the flesh and the relative rarity of specimens in deep-set trawls preclude any potential fisheries.

#### SIZE

Moderate to large-sized squids; maximum ML ~600 mm.

#### REMARKS

The family Cycloteuthidae contains two rather dissimilar genera: *Cycloteuthis* and *Discoteuthis*. These two genera are united principally by the single character of the subtriangular funnel-locking component. Otherwise there is little morphological evidence for a close relationship. Molecular analysis, however, confirms the close relationship of these genera in the Cycloteuthidae (Carlini 1998). Although four nominal species are currently recognised, other species are likely to exist.

#### DISTRIBUTION

Cosmopolitan in tropical and subtropical waters.

#### REFERENCE

Roper and Jereb (2010i).

### Key to Cycloteuthidae genera (after Roper and Jereb 2010i)

1a. Mantle in adults drawn out posteriorly into 'tail'. Fins subtriangular, length in adults to 75% ML, including 'tail'. Suckers tetraserial on manus of club of equal/subequal size. A single photophore present on ink sac; a series of 15–30 very small, irregularly shaped photophores on iris of eye around pupil; no photophores embedded on surface of body. Gladius thin, narrow, with long, terminal cone .........*Cycloteuthis* (*C. sirventi*)

1b. Mantle in adults not drawn out into a 'tail'. Fins large and rounded, length equals or slightly exceeds ML in adults. Suckers in central two series on clubs greatly enlarged, suckers in the two marginal rows greatly reduced in size. No photophore on ink sac; one to several photophores embedded on surface of body. Vane of gladius wide, heavily thickened, posteriorly rounded, terminal cone absent or minimal ...................*Discoteuthis*

## *Cycloteuthis* Joubin, 1919

Two species are recognised in the genus, but they are poorly known because they are relatively rare in collections. One of these, *Cycloteuthis sirventi* Joubin, 1919, is found in Australian waters.

## *Cycloteuthis sirventi* Joubin, 1919

*Cycloteuthis sirventi* Joubin, L. (1919). Estudes preliminaires dur les Cephalopodes recueillis au cours des croisieres de S.A.S. le Prince de Monaco. 7e Note: *Cycloteuthis sirventi* nov. gen. et sp. *Bulletin de l'Institut Oceanographique 351*, 1–7. [1, figs 1, 2].

Type data: holotype MOM.

Type locality: eastern North Atlantic Ocean, off Madeira, 30°45′40″N, 25°47′W.

#### COMMON NAME

Sirvent's Disc-fin Squid

#### DISTINGUISHING CHARACTERS

Easily distinguished from *Discoteuthis* by its general shape, which includes a 'tail'.

#### DIAGNOSTIC FEATURES

Distinct 'tail' (i.e. mantle extends posteriorly beyond fins). Fins large, oval; ~80% ML in adults. Large suckers

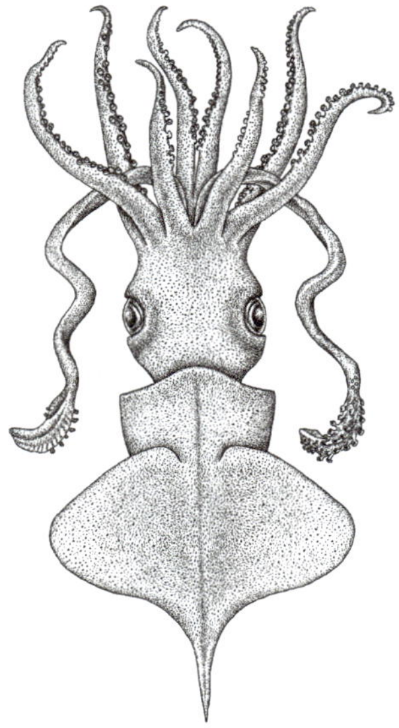

Dorsal view (ABRS, K. Hollis).

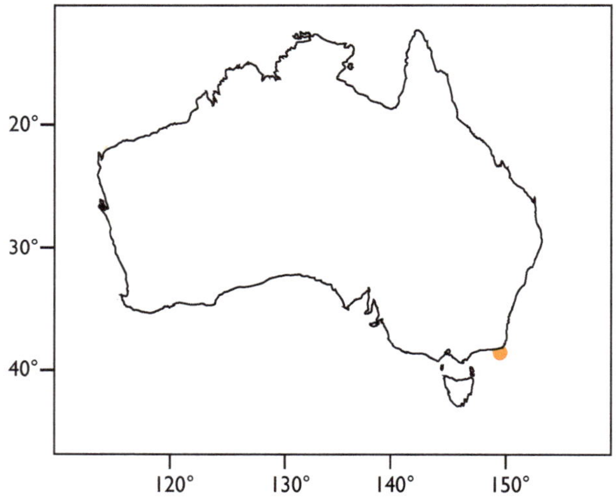


*Cycloteuthis sirventi* distribution. (Australian records are based on few specimens, but the species is likely to occur more widely.)

from arm 3 with ~20 low, truncate teeth (largest distally) around entire margin. Club suckers almost equal in size across manus, with ~20 teeth pointed distally and truncated proximally. Gladius with secondary conus.

Young specimens: mantle elongate with small terminal fins; posterior tail a small bump between posterior margin of fins; visceral photophore barely visible; eyes protrude noticeably, not stalked; arm formula $2 = 3 \geq 4 > 1$.

*Photophores.* Eyeball in adults with large photophore on ventral surface and irregular series of small photophores at edge of iris. Large visceral photophore on ink sac.

#### HABITAT AND BIOLOGY

Little is known about the biology of this species. The paralarvae of this species are very distinctive (see above).

#### SIZE

Mantle length to ~480 mm.

#### DISTRIBUTION

Tropical, subtropical, temperate waters of the Atlantic, Mediterranean, Indo-West Pacific. Lower epipelagic, mesopelagic, rarely bathybenthic.

#### REFERENCE

Roper and Jereb (2010i).

## *Discoteuthis* Young & Roper, 1969

#### DIAGNOSTIC FEATURES

See family key. Large rounded fins extend length of mantle. Small tentacular clubs and suckers, but no hooks on arms.

#### HABITAT AND BIOLOGY

These squid have not been observed alive and little is known about their biology and ecology.

#### REMARKS

*Discoteuthis* is currently represented by two species: *D. discus* and *D. laciniosa*, both of which are rare in collections. However, more species exist but are yet to be formally described. The photophore arrangement varies among species.

#### DISTRIBUTION

Tropical and subtropical waters of all oceans.

#### REFERENCES

Young (1999), Roper and Jereb (2010i).

### Key to *Discoteuthis* species

1a. No flaplike papillae on free anterior mantle margin. Funnel-locking cartilage a diagonal groove. Photophores at posterioventral tip of mantle; brachial crown photophores absent ............................ *D. discus*

1b. Flap-like papillae on free anterior mantle margin. Funnel-locking cartilage a V-shaped groove. Pair of photophores near anterioventral mantle margin; brachial crown with photophores ........... *D. laciniosa*

## *Discoteuthis discus* Young & Roper, 1969

*Discoteuthis discus* Young, R. E. and Roper, C. F. E. (1969). A monograph of the Cephalopoda of the North Atlantic: the family Cycloteuthidae. *Smithsonian Contributions to Zoology 5*, 1–24 [5, pls 3, 4, 7–9].

Type data: holotype NMNH 576214.

Type locality: eastern Atlantic Ocean, 3°50′N, 2°37′W.

### COMMON NAME

Rounded Disc-fin Squid

### DISTINGUISHING CHARACTERS

Without tissue flaps on anterior mantle margin and without brachial photophores (both present in *Discoteuthis laciniosa*).

### DIAGNOSTIC FEATURES

Mantle without 'tail'. Anterior mantle margin smooth, without flaplike papillae. Fins large, oval; length equal to ML in subadults. At 16 mm ML fins ~two-thirds ML and reach ~98% ML in squid of 45 mm ML. Groove of funnel-locking apparatus diagonal. Arms weak, approximately subequal, 60–85% ML. Large arm suckers with few, low, broad, round teeth, mostly on distal margin. Large suckers on club manus with numerous small, flattened to rounded teeth around entire margin of inner ring. Medial manal suckers greatly enlarged. Suckers on dactylus tetraserial.

*Photophores.* At least four photophores on margin of eyelid. A single photophore on ventroposterior tip of mantle. No photophores on brachial crown.

### HABITAT AND BIOLOGY

This squid has never been observed alive and little is known about its biology and ecology.

### SIZE

Mantle length to 90 mm, but mature individuals are unknown.

### REMARKS

The type locality is off Ghana, West Africa. Further work is needed to determine whether the specimens from Australian waters that are attributed to this species are actually *D. discus*. Characters listed above refer to *D. discus*.

### DISTRIBUTION

Tropical and subtropical Atlantic, Pacific and Indian Oceans.

### REFERENCES

Young (2009a), Roper and Jereb (2010i).

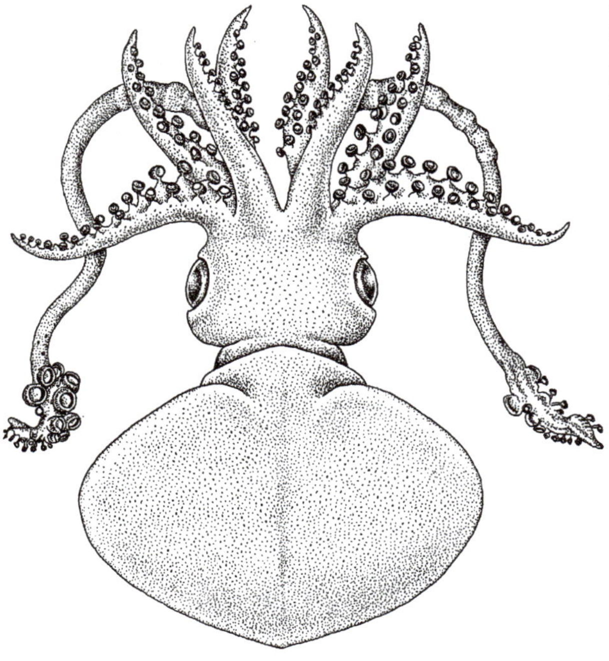

Dorsal view (ABRS, K. Hollis).

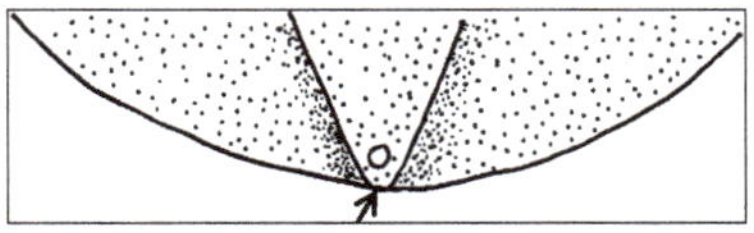

Ventral view of posterior tip of mantle, showing photophore (arrow) (A. Reid, redrawn from Young 2009a).

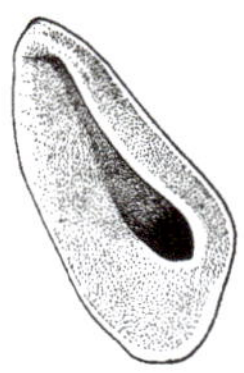

Funnel-locking cartilage (ABRS, K. Hollis).

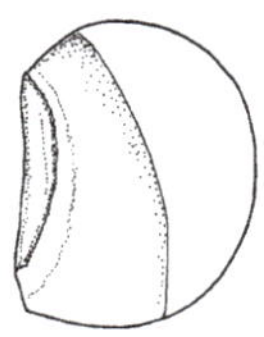

Side view of large manus sucker (A. Reid, redrawn from Young 2009a).

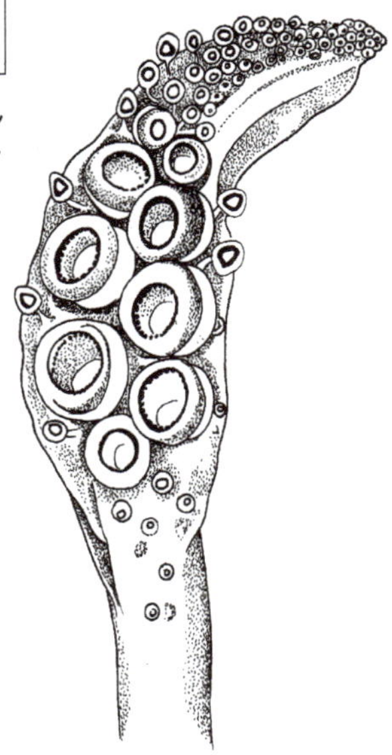

Tentacular club (ABRS, K. Hollis).

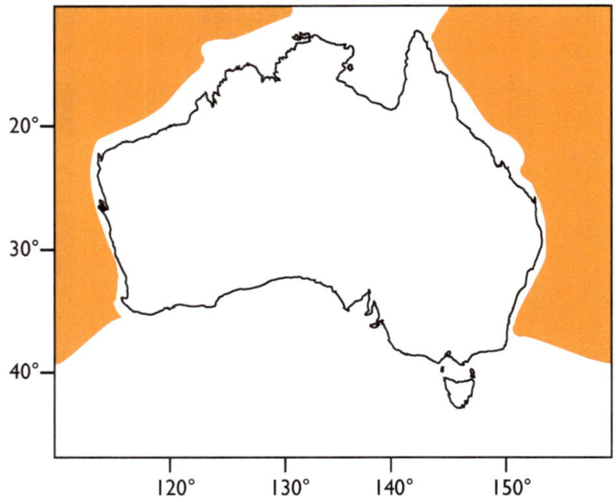


*Discoteuthis discus* distribution.

## SIZE
Up to 70 mm ML.

## REMARKS
*Discoteuthis laciniosa* can be distinguished from *D. discus* by the arrangement of the photophores. The paralarvae of this species are easily recognised by the wide separation of the fins due to the broad gladius and by the small flaps on the mantle margin. Mature squid are unknown.

## DISTRIBUTION
Tropical, subtropical Atlantic, Pacific and Indian Oceans.

## REFERENCES
Young (2006), Young (2009a), Roper and Jereb (2010i).

# *Discoteuthis laciniosa* Young & Roper, 1969

*Discoteuthis laciniosa* Young, R. E. and Roper, C. F. E. (1969). A monograph of the Cephalopoda of the North Atlantic: the family Cycloteuthidae. *Smithsonian Contributions to Zoology* 5, 1–24 [9, pls 5–9].

Type data: holotype NMNH 576215.

Type locality: eastern Atlantic Ocean, 33°04′N, 64°37′W.

## DISTINGUISHING CHARACTERS
Flaplike papillae on free anterior margin of mantle (unlike *D. discus*). Medial suckers on manus of club flat relative to *D. discus* (inner ring height ~35% of diameter compared to 60% of diameter in *D. discus*). Photophores on brachial crown.

## DIAGNOSTIC FEATURES
Funnel-locking apparatus with V-shaped groove. Arm 3 suckers large with 10–12 low, scalloped teeth on distal margin. Tentacular club with three series of suckers on distal half of club dactylus. Largest club suckers with numerous (50–60) minute, irregular, low, scalloped teeth around entire ring. Gladius with conus.

*Photophores.* Small photophores on brachial crown on membrane connecting arms 3 and 4. Small pair of photophores on anterioventral margin of mantle.

## HABITAT AND BIOLOGY
In Hawaiian waters, *D. laciniosa* is known to occupy mesopelagic depths during the day and migrates to epipelagic water at night.

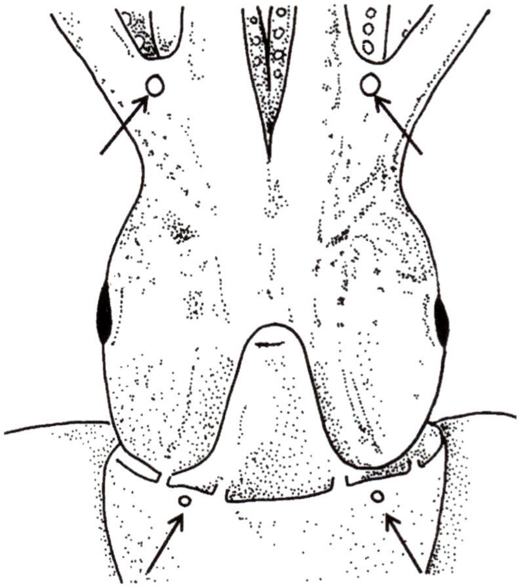

Ventral view of head, anterior mantle and arm webs showing photophores (arrows) and small flaps on mantle margin (A. Reid, redrawn from Young 2006).

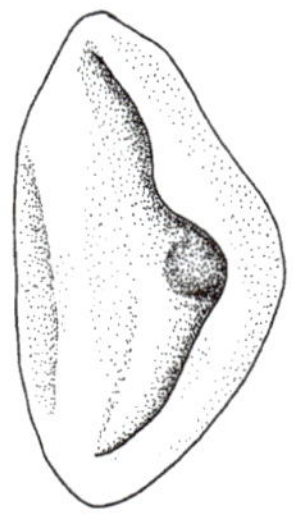

Funnel-locking cartilage (A. Reid, redrawn from Young 2006).

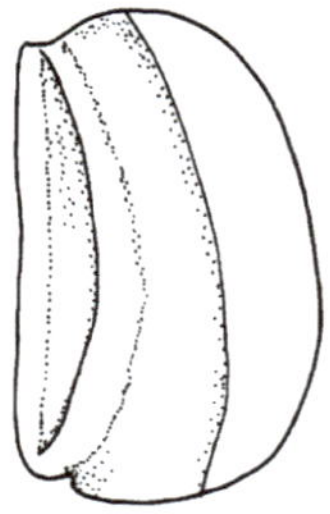

Side view of large manus sucker (A. Reid, redrawn from Young 2006).

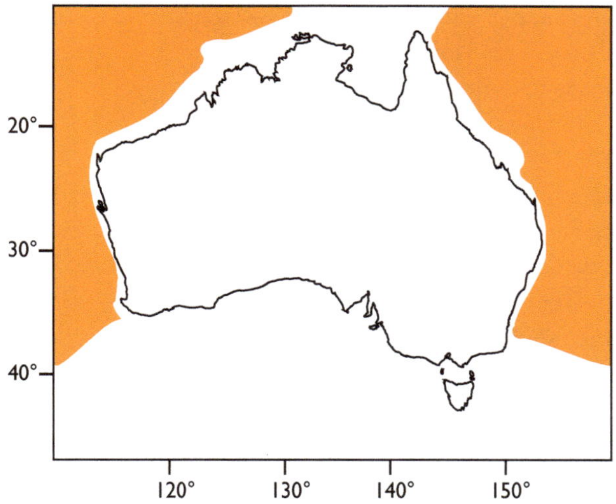


*Discoteuthis laciniosa* distribution.

## *Enoploteuthidae* Pfeffer, 1900

### COMMON NAME(S)

Firefly Squids, Enope Squids, Armed Squids

### DIAGNOSTIC FEATURES

Small- to medium-sized squids with only moderately muscular or gelatinous mantles. Fins usually terminal; shape varies; always lack posterior lobes. Tail fleshy, extends well beyond conus of gladius. Nuchal folds present. Funnel-locking apparatus simple, straight. Arm suckers biserial with some strongly developed hooks on all arms. One or both ventral arms hectocotylised in males (with considerable variation between species). Clubs with one or two series of hooks on manus (arrangement varies with genus); manus with biserial, or triserial suckers. Buccal crown with eight supports. Buccal connectives attach to dorsal border of ventral arms. No secondary buccal connectives attached to ventral margins of arms 1 and 2. Nidamental glands absent; oviducal glands enlarged in some genera (*Abralia* and *Abraliopsis*); oviducts unpaired. Ink sac not embedded in digestive gland.

*Photophores.* Present in adults of all genera (arrangement, size and number of photophores are generic and species-specific traits). Photophores absent from tentacles, viscera, and most of fins. Photosensitive vesicles present.

### HABITAT AND BIOLOGY

Enoploteuthids are epi- to mesopelagic tropical and subtropical oceanic squids that undergo extensive diel vertical migrations. Paralarvae and juveniles are epipelagic and inhabit the upper 200 m.

Some females spawn individual eggs into the plankton in continental slope and deeper continental shelf waters and lack the jelly-like egg masses characteristic of some other oegopsids. Female *Abralia* and *Abraliopsis* that mature at ~50 mm ML are capable of producing between 10 000 and 20 000 eggs, typically with a maximum length of 1.2 mm. Ripe eggs of mature *Ancistrocheirus* are large, typically reaching 3 mm long.

*Interest to fisheries.* They are the most commonly caught squids in midwater trawls, but enoploteuthids are not the subject of targeted fisheries in Australia.

### SIZE

Generally < 150 mm ML.

### REMARKS

The family is represented in Australian waters by an abundance of small- to medium-sized (< 150 mm ML) mesopelagic species. Together with Ancistrocheiridae and Pyroteuthidae, the Enoploteuthidae form a group known as the 'Enoploteuthid group' (see 'Remarks' under Ancistrocheiridae).

The three enoploteuthid genera are found in Australian waters: *Abralia*, *Abraliopsis* and *Enoploteuthis*. Young *Abralia* and *Enoploteuthis* are not easily distinguished: the presence of two photophores on the developing eyes of *Enoploteuthis v.* three enlarged photophores on *Abralia* seems to apply for most species but not necessarily all. *Abraliopsis* species, from a very small size, generally differ from other related genera due the development of the bulbous photophores on the tips of the ventral arms and by relatively longer arms (greater than or equal to the ML) and tentacles. Characters useful for species discrimination include chromatophore patterns, size of largest club suckers *v.* arm suckers, relative sizes among club suckers, number of club suckers and hooks, photophore patterns, photophore sizes and hectocotylus structure.

### DISTRIBUTION

Inhabitants of the tropical, subtropical and temperate oceans of the world.

### REFERENCE

Roper and Jereb (2010j).

### Key to genera of the Enoploteuthidae (after Dunning 1998a)

1a. Eight to 12 photophores on ventral side of eyeball, two marginal ones large, others small and abutting one another. Posterior end of mantle in adults developed into soft, conical tail extending beyond edge of fins............................................................ *Enoploteuthis*

1b. Five to eight unequal-sized, separate light organs on ventral side of eyeball. Fins reach to posterior end of mantle............................................................................ 2

2a. One or more enlarged photophores on distal tip of arms 4 (> 3 to 5 mm ML specimens); appear like small black balls in preserved specimens. Five round light organs on ventral side of eyeball, both marginal light organs larger than median ones, which are equal-sized. Buccal membrane dark violet. Right ventral arm 4 hectocotylised. Tentacular club with two rows of hooks ...................................... *Abraliopsis*

2b. No large, globular photophores on distal tips of arms 4 (silvery light organs may be present). Five to 12 light organs of different size and structure on ventral side of eyeball (> 2.0 to 2.5 mm ML). Buccal membrane pink. Left (sometimes right) ventral arm 4 hectocotylised. Tentacular clubs with a single row of hooks ............................................................. *Abralia*

## *Abralia* Gray, 1849

### DISTINGUISHING CHARACTERS

Enoploteuthid without large black photophores at the tips of arms 4 (except in *Abralia astrolineata*; however, they are smaller than those of *Abraliopsis*). Fins extend to posterior tip of tail.

### DIAGNOSTIC FEATURES

Fins terminal. Suckers present distally on arms 4. Large arm hooks with closed apertures. Right or left arm 4 hectocotylised. Manus of tentacular club with one series of hooks, and biserial suckers. Buccal crown with typical chromatophores on oral surfaces, no other pigmentation present, only occasionally with silvery photophores. Spermatophore receptacle(s) located under collar, in median pocket at posterior end of nuchal cartilage, or between stellate ganglia. Radula with homodont teeth.

*Photophores.* Five to 12 photophores of various sizes on eyeball. Enlarged photophores on tips of arms 4 generally absent; when present, photophores not covered by black chromatophores (except in *A. astrolineata*). Complex photophores of integument without red colour filters. In live animals, integumental photophores are of three types: (1) 'lensed' with blue centre ringed by white; (2) 'simple', small and violet coloured; (3) complex photophores with green or blue centres and small green points in darker outside ring.

### HABITAT AND BIOLOGY

*Abralia* are epipelagic to mesopelagic, associated with shallow bottoms on steep slopes. They ascend from daytime depths of ~600 m to less than 200 m at night. Some species may occur at shallow depths when they inhabit waters where the mesopelagic zone intersects slopes of land masses – for example, of oceanic islands (the mesopelagic-boundary zone).

### REMARKS

*Abralia* is the most speciose genus in the Enoploteuthidae. Species are generally placed into one of six subgenera (which may not all be natural groups so the subgeneric names are not listed with the taxon names, but are referred to in the remarks for each species).

### DISTRIBUTION

Worldwide in tropical and subtropical waters.

### REFERENCES

Young and Tsuchiya (2009), Roper and Jereb (2010j).

### Key to *Abralia* species

1a. One type of ocular photophore and large, unpigmented caudal and arm tip 4 photophores............... 2

1b. Two types of ocular photophores and no large, unpigmented caudal tip photophores........................ 3

2a. Distinct longitudinal stripes of photophores on ventral mantle and head. Photophores on tips of arms 4 pigmented in young animals ........................... .........................................*A. (Astrabralia) astrolineata*

2b. Indistinct longitudinal stripes of photophores on ventral mantle and head. Photophores on tips of arms 4 not pigmented in young animals.................................. ............................................. *A. (Astrabralia) astrostricta*

3a. Left arm 4 hectocotylised in males. Tentacular club with < 5 hooks............. *A. (Heterabralia) andamanica*

3b. Right arm 4 hectocotylised in males. Tentacular club with 5–7 hooks .................................................... 4

4a. One large isolated white photophore in funnel groove on each side of funnel near base of first nuchal fold .................................. *A. (Abralia) spaercki*

4b. No large white photophores in funnel groove ........... ....................................................... *A. (Abralia) armata*

## *Abralia andamanica* Goodrich, 1896

*Abralia andamanica* Goodrich, E. S. (1896). Report on a collection of Cephalopoda from the Calcutta Museum. *Transactions of the Linnaean Society of London, (series 2, Zoology)* 7(1), 1–24 [9, pl 2, figs 38–45].

Type data: syntypes (3)?, holotype ZSI M386/1.

Type locality: eastern Indian Ocean, Andaman Sea.

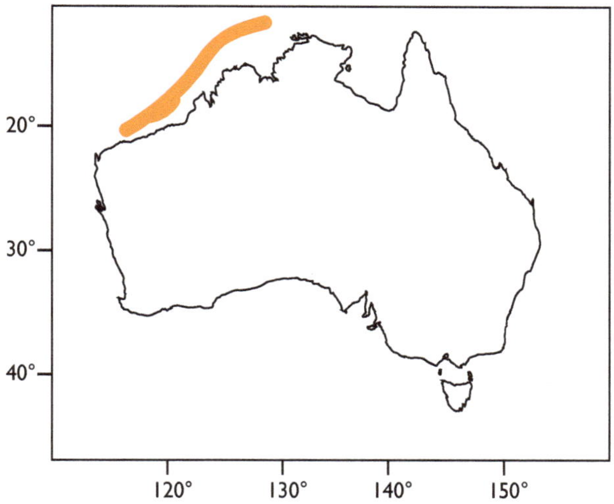


*Abralia andamanica* distribution. (Australian records are based on few specimens, but the species is likely to occur more widely.)

### DISTINGUISHING CHARACTERS

Tentacular club with <5 hooks.

### DIAGNOSTIC FEATURES

Posterior mantle with very long conical tail. Two types of ocular photophores. Distal suckers on arms 1–3 biserial. Left ventral arm of male hectocotylised: two differently sized offset flaps. Tentacular clubs with two or three hooks on ventral side; two rows of large suckers on dorsal side of manus.

*Photophores.* Five complex photophores around eyes: two large terminal opaque organs and three intermediate silvery organs. Ventral mantle and head integument with numerous scattered photophores. No large, unpigmented caudal and arm 4-tip photophores. Ventral attachment of buccal connective to arm 3.

### HABITAT AND BIOLOGY

A mesopelagic-boundary species that ascends from the mesopelagic zone to 100–300 m at night.

### SIZE

Mantle length to 60 mm.

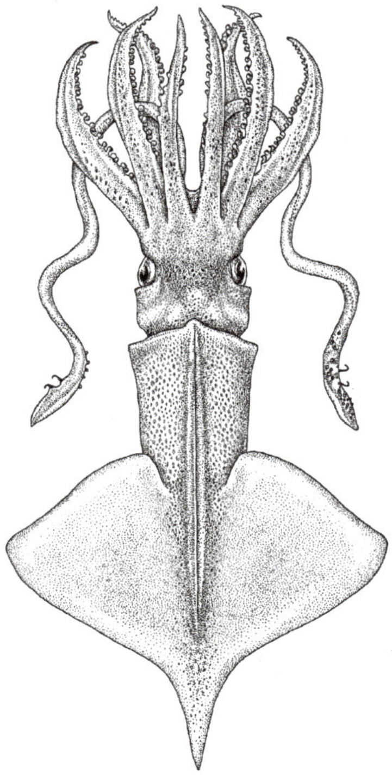

Dorsal view (ABRS, K. Hollis).

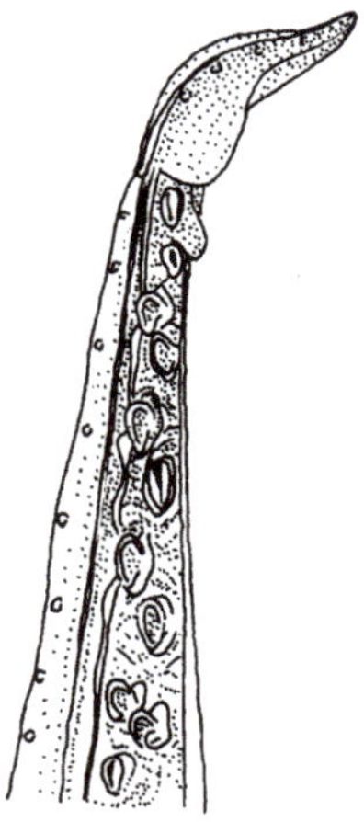

Left ventral arm (M. Shea, redrawn from Voss 1963).

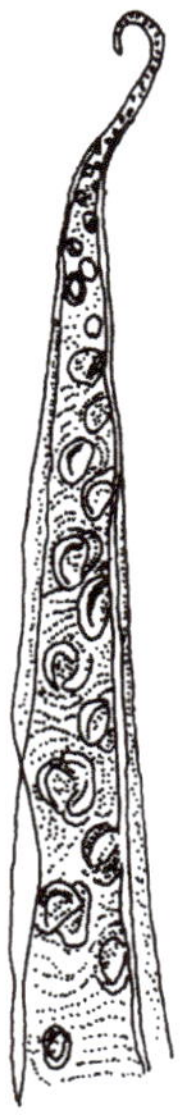

Right arm 2 (M. Shea, redrawn from Voss 1963).

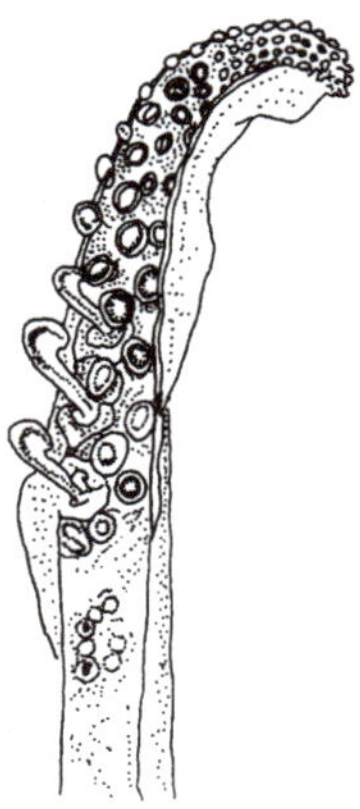

Tentacular club (M. Shea, redrawn from Voss 1963).

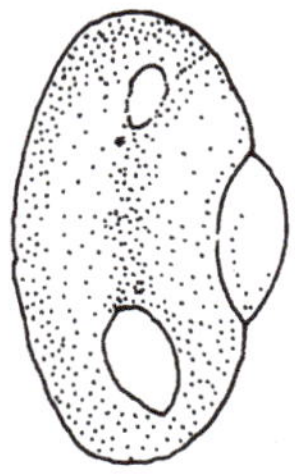

Ventral view of left eyeball (M. Shea, redrawn from Voss 1963).

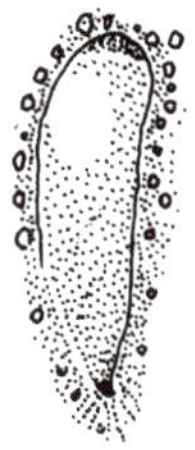

Left eyelid (below) (M. Shea, redrawn from Voss 1963).

#### REMARKS

This taxon as currently recognised may, in fact, represent several closely related species. It is placed in the subgenus *Heterabralia*.

#### DISTRIBUTION

Indian and western Pacific Oceans: off Australian North West Shelf, Arabian Sea and Seychelles, Philippines, Indonesia, off Japan to Tasman Sea, East China Sea, South China Sea, Yellow Sea.

#### REFERENCES

Tsuchiya (2009a), Roper and Jereb (2010j).

### *Abralia armata* (Quoy & Gaimard, 1832)

*Onychoteuthis armatus* Quoy, J. R. and Gaimard, J. P. (1832). Mollusques. *Voyage de decouvertes de l'Astrolabe pendant les annees 1826-1827-1828-1829, Zoologie 2*(1), 1–320 [84, pl 5 figs 14–22].

Type data: holotype MNHN 2-14-1943 (2-14-607 gladius).

Type locality: Indonesia, near Sulawesi [Celebes] Island.

#### COMMON NAME

Armed Enope Squid

#### DIAGNOSTIC FEATURES

Fin length 50–67% ML, width >83% ML. Tentacular club with six hooks. Right ventral arm hectocotylised in males.

*Photophores.* Photophores on ventral side of mantle distributed diffusely, without photophore-free median longitudinal band. Six (rarely five or seven) photophores on ventral side of eyeball. Photophore on ink sac.

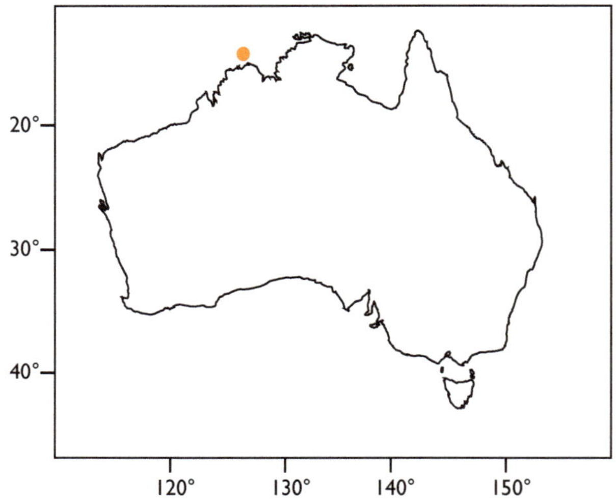


*Abralia armata* distribution. (Australian records are based on few specimens, but the species is likely to occur more widely.)

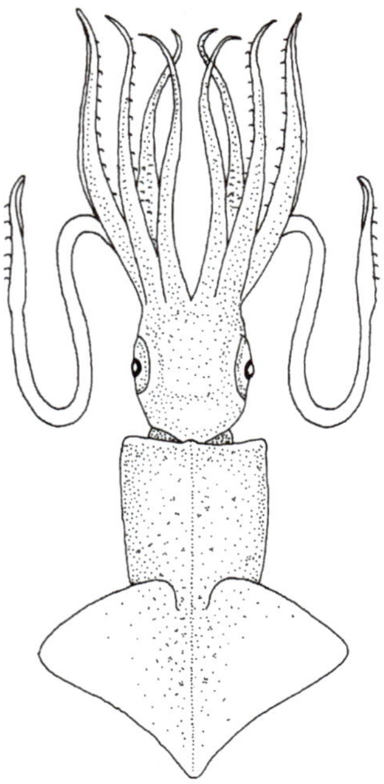

Dorsal view (M. Shea, redrawn from Voss, 1963).

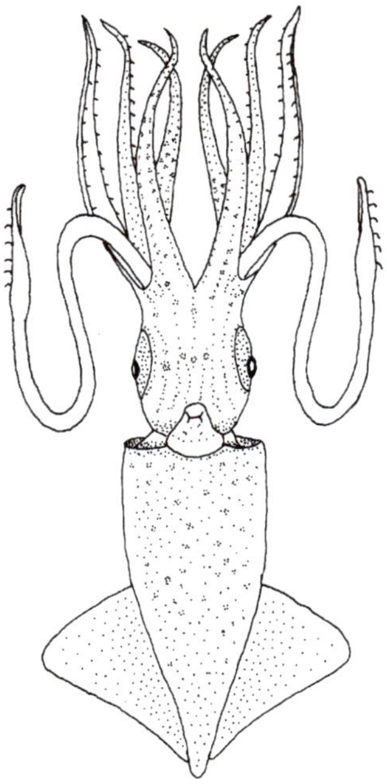

Ventral view (M. Shea, redrawn from Voss, 1963).

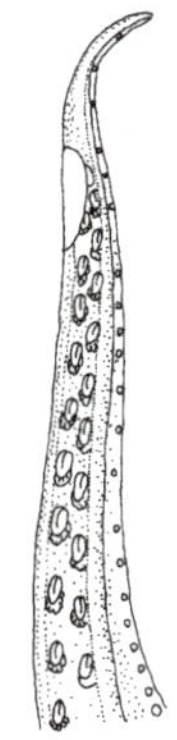

Right ventral arm of male (M. Shea, redrawn from Voss, 1963).

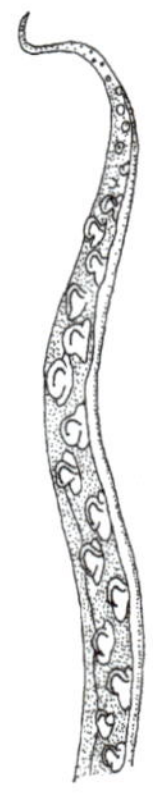

Dorsolateral arm of female (M. Shea, redrawn from Voss, 1963).

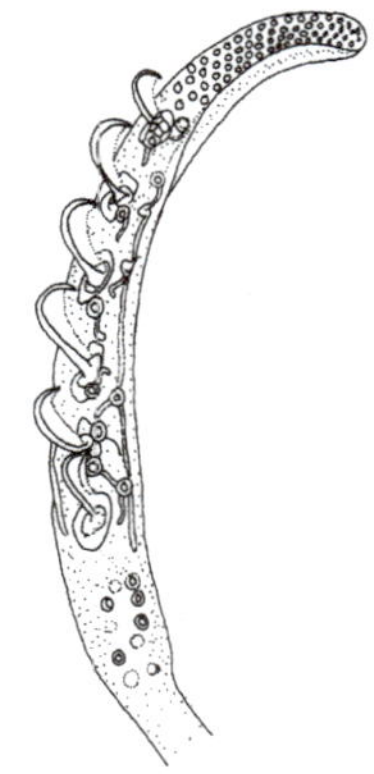

Tentacular club (M. Shea, redrawn from Voss, 1963).

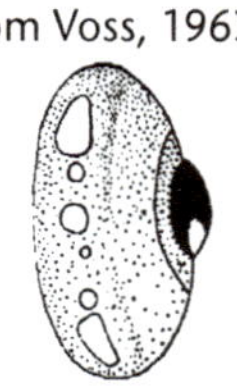

Ventral view of right eyeball with photophores (M. Shea, redrawn from Voss, 1963).

**SIZE**
Mantle length to ~20 mm.

**DISTRIBUTION**
Indo-Pacific Ocean: northern Australia, Indonesia, Philippines, Moluccas Islands, South China Sea.

**REFERENCES**
Nesis (1982/1987), Roper and Jereb (2010j).

## *Abralia astrolineata* (Berry, 1914)

*Abralia astrolineata* Berry, S. S. (1914). Notes on a collection of cephalopods from the Kermadec Islands. *Transactions and Proceedings of the New Zealand Institute 46*(24), 134–149 [145, pl 10 figs 1, 2].

Type data: holotype NMNH 816428.

Type locality: western South Pacific Ocean, Kermadec Islands.

**DISTINGUISHING CHARACTERS**
Four distinct longitudinal stripes of photophores with broad spaces between them. Large photophores present on tips of arms 4 similar to those seen in *Abralia. astrostricta*. The arrangement is similar to that seen in *Abraliopsis* except they are smaller. In young *A. astrolineata* the photophores on the tips of arms 4 are pigmented, unlike those of *A. astrostricta*.

**DIAGNOSTIC FEATURES**
Suckers absent from distal tips of arms 4 opposite photophores. Right ventral arm of male hectocotylised: subequal protective membranes, flap on ventral margin wider than that on dorsal margin, widening begins in a more proximal position on ventral than dorsal side of arm. Tentacular clubs with approximately four hooks on ventral side; two rows of large suckers on dorsal side of manus.

*Photophores.* Eyes with five large silvery organs of subequal size. Ventral mantle with four distinct

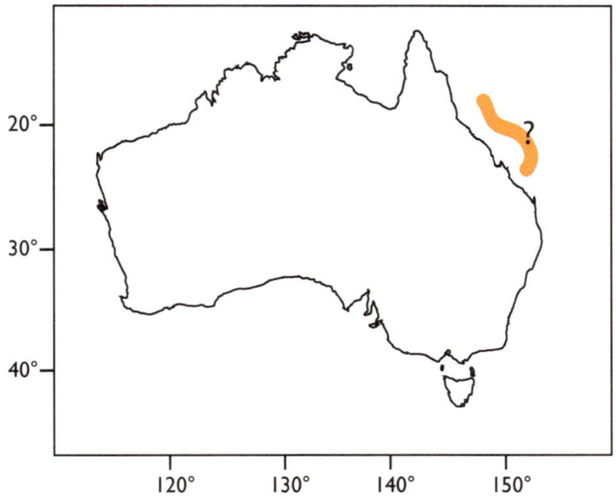


*Abralia astrolineata* distribution. (Australian records are based on few specimens, but the species is likely to occur more widely.)

Dorsal view (K. Graham).

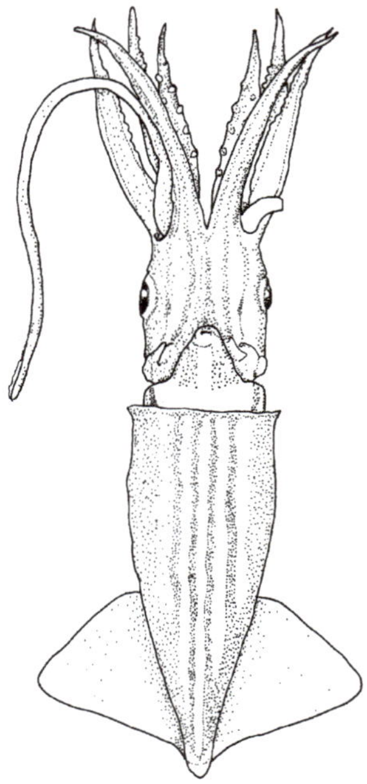

Ventral view (A. Reid, redrawn from Riddell 1985).

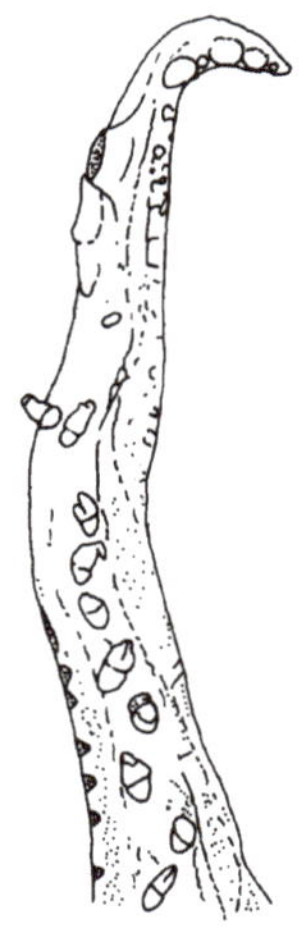

Hectocotylus (M. Shea, redrawn from Tsuchiya 2009b).

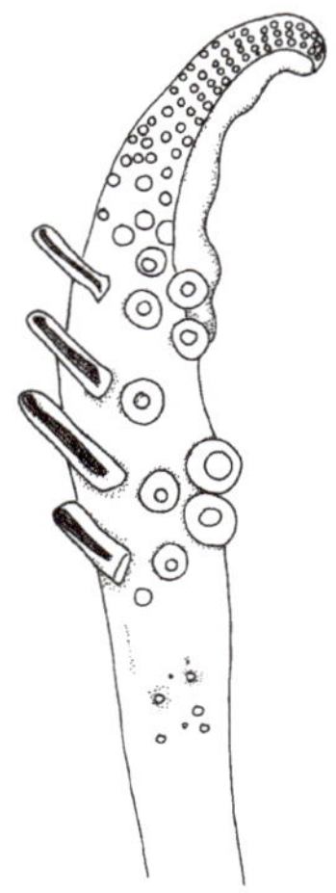

Tentacular club (A Reid, redrawn from Tsuchiya 2009b).

longitudinal stripes of photophores with broad spaces between them; ventral side of head with three distinct longitudinal stripes of photophores. Distal tips of arms 4 with four large, ovoid photophores. Ventrolateral posterior tip of mantle with pair of ovoid photophores.

### HABITAT AND BIOLOGY

A presumed mesopelagic-boundary species, with three discrete spawning periods during the year. Mature females have been caught off the Australian North West Shelf in late winter.

### SIZE

Mantle length to 100 mm.

### REMARKS

This species, together with *A. astrostricta*, belongs in the subgenus *Astrabralia.*

### DISTRIBUTION

South-western Pacific Ocean; eastern Australia (off Queensland); along slope of the Kermadec Islands.

### REFERENCES

Nesis (1982/1987), Lu and Dunning (1998), Tsuchiya (2009b), Roper and Jereb (2010j).

## *Abralia astrostricta* Berry, 1909

*Abralia astrostricta* Berry, S. S. (1909). Diagnosis of new cephalopods from the Hawaiian Islands. *Proceedings of the United States National Museum, 37*(1713), 407–419 [412, figs 4–7].

Type data: holotype NMNH 214313.

Type locality: Hawaiian Islands, off Oahu Island.

### DISTINGUISHING CHARACTERS

Large photophores present on tips of arms 4 similar to those seen in *A. astrolineata*; arrangement similar to that seen in *Abraliopsis* except they are smaller and lack the black chromatophores that surround the photophores in the latter genus.

### DIAGNOSTIC FEATURES

Relatively large species in genus. Arm suckers absent from distal tips of arms 4 opposite photophores in both sexes. Right ventral arm of male hectocotylised; protective membranes well developed (much more pronounced on ventral side than dorsal side of arm). Tentacular clubs with about four hooks on ventral side; manus with two rows large suckers on dorsal side.

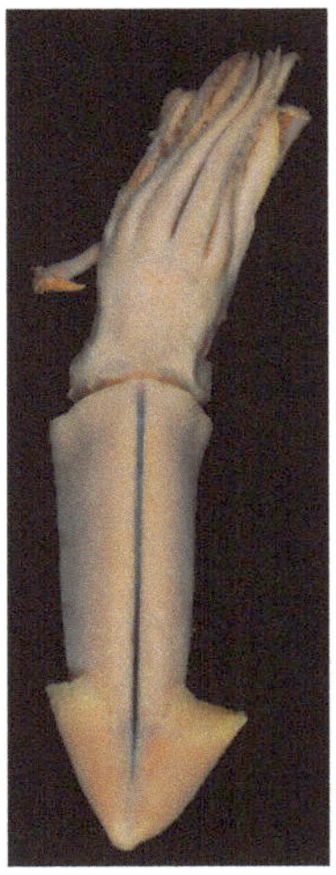

Preserved museum specimen, dorsal view (AM C.131811, A. Reid).

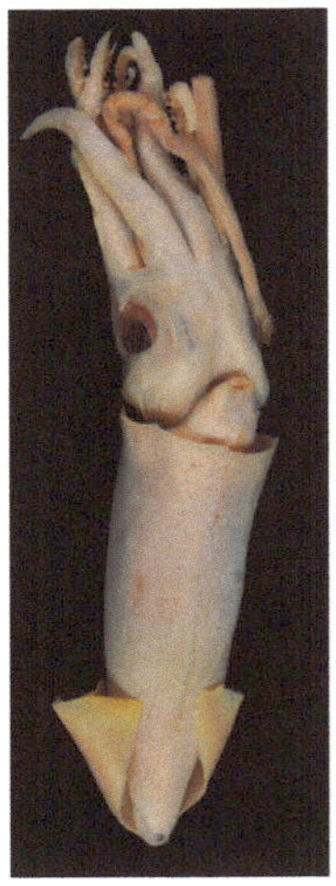

Preserved museum specimen, ventral view (AM C.131811, A. Reid).

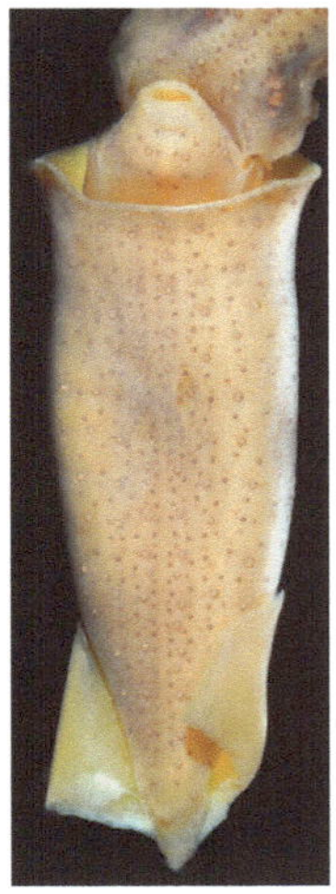

Preserved specimen showing ventral photophores © Richard E. Young (Tsuchiya and Young 2009).

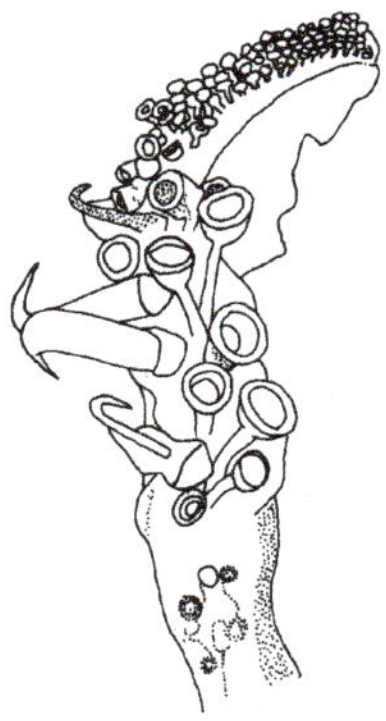

Tentacular club, oral view (M. Shea, redrawn from Tsuchiya and Okutani 1988).

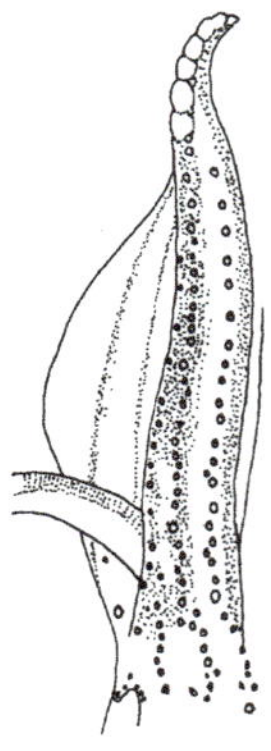

Arm 4 showing terminal photophores, ventral view (M. Shea, redrawn from Tsuchiya and Okutani 1988).

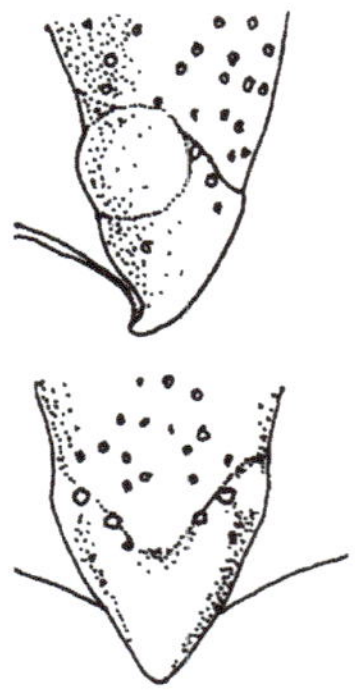

Caudal photophores (M. Shea, redrawn from Tsuchiya and Okutani 1988).

*Photophores.* Indistinct longitudinal stripes of photophores on mantle (scattered arrangement) and head (three indistinct longitudinal stripes); eyes with five large, subequal, silvery photophores. Arm 4 with four ovoid photophores at distal tip that are not encased in chromatophores. Ventrolateral posterior tip of mantle with pair of ovoid photophores buried in tissue. Paralarvae with very large integumental photophores. Caudal photophores visible in paralarvae.

### HABITAT AND BIOLOGY

This is a widely distributed tropical, subtropical mesopelagic-boundary species. The species is thought to inhabit near-bottom waters during the day at upper mesopelagic

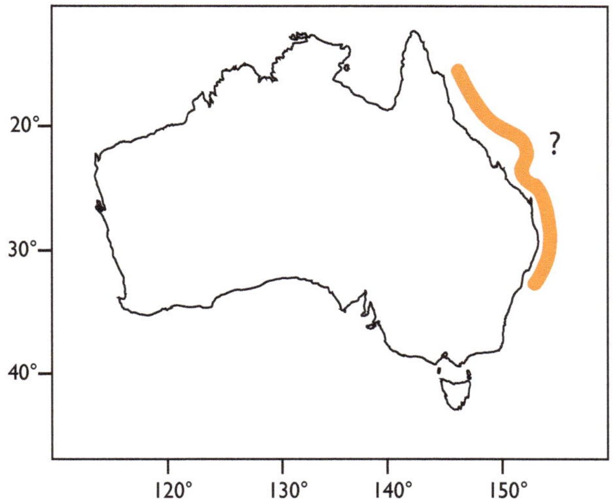


*Abralia astrostricta* distribution. (Australian records are based on few specimens, but the species is likely to occur more widely.)

or shallower depths, moving into shallower waters at night.

#### SIZE

Mantle length to 70 mm.

#### REMARKS

This species, together with *A. astrolineata*, belongs in the subgenus *Astrabralia.*

#### DISTRIBUTION

North-eastern and north-western Australia. Widely distributed throughout the central and western Pacific Ocean including New Zealand, Hawaiian Islands to Philippine Sea, off Japan, Kuroshio Current, South China Sea.

#### REFERENCES

Tsuchiya and Young (2009), Roper and Jereb (2010j).

### *Abralia spaercki* Grimpe, 1931

*Abralia spaercki* Grimpe, G. (1931). Teuthologische Mitteilungen. XIII. Uber die Cephalopoden der Sunda-Expedition Rensch. *Zoologischer Anzeiger* 95(5/8), 149–174 [150].

Type data: holotype, ZMUC.

Type locality: Indonesia, Banda Sea, near Amboina.

#### DISTINGUISHING CHARACTERS

Unusual photophores in funnel groove.

#### DIAGNOSTIC FEATURES

Right ventral arm of male hectocotylised; protective membranes flap-like, offset proximally, differ in size; proximal part of modified portion with hooks. Tentacular club with five to seven hooks on ventral side; two rows of large suckers on dorsal side of manus.

*Photophores.* Eyes with five complex organs: two large opaque organs and three intermediate silvery organs; some minute additional organs present. Ventral mantle and head with sparse, scattered photophores. Funnel groove with large white photophores posteriomedial to inner nuchal fold.

#### HABITAT AND BIOLOGY

Demersal mesopelagic boundary species associated with continental or island shelves.

#### SIZE

Mantle length to 50 mm.

#### REMARKS

Based on the drawings in Tsuchiya (2009c), the arrangement of suckers on the tentacular club appears to differ among specimens (animals from the Philippines and northern Australia appear to differ). In addition, the number and arrangement of photophores on the eye in northern Australian specimens appears to differ from the

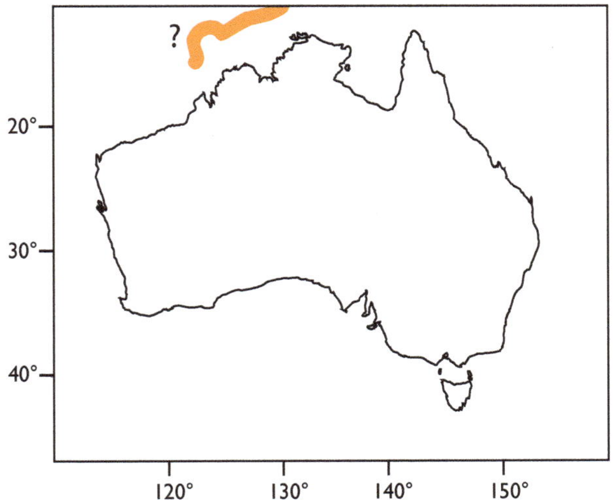


*Abralia spaercki* distribution. (Australian records are based on few specimens, but the species is likely to occur more widely.)

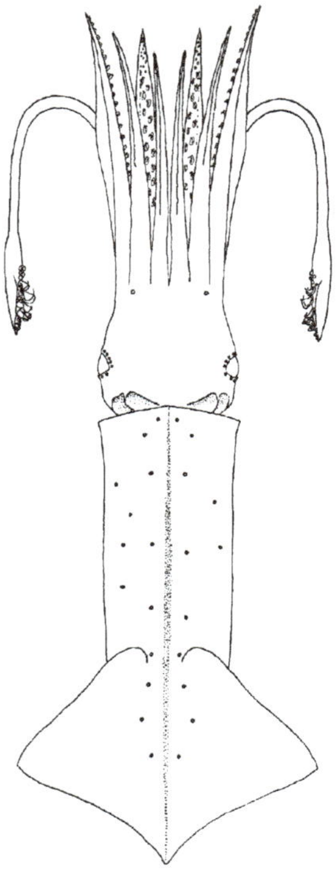

Dorsal view (M. Shea, redrawn from Voss 1963).

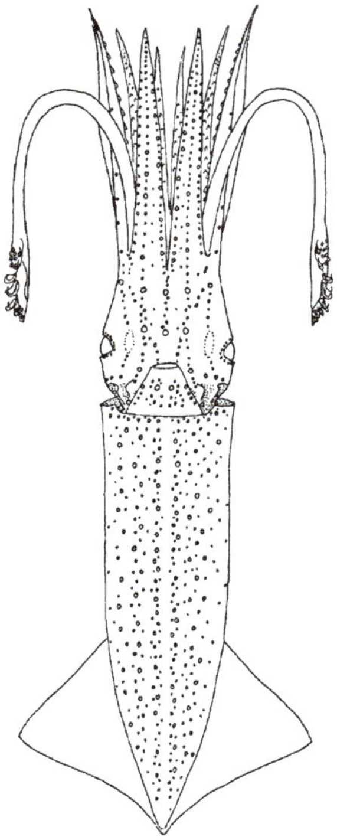

Ventral view (M. Shea, redrawn from Voss 1963).

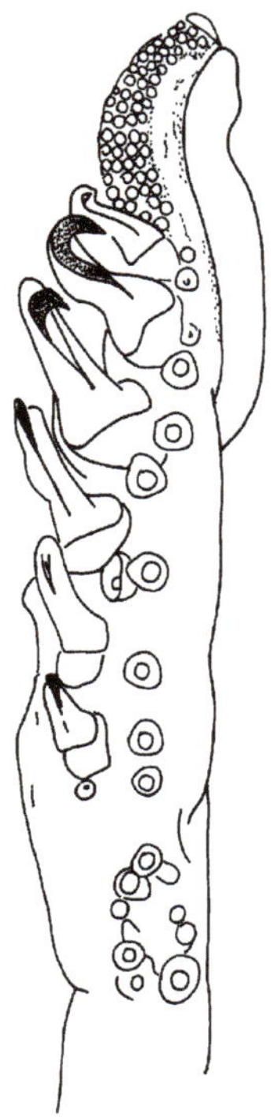

Tentacular club (A. Reid, redrawn from Tsuchiya 2009c).

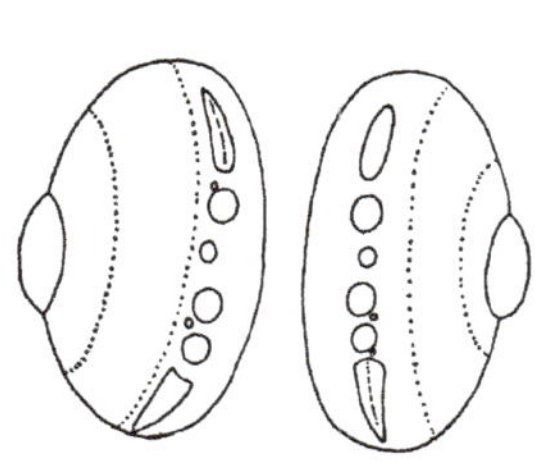

Eyes showing number and arrangement of photophores (M. Shea, redrawn from Voss 1963).

*Abralia spaercki*

holotype (from Indonesia). This may be worthy of more detailed investigation.

#### REFERENCES

Tsuchiya (2009c), Roper and Jereb (2010j).

## *Abraliopsis* Joubin, 1896

#### DISTINGUISHING CHARACTERS

Buccal membrane darkly pigmented on oral surface (rather than typical chromatophores). Hooks on manus of club biserial. Most easily recognised by the spherical black photophores (usually three) at the tips of arms 4.

#### DIAGNOSTIC FEATURES

Fins extend to posterior end of tail. Suckers at arm tips biserial. Large arm hooks with closed apertures. Suckers absent from arms 4 (only hooks present). Right arm 4 hectocotylised. Manus of club with biserial hooks, and one (usually incomplete) series of suckers. Spermatophore receptacle(s) located under collar and/or between stellate ganglia. Attachment of buccal connectives to arms DDVD (usually connected to ventral borders of arms 3). Radula with homodont teeth.

*Photophores.* Eyeball with five photophores ventrally, anterior and posterior photophores enlarged, three remaining photophores small, subequal. Two to four (mostly three) large photophores covered by black chromatophores on tips of arms 4. Three types of common integumental photophores (in life), covering arms, head, funnel and mantle. Most complex of these with red colour filters are called 'complex' or 'red photophores', other two common or 'non-complex' integumental photophores are 'blue photophores'. Largest of blue 'types' has large lens and other does not. Fourth but uncommon type of photophore is white. Photophores, such as are found in funnel groove on heads are white.

*Colour.* Buccal crown without typical chromatophores, with dark epithelial pigmentation on oral surface.

#### HABITAT AND BIOLOGY

*Abraliopsis* species occupy similar habitats as *Abralia* species, both geographically and vertically.

#### REMARKS

The black appearance of the photophores on the arm tips is due to black chromatophores that tightly cover the photophores but when they luminesce, the chromatophores withdraw from the photophores.

*Abraliopsis* species can be placed into one of four subgenera (*Abraliopsis*, *Boreabraliopsis*, *Micrabralia* and *Pfefferiteuthis*) but these may not all be natural groups. Two of these subgenera, *Abraliopsis* and *Micrabralia*, are represented in Australian waters.

Most of the information on this genus has been obtained from the most up-to-date reference at the time of writing, Young and Tsuchiya (2013) recently revised by Richard Young, based largely on Hawaiian material. The Australian representatives of this group were recently reviewed by K. Li (Wollongong University) based on specimens housed in Australian Museum and Museum Victoria collections. Three species in three subgenera were previously thought to occur in Australian waters: *A.* (*Micrabralia*) *gilchristi* (Robson, 1924), *A.* (*Abraliopsis*) *hoylei* (Pfeffer, 1884) and *A.* (*A.*) *tui* Riddell, 1895. It appears that the Australian records of *A. hoylei* (originally described from the Mascarene Islands, south-western Indian Ocean) are based only on the misidentification of some larval *Abraliopsis* collected off south eastern Australia (Allan 1945). Rexamination of the specimens cited in that paper have shown them to be *A.* (*M.*) *gilchristi*, so it does not appear that *A. hoylei* occurs in Australian waters. A third species, *A. lineata* from the Coral Sea (off far north-east Queensland) was recently found among the collections of the Australian Museum (Li and Reid, unpublished data), and this is the first record of the occurrence of this species in Australia.

#### REFERENCES

Roper and Jereb (2010j), Young and Tsuchiya (2013), Li and Reid (unpublished data).

### Key to *Abraliopsis* species

1a. Photophores scattered over entire ventral mantle........................................................................ *A. gilchristi*

1b. Mid-ventral longitudinal gap devoid of photophores, extends entire length of mantle.................................2

2a. Large photophores on ventral mantle arranged in six longitudinal series; large photophores on head arranged in three longitudinal series. Tentacular club without keel and carpal flap................ *A. lineata*

2b. Large photophores on ventral mantle, scattered, not arranged in clear longitudinal series; large photophores on head in scattered arrangement, not in linear series. Tentacular club with keel and carpal flap ........................................................................ *A. tui*

### *Abraliopsis (Micrabralia)*

#### DISTINGUISHING CHARACTERS

*Photophores.* Ventral surface of head with red and blue integumental photophores in linear pattern (thus differing from *A.* (*Abraliopsis*)), with, or without scattered blue photophores; funnel groove photophores in 'moderate' 'complex-1' pattern (see image *A.* (*M.*) *gilchristi*).

#### DIAGNOSTIC FEATURES

Arm hooks differ greatly in size between and within series. Right arm 4 of males with flaps (one large trapezoidal flap, distally offset by smaller dorsal flap, and low ventral membrane extends from ventral flap to terminal large photophore); dorsal flap of hectocotylus (right arm 4) about one-third to half the length of the ventral flap; characteristic pattern of hooks that differ greatly in size between and within series, small mostly free trabeculae with tubercles. Ventral protective membrane of left arm 4 of male variable but not greatly enlarged.

#### REMARKS

This subgenus is a 'catch-bag' of species that don't fit the other two subgenera.

#### REFERENCES

Roper and Jereb (2010j), Young and Tsuchiya (2013).

### *Abraliopsis (Micrabralia) gilchristi* (Robson, 1924)

*Abralia gilchristi* Robson, G. C. (1924). Preliminary report on the Cephalopoda (Decapoda) procured by the S.S. 'Pickle.' *Report of the Fisheries and Marine Biological Survey of the Union of South Africa 3*, 1–14 [3].

Type data: syntypes (2) BMNH 1924.9.9.41-2.

Type locality: South Africa: off Cape Town.

#### DISTINGUISHING CHARACTERS

Arms 1 and 2 (mostly arm 2) with tubercles. Distal third to half of arms 4 not bare. Arms 4 with 23–28 hooks in two rows. Similar to *Abraliopsis* (*Abraliopsis*) in presence of complex integumental photophores in linear series on the head, and similar to *A.* (*Abraliopsis*) in the absence of arm 4 modifications.

*Abraliopsis (M.) gilchristi* occurs sympatrically with *A. (A.) tui* off eastern and south-eastern Australia. The two species can be distinguished by the arrangement of photophores on the ventral side of the head, with *A. (M.) gilchristi* having photophores arranged in four rows, with a few scattered organs between the median rows but none between each median row and the row lateral to it. In contrast, *A. tui* has a uniform, scattered photophore distribution on the ventral surface of the head.

Juvenile *A. (A.) tui* first develop the photophores of the head in four rows, as in *A. (M.) gilchristi*, but the rows are never as distinct as in the latter species and there are always additional photophores between all rows, rather than just between the middle pair as in *A. (A.) tui*. The ventral mantle photophores are in broad fields, rather than distinct rows as in *A. (M.) gilchristi*.

### DIAGNOSTIC FEATURES

Median mantle sector with bare, unpigmented strip indistinct in anterior fifth of mantle. Fin width 95% ML. Arms 4 relatively short, 45–65% ML. Arms 1–4 with 17–28 hooks; only arms 1–3 with distal suckers. Male arms 1 dorsal border with small lappets, ventral border with large membrane bearing long, broad trabeculae; arms 2 dorsal margin with short trabeculae with tubercles but membranes nearly absent, ventral margin with enlarged membrane and long, flat, trabeculae, both trabeculae and membranes with tubercles; arms 3 dorsal margin with lappets only, ventral margin with narrow trabeculae and deep membrane but no papillae. In females, dorsal margins of arms 1–3 virtually without trabeculate protective membranes, membranes absent on dorsal and ventral sides of arms 4; protective membranes well developed on ventral margins of arms 1–3. Tubercles absent from trabeculae and arms. Hectocotylus (right arm 4): long ventral flap and shorter offset dorsal flap; modified portion with armature, hooks about equal size on proximal margins, medially hooks of ventral margin abruptly smaller to end

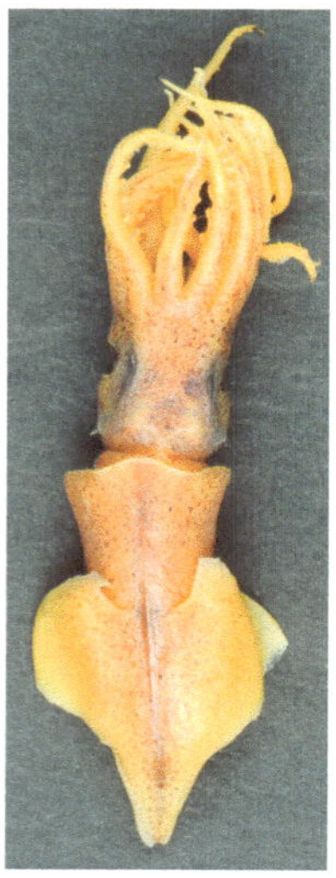

Preserved museum specimen, dorsal view (AM C.398570, L. Mott).

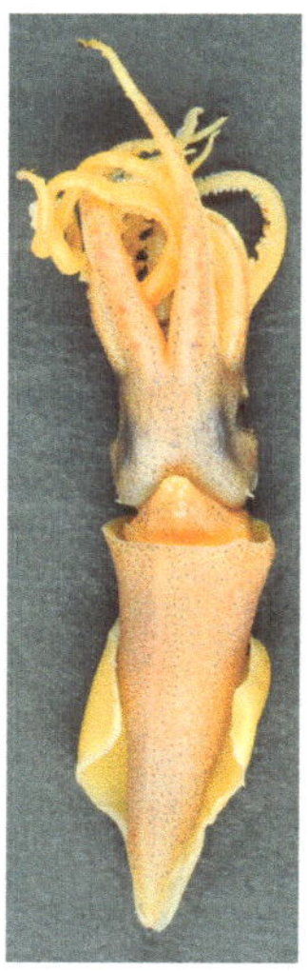

Preserved museum specimen, ventral view (AM C.398570, L. Mott).

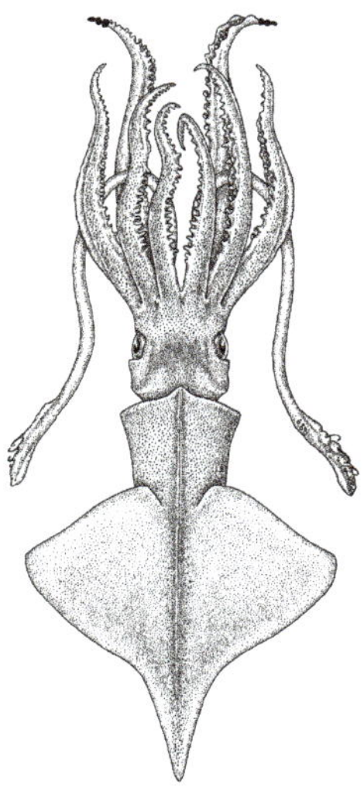

Dorsal view (ABRS, K. Hollis).

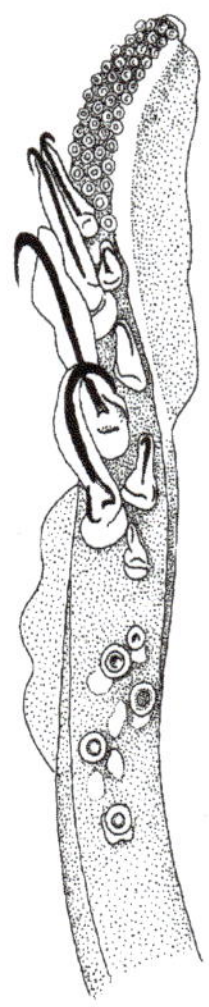

Tentacular club, oral view (M. Shea, redrawn from Voss 1967).

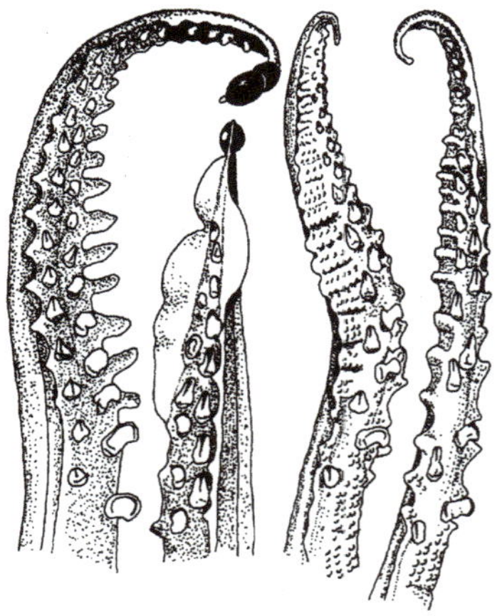

Arms, oral view, including the hectocotylised right arm 4 (M. Shea, redrawn from Voss 1967).

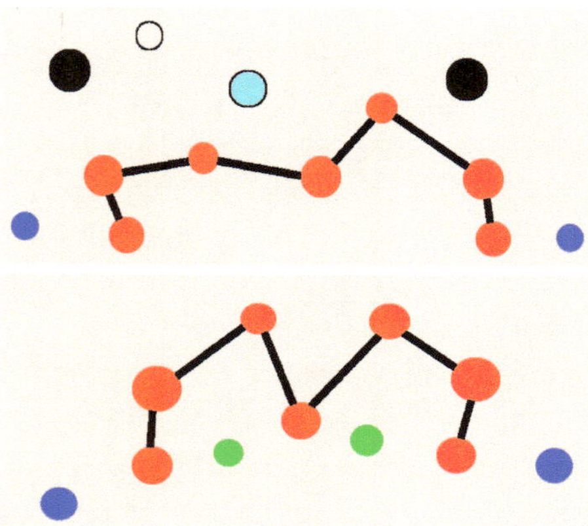

Variable arrangement of photophores in funnel groove (females).

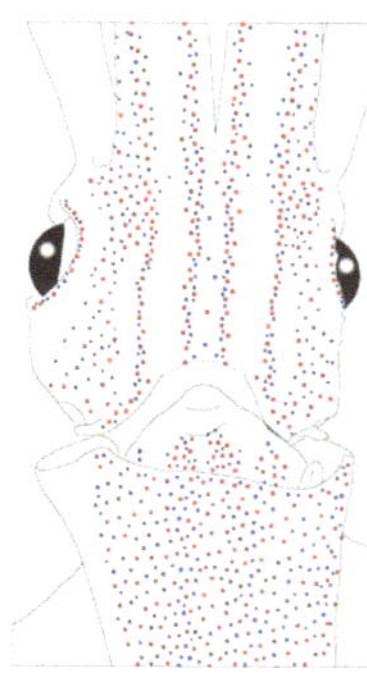

Arrangement of complex photophores (red) and non-complex photophores (blue) on ventral side of head (Young and Tsuchiya 2013).

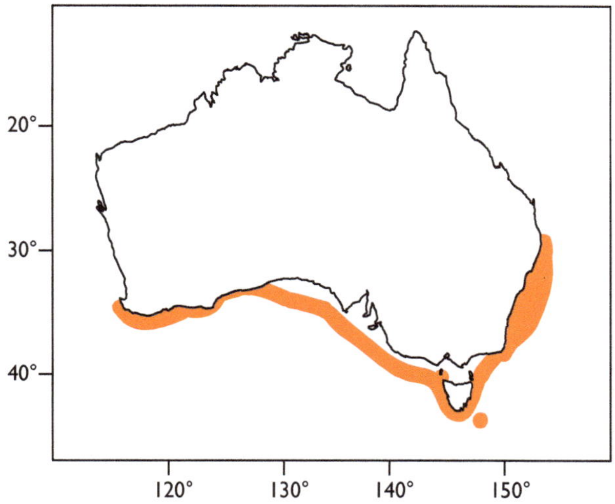


*Abraliopsis gilchristi* distribution. (Australian records are based on few specimens, but the species is likely to occur more widely.)

of arm; spatulate trabeculae on protective membrane of left arm 4; proximally dorsal hooks larger than ventral counterparts. Club keel large; carpal flap large. Large ventral club hooks > 3 × size of dorsal club hooks.

*Photophores.* Mantle with scattered red photophores. Ventral head with four series of photophores. No or few photophores in medial head sector. Five ocular photophores with end member ~2 × larger than middle member.

### HABITAT AND BIOLOGY

Probable diel vertical migrator. This species is thought to have a protracted summer spawning season. *Abraliopsis (M.) gilchristi* was shown by Brandt (1983) to be associated with a warm core eddy of the East Australian Current. The length–frequency distributions of this species were larger outside than inside the eddy. This may reflect the effect of eddies on growth, spawning times, reproduction and survival.

### SIZE

Mantle length to 40 mm.

### REMARKS

*Abraliopsis (Micrabralia) gilchristi* is one of the most easily recognised species of *Abraliopsis*. The arm features of the male are very similar to those of the subgenus *Pfefferiteuthis* and the scattered arrangement of the red photophores on the head are similar members of the subgenus *Abraliopsis*.

### DISTRIBUTION

Eastern South Atlantic Ocean; Indian Ocean, off Cape Town; off New Zealand, central South Pacific Ocean (possibly circumglobal in southern subtropical belt). Circum-southern temperate waters between 20–45°S. Southern border of the Sub-Antarctic Frontal Zone in the central South Pacific.

### REFERENCES

Roper and Jereb (2010j), Young and Tsuchiya (2013).

## *Abraliopsis (Micrabralia) lineata* (Goodrich, 1896)

*Abralia lineata* Goodrich, E. S. (1896). Report on a collection of Cephalopoda from the Calcutta Museum. *Transactions of the Linnaean Society of London, (series 2, Zoology)* 7(1), 1–24, 5 plates. [pl. 3, figs 46–50].

Type data: syntypes (2), holotype ZSI M389/1.

Type locality: Andaman Sea and off Ganjam coast.

### DISTINGUISHING CHARACTERS

Tentacular clubs without keels and carpal flaps. Largest club hook of ventral series < 1.5 × height of dorsal counterpart. Arms 1 of males with enlarged ventral trabeculae bearing tubercles. Bases of arms 2 swollen in males.

### DIAGNOSTIC FEATURES

Horn-like membrane on aboral side of eyes. Two rows of subequal sized hooks on manus of tentacular club; eight hooks in total. Dactylus with 12–14 transverse rows of suckers. Arms 4 longer than other arms in both sexes. Arms 1–3 with 14–18 hooks in females and 19–25 hooks in males; arms 4 with 10–13 hooks. Arms 1–3 with 11–23 suckers on distal tips; arms 4 without suckers. Half of inner ring of arm suckers with six or seven rectangular teeth. Arms 1 and 2 with narrow aboral keel on distal half of length; arms 3 with broad aboral keel along entire length. Arms 1–3 with protective membrane with trabeculae along ventral side only; arms 4 without trabeculae. Right arm 4 of males hectocotylised with three different types of crest: proximal crest long, quadrate and wavy, extending from distalmost hook on ventral margin to about 70% of arm length; second crest triangular, extending from middle of proximal crest to about 80% of arm length along its dorsal margin; distal crest laterally lunate, extending from distal end of proximal crest to terminal photophore along ventral margin of arm; no armature in region of flaps. Protective membrane of both arms 1 in male well-developed dorsally and ventrally; trabeculae swollen and flattened; oral surface ornamented with numerous small, granular papillae. Gills with 19–21 lamellae on each demibranch.

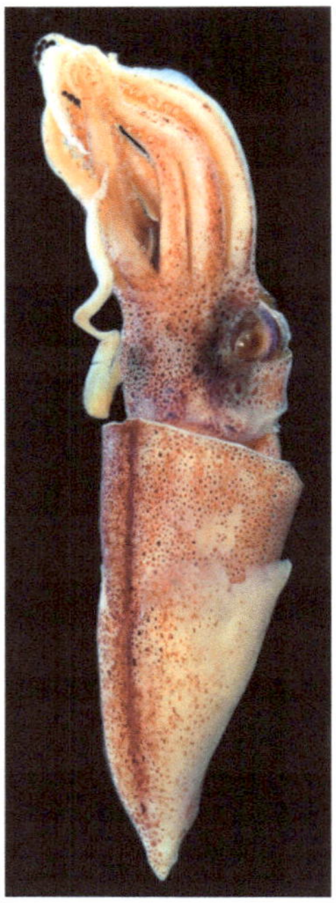

Preserved museum specimen, dorsal view (AM C.495718, S. Lindsay).

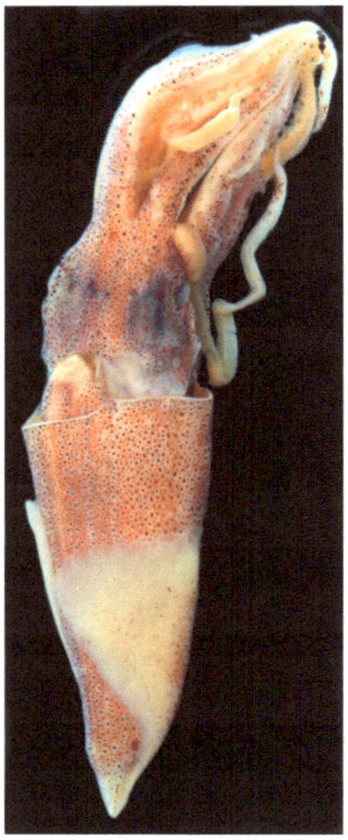

Preserved museum specimen, ventral view (AM C.495718, S. Lindsay).

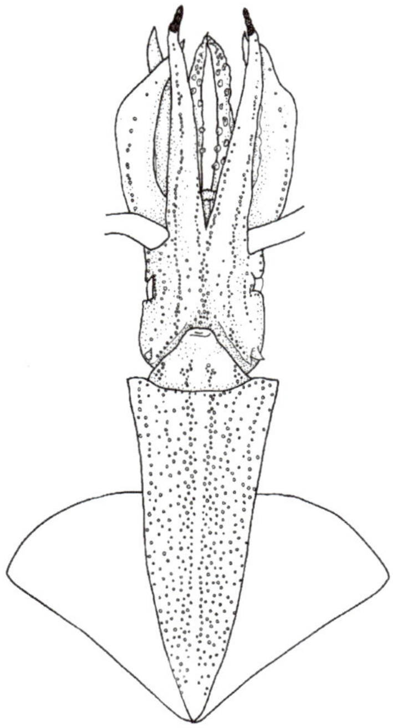

Ventral view showing photophore arrangement (A. Reid, redrawn from Tsuchiya *et al.* 1991).

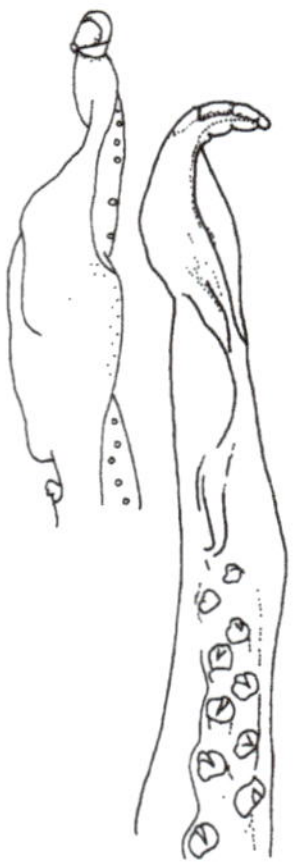

Hectocotylised arm 4, left, lateral view, right, oral view (A. Reid, redrawn from Tsuchiya *et al.* 1991).

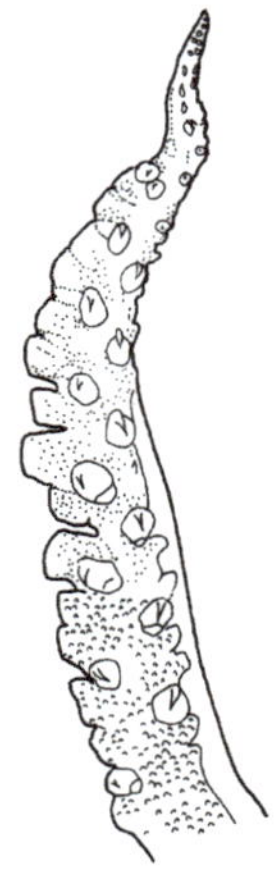

Right arm 1, oral view, male (A. Reid, redrawn from Tsuchiya *et al.* 1991).

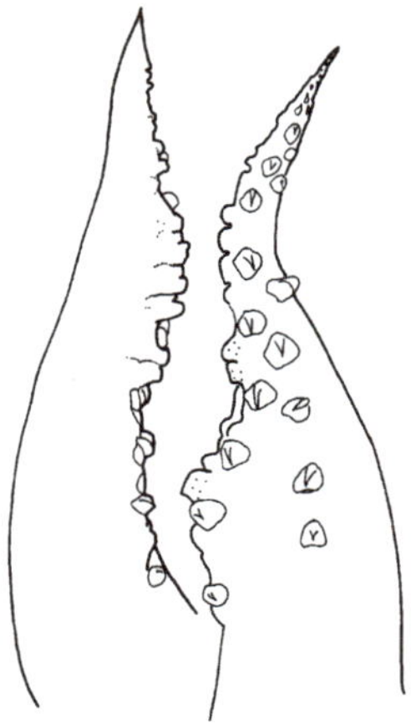

Right arm 2, left, lateral view, right oral view, male (A. Reid, redrawn from Tsuchiya *et al.* 1991).

Right arm 3, lateral view, female (A. Reid, redrawn from Tsuchiya *et al.* 1991).

Left tentacular club, male, oral view (A. Reid, redrawn from Tsuchiya *et al.* 1991).

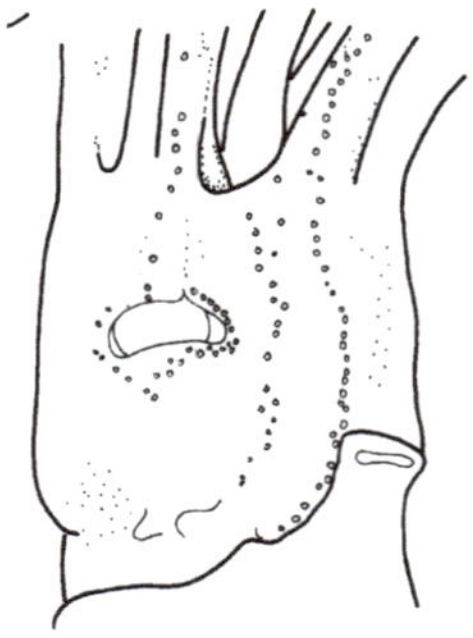

Lateral view of head showing photophore arrangement (A. Reid, redrawn from Tsuchiya *et al.* 1991).

*Photophores.* Ventral side of head with five longitudinal rows of large photophores; middle three rows extend to arms 4, those on ventral side continue to arm tips. Arms 3 with photophores along entire length. Eyes with photophores around entire margin. Ventral mantle with six longitudinal rows of large photophores, some scattered photophores between rows; narrow median longitudinal strip without photophores. Tips of arms 4 with three or four large black photophores. Funnel with four longitudinal rows of photophores.

## HABITAT AND BIOLOGY

The Australian specimens were collected in plankton trawls between 12 m and 200 m at night in water with

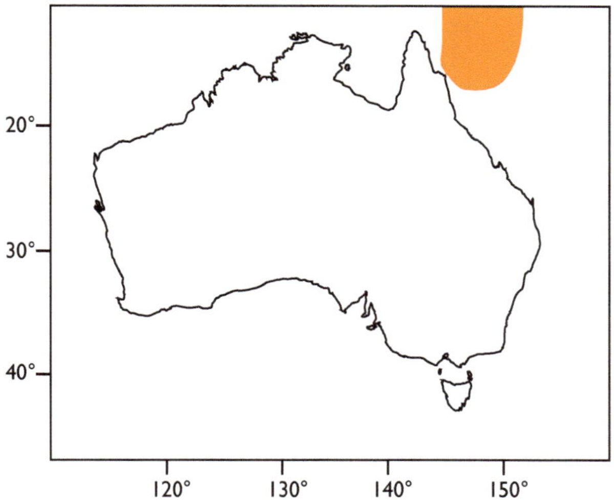


*Abraliopsis lineata* distribution. (Australian records are based on few specimens, but the species is likely to occur more widely.)

temperatures ranging from approximately 26°C to 27°C. The actual capture depth was probably shallow surface waters.

#### SIZE

Mantle length to 30 mm.

#### REMARKS

This is the first record of *A. lineata* in Australian waters.

#### DISTRIBUTION

The species is widely distributed in the Indowest Pacific. In Australian waters the species is known to occur in the Coral Sea from 9°19′S 145°20′E to 15°12′S 149°34′E, but is probably more widespread off northern Australia.

#### REFERENCES

Tsuchiya *et al.* (1991), Tsuchiya (2013c).

### *Abraliopsis (Abraliopsis)*

#### DISTINGUISHING CHARACTERS

Complex and non-complex photophores scattered, not in a regular arrangement (the exception to this is in a species that is not found in Australian waters).

#### REMARKS

The identification of species in this subgenus requires the examination of characters that are not related to the photophores.

### *Abraliopsis (Abraliopsis) tui* Riddell, 1895

*Abraliopsis tui* Riddell (1985). The Enoploteuthidae (Cephalopoda: Oegopsida) of the New Zealand region. *Fisheries Research Bulletin (N. Z. Ministry of Agriculture and Fisheries)*, *27*, 1–52, 33 figures. [37, figs 28–29].

Type data: holotype, NMNZ M.89787.

Type locality: New Zealand, Kermadec Islands.

#### COMMON NAME

Tui Squid

#### DISTINGUISHING CHARACTERS

Distinct posterior crescentric ridge connecting occipital folds 3 and 4. Ventral mantle with distinct bare, pale strip that lacks photophores. Hectocotylus lacks a dorsal flap.

#### DIAGNOSTIC FEATURES

Fins very large, length 73–79% mantle length, width 101–111% ML. Arms 4 long, 100–110% mantle length. Arms 1–4 with 17–23 hooks, arms 1–3 with ~30 suckers distally. Hectocotylus with a single long, narrow flap along ventral margin extending from level of hooks 10–15; dorsal flap absent; modified portion with armature. Tentacular club with two series of hooks on manus that differ in size; five suckers in carpal cluster; dactylus with four series of minute suckers. Aboral keel large, extends from tip of dactylus to level of second pair of hooks. Large membranous flap aborally from level of carpal cluster to level of first or second pair of hooks.

*Colour.* Buccal membrane dark, purplish.

*Photophores.* Ventral mantle with many scattered photophores that are absent from mid-ventral strip except for a single row around anterior mantle margin. Indistinct line of large organs extend posteriorly from locking cartilages; line obvious in juveniles, separated from other ventral photophores by clear band. Ventral head with scattered photophores; ventrolateral region devoid of photophores; dorsal to this are two indistinct rows that merge with other photophores posteriorly and continue to outer edge of ventral arm web anteriorly. Series of ~50 photophores around ventral eyelid.

#### HABITAT AND BIOLOGY

Diel vertical migrator.

#### SIZE

Mantle length to 35 mm.

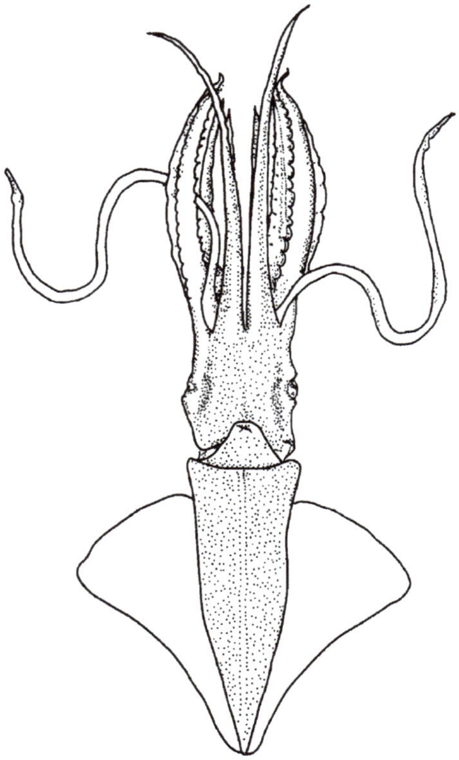

Ventral view (A. Reid, redrawn from Riddell 1985).

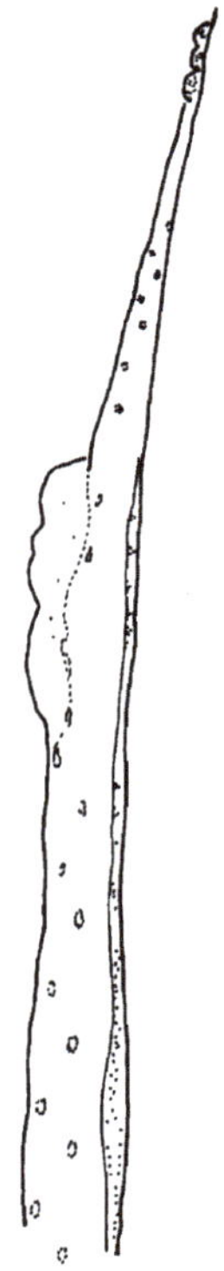

Hectocotylus, oral view (A. Reid, redrawn from Riddell 1985).

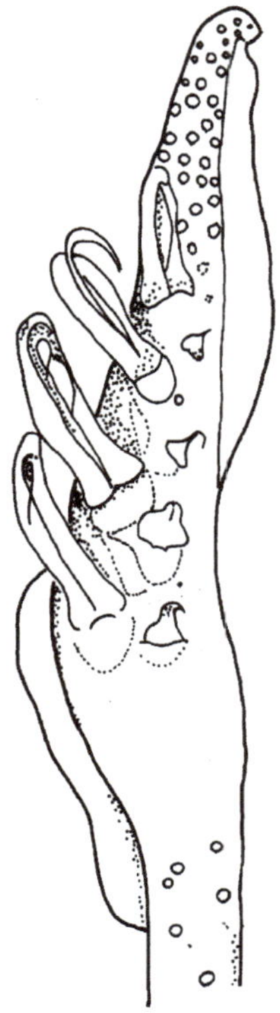

Tentacular club, oral view (A. Reid, redrawn from Riddell 1985).

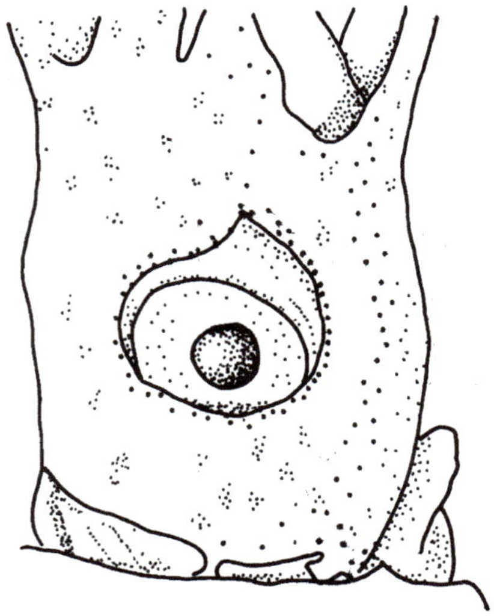

Lateral view of head with photophores (A. Reid, redrawn from Riddell 1985).

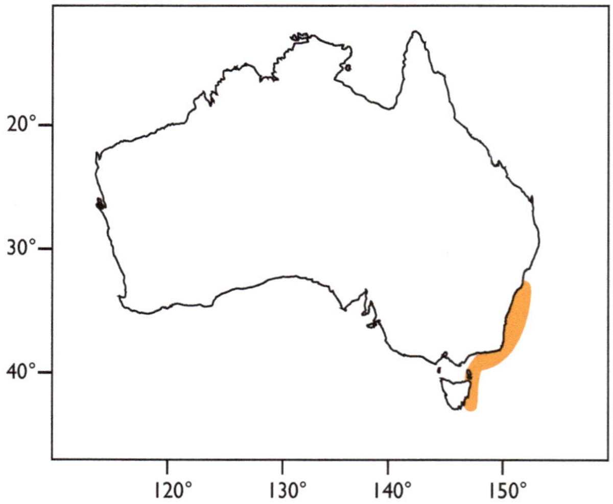


*Abraliopsis tui* distribution. (Australian records are based on few specimens, but the species is likely to occur more widely.)

## REMARKS

*Abraliopsis tui* has previously been reported as a New Zealand endemic. However, specimens identified as *A. tui* from off south-eastern Australia have been found among the collections of Museum Victoria and the Australian Museum (C. C. Lu and K. Li).

## DISTRIBUTION

New Zealand, Kermadec Island waters, south-eastern Australia.

## REFERENCES

Roper and Jereb (2010j), Tsuchiya, K. (2013c).

## *Enoploteuthis* d'Orbigny, 1848

### DIAGNOSTIC FEATURES

Mantle with tail-like projection beyond posterior end of fin insertion. Fins terminate well anterior to posterior tip of mantle. Fins subterminal. Large arm hooks with open apertures; suckers present distally on arms 4. Right arm 4 hectocotylised. Manus of club with biserial hooks; marginal suckers absent. Radula with heterodont rhachidian tooth. Spermatophore receptacles at posterior junction of retractor muscles of funnel and head. Buccal crown with typical chromatophores on aboral surface, possibly with light epithelial pigmentation on oral surface.

*Photophores.* Two types of integumental photophores: simple photophores with dark blue centres (in life) and a narrow white ring, and slightly larger 'complex' filtered photophores. No enlarged photophores at tips of arms 4. Nine or 10 photophores on eyeball. Anterior and posterior ocular photophores small, subequal. Complex photophores of integument without red colour filters.

### HABITAT AND BIOLOGY

Mesopelagic boundary species.

### SIZE

Up to 110 mm ML.

### REMARKS

*Enoploteuthis* species in Australian waters have little commercial value at the present time. Two subgenera, *Enoploteuthis* (*Enoploteuthis*) and *Enoploteuthis* (*Paraenoploteuthis*), are recognised. One representative of each of the two subgenera is found in Australian waters. Differences between the two subgenera are discussed under the species accounts below for *E.* (*P.*) *galaxias* Berry, 1918 and *E.* (*E.*) *leptura* (Leach, 1817) respectively.

The arrangement of the photophores is an important diagnostic trait, but care needs to be taken with identification because the photophore arrangement changes with growth.

### REFERENCES

Roper and Jereb (2010j), Tsuchiya (2013c).

### Key to *Enoploteuthis* species

1a. Tentacles short, thin; clubs narrow; carpal-locking apparatus elongate; manus hooks all similar-sized; dactylus suckers biserial. Males with numerous small papillae on arms 1–3........... *E. leptura* (*?magnoceani*)

1b. Tentacles long, thick; clubs robust; carpal-locking apparatus circular; manus hooks unequal in size; dactylus suckers tetraserial. Males without numerous papillae on arms 1–3 ............... *E. galaxias*

## *Enoploteuthis* (*Paraenoploteuthis*) *galaxias* Berry, 1918

*Enoploteuthis galaxias* Berry, S. S. (1918). Report on the Cephalopoda obtained by the F.I.S. 'Endeavour' in the Great Australian Bight and other southern Australian localities. *Biological Results of the Fishing Experiments carried on by the F.I.S. 'Endeavour', 1909–14*, *4*(5), 201–298 [211, text figs 1–4, pls 59, 60].

Type data: holotype, AM C.148251.

Type locality: Australia: Victoria, Tasman Sea, western South Pacific Ocean.

### DIAGNOSTIC FEATURES

Small, firm, fleshy mantle widest anteriorly, tapering past fins to an attenuate posterior point. Fins large, triangular, length about half ML. Funnel large, broad and swollen ventrally. Arm formula 4 > 3 > 2 > 1 or 4 > 3 = 2 > 1, dorsal pair much shorter than rest. Arms 3 with distinct keel. Tentacular sheath continues as broad membranous fold along ventral outer angles of arms to their tips. All arms with two regularly alternating series of well spaced hooks, arm tips with minute suckers. Basal sixth of arms devoid of suckers or hooks. Males without numerous, small papillae on oral surfaces of arms 1–3. Right ventral arm of male hectocotylised: distal to eighth ventral hook, protective membrane thickened and widened to form elongate flap, narrows abruptly distally at level of thirteenth hook. Hectocotylus with flaps on both margins; enlargement of dorsal membrane not as great as ventral membrane, enlargement beginning more distally; modified portion with armature. Relatively long, thick tentacles with robust clubs. Tentacular club narrower than carpus, tapering distally; club carpal-locking apparatus nearly circular; carpus with four to six suckers and four or five pads on oral side. Club manus with two alternating series of very unequal-sized hooks; ventral series with small proximal hook succeeded by five larger, longer slender hooks, the second and third largest of series, remainder gradually diminishing in size; dorsal series with five smaller hooks; no suckers proximal to hooks in either series or between hooks and carpus. Ventral

Fresh-caught specimen (K. Graham).

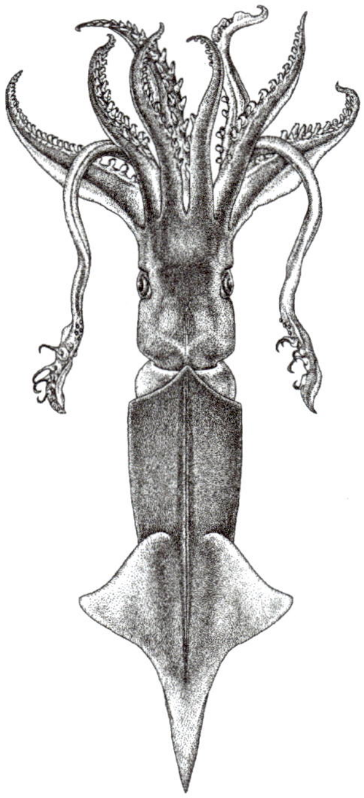

Dorsal view (ABRS, K. Hollis).

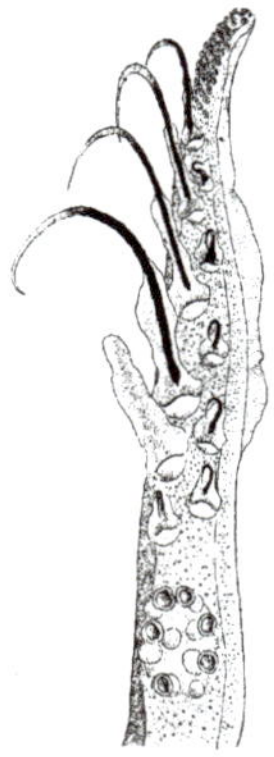

Tentacular club, ventral view (A. Reid, redrawn from Berry 1918).

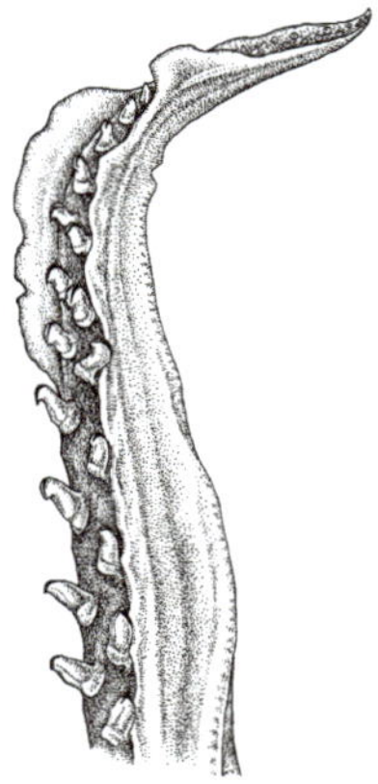

Hectocotylus (ABRS, K. Hollis).

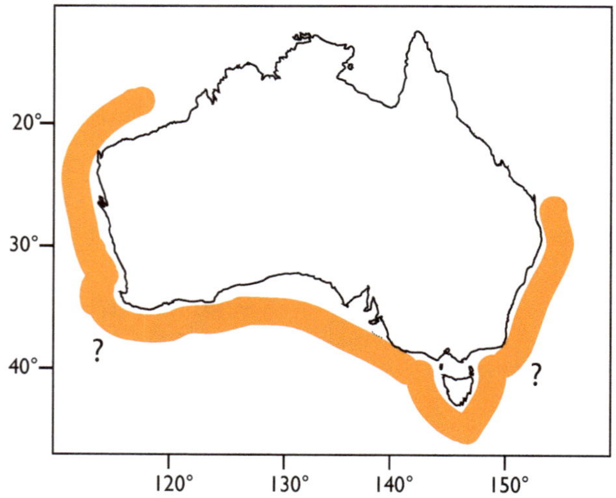


*Enoploteuthis galaxias* distribution. (Australian records are based on few specimens, but the species is likely to occur more widely.)

protective membrane on manus. Dactylus with tetraserial closely packed suckers.

*Photophores.* Ventral head with three wide, longitudinal stripes of photophores. Photophores numerous over ventral integument; anteriorly with six longitudinal stripes of photophores that are interconnected with additional photophores leaving scattered pale patches, without photophores. Ventral side of arms 3 with series of photophores along length.

*Colour.* Preserved specimens dark slate mottled with brown. Photophores purplish. Buccal membrane dark due to dense chromatophores.

## HABITAT AND BIOLOGY

A southern subtropical, nerito-oceanic mesopelagic and benthic-bathyal species.

*Interest to fisheries.* This species is occasionally abundant in demersal trawl catches off the eastern Australian coast and is of a similar size to the commercially important Japanese enoploteuthid, *Watasenia scintillans* (Berry, 1911).

## SIZE

Mantle length to 110 mm.

## REMARKS

The fenestrate pattern of photophores on the ventral mantle is obscured during growth as more photophores are added.

## DISTRIBUTION

South-western South Pacific Ocean: southern and eastern Australia and northern New Zealand.

## REFERENCE

Tsuchiya (2009d).

## *Enoploteuthis (Enoploteuthis) leptura* (Leach, 1817)

### *?E. (E.) magnoceani* Nesis, 1982

*Loligo lepturo* Leach, W. E. (1817). Synopsis of the orders, families, and genera of the class Cephalopoda. *The Zoological Miscellany; being descriptions of new or interesting animals* 3(30), 137–141 [141].

Type data: holotype BMNH 1993017.

Type locality: eastern Atlantic Ocean, Gulf of Guinea, 01°08′N, 07°26′E.

#### COMMON NAME

Hooked Enope Squid

#### DIAGNOSTIC FEATURES

(See 'Remarks' regarding italicised characters.) Two rows of hooks extend along arms. Arms 1–3 with ~22–27 hooks; arms 4 with 32–39 hooks. Males with numerous minute papillae over oral surfaces of arms 1–3. *Right arm 4 in males hectocotylised: ventral protective membrane expanded in distal third of arm* [*with armature*] opposite about six pairs of hooks. Dorsal protective membrane expands slightly into series of small flaps opposite large flap of ventral membrane. Tentacles short, narrow; clubs not expanded, thin; *manus with biserial [subequal] hooks*, ventral series with six or seven large hooks and dorsal series with four or five smaller hooks, ventral protective membrane absent; *dactylus with 10–15 biserial suckers; carpal cluster very long and narrow.*

*Photophores.* Ventral head with six longitudinal series of photophores, lateralmost broken by ventral 'window'. Six complex photophore rows on anterioventral side of mantle; seventh series does not extend to anterior mantle margin but curves to join one nearby medial series to the right or left. Anterioventral mantle margin with discontinuous row of photophores; gaps correspond to spaces between photophore series on mantle. Ventral mantle with five longitudinal but weakly oblique stripes of photophores. Median stripe composed of two closely adjacent longitudinal rows that converge anteriorly then join and terminate just short of mantle margin. Ventralmost pair of stripes is obscure and fused to lateral rows. Ventral side of arm 3 with a row of photophores along *almost entire arm.*

#### HABITAT AND BIOLOGY

Mesopelagic to epipelagic in tropical, subtropical (rarely temperate) waters. Diel vertical migrators.

#### SIZE

Maximum ML to 92 mm.

#### REMARKS

According to Roper and Jereb (2010j), records from Pacific and Indian Oceans probably refer to *Enoploteuthis leptura*

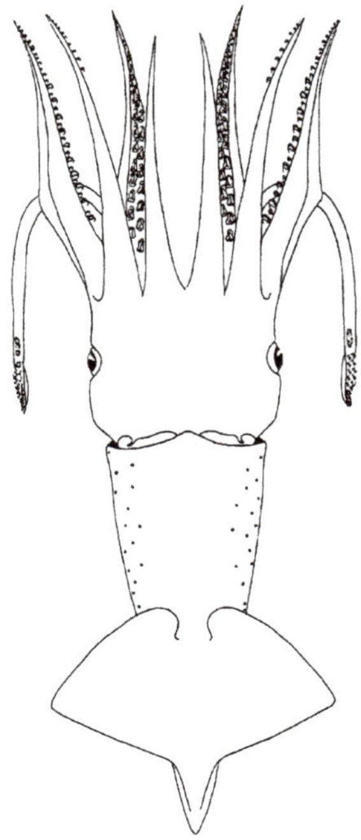

Dorsal view (FAO, E. D'Antoni).

Ventral view (Plate XI, Chun 1910).

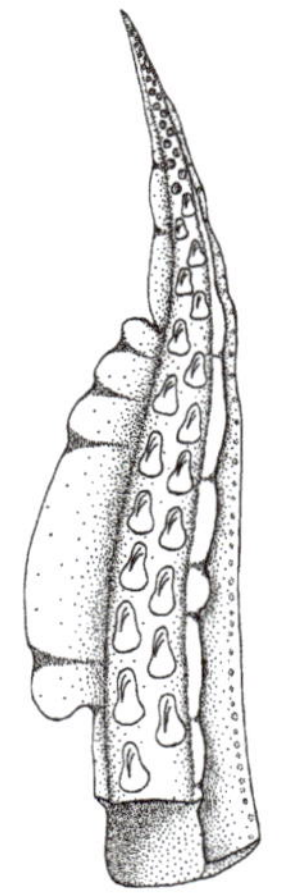

Hectocotylus, oral view (FAO, E. D'Antoni).

Tentacular club (Plate XI, Chun 1910).

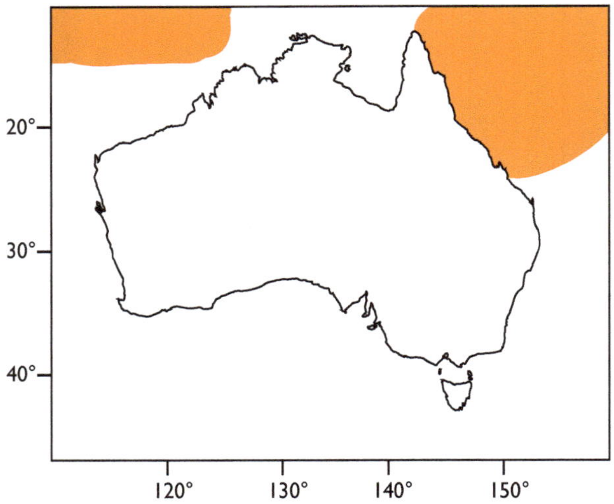


*Enoploteuthis (Enoploteuthis) leptura* distribution. (Australian records are based on few specimens, but the species is likely to occur more widely.)

*magnoceani* Nesis, 1982. This taxon has never been fully described; however, Tsuchiya and Young (2014) consider the evidence is adequate to consider it as a valid species. It is possible then that the species recognised as *E. leptura* in Australian waters is actually *E. magnoceani* Nesis, 1982, but until the latter species is fully described and Australian material examined in detail, the name *E. leptura* is retained here. (Unfortunately, the type locality for *E. magnoceani* is given only as 'Pacific Ocean'.) The description above refers to *E. leptura*. Based on the information given in Tsuchiya and Young (2014) the traits listed for *E. magnoceani* are shown above in italics. The arrangement of photophores on the ventral mantle differs between the two taxa.

### DISTRIBUTION

Gulf of Guinea, eastern and western central Atlantic Ocean, Indo-West Pacific Ocean.

### REFERENCES

Roper and Jereb (2010j), Tsuchiya and Young (2014).

## Gonatidae Hoyle, 1886

### COMMON NAME

Gonate Squids

### DIAGNOSTIC FEATURES

Body narrow, muscular. Fins medium to large, rhomboid or heart-shaped. In some species, muscle tissues degenerate upon maturation and become watery and flaccid, associated with post-spawning egg brooding. Funnel-locking cartilage simple, straight, may flare slightly at posterior end. Most species without true hectocotylus. Mature male *Gonatus* have a long penis that is used to transfer spermatophores onto the female's buccal membrane. Tentacles with well-developed fixing apparatus along dorsal margin. Buccal connectives DDVV.

### HABITAT AND BIOLOGY

Most species inhabit the upper water levels as paralarvae and juveniles and then they undergo significant ontogenetic descent. Adults descend to great depths in the midwaters (mesopelagic) or to the deep-sea floor where they mate and spawn. Post-spawing egg brooding is a unique feature among the teuthids. Paralarvae and juveniles ascend to food-rich epipelagic waters to feed until ontogenetic growth begins. Spent female *Gonatus* become degenerate and gelatinous, and float, suspended in the meso- and bathypelagic layers or even in the surface waters. Some species also undergo diel vertical migrations during which they descend to the mesopelagic zone during the day and then ascend to the surface layers at night.

### SIZE

Small- to medium- and large-sized squids; maximum ML to 550 mm.

### REMARKS

Species in this family belong to four genera (*Berryteuthis*, *Eogonatus*, *Gonatopsis* and *Gonatus*), of which only *Gonatus* is found in Australian waters. All genera inhabit the North Pacific, and only *Gonatus* has expanded its range far beyond the North Pacific region into the Arctic, North Atlantic and Southern Oceans. All species of the Gonatidae are cold-water forms. Of the 19 currently described species in the family, only one species dwells in the Southern Ocean.

### DISTRIBUTION

The family has a bipolar distribution: the western arctic-boreal region, the boreal North Atlantic Ocean, the boreal (sub-Arctic) North Pacific Ocean and the whole sub-Antarctic belt.

### REFERENCE

Roper *et al.* (2010b).

## *Gonatus* Gray, 1849

### DIAGNOSTIC FEATURES

Fins heart shaped, extend posteriorly beyond end of mantle, attenuate into tail supported by cartilaginous rod. Hooks well developed in two medial series on arms 1–3 from paralarval or immediate post-paralarval stages. Tentacular club with at least one large central hook on manus and generally with additional hooks (large distally and small proximally) along midline. Club with distinctive locking apparatus on dorsal border of manus consisting of series of four to six large, thick, transverse pads with medial alternating oval pads and suckers. Radula with five longitudinal rows of teeth.

### HABITAT AND BIOLOGY

As adults, *Gonatus* species are typically mesopelagic and upper bathypelagic. Cannibalism is common in this genus.

*Interest to fisheries.* *Gonatus* species do not support any significant commercial fishery, despite having a good flesh consistency and the large size of some species. Potential exists for future exploitation.

### REFERENCE

Roper *et al.* (2010b).

## *Gonatus antarcticus* Lönnberg, 1898

*Gonatus antarcticus* Lönnberg, E. (1898). On the cephalopods collected by the Swedish Expedition to Tierra del Fuego, 1895–96. *Svenska Expeditionen till Magellansländerna* 2(4), 49–64 [51, pl. 5. Figs 4, 5].

Type data: holotype NRM?

Type locality: Chile, Punta Arenas, on the beach.

### DIAGNOSTIC FEATURES

Mantle elongate, slender, with long 'tail', body soft. Head squarish in shape. Fins long, narrow, 50% ML, saggitate with roundish margins. Funnel-locking cartilage lanceolate in shape; nuchal cartilage rectangular with three grooves. Arms short (40–50% ML), thick and muscular, all approximately similar in length. Arms 4 with thin lateral keels and arm 3 with keel along entire length. Arm suckers tetraserial (arm tips may have more than four series of minute suckers); two medial series on arms 1–3 typically hooks along nearly entire arm length, except for basal-most and distal arm tips. Median and marginal arm 4 suckers similar in size and 1.5 × larger than marginal suckers of other arms. Large marginal suckers of arms 1–3 with six to nine pointed teeth on distal margins of inner rings, proximal margins smooth; marginal and medial suckers of arms 4 with seven or eight pointed teeth. Tentacle long, stout with small club (16 or 17% ML); large central hook and medium-sized hook distal to this; three or four hooks proximal to central large hook, distal one largest and decreasing in size proximally; numerous crowded small suckers on median portion of stalk. Ventro-marginal part of manus with 14 or 15 trabeculae, three or four suckers and six or seven blunt teeth. Carpus with five or six large, smooth-ringed suckers alternately positioned on thick ridges with fleshy knobs.

*Photophores.* Absent.

*Colour.* Skin densely spotted with purplish chromatophores.

### HABITAT AND BIOLOGY

Paralarvae and juveniles of *G. antarcticus* live in epipelagic and mesopelagic zones, adults occupy mesopelagic and bathypelagic zones. It is a non-ammoniacal species.

### SIZE

This species grows to a maximum ML of 350 mm.

### DISTRIBUTION

*Gonatus antarcticus* is a notalian circumpolar species in the southern hemisphere in sub-Antarctic waters north of the Antarctic Polar Front, south of the Subtropical Front reaching Antarctic waters (south of 40°S). It is also distributed around South Georgia and along the Scotia

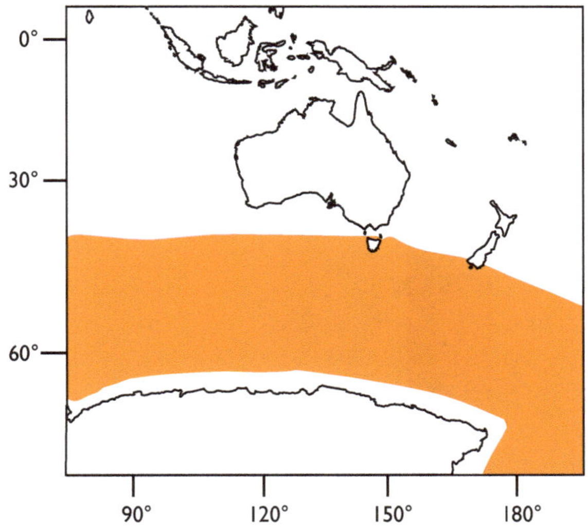


*Gonatus antarcticus* distribution.

Dorsal view (Darren Stevens, NIWA, NZ IPY-CAML (voyage TAN0802)).

Ventral view (Darren Stevens, NIWA, NZ IPY-CAML (voyage TAN0802)).

Left tentacular club, © Richard E. Young (Kubodera 2006).

*Gonatus antarcticus*

Arc, north to southernmost South Africa, the Cook Strait and Peru.

#### REFERENCES

Kubodera (2006), Roper *et al.* (2010b).

## Histioteuthidae Verrill, 1881

#### COMMON NAME(S)

Jewel Squids, Umbrella Squids

#### DISTINGUISHING CHARACTERS

Unique among oegopsid squids, adults usually with asymmetrically developed eyes, the left eye usually considerably larger than the right, often tubular.

#### DIAGNOSTIC FEATURES

Medium-sized to large squids. Most species with short, broad, conical mantle. Head broad. Body texture gelatinous. Head large in adults; left eye considerably larger than right eye. Fins terminal, medium to large, together transversely oval in outline, unite posteriorly with median notch; posterior margin of fins may extend posteriorly beyond mantle. Funnel–mantle locking cartilage simple, straight, or slightly curved. Arms long, thick proximally. Arm suckers biserial, unmodified, toothed. Inner web connects arms and extends to greater than two-thirds of arm length. Club suckers tetraserial, or in eight rows that vary in size. Both dorsal arms usually modified in males: terminal suckers abruptly reduced in size on elongate, palisaded pedestals; basal (or all normal) suckers enlarged, with swollen, fleshy collars; dorsal arms often more elongate and robust than remaining arms. Tentacles long with expanded club; distinct dactylus, manus, and carpal adhesive apparatus, with alternating suckers and pads in a single series that extends proximally on tentacular stalk for 1–3 club lengths. Sucker rings on manus toothed around entire circumference. Buccal connectives attached to dorsal border of ventral arms. Primary and secondary efferent blood vessels highly pigmented. Gladius with ventrally inrolled, cupped cone at posterior end.

*Photophores.* Many anteriorly directed photophores over surface of mantle, head and arms, and ring around eyelid; more concentrated on ventral surface; arrangement species-specific. Internal photophores absent.

#### HABITAT AND BIOLOGY

Inhabits open ocean and slope. Members of the cosmopolitan family Histioteuthidae are very broadly distributed throughout the world's oceans and seas in the mesopelagic, bathypelagic and bathybenthic zones. They undergo diel vertical migration into the epipelagic at night. Although the family is considered to be primarily mesopelagic, *Histioteuthis pacifica* (G. Voss, 1962) and *H. miranda* (Berry, 1918) have been caught frequently in deep-water lobster trawls at depths of 300–600 m off northern Australia and in demersal trawls elsewhere in their range suggesting they are associated with the bottom during part of their life cycles. Species range from the tropics to the polar seas. Some species are extremely abundant. Some species are quite large and attain mantle lengths of at least 350 mm, but the total length can be 4 or 5 × longer than the ML.

The number and size of beaks in sperm whale stomachs suggests that histioteuthids aggregate in 'schools', even at great depths. Most species are major components in the diets of sperm whales (up to 62% in some regions), other odontocete cetaceans, pinnipeds, albatrosses, penguins, other seabirds, sharks, tunas, lancetfishes and other squids.

The large posteriodorsally directed eye is thought to detect down-welling light during the day, primarily at depth. The photophores probably function in ventral countershading.

*Interest to fisheries.* While there are no large-scale directed fisheries for any species of *Histioteuthis*, it is possible that a small by-catch interest could develop as commercial fishery trawling probes increasingly deeper, to 1500 m and beyond. The flesh on the mantle and arms of histioteuthids is firm and thick in adults of pre-spawning condition. At full maturation of both males and females, however, the tissue of the mantle, head and arms begins to soften, and by full spawning and post-spawning stages, the tissues of the mantle, head and arms soften to a semi-gelatinous consistency. A further impediment to development of a consistent fishery would be the suspected ammoniacal composition of the flesh.

### SIZE

Small- to medium-sized squid; maximum ML 350 mm.

### REMARKS

Young and Vecchione (2007a) consider members of the Histioteuthidae with paired secondary reproductive organs (penis, spermatophore gland complex, spermatophore storage sac) distinctive and place them in a separate clade. *Histioteuthis hoylei* (Goodrich, 1896) and *H. arcturi* (Robson, 1948) are placed in this separate clade. The males that originally represented *H. hoylei* came from north temperate waters off California and Japan. After the examination of material from off Hawaii, Young and Vecchione (2008) consider the Pacific forms as divided into two different species, a temperate North Pacific one, for which they keep the name *H. dofleini* and a more southern species, *H. hoylei*. The same authors then consider the generic name *Stigmatoteuthis* Pfeffer, 1900 as the valid designated genus name for this separate clade of histioteuthids and this is the classification followed here.

Species are easily distinguished by numerous (usually), anteriorly directed, complex photophores with red colour filters over the surface of the mantle, head and arms, especially on the ventral and ventrolateral surface (excluding the funnel). The pattern of photophores is important in identifying species and species-groups. Paralarvae are easily recognised by the long arms in combination with the very large arm suckers relative to the small club suckers; however, paralarvae have not been identified to species due to their pronounced morphological changes as they transition to adults.

### DISTRIBUTION

Histioteuthids occur in all oceans of the world except the Arctic and Antarctic. Known depth range 200–2000 m.

### REFERENCES

Young and Vecchione (2007a), Roper and Jereb (2010k).

## Key to genera of the Histioteuthidae

1a. Arms 160–250% ML. Male arms 1 <400% ML. Males without paired secondary reproductive organs ........ *Histioteuthis*

1b. Arms usually < 150% ML (exceptions *Histioteuthis bonnellii* up to 300% ML, and *H. atlantica* up to 176% ML). Male arms 1 very elongate > 400% ML. Males with paired secondary reproductive organs (penis, spermatophore gland complex and spermatophore storage sac) ........ *Stigmatoteuthis (S. hoylei)*

## *Histioteuthis* d'Orbigny in d'Orbigny & Ferussac, 1841

### COMMON NAME(S)

Strawberry Squid, Jewel Squid, Violet Squid

### DISTINGUISHING CHARACTERS

Arms not very long relative to ML. Males without paired secondary reproductive organs.

### Key to *Histioteuthis* species (after Voss *et al.* 1998)

1a. Median row of tubercles on dorsal surface of mantle and basal portions of arms 1–3 .......... 2

1b. Median row of tubercles on mantle and arms absent... 3

2a. Photophores uniformly small, arranged in dense pattern on ventral surfaces of mantle and head, in eight or nine longitudinal rows on basal portions of arms 4, in circlet of 19–22 around right eye .......... *H. meleagroteuthis*

2b. Photophores of uniformly medium size, arranged in moderately dense pattern on ventral surfaces of mantle and head, in four to six longitudinal rows on basal portions of arms 4, in circlet of 16 or 17 (rarely 15 or 18) around right eye .......... *H. miranda*

3a. A single, large terminal photophore present on arms 1–3 or 1–4; inner web connecting basal portions of arms deep: 50% or more of arm length in adults, subadults and large juveniles .......... 4

3b. A single, large terminal photophore absent on arms; inner web vestigial to moderate in depth: one-third or less of arm length in adults, subadults and juveniles .......... 5

4a. Buccal membrane with six lappets and one connective to arms 4; segment of inner web absent between right and left junctures of web segments from arms 3 and 4 .......... *H. bonnellii*

4b. Buccal membrane with seven lappets and two connectives to arms 4; segment of inner web present between right and left junctures of web segments from arms 3 and 4 .......... *H. macrohista*

5a. Photophores large, arranged in widely to moderately widely spaced pattern on anterior one-third to half of ventral surface of mantle; circlet around right eye composed of 16 or 17 (rarely 15 or 18) large photophores .......... *H. pacifica*

5b. Photophores intermixed large and small, arranged in moderately dense pattern on ventral surface of mantle; circlet around right eye composed of 17 large and one small photophore .......... 6

6a. Three longitudinal series of large photophores on arms 4 .......... *H. eltaninae*

6b. Four longitudinal series of photophores on arms 4 (three rows of large organs, one row of small or mixed-size organs) .......... *H. atlantica*

## *Histioteuthis atlantica* (Hoyle, 1885)

*Histiopsis atlantica* Hoyle, W. E. (1885). Diagnosis of new species of Cephalopoda collected during the cruise of H.M.S. Challenger. – II. The Decapoda. *Annals and Magazine of Natural History (series 5) 16*, 181–203 [201].

Type data: holotype BMNH 1890.1.24.18.

Type locality: Atlantic Ocean, 35°36′S, 21°12′W.

### COMMON NAME

Atlantic Jewel Squid

### DIAGNOSTIC FEATURES

Tubercles absent. Inner arm web shallow; web depth 17–30% of length of longest arm.

*Photophores.* Large compound photophores with posteriorly directed reflectors. Ventral surface of mantle with large and small photophores intermixed. At maturity, numerous simple photophores on posterior end of mantle in males, becoming larger and denser with age. Eighteen photophores (17 large and one small) around right eyelid. Distal third of arms 1–3 with series of several elliptical black photophores, largest of which is 2 or 3 × larger than ordinary photophores. Gap between arm photophores absent. Four longitudinal rows of photophores (not including small ones) on ventral arms; dorsal series with photophores of various sizes.

### HABITAT AND BIOLOGY

Maximum known depth range ~40–2000 m. Juveniles are taken in open nets at 100–200 m at night and at 500–1000 m during both day and night. Adults and subadults have been captured with open nets between 300 and 2000 m at night and between 900 and 1000 m during the day; subadults have been caught on the bottom at night at 700–850 m.

### SIZE

The maximum ML reaches 260 mm in mature males and 150 mm in females.

### REMARKS

Mature females are unknown. The distribution of *Histioteuthis atlantica* (Hoyle, 1885) overlaps with that of *H. eltaninae* N. Voss, 1969.

### DISTRIBUTION

*Histioteuthis atlantica* is circumglobal in distribution between ~30–33°S and 50°S, normally in the southern

Preserved museum specimen, dorsal view (AM C.391830, A. Reid).

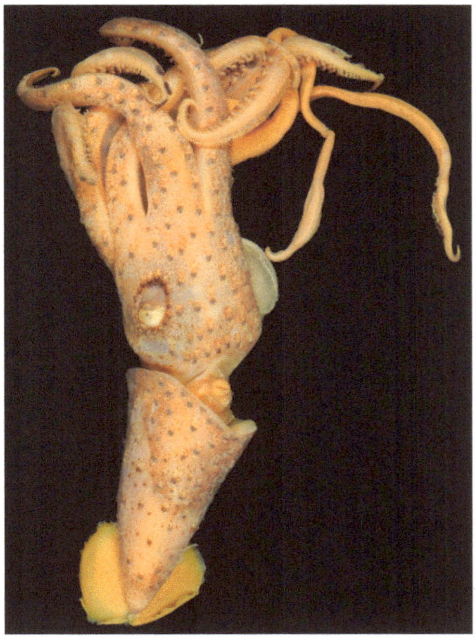

Preserved museum specimen, ventral view (AM C.391830, A. Reid).

Subtropical Front, as well as in the fringes of the subtropical and the sub-Antarctic waters, over oceanic basins, plateaus and shelf areas, extending south to the Antarctic Polar Front.

## REFERENCES

Young and Vecchione (2000a), Young and Vecchione (2007a).

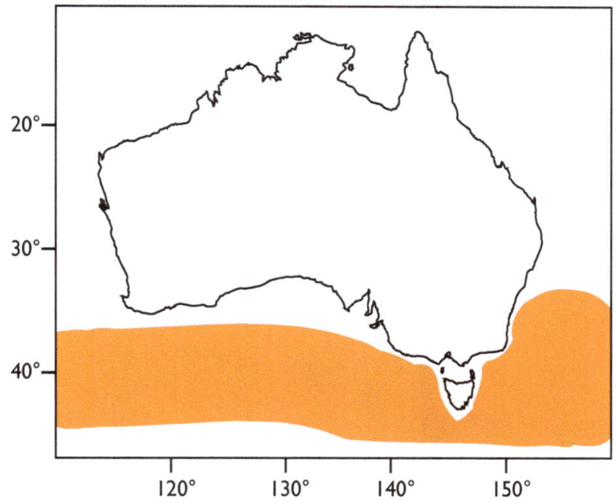


*Histioteuthis atlantica* distribution.

## *Histioteuthis bonnellii* (Ferussac, 1834)

*Cranchia bonnellii* Ferussac, A. E. de (1834). Notice sur deux nouvelles especes des Cephalopodes appartenant aux genres Calmaret et Cranchie. *L'Institut, Journal General des Societes et Travaux Scientifiques de la France et de l'Etranger* 2(77), 355 [355].

Type data: syntype MNHN 3-3-663.

Type locality: Mediterranean Sea, off Nice.

### COMMON NAME

Jewel Squid

### DISTINGUISHING CHARACTERS

See *Histioteuthis macrohista* N. Voss, 1969 for differences between these two closely related species.

### DIAGNOSTIC FEATURES

Mantle conical, relatively short, broad. Fins oval, medium to large, length ~40–60% ML, width ~70–90% ML. Head large, with one nuchal fold. Arms long, 130–300% ML, connected with very broad, deep, maroon-coloured web extending > 50% of arm length; arm formula typically 2 = 3 > 1 = 4. Buccal membrane with six lappets.

*Photophores.* Mantle covered with uniform-sized, dark, elongate, compound photophores in seven or eight diagonal rows on ventral surface (lateral surfaces of mantle, ventral and lateral surfaces of head and arms also covered with these photophores). Three series of photophores on arms 4. A distinct, single, enlarged, elongate, dark, simple photophore on each arm tip; no gap between arm photophores. Seventeen (rarely 16 or 18) small, oval photophores around periphery of right eyelid.

*Sculpture.* Tubercles absent.

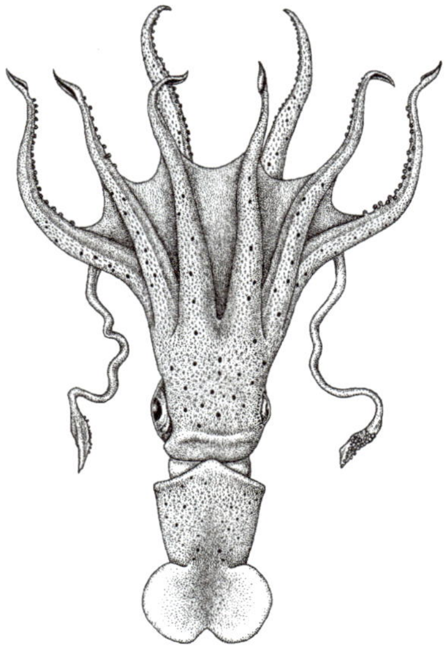

Dorsal view (K. Hollis, ABRS).

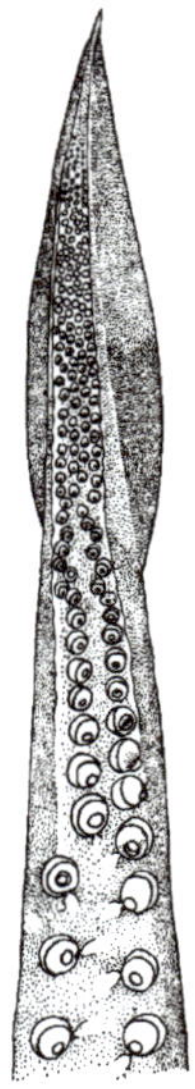

Hectocotylus (C. K. S., FAO).

Tentacular club (C. K. S., FAO).

Arrangement of photophores around right eye (C. K. S., FAO).

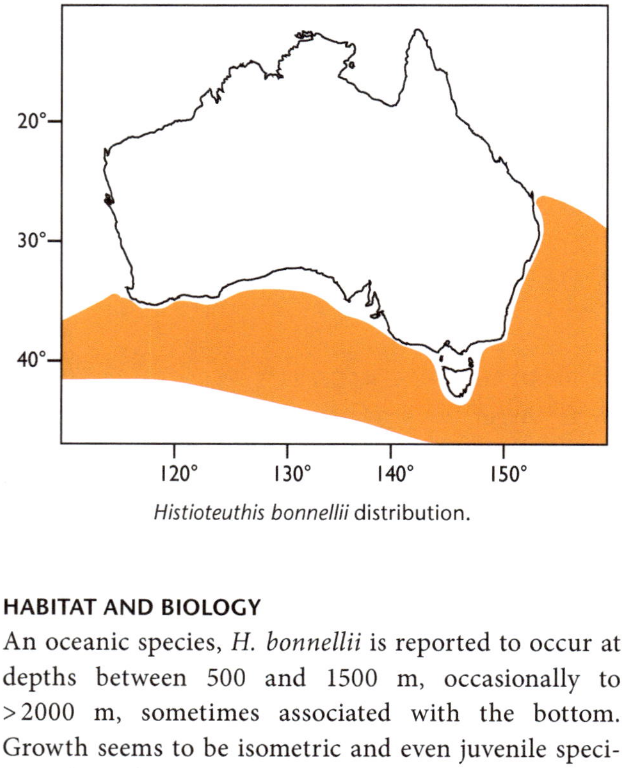


*Histioteuthis bonnellii* distribution.

#### HABITAT AND BIOLOGY

An oceanic species, *H. bonnellii* is reported to occur at depths between 500 and 1500 m, occasionally to >2000 m, sometimes associated with the bottom. Growth seems to be isometric and even juvenile specimens have the proportions of adults. Although they are usually collected singly or in pairs in sampling nets, it is suspected, from evidence in sperm whale stomachs and from very large midwater trawls, that it is a schooling species.

*Interest to fisheries.* None currently.

#### SIZE

This species attains a maximum ML of 330 mm.

#### REMARKS

This is the largest *Histioteuthis* species. It is probably a member of a species-complex that has yet to be resolved.

#### DISTRIBUTION

Very broadly distributed: south-western Pacific Ocean between Australia and New Zealand and the southern Indian Ocean, eastern central North Atlantic Ocean and western Mediterranean Sea to the south-eastern Atlantic Ocean.

#### REFERENCES

Young and Vecchione (2007a), Roper and Jereb (2010k).

### *Histioteuthis eltaninae* N. Voss, 1969

*Histioteuthis eltaninae* Voss, N. A. (1969). A monograph of the Cephalopoda of the North Atlantic. The family Histioteuthidae. *Bulletin of Marine Science 19*(4), 713–867 [755, figs 3d, e, 5c, 7d, 12, 13, 14i–m].

Type data: holotype NMNH 576164.

Type locality: South Pacific Ocean, 40°05′S, 149°55′W.

#### COMMON NAME

Jewel Squid

#### DIAGNOSTIC FEATURES

Fin length 29–37% ML; width 50–55% ML. Funnel organ inverted V-shape with low ridges medially on each arm. Arms 100–127% ML. Inner arm web low to vestigial; outer web absent. Arms 4 with three longitudinal series of suckers basally; suckers with variable number of low, triangular teeth around margin. Mature males with swollen suckers

Fresh-caught specimen, ventral view (K. Graham).

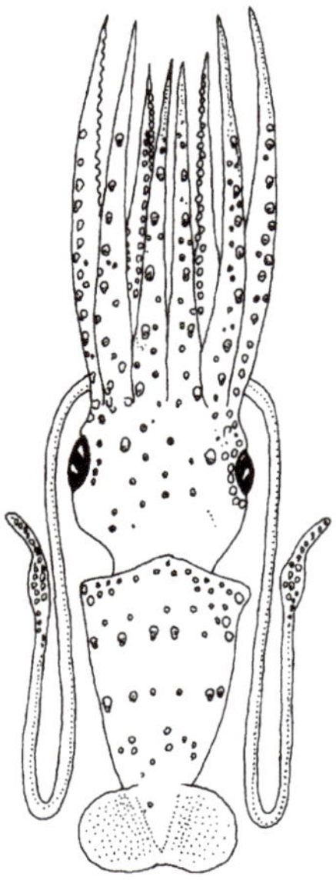

Dorsal view (M. Shea, redrawn from Nesis 1982/1987).

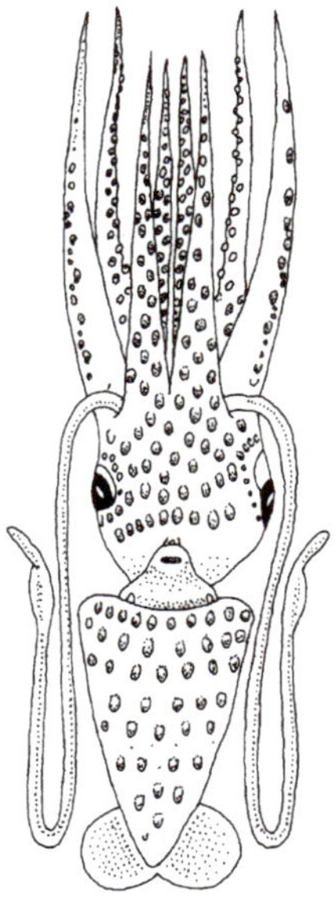

Ventral view (M. Shea, redrawn from Nesis 1982/1987).

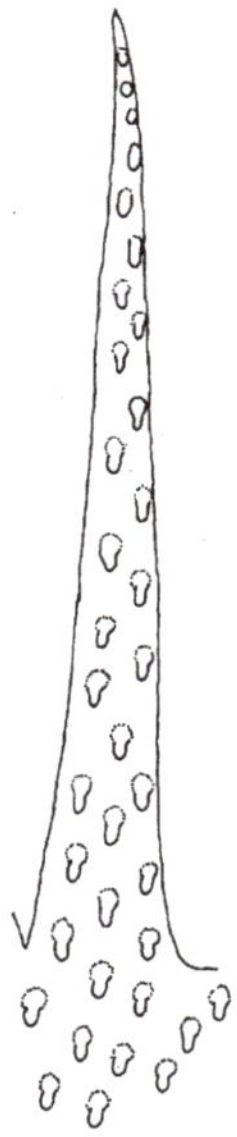

Arm 4, ventral view (M. Shea, redrawn from Nesis 1982/1987).

Tentacular club (M. Shea, redrawn from Nesis 1982/1987).

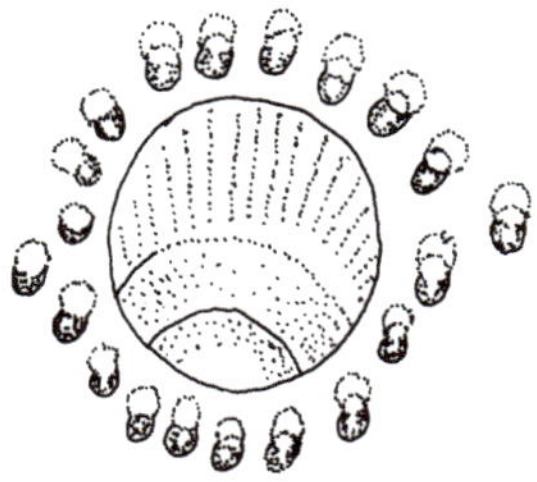

Photophores around right eye (M. Shea, redrawn from Nesis 1982/1987).

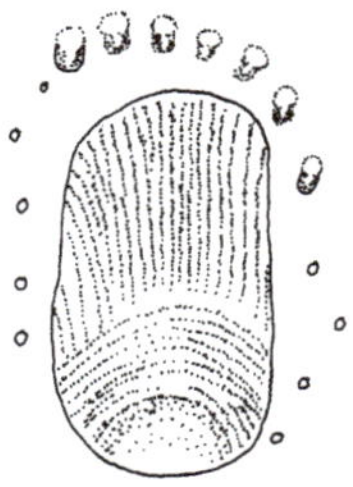

Photophores around left eye (M. Shea, redrawn from Nesis 1982/1987).

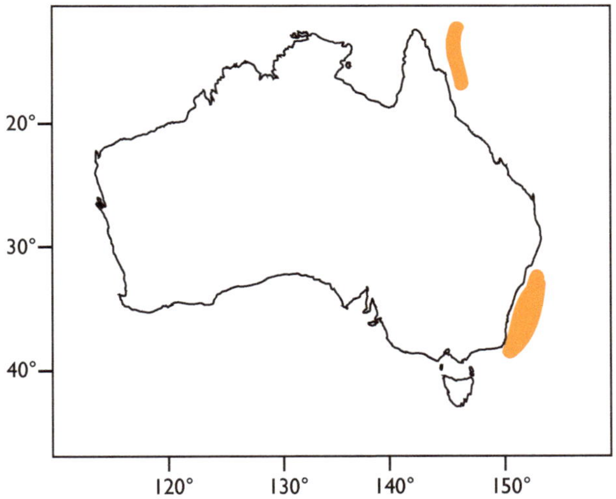


*Histioteuthis eltaninae* distribution. (Australian records are based on few specimens, but the species is likely to occur more widely.)

proximally on dorsal arms (no other hectocotylisation known). Tentacular club with five or six suckers in each transverse row; four median suckers on manus 3 or 4 × larger than ventral marginal suckers. Largest suckers on manus with 36–52 blunt triangular teeth around margin. Buccal crown with seven supports. Male copulatory organ, spermatophore storage sac and spermatophore glands and ducts not paired.

*Photophores.* Eighteen photophores around right eyelid (17 large, one small). Large and small photophores on mantle intermixed. Ventral mantle with large compound photophores interspersed with small compound photophores. Large compound photophores appear to be without posteriorly directed reflectors. No terminal group of large photophores on arms 1–3; gap between arm photophores absent. Three series of photophores on ventral arms.

*Sculpture.* Tubercles absent.

## HABITAT AND BIOLOGY

While the species occurs over oceanic basins, its greatest abundance is in the higher-productivity waters associated with submarine ridges and continental shelves. It is found from the upper 100 m to in excess of 1000 m, perhaps to 2000 m. It is an ammoniacal species.

### SIZE

This small species attains a maximum ML of 110 mm.

### REMARKS

The distribution of *H. eltaninae* overlaps that of *H. atlantica* (Hoyle, 1885).

### DISTRIBUTION

*Histioteuthis eltaninae* is distributed circumglobally in sub-Antarctic waters. Its normal northern boundary is the southern Subtropical Front and its normal southern boundary is the Antarctic Polar Front. A few captures have been made around the Antarctic Peninsula and Scotia Sea. It also has been taken in the East Australian Current to ~33°S, and in the north-eastern Tasman Sea.

### REFERENCES

Young and Vecchione (2000b), Young and Vecchione (2007a), Roper and Jereb (2010k).

## *Histioteuthis macrohista* N. Voss, 1969

*Histioteuthis macrohista* Voss, N. A. (1969). A monograph of the Cephalopoda of the North Atlantic. The family Histioteuthidae. *Bulletin of Marine Science 19*(4), 713–867 [845, figs 4a, b, 6f, 8g, 33, 36–37].

Type data: holotype NMNH 576761.

Type locality: Tasman Sea, 45°10′S, 160°10′E.

### COMMON NAME

Jewel Squid

### DISTINGUISHING CHARACTERS

Most easily separated from *Histioteuthis bonnellii* (Ferrussac, 1834) by the presence of seven rather than six supports in the buccal crown, 16 rather than 17 photophores on the right eyelid, and the presence of two rather than three conspicuous, dark, round photophores on the posterioventral margin of the head on the left side. Terminal photophores never develop on arms 4 in *H. macrohista* (in contrast to large photophores in *H. bonnellii*). In *H. macrohista* the two junctions of the inner webs between arms three are widely separated where they join those of the ventral arms, while in *H. bonnellii* they nearly form a common junction.

### DIAGNOSTIC FEATURES

Fins 52–56% ML, width 80–89% ML. Heavy, swollen ridge on each lateral arm of funnel organ. Inner arm web deep; webbing extends >50% of arms 1–3. Arms 80–130% ML. Both arms 1 in males modified in distal third; suckers abruptly reduced in size, and on elongated palisade pedicels. Arm suckers on inner rings with 4–10 low, rounded or square teeth on distal and lateral margins. Suckers on manus of tentacular club in six irregular series; median suckers moderately enlarged (1.5–2 × ventral marginal suckers); enlarged suckers with numerous sharp teeth. Buccal crown with seven supports. Multiple attachments of fourth ventral supports of the buccal crown (one each to sides of ventral arms and to junctures of web segments from arms 3 and 4). Ejaculatory apparatus of spermatophores with a single loop.

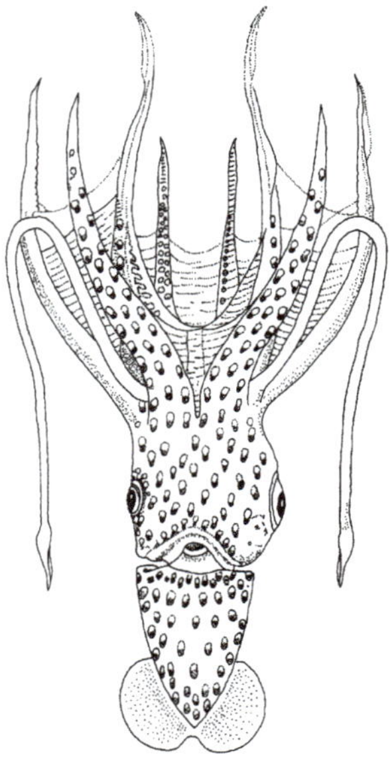

Ventral view (M. Shea, redrawn from Nesis 1982/1987).

Tentacular club (M. Shea, redrawn from Nesis 1982/1987).

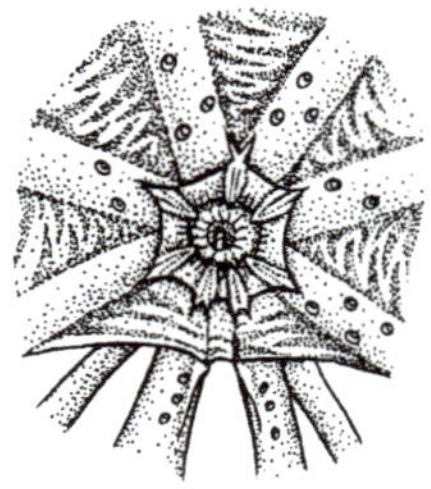

Buccal membrane (M. Shea, redrawn from Nesis 1982/1987).

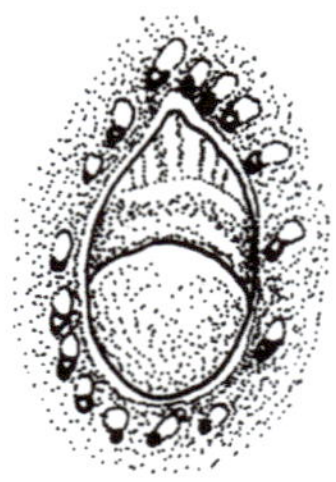

Photophores around right eye (M. Shea, redrawn from Nesis 1982/1987).

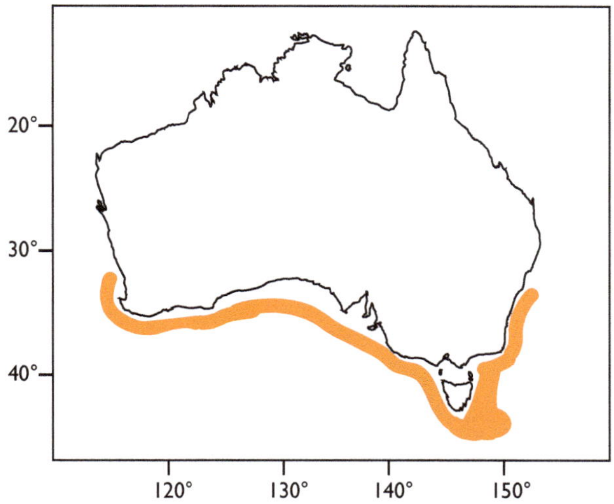

*Histioteuthis macrohista* distribution.

*Photophores.* Photophores on mantle of uniform size. Two or three large, round, broadly separated, dark photophores on left posterior margin of ventral surface of head. Sixteen (rarely 15) compound photophores around right eye. Compound photophores of large, uniform size on anterior half of ventral mantle. Arms 4 with three longitudinal series of photophores proximally; no gap between proximal and distal photophores. A single, elongate simple photophores at distal tip of each arm 1–3; gap between arm photophores absent.

*Sculpture.* Tubercles absent.

## HABITAT AND BIOLOGY

This species occurs close to continental slopes as well as in oceanic basins. The vertical distribution of the species extends between the surface 100 m to >1000 m; both juveniles and subadults have been caught over the entire known vertical range at night, so a diel vertical migration is not confirmed. Females mature at 49 mm to >65 mm ML; males mature at 40–55 mm ML.

## SIZE

The maximum ML is 75 mm in females and 55 mm in males.

## REMARKS

This species is closely related to *H. bonnellii.*

## DISTRIBUTION

*Histioteuthis macrohista* is broadly distributed in the Atlantic, Indian and western Pacific Oceans; it occurs primarily in fringing and transitional waters of the southern Subtropical Front between ~33°S and 47°S (where it is sympatric with *H. atlantica* (Hoyle, 1885)). It is absent from the eastern Pacific.

## REFERENCES

Young and Vecchione (2000c), Young and Vecchione (2007a), Roper and Jereb (2010k).

### *Histioteuthis meleagroteuthis* (Chun, 1910)

*Calliteuthis meleagroteuthis* Chun, C. (1910). Die Cephalopoden. Oegopsida. *Wissenschaftliche Ergebnisse der Deutschen Tiefsee-Expedition dem Dampfer Valdivia, 1898–1899 18*, 1–401. [170].

Type data: repository unresolved ?AIM neotype 057.13; NMNZ neotype M.22403.

Type locality: western South Pacific Ocean, 35°45′S, 176°20′E.

## COMMON NAME

Jewel Squid

## DIAGNOSTIC FEATURES

Inner arm web shallow.

*Photophores.* Photophores on mantle uniform in size. Simple photophores on arm tips absent. Gap between arm photophores present. Eight or nine series of photophores on arms 4.

*Sculpture.* Tubercles present. Row of small tubercles below epithelium of midline of anterior mantle midline and basal half of first three arms. These tubercles become less apparent with growth.

## HABITAT AND BIOLOGY

The species appears to reach greatest abundance in zones of high productivity, particularly those waters associated with bottom slopes and submarine ridges. The vertical distribution extends from the surface to in excess of 1000 m. Closing nets have recorded juveniles at 210–300 m during the day, with larger subadults at 610–700 m. At night, juveniles have been captured at 100–360 m. Open nets have captured specimens to 1950 m in midwaters and to 1250 m near the bottom.

## SIZE

The maximum recorded ML is 114 mm in mature females and 102 mm in mature males.

Preserved museum specimen, dorsolateral view (AM C.391848, A. Reid).

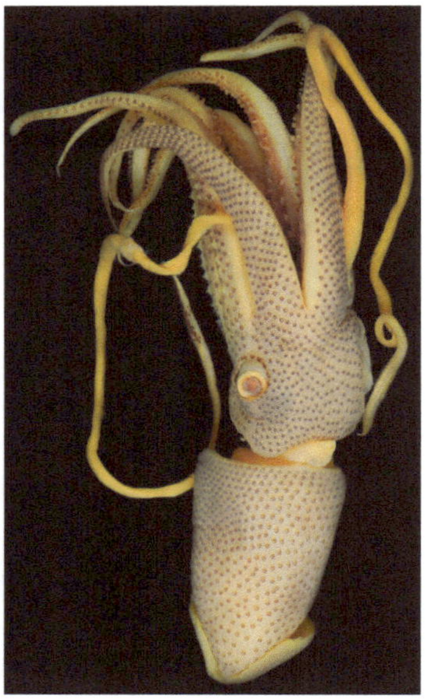

Preserved museum specimen, ventrolateral view (AM C.391848, A. Reid).

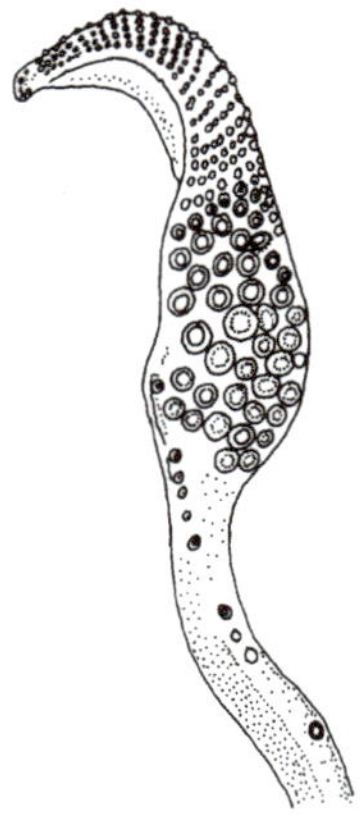

Tentacular club (M. Shea, redrawn from Voss 1963).

Enlargement of arms, dorsal view (AM C.391848, A. Reid).

*Histioteuthis meleagroteuthis*

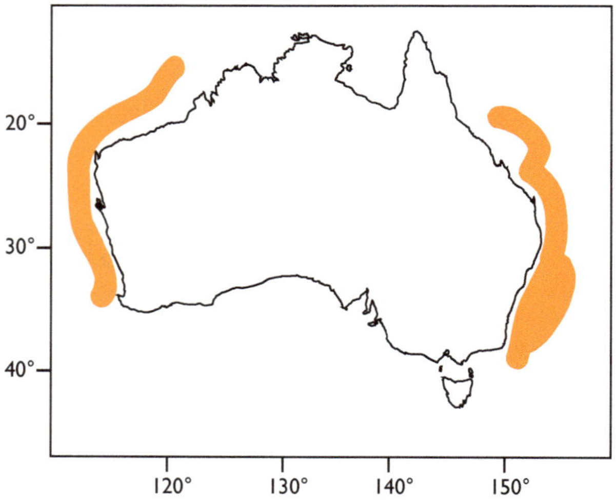


*Histioteuthis melagroteuthis* distribution.

## DISTRIBUTION

*Histioteuthis meleagroteuthis* has a circumglobal distribution in tropical and subtropical waters. The southern limit of occurrence is sharply defined by the southern Subtropical Front; the species does not normally inhabit the convergence transition waters. It is a pan-Pacific Ocean species with many records concentrated in New Zealand and Australian warm waters, northward through the islands to southern Japan.

## REFERENCES

Young and Vecchione (2007a), Roper and Jereb (2010k).

# *Histioteuthis miranda* (Berry, 1918)

*Calliteuthis miranda* Berry, S. S. (1918). Report on the Cephalopoda obtained by the F.I.S. 'Endeavour' in the Great Australian Bight and other southern Australian localities. *Biological Results of the Fishing Experiments Carried on by the F.I.S. Endeavor, 1909–1914* 4(5), 201–298 [221, text figs 5–9, pls 61–62].

Type data: neotype NMNH 576165.

Type locality: western South Pacific Ocean, 37°54′S, 178°56′E.

## COMMON NAME

Wondrous Jewel Squid

## DIAGNOSTIC FEATURES

Mantle robust, moderately elongate; wall thick. Fins medium-sized, oblong, length ~31–44% ML, width ~44–57% ML. Head relatively large; one nuchal fold present in juveniles, absent in adults. Arms of moderate length, ~100–150% ML; arm formula 1 = 2 = 3 >4. Inner web shallow, connects basal 16–25% of arms 1–3; outer web only slightly developed. Tentacles long, 150–260% ML. Tentacular club without longitudinal cleft on aboral surface. Suckers on manus closely packed in six or seven series; median suckers enlarged, twice diameter of ventral marginal suckers. Buccal membrane with six lappets. Mature egg diameter 0.8 mm.

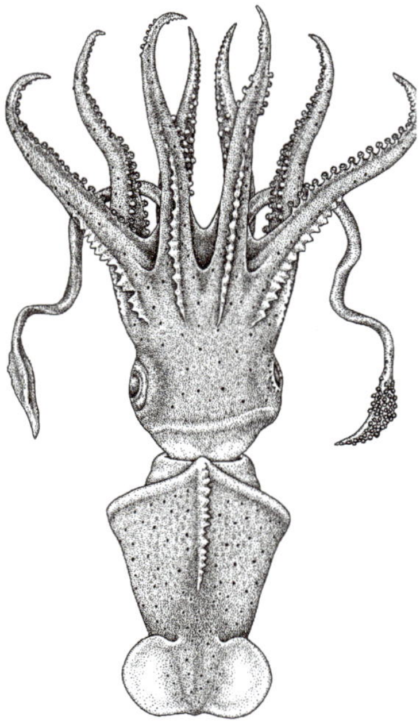

Dorsal view (ABRS, K. Hollis).

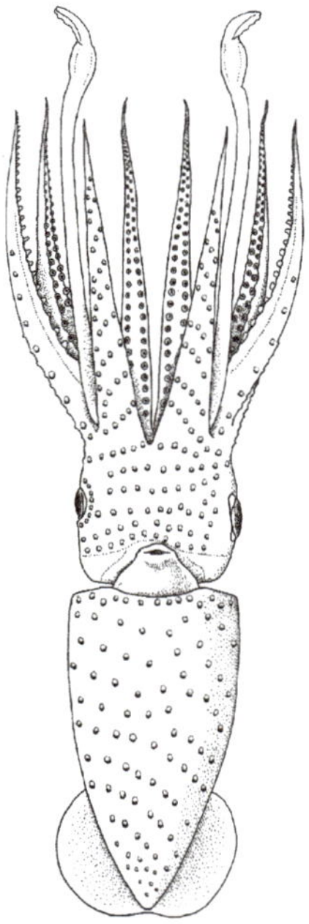

Ventral view (FAO, E. D'Antoni).

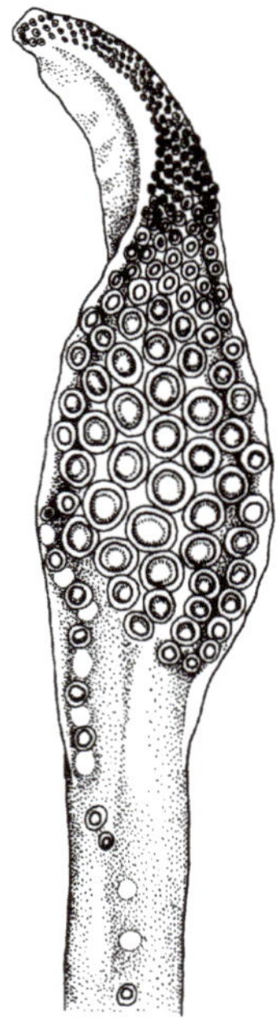

Tentacular club (FAO, E. D'Antoni).

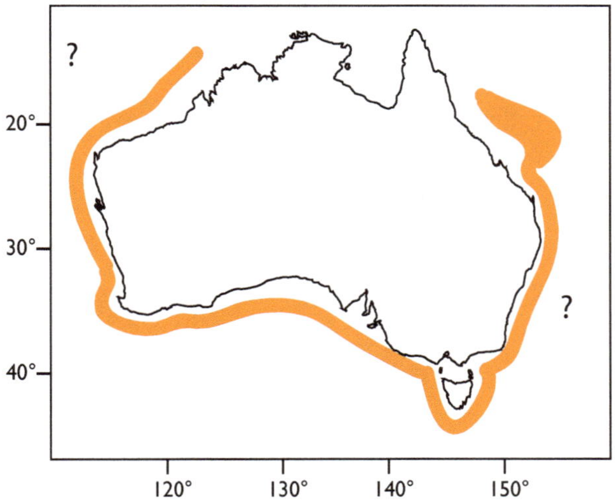


*Histioteuthis miranda* distribution.

*Photophores.* Large and small photophores on mantle slightly intermixed. Compound photophores numerous, uniform medium size, moderately widely spaced, arranged in nine or 10 diagonal rows across ventral mantle; 16 or 17 (rarely 15) moderately large photophores encircle the right eyelid. Five series of photophores on arms 4. Simple photophores on arm tips absent; gap between arm photophores absent.

*Colour.* Skin colour a dark grey-blue to purple.

*Sculpture.* Tubercles present: ridge of small tubercles extends along dorsal midline of mantle below epithelium on anterior half, well developed in medium-sized and large juveniles, but hidden beneath mantle tissue in large subadults and adults. Basal one-third to half of arms 1–3 with median row of low tubercles.

## HABITAT AND BIOLOGY

This species occurs in close association with the slopes of continental masses and their offshore islands and submarine rises. Capture data suggest that this is the most abundant histioteuthid species encountered throughout most of its normal range. It ranges from subsurface waters to at least 1200 m, and appears to undergo ontogenetic descent. Many specimens of *Histioteuthis miranda* have been frequently caught in deep-water lobster trawls at 300–600 m off the North West Shelf of Australia and the Great Barrier Reef, suggesting an association with the sea floor at some time during their life cycles. Eggs 0.8 mm diameter.

Off the eastern Australian coast maturity is reached at mantle lengths of ~200–250 mm.

*Interest to fisheries.* There is no direct fisheries interest, but the species is important as prey of commercially harvested fishes.

## SIZE

A large- to medium-sized squid. The maximum ML of mature females is 270 mm, of mature males is 260 mm.

### DISTRIBUTION

*Histioteuthis miranda* occurs in the Pacific and Indian Oceans but it is absent from the Atlantic Ocean, except off extreme southern South Africa in the Agulhas Current (~33°S). It occurs in the Indian Ocean southern subtropical waters from south-eastern South Africa eastward to Australian/New Zealand waters.

### REFERENCES

Young and Vecchione (2007a), Roper and Jereb (2010k).

## *Histioteuthis pacifica* (G. Voss, 1962)

*Calliteuthis celetaria pacifica* Voss, G. L. (1962). Six new species and two new subspecies of cephalopods from the Philippine Islands. *Proceedings of the Biological Society of Washington 75*, 169–176 [174].

Type data: holotype, NMNH 575453.

Type locality: Philippine Islands, between Jolo and Tawi Tawi, Dammi Island.

Preserved museum specimen, dorsal view (AM C.91869, A. Reid).

### COMMON NAME

Jewel Squid

### DIAGNOSTIC FEATURES

Funnel organ simple inverted V-shape. Fins 40–60% ML; width 63–75% ML. Arms 80–130% ML. Inner arm web vestigial, outer arm web absent. Male arms 1 hectocotylised: distal tips with equal-sized suckers on long pedicels; in large mature males, basal suckers on all arms enlarged and with swollen, fleshy collars. Manus of tentacular club with six or seven suckers in each transverse row. Two ventro-marginal series of club suckers with asymmetrical outer sucker rings. Median manal suckers 1.5 × marginal suckers. Large manal suckers with 28–32 teeth on margin. Buccal crown with seven supports.

*Photophores.* No large simple photophores on head; nine photophores on posterioventral margin of head. Photophores on mantle uniform in size, compound photophores large. Ventral mantle with large, uniformly sized and evenly spaced photophores. Arms 4 with three longitudinal series of photophores basally. Distal tip of arms 4 with group of small, compound photophores, separated by gap from proximal arm photophores. Simple photophores on arm tips absent; gap between arm photophores and those on distal tip present. Three series of photophores on arms 4. (In preserved specimens, small, simple photophores are not easy to recognise and may be variable.)

Preserved museum specimen, ventral view (AM C.91869, A. Reid).

*Sculpture.* Tubercles absent.

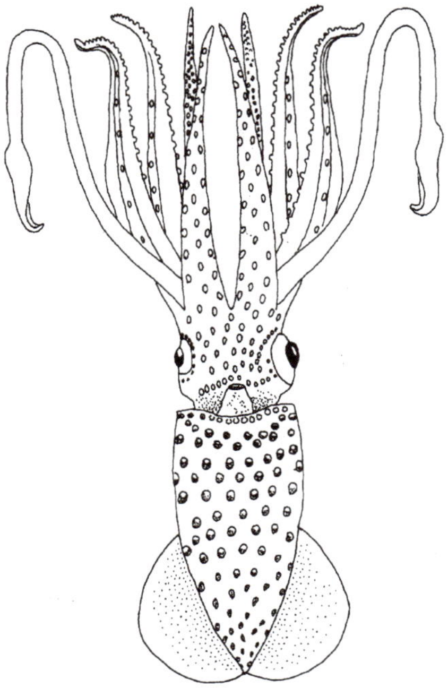

Ventral view (M. Shea, redrawn from Voss, 1963).

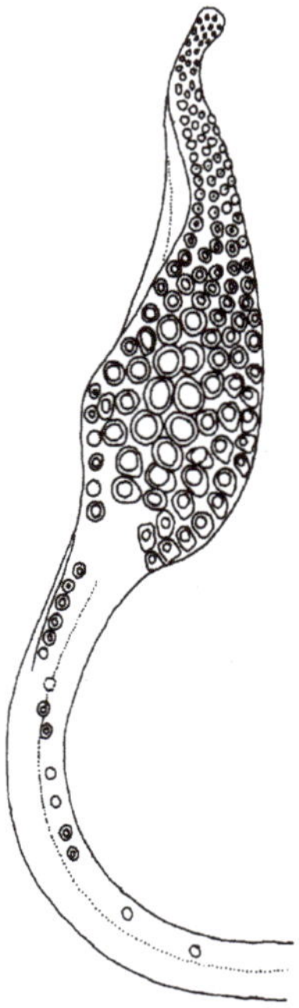

Tentacular club (M. Shea, redrawn from Voss, 1963).

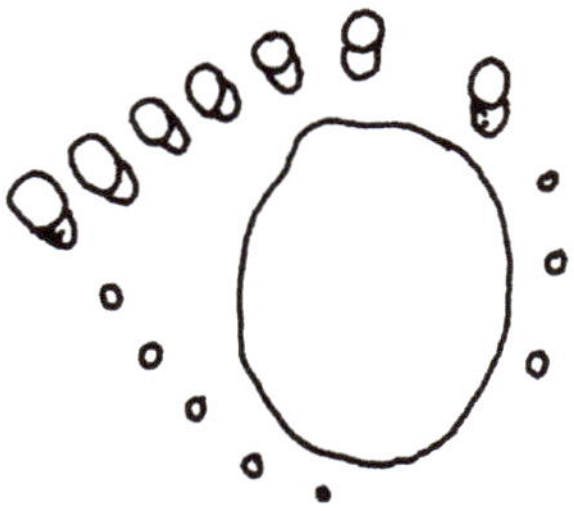

Photophores around left eye (M. Shea, redrawn from Voss, 1963).

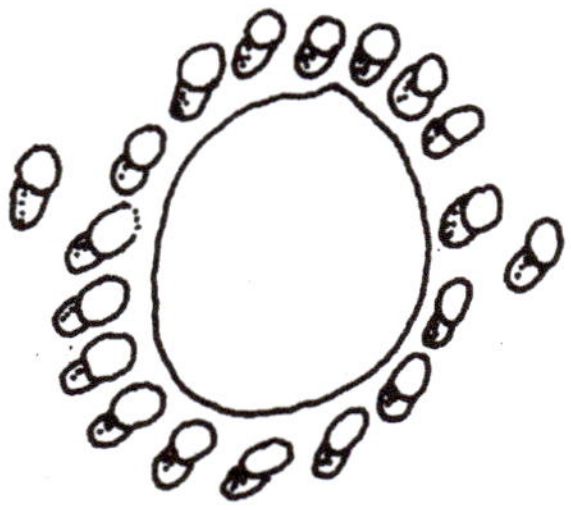

Photophores around right eye (M. Shea, redrawn from Voss, 1963).

*Histioteuthis pacifica*

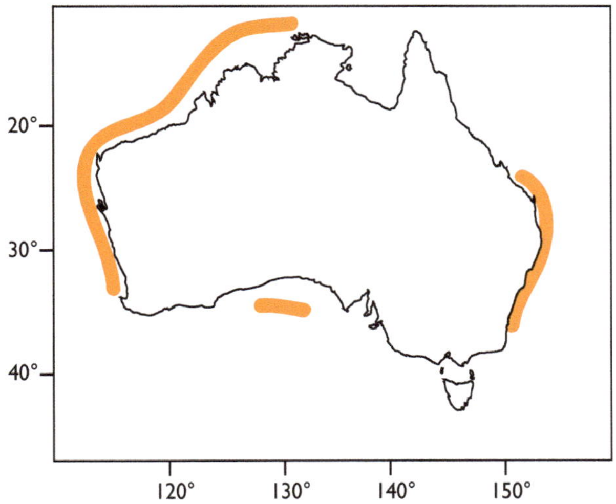


*Histioteuthis pacifica* distribution.

## HABITAT AND BIOLOGY

In other parts of the world (and unknown for Australia), juveniles occur in the upper 200 m but are capable of descending with growth to increasingly deeper depths, possibly to 1000 m. Subadult and adult males have been found during the day and at night in midwaters and near the bottom in shelf and slope waters at 250–1000 m. Mature females are unkown. The largest immature female recorded to date is 234 mm ML. Males mature between 60–280 mm ML.

## SIZE

The maximum ML reaches 280 mm in mature males.

## DISTRIBUTION

*Histioteuthis pacifica* is widely distributed in the tropical and subtropical regions of the Pacific and Indian Oceans.

## REFERENCES

Young and Vecchione (2000d), Young and Vecchione (2007a), Roper and Jereb (2010k).

# *Stigmatoteuthis* Pfeffer, 1900

## DIAGNOSTIC FEATURES

Large squids with dermal warts beneath outer epithelium. Tubercles absent. Arms very long relative to ML. Medial suckers of tentacular clubs greatly enlarged, more than 4 × diameter of minute marginal suckers. Secondary reproductive organs (penis, spermatophores gland complex, spermatophore storage sac) paired.

*Photophores.* Compound photophores large, uniform size and evenly spaced on anterior half of ventral mantle. Sawtooth arrangement of photophores on posterioventral margin of head (eight photophores and three sawteeth). Tips of arms 4 without gap between distal and proximal photophores.

## REFERENCE

Voss *et al.* (1998).

## *Stigmatoteuthis hoylei* (Goodrich, 1896)

*Histiopsis hoylei* Goodrich, E. S. (1896). Report on a collection of Cephalopoda from the Calcutta Museum. *Transactions of the Linnean Society of London, Zoology* 7(1), 1–24 [15, pl. 4 figs 62–71].

Type data: holotype NMGW.

Type locality: Indian Ocean, Andaman Sea, near Andaman Islands.

### COMMON NAME

Flowervase Jewel Squid

### DIAGNOSTIC FEATURES

Mantle conical, relatively thick-walled, relatively short, stout; skin of whole body (mantle, head, arms and basal half of fins) covered with thickly spaced, low, fleshy papillae, giving a rough, textured appearance. Head large, without nuchal folds. Fins medium-sized, round, 30–40% ML, width ~45–70% ML. Dorsal pad of funnel organ with strong median ridge from apical papilla posteriorly along each lateral arm; expands posteriorly to form two broad flaps.

Arms long, 160–250% ML; mature males with arms 1 very elongate, >400% ML. Outer web between arms very low, indistinct; inner web virtually absent. Hectocotylus with multiple sucker series, enlarged distal suckers and no tubercles. Both arms 1 modified on distal two-thirds in mature males: suckers palisade on elongated pedicels except near tip; suckers in two broadly separated series proximally, scattered suckers between marginal series over middle section of modified region; without papillated skin on ventral surface. Suckers on modified ends of arms 1 with short, sharply pointed teeth on distal and lateral or entire margins, suckers enlarged near tip. Arm sucker rings with 6–16 round or truncate teeth on distal and lateral margins, sometimes with numerous truncate teeth around entire margins on arms 4. Club with suckers in five to seven irregular rows on manus. Median suckers more than 4 × marginal suckers. Enlarged manus suckers with ~38 teeth in large animals. Spermatophore ejaculatory apparatus highly coiled, connective complex appears to be absent. Skin covered with low fleshy pads or 'warts'.

*Photophores.* Ventral and lateral surfaces of mantle, head, arms covered with elongate, compound photophores; widely spaced on ventral mantle and arranged in about six diagonal series; photophores uniform in size on anterior half of ventral mantle, progressively decrease in size on posterior two-thirds of mantle. Seventeen large photophores around right eyelid. No photophores on arm tips. Arms 4 with three longitudinal series of photophores proximally, arms 1–3 with one longitudinal series proximally. Photophores on mantle of uniform size. Simple photophores on arm tips absent, gap between arm photophores absent. Three series of photophores on arms 4.

*Sculpture.* Tubercles absent.

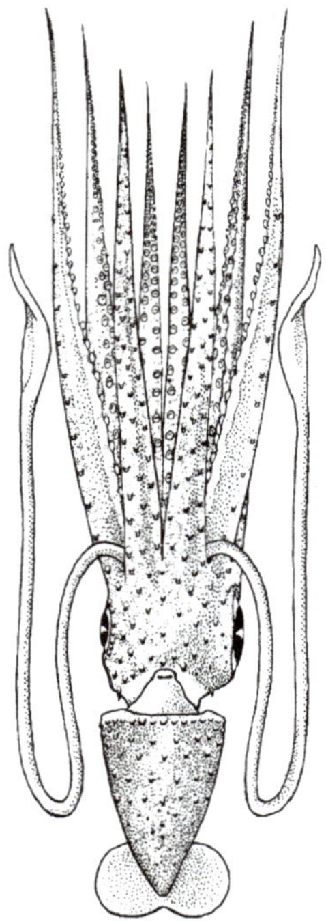

Ventral view (FAO, E. D'Antoni).

Tentacular club (FAO, E. D'Antoni).

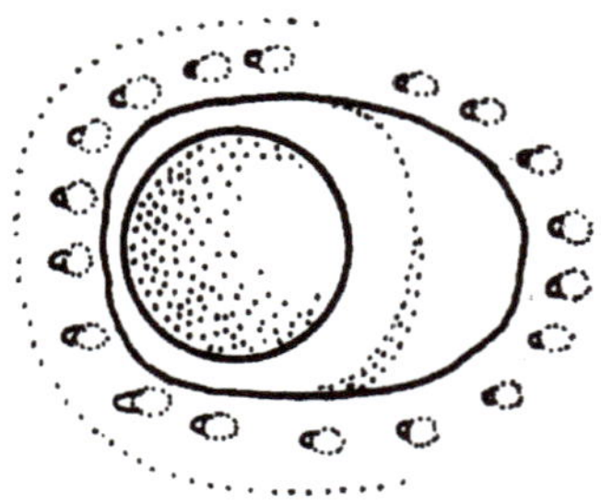

Photophores around right eye (FAO, E. D'Antoni).

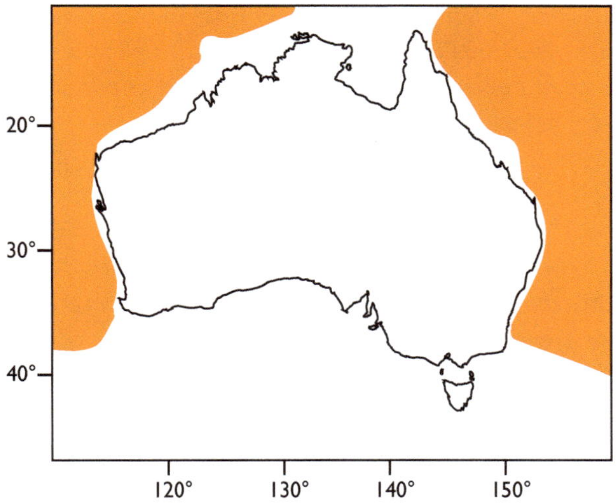

*Stigmatoteuthis hoylei* distribution.

#### HABITAT AND BIOLOGY

Analysis of the dimorphic eyes suggests that the large left eye points upward and the small right eye points ventrolaterally when the animal is in its normal oblique, arms downward, position. This enables the squid, in its daytime depth of 500–700 m, to use its large eye to utilise the dim daylight from above, while the small right eye perceives bioluminescent light from the side and below. Off Hawaii, *S. hoylei* has been captured at depths of 375–850 m during the day (with most captures >500 m). At night, captures were made at depths of 100–500 m. Larger squids tend to be caught at greater depths.

*Interest to fisheries.* None at present.

#### SIZE

The maximum ML attained is 240 mm in females, and to 210 mm in males.

#### DISTRIBUTION

*Stigmatoteuthis hoylei* is widely distributed in the tropical and subtropical waters of the Pacific Ocean between ~45°N and 45°S and in the Indian Ocean between ~10°N and the southern Subtropical Front.

#### REFERENCES

Young and Vecchione (2008), Roper and Jereb (2010k).

## Joubiniteuthidae Naef, 1922

The family is monotypic, represented by *Joubiniteuthis portieri* (Joubin, 1916).

### *Joubiniteuthis* Berry, 1920

As for family.

### *Joubiniteuthis portieri* (Joubin, 1916)

*Chiroteuthis portieri* Joubin, L. (1916). Estudes preliminaires sur les Cephalopodes recueillis au cours des croisieres de S.A.S. le Prince de Monaco. 4e Note: *Chiroteuthis portieri* nov. sp. *Bulletin de l'Institut Océanographique, Monaco 317,* 1–10 [1, text figs 1–3].

Type data: holotype MOM [station 3100].

Type locality: eastern Central Atlantic Ocean, 29°03′N, 16°08′30″W.

#### COMMON NAME

Joubin's Squid

#### DIAGNOSTIC FEATURES

Mantle long, narrow, conical posteriorly, semigelatinous. Head narrow, eyes small, neck long. 'Tail' (not shown in the figure) long (longer than the mantle), very slender, needle-like; composed of conus that is elongate with ventral fusion, and covered only by soft, thickened integument with membranous non-motile appendages. Fins relatively very small, ovoid, width nearly equal to length, ~30% ML; thick and gelatinous medially, without anterior or posterior lobes. Funnel-locking apparatus with oval depression. Arm pairs 1–3 extremely long, thin, >2 × ML; three or more times longer than ventral arms; joined by low web; very small suckers in six transverse rows. Ventral arms short (length one-third of, or less, than length of arms 1–3) with tetraserial suckers. Suckers in six series on arms 1–3; suckers in four series on arms 4. Basal-most suckers with smooth rings, more distal suckers with 6–10 teeth on distal half of inner ring. Arms 4 with broad lateral membranes. Tentacles in juveniles shorter and much thinner than arms 1–3; adult tentacles, thread-like; tentacular clubs laterally compressed, with minute suckers in five or six transverse rows, increasing to 8–12 transverse rows towards distal third of club; protective (or other) membranes present only on distal tip of club; no fixing apparatus (carpus), club keels absent. Club suckers with smooth inner rings. Hectocotylus absent. Buccal membrane with six lappets, oral face of membrane with numerous finger-like papillae; buccal connectives attach to ventral borders of arms 4.

Paralarvae with semi-transparent, spindle-shaped mantle, very small leaf-like fins, needle-like 'tail' longer than mantle, possibly a small secondary fin at posterior one-third

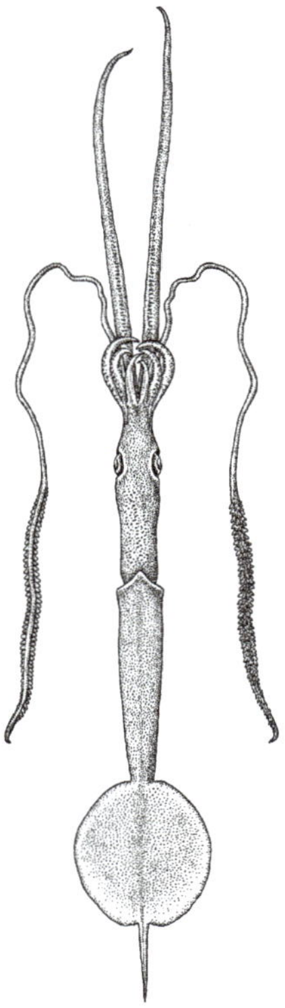

Dorsal view (ABRS, K. Hollis).

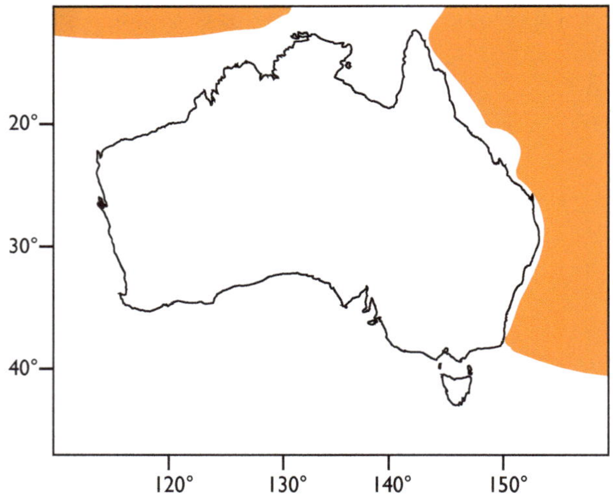


*Joubiniteuthis portieri* distribution.

of needle-like 'tail'. Neck short without septa. Arm-crown pillar absent. Arms and tentacles proportionally very short at mantle lengths < 10 mm; tentacles short and thick, club very short, with tetraserial suckers at 6–9 mm ML, 8 transverse rows at 18 mm ML. Thick tentacles, long arms, and very long 'tail' are characteristic in adults.

*Colour.* Surfaces of mantle, fins, funnel, head and arms with small reddish brown chromatophores. Chromatophores closely packed on oral base of arms and both sides of buccal membrane.

*Photophores.* Absent.

#### HABITAT AND BIOLOGY

A mesopelagic to bathypelagic, rarely encountered squid. Little is known about its biology. Depths of capture in open nets range from 300–500 m at night and 800–2500 m during the day. Recent ROV (remotely operated vehicle) observations have shown that, aside from constant beating of the 'tail' fin, the squid moves very slowly. The squid appears to extend its tentacles in a similar fashion to that seen in *Mastigoteuthis* and *Chiroteuthis* with the lateral membranes of the ventral arms forming tentacle sheaths. The arms are often spread apart and curled or bent in the middle, perhaps preventing their entanglement.

*Interest to fisheries.* None.

#### SIZE

A small-sized squid; maximum ML to 105 mm; 'tail' length an additional 155 mm; females clearly developing but immature at 85 mm ML, with paired nidamental glands.

#### REMARKS

This is a very rarely captured species with only a few dozen specimens known worldwide.

#### DISTRIBUTION

Circumglobal, in tropical and subtropical waters, and even in temperate waters. It is found off eastern Australia, the Tasman Sea, New Zealand as well as the Atlantic Ocean (40°N to 30°S), including the Caribbean Sea; Pacific Ocean, Hawaii and Japan. Subersibles have found these squids at ~1100 m during the day.

#### REFERENCES

Young (2009b), Roper and Jereb (2010k).

### Lepidoteuthidae Pfeffer, 1912

The family Lepidoteuithidae is monotypic, represented by *Lepidoteuthis grimaldii* Joubin, 1895.

## *Lepidoteuthis* Joubin, 1895

### *Lepidoteuthis grimaldii* Joubin, 1895

*Lepidoteuthis grimaldii* Joubin, L. (1895). Cephalopodes recueillis dans l'estomac d'cachalot, capture aux isles Acores. *Compte Rendu des Sèances de l'Acadèmie des Sciences 121*, 1172–1174 [1172, text fig.].

Type data: syntypes (2) MOM.

Type locality: central North Atlantic Ocean, Azores Islands, near Teceira, 38°34′45″N, 29°37′W (from sperm whale stomach).

Dorsal view (K. Hollis, ABRS).

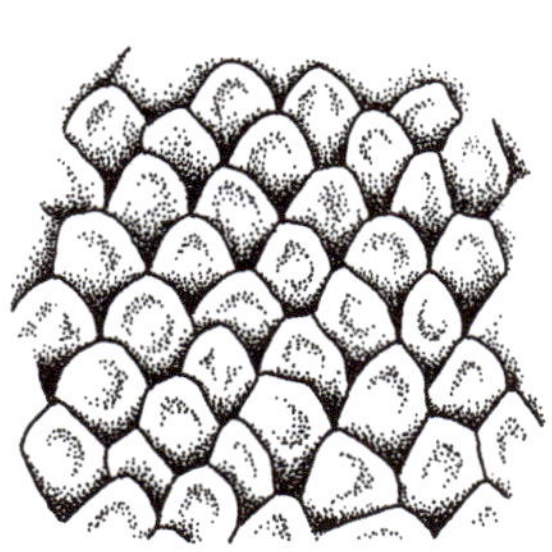

Enlargement of scales (K. Hollis, ABRS).

#### COMMON NAME

Soft-scaled Squid

#### DIAGNOSTIC FEATURES

Posterior end of mantle bulges slightly. Fins large, terminal, not lobed or scaled, together oval-shaped. Funnel-locking cartilage a straight, simple groove, deeper anteriorly. Adult arms thickened, subequal in length; suckers biserial with long, sharply pointed teeth on distal half of inner ring. A single saber-like hook present near base of each arm 2 in males. Hectocotylus absent. Large, thick, tentacles fail to grow much beyond paralarval stage and are lost in subadults and adults (from ~85 mm ML); few (six to eight) variously sized suckers in two longitudinal series on small, slightly expanded tentacular clubs of paralarvae and juveniles. Buccal connectives attach to ventral borders of arms 4.

*Photophores.* Absent.

*Sculpture.* Mantle covered with overlapping, relatively large (at least 10 mm × 12 mm in adults), dermal cushions; dermal cushions diamond- to hexagonal-shaped structures that cover all of mantle (except posterioventrally under posterior half of fins); they resemble scales of some fishes in outline and in tightly overlapping arrangement; internal structure highly vacuolate with loose connective tissue forming 'walls' of chambers. This layer ends in anterior half of fins. Mantle of paralarvae and juveniles covered with small papillae – precursors of dermal cushions – with two to four minute 'cartilaginous' points or tips.

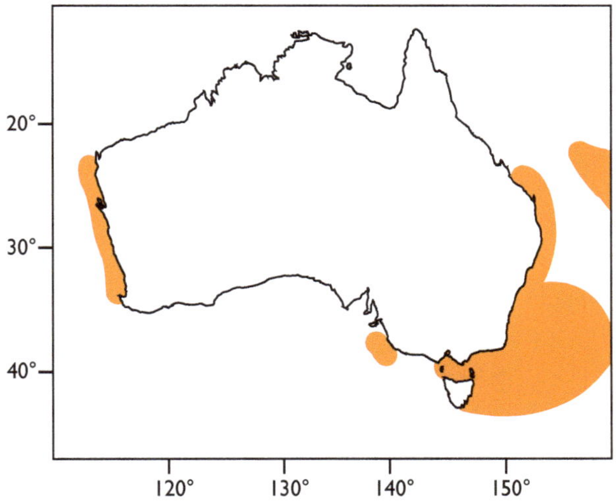


*Lepidoteuthis grimaldii* distribution.

#### HABITAT AND BIOLOGY

This squid is rarely captured so little is known of its biology. The extent of its vertical distribution not clear but it appears to be a mesopelagic to bathypelagic, benthic-bathyal species; paralarvae have been taken in upper 100 m at night; juveniles and subadults have been taken in open nets to 700 m day and night. All other records of larger subadults and adults come from the stomachs of predatory deep-sea fishes and whales that normally feed from a few hundred metres to 2000 m. Paralarvae attain a relatively large size (10 mm ML or larger) before becoming juveniles.

The unusual dermal cushions with very numerous internal vacuoles might function as a buoyancy mechanism if the vacuoles contain a light-weight fluid such as ammonium chloride.

*Interest to fisheries.* None currently, but it attains a large size and its mantle musculature is relatively firm, not gelatinous. Its deep-sea habitat and apparently low population density may deter developing fisheries.

#### SIZE

A large-sized squid; maximum ML to 1 m.

#### DISTRIBUTION

Tropical and subtropical regions of the Atlantic, Pacific and Indian Oceans. It is a cosmopolitan tropical and subtropical species with broadly disjunct records, for example in the Pacific Ocean from Hawaii, Honshu, Japan, New Caledonia, eastern Australia, Tasman Sea; western Australia, southern Africa; broadly in the Atlantic Ocean.

#### REFERENCE

Roper and Jereb (2010m).

## Lycoteuthidae Pfeffer, 1908

#### COMMON NAME

Firefly Squids

#### DIAGNOSTIC FEATURES

Mantle conical. Fins broad, rhomboidal, sometimes posteriorly extended into a tail. Funnel-locking apparatus a straight, simple groove. Dorsolateral arms elongate in males. Arm suckers biserial, sharp-toothed. Club suckers tetraserial; hooks absent; carpal-locking apparatus with few suckers and knobs, often compact, in circular arrangement. Buccal membrane with eight lappets and supports; connectives to arms 4 attach to dorsal margins. Spermatangia attach to modified tissue in nuchal region.

*Photophores.* Photophores on viscera, apex of mantle and ventral surface of eyes in some genera. Four or five oval photophores on ventral surface of eyeball. Visceral photophores on anal, branchial, abdominal and posterio-abdominal organs. Spherical photophores embedded in tentacular stalks. Sexual dimorphism is pronounced in some genera in which males have branchial photophores and females have larger, more numerous photophores.

#### HABITAT AND BIOLOGY

These squids occupy mesopelagic to bathyal/mesopelagic depths during the day and migrate into near-surface waters at night. They possess a large variety of luminous organs.

#### SIZE

Mostly small-, rarely medium-sized squids; maximum recorded ML 194 mm.

#### REMARKS

Two subfamilies are recognised: Lycoteuthinae and Lampadioteuthinae. Only one representative of the Lycoteuthinae is recognised as occurring in Australian waters: *Lycoteuthis lorigera* (Steenstrup, 1875).

#### DISTRIBUTION

Circumglobal, mostly in tropical and subtropical waters, but not known from the North Pacific Ocean.

#### REFERENCES

Arocha (2003), Roper and Jereb (2010n).

## Lycoteuthinae Pfeffer, 1908

### *Lycoteuthis* Pfeffer, 1900

#### DIAGNOSTIC FEATURES

Arms 2 greatly elongate in males with series of regularly spaced photophores on aboral surface. Two tentacular photophores in females (more in males); three abdominal photophores; one transverse tail photophore (at least in *Lycoteuthis springeri* Voss, 1956); males with additional photophores on arms 3, head and mantle. Male genitalia paired.

#### REMARKS

Sexual dimorphism in body shape and photophore arrangement suggests bioluminescent courtship behaviour, but this has never been observed. These are thought to be mesopelgic squids located over island and continental slopes (Nesis 1982/1987). The genus is represented by two species, one of which, *L. lorigera* (Steenstrup, 1875), is found in Australian waters.

### *Lycoteuthis lorigera* (Steenstrup, 1875)

*Onychoteuthis lorigera* Steenstrup, J. (1875). *Hemisepius*, en ny Slaegt af Sepia-Blacksperutternes Familie, med Bemaerkninger om Sepia-Formerne i Almindelighed. *Danske Videnskabernes Selskabs Skrifter, 5 Raekke, Naturvidenskabelig og Mathematisk 10*(7), 465–482 [473].

Type data: holotype ZMUC.

Type locality: Pacific Ocean, from sperm whale stomach.

#### COMMON NAME

Crowned Firefly Squid

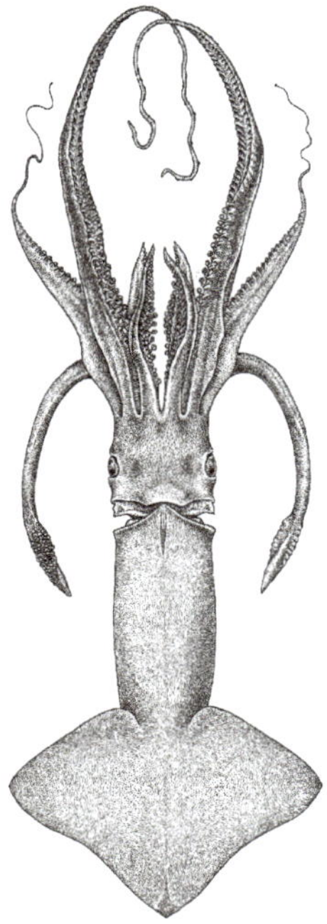

Male, dorsal view (ABRS, K. Hollis).

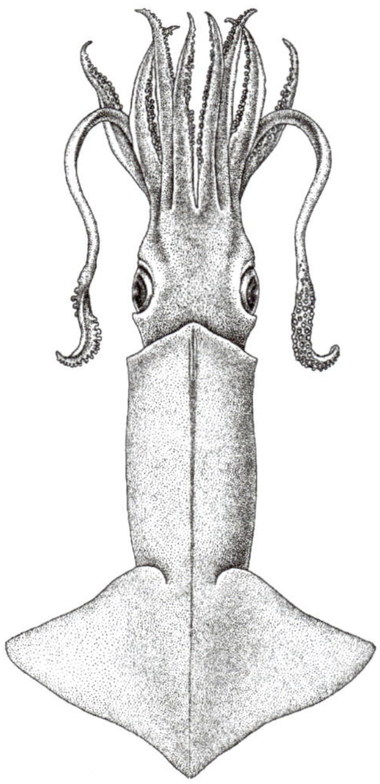

Female, dorsal view (ABRS, K. Hollis).

### DIAGNOSTIC FEATURES

Males lack elongate tail. Arms 2 greatly elongated in males. Male genitalia paired.

*Photophores.* Live and fresh-caught specimens with bright blue mid-ventral ocular photophores on each eye. Series of regularly spaced photophores on aboral surface of arms 2 in males. Three groups of ventral mantle photophores present. Two widely spaced photophores anteriorly, median transverse row of about six photophores, with the middle three grouped together; posterior abdominal photophores fused into a single, broad organ in males. Males with additional photophores on arms 3, head and mantle.

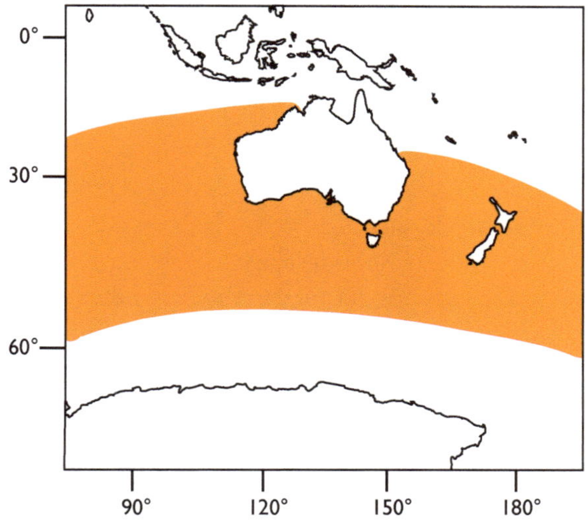


*Lycoteuthis lorigera* distribution.

### HABITAT AND BIOLOGY

Bathyal- and mesopelagic over slopes and seamounts. Females smaller than males, ~110 mm ML maximum. Paired spermatophoric and terminal organs in males are both functional.

*Interest to fisheries.* No commercial fishery currently exists.

### SIZE

Mantle length maximum to ~190 mm in males.

### REMARKS

Females of this species are currently indistinguishable from those of *Lycoteuthis springeri*. It has been suggested that the very elongated and modified arms 2 and 3 in males may play a role in spermatophore transfer during mating (Villanueva and Sanchez 1993, Hoving *et al.* 2007).

### DISTRIBUTION

Circumglobal in southern, subtropical and notalian zones. In Australia known from southern New South Wales and eastern Tasmania. Depth to at least 400 m.

### REFERENCES

Vecchione and Young (1999), Roper and Jereb (2010n).

## Mastigoteuthidae Verrill, 1881

### COMMON NAME(S)

Mastigoteuthid Squids, Whiplash Squids

### DISTINGUISHING CHARACTERS

Differ from chiroteuthids (also with enlarged ventral arms) in having poorly developed knobs on the funnel-locking cartilage and having hundreds of minute suckers on the tentacular clubs (rather than fewer arranged tetraserially in chiroteuthids, apart from *Idioteuthis cordiformis* Chun, 1908).

### DIAGNOSTIC FEATURES

Medium-sized to large squids. Mantle weakly muscular to semigelatinous; anterior portion cylindrical, posterior part tapered, conical; ventral anterior margin produced into two points in position of underlying mantle-locking cartilage. Some species with minute dermal tubercles covering surface of mantle, fins, arms, head and funnel. Fins large, with half or more of their length posterior to muscular part of mantle; heart-shaped, circular, elliptical or lanceolate in outline, usually longer than 50% ML. Funnel-locking cartilage oval, with inward projecting knobs or ridges; antitragus (generally) and tragus (occasionally). Mantle-locking cartilage ovate, flask, C-shaped or nose-shaped. Ventral arms elongate, thick with expanded lateral membranes; arm suckers biserial, without hooks. Hectocotylus absent. Tentacles long, whip-like, weakly muscled (often lost during capture; transverse muscles poorly developed), with hundreds of densely set minute suckers. Clubs not expanded or only slightly expanded; elongate with very numerous (often thousands) of small to microscopic suckers (except in *I. cordiformis*) arranged in many irregular series (usually more than 15, even more than 30 in some species); club without keels or locking (fixing) apparatus. Buccal connectives with seven lappets; attached to ventral border of ventral arms. Buccal formula DDVV. Gladius narrow, with long conus that forms a needle-like 'tail' spike that extends posterior to terminus of fins.

*Photophores.* Photophores variably present (depending on species) on surface of mantle, ventral surfaces of head, ventral arms, eyeball and eye-sinus. Photophores variously present in most species: some species with numerous minute, round cutaneous organs embedded in ventral surface of mantle, head and arms.

*Colour.* Body pink or brick reddish coloured; much pigmentation not incorporated in chromatophores but dispersed in other integumental cells; many typical chromatophores usually present.

### HABITAT AND BIOLOGY

Mastigoteuthids are deep-living pelagic, or benthopelagic oceanic squids occurring from tropical waters to polar regions. They appear to live between 500 and 1000 m during the day and ascend into shallower water at night.

Observations from submersibles at 700–1000 m recorded mastigoteuthids in close proximity to the bottom, oriented vertically with head down and tentacles extended towards the sediment, presumably to capture copepods and other small plankton of the epibenthic zooplankton; position was maintained by gentle undulations of the very large fins. The tentacles are supported proximally by the tentacular sheath of arms 4 and are held apart in the 'tuning fork' posture. Paralarvae do not exhibit a high degree of sudden morphological change during early ontogeny, but tentacle length increases disproportionately rapidly, probably as a support for the 'dangling tentacle' feeding technique observed in adults. Tentacular suckers have a weak release mechanism (readily adhering to anything they touch) and retain a sensory function; these morphologies enable the species to forage in the relatively food-rich zone on and immediately above the bottom. They appear to lack a strong, rapid, jet propulsion capability. Tissues of the mantle, head and especially the large ventral arms are vesiculate and contain ammonium ions in the tissue fluid, providing near-neutral buoyancy; the positive buoyancy of the anterior portion is overcome by complex rhythmic wave-like undulations of the fins to maintain the head-down posture during feeding.

*Interest to fisheries.* None, due to their gelatinous bodies.

### SIZE

Medium to large-sized squids; ML up to 1 m.

### REMARKS

The Mastigoteuthidae is among the most taxonomically confused of all deep-sea squid families, principally because many characters are based on structures frequently lost during capture, such as the tentacles and skin photophores. Further, the family includes many poorly known species, some described only from one or very few, often badly damaged, specimens, or from paralarvae. The systematics of the family has been in a state of flux for many years. The most recent revision by Braid *et al.* (2014) was based on the phylogenetic analyses of three mitochondrial genes (COI, 16S rRNA and 12S) in addition to comparing morphology among species. This analysis, the results of which are followed here, supports the division of the family into five genera: *Mastigoteuthis*, *Idioteuthis*, *Mastigopsis*, *Echinoteuthis* and *Magnoteuthis*. Four genera occur in Australian waters, each represented by a single species.

### REFERENCES

Roper and Jereb (2010o), Braid *et al.* (2014).

## Key to genera and species of the Mastigoteuthidae

1a. Arm lengths subequal; arms 2 longer than arms 4 Tentacular club with very large suckers proximally. .... ...................................................... *Idioteuthis cordiformis*

1b. Arms 4 much longer than remaining arms. Tentacular club with minute suckers....................................... 2

2a. Funnel-locking cartilage ear-shaped; antitragus present. Photophores on integument and eye sinus ...... .............................................. *Mastigoteuthis psychrophila*

2b. Funnel-locking cartilage oval or flask-shaped; antitragus absent. No photophores on integument and eye sinus........................................................................ 3

3a. Fins heart-shaped. Funnel-locking cartilage oval; tragus absent. Skin tubercles present. Eye with two photophores..................................... *Mastigopsis hjorti*

3b. Fins elliptical. Funnel-locking cartilage flask-shaped; tragus present. Skin tubercles absent. Eye photophores absent .................. *Magnoteuthis magna*

## *Idioteuthis* Sasaki, 1916

### DIAGNOSTIC FEATURES

Fins heart-shaped or elliptical in outline; fin length ~81% ML. Funnel bulbous posteriorly, cylindrical anteriorly. Funnel-locking cartilage ear-shaped, tragus and antitragus present; funnel pocket absent. Mantle-locking cartilage nose-shaped, posteriorly undercut in profile. Arms suckers biserial, with blunt teeth, or teeth absent. Arm length subequal, arms 2 often slightly longer than arms 4, trabeculae present on protective membranes. Tentacular club suckers large proximally, minute distally; club slightly expanded with trabeculate membrane.

*Photophores.* Integumental photophores absent. Eye sinus photophore absent. A single crescent-shaped photophore present on ventral surface of eye.

*Sculpture.* Skin tubercles present.

### SIZE

Mantle length at maturity ~450 mm to > 1 m.

### REFERENCE

Braid *et al.* (2014).

## *Idioteuthis cordiformis* Chun, 1908

*Mastigoteuthis cordiformis* Chun, 1908, Über Cephalopoden der Deutschen Tiefsee-Expedition. *Zoologischer Anzeiger 33*(2), 86–89 [88].

Type data: holotype ZMB Moll.-110006.

Type locality: eastern Indian Ocean, Indonesia, Sumatra, off Nias Island, 0°15′N, 98°08′E.

### DIAGNOSTIC FEATURES

Fins heart-shaped or elliptical; fins 72–*81*–92% ML. Funnel-locking cartilage ear-shaped; tragus present, antitragus present, weak; mantle-locking cartilage nose-shaped. Arms 4 64–*78*–99% ML (min.–*mean*–max.), arm sucker teeth blunt or absent. Tentacular club suckers large to minute.

*Photophores.* A single photophore present on each eye, none on integument or eye sinus.

*Sculpture.* Skin tubercles present.

### HABITAT AND BIOLOGY

This wide-ranging species lives in tropical and subtropical waters to depths of 1000 m and undergoes diel vertical migration.

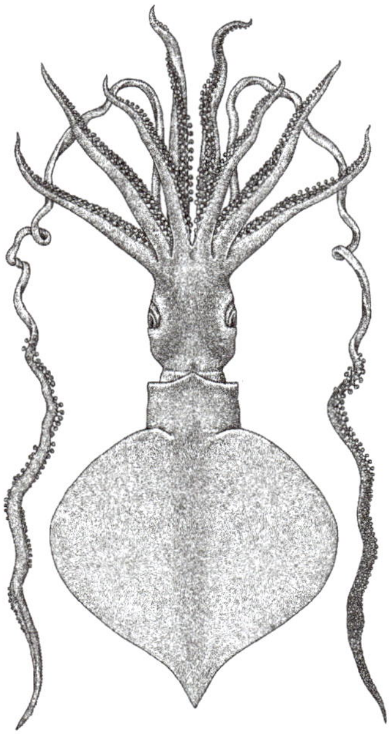

Dorsal view (ABRS, K. Hollis).

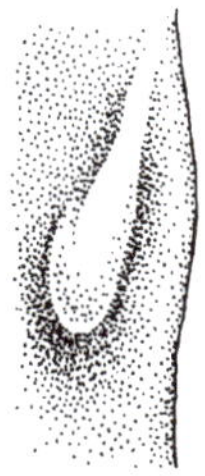

Mantle-locking cartilage (ABRS, K. Hollis).

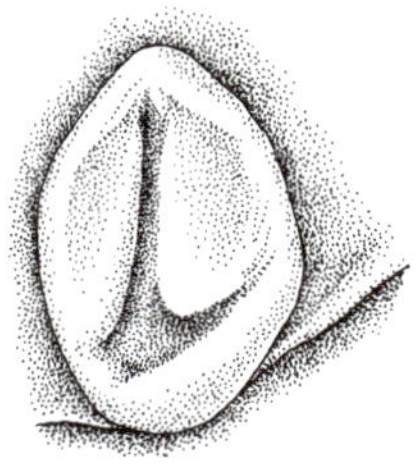

Funnel-locking cartilage (ABRS, K. Hollis).

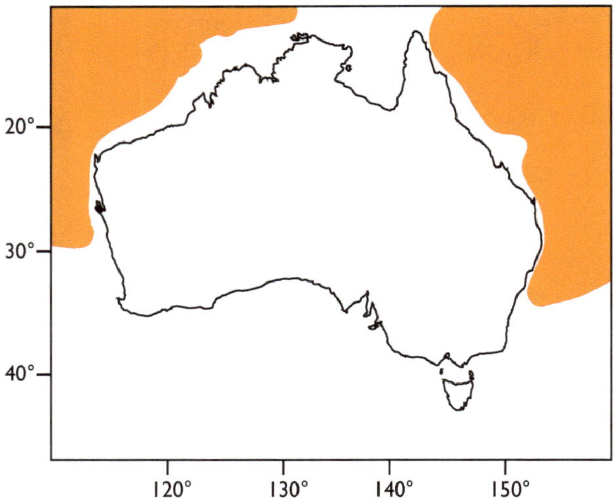


*Idioteuthis cordiformis* distribution.

### SIZE

Maximum ML known to reach 700 mm. (Unpublished reports record it to 1 m or more.)

### REMARKS

According to Braid *et al.* (2014), the status of *Idioteuthis cordiformis* is unclear. It shares several traits with other mastigoteuthids, such as the ear-shaped funnel-locking cartilage and whip-like tentacles with minute distal suckers. However, in contrast to other mastigoteuthids, the tentacular clubs of *I. cordiformis* are slightly expanded and the proximal suckers are large and have similar morphology to the arm suckers. Additionally, the arms are subequal, with arms 2 often longer than arms 4. In other mastigoteuthids, arms 4 are much longer than the other arms. The ML of large *I. cordiformis* individuals is at least five times greater than any other mastigoteuthid at maturity.

### DISTRIBUTION

Broadly distributed on the continental slope in the Indo-West Pacific, Indonesia, the Philippines, southern Japanese waters and of the north-west slope, Australia, the Tasman Sea and New Zealand waters.

### REFERENCES

Roper and Jereb (2010o), Braid *et al.* (2014).

## *Magnoteuthis* Salcedo-Vargas & Okutani, 1994

### DIAGNOSTIC FEATURES

Fins heart-shaped or elliptical in outline; fin length ~66% ML. Funnel bulbous posteriorly, cylindrical anteriorly. Funnel-locking cartilage flask-shaped, tragus present, antitragus absent, convex protrusion present along lateral margin; funnel pocket absent. Mantle-locking cartilage flask-shaped, posteriorly undercut in profile. Arm suckers biserial but merge into a single series midway on arms 4. Arm suckers adentate, or with blunt teeth; arms 4 ~110% ML; protective membrane without trabeculae. Tentacular suckers microscopic, club not expanded, protective membranes without trabeculae.

*Photophores.* Integumental photophores minute or absent. Two round photophores present on each eye.

### SIZE

Mantle length at maturity ~220 mm.

### REFERENCE

Braid *et al.* (2014).

## *Magnoteuthis magna* Joubin, 1913

*Mastigoteuthis magna* Joubin, L. (1913). Estudes preliminaires sur les Cephalopodes recueillis au cours des croisieres de S.A.S. le Prince de Monaco. 3e Note: *Mastigoteuthis magna* nov. sp. *Bulletin de l'Institut Océanographique, Monaco 275*, 1–11, Fig. 1. [2, fig. 1].

Type data: holotype MOM [station 2108].

Type locality: Atlantic Ocean, Sargasso Sea, 31°44′30″N, 42°39′W.

### DIAGNOSTIC FEATURES

Fins elliptical; fin length index 55–*67*–77 (min.–*mean*–max.). Funnel-locking cartilage flask-shaped; tragus present, antitragus absent; mantle-locking cartilage flask-shaped. Arms 4 length index 91–*103*–75, arm sucker teeth absent. Tentacular club suckers minute, microscopic.

*Photophores.* Absent from integument, eye sinus and eye.

*Sculpture.* Skin tubercles absent.

### SIZE

Maximum ML known 160 mm.

### DISTRIBUTION

*Mastigoteuthis magna* appears to have a disjunct distribution. It occurs commonly throughout the western central North Atlantic, from the Bahamas, off Cape Hatteras, North Carolina, and northward, as well as along a transect in the eastern central Atlantic at 30°W. The species also occurs over deep water in the western North Pacific off Japan. Recorded from off eastern Australia.

Ventral view (Joubin 1920, Plate X).

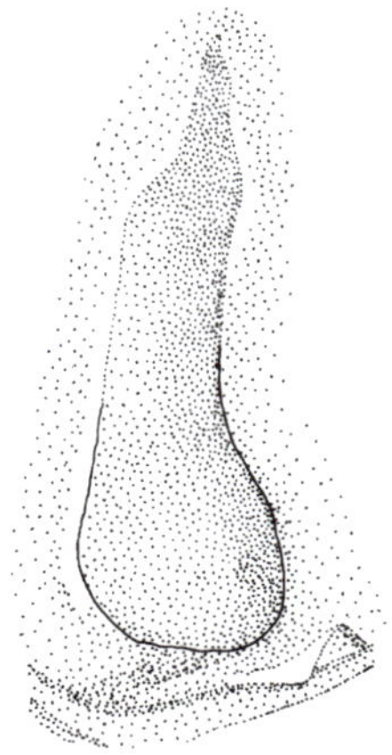
Funnel-locking cartilage (M. Shea, redrawn from Vecchione and Young 2010).

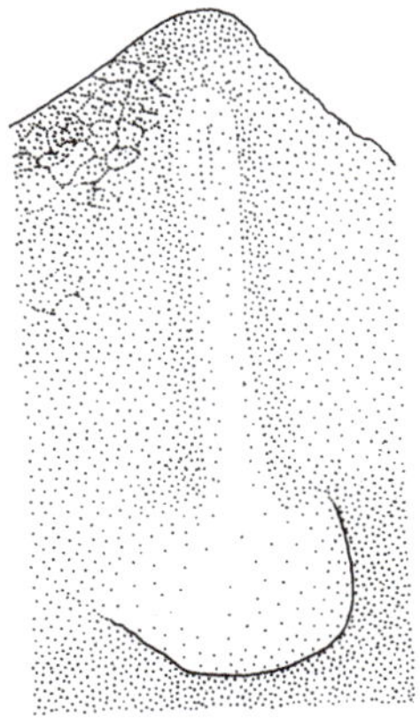
Mantle-locking cartilage (M. Shea, redrawn from Vecchione and Young 2010).

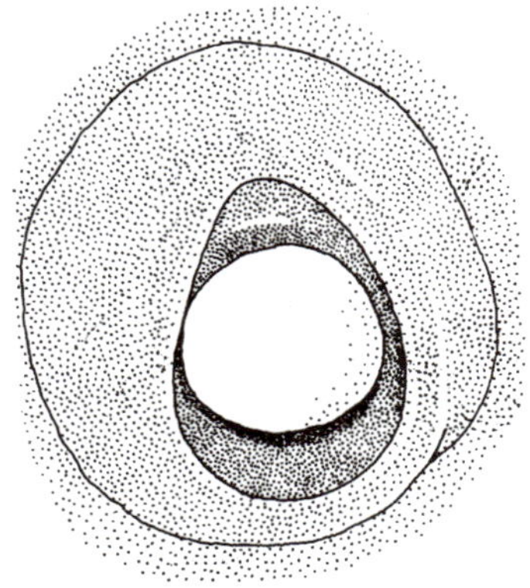
Arm sucker ring (M. Shea, redrawn from Vecchione and Young 2010).

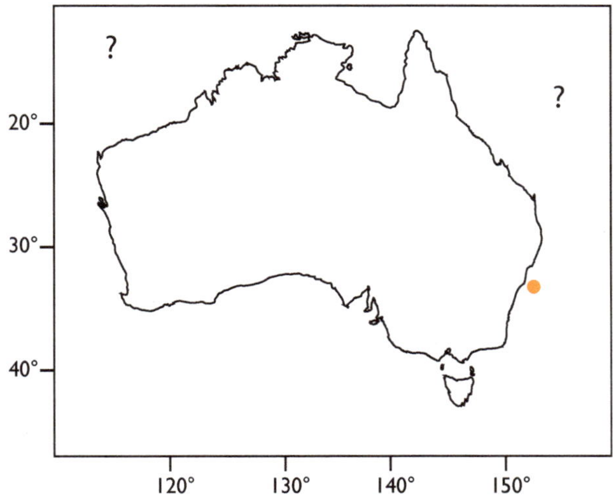

*Mastigoteuthis magna* distribution. (Australian records are based on few specimens, but the species is likely to occur more widely.)

#### REFERENCES

Roper and Jereb (2010o), Braid *et al.* (2014).

## *Mastigopsis* Grimpe, 1922

#### DIAGNOSTIC FEATURES

Skin tubercles present. Fins heart-shaped in outline, fin length ~89% ML. Funnel narrowly conical. Funnel-locking cartilage oval with convex protrusion along outer margin; tragus and antitragus absent. Mantle-locking cartilage ovate, posteriorly undercut in outline. Arm suckers biserial, but appearing to merge into a single series halfway along arms 4; arm suckers with blunt teeth. Arms 4 ~78% ML; protective membranes without trabeculae. Tentacular club suckers minute, club not expanded, with trabeculate protective membrane.

*Photophores.* Integumental photophores absent. Two round photophores on each eye (one posterioventral, one anterioventral). Eye sinus photophore absent.

#### SIZE

Mantle length at maturity ~180–235 mm.

#### REFERENCE

Braid *et al.* (2014).

### *Mastigopsis hjorti* Chun, 1913

*Mastigoteuthis hjorti* Chun, C. (1913). *Report on the Scientific Results of the 'Michael Sars' North Atlantic Deep-sea Expedition 1910, 3*(1), 1–28 [8, text-fig. 1, pl. 2, fig. 1].

Type data: syntypes (5) ZMUB.

Type locality: tropical central North Atlantic Ocean, 31°24′N, 34°47′W.

#### DIAGNOSTIC FEATURES

Fins heart-shaped; fin length index ~89. Funnel-locking cartilage oval; tragus and antitragus absent; mantle-locking cartilage oval. Arms 4 length index ~78, arm sucker teeth long, blunt. Tentacular club suckers minute.

*Photophores.* Two photophores present on each eye, none on integument or eye sinus.

*Sculpture.* Skin tubercles present.

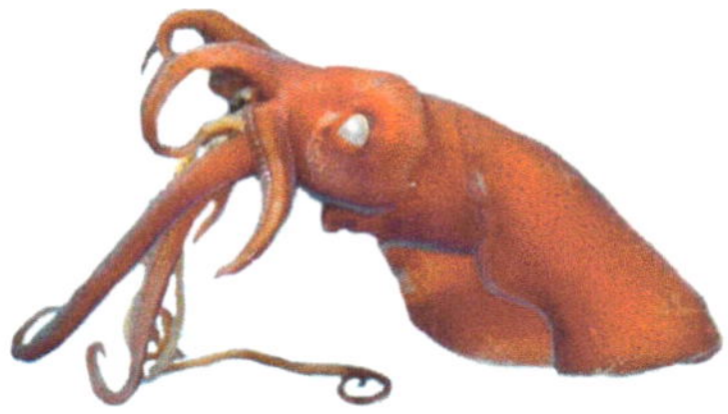

Specimen photographed using slurp gun from the *Johnson Sea Link II* submersible in the western North Atlantic (R. Young).

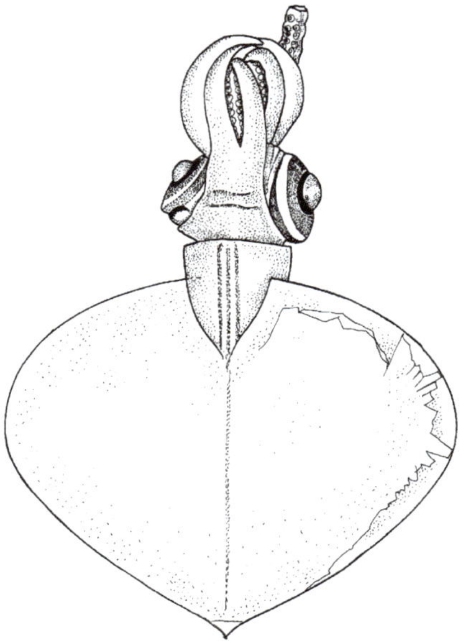

Dorsal view (FAO, E. D'Antoni).

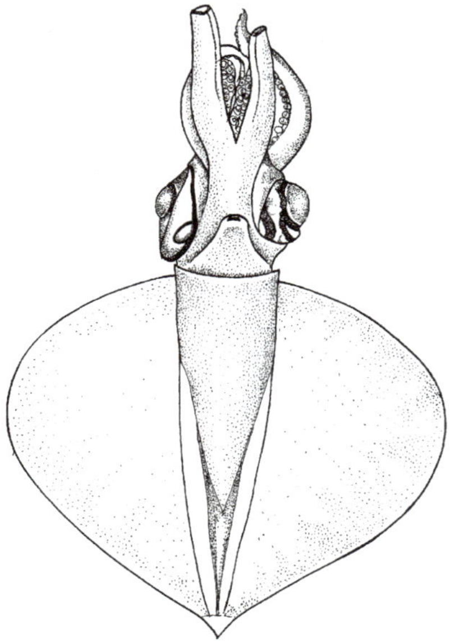

Ventral view (FAO, E. D'Antoni).

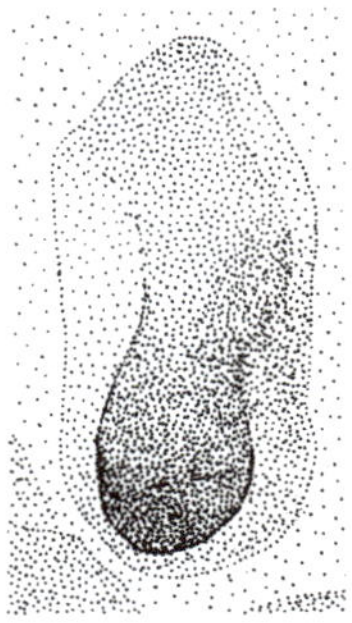

Funnel-locking cartilage (M. Shea, redrawn from Vecchione and Young 2007).

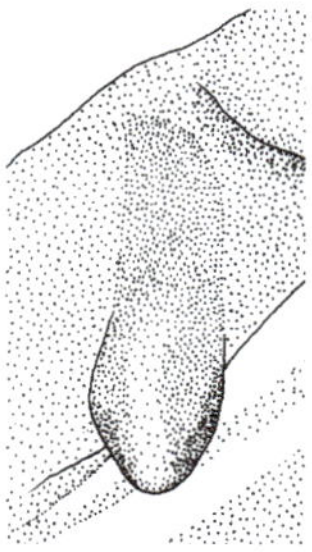

Mantle-locking cartilage (M. Shea, redrawn from Vecchione and Young 2007).

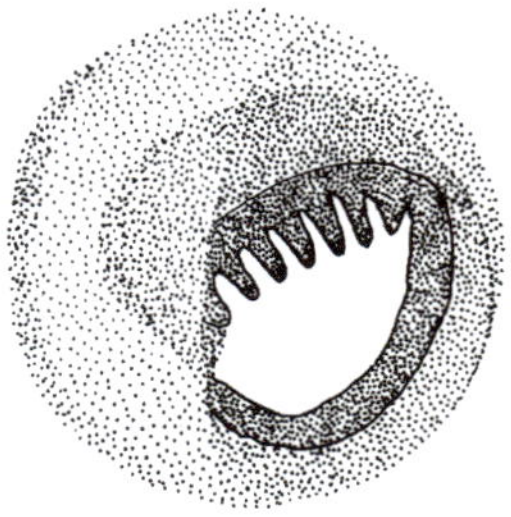

Arm sucker ring (M. Shea, redrawn from Vecchione and Young 2007).

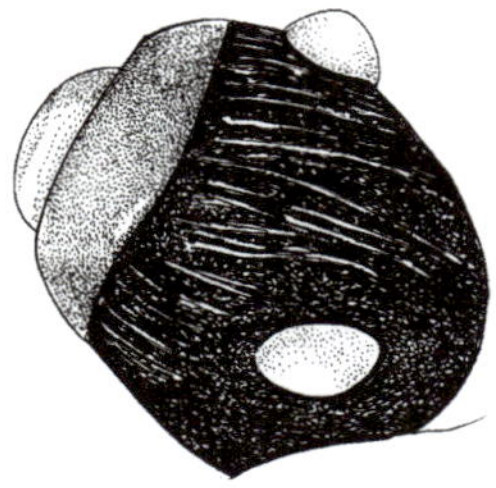

Ventral view of eye showing photophores (FAO, E. D'Antoni).

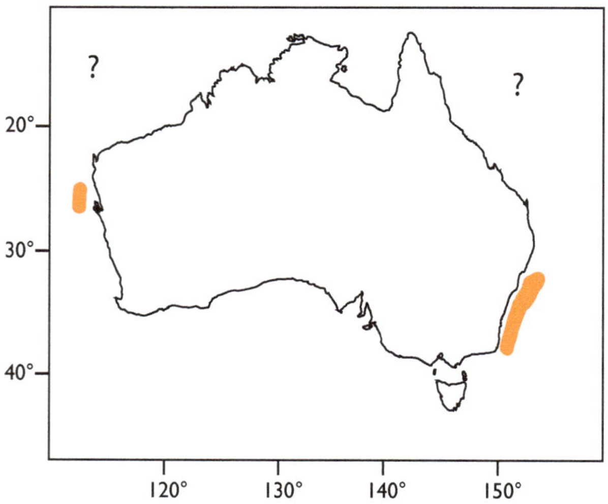


*Mastigopsis hjorti* distribution. (Australian records are based on few specimens, but the species is likely to occur more widely.)

## HABITAT AND BIOLOGY

Observations of *Mastigopsis hjorti* that were made from the manned submersible *Johnson Sea-Link II* in the Gulf of Mexico showed that it oriented vertically with the head down and tentacles extending to the sediment. The orientation is maintained by gentle, complex undulation of the fins.

## SIZE

Maximum ML reaches 100 mm.

## DISTRIBUTION

Very broadly distributed, possibly as a tropical and subtropical circumglobal species with northern limits of ~40°N and southern boundaries undetermined. It occurs in the North Atlantic Ocean from eastern regions of the Gulf of Mexico, northward to Bermuda and to the Gulf of

Maine in the western North Atlantic Ocean, and in the eastern Atlantic Ocean, off South Africa, the central Pacific Ocean and the Indian Ocean.

#### REFERENCES

Roper and Jereb (2010o), Braid *et al.* (2014).

## *Mastigoteuthis* Verrill, 1881

#### DIAGNOSTIC FEATURES

Skin tubercles absent. Fins circular or elliptical in outline; fin length ~64% ML. Funnel conical, broad, with recurved tip. Funnel-locking cartilage ear-shaped with tragus and antitragus present; funnel pocket present. Mantle-locking cartilage nose-shaped, posteriorly undercut in profile. Arm suckers biserial, appearing to merge into a single series midway along arms 4; suckers with sharp teeth. Arms 4 ~137% ML with minute protective membrane. Tentacular club suckers minute, club not expanded, protective membranes without trabeculae.

*Photophores.* Integumental photophores present. Small eye-sinus photophore present, diameter ~9% eye diameter. Eye photophore absent in most species.

#### SIZE

Medium to large-bodied (ML ~130–200 mm).

#### REFERENCE

Braid *et al.* (2014).

### *Mastigoteuthis psychrophila* Nesis, 1977

*Mastigoteuthis psychrophila* Nesis, K. N. (1977). *Mastigoteuthis psychophila* sp. n. (Cephalopoda, Mastigoteuthidae) from the Southern Ocean. *Zoologicheskij Zhurnal 65*(6), 835–842 [835, text figs a–e].

Type data: holotype ZIN.

Type locality: Antarctic Ocean waters south of eastern Australia, 59°26′S, 158°36′E.

#### DIAGNOSTIC FEATURES

Fins elliptical; fin length index ~63. Funnel-locking cartilage ear-shaped; tragus and antitragus present, antitragus strong; mantle-locking cartilage nose-shaped. Arms 4 length index 125–*153*–185 (min-*mean*-maximum), arm sucker teeth sharp. Tentacular club suckers minute. Skin tubercles absent. Photophores present on integument, eye sinus and eye. Ammoniacal species.

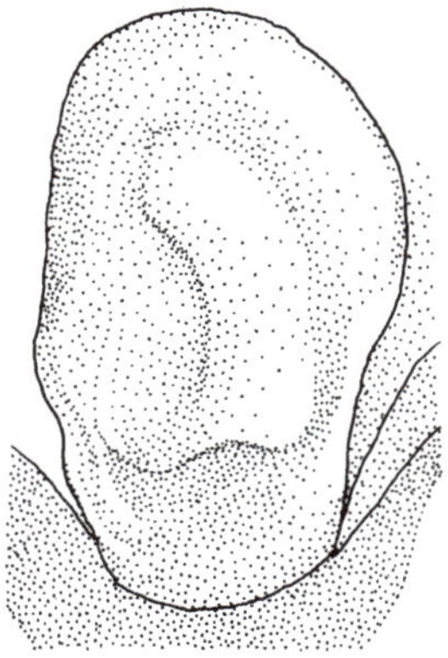

Funnel-locking cartilage (M. Shea, redrawn from Young and Vecchione 2014a).

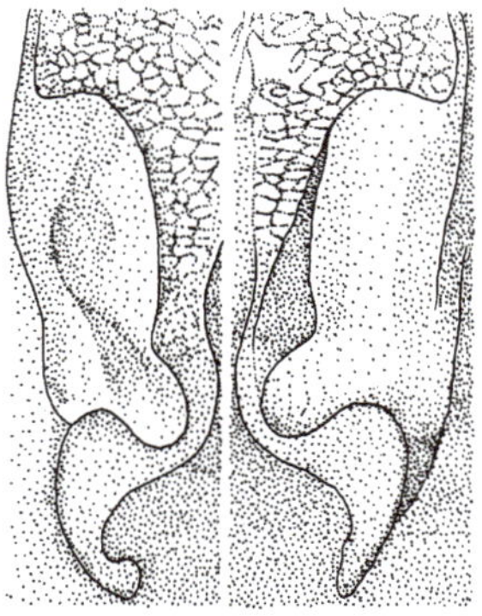

Mantle-locking cartilage cut along midline to show longitudinal sections of the lock (M. Shea, redrawn from Young and Vecchione 2014a).

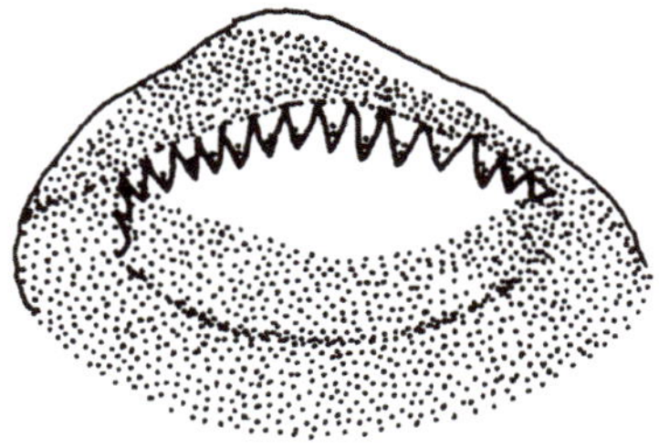

Right, arm 4 sucker ring (M. Shea, redrawn from Young and Vecchione 2014a).

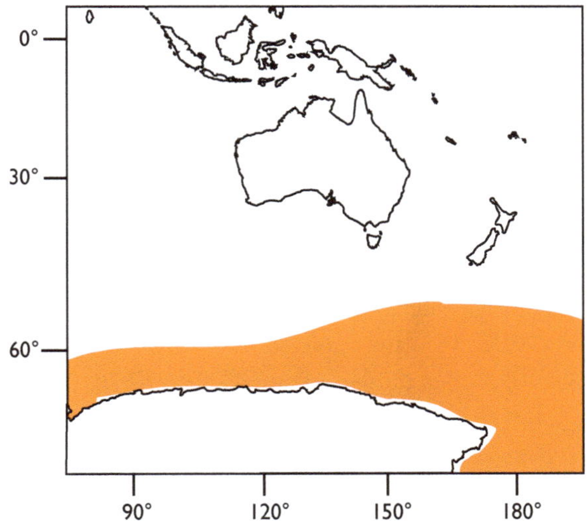


*Mastigoteuthis psychrophila* distribution.

#### SIZE

Maximum ML is 150 mm.

#### DISTRIBUTION

Circumpolar in sub-Antarctic and Antarctic waters, with records from mesopelagic to bathypelagic depths over sub-Antarctic and Antarctic island trenches, such as South Sandwich, Macquarie and Hjort trenches and off South Georgia's north-west slope. This species is an Antarctic endemic, extending north to the Antarctic Polar Front.

#### REFERENCES

Roper and Jereb (2010o), Braid *et al.* (2014).

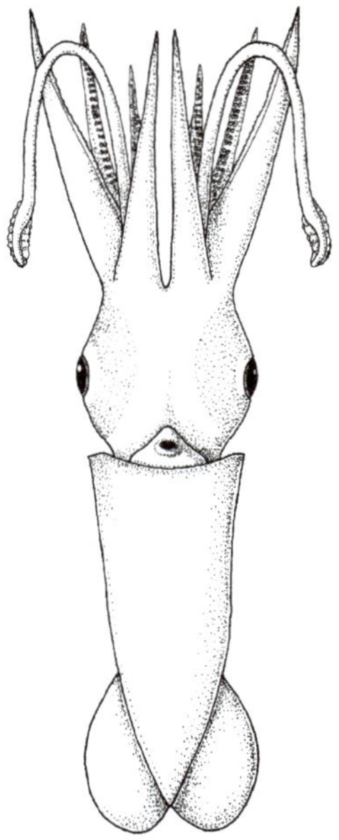

Ventral view (FAO, E. D. Antoni).

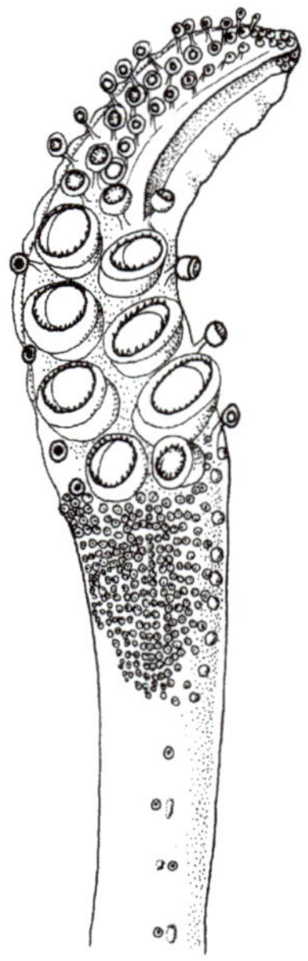

Tentacular club (FAO, E. D. Antoni).

## Neoteuthidae

Four genera, each represented by a single species, are found in this family. *Alluroteuthis* is the only known Australian representative.

### *Alluroteuthis* Odhner, 1923

*Alluroteuthis* is currently monotypic, represented by *Alluroteuthis antarcticus* Odhner, 1923.

### *Alluroteuthis antarcticus* Odhner, 1923

*Alluroteuthis antarcticus* Odhner, N. H. (1923). Die Cephalopoden. *Further Zoological Results of the Swedish Antarctic Expedition, 1901–1903* 1(4), 1–7 [2, pl. 1].

Type data: syntypes NRM 1080.

Type locality: 63°25′S, 45°39′W, Weddell Sea, Antarctic Ocean.

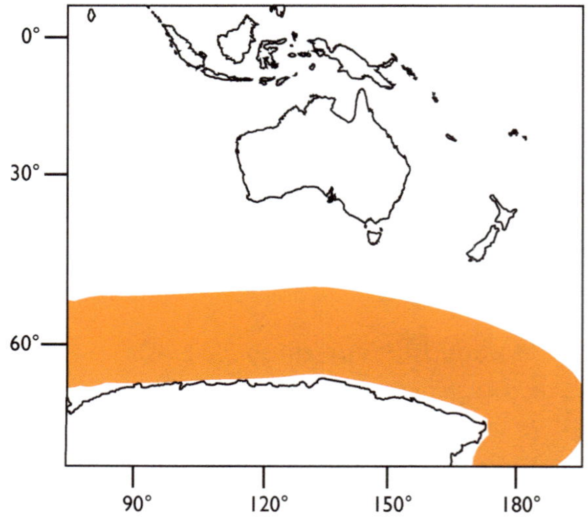


*Alluroteuthis antarcticus* distribution.

#### COMMON NAME

Antarctic New Squid

#### DIAGNOSTIC FEATURES

Weakly muscled. Fins short (35–40% ML), attached laterally on mantle muscle, distinctive in having free posterior fin lobes and anterior lobes are absent. Tentacular clubs short (33% ML). Manus divided into two distinct regions: proximally with numerous small suckers in irregular series and distal manus with four regular series of greatly enlarged suckers (may be more than four series on proximal end of this series). Circular proximal manal suckers with toothed inner rings. Enlarged tooth/hook on proximal arm suckers.

#### HABITAT AND BIOLOGY

Paralarvae and juveniles are mesopelagic; adults are mesopelagic to bathypelagic at 750–2800 m. Adults concentrate at 800–900 m during the day, spread out vertically at night; ontogenetic descent occurs with juveniles and subadults. It is an ammoniacal species that feeds in midwater on Antarctic krill, fishes and squids.

#### SIZE

The maximum ML extends to 270 mm.

## DISTRIBUTION

This species is an Antarctic endemic, extending north to the Antarctic Polar Front (i.e. occurs south of 50°S).

## REFERENCES

Roper and Jereb (2010p), Rodhouse (2013).

# Octopoteuthidae Berry, 1912

## DIAGNOSTIC FEATURES

Medium to large squids. Very broad fins extending for length of mantle. Broad, straight funnel-locking apparatus. Arms with biserial hooks and biserial suckers sometimes present on distal arm tips. Tentacles lost in adults. Larve with few distal club suckers. Hectocotylisation unknown. Buccal connectives attach to ventral borders of ventral arms. Several small, spindle-shaped photophores on tips of arms in *Octopoteuthis*. A single large photophore at tip of each dorsolateral arm in *Taningia*.

## HABITAT AND BIOLOGY

Unknown. Nothing known of life history. Octopoteuthids are a significant part of the diet of sperm whales, lancetfishes and sharks.

*Interest to fisheries.* No commercial fisheries potential due to their soft, gelatinous bodies. Readily damaged in nets, so unsuitable for laboratory study.

## SIZE

Medium to large, up to 1700 mm ML.

## REMARKS

The family comprises two genera: *Octopoteuthis* and *Taningia*. Species in *Octopoteuthis* are poorly delineated.

## DISTRIBUTION

Until their taxonomy is clarified, the distribution of species in Australian waters will remain poorly known.

### Key to genera of the Octopoteuthidae (after Roper and Jereb 2010q)

1a. Small, narrow, spindle-shaped photophore on all arm tips; photophores embedded in tissue of mantle, head and arms. Paralarvae with gelatinous tentacles that are lost when they reach ~12 mm ML. Size up to 500 mm ML (usually < 200 mm)......... *Octopoteuthis*

1b. Large oval composite photophore on tips of arms 2; no photophores in tissue of mantle, head and arms. Paralarve with robust tentacles that diminish and disappear when they exceed 40 mm ML. Mantle length up to 1.5 m........................*Taningia* (*T. danae*)

# *Octopoteuthis* Ruppell, 1844

## DIAGNOSTIC FEATURES

Mantle of some species with thick, gelatinous layer over pigmented skin-like layer; particularly thick over 'tail'. Longitudinal grooving in anterior mantle jelly (this is where spermatophores are lodged) in at least one species. Body proportions of larvae change rapidly during growth up to 15 mm ML. Eyes initially stalked, become sessile at ML of 10–15 mm. By this size, tentacles are usually lost; subadult and adult *Octopoteuthis* lack tentacles.

*Photophores.* Slender, small, spindle-shaped dark photophores present at tips of all arms. One or two large 'tail' photophores. Arm photophores lack muscular lids; they have mobile chromatophores that conceal photogenic tissue.

## HABITAT AND BIOLOGY

The arm tips are frequently broken during capture. This fragility is probably a normal defensive weapon (Bush *et al.* 2009).

## REMARKS

Species within the genus are poorly delineated and the genus is badly in need of revision. The hectocotylus is unknown. According to Young (1972) the systematics within the genus currently rests on three traits: (1) the number of 'tail' photophores, (2) the size of the 'tail' in small specimens, and (3) the presence of accessory cusps on the hooks in small specimens.

Unusual semicircular 'tail' flaps have been seen in live *Octopoteuthis deletron* Young, 1972 filmed *in situ* by the MBARI ROV. It is possible that these 'tail' flaps may occur also in other species.

Live specimens of one species, *O. neilseni* (Robson, 1948), have been seen to emit bright bioluminescent flashes from the caudal area. In addition, it appears that the arm tips can be autotomised or are easily broken by predators. When this occurs, the wriggling arm flashes and wriggles, possibly acting as a lure to detract a predator while the squid escapes. The arms seem to regenerate a new tip when this occurs.

#### REFERENCES

Roper and Jereb (2010q), Young and Vecchione (2013).

### Key to *Octopoteuthis* species

1a. No photophores on posterioventral mantle ....*O. rugosa*
1b. Photophores on posterioventral mantle...................2
2a. Photophores at base of arms 3 and 4 ...........*O. sicula*
2b. No photophores at base of arms 3 and 4.....................
.................................................................... *O. megaptera*

## *Octopoteuthis megaptera* (Verrill, 1885)

*Ancistrocheirus megaptera* (Verrill, 1885). Third catalogue of Mollusca, recently added to the fauna of the New England coast and the adjacent parts of the Atlantic, consisting mostly of deep-sea species, with notes on others previously recorded. *Transactions of the Connecticut Academy of Sciences* 6(2), 395–452 [399, pl. 42 fig 1–1a].

Type data: holotype NMNH 40128.

Type locality: 39°12′N, 72°03′W, off New Jersey, western North Atlantic Ocean.

Live animal. *Note*: The many broken arm tips, the tips perhaps broken off as flashing predator decoys (D. Shale).

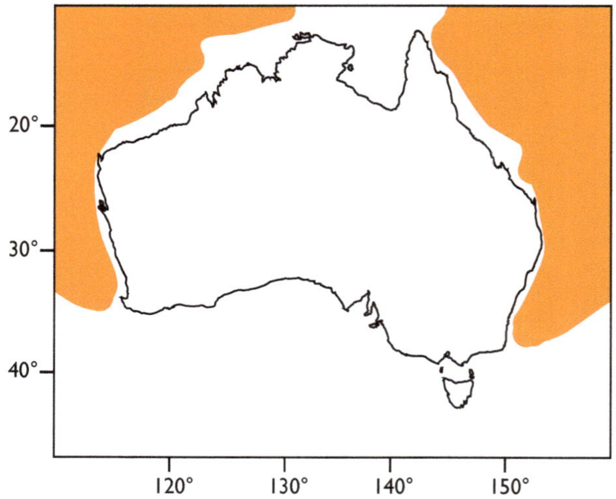


*Octopoteuthis megaptera* distribution.

#### DIAGNOSTIC FEATURES

Posterior mantle tail-like, extends beyond posterior junction of fins and mantle. Fins ~75% ML. Tail long, acuminate, flattened from above, with narrow lateral fringes.

*Photophores*. Two photophores on posterioventral mantle. Photophores at base of arms 3 and 4 absent.

#### HABITAT AND BIOLOGY

Observations of *Octopus megaptera* from the manned submersible *Johnson Sea-Link II* in the Gulf of Mexico showed that this species swims with its body oriented obliquely with the head upward and arms flexed laterally and dorsally. The fins flap in the arc between overlapping ventrally and extending laterally, never traversing dorsally above the body plane. The vertical distribution of this species is mesopelagic, bathypelagic and bathyal with vertical ascent into epipelagic zone at night.

#### SIZE

Mantle length up to 200 mm.

#### DISTRIBUTION

Subtropical to tropical cosmopolitan: western tropical Pacific Ocean to 36°S off Australian east coast and to Cape Leeuwin off Australian west coast, Indian Ocean, western, central and eastern Atlantic Ocean.

#### REFERENCE

Roper and Jereb (2010q).

## *Octopoteuthis rugosa* Clarke, 1980

*Octopoteuthis rugosa* Clarke, M. (1980). Cephalopoda in the diet of sperm whales of the southern hemisphere and their bearing on sperm whale biology. *Discovery Reports 37*, 1–324 [156, text figs 122–132].

Type data: holotype BMNH 1973100.

Type locality: off southeastern South Africa, western South Indian Ocean (sperm whale stomach).

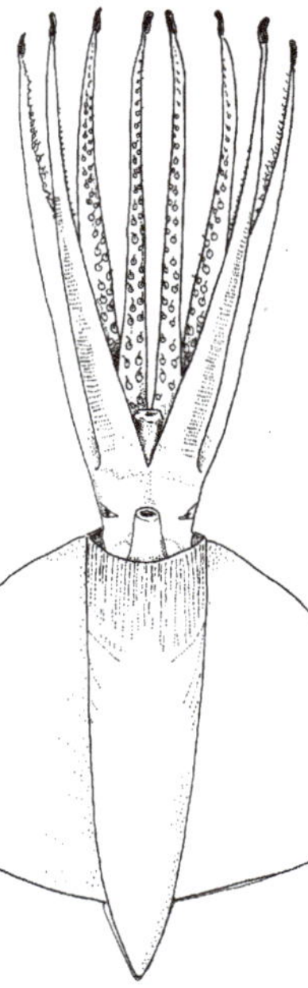

Ventral view (A. Reid, redrawn from original description).

### DIAGNOSTIC FEATURES

Mantle, head and arms with thick, gelatinous layer external to pigmented skin-like layer. Jelly very thick over tail region. Anterior mantle of female with longitudinal grooves in gelatinous layer. Body extends well beyond fins; in juveniles, fin reaches posterior end of mantle; adults with tail. Fin width approximately equal to ML. Arms muscular, subequal in length. Tentacles absent. All arms with two rows of hooks, each with two small, ancillary hooks basally. Buccal membrane with six lappets. Radula with seven teeth in each transverse row. Median teeth with small cusp on either side, first lateral teeth with one small cusp.

*Photophores.* Distinct elongate black photophores on all arm tips. No photophores on tail, head or mantle cavity.

*Colour.* Purple in life. Aboral surface of arm 4 with transverse pigment stripes.

### HABITAT AND BIOLOGY

This is a mesopelagic, probably schooling species (as indicated following examination of the stomach contents of sperm whales). The type specimen and additional material have spermatophores embedded in the mantle and head jelly.

### SIZE

Mantle length to 230 mm.

### REMARKS

This species differs from *Octopoteuthis sicula* Rüppell, 1844 in the head size, tentacular hook shape, arm tip and radula, and from *O. megaptera* in proportions and hook shape. In juveniles, the fin reaches the posterior end of the mantle but adults have a short tail.

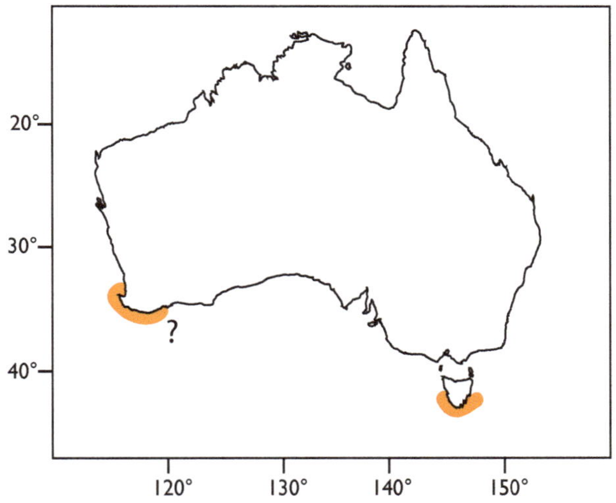


*Octopoteuthis rugosa* distribution. (Australian records are based on few specimens, but the species is likely to occur more widely.)

### DISTRIBUTION

Tropical and subtropical Atlantic Ocean and the Indo-West Pacific Ocean, for example Mauritania, and the equatorial Atlantic Ocean, southern Africa, south-western and eastern Australia, Sulu and Flores Seas.

### REFERENCE

Roper and Jereb (2010q).

## *Octopoteuthis sicula* Rüppell, 1844

*Octopoteuthis sicula* Rüppell, E. (1844). Intorno ad alcuni cefalopodi del mare di Messina: lettera del Dr. Eduardo Ruppell di Frankfort sul Meno al Prof. Anastasio Cocco. *Giornale del Gabinetto Letterario di Messina* 5(27–28), 129–135 [134].

Type data: holotype BMNH 1845.9.8.13.

Type locality: Sicily.

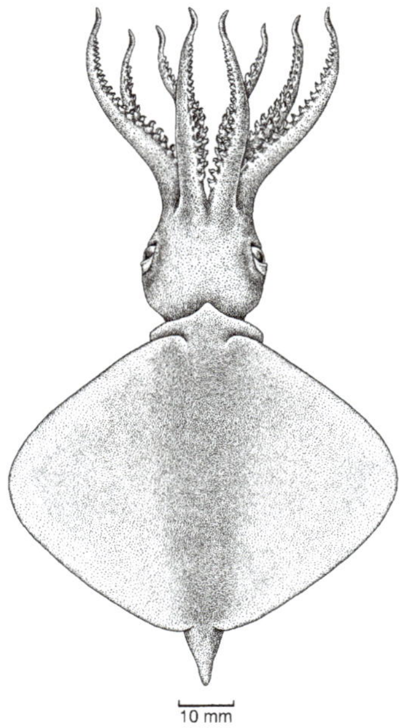


Dorsal view (ABRS, K. Hollis).

### COMMON NAME

Rüppell's Octopus Squid

### DIAGNOSTIC FEATURES

Mantle conical, short; acuminate posteriorly; musculature flabby with gelatinous integumentary layer and rugosities. 'Tail' short, indistinct in juveniles. Fins large, length ~90% ML, width 115% ML; extend to posterior tip of mantle.

*Photophores.* Pair of photophores embedded in posterior mantle, ~20–25% ML from posterior tip; three pairs of embedded photophores on lateral sides of funnel groove and near neck; photophores embedded at bases of arms 3 and 4.

### HABITAT AND BIOLOGY

The vertical distribution of *Octopoteuthis sicula* is mesopelagic to bathypelagic (to 2000 m) during the day, with nocturnal ascent into the epipelagic zone. In the eastern North Atlantic Ocean juveniles are caught from May to September as planktonic vertical migrators.

*Interest to fisheries.* None because of the semi-gelatinous consistency of the flesh.

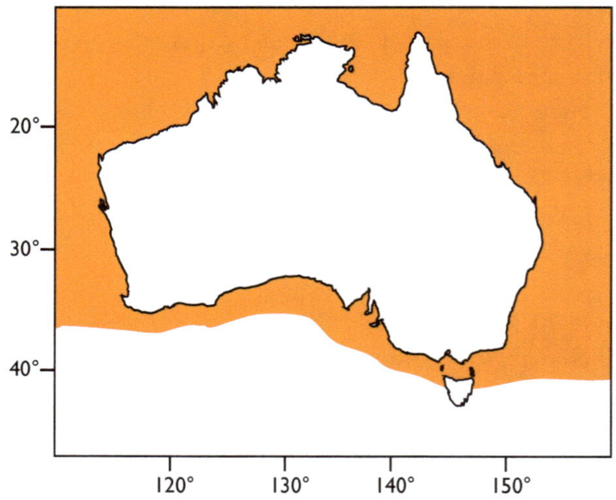


*Octopoteuthis sicula* distribution.

### SIZE

Grows to 200 mm ML (possibly to 500 mm).

### DISTRIBUTION

This species occurs in tropical, subtropical, and subtemperate waters in the Atlantic Ocean to South Africa and in the Indo-West Pacific Ocean; Mediterranean Sea; and possibly tropical eastern Pacific Ocean from Mexico to Chile, westward to Hawaii.

### REFERENCE

Roper and Jereb (2010q).

## *Taningia* Joubin, 1931

*Taningia* is currently monotypic, represented by *Taningia danae* Joubin, 1931, although at least one additional species exists (J. Kelly in prep.), mentioned in Hoving *et al.* (2014).

## *Taningia danae* Joubin, 1931

*Taningia danae* Joubin, L. (1931). Notes preliminaires sur les Cephalopodes des croisieres du DANA (1921-1922), 3e Partie. *Annals de l'Institut Oceanographique, Monaco (new series) 10*(7), 169–211 [181, text figs 11–16].

Type data: holotype ZMUC.

Type locality: Central Atlantic Ocean, 14°52′N, 28°04′W.

### COMMON NAME

Taning's Octopus Squid

### DIAGNOSTIC FEATURES

Mantle broad, robust, muscular. From ~3 mm ML paralarvae, fins very large, occupy nearly entire ML; fins very broad, width exceeds ML (up to 130% ML). Arms with large hooks in two series to tips, where minute suckers may occur as incipient hooks; hooks begin to develop in animals of ~5 mm ML. Tentacles robust on paralarvae, with some suckers considerably enlarged; tentacles cease to develop and are lost at ML of 40–45 mm (although stumps may remain for some time).

*Photophores.* Tips of arms 2 with large, oval, swollen, composite photophore with creamy white, crenulate bioluminescence-producing surface, exposed when black

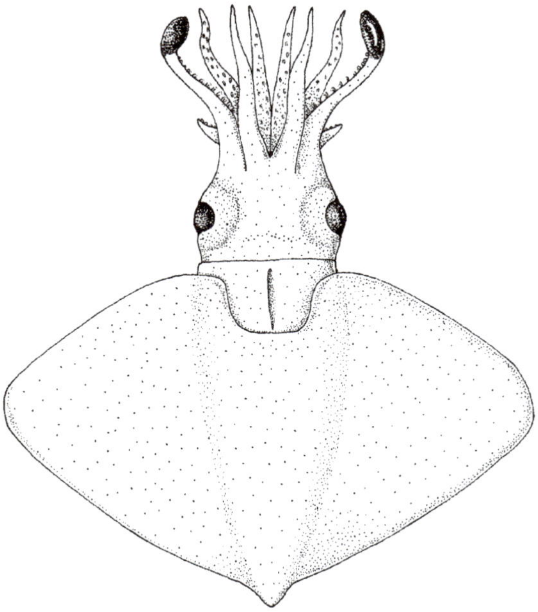

Dorsal view (FAO, E. D'Antoni).

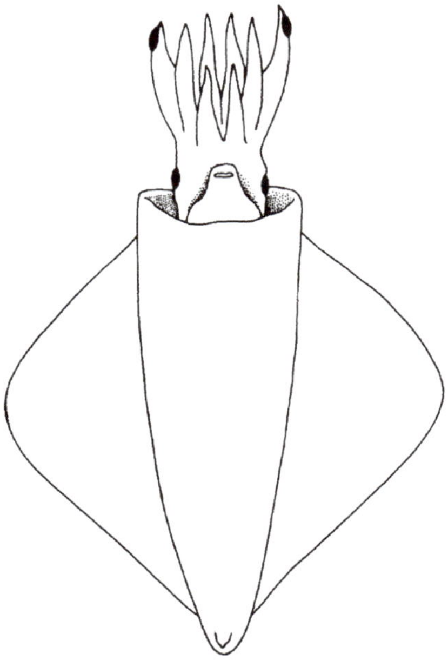

Ventral view (FAO, E. D'Antoni).

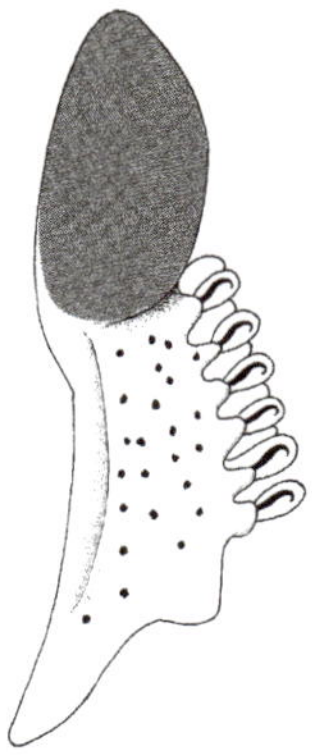

Photophores, arms 2, ventral view (FAO, E. D'Antoni).

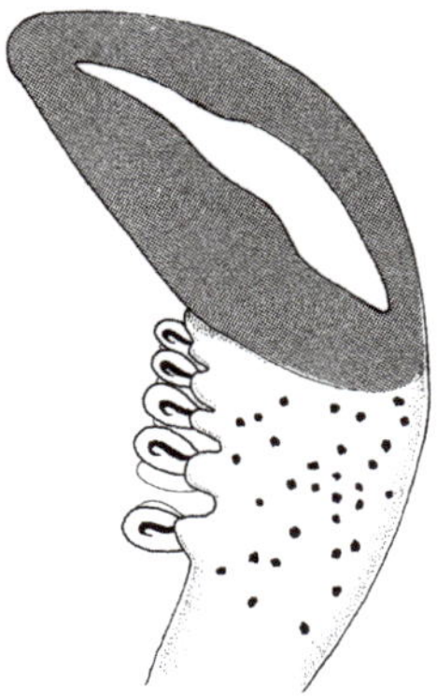

Photophores, arms 2, dorsal view (FAO, E. D'Antoni).

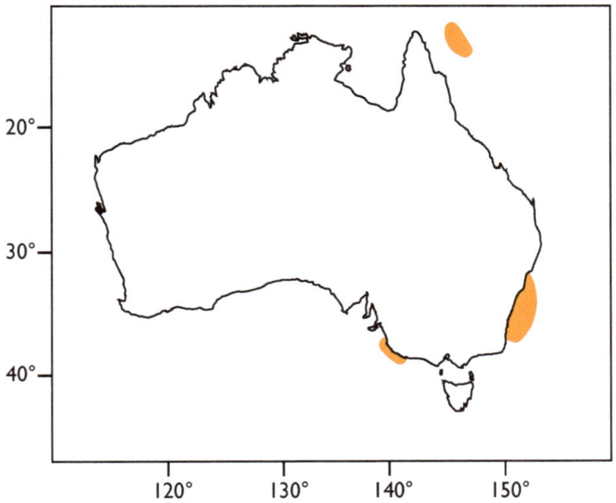


*Taningia danae* distribution. (Australian records are based on few specimens, but the species is likely to occur more widely.)

eyelid-like skin folds are contracted open (no photophores on tips of any other arms); photophores begin to develop at ~4.5–5.0 mm ML. No photophores (known to) exist embedded in the tissues of mantle, head or arms; one photophore embedded on each side of intestine ventral to ink sac.

*Colour.* Intense maroon.

#### HABITAT AND BIOLOGY

*Taningia danae* is an oceanic species; paralarvae and juveniles are epipelagic to upper mesopelagic and then undergo ontogenetic descent into lower mesopelagic, bathypelagic and bathyal zones. It may be solitary or form only small groups. Adults undergo diel vertical migration and ascend into the epipelagic zone at night. Juveniles have been captured at night within 175 m of the surface, while the remains of a large specimen (50 kg) have been taken from the stomach of the bottom-living shark, *Centroscymnus coelolepis* (Saemundsson, 1922), caught on a bottom-set longline at 1246 m. A mature female of 1.4 m ML had 250 000 ovarian eggs. Observations on bioluminescence from living animals have shown a rapid flash response from the arm tip photophores, probably used as a startling or warning function, and a steady glow from the visceral photophores as is commonly used for protective counter illumination. *In situ* video footage taken in the northern hemisphere reveals that *T. danae* is a strong swimmer, both forwards and backwards, propelled by its huge fins, and it changes directions quickly by bending its body. It attained speeds of 7.2–9.0 km hour$^{-1}$ when it attacked a bait rig, emitting short, bright flashes from its two arm-tip photophores before it attacked. It also emitted long and short intermittent glows when swimming around the lighted bait rig.

*Interest to fisheries.* *Taningia danae* grows to a very large size with thick mantle and fin musculature. Potential exists for the development of a significant fishery for this species.

#### SIZE

The maximum ML recorded is 170 cm; maximum weight 161 kg.

#### DISTRIBUTION

Occurs as far south as 44°S. *Taningia danae* is circumglobal in tropical and subtropical waters, including the Pacific Ocean (Australian, New Zealand, Hawaiian, and Japanese waters, and across the Indian Ocean). It is found from 50°N to 55°N and 40°S to 45°S throughout the Atlantic Ocean.

#### REFERENCE

Roper and Jereb (2010q).

## Ommastrephidae Steenstrup, 1857

#### COMMON NAME(S)

Arrow Squid, Flying Squid

#### DISTINGUISHING CHARACTERS

Ommastrephids can be distinguished from onychoteuthids (also large, muscular squids) by the following traits: four rows of club suckers (Australian species) without hooks (onychoteuthids have biserially arranged hooks and rarely marginal suckers); ⊥-shaped funnel–mantle locking cartilage in ommastrephids (rather than simple, straight as in onychoteuthids), and buccal connectives attached to the dorsal (rather than the ventral) border of the arms in ommastrephids.

#### DIAGNOSTIC FEATURES

Medium to large oegopsid squids with strong, muscular, torpedo-shaped mantle. Head and eyes large. Fins terminal, muscular, transversely rhomboidal, or heart-shaped and strongly attenuate posteriorly (*Ornithoteuthis*). Fins wide, short (< two-thirds ML). Funnel groove deeply recessed, distinct, triangular, either smooth, without foveola (*Todaropsis*), or with foveola, with either: (1) an arched transverse fold (pocket) of skin (other Todarodinae), or (2) foveola with additional small, arched, cutaneous pockets situated on each side of funnel groove (Ommastrephinae). Funnel–mantle locking apparatus ⊥-shaped, sometimes fused with mantle and funnel elements, present and stable throughout ontogenesis from earliest paralarval stage. Funnel- and mantle-locking cartilages fused in *Sthenoteuthis oualaniensis* (Lesson, 1830) and *Eucleoteuthis luminosa* (Sasaki, 1915), beginning at end of paralarval stage. Nuchal cartilage a relatively long, narrow ridge and groove plate with rounded ends; mantle component corresponds to fit into nuchal component. Lateral arms (2, 3) strongly keeled in many species; protective membranes well developed. Arm suckers biserial with denticulate rings. One or both ventral arms hectocotylised in males. If one arm hectocotylised, it may be longer than adjacent non-modified ventral arm, or both arms are equal-sized. Modifications vary from elaborate development and sculpturing of sucker bases and trabeculae to loss of suckers and stalks and development of protective membranes. Differences in distal protective membranes of hectocotylus distinguish subfamilies. Tentacle with well-defined stalk, carpus, manus and dactylus. Manus and dactylus with tetraserial suckers; suckers of two median series on manus larger than those on marginal series; large sucker rings on manus have sharp teeth, low broad plates, or are smooth. Fixing apparatus with knobs and opposing smooth-ringed suckers on the carpus only in Ommastrephinae. Buccal membranes attach to dorsal border of ventral arms. Radula with seven series of teeth: central (rachidian) row of trident-shaped teeth, two lateral series of bicuspid teeth flanking it on each side, and then two rows of canine-like unicuspid marginal teeth on each side. Marginal plates present, reduced.

*Photophores.* Photophores present in some species on dorsal and ventral mantle, viscera, and ventrally on eyes and head. Photophores absent in Todarodinae, present in Ommastrephinae, either as small, subcutaneous granules, or as round, oval, or longitudinal, reflective, cutaneous structures.

### HABITAT AND BIOLOGY

After Roper *et al.* 2010a.

#### Life history and growth

*Larvae.* Unlike most teuthoids that hatch as miniature adults ('paralarvae'), ommastrephids have a distinct larval form, the rhynchoteuthion, in which the precursors to the tentacles remain fused along their length into a proboscis with a terminal disc of suckers until the larva reaches ~5 mm ML. By the end of the paralarval stage (between 6.5 and 10 mm ML) the proboscis progressively splits, starting from the proximal base, and soon the tentacles are separated.

*Life cycle.* Based on studies undertaken primarily in the northern hemisphere (but included here because of the importance of this group both ecologically and as a fisheries resource elsewhere) it appears that ommastrephids are among the fastest growing squids. The maximum relative daily growth rates have been observed in paralarvae at 15–20% body weight, juveniles up to 10–12% body weight, and immature and maturing subadults 2 or 3% body weight. Somatic growth in mature shelf-slope ommastrephids (such as *Todaropsis* and *Todarodes*) sharply decreases to 0.3–0.6% body weight and then ceases during the spawning period when feeding activity stops completely. Mature and spawning oceanic squids (e.g. *Ommastrephes*, *Sthenoteuthis*) demonstrate slow growth at a level of 1.5–2% body weight while active feeding continues. There is considerable variation in adult size among ommastrephids that is basically dependent on the differences of relative daily growth rates at the larval and juvenile stages. Usually males are smaller than females due to earlier maturation and consequent earlier decrease of somatic growth. Individual and group phenotypic variability in growth and maturation rates and year-round spawning with high fecundity enables ommastrephids to maintain population stability under considerable variations in numbers. In spite of differences in growth rates, the life cycles of almost all ommastrephids take place over 1 year; however, some forms have a 6 month life cycle.

Male fecundity ranges from 100–2500 spermatophores. Spermatophores are transferred to the buccal membrane where the seminal receptacles are situated. Ommastrephids mate in the 'male-parallel' or 'head to head' positions. Mating is quick (not more than 1–2 minutes) and promiscuous, with both sexes accepting multiple partners. Sperm bulbs have been found in the spermathecae of immature females in some species, suggesting that copulation may precede spawning by some time. The oocyte stock development is asynchronous throughout ontogenesis. The ripe egg size varies in different species from 0.75 to 1.6 mm. Maximum egg number varies from species to species from 50 up to 1.2 million eggs and they are intermittent spawners. Two strategies of spawning exist: the coastal type and the oceanic type. In coastal species (e.g. Todarodinae), after maturation and initiation of spawning activity (i.e. the oocyte maturation stage followed by periodic laying of egg masses), females cease to feed. During the spawning period the main source of energy is the digestive gland and mantle tissue. This is called the descending type of spawning, during which the initial egg masses laid are the largest with the greatest number of eggs, and then in subsequent egg masses the number of eggs gradually decreases due to the reduction in female energetic reserves. Consequently, somatic growth becomes negative and by the end of the spawning period the female's body proportions are significantly changed. In oceanic species (characteristic of the Ommastrephinae), during the spawning period a significant number of oocytes mature, periodical laying of egg masses of approximately the same size occurs, and active feeding between egg-laying events continues, accompanied by appreciable somatic growth. However, total post-spawning mortality occurs in all ommastrephids. Large, pelagic, gelatinous egg masses that are slightly denser than seawater float either near the surface or near the bottom, depending on the species. The duration of embryonic development (in species for which data are available, with egg sizes of 0.75–0.90 mm) varies from 3 to 22 days and depends mostly on water temperature. Ommastrephids exhibit a typical 'r-strategy' biological model.[1] The most important factor that determines the pattern of abundance, dynamics and distribution of exploitable ommastrephid populations is the availability of planktonic paralarvae and juveniles that inhabit subsurface depths. These stages are characterised by a high mortality rate. Therefore, spawning success and paralarval survival rates constitute the principal factor that determines recruitment levels. Another important ecological-population feature of these squids is the length of their life cycles.

Collection and laboratory maintenance of these large and mobile cephalopods presents major difficulties so they have not yet been subjected to laboratory studies to further elucidate features of their life cycles and behaviour.

### Depth distribution and habitat

Larvae and juveniles are found in shallow coastal waters, while adults occur in deeper continental shelf, slope and open ocean waters to depths of 2000 m, forming schools that decrease in size as the animals grow.

Almost all species, particularly the oceanic ones, are abundant and widely distributed. The main taxonomic groups inhabit different but loosely defined reproductive zones and spatial feeding areas. Three main groups can be distinguished by characteristics of habitat, especially spawning habitat, which includes the degree of relationship with the bottom: (1) the slope-shelf group mainly inhabiting the waters of the continental slope and shelf, for example *Nototodarus gouldi* (McCoy, 1888), *Todarodes pusillus* Dunning, 1988 and *T. eblanae* (Ball, 1841); (2) the nerito-oceanic group mainly inhabiting the waters of the continental slope, oceanic seamounts and islands, and adjacent oceanic waters, for example *Martialia hyadesi* Rochebrune & Mabille, 1889, *N. hawaiiensis* (Berry, 1912), *Ornithoteuthis volatilis* (Sasaki, 1915) and *T. filippovae* Adam, 1975; and (3) the oceanic group, in which the entire life cycle takes place only in oceanic waters and no connections or relationships ever exist with the bottom, for example *Eucleoteuthis luminosa* (Sasaki, 1915), *Hyaloteuthis pelagica* (Bosc, 1802), *Ommastrephes bartramii* (Lesueur, 1821) and *S. oualaniensis* (Lesson, 1830).

[1] An 'r-strategy' model refers to animals that reproduce rapidy (exponentially). They have a high fecundity, or produce large numbers of young, have short lifespans, are quick to mature with little investment in individuals, and the strategic intent is to flood the habitat with progeny so that, regardless of predation, some will survive to reproduce.

### Behaviour

*Flying.* Juveniles of several species are able to glide out of the water to escape predators. Tsunemi Kubodera presented a paper at CIAC 2012 (Kubodera 2012) in which he had calculated from high definition video observations that *S. oualaniensis* with body lengths of 24 mm were able to fly distances of up to 50 m during flights lasting 7 or 8 seconds.

*Migration.* Ommastrephids can be found at all depths and are capable of extensive vertical and horizontal migration. Evidence from commercial fishing operations suggest that they congregate close to the bottom during the day and move up through the water column at night where they feed close to the surface. Long distance migration may be associated with spawning in some species.

### Predators and prey

Throughout the life cycle, ommastrephids 'traverse' the trophic pyramid, changing the taxonomic and ecological spectrum of their food organisms (prey), enemies (predators) and parasites. In some oceanic communities, large-sized adult ommastrephids are actually the top-level predators, and this effect is becoming more prevalent due to the collape of many finfish fisheries worldwide. Ommastrephids are strongly active predators. Their food composition depends on their ML: in young small-sized squid, prey consist predominantly of meso- and macroplanktonic crustacea and juvenile fishes; in middle-sized squid prey consists of micronektonic fishes (mainly myctophids) and marcroplanktonic euphausiids and shrimps; and in large-sized individuals prey are only fishes, squids and to a lesser degree, shrimps. Cannibalism is common. They have high general and active metabolisms, and a daily food consumption rate of 6–12% of the body weights of adult squids. Protein is the main substrate for energy metabolism. Predators include seabirds, teleosts, sharks, whales, dolphins and seals.

*Interest to fisheries.* Worldwide, the Ommastrephidae are the most important commercial fishery group among cephalopods. No target fisheries for these species exist in Australian waters; however, they are caught incidentally in demersal trawls in shelf and slope waters. They can also be caught using baited or artificial lures and jigs. *Nototodarus hawaiiensis* (Berry, 1912) and *T. pusillus* occur off north-eastern Australia and represent an underutilised resource. More information regarding fisheries potential generally and problems that need to be overcome can be found in Roper *et al.* (2010a).

#### REMARKS

The ommastrephids are the most well-studied and, therefore, well-known oegopsid squid family due to their fisheries importance.

Most adult ommastrephids are easily identified; however, ontogenetic development or loss of photophores and variability in sucker dentition with growth may make identification of small specimens (less than ~80 mm ML) more difficult.

Since the mid-twentieth century the traditional classification of the Ommastrephidae consisted of three subfamilies: Illicinae, Todarodinae and Ommastrephinae (Illicinae are not found in Australian waters). Recently, the phylogenetic relationships in the Ommastrephidae were analysed using mitochondrial DNA sequencing (Wakabayashi *et al.* 2006) and the monophyletic status of the three subfamilies was supported (Illicinae, Ommastrephinae, Todarodinae). However, this study was not able to resolve the phylogenetic relationships within the Ommastrephidae because of the significant genetic divergence among species. The subfamilial classification has been further subdivided with the addition of two new subfamilies, Todaropsinae and Ornithoteuthinae, in the Todarodinae. Following Roper *et al.* (2010a) the taxa are divided only into three previously recognised subfamilies here.

#### DISTRIBUTION

Ommastrephid squids are distributed worldwide. They inhabit waters of the shelf, slope and open ocean, from the surface to depths that range to 2000 m. Considerable distributional overlap occurs for many species. In East Australian Current waters, juveniles of up to five genera have been taken in the same trawl catch (Dunning and Brandt 1985). Ecologically, the Todarodinae are nerito-oceanic and the Ommastrephinae are truly oceanic.

#### REFERENCE

Roper *et al.* (2010a).

### Key to subfamilies and genera of the Ommastrephidae (modified after Lu and Dunning 1998 and Roper *et al.* 2010a)

1a. Photophores present. Funnel-groove side pockets usually present. Ventral protective membranes on distal end of hectocotylus enlarged ............................ ....................................Subfamily Ommastrephinae 2

1b. Photophores absent. Funnel-groove side pockets usually absent. Ventral protective membranes on distal end of hectocotylus with thickened trabeculae ............................................. Subfamily Todarodinae 6

2a. Mantle narrow, elongate, posteriorly drawn out into a pointed 'tail'. Funnel groove side pockets often obscure. Tentacle fixing apparatus without tubercle-like knobs; rings of large club suckers with 15–25 equal-sized pointed teeth. No external photophores; no small subcutaneous photophores; one photophore on each eyeball, two intestinal photophores; in adults a narrow pinkish photogenic band extends along viscera from rectum to near posterior end of mantle cavity with two gaps ....................................... .........................................*Ornithoteuthis (O. volatilis)*

2b. Mantle wide, end of mantle without pronounced 'tail'. Funnel groove side pockets clearly defined; tentacle fixing apparatus with one or several knobs alternating with smooth ringed suckers; rings of large club suckers with either one large median tooth in distal part of ring, or with four large teeth, one at each quadrant. Small, yellow subcutaneous photophores (like short rice grains) on head, mantle, fins and arms; some species may have arrangement of different form (large mantle photophores) with reflective subjacent layer; one ocular and one or two intestinal photophores present................................. 3

3a. Rings of large club suckers with four large, evenly spaced teeth. No large cutaneous photophores in form of spots and stripes on ventral surface of mantle; small subcutaneous photophores present; no or two intestinal photophores............................. 4

3b. Rings of large club suckers with one large medial tooth in distal part of ring. Large cutaneous spot or stripe-like photophores on ventral surface of mantle; one intestinal photophore ......................................... 5

4a. Funnel–mantle locking cartilage not fused. End of fin slightly attenuate posteriorly. Four to seven suckers with denticulate rings in carpal part of club proximal to first knob. Large, compound, subcutaneous photophore absent on anteriodorsal part of mantle; no ocular or intestinal photophores; one wide gold or silvery longitudinal, cutaneous, bioluminescent stripe extends along ventral midline of mantle from near mantle edge to anterior margin of fin................................ *Ommastrephes (O. bartramii)*

4b. Funnel–mantle locking cartilage fused. Fin rhomboidal, not attenuate posteriorly. Not more than two suckers with denticulate rings in carpal part of club proximal to first knob. A single, large, oval, compound, subcutaneous photophore on anteriodorsal part of mantle; one ocular and two intestinal photophores present; no longitudinal, cutaneous, bioluminescent stripe on ventral side of mantle........ .....................................*Sthenoteuthis (S. oualaniensis)*

5a. Mantle–funnel locking cartilage fused (after paralarval stage). Pair of long stripes of cutaneous, bioluminescent tissue (large elongate photophores) extends along ventral side of mantle, one on each side of midline, each stripe broken posteriorly by gap at level of maximum width of fin; two oval bioluminescent patches on head near anterior margin of mantle aligned with base of each ventral arm....... .......................................... *Eucleoteuthis (E. luminosa)*

5b. Mantle–funnel locking cartilage not fused. Nineteen round, cutaneous, bioluminescent patches arranged in a fixed pattern on ventral side of mantle that form converging or separated pairs; cutaneous, bioluminescent patches at base, in middle, and near tip of each ventral arm................*Hyaloteuthis (H. pelagica)*

6a. Funnel groove smooth, without foveola. Largest manus sucker teeth without intermediate ridges or plates...................................... *Todaropsis (T. eblanae)*

6b. Funnel groove with foveola. Largest manus sucker teeth linked by intermediate ridges or plates...........7

7a. Finger-like cirri project beyond protective membranes along both sides of arms and tentacular clubs .................................................. *Martialia (M. hyadesi)*

7b. Finger-like projections and cirri on arms and tentacular clubs absent.....................................................8

8a. One ventral arm hectocotylised (usually right, rarely left).................................................... *Todarodes*

8b. Both ventral arms hectocotylised: one arm modified in proximal and distal parts, other arm only in proximal part.......................................... *Nototodarus*

## Subfamily Ommastrephinae Posselt, 1891

Foveola on funnel groove; side pocket on funnel groove. Light organs present.

### *Eucleoteuthis* Berry, 1916

*Eucleoteuthis* is monotypic, represented by *Eucleoteuthis luminosa* (Sasaki, 1915).

### *Eucleoteuthis luminosa* (Sasaki, 1915)

*Symplectoteuthis luminosa* Sasaki, M. (1915). On three interesting new oegopsids from the Bay of Sagami. *Journal of College of Agriculture, Tohoku Imperial University, Sapporo* 6(6), 131–150 [144, text fig. 4, pl. 4 figs 7–13].

Type data: syntypes MSUT[TIU] 997.3.2 (9 of 10).

Type locality: western Pacific Ocean, Japan, Sagami Bay.

#### COMMON NAME(S)

Striped Squid, Luminous Flying Squid

#### DIAGNOSTIC FEATURES

Mantle–funnel locking cartilage fused after paralarval stage. Fins heart-shaped with attenuate 'tail'; fin length 41–50% and width 50–65% ML; fin angle 35–50°. Funnel groove foveola with four to nine longitudinal folds and one to three side pockets (often obscure). Arms with 19–32 pairs of suckers. Left ventral arm hectocotylised: suckers absent on distal third of arm, only series of small paired sucker stalks. Tips of all arms not attenuate. Dactylus of tentacular club with tetraserial suckers; carpal-locking apparatus on tentacular stalk with one to four knobs and one to three smooth-ringed suckers. Two to four suckers with denticulate rings located on carpus proximal to first knob. Largest medial suckers on manus of club with one large, pointed tooth at distal margin of ring. Spermatozoa with one flagellum.

*Photophores.* External photophores include small, scattered, subcutaneous structures on mantle, head, arms and tentacles and large cutaneous mantle photophores that consist of pair of long yellowish strips that extend along ventral surface of mantle, one on each side of midline; each strip interrupted posteriorly by gap at level of maximum fin width; two oval bioluminescent patches near anterior margin of mantle anterior to each strip; two patch-like photophores at base and in middle on ventral surface of each ventral arm. A single, round photophore

Fresh-caught specimen, dorsal view (K. Graham).

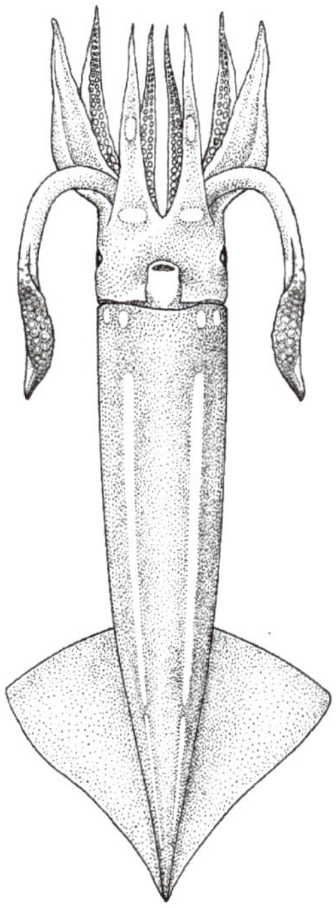

Ventral view (FAO, E. D'Antoni).

Tentacular club (FAO, E. D'Antoni).

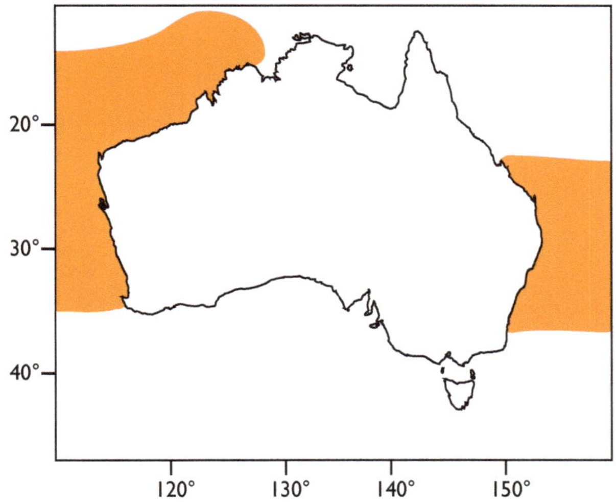


*Eucleoteuthis luminosa* distribution.

on ventral surface of each eye and one round photophore on intestine. Juveniles with oval light organs ventrally on eyes and between ink sac and intestine.

*Colour.* Two long luminous stripes on ventral mantle extending nearly entire length of mantle.

## HABITAT AND BIOLOGY

*Eucleoteuthis luminosa* is the least studied species among ommastrephids. It is an epipelagic and upper mesopelagic species distributed from the surface to 400 m. It inhabits open waters mostly over great depths of more than 200 m, and it is not associated with the bottom or slope habitats. Subadult and adult squids can be observed at the surface at night, but only infrequently. During the day they inhabit depths >50–100 m. The abundance is low or moderate locally (especially in the South Pacific Ocean between 30°S and 40°S). Males reach maturity at 95 mm to 185 mm ML (age 180–270 days), and females mature at 102–200 mm ML (age 220–240 days). Geographical variability of size at maturity and size morphology is pronounced. Maximum size and size at maturity increases from the low latitudes to the high latitudes. Lifespan is a maximum of 1 year. In the lower latitudes spawning occurs throughout most of the year, and in the higher latitudes spawning occurs in the spring and summer seasons. Spermatophore lengths are 11.2–15.8 mm, and their numbers in the spermatophore storage sac of mature males reach up to 150 (mostly between 70 and 100). Egg size is 0.8–1.0 mm, and the potential fecundity of mature females varies between 300 000 and 625 000 ova. The oviducts of mature females contain ~17 000 ripe eggs, probably more. This species is an intermittent, multibatch spawner. Paralarvae can be distinguished from other confamilial paralarvae by their specific chromatophore pattern on the surface of the mantle and head.

*Interest to fisheries.* The species is not fished commercially due to its low abundance and widespread population structure.

### SIZE

Maximum ML in males 207 mm; females 227 mm, body weight ~235 g. Up to 170 mm ML.

### DISTRIBUTION

This oceanic species has ranges in the North Pacific Ocean and the southern hemisphere. It is found in the southern South Pacific Ocean between 20°S and 35°S in the west and between 13°S and 43°S in the east. In the southern Indian Ocean it is found between 15°S and 34°S and in the southern Atlantic Ocean between 10°S and 36°S. It is absent from the North Atlantic Ocean.

### REFERENCE

Jereb and Roper (2010).

## *Hyaloteuthis* Gray, 1849

*Hyaloteuthis* is monotypic, represented by *Hyaloteuthis pelagica* (Bosc, 1802).

## *Hyaloteuthis pelagica* (Bosc, 1802)

*Sepia pelagica* Bosc, L. A. G. (1802). Histoire naturelle de vers contenant leur description et leurs moeurs avec figures dessinees d'apres nature. Volume 1. *Paris: Chez Deterville. de l'imprimerie de Guilleminet. I*: pp. 324. [46, (vol. 1), pl 1 figs 1, 2].

Type data: type repository unresolved.

Type locality: 'Trouvee en pleine mer, par Bosc dans l'estomac d'une dorade' [Found these on the open sea by Bosc in the stomach of a sea bream.]

### COMMON NAME

Glassy Flying Squid

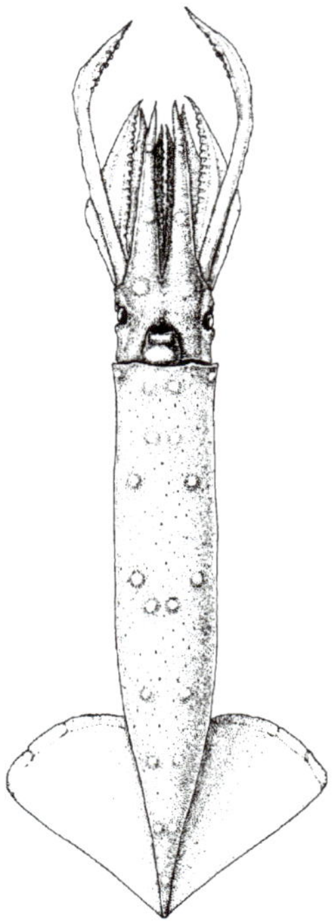

Ventral view (FAO, E. D'Antoni).

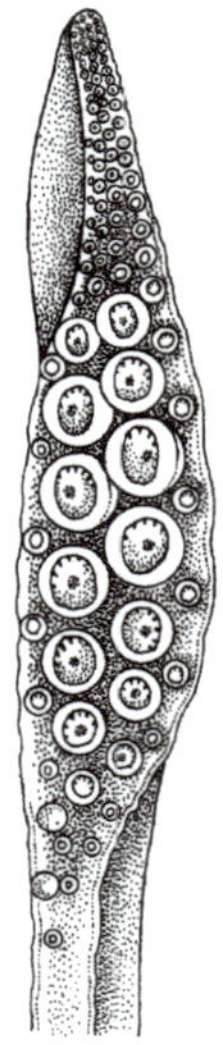

Tentacular club (FAO, E. D'Antoni).

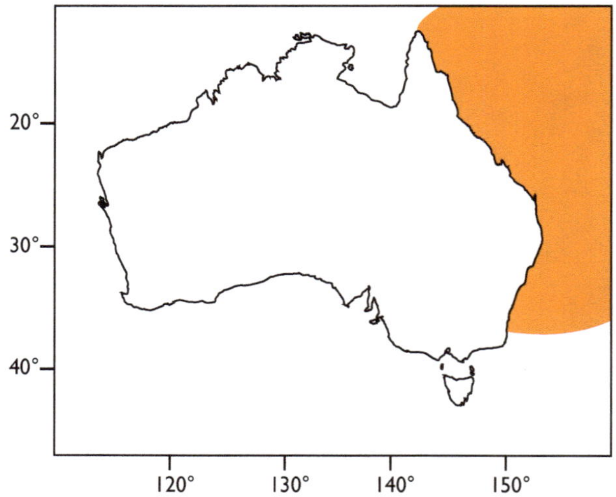


*Hyaloteuthis pelagica* distribution.

### DIAGNOSTIC FEATURES

Fins rhomboidal with straight or very slightly convex posterior edges. Fin length 35–40% and fin width 55–62% ML; fin angle 45–55°. Funnel groove has foveola with seven to nine longitudinal folds and zero to three side pockets (often obscure). Mantle–funnel locking cartilage not fused. Arms 1 with 15–32 pairs of suckers; ventral protective membrane arms 2 and 3 expanded, most prominantly on arms 3, where membrane is 2 or 3 × height of sucker level, thin with prominent trabeculae. Females with suckers in ventral series of arms 2 in rows 4–6 ~2–3 × larger in diameter than dorsal suckers on arms 2 and other arms. Dactylus of tentacular club with four series of small suckers. Carpal-locking apparatus on tentacular stalk with one knob and one smooth-ringed sucker. Largest medial suckers on manus of club with smooth ring and may have one large, pointed tooth at distal margin. Protective membrane of tentacular club very poorly developed. Right, or rarely left, ventral arm hectocotylised: 12–15 suckers; distal 2–4 very small;

dorsal protective membrane opposite last dorsal sucker expands to form a flap and then disappears; in same region ventral membrane expanded and thickened to arm tip without detectable trabeculae.

*Photophores.* Animals larger than 9 mm in length have distinctive pattern of 19 light organs on ventral surface of mantle.

Two round photophores on ventral surface of head, one at base of each arm 4 and two round photophores on intestine. Two spherical photophores in core of each arm 4 at one-third and two-thirds of arm length (difficult to see without dissection). Large cutaneous photophores present on ventral surface of mantle and arms in form of round, yellow, reflective patches: three situated along each ventral arm and 19 arranged in fixed pattern on ventral surface of mantle, where they form pairs, either closely set or well separated. Juveniles with oval light organs ventrally on eyes and between ink sac and intestine. Spermatozoa with one flagellum.

#### HABITAT AND BIOLOGY

This epipelagic to mesopelagic and upper bathypelagic species inhabits open waters over great depths of more than 400 m, and it is not related in any biological or oceanographic way to bottom and slope waters. Paralarvae and juveniles inhabit the upper 50 m at night and 100–200 m during the day. Subadult and adult squids inhabit subsurface layers from 15–20 to 150 m at night. They also have been observed at night at the surface but only infrequently. During daytime they inhabit depths from 200 to at least 800 m. The population numbers are low or moderate locally. Males reach maturity at 50–65 mm ML (age 80–100 days), and females mature at 50–90 mm ML (age 80–135 days). No pronounced geographical variability exists for size at maturity or size morphology. The lifespan is 6 months. Spawning occurs throughout the year with some seasonal variability in activity. Spermatophores attain lengths of 5.0–11.5 mm, and their numbers in the spermatophore storage sac of mature males reach a maximum of 120, but normally they vary between 50 and 80 total. Egg size is 0.9–1.0 mm, and female potential fecundity varies between 30 000 and 100 000 ova. The oviducts of mature females contain a maximum of 500 ripe eggs. The species is an intermittent, multibatch spawner.

A hectocotylus may be present by 45 mm ML; females can mature at least by 55 mm ML.

*Interest to fisheries.* The species is not fished commercially due to its small size and low local area abundance.

#### SIZE

The maximum male ML is 81 mm and the maximum female ML is 105 mm; body weight 35 g.

#### REMARKS

*Hyaloteuthis pelagica* is one of the smallest ommastrephid species and may mature earlier than any other ommastrephid. The restriction profiles of two endonucleases, AluI and Tsp509I, are diagnostic for species identification. It is a small species that is rarely encountered.

#### DISTRIBUTION

Tropical to warm temperate waters in the Atlantic and Pacific Oceans. In the south Pacific Ocean *H. pelagica* is found in oceanic waters from 0°S to 36°S off Australia and to 20°S off southern Chile. It mostly absent from the equatorial zones of both oceans, and has not been found in the open waters of the Indian Ocean. This is a bi-subtropical (anti-tropical) species.

#### REFERENCE

Jereb and Roper (2010).

### *Ommastrephes* d'Orbigny, 1835

The genus is monotypic, represented by *Ommastrephes bartramii* (Lesueur, 1821).

### *Ommastrephes bartramii* (Lesueur, 1821)

*Loligo bartramii* Lesueur, C. A. (1821). Description of several new species of Cuttle-fish. *Journal of the Academy of Natural Sciences of Philadelphia* 2(1), 86–101 [90, pl. 7].

Type data: ANSP type not extant.

Type locality: Not designated.

#### COMMON NAME(S)

Neon Flying Squid, Red Ocean Squid

#### DIAGNOSTIC FEATURES

Mantle–funnel locking apparatus not fused. Fins rhomboidal, slightly attenuate posteriorly; fin length 40–50% and width 60–85% ML; fin angle 46–65°. Funnel groove has foveola with five to eight, occasionally nine, longitudinal folds and two to five (mainly three or four) distinct side pockets. Tips of all arms not attenuate. Arms with 24–35 pairs of suckers. Tips of trabeculae of protective membranes do not project beyond edge of membrane; ventral protective membranes of arms 3 very wide, and in adult females expand into large, triangular, membranous lobe. Right or left ventral arm hectocotylised, smooth, without suckers and sucker pedicels distally in mature males; laterally this area is demarcated by ridge-like continuation of protective membrane. Dactylus of tentacular club with small tetraserial suckers. Carpal-locking apparatus on tentacular stalk with two to five knobs and two to four smooth-ringed suckers; four to seven suckers with denticulate rings on carpus proximal to first knob. Large tentacular suckers with four equally spaced teeth, interspersed with 7–10 smaller teeth. Medial manus sucker rings with one tooth in each transverse row greatly enlarged.

*Photophores.* Small, dense, irregularly shaped, subcutaneous photophores present in adults, embedded in skin of ventral mantle and ventrally on head and ventral arms. When mantle cut, appear as small, yellow spots ~1–2 mm long embedded in ventral mantle tissue just below its outer surface. No large dorsal mantle photophores; ocular and intestinal photophores absent. No photophores in larvae.

*Colour.* A long, wide, silvery or golden opalescent stripe extends along ventral midline from just posterior to the mantle opening to level of anterior edge of fins. Similar silvery or golden opalescent tissue on ventral surface of head and ventral arms. In life, a broad bluish purple longitudinal band on dorsal surface; ventral surface bronze, with golden midventral patch.

## HABITAT AND BIOLOGY

*Ommastrephes bartramii* primarily inhabits open ocean subtropical water masses. In temperate and partly tropical zones it penetrates warm and cold boundary currents respectively. For example, it inhabits the unique frontal areas between the Southern Ocean and the areas to the north, where it effectively occupies the ecological niche normally associated with fishes in the epipelagic zone.

It occurs in areas with sea-surface temperatures from 10 to 25°C, generally over bottom depths >200–400 m, and it is not associated in any way with the bottom, appearing, for example, to actively avoid areas over seamounts. It inhabits the entire water column through epipelagic, mesopelagic and upper bathypelagic zones from 0 to 1500 m and it has a true oceanic distribution and life style strategy, which differs from the squids more associated with the continental shelf.

The planktonic paralarvae, post-paralarvae and juveniles up to 100–150 mm ML inhabit 0–100 m in surface waters during day and night. At night, most juveniles of 60–150 mm ML descend from the surface layers to depths of ~40–100 m. Subadults and adult squids up to 600–650 mm ML undergo daily vertical migrations through a significant depth range. At night they concentrate for active feeding mostly in surface and subsurface layers between 0 and –150 m (mainly 0–40 m or 30–70 m), but some portion of the population may remain in the deep layers of the daytime habitat. About an hour before sunrise they begin to migrate from surface waters to depths of 150–600 m in the higher latitudes of their geographical range and to depths from 400–1500 m (mainly 600–800 m) in the lower latitudes; both populations remain at their depths during the day in a state of low activity. Large females of more than 650 mm ML do not migrate to the surface at all, but permanently inhabit depths of 300–800 m.

In the southern hemisphere the latitudinal, feeding and spawning structure of the population is probably similar to that in the northern hemisphere: the reproductive/spawning area is in subtropical waters (mainly between 20°S and 30–35°S) and the foraging area occurs in temperate waters (35–50°S) that are used in the southern summer. The size at maturity varies considerably throughout the range of this species. The North Atlantic and southern hemisphere males start to mature at 270–300 mm ML and all males larger than 320 mm ML are mature. Two groups are distinguished among females by their size at maturation: (1) females mature at 360–600 mm ML, and (2) females mature at >650–700 mm ML. The lifespan of males and group 1 females is 1 year; the lifespan of group 2 females is not known.

The size of ripe eggs is 0.9–1.2 mm, with the maximum number of eggs in the oviducts of mature females varying between 0.5 and 2 million, depending on the females' sizes. The number of spermatophores in the spermatophore storage sac of mature males from the southern hemisphere is 900–1700. Mass mating occurs in surface layers at night in the 'head to head' position, and probably promiscuity occurs. The duration of copulation is 1–2 minutes. Spermatangia are attached to the buccal cone and membrane of the female. Between 115 and 160 seminal receptacles are found on the buccal membrane. Spawning occurs throughout the year, with some seasonal activity; it occurs off the central eastern Australian coast in spring. They are intermittent, multibatch spawners. After the first spawning, they continue to feed and grow in order to mature the next batch of oocytes for subsequent spawnings.

As with its other family members *O. bartramii* is a high speed, manoeuvrable and powerful squid that reacts very quickly to any change in its surroundings. It is a powerful, epipelagic, jet-propelled swimmer equipped with a strongly muscular mantle for maximum thrust and short, muscular,

Fresh-caught specimen, dorsal view (K. Graham).

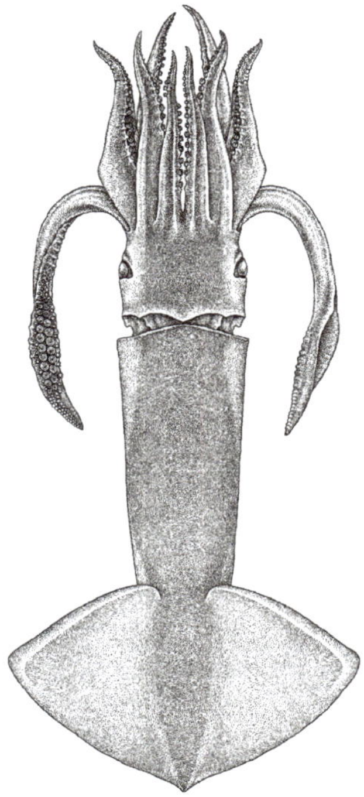

Dorsal view (ABRS, K. Hollis).

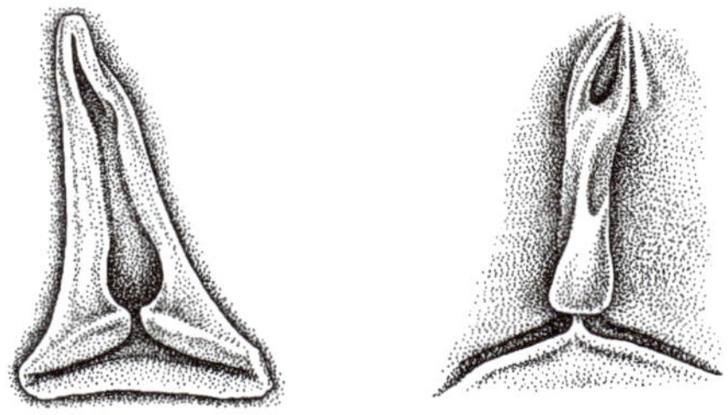

Funnel- (left) and mantle-locking cartilage (right) (ABRS, K. Hollis).

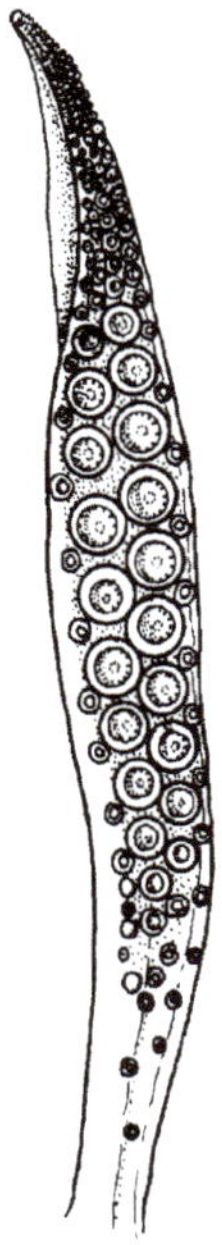

Hectocotylus (ABRS, K. Hollis)

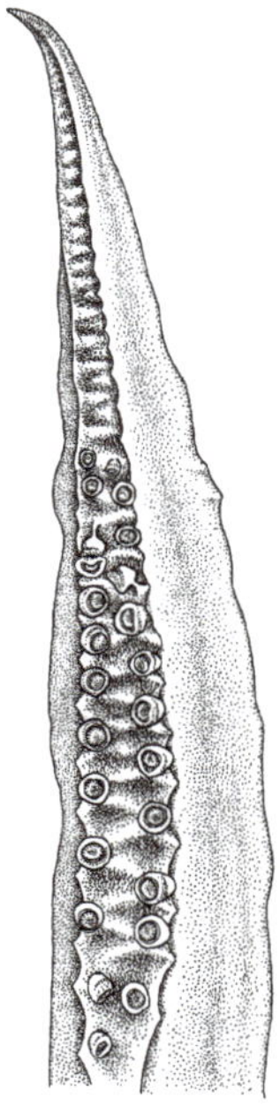

Tentacular club (FAO, E. D'Antoni).

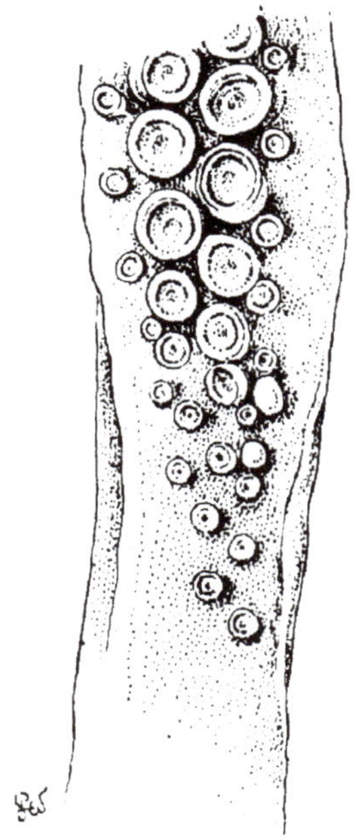

Enlargement of base of tentacular club (FAO, E. D'Antoni).

triangular fins for stabilisation and manoeuvring. Squids of 60–150 mm ML may burst through the surface and glide through the air when they are threatened and pursued by predators. The measured horizontal sustained speed of adult squid (370–530 mm ML) in the sea varies between 6 and 17.6 km $h^{-1}$, while the burst speed is greater. During the evening and morning vertical migrations the swimming speed is less and varies on average from 3–5 km $h^{-1}$.

It is an obligate shoaling squid throughout ontogenesis. Distinct shoals, or schools, are composed of uniform-sized squids. Shoal size varies from two to ~150 or more individuals. *Ommastrephes bartramii* is an active predator throughout all ontogenetic stages, requiring periodic changes in prey selection throughout its life. Feeding of subadult and adult squid of 500–600 mm ML takes place during night in surface layers from 0–70 m, and

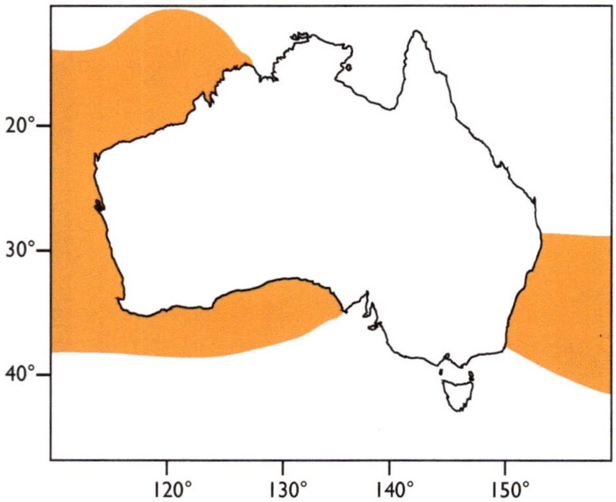

*Ommastrephes bartramii* distribution.

feeding activity is significantly reduced at the greater depths of the mesopelagic zone. The broad spectrum of prey for this adult size group of squid is relatively similar in the different parts of the vast geographical range of the species. Males mature at ~300 mm ML; females mature at ~400 mm ML.

*Interest to fisheries.* This species is not commercially exploited in Australia, but it is caught in continental slope waters around the southern Australian coast. The flesh is tasty and edible, but may be tough in large individuals. Experimental fisheries were carried out in the Tasman Sea without successful commercial results. However, continued demand for squid products, and increased knowledge about the biology and population structure of the species, may lead to geographically expanded fishery exploitation. Regions of high biological productivity are conducive for concentrations of feeding *O. bartramii* and consequently for fishing effort. Such areas where warm-core rings form and *O. bartramii* fisheries exist include the East Australia Current off south-eastern Australia.

### SIZE

The maximum ML of females is 800–900 mm (body weight 20–25 kg) and of males it is 400–420 mm (body weight 2.0–2.2 kg). Males mature at ~400 mm ML; females at ~600 mm ML.

### REMARKS

The genus is monotypic: a single oceanic species, *O. bartramii*, is recognised. Restriction profiles of two endonucleases, Alu I and Tsp509 I, are diagnostic for species identification.

Three populations (perhaps subspecies) of *O. bartramii* inhabit the North Atlantic, the North Pacific and the circumglobal southern hemisphere, but they have not yet been formally described. These three geographically isolated populations are differentiated by spermatophore morphology, by their size structure and by substrate-inhibitory traits of optic ganglia cholinesterases. The results of a mitochondrial DNA sequence analysis (mtDNA ND1–16S rRNA) by Kurosaka *et al.* (2012) also indicated significant differences among the North Pacific and Indian Ocean samples, but no clear genetic differences were observed among North Pacific sampling locations.

A revision of this taxon is critically needed. This is necessary not only for systematic and nomenclatural purposes, but it also is important for biological and, especially, fisheries management decisions.

### DISTRIBUTION

*Ommastrephes bartramii* is found circumglobally in subtropical and temperate oceanic waters. It is an extremely widely distributed ommastrephid species with an oceanic bi-subtropical (anti-tropical) worldwide distribution and is rarely encountered in continental slope waters and tropical latitudes. It inhabits the great subtropical ocean gyres in subtropical and partly temperate oceanic waters. In Australia, the species occurs in southern waters, south of the Tropic of Capricorn, including eastern Tasmania and the western Great Australian Bight. *Ommastrephes bartramii* have been caught between 23°42′S and 45°45′S, and larvae as far north as 24°S in eastern Australia. It is the dominant ommastrephid in the south-eastern Indian Ocean and Tasman Sea north of 40°S. It is replaced in the south by *Todarodes filippovae* Adam, 1975 (Dunning, 1993).

### REFERENCES

Lu and Dunning (1982), Dunning (1998b), Roper *et al.* (2010a).

## *Ornithoteuthis* Okada, 1927

### DIAGNOSTIC FEATURES

Mantle narrow, drawn out posteriorly into long, pointed 'tail'. Fins long, narrow, strongly pointed posteriorly, length 50–60% and width ~45–52% ML; fin angle 20–35°; funnel groove smooth with foveola with seven to nine longitudinal folds and a few side pockets (often obscure). Protective membranes of arms and tentacles normally developed (not greatly expanded). Tentacular club moderately expanded but without carpal-locking apparatus. Small tetraserial suckers on dactylus; carpal sucker area reduced, tubercle-like knobs absent; largest medial suckers on manus with 15–25 equal-sized pointed teeth; no enlarged teeth. Right (or very rarely left) arm 4 hectocotylised in mature males: distal modified section bears papillae formed from sucker stalks and with honeycomb sculptural formation along midpart of ventral surface of arm. Spermatozoa with one flagellum.

*Photophores.* External photophores and subcutaneous photophores absent; a single, round photophore on ventral surface of each eye and two round photophores on ventral surface of intestine in paralarvae and narrow, luminous pinkish photogenic stripe, with two interruptions, extends along ventral midline of viscera in adults.

#### HABITAT AND BIOLOGY

Epi-mesopelagic, lower sublittoral and bathyal squids. Adults occur in the lower epipelagic and mesopelagic zones and to the bottom in the bathyal realm. Paralarvae and juveniles occur in midwater and into the near-surface epipelagic waters.

#### SIZE

Maximum reported ML 310 mm.

#### REMARKS

Historically, the distributional ranges of the two nominal species in this genus, the Indo-West Pacific (*Ornithoteuthis volatilis* (Sasaki, 1915)) and the tropical and subtropical Atlantic (*O. antillarum* Adam, 1957) were probably continuous, but biogeographic divergence resulted in the formation of the separate, closely related species recognised currently.

#### DISTRIBUTION

Indo-West Pacific and tropical and subtropical Atlantic Ocean.

#### REFERENCE

Jereb and Roper (2010).

### *Ornithoteuthis volatilis* (Sasaki, 1915)

*Ommastrephes volatilis* Sasaki, M. (1915). On three interesting new oegopsids from the Bay of Sagami. *Journal of the College of Agriculture, Tohoku Imperial University, Sapporo* 6(6), 131–150 [138, text fig. 3, pl 4 figs 1–6].

Type data: syntypes MSUT.

Type locality: western Pacific Ocean. Japan, Sagami Province off Atami.

#### COMMON NAME(S)

Shiny Bird Squid, Long-tailed Flying Squid

#### DIAGNOSTIC FEATURES

Mantle very narrow, muscular, drawn out posteriorly into a long pointed 'tail'. Fins long, saggitate, sharply lanceolate posteriorly, posterior margins concave; fin length 55% (51–59%) of ML; fin width 47% (45–51%) of ML; single fin angle 27% (20–35%). Head broad, considerably broader than mantle width. Funnel groove with folds forming a pocket. Arm sucker rings with 10–14 sharp teeth on distal margin and with smooth proximal rim. Right arm 4 hectocotylised: distal half with reduced-sized suckers, sucker stalks modified into papillae, particularly in dorsal row, and honeycomb sculpturing with lateral pores along midventral surface with two or three longitudinal columns of depressions and swollen ridges, and with 10–15 depressions (pits or pores) in each column. Tentacular club expanded with very large medial manal suckers, their rings with 18–22 evenly spaced, equal-sized (not enlarged), sharply pointed teeth, occasionally interspersed with low, flat platelets. Proboscis typical of ommastrephid rhynchoteuthion-stage with a length of 50–75% ML at <4.0 mm ML; proboscis length approximately equal to longest arms; two lateral distal suckers on tip of proboscis up to 150% larger than remaining six suckers. Dorsal mantle with five oblong chromatophores in a + shape and a single small chromatophore between minute fins; ventral mantle chromatophore pattern forms side-to-side oblong of six rectangular chromatophores and two very small, roundish chromatophores at posterior tip of mantle.

*Photophores.* A single, round photophore patch on ventral surface of each eye. On viscera one large, round photophore near anus and one small oval photophore at posterior end of intestine; narrow strip of pinkish bioluminescent tissue extends from small photophore to posterior tip of mantle cavity, interrupted by two gaps where

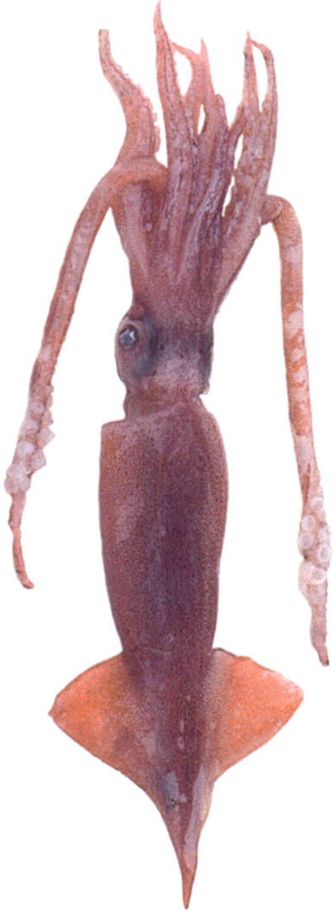

Fresh-caught specimen, dorsal view (K. Graham).

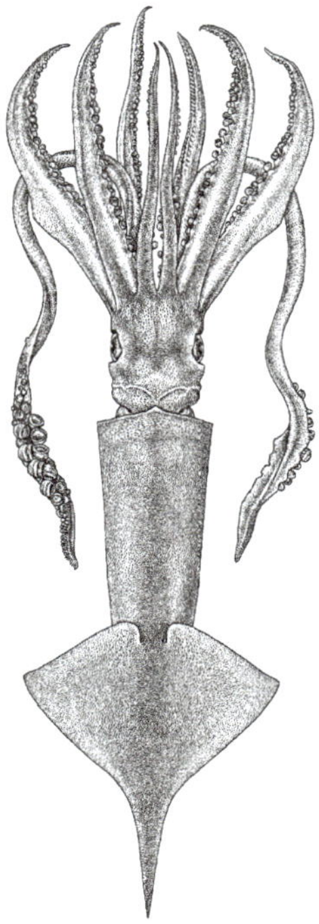

Dorsal view (ABRS, K. Hollis).

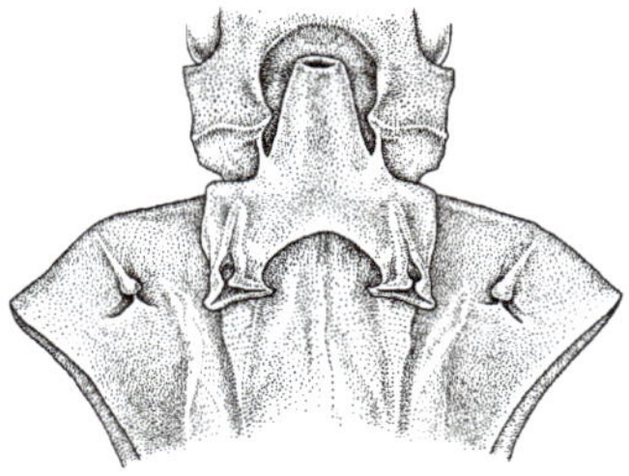

Funnel- and mantle-locking cartilages (ABRS, K. Hollis).

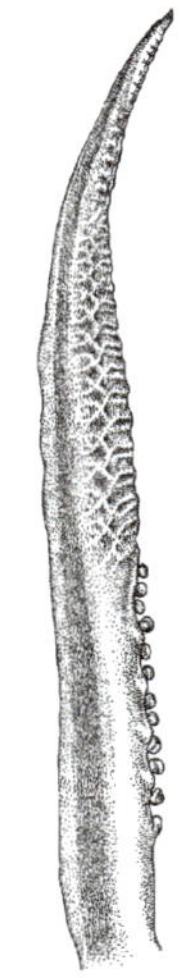

Hectocotylus (ABRS, K. Hollis).

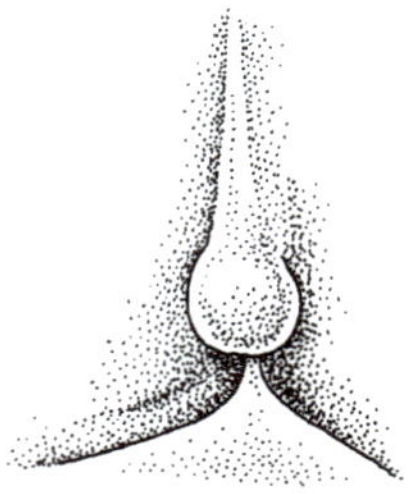

Mantle-locking cartilage (ABRS, K. Hollis).

Funnel-locking cartilage (ABRS, K. Hollis).

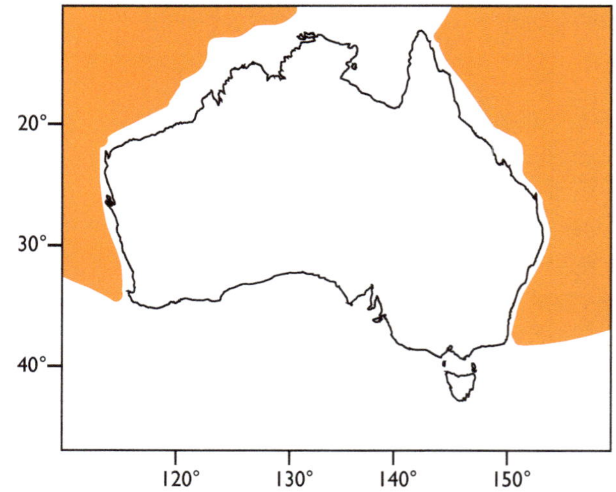


*Ornithoteuthis volatilis* distribution.

arteries intervene. Round photophores persist from paralarval stage to fully mature adults. Paralarvae with two round, unequal-sized intestinal photophores that are evident at ~3.5 mm ML; larger, anterior one ventral to intestine near the anus; smaller, posterior one ventral to intestine just anterior to its descent from the caecum; a single, ovoid photophore patch on ventral surface of each eye at ~4.0 mm ML.

*Colour.* In life mantle dark red-brown. Ventrally, bright strip and large photophores visible through mantle.

## HABITAT AND BIOLOGY

*Ornithoteuthis volatilis* is a tropical/subtropical species, widely distributed in the Indo-Pacific Ocean. It occupies tropical slope and oceanic waters from the surface (paralarvae; adults at night) to moderate depths, at the bottom in the bathyal zone and in midwater above the slope (lower epipelagic, mesopelagic) as well as above the tops and slopes of seamounts and midocean ridges. Paralarvae and juveniles also occur in the equatorial zone in midwater above great oceanic depths. *Ornithoteuthis volatilis* is commonly captured in trawl catches in

Australian slope waters. It is captured mainly in midwater and demersal trawls in continental slope waters and near seamounts but are rarely caught at the surface. It has been suggested that *O. volatilis* undergoes diel vertical migrations.

Size at maturity varies significantly in both sexes, depending on season and locality. Mature males were caught in south-eastern Australia in summer, at 160–240 mm ML, while other males were still immature at 180–200 mm ML. They have very small eggs (0.7–1.0 mm, mostly 0.75–0.85 mm), and spawning is intermittent with several egg masses produced over an extended period. They spawn near the bottom around seamounts and ridges, some after long migrations to these localities. Spawning of *O. volatilis* off south-eastern Australia occurs in the deeper shelf and upper continental slope waters that comprise the warm East Australian Current. Size distribution analysis of paralarvae and adults suggests that spawning in this region occurs year-round.

*Interest to fisheries.* While *O. volatilis* is a widely distributed species in tropical/subtropical Indo-West Pacific waters, to date adults have not been captured in large numbers. They are occasionally caught in surface waters on jigs and in nets. If concentrations of this species can be found, they will provide a potentially desirable product because of their size, texture and nutrient composition.

### SIZE

The maximum ML recorded from the Western Deepwater Trawl Fishery is 140 mm for a female. Worldwide, the maximum size reported is 250 mm in females and 310 mm in males.

### DISTRIBUTION

Indo-West Pacific, tropical to subtropical western Pacific and Indian Oceans. Predominantly in northern and eastern waters in the Timor, Coral and Tasman Seas. Off the eastern Australian coast its distribution extends to 40°S, off the north-western Australian coast it occurs in continental slope waters to approximately 35°S.

### REFERENCES

Lu and Dunning (1982), Roper *et al.* (2010a).

## *Sthenoteuthis* Verrill, 1880

### DIAGNOSTIC FEATURES

Mantle–funnel locking cartilage fused or not fused; if fused, occurs at post-paralarval stages. Funnel groove has foveola with five to nine longitudinal folds and four to six distinct side pockets. Fins rhomboidal, not attenuate posteriorly. Fin length 40–50% and fin width 70–88% ML; fin angle 58–72°. Arm tips not attenuate. Arms have 26–35 pairs of suckers. Tips of trabeculae of protective membranes do not project beyond edge of membrane. Ventral protective membranes of arms 3 subequal to arm width. Right or left ventral arm hectocotylised, smooth, with six to eight pairs of suckers. Dactylus of tentacular club with tetraserial suckers. Carpal-locking apparatus on tentacular stalk with two to five knobs and an equivalent number of alternating smooth-ringed suckers; zero to two suckers with denticulate rings on carpus proximal to first knob. Largest medial suckers on manus with four large, pointed teeth.

*Photophores.* External photophores present, including small subcutaneous photophores scattered on mantle, head and bases of ventral arms. Large, yellow, oval photophore located anteriorly on dorsal part of mantle in squid > 140 mm ML; photophores consist of numerous, densely packed, small, yellow, subcutaneous organs, 1.0–2.5 mm in diameter, embedded in mantle musculature, beneath which is white reflective layer. A single round photophore on ventral surface of each eye and two round photophores on ventral surface of intestine. Juveniles with oval light organs ventrally on eyes and between ink sac and intestine.

### SIZE

A large-sized squid; ML up to 650 mm.

### REMARKS

*Sthenoteuthis* includes two oceanic species, one of which, *S. oualaniensis* (Lesson, 1830), is found in Australian waters. Two endonucleases, AluI and Tsp509I, are diagnostic for species identification.

### DISTRIBUTION

Tropical and partly subtropical waters of the Atlantic, Indian and Pacific Oceans.

### REFERENCE

See species account.

## *Sthenoteuthis oualaniensis* (Lesson, 1830)

*Loligo oualaniensis* Lesson, R. P. (1830) in Lesson, R. P. (1830–1831). Mollusques, Annelides et Vers. Voyage Autour du Monde, sur la Corvette de la Majeste, La Coquille, pendant les annes 1822, 1823, 1824 et 1825, sous le commandement du capitaine Duperrey. Paris, *Zoologie 2*(1), 471 pages [240, pl. 1 fig 2].

Type data: type not extant MNHN.

Type locality: western Central Pacific Ocean, Caroline Archipelago, Micronesia, Oualan Island.

### COMMON NAME(S)

Purpleback Flying Squid, Yellow-backed Squid, Flying Squid, Purple Squid

### DIAGNOSTIC FEATURES

Mantle cylindrical, muscular, tapered caudally. Head large, slightly narrower than mantle. Fins rhomboidal, about twice as wide as long, fin angle 60–70°. Males have wider fins than females. Funnel and mantle-locking cartilage fused, shaped like an inverted 'T'. Mantle element of inverted T-shaped locking apparatus curved with anterior bifurcation, fused to funnel element at post-paralarval stage by 9–12 mm ML along posterior third of longitudinal groove. Funnel groove with foveola and side pockets. Either left or right ventral arm 4 hectocotylised in mature males: six to eight pairs of normal suckers proximally, sucker stalks and bases lost distally and about six pores developed in thickened dorsal and ventral protective membranes basally and medially. Large tentacular suckers with four equally spaced large, sharp teeth interspersed with five or six smaller sharp teeth. Dactylus of tentacular club with tetraserial suckers; medial manal sucker rings with one tooth in each quadrant greatly enlarged, five to seven subequal teeth between these; carpal fixing apparatus consisting of smooth-ringed suckers and knobs on tentacular stalk.

*Photophores.* Luminous oval patch on anteriodorsal mantle. Large dorsal photophore may be present in larger individuals anteriorly on the mantle; this begins to develop in immature squid of 100–120 mm ML; single, oval photophore present ventrally on each eye in larvae and juveniles. Two approximately equal-sized photophores present between intestine and ink sac in larvae and juveniles, one anteriorly near anus and another posteriorly.

*Colour.* Live animals with dark red-brown mantle, slightly darker along dorsal midline; bright yellow dorsal luminous patch may light up following capture.

### HABITAT AND BIOLOGY

*Sthenoteuthis oualaniensis* is a widely distributed, abundant and highly productive, epipelagic to mesopelagic and upper bathypelagic species that inhabits open waters over great depths of more than 200–400 m, at sea-surface temperatures from 16 to 32°C, but usually above 20–22°C. Dunning (1998b) captured juveniles between 22and 39°S where sea-surface temperatures were 26.7°C–20.8°C. It is the most abundant large squid in the tropical and subtropical waters of the Indo-Pacific region and has a patchy distribution concentrated in areas of high productivity.

It is truly pelagic, not associated with the bottom. Diurnal vertical migrations by paralarvae and juveniles are absent or very short in distance. During daytime the main bulk of juveniles are found near the surface from 0 to 30 m, while at night they descend deeper, between 25 and 75 m, moving out of the zone of maximum numbers of conspecific predatory adult squids. Subadult and adult squids are interzonal animals. At night they inhabit surface and subsurface layers from 0 to 150 m to feed, with maximum numbers between the surface and ~25 m. In the morning they descend to depths of 200–1100 m and remain there during the day.

In northern Australian waters, males reach maturity from ~160 mm ML and females from 250 mm ML. In the eastern tropical Pacific, Dunning (1998b) found males mature > 110 mm ML and females at 180–190 mm ML. Mating occurs in surface waters at night in the 'head to head' position without preliminary rituals; probably promiscuity is common. The duration of copulation is short, lasting only 0.5–2 minutes, and results in up to 150 spermatangia being attached to the buccal cone and membrane of females, and then the viable sperm are stored in 70–120 seminal receptacles that are found on the buccal membrane. At the species level, spawning takes place all year round, with the duration of the spawning season for individual females up to 1–3 months. Data from Australia suggest that *S. oualaniensis* spawns in summer along the continental shelf. This squid is an intermittent, multi-batch spawner. After spawning once, females continue to feed and grow, while the next batch of oocytes matures;

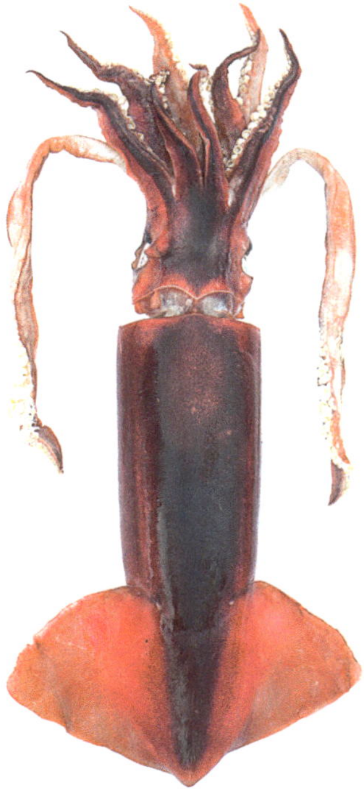

Fresh-caught specimen, dorsal view (K. Graham).

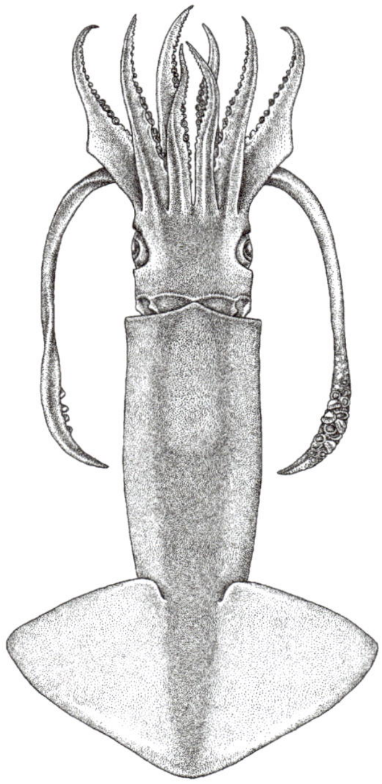

Dorsal view (ABRS, K. Hollis).

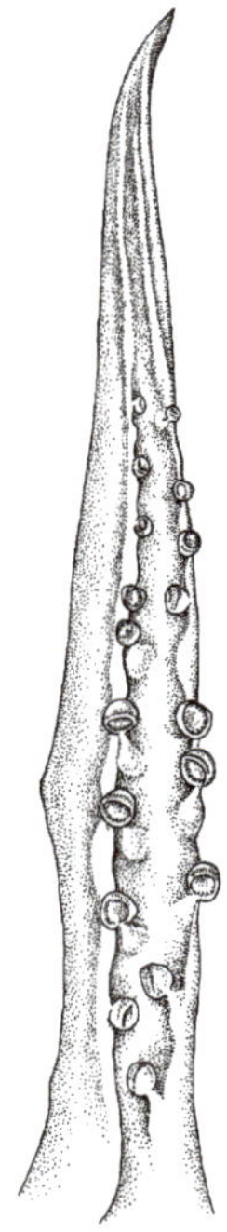

Hectocotylus (ABRS, K. Hollis).

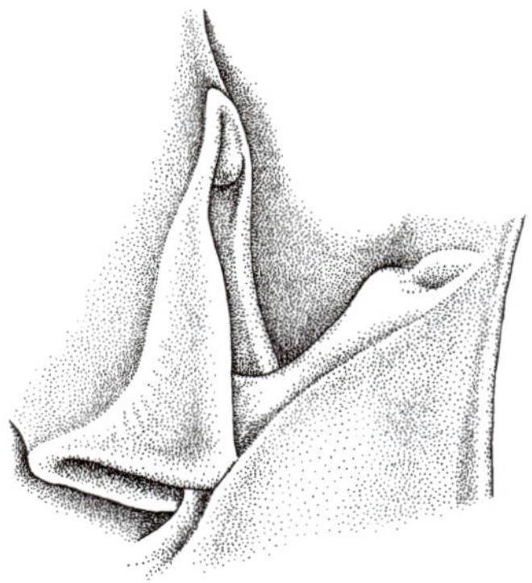

Funnel- and mantle-locking cartilages (ABRS, K. Hollis).

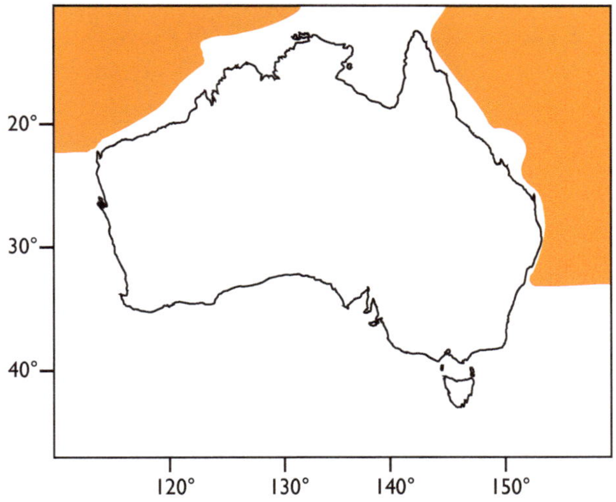


*Sthenoteuthis oualaniensis* distribution.

then spawning begins again. Egg masses (of typical ommastrephid form: a gelatinous mass containing thousands of eggs) are pelagic, and they float above the upper pycnocline layer. The duration of embryogenesis in water temperatures of 20–25°C is 3–6 days. Hatching size is ~1.0 mm ML. Paralarvae have the typical ommastrephid rhynchoteuthion-stage defined by the tentacles being fused into a trunk-like proboscis. During growth the proboscis gradually separates and in *S. oualaniensis* this separation is complete by 7.0–8.0 mm ML, signalling the end of this stage.

*Sthenoteuthis oualaniensis* is a fast-swimming, highly manoeuvrable, nektonic squid. Juveniles are capable of 'flying' during daytime. Under threat from pursuit of predators they can reach a high speed and then glide above the surface of the sea some tens of metres. The cruising speed of adult squid is ~3–10 km $h^{-1}$, but the burst speed reaches >25–35 km $h^{-1}$. During the day at depth, squids are nearly non-active, and either 'hang' in the water or move slowly with a speed of only

1–1.5 km $h^{-1}$, propelled by the undulating fins. The species is an obligate shoaling squid throughout ontogenesis. Usually, shoals consist of squid of uniform size. Shoal size is highly variable and depends on the productivity of the area and the density of the squid population: it varies from two to ~800 individuals, occasionally more. At the periphery of its geographical range, this species may form joint schools with same-sized *Dosidicus gigas* (d'Orbigny, 1835) and *Ommastrephes bartramii* (Lesueur, 1821).

*Interest to fisheries.* *Sthenoteuthis oualaniensis* is not fished commercially on an industrial level anywhere in the world, mainly due to the absence of effective fishery methods and to the relatively poor quality of the meat. The catches are used mainly as bait for the tuna fishery. Many attempts to use jigging machines to fish for the species in different parts of the species range have not been effective. However, this species has great potential for a future significant fishery. Two representatives of nematodes found in the species that have significant negative impact on the commercial importance of the species. Larvae of *Anisakis* sp. are a life-threatening danger to humans, but they are localised in the ovaries and encysted in the coelomic membranes, neither of which is used for human food. Larvae of *Porrocaecum* sp. are encysted on the internal mantle wall and they may significantly degrade the quality of the marketable components of the infested squid.

#### SIZE

Maximum ML ~300 mm in females, 250 mm in males (commonly to ~200 mm ML).

#### REMARKS

On the basis of size differences of mature squid, as well as dorsal photophore and gladius morphology, at least five forms of undetermined status are distinguishable. This suggests a complex population structure. The extent of subpopulation mixing and genetic isolation is unknown. Several size classes are known around Australian waters.

*Sthenoteuthis oualaniensis* as is currently recognised includes two forms in the western central Pacific, one with and the other without a large dorsal photophore patch on the anterior mantle in adults. The former reaches in excess of 300 mm ML, while the latter reaches maturity at <200 mm ML. The dorsal photophore patch develops in specimens from 65 mm ML and is readily visible at 100 mm ML.

#### DISTRIBUTION

In Australia, North West Shelf and slope; eastern coast from Queensland to southern New South Wales. Indo-Pacific, tropical, oceanic species; its range extends in a broad band that stretches throughout the open waters of the tropical zone of the Indian and Pacific Oceans (from ~40°N to 40°S) generally where bottom depths exceed 200 m. It enters into subtropical zones where warm currents occur near Japan, the eastern Australian coast and South Africa.

The southern boundary of the species reaches to the southernmost point of Africa to ~35°S and then it continues eastward approximately along the longitudes 25°S to 30°S, and afterwards to 35°S near the western Australia coast. It extends from the southern part of the Australian coast (38°40′S) eastward to northern Chile (18°S to 22°S). Squid migrate to high latitudes with seasonal warming during summer in their appropriate hemisphere along the shift of surface isotherms of 20–22°C; then they move to lower latitudes with the seasonal autumn of water temperature.

#### REFERENCES

Jereb and Roper (2010), Chen *et al.* (2013).

## Subfamily Todarodinae Adam, 1960

#### DIAGNOSTIC FEATURES

Ommastrephids with funnel groove and foveola present (except *Todaropsis*), side pockets absent. Right or left or both arms 4 hectocotylised in mature males; ocular, visceral and mantle photophores absent. Left tentacular club with two pairs of carpal suckers; tentacles without carpal-locking apparatus; suckers on dactylus of tentacular club tetraserial; medial manus sucker rings with 20 or fewer long, pointed teeth, with single, large, pointed tooth distally (in some species).

#### REMARKS

Within the Todarodinae, using 16S rRNA and COI partial sequences, Wakabayashi *et al.* (2012) found *Todarodes* and *Nototodarus* to be polyphyletic.

## *Martialia* Rochebrune & Mabille, 1889

*Martalia* is montypic, represented by *Martialia hyadesi* Rochebrune & Mabille, 1889.

### *Martialia hyadesi* Rochebrune & Mabille, 1889

*Martialia hyadesi* Rochebrune, A. and Mabille, J. (1889). *Mollusques. Mission Scientifique du Cap Horn, 1882–1883 6*(2) (Zoologie), H1–H143 [H9], 8 plates. [9, pl. 1 figs 1–5].

Type data: syntypes MNHN 7-3-512 and 7-3-513 (gladius).

Type locality: eastern South Pacific Ocean, Chile, Orange Bay (Isla Navarino).

#### COMMON NAME

Sevenstar Flying Squid

#### DIAGNOSTIC FEATURES

Mantle robust, tapers to somewhat elongated 'tail'. Funnel groove with foveola possesses seven longitudinal folds; no side pockets. Fins rhomboidal, elongated posteriorly forming small 'tail'; fin length 38–46% and fin width 54–66% ML; fin angle 45–55°. Protective membranes of arms and tentacular clubs low, poorly developed; trabeculae strongly developed, prominent, like cirri. Arm sucker stalks attached at bases of trabeculae. Suckers of distal third of left arm 4 sometimes with elongated stalks. Arm suckers proportionally small; rings with five pointed teeth in distal half, central one conical, lateral ones become truncated; proximal half of ring smooth. Right ventral arm 4 hectocotylised, left arm 4 slightly modified: trabeculae on distal one-third of arm modified as rounded flaps on ventral row; flaps reduced and then absent distally on dorsal row. Tentacular clubs occupy almost entire length of tentacles; not expanded. Carpal-locking apparatus absent on tentacular club/stalk. Dactylus of tentacular club with four rows of small suckers. Largest medial suckers on manus with 15 relatively large, sharp teeth (one conical tooth usually larger than others) alternating with 15 small flat plates or blunt teeth. Tentacles like enlarged arms, most of tentacle occupied by club; not differentiated into carpus, manus and dactylus. Tentacular stalks with three to five free trabeculae proximal to first suckers on each margin.

*Photophores.* Absent throughout life cycle.

*Colour.* In live animals, dorsal surface of mantle very dark purple, lateral mantle red-brown, ventral mantle silver-white.

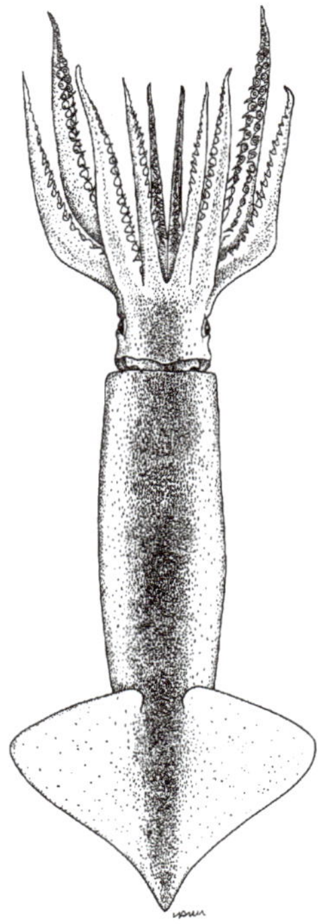

Dorsal view (FAO, E. D' Antoni).

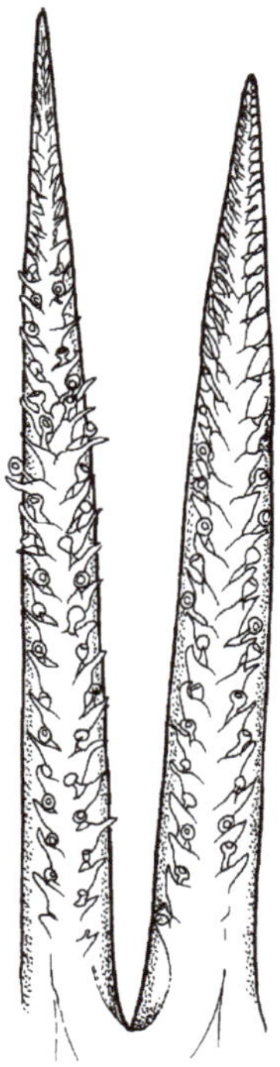

Hectocotylus (FAO, E. D' Antoni).

#### HABITAT AND BIOLOGY

Adults are oceanic-neritic in slope waters, seldom penetrating waters where bottom depths are < 200 m. *Martialia hyadesi* is classified as a notalian-Antarctic species, endemic to the circumpolar zone bounded roughly by

Enlargement of sucker (FAO, E. D' Antoni).

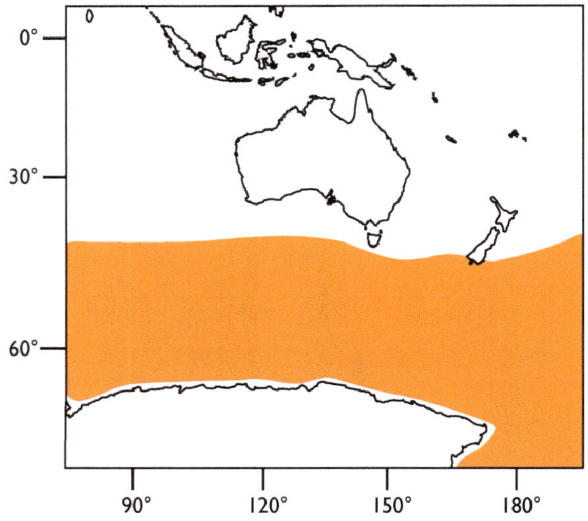


*Martialia hyadesi* distribution.

the subtropical southern boundary and the Antarctic Polar Front. Small juveniles, < 10 mm ML, have been found in southern New Zealand waters south of the Subtropical Front, indicating spawning in the region of the Auckland Island shelf. The very distinctive tentacular club is evident on juveniles from 19 mm ML, while the distinctive arm trabeculae first appear at 40–50 mm ML. The size of ripe eggs is 0.7–2.4 mm and the potential fecundity is 20 000–2.5 million eggs. Mature males of 233–260 mm have spermatophores 25 mm long with a fecundity of ~500. Growth in *M. hyadesi* is characterised by slow juvenile growth (< 100 mm ML), rapid growth in immature squids, and a sharp decrease in growth rate during maturation. The species matures later than other temperate ommastrephids (> 270 days), but final maturation is quite rapid: 2–3 months. Statolith analysis indicates an age of 12–13 months, but other life cycle data suggest a lifespan that approaches 2 years. Unike most other Southern Ocean species (except *Todares filippovae* Adam, 1975) *M. hyadesi* forms schools. In spite of its oceanic habitat, occasionally this species will undergo mass strandings, as occurred in the Falkland Islands in 1997, when more than 3000 immature *M. hyadesi* (220–270 mm ML) forcibly drove ashore; 61% were females. It has also occurred after storms at Macquarie Island. Such strandings are believed to be related to temporal shifts in frontal zones that affect the behaviour, range and environment of foraging squids.

*Martialia hyadesi* is a non-ammoniacal species.

*Interest to fisheries.* Accumulated data on the circum-Antarctic population of *M. hyadesi* indicate a very significant biomass of this species, with high potential for supporting a managed fishery.

### SIZE

Maximum ML to 400 mm and body weight to 820 g.

### REMARKS

Genetic studies indicate considerable genetic differentiation across the circum-Antarctic range of the species, suggesting the possibility that sibling or cryptic congeneric taxa might exist (Brierley *et al.* 1993).

### DISTRIBUTION

Circumglobal in cool temperate waters of the Southern Ocean; a nerito-oceanic slope species associated with high velocity, cool waters of the Antarctic Circumpolar Current, especially along the Antarctic Polar Frontal Zone south of the Subtropical Front; southern Patagonian Shelf edge, Scotia Sea, New Zealand waters south of the Subtropical Front and the sub-Antarctic islands, sub-Antarctic waters north of the Antarctic Polar Front.

### REFERENCE

Jereb and Roper (2010).

## *Nototodarus* Pfeffer, 1912

### DIAGNOSTIC FEATURES

Fins rhomboidal, slightly attenuate posteriorly; fin length varies between 36% and 50% ML; fin width between 50 and 61% ML; fin angle 40–57°. Funnel groove has foveola, but lacks side pockets. Both ventral arms hectocotylised: right arm modified completely and left arm partially; distal part of right arm with papillae; proximal part of both arms with very unusual sculptured structure of hyalineous integuments. Carpal-locking apparatus absent from tentacular stalk; dactylus of tentacular club with small suckers in four rows; largest medial suckers on manus have teeth of equal size, or one sharply pointed tooth larger than the other teeth (*Nototodarus hawaiiensis* Berry, 1912). Protective membranes on arms and tentacles normally developed, without exposed trabeculae.

*Photophores.* Absent.

*Colour.* Typically with dark, narrow mid-dorsal stripe on mantle; oceanic species dark over entire dorsum.

#### HABITAT AND BIOLOGY

*Nototodarus* species comprise major populations in the shelf-break zone, where enriched food supplies are created by eastern and western boundary currents and their associated mesoscale and downwelling features, such as upwelling cells and eddies. Young of the species congregate seasonally over the continental shelf in large schools.

#### SIZE

A medium-sized squid; maximum recorded ML ~420 mm.

#### REMARKS

After lengthy taxonomic confusion over many years, Dunning and Förch (1998) concluded that the valid species within *Nototodarus* are *Nototodarus sloanii* (Gray, 1849), *N. gouldi* (McCoy, 1888) and *N. hawaiiensis* (Berry, 1912) and reviewed the systematics, distribution and biology of the three species. Recent mitochondrial DNA analyses indicate that *Nototodarus* is paraphyletic – *N. gouldi* and *N. sloani* are sister species and are more closely related to *Todarodes pacificus* (Steenstrup, 1880) than to *N. hawaiiensis*. These results suggest that the form of the hectocotylus, currently a major morphological distinguishing character in *Nototodarus*, may not be a reliable character (Wakabayashi *et al.* 2006). (Also see 'Remarks' under *T. filippovae* below.)

#### DISTRIBUTION

Distributed in tropical, subtropical and temperate waters of the Indian and Pacific Oceans.

#### REFERENCE

Jereb and Roper (2010).

### Key to *Nototodarus* species

1a. Smooth skin. Arms with more than 35 pairs of suckers. Maximum size up to 420 mm ML.................. *N. gouldi*

1b. Rough textured skin. Arms with fewer than 30 pairs of biserial suckers. Maximum size (Australian waters) up to 250 mm ML ....................*N. hawaiiensis*

### *Nototodarus gouldi* (McCoy, 1888)

*Ommastrephes gouldi* McCoy, F. (1888). Natural History of Victoria. *Ommastrephes gouldi* (McCoy). *Prodomous of the Zoology of Victoria, Decade 17*, 255–257 [255, pls 169, 170].

Type data: holotype MV F.5104.

Type locality: Australia: southern Australia, Victoria, Hobson's Bay.

#### COMMON NAME(S)

Arrow Squid, Gould's Flying Squid, Gould's Squid; Gould's Arrow Squid, Red Arrow Squid

#### DIAGNOSTIC FEATURES

Mantle robust, heavily muscled, tapers gradually then very acutely to a point posteriorly. Fins ~45% as long as ML; angle broad, 90–110° (45–55° single fin). Funnel–mantle locking cartilage inverted T-shape. Funnel groove with foveola and six to eight low longitudinal ridges anteriorly; folds of skin forming pockets. Arms 1 with 28–50 pairs of suckers. Arm sucker rings with 12 or 13 sharp, triangular teeth graded to a single large central tooth distally; proximal third smooth, without teeth or denticles. Both ventral arms (4) in males hectocotylised: right arm with heavily tuberculate basal region, narrow sucker-bearing midpart, and distal part with very broadly expanded, pectinate protective membrane; left ventral arm one-quarter longer than right 4, about five proximal-most pairs of suckers modified into large, swollen tubercles, remaining suckers normal. Tentacular club long, 60–70% of tentacle length; about eight pairs of enlarged manal suckers in median rows, marginal suckers small, subequal; largest sucker rings toothed all around with 14 or 15 separated, stout, triangular, pointed teeth interspersed with low, side platelets. Carpus of tentacle with six or seven pairs of suckers.

*Colour.* Live animals light brownish pink with bluish purple dorsal stripe.

#### HABITAT AND BIOLOGY

*Nototodarus gouldi* is a neritic and oceanic, mostly epipelagic species, that inhabits inshore waters over the continental shelf to ~500 m depth. Its abundance in surface waters is correlated with the lunar cycle; on full moon nights the squids tend to remain in deeper waters where they are less vulnerable to fishing gear, and presumably some predators. The principal oceanographic feature that supports and sustains *N. gouldi* populations is the productivity of the Antarctic Circumpolar Current through the distribution and dispersal of energy flow into the region of secondary producers and predators that form their prey. While its greatest abundance is over shelf waters at 50–200 m (sometimes 300 m), *N. gouldi* is occasionally abundant in shallow coastal waters in the summer in Australia.

*Growth.* Across the range of the species in south-eastern Australian waters considerable variation exists in size, growth rates and maturity between sites, seasons and years. The age at sexual maturation has been determined using statolith ageing techniques. However, over its range, the age of mature squid varies by > 100 days from youngest to oldest. Around Tasmania, males start to mature at ~200 days, and the testis somatic index (TSI) reaches its maximum at ~270 days. Females begin to mature as the TSI reaches its maximum in males. The rapid development of the ovary, oviducts and nidamental glands occurs simultaneously, and copulation takes place during this period as well. Spawning occurs about 2 months after copulation, completing the slightly < 1 year life cycle, based on statolith analysis. According to Jackson *et al.* (2005), males exhibit no seasonal differences in growth rates, but females do: females hatched in summer have significantly higher growth rates than those hatched in winter or spring. During cool months, females are larger and have slower lifetime growth, lower gonadic investment and better somatic condition than those caught in warm months, representing a trade-off between gonadic investment and somatic condition. Males have a tight coupling between gonadic investment and somatic condition throughout the seasons, with gonadic and somatic tissues growing concurrently. This strategy of 'using' monthly variations in ambient conditions to govern egg production and spawning would have positive effects to maximise reproduction and survival. Females grow larger than males, but males tend to be heavier at a given ML, until females surpass them at maturity.

*Spawning.* The species appears to be a multiple spawner with eggs being released over time in discrete batches (McGrath and Jackson 2002). O'Shea and Bolstad (2004) have described the egg mass of *N. gouldi* (identified using 16S rDNA sequence determination) to be a gelatinous, free-floating sphere of at least 1.5 m diameter and containing several thousand eggs. Such egg masses are likely to be damaged by trawling and other shipping activities.

Based on hatch dates calculated from statolith age estimates, Green (2011) found that *N. gouldi* collected in Victoria and the Great Australian Bight during 2007–09 hatched year-round, with peaks of greater recruitment.

*Egg laying.* *Nototodarus gouldi* exhibits the multiple-spawning strategy whereby eggs are laid in discrete batches during the course of the spawning period. Mature ovaries contain oocytes with a range of sizes with discrete peaks. The life cycle generally is related to warm currents and eggs, paralarvae and occasionally juveniles are transported by the current from the spawning grounds to the feeding grounds, where they become available to the fishery. Following mating and spawning, the muscular tissues of spent individuals begin to degrade and become weakened and flaccid. Probably the very rapid disposition of somatic protein into protein-rich yolk during vitellogenesis in the ovary, coupled with reduced or ceased feeding, leads to the somatic degradation after spawning, soon followed by death.

*Stock structure.* An allozyme investigation of the stock structure of *N. gouldi* compared the variation at 48 loci from samples collected from six localities across southern Australia separated by distances of between 700–4300 km. The data revealed no evidence for more than a single species among 203 squid examined. There was some evidence of stock structure, with animals from Iluka (northern New South Wales) displaying significant differences in allelic structure from Ulladulla (southern New South Wales) and Tasmania. Thus, observed regional differences in growth appear to be largely due to responses to environmental differences rather than genetic segregation.

Interannual variations in the Southern Oscillation Index (SOI) and the Trans Polar Index (TPI) affect the environment and abundance of *N. gouldi*: juveniles are affected mainly by the SOI at lower latitudes, while the adults are affected by the TPI in higher latitudes. This mechanism influences recruitment and stock abundance.

*Movement.* Stark *et al.* (2005) used an acoustic telemetry system to follow the movement and activity of tagged squid in and out of Storm Bay and the Derwent River in south-east Tasmania (known depth range up to 85 m). They showed that the population was highly dynamic and moved widely between the bay and the river

Fresh-caught specimen (CSIRO ScienceImage AF9591).

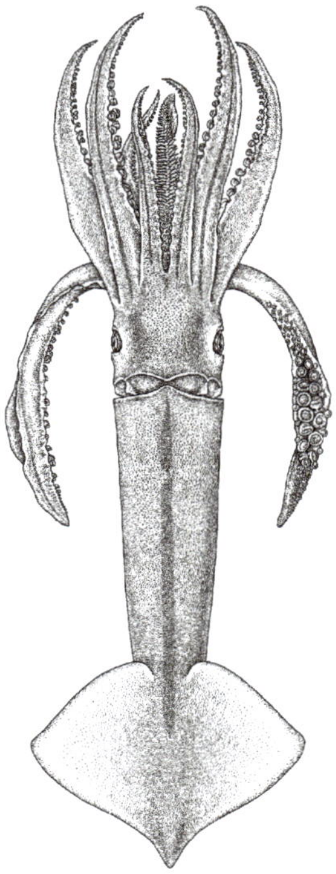

Dorsal view (ABRS, K. Hollis).

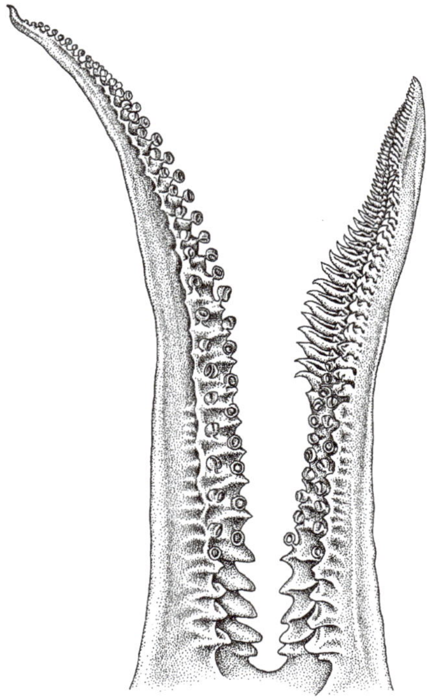

Hectocotylus (ABRS, K. Hollis).

during the sampling period (summer 2003), and was more active at night.

*Feeding.* *Nototodarus gouldi* is an opportunistic predator with feeding occurring mainly at night and at dawn and digestion is rapid. The digestive gland of *N. gouldi* accumulates high levels of trace metals (cadmium, copper, zinc and polonium-210). Pethybridge *et al.* (2012) examined the feeding ecology of *N. gouldi* collected off south-eastern Australia and found clear seasonal dietary shifts in prey composition. Teleost remains (mainly otoliths) were found in 67% of stomach contents with the mesoplanktonic planktivorous fishes *Lampanyctodes hectoris* (Günther, 1876) and *Maurolicus muelleri* (Gmelin, 1789) dominating. Cephalopods and crustaceans were supplementary dietary components, with an increased representation over winter. Total lipid content and fatty acid composition in the squid stomachs were related to summer upwelling and local changes in productivity, and seasonal differences in prey consumption, respectively.

*Interest to fisheries.* Prior to the development of a directed fishery, the species was taken only as by-catch in demersal trawling in the Australian prawn fishery, but the main fishing method is jigging under lights. Feasibility fishing started in the summer 1978–79, with 19 Japanese jigger vessels operating in south-east Australian waters.

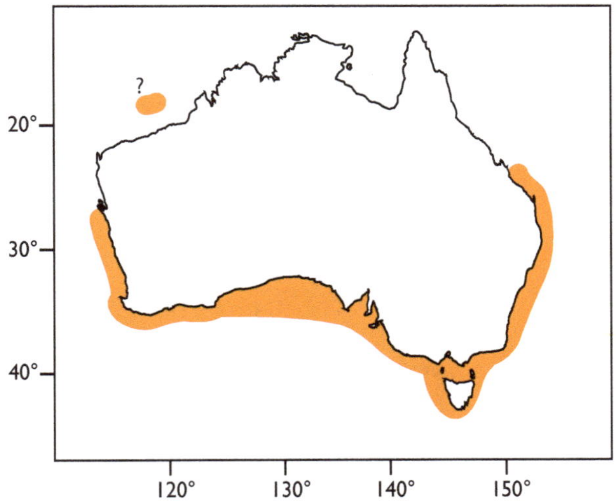


*Nototodarus gouldi* distribution.

In the light of promising catch rates, exploratory fishing was subsequently expanded. Chinese jigging vessels (from Taiwan) operate intermittently in the area. Catches are best in the waters around Tasmania and western Bass Strait, particularly from December to March. In early years, the annual catch fluctuated between 4000 and 5000 tonnes. An active fishery then developed for *N. gouldi*, principally around southern Australia and Tasmania in temperate waters and then in subtropical waters north to 27°S off the east coast, and to 25°S along the west coast. The instantaneous biomass of *Nototodarus* species throughout their ranges has been estimated at 2–3 million tonnes of which 0.5–1 million tonnes are *N. gouldi*. The principal goal in management of a fishery on squid with an annual life cycle is to enable a sufficient number of spawning individuals to escape. In the case of *N. gouldi*, regulation of effort is the best tool to manage the fishery, regardless of the fishing area. Clearly, knowledge of the timing of spawning, the ecology of hatchlings, the larval distribution and the growth dynamics of juveniles is critical to understand the location and life cycle of the adults in order to sustain a well-managed, consistent fishery.

Echo sounding and sonar surveys have been conducted to determine the suitability of alternative fishing techniques for *N. gouldi*. The species forms demersal aggregations during daylight hours. Around dusk, squid leave the bottom and disperse into the water column to feed at night. This behaviour is conducive for bottom and midwater trawling. Purse seining was suggested as an alternative technique for capturing squid attracted by lights at night and clustered in demersal aggregations during daylight hours. During the week of the full moon, catch rates are ~25–50% lower than during darker weeks. Catch rates generally are greatest during the days following a new moon. Commercial jigging operations are conducted both during the day and at night, but night-time catches are greater, probably because the diel vertical migration pattern brings the squid off the bottom layer and closer to the surface at night. In the developing squid fishery off New South Wales, *N. gouldi* is the most abundantly caught species. The fishery for *N. gouldi* at present is managed as one stock. Given the results of the allozyme studies described above, further studies are needed to delineate the degree of stock structure for fisheries management purposes.

#### SIZE

Females up to 420 mm ML, up to ~1.7 kg in weight; lifespan ~360 days; males up to 366 mm ML and weigh up to 1.0 kg.

#### REMARKS

*Nototodarus gouldi* was the first ommastrephid described from Australian waters (as *Ommastrephes gouldi* McCoy, 1888).

#### DISTRIBUTION

Found in waters south of 27°S and the northern and central coasts of New Zealand. Temperate to subtropical areas off Australia: from south of 27°S off the east coast (Queensland), southward around Tasmania through the Bass Strait, westward along the south coast, and then northward off Western Australia to ~25°S off Shark Bay; off the northern and central coasts of New Zealand, principally off North Island.

#### REFERENCES

Triantafillos *et al.* (2004), Roper *et al.* (2010a).

### *Nototodarus hawaiiensis* (Berry, 1912)

*Ommastrephes hawaiiensis* Berry, S. S. (1912). A catalogue of Japanese Cephalopoda. *Proceedings of the Academy of Natural Sciences of Philadelphia 64*(2), 380–444 [434].

Type data: holotype NMNH 214382.

Type locality: North Pacific Ocean, Hawaiian Islands, off Oahu Island.

#### COMMON NAME

Hawaiian Arrow Squid

#### DISTINGUISHING CHARACTERS

Sometimes confused with *Nototodarus gouldi* (McCoy, 1888), this species is most easily differentiated by the rough texture of the skin.

#### DIAGNOSTIC FEATURES

Mantle cylindrical, moderately muscular, tapering gradually to fins and then abruptly to a narrow, conical posterior tip. Texture of mantle rough or 'chamois-like' due to regularly spaced skin thickenings. Fins short, broad, rhombic; fin length 35–43% ML; single fin angle 50–57°. Funnel groove with foveola, without side pockets. Head large, slightly wider than mantle. Mantle element of T-shaped locking apparatus with straight ridge, no muscular fusion to funnel element; funnel groove with foveola but no side pockets. Arms subequal and large, longest slightly < half ML; swimming keels well developed on arms 3. Arms with 18–28 transverse rows of suckers. Largest arm suckers with 9–21 conical (distally) to rounded (proximally) teeth interspersed with low, flat, truncated teeth, distal tooth significantly larger. Small denticles between some distal teeth. A single large distal tooth plus 16 or 17 smaller teeth, most obvious in males and in largest suckers close to body.

Both arms 4 hectocotylised in males, right arm with greater modification, longer and larger than left and has double row of slender, conical papillae and expanded

Fresh-caught specimen, dorsal view (K. Graham).

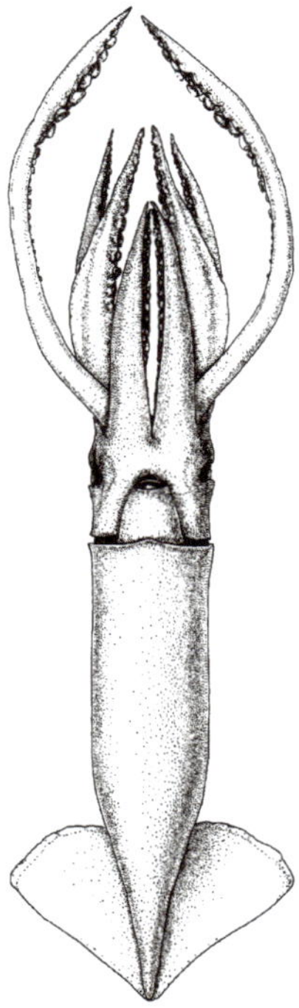

Ventral view (FAO, E. D'Antoni).

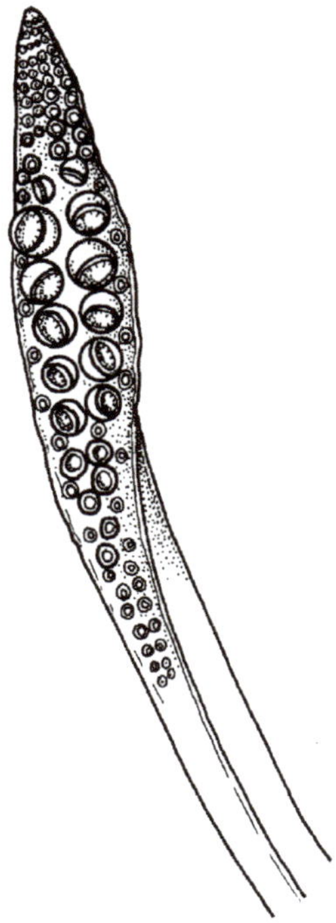

Tentacular club (FAO, E. D'Antoni).

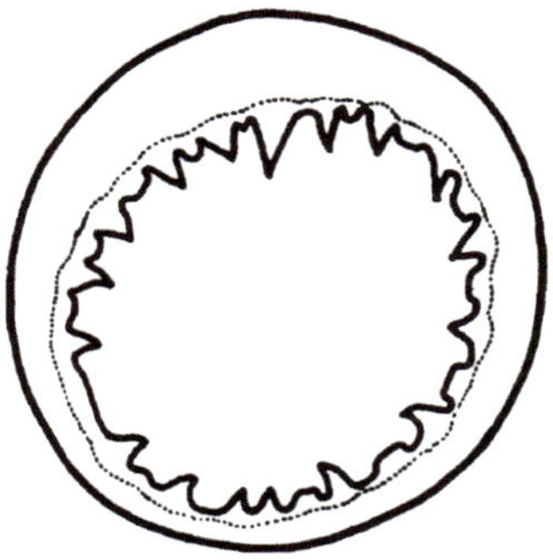

Enlargement of arm 2 sucker ring (FAO, K. Hollis).

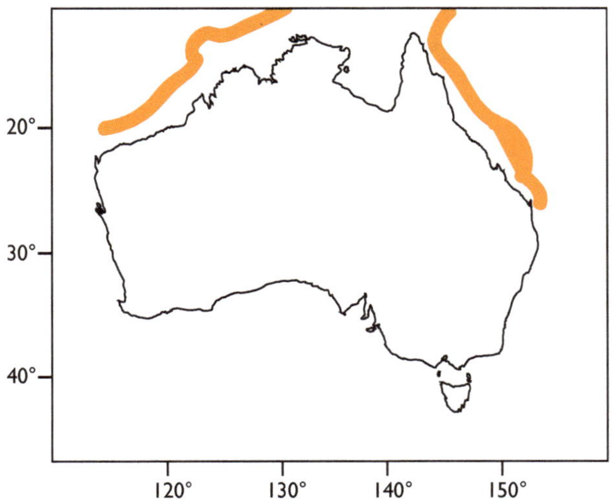


*Nototodarus hawaiiensis* distribution.

protective membrane distally; proximal one-quarter of each arm 4 has four or five enlarged cushion-like trabeculae and wide protective membranes (saw-like or comb-like appearance) that have attenuated supports that extend to arm tip; dorsal and ventral distal right arm trabeculae, sucker stalks and suckers modified by enlargements and reductions.

Tentacular manus with 9–11 tetraserial rows of suckers; manus suckers 3 or 4 × larger than lateral suckers; largest medial manus sucker ring with 14–18 moderately large, conical teeth interspersed with low, horny plates; one distal conical tooth significantly larger than rest. Fixing apparatus absent. Four series of suckers on dactylus; carpal area occupies most of tentacle length along stalk (~70%). Photophores absent.

*Colour.* Mantle and arms chestnut brown with distinctly darker brown dorsal longitudinal stripe along mantle midline, broadening over head and extending anteriorly as thin stripes along arms; dark stripe of thickly set chromatophores along dorsal midline of fins.

## HABITAT AND BIOLOGY

*Nototodarus hawaiiensis* is a pan-tropical nerito-oceanic species associated with continental shelf and slope waters, oceanic seamounts, ridges and banks. Distributed down to depths of 650 m, it also occurs in shallower waters where it is an important food item in the diet of seabirds and fishes. During the day, *N. hawaiiensis* 'sits' on the bottom on its arm tips and posterior mantle tip, a common behaviour in ommastrephids, and then it disperses into midwaters at night to forage. Paralarvae, juveniles and subadults inhabit midwaters in the epi- to upper mesopelagic zones over the tops of seamounts, slopes and oceanic depths, and the adults spawn near or on the bottom. Adults are not found in the open ocean, but they always remain near oceanic seamounts, ridges and banks. Its biology is poorly known. A survey of egg size and fecundity in selected species in the Todarodinae revealed that *N. hawaiiensis* is characterised in the group that has ripe egg sizes in the order of 0.7–2.4 mm diameter and a low to medium fecundity of 20 000–2.5 million oocytes. The spermatophoric organ in mature males of ~150 mm ML contains ~2700 developing spermatophores each with an average length of 18 mm. About 1000 spermatophores are fertile. Size at reproductive maturity varies seasonally and with locality. In northern Australian populations, males generally mature from 130 mm and females from 150 mm ML; occurrence of mature females and larvae suggests that spawning occurs throughout the year. Off western Australia males of 98 mm ML and females of 100 mm are mature.

*Interest to fisheries.* *Nototodarus hawaiiensis* is the predominant cephalopod species caught as by-catch in the shrimp and prawn demersal trawl fishery on the north-west slope off Western Australia, principally in the Ramkin Bank area. Its abundance and flesh quality indicate a high potential as a fishery resource, even though its occurrence in relatively deep waters may limit access by common trawling techniques.

### SIZE

Up to 250 mm ML, commonly to 150 mm ML. Maximum ML varies according to locality: off southern Queensland, Australia, males are up to 215 mm ML, and females up to 248 mm ML.

### REMARKS

This species occurs predominantly in continental slope waters from depths of more than 200 m to at least 500 m, although small juveniles and larvae have been caught in adjacent shelf waters. It is only occasionally taken on jigs, suggesting it is predominantly demersal rather than pelagic.

*Interest to fisheries.* Caught incidentally in demersal trawling in deep water and occasionally abundant in catches, but no targeted fisheries exist. This species is commonly caught by North West Shelf demersal trawlers.

### DISTRIBUTION

*Nototodarus hawaiiensis* has a very broad, disjunct distribution in tropical to subtropical waters of the Pacific and Indian Oceans: Hawaiian Islands to Midway Island, South China Sea, Japanese waters (south of 35°N), Hong Kong, Philippine and Vietnamese waters, north-eastern Australia, North West Shelf of Western Australia, Andaman Sea, Bay of Bengal, Hawaiian Ridge, Sala y Gomez Ridge to Nazka Ridge, Saya de Malha Bank, off Somalia, Madagascar, Seychelle Islands, Mozambique, KwaZulu-Natal, Southern African waters. It also has been reliably reported from a seamount in the far south-eastern South Pacific Ocean off Chile, 35°39′S, 85°31′W.

Within Australia it is found in northern slope waters off the west coast, along the North West Shelf and the north-eastern and eastern slope to southern Queensland.

### REFERENCE

Jereb and Roper (2010).

## *Todarodes* Steenstrup, 1880

### DIAGNOSTIC FEATURES

Funnel groove with foveola, side pockets absent. Tentacular club with two pairs of carpal suckers, tentacles without fixing apparatus; suckers on dactylus of tentacular club tetraserial; medial manus sucker rings with 20 or fewer long, pointed teeth with a single larger tooth distally in some species. Only right arm 4 hectocotylised in mature males. Ocular, visceral and mantle photophores absent.

*Colour.* Typically dark over entire dorsum, with narrow mid-dorsal stripe on mantle.

*Photophores.* Absent.

### REMARKS

The genus *Todarodes*, represented by five species, is widely represented in the world's oceans. Some species of *Todarodes*, particularly *T. pacificus* (Steenstrup, 1880) from the north-west and northern Pacific Ocean and, less so, *T. sagittatus* (Lamarck, 1798) from the eastern Atlantic, support among the largest fisheries in the world.

Wakabayashi *et al.*'s (2012) study based on 16S rRNA and COI partial sequences, found *Todarodes* to be polyphyletic. Further studies and observations are necessary to help clarify the status and the distribution of *Todarodes* species in the southern hemisphere.

### Key to Todarodes species

1a. Manus of tentacular club with six to eight transverse rows of tetraserial suckers .......................... *T. pusillus*
1b. Manus of tentacular club with 12 or 13 transverse rows of tetraserial suckers ....................... *T. filippovae*

## *Todarodes filippovae* Adam, 1975

*Todarodes filippovae*, Adam, W. (1975). Notes sur sur les Cephalopodes XXVI. – Une nouvelle espece de *Todarodes* (*Todarodes filippovae* sp. nov.) de l'Ocean Indian. *Bulletin de l'Institut royal des Sciences.naturelles de Belgique 50*(9), 1–10. [3, pls 1, 2].

Type data: holotype IRSNB.

Type locality: southern Indian Ocean, 35–38°S, 66–77°E.

#### COMMON NAME(S)

Antarctic Flying Squid; Southern Ocean Arrow Squid

#### DISTINGUISHING CHARACTERS

Clubs very expanded occupying nearly entire length of tentacles; only two pairs of carpal suckers at base of club.

#### DIAGNOSTIC FEATURES

Mantle muscular, long, narrow (width 16–24% ML); tapers to a pointed 'tail'. Fins saggitate; length and width ~50% ML; single fin angle ~30–35° (60–70° both fins); attenuated posteriorly. Right arm 4 hectocotylised along the distal 21–36% (up to distal two-thirds, Dunning 1993) of arm, with suckers transformed to papillae and tubercles, with ventral protective membrane and trabeculae very expansively developed. Tentacles very large and robust. Clubs very expanded occupying nearly entire length of tentacles; two pairs of carpal suckers at base of club; medial manus sucker rings with 7–13 long, pointed teeth, evenly spaced, alternating with low plates; largest sucker diameter 2.7–4.5% ML; manus with 12–14 tetraserial sucker rows. Arms relatively short; sucker rings with 10 sharp teeth.

*Colour.* Live and fresh-caught animals deep red or carmine; no distinct medial stripe on dorsal mantle.

#### HABITAT AND BIOLOGY

*Todarodes filippovae* is an oceanic species found from the surface to ~1200 m depth. It inhabits open waters or slope locations, very rarely, if ever, extending over the continental shelf, and is a non-ammoniacal species. As in other parts of the world *T. filippovae* is associated with a high velocity current. Links between the life cycle and high velocity systems are likely, both in the transport of planktonic eggs and paralarval stages, as well as in adult feeding ecology. No clear relationship is apparent between the species' distribution and water temperature: *T. filippovae* is found associated with cold Antarctic Intermediate Water (3–3.8°C), as well as with warm superficial streams (up to 24°C). These observations suggest the presence of either a cryptic complex of species with different temperature requirements or a single, eurythermal species. While the species is widely distributed in the southern hemisphere, most information on its biology comes from the Tasman Sea (southern Australia) and South Africa. In Australian waters during summer *T. filippovae* occurs from central New South Wales to south-western Western Australia and it is the most abundant ommastrephid squid in the Subtropical Front zone of the Tasman Sea and in the slope waters along the south-eastern Australian coast. Juveniles as well as large adult individuals are most abundant in the upper, surface layer of the water column, where they are caught by jigs and driftnets in surface water temperatures between 11.5°C and 24.7°C and at 250 m between 9.7°C and 11.9°C. No mass migration in a north–south direction nor across the Tasman Sea has been observed. However, mature specimens of both sexes move to the continental slope waters during summer months, before spawning. The spawning season occurs between December and August. Mature males represent a higher proportion of the population off southern New South Wales (38°S to 32°S), suggesting that mating and spawning probably occur towards the northern boundary of the species' distribution in this area. Spermatophores are evident in males at 260 mm ML, and at 320 mm ML all males have fully formed spermatophores. The smallest mature female observed with eggs in the oviduct was

Fresh-caught specimen, dorsal view (K. Graham).

Fresh-caught specimen, ventral view, anterior end (K. Graham).

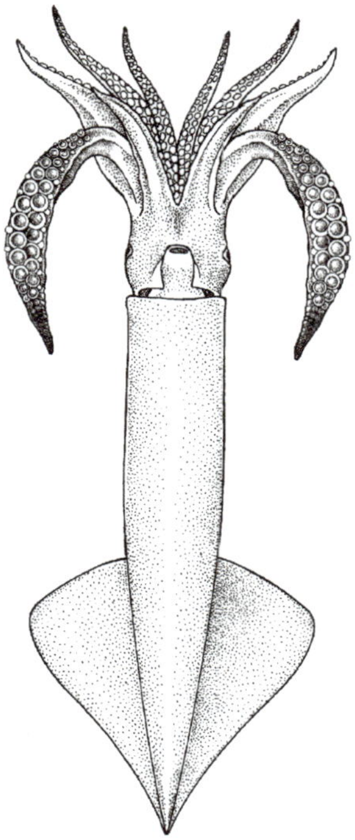

Ventral view (FAO, E. D'Antoni).

Hectocotylus (FAO, E. D'Antoni).

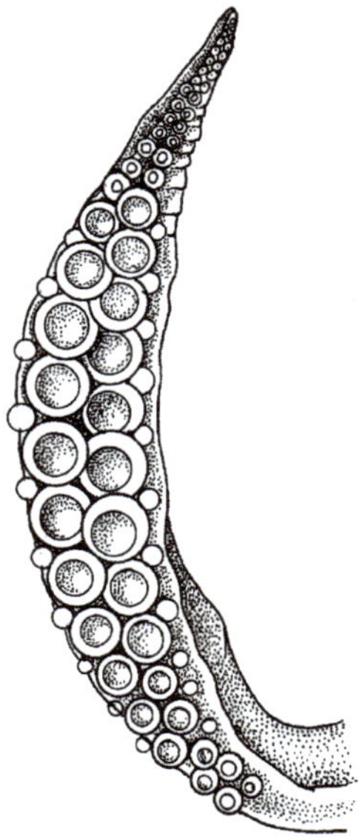

Tentacular club (FAO, E. D'Antoni).

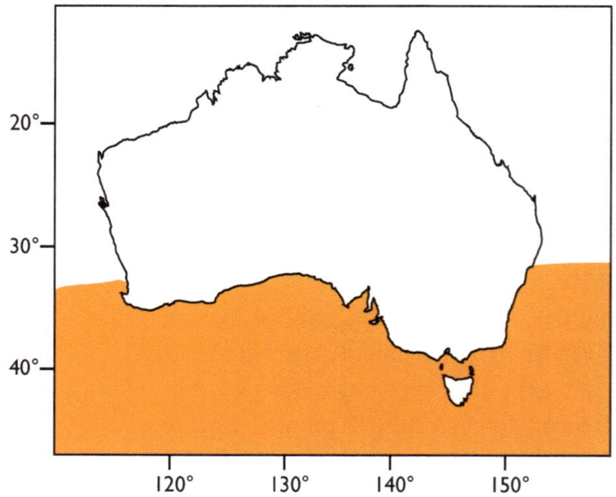


*Todarodes filippovae* distribution.

380 mm ML. A few juveniles between 40 mm and 80 mm ML have been caught off eastern Tasmania during midsummer and juveniles of unreported length have been collected in the southern Pacific in autumn. Growth is rapid, as in most ommastrephids, with females generally growing faster than males and reaching larger sizes, while males mature earlier. Based on estimated growth rates of the summer Tasman Sea population, as well as on assumed daily periodicity in statolith increments, Jackson *et al.* (2005) determined the life cycle is about one year or slightly more. In contrast with other ommastrephid species, *T. filippovae* appears to grow faster in cooler oceanographic conditions, with growth rates of autumn- and winter-hatched squids significantly faster than summer-hatched squids. Unlike most other Southern Ocean species (except *Martialia hyadesi* Rochebrune & Mabille, 1889) this species forms schools. Adults of this species may be infested

with large numbers of encysted nematodes and cestodes in their mantle tissue.

*Interest to fisheries.* This species is taken as by-catch to the jig fishery for *Nototodarus sloanii* (Gray, 1849) off New Zealand and southern Australia. It is caught as by-catch by trawlers around most of its distributional range. *Todarodes filippovae* is a likely candidate for future exploitation, so it is important that the taxonomy of this species is resolved (i.e. whether it represents one or more species).

### SIZE

A large-sized squid; maximum reported size 540 mm ML, 4.7 kg for a female, ~400 mm ML for males; commonly between 200 mm and 400 mm ML.

### REMARKS

This species is the most abundant ommastrephid in the Subtropical Front zone and in slope waters off south-eastern Australia. It is a large and muscular squid and is likely to be an important part of the ecosystem. It displays rapid growth, a short lifespan and early maturity, together with intra- and interannual plasticity in size and growth parameters (Jackson *et al.* 2007). This interannual plasticity in growth dynamics, driven by environmental variability, appears to be a common phenomenon in squids. Growth is more rapid in cooler oceanographic conditions (in contrast to the other common ommastrephid in southern Australian waters, *Nototodarus gouldi* (McCoy, 1888)).

Nesis (1979) synonymised *T. filippovae* with *T. angolensis*, and most Russian scientists then referred all *Todarodes* (adults and paralarvae) caught in Southern hemisphere waters to *T. angolensis*. Therefore, many records of *Todarodes* squid off New Zealand, the Auckland and Campbell Islands and in the area west and south of Tasmania were considered to belong to *T. angolensis*. This synonymy is not widely accepted by the cephalopod community and is not followed here.

This species may prove to be important in the circumpolar Antarctic ecosystem, both because of its abundance and its trophic links with other organisms. There is a possibility that *T. filippovae* represents a species-complex (Rodhouse 1998). Detailed genetic work throughout its range would help resolve the genetic status of *T. filippovae* across its distribution and is much needed for fisheries management.

Wakabayashi *et al.* (2012) found in a phylogenetic analysis of the ommastrephids that *Todarodes* is polyphyletic. *Todarodes filippovae* was consistently the sister taxon to *M. hyadesi* and distant from *T. pacificus* (Steenstrup, 1880), with *T. pacificus* consistently placed in a clade among *Nototodarus* species.

### DISTRIBUTION

Circumpolar in the Southern Ocean, particularly in association with high velocity current systems; south of ~35°S; common in the Antarctic Polar Front zone. Inhabits deep offshore continental slope waters and seamounts.

### REFERENCE

Jereb and Roper (2010).

## *Todarodes pusillus* Dunning, 1988

*Todarodes pacificus pusillus* Dunning, M. (1988). *Todarodes pacificus pusillus*, new subspecies (Cephalopoda: Ommastrephidae) from northern Australia. *Memoirs of the Museum of Victoria 49*(1), 149–157 [149].

Type data: holotype MV F.31571.

Type locality: north-east Indian Ocean 13°48′S, 124°19′E.

### COMMON NAME

Little Flying Squid

### DISTINGUISHING CHARACTERS

Differs from *Todarodes pacificus* (Steenstrup, 1880) from Japan in having a greater proportion of the hectocotylised

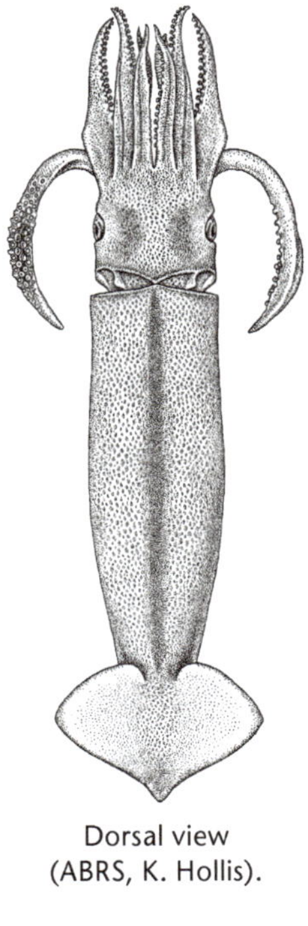

Dorsal view (ABRS, K. Hollis).

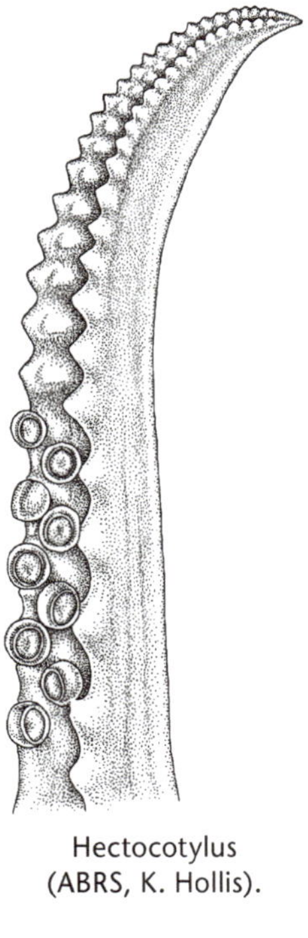

Hectocotylus (ABRS, K. Hollis).

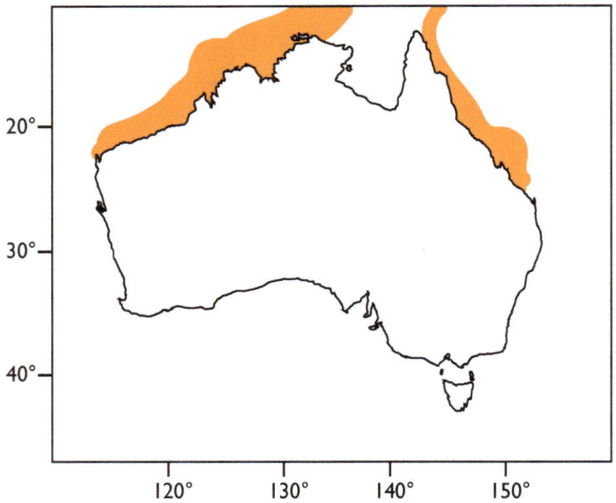


*Todarodes pusillus* distribution.

arm modified (45–57% in *T. pusillus* compared with 30% of total arm length in *T. pacificus*) and in having fewer proximal suckers on the hectocotylus (11–13 in *T. pusillus v.* 44–48 in *T. pacificus*).

#### DIAGNOSTIC FEATURES

Mantle slender, moderately muscular, tapering abruptly anterior to fin junction. Fins short, small, rhomboidal; fin length 25–31% ML. Mantle element of T-shaped locking apparatus with straight ridge, no muscular fusion to funnel element; funnel groove with foveolla, but no side pockets. Arms subequal, large, longest slightly < 50% ML with well-developed swimming keels. Arms 1 with 12–17 suckers. Largest arm sucker rings with 9–11 subequal, slightly flattened teeth on distal half, smooth proximally. Right ventral arm 4 in males hectocotylised, slightly shorter and thicker than left ventral arm 4; 11–13 normal arm suckers proximally; ~20 pairs of trabeculae on modified distal section (in dorsal column, sucker stalks and bases disappear immediately but ventrally they are still evident for at least 10 rows before fusing with trabeculae; hectocotylised portion of arm occupies ~45–57% of arm length. Largest medial manus sucker rings with 16–18 long, pointed teeth interspersed with low chitinous plates; manus with six to eight transverse rows of tetraserial suckers; largest median manus sucker rings with 16–18 moderately large subequal conical teeth interspersed with horny plates; by fourth manus row, they possess complete ring of 18–20 sharp teeth; carpal fixing apparatus absent.

*Photophores.* Absent.

*Colour.* Dorsal mantle dark brown with distinct blue-black stripe along dorsal mantle midline that broadens on head and then extends anteriorly along dorsal surfaces of arms 1 and 2.

#### HABITAT AND BIOLOGY

This species reaches maturity at < 80 mm ML. All females larger than 60 mm ML are sexually mature with eggs in the oviducts; males larger than 50 mm ML are mature with fully developed spermatophores (although variation in size at maturity has been observed in northern hemisphere populations of the closely related *T. pacificus*). Many individuals may be captured in a trawl, so this species appears to be a schooling form.

*Interest to fisheries.* None at present.

#### DISTRIBUTION

Continental shelf and upper slope of northern Australia from the Timor Sea between 8°53′S and 19°31′S, on the North West Shelf, and off the east coast from Torres Straight to approximately off Brisbane. Also occurs in Indonesia. Known depth range 78–357 m.

#### REFERENCE

Roper *et al.* (2010b).

### *Todaropsis* Girard, 1889/1890

Oceanic species dark over entire dorsum. Typically with dark, narrow mid-dorsal stripe on mantle.

### *Todaropsis eblanae* (Ball, 1841)

*Loligo eblanae* Ball, R. (1841). On a species of *Loligo*, found on the shore of Dublin Bay. *Proceedings of the Royal Irish Academy* 1(19), 362–364 [364, figs 1–7].

Type data: holotype NMI.

Type locality: North Atlantic Ocean. Irish Sea, Ireland, Dublin Bay.

#### COMMON NAME(S)

Golden Arrow Squid, Lesser Flying Squid

#### DIAGNOSTIC FEATURES

Mantle cylindrical, moderately muscular, relatively short with abrupt caudal taper; fins < 50% ML but broad (90% ML), rhombic. Mantle robust, stout; head large, broad, four nuchal folds on posterior of head (neck). Mantle element of inverted T-shaped locking apparatus with straight ridge, no muscular fusion to funnel element; funnel groove without foveola or side pockets. Fins broad, anterior border more convex than posterior border; fin length < 50% of dorsal ML. Head large, slightly wider than mantle. Arms subequal and large, longest more than half the ML; swimming keels show greatest development on arms 3; largest arm suckers with ~10 conical teeth on distal margin; distal tooth only marginally larger. Protective membranes and their supports of uniform height, not higher than suckers. Both ventral arms (4) hectocotylised in mature males; suckers on proximal 20–30% of both arms lost and trabeculae modified hard, bracket-like lappets; distally to this modified area, on left arm suckers normal, in two rows, and on right arm, suckers absent, trabeculae modified into series of papillae; dorsal protective membrane missing, ventral protective membrane larger than its counterpart on left ventral arm; right ventral arm slightly shorter than left. Dactylus of tentacular club with tetraserial small suckers; manus of club with six transverse rows of tetraserial suckers, each with median pairs up to 4 × larger in diameter than lateral suckers. Sucker rings of largest median club suckers with ~30 short, pointed teeth, occasionally alternating with minute teeth.

*Colour.* In life, mantle and arms golden brown with darker brown dorsal stripe along mantle midline.

## HABITAT AND BIOLOGY

This demersal species is usually associated with sandy and muddy bottoms, in depths between ~20 m and 850 m. Females grow larger than males, but no other sexual dimorphism is evident. The Australian populations begin to mature at a slightly larger size than their counterparts in other parts if the world (males at 160 mm ML, and females at 120–150 mm ML). In eastern Australian waters, spawning occurs from late summer until early winter. Fecundity varies greatly, depending on size of females and specific populations; total fecundity ranges between a few thousand to > 250 000 ova. The number and the size of spermatophores increase with the size of the males; spermatophore length ranges from 11–30 mm, and their numbers range between 10 to in excess of 250. The final maturation phase is rapid, and spawning constitutes the terminal phase of the life cycle. Intermittent terminal spawning is possible, but the degeneration process after the last spawning event must be rapid, since spent females are seldom found. Immature oocytes are pale yellow in colour, while mature oocytes are bright orange and measure from 0.8 to 2.5 mm in diameter. Eggs are laid in a large capsule secreted by the greatly enlarged nidamental glands. No *in situ* observations on egg deposition are available, although they are probably laid in floating masses, as observed for many other ommastrephids in the natural environment and in captivity. No information exists on embryonic development or on the newly hatched squids; however, as in all ommastrephids *Todaropsis eblanae* has a distinct paralarval form, the 'rhynchoteuthion', characterised by the fusion of the tentacles into a proboscis. The paralarvae have yet to be definitively identified and described. The lesser flying squid is a typical opportunistic predator that feeds on its principal available prey species depending on geographical area and other environmental factors. The absence of sympatric benthic organisms in stomach contents indicates that *T. eblanae* feeds mainly in the water column. Cannibalism also occurs, but it is mainly an occasional phenomenon associated with anomalous concentrations of conspecific squid and/or scarcity of commonly available prey. Recent observations indicate that *T. eblanae* is an important host for anisakid nematodes.

*Interest to fisheries.* *Todaropsis eblanae* is taken mainly as by-catch in otter trawl fisheries throughout its distributional range, but is not fished in Australian waters. It is not taken on jigs. It is less abundant than *Nototodarus gouldi* (McCoy, 1888).

## SIZE

Maximum ML to ~160 mm for males and 270 mm, females (data from North West Shelf trawl fishery).

## REMARKS

*Todaropsis eblanae* occurs predominantly in continental slope waters from depths of more than 200 m to at least 800 m (temperatures 9–17.5°C) (where its distribution overlaps with that of *N. gouldi*) although small juveniles have also been caught in adjacent deeper shelf waters. Adults are demersal rather than pelagic. Males generally mature from 90 mm ML and females from 150 mm ML in northern Australian waters; the spawning season is apparently protracted from midsummer to winter.

Guerra *et al.* (2013) present a review of the biology, ecology and fisheries of this ommastrephid squid. In the northern hemisphere, estimations from growth increments in statoliths show average growth rates of 0.30 mm ML per day and 0.67 mm ML per day in females

Fresh-caught specimen, dorsal view (K. Graham).

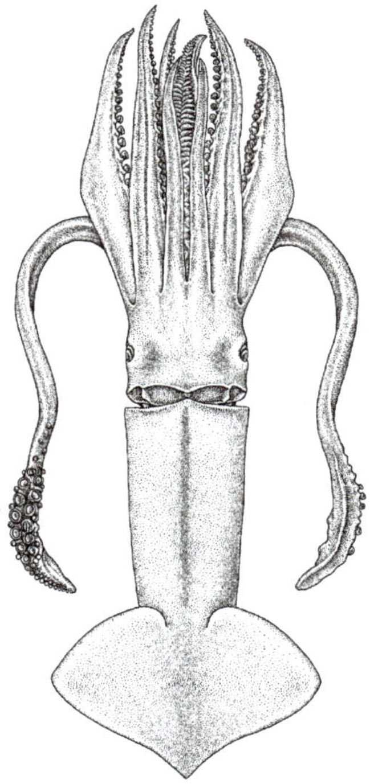

Dorsal view (ABRS, K. Hollis).

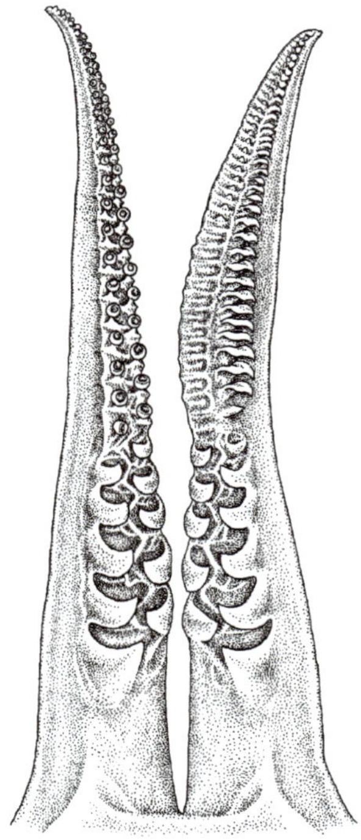

Hectocotylus (ABRS, K. Hollis).

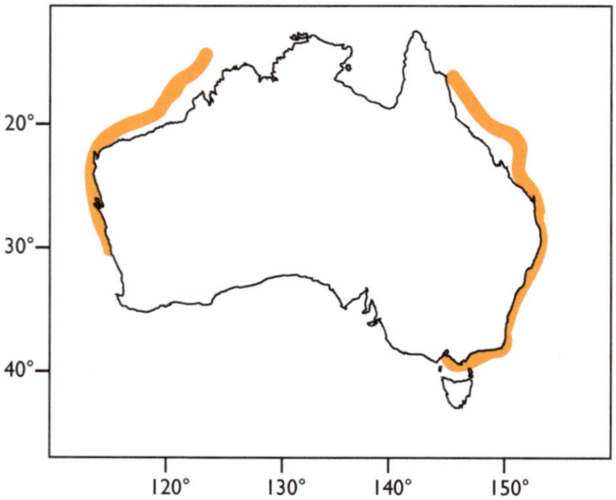


*Todaropsis eblanae* distribution.

and a lifespan of 1 year. The species is an intermittent terminal spawner. The hatching season extends year-round with peaks according to latitude.

Further detailed studies on the identity and systematic status of this species throughout its worldwide range are needed. Its distribution is quite disjunct and it may be that it is actually a species-complex. The existence of at least three genetically isolated populations in the eastern Atlantic has been shown: South African and Mauritanian populations are genetically distinct both from one another and from European populations.

#### DISTRIBUTION

This species is found in the eastern Atlantic and Mediterranean, the western Indian Ocean and Timor Sea, South China Sea and Australia waters. It occurs off the northern Australian coast, along the western and eastern Australian coasts, and to Tasmania on the eastern side. Known depth range 30–850 m (over entire range). (Map shows museum records only.)

#### REFERENCES

Jereb and Roper (2010), Guerra *et al.* (2013).

## Onychoteuthidae Gray, 1849

#### COMMON NAME

Hooked Squids, Clubhook Squids

#### DISTINGUISHING CHARACTERS

Onychoteuthids differ from ommastrephids in having a simple, straight funnel-locking apparatus rather than a ⊥-shaped mantle-locking apparatus. The sucker rings in onychoteuthids are smooth rather than toothed, and the buccal connectives are attached to the ventral border, rather than the dorsal border of the ventral arms. The lower beak in onychoteuthids has a 'step' or ridge at the base of the jaw angle. This is otherwise only seen in *Pholidoteuthis massyae* (Pfeffer, 1912) and some *Enoploteuthis*.

### DIAGNOSTIC FEATURES

Head with three primary ventrolateral nuchal folds in all genera and three to eight secondary nuchal folds in *Onychoteuthis*, *Ancistroteuthis* and *Notonykia*. Funnel-locking apparatus, simple, straight groove. Arms suckers biserial with smooth rings. Hectocotylisation absent. Tentacular clubs with biserial hooks (marginal suckers may be present in juveniles but usually lost by subadult stage); largest hooks in ventromedial series; suckers usually limited to carpal-locking apparatus and terminal pad in subadults; dactylus present in *Notonykia* and possibly *Kondakovia*; carpal-locking apparatus circular to oval in shape and usually defined by ridge or elevation. Carpal-locking apparatus consisting of smooth-ringed suckers and corresponding knobs at base of the tentacular clubs. Buccal connectives attach to ventral borders of arms 4. Lower beak with 'step' at base of jaw angle (ridge along interior surface of lateral wall).

*Photophores.* Intestinal and eye photophores only in *Onychoteuthis*. *Ancistroteuthis* may have photophores on ventral surface of eye (nature of tissue not determined).

*Colour.* Maroon to brick red, darker dorsally. *Onykia* dark blue to deep purple, especially on dorsal surface, with reflective sheen. Skin smooth to rugose. Some immature species, like other epipelagic cephalopods, deep grey colour during the day.

### HABITAT AND BIOLOGY

Members of this family are found throughout the world's oceans, with the exception of Arctic waters. Some species are common in the open ocean, and others live in the bathyal region near the ocean floor along continental or island slopes where they are caught in demersal trawls. They can be very abundant and form schools. Juveniles are able to leap above the surface to escape predators as do ommastrephids. Some species undergo seasonal migrations for feeding and then spawning.

*Interest to fisheries.* Onychoteuthids are not fished commercially in Australian waters, although they are sometimes caught incidentally in demersal trawls in continental slope waters (300–600 m) of northern Australia. Several species have significant fisheries potential. They are staples in the diets of many teuthophagous marine predators, including seals, sharks and whales.

### SIZE

Medium-sized to very large, size varies among species ranging up to 230 cm ML.

### REMARKS

This family was revised by Bolstad (2010), who recognised seven genera: *Ancistroteuthis*, *Callimachus*, *Filippovia*, *Kondakovia*, *Notonykia*, *Onychoteuthis* and *Onykia* with ~25 species. The status of the genus *Walvisteuthis*, while retained here, is formally unresolved. Bolstad (2010) gives a summary of the status of this taxon. It has been suggested by some researchers (Young *et al.* 2003) that *W. virilis* may be a junior synonym of *C. rancureli* (Okutani, 1981). Nesis and Nikitina (1986) report that it should be placed in its own family, Walvisteuthidae (the option followed in Bolstad 2010). According to Roper *et al.* (2010a), the status of the family has also been reviewed by Tsuchiya (unpublished data), who has also concluded, as did earlier researchers, that *W. virilis* is a synonym of *C. rancureli*. Until these nomenclatural tangles are formally unravelled, the name *Walvisteuthis virilis* is retained here.

Lindgren (2010b) investigated relationships within this family using five genes (18S rRNA, 28S rRNA, histone H3, 16 rRNA and COI). Based on this study, *Onykia* was found to be polyphyletic.

The Australian representatives of this family need to be examined in the light of recent reviews (Bolstad 2007, 2008, 2010). Several specimens in museum collections have been identified as *O. banksii*, a species that occurs only in the central North Atlantic and the Gulf of Mexico. These specimens need to be re-examined to determine their identity and the distribution of this/these taxa.

### REFERENCES

Bolstad (2007), Bolstad *et al.* (2010a), Roper and Jereb (2010r).

## Key to genera of the Onychoteuthidae

1a. Warts or longitudinal ridges on the skin ................. 2
1b. Skin smooth, no warts or longitudinal ridges ........ 3
2a. Marginal suckers on manus of tentacular club absent in mature animals; carpus with 9–18 suckers. Radula rhachidian tooth tricuspid ................ *Onykia*
2b. Full series of marginal suckers on manus of tentacular club present in mature animals; carpus with 30–36 suckers. Radula rhachidian tooth unicuspid to weakly tricuspid......... *Kondakovia* (*K. longimana*)
3a. Two intestinal photophores ................ *Onychoteuthis*
3b. No intestinal photophores ........................................ 4
4a. Three secondary nuchal folds present...... *Notonykia*
4b. Secondary nuchal folds absent ................................. 5
5a. Mantle robust, muscular. Fins rhombic..................... ........................................... *Filippovia* (*F. knipovitchi*)

5b. Mantle muscle soft, fleshy. Fins ovate or paddle-shaped in adults .................. *Walvisteuthis* (*W. virilis*)

## *Filippovia* Bolstad, 2010

### DIAGNOSTIC FEATURES

Skin of mantle, head and arms smooth. Head much narrower than mantle opening in subadults to adults. Secondary nuchal folds absent. Fins rhombic, length ~two-thirds ML, width ~74% ML. Funnel groove U-shaped, without Y-shaped ridge; arms subequal, ~50% ML. Bases of large ventral hooks of tentacular club asymmetrical. Radula rhachidian tooth unicuspid. Photophores absent.

### *Filippovia knipovitchi* (Filippova, 1972)

*Moroteuthis knipovitchi* Filippova, Y. A. (1972). New data on the squids (Cephalopoda: Oegopsida) from the Scotia Sea. *Malacologia 11*(2), 391–406 [392, figs 2, 3].

Type data: holotype ZMMGU.

Type locality South Atlantic Ocean, sub-Antarctic waters near South Georgia Island.

### DIAGNOSTIC FEATURES

Mantle moderately broad, stout, not drawn out into prominent 'tail'; ammoniacal; thin, smooth-textured skin surface. No secondary nuchal folds. Funnel groove appears to lack inverted Y-shaped ridge. Fins large, broad, rhomboidal, not attenuate; fin angles 45–50° each; fin length 50–60% ML, width ~70% ML. Funnel groove U-shaped (no Y-shaped ridge). Arms subequal in length, or arm pair 1 slightly shorter than rest. Tentacular club 35% ML; manus long, slender, with 20–30 long, narrow hooks in two medial series, those of ventral series larger than hooks on dorsal series; hooks in ventral series have asymetrical bases; second to fourth hooks on dorsal series and sixth to eighth hooks on ventral series are largest in respective series; 9–13 small suckers plus their accompanying knobs on carpus; 14–16 minute suckers on dactylus. Radula homodont with unicuspid rhachidian and first lateral teeth.

*Photophores.* Absent.

*Colour.* Light violet.

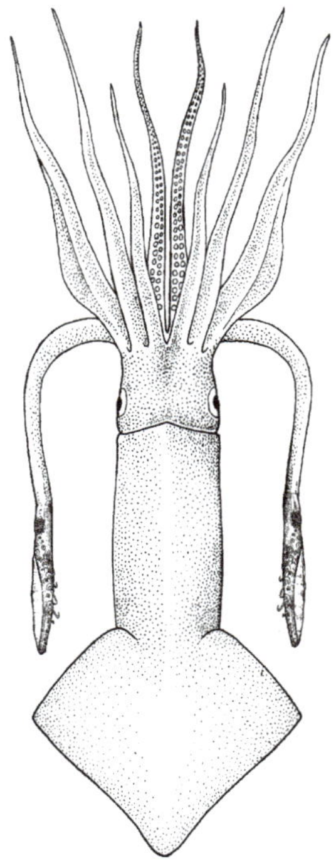

Dorsal view (FAO, E. D'Antoni).

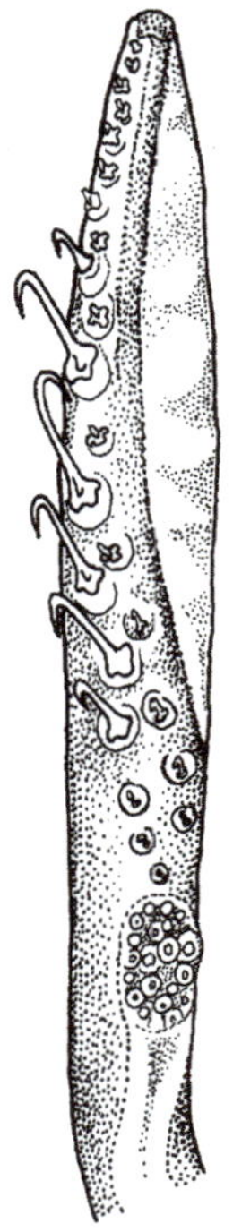

Tentacular club (FAO, E. D'Antoni).

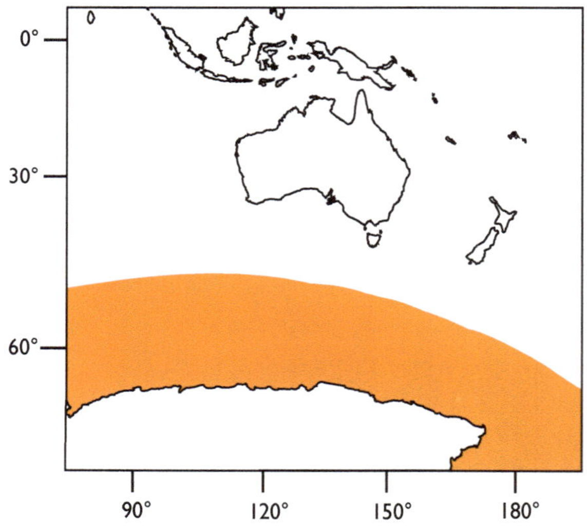


*Filippovia knipovitchi* distribution.

### HABITAT AND BIOLOGY

*Filippovia knipovitchi* is an oceanic, mesopelagic, bathyal, ammoniacal species. It feeds on mesopelagic fishes and Antarctic krill. It appears to be a deep living, midwater feeder that approaches the surface at night. The potential fecundity of females is 100 000 oocytes.

*Interest to fisheries.* Its abundance, size and consistency before spawning make this species a potential target for a significant fishery. However, as it is an ammoniacal species, *F. knipovitchi* is unsuitable for human consumption until a neutralising process becomes available.

#### SIZE

Large-sized species: maximum ML 450 mm.

#### REMARKS

This species was first placed in the genus *Moroteuthis*. Wakabayashi *et al.* (2007) showed that *Onykia* was synonymous with *Moroteuthis*, but suggested that *M. knipovitchi* was genetically distant and morphologically distinct. Bolstad (2010) later erected *Filipovia* to accommodate this species.

#### DISTRIBUTION

This species occurs circumglobally in the Southern Ocean. It occurs in the Scotia Sea, Argentine Basin, Drake Passage and around South Georgia, Kerguelen, Crozet and Prince Edward Islands. It is an Antarctic endemic, extending north to the Antarctic Polar Front.

#### REFERENCES

Bolstad *et al.* (2010c), Roper and Jereb (2010r), Rodhouse (2013).

### *Kondakovia* Filippova, 1972

#### DIAGNOSTIC FEATURES

Skin with reticulate or longitudinal folds. Dorsolateral nuchal folds absent. Two complete rows of marginal suckers present on adult tentacular club (at least in *Kondakovia longimana* Filippova, 1972). Club with 23–40 hooks. Photophores absent.

#### REMARKS

Only two species are presently known to occur in this genus, *K. longimana* and *K. nigmatullini* Laptikhovsky *et al.* (2009) from the Falklands. Laptikhovsky *et al.* (2009) provide a revision of the genus.

### *Kondakovia longimana* Filippova, 1972

*Kondakovia longimana* Filippova, Y. A. (1972). New data on the squids (Cephalopoda: Oegopsida) from the Scotia Sea. *Malacologia 11*(2), 391–406 [395, figs 4, 5].

Type data: holotype ZMMGU.

Type locality: Antarctic Ocean, South Atlantic sector, north of the South Orkney Islands.

#### COMMON NAME(S)

Antarctic Clubhook Squid, Giant Warty Squid

#### DIAGNOSTIC FEATURES

Mantle soft, fleshy, ammoniacal, broadly cylindrical, slightly tapered posteriorly; numerous soft longitudinal ridges in outer integument external to thick, soft muscle layer, especially in young stages. Head and arms more massive than in *Onykia* (formerly '*Moroteuthis*'), and together they are longer than mantle; thick, rugose, tuberculate in young. Three nuchal folds; secondary folds absent. Funnel groove rounded anteriorly; mantle component of locking apparatus longer than funnel component. Fins rhomboidal, but weak, short, broadly heart-shaped, length ~42% ML, width ~60% ML. Arms become very attenuate at tips. Arms long, ~110–120% ML, arm pair 1 slightly shorter than rest; protective membrane undulating with muscular supports; arm 4 suckers smaller than rest. Tentacular club elongate, with two medial series of a total of 27–38 hooks, and always two marginal series of small suckers on manus throughout development to adulthood (~33 small marginal suckers); carpus well developed with 9–13 small suckers plus their associated knobs; dactylus with 17–40 minute, closely set suckers.

*Gladius.* Not visible beneath skin in dorsal midline.

*Photophores.* Absent.

*Colour.* Squids in type series purple-brown.

*Sculpture.* Longitudinal dermal ridges on surface of mantle in large specimens.

#### HABITAT AND BIOLOGY

*Kondakovia longimana* is a large, very broadly distributed, abundant epipelagic to mesopelagic and benthic-bathyal species. Because it is an ammoniacal species, it is considered to be neutrally buoyant and to exhibit a rather passive life mode, feeding on massive, almost stationary, shoals of krill as well as fishes, sponges and squids. It appears to mate and spawn in the austral winter around Crozet Island, and presumably around the other sub-Antarctic islands. Apparently this species is most abundant near the Antarctic Polar Front.

This species is an important element of the Antarctic ecosystem and is a common prey item for whales and seabirds.

*Interest to fisheries.* Because this large species is a soft-bodied, ammoniacal squid, its potential as a fishery target for human consumption currently seems questionable. However, its value as a major food source for a broad cross-section of predators is massive.

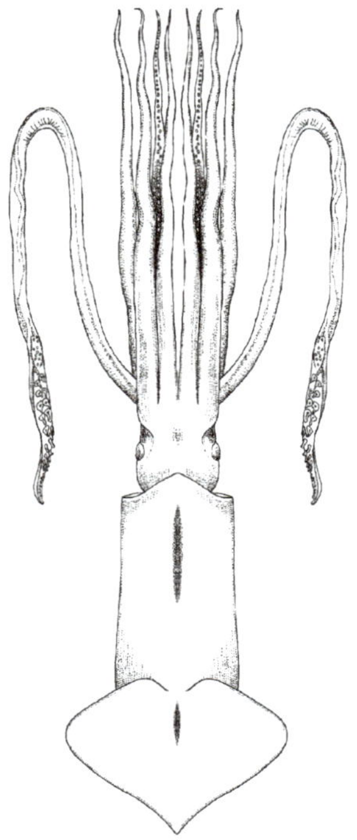

Dorsal view (FAO, E. D'Antoni).

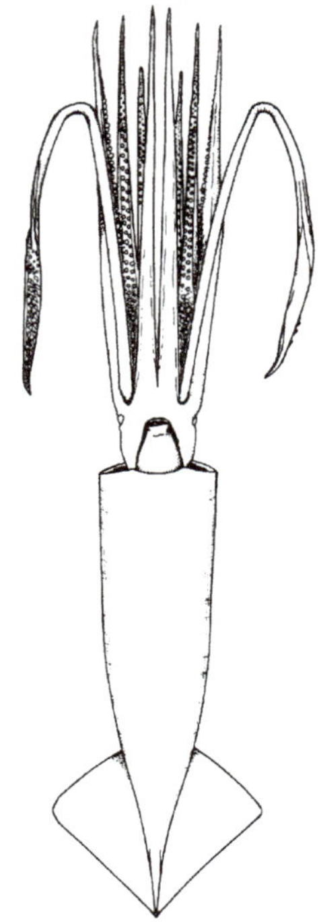

Ventral view (FAO, E. D'Antoni).

Tentacular club (FAO, E. D'Antoni).

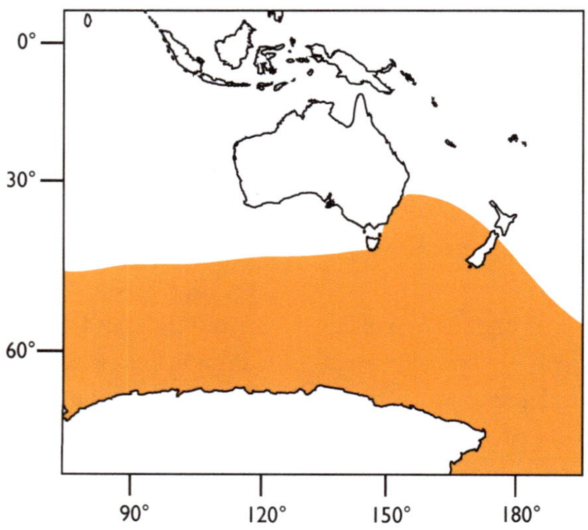


*Kondakovia longimana* distribution.

## SIZE

This is a large squid, perhaps second in size to the largest known onychoteuthid, *Onykia robusta* (Verrill, 1876) from the North Pacific. It has a maximum reported ML of 108 cm and a total length of 225 cm.

## REMARKS

Laptikhovsky *et al.* (2009) lists the following features that differ beween *K. longimana* and its congener *K. nigmatullini* Laptikhovsky *et al.* (2009) from the Falklands: (1) the fins are narrower in *K. longimana*, (2) the rostum of the gladius is shorter in *K. longimana* (5–10% ML) than in *K. nigmatullini* (~13%), (3) the mantle component of the funnel–mantle locking apparatus is much longer than the funnel component in *K. longimana* (funnel component 45–70% of mantle component *v.* funnel component ~90% of the mantle component in *K. nigmatullini*), and (4) *K. nigmatullini* has a tricuspid rachidian tooth, while the rachidian tooth is unicuspid in *K. longimana.*

Laptikhovsky *et al.* (2009) suggest that the development of the Southern Ocean Polar Front may have acted as an effective geographic boundary resulting in allopatric speciation in *Kondakovia*.

## DISTRIBUTION

This species is found circumglobally in the Antarctic Ocean. It is an Antarctic endemic extending north to the Sub-Antarctic Front. It occurs in the Tasman Sea, and around South Georgia, Falkland, South Orkney (Scotia Sea), Prince Edward, Marion, Gough, Auckland, Antipodes and Macquarie Islands, as well as off Chile. Known depth range to at least 1370 m.

## REFERENCES

Bolstad *et al.* (2010d), Roper and Jereb (2010r), Laptikhovsky *et al.* (2013).

## *Notonykia* Nesis, Roeleveld & Nikitina, 1998

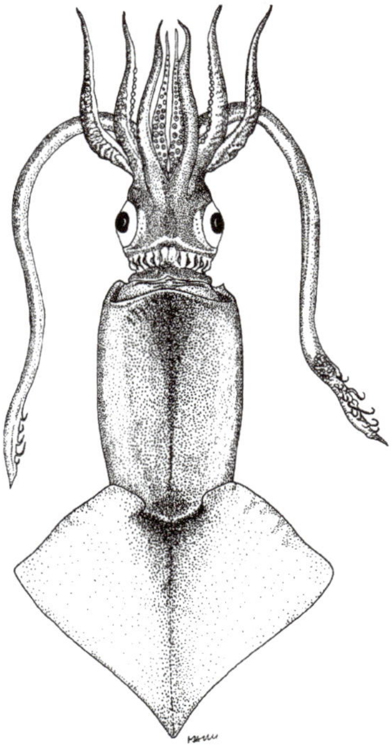

Dorsal view (FAO, E. D'Antoni).

### DIAGNOSTIC FEATURES

Mantle smooth without wrinkles or warts. Fins rhomboidal. Six or seven nuchal folds on each side of head. Marginal tentacular club suckers absent proximally; large number of suckers (30–40) on distal portion of club; suckers cover terminal pad and distal portion of club; manus with 16–18 hooks.

*Gladius.* Gladius visible along dorsal midline of mantle only between fins.

*Photophores.* Photophores absent.

### REFERENCES

Bolstad *et al.* (2007), Bolstad *et al.* (2010a).

### Key to *Notonykia* species

1a. Tentacular club carpus with 10–12 suckers. Largest ventral hooks robust, C-shaped............. *N. africanae*

1b. Tentacular club carpus with 7–9 suckers. Largest ventral hooks slender, J-shaped.................... *N. nesisi*

## *Notonykia africanae* Nesis, Roeleveld & Nikitina, 1998

*Notonykia africanae* Nesis, K. N., Roeleveld, M. A. C. & Nikitina, I. V. (1998). A new genus and species of onychoteuthid squid (Cephalopoda, Oegopsida) from the Southern Ocean. *Ruthenica*, *8*(2), 153–168 [154, figs 1–4].

Type data: holotype SAMC S3414.

Type locality: South Africa, off South-Western Cape, 33°42′S, 17°20′E.

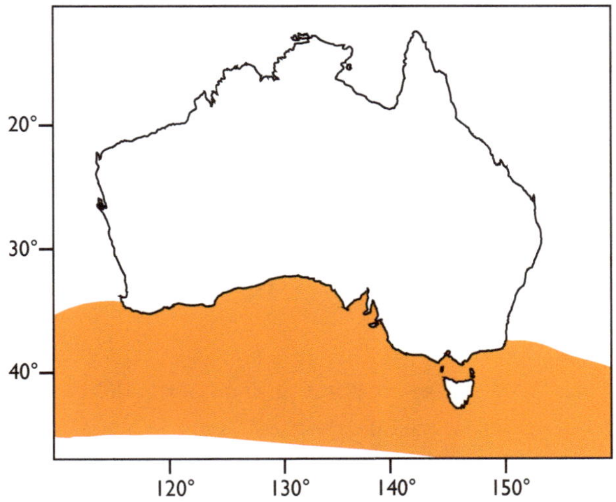


*Notonykia africanae* distribution.

### COMMON NAME

African Notalian Clubhook Squid

### DISTINGUISHING CHARACTERS

In specimens >18 mm ML the gladius is visible dorsally only between the fins. The gladius, in contrast to species of *Onychoteuthis*, is not needle-like and does not appear to pierce the skin. Positive identification of juveniles depends on the appearance of nuchal folds; only three folds are visible by 18 mm ML. The chromatophore pattern on the ventral mantle surface and lack of photophores can also be used to distinguish juveniles.

### DIAGNOSTIC FEATURES

Mantle densely muscular (non-ammoniacal), broad, cylindrical anteriorly, tapers to moderately pointed, conical 'tail'. Fins large, muscular, rhomboidal; length 58–66% ML (average 59%), width 66–84% ML (average 72%). Anterior margins of fins slightly convex, posterior margins nearly straight, very slightly drawn out into a short 'tail'. Nuchal folds present, four or five per side. Tentacles ~70% ML. Tentacular club narrow, with two medial series of 14–20, usually 17 or 18, hooks; fifth to sixth hooks in ventral row largest; largest ventral hooks robust C-shaped to nearly semicircular in lateral profile; hooks on dorsal row small. No marginal series of suckers. Dactylus with 20–38 minute suckers; carpus well defined, with 10–12 suckers plus their accompanying knobs. Buccal connectives attach dorsally to arms 1 and 2, ventrally to 3 and 4 (type DDVV). Hectocotylus absent.

*Photophores.* Photophores absent.

*Sculpture.* Skin smooth, not rugose or tubercular.

### HABITAT AND BIOLOGY

*Notonykia africanae* has a distributional range intermediate between a Southern Subtropical Convergence distributional pattern and the true notalian type of distribution. The species has a very broad vertical distributional range, also, as it occurs in epi-, meso- and bathypelagic layers, between 0 and 1200 m, mostly 30–200 m, and ascends to the surface at night. Subadults and the earliest stages of maturity are broadly eurybathic, from ~0 to 1000 m, but they are primarily mesopelagic with records from ~500 to 900 m. Maturation in males begins at 90–100 mm ML and in females at 100–130 mm ML. The size at full maturity for males seems to be >130 mm and for females >150 mm.

*Interest to fisheries.* No interest in a fishery exists at the present time. But, because it is a member of the muscular, non-ammoniacal group of onychoteuthids, it seems possible that a fishery could develop in the future if large concentrations of this small squid are discovered.

### SIZE

Small-sized species; maximum known ML 180 mm (but mature animals are unknown).

### DISTRIBUTION

This species is circumpolar in a sub-Antarctic belt between 30°S (off Western Cape, South Africa) and 53°S (south of New Zealand): it occurs off South Africa, off southern Chile, off southern Australia, off eastern North Island, New Zealand to ~145°W.

### REFERENCE

Roper and Jereb (2010r).

## *Notonykia nesisi* Bolstad, 2007

*Notonykia nesisi* Bolstad, K. S. (2007). Systematics and distribution of the New Zealand onychoteuthid fauna (Cephalopoda: Oegopsida), including a new species, *Notonykia nesisi* sp. nov. *Reviews in Fish Biology and Fisheries 17*, 305–335 [327, figs 51–60].

Type data: holotype NMNZ M.91597.

Type locality: Chatham Rise, New Zealand, 46°06.80′S, 171°52.00′E.

### COMMON NAME

Nesis' Notalian Clubhook Squid

### DIAGNOSTIC FEATURES

Mantle muscular, cylindrical, very narrow posteriorly. Fins rhombic. Funnel set in deep, well-defined V-shaped groove bordered by low muscular ridge; posterior end of ridge widened into ventral-most primary nuchal fold. Arms 1 and 4 slightly shorter than arms 2 and 3, arms 1 shortest; each with ~70 suckers. Tentacular club one-third ML, slightly expanded. Club manus with 6–18 hooks. Neck (shaft) of largest ventral tentacular hook straight in lateral profile; profile of hook forms an overall J-shape in lateral profile; dorsal tentacular hooks larger than ventral tentacular hooks 1 and 2 (counting from the proximal end of the club); carpus with seven to nine suckers; dactylus with 17–24 suckers; terminal pad with 11–14 suckers. Radula teeth unicuspid.

*Colour.* Dorsal surface of mantle and fins in preserved specimens purple, with pink tissue overlaid by many small, dark purple chromatophores, ventral surface paler, fins darkest medially, paler laterally; ventral surface with small, pale chromatophores over lateral fin margins. Dorsal surface of head slightly iridescent, with layers of dark chromatophores. Arms darkest aborally. Tentacles and clubs with chromatophores on all surfaces, apart from oral face of lateral membranes.

### HABITAT AND BIOLOGY

*Interest to fisheries.* No interest in a fishery exists at the present time.

### SIZE

Small, up to 105 mm ML.

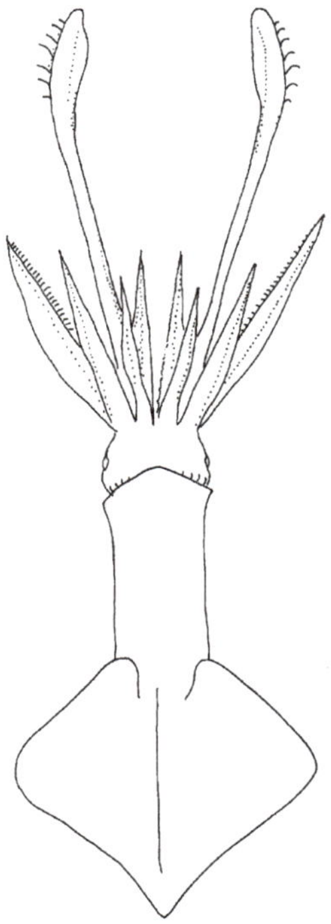

Dorsal view (A. Reid, redrawn from Bolstad 2010).

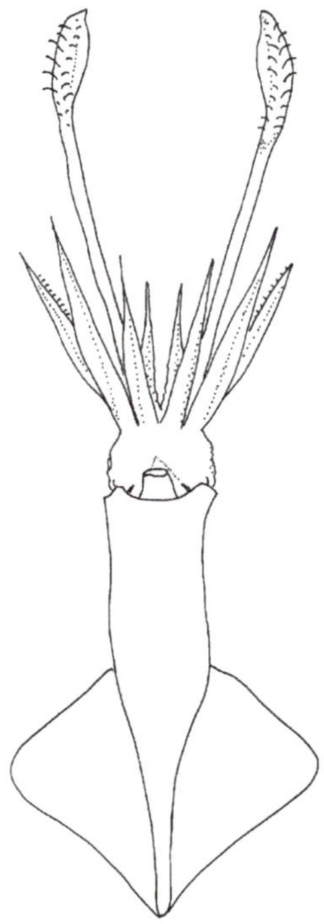

Ventral view (A. Reid, redrawn from Bolstad 2010).

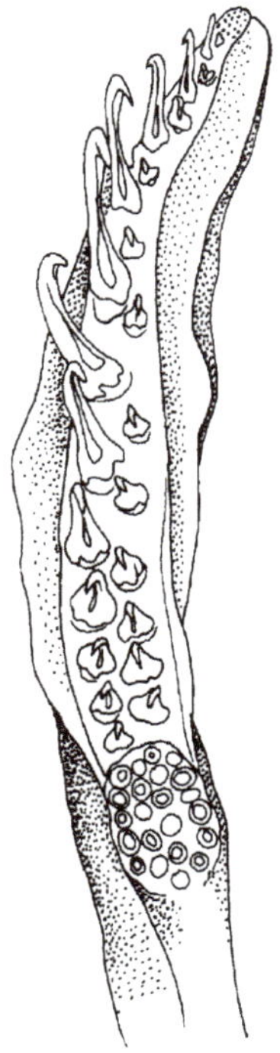

Tentacular club (FAO, E. D'Antoni).

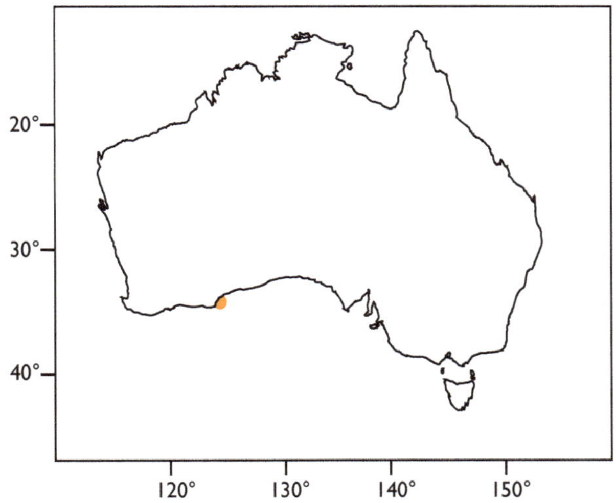


*Notonykia nesisi* distribution. (Australian records are based on few specimens, but the species is likely to occur more widely.)

#### DISTRIBUTION

Southern waters of New Zealand, south of the Subtropical Convergence, southern Australia, known depth range 0–950 m, mostly the upper 300 m.

#### REFERENCES

Bolstad and Young (2010), Roper and Jereb (2010r).

## *Onychoteuthis* Lichtenstein, 1818

#### DISTINGUISHING CHARACTERS

Eight to 13 nuchal folds (three primary/ventrolateral and five to eight secondary/dorsolateral). Gladius visible beneath skin in dorsal midline. Photophores present. Paralarvae can be recognised by the pointed rostrum on the gladius.

#### DIAGNOSTIC FEATURES

Mantle densely muscular with smooth epidermis, cylindrical, tapered posteriorly to pointed 'tail'; non-ammoniacal. Most species with gladius clearly visible through dorsal mantle musculature, rostrum of gladius visible through distal 5–10% ML, and ventral surface of mantle with small, firm, posterior keel. Funnel groove with distinct ridges along margin, margins convex anteriorly. Fins rhombic, to slightly heart-shaped, saggitate, groove broadly V-shaped posteriorly, with nearly parallel margins anteriorly. Eight to 13 prominent nuchal folds (three primary/ventrolateral and five to eight secondary/dorsolateral). Olfactory organ at posteriodorsal end of nuchal fold 2. Arms 1 shortest, arms 2–4 42–56% ML, with 50–110 suckers; arm suckers with distal, fleshy thickening in most species (not visible on all suckers and easily damaged during capture); protective membrane low and trabeculae fused to sucker base. Tentacle stalk with low, longitudinal aboral keel. Tentacular club with two medial series of large strong hooks on manus; bases of largest ventral hooks asymmetrical, variably produced into spike; length of longest ventral hook 16–25% club length; median dorsal hooks small, flanked on each side by larger hooks; distal suckers on club restricted to terminal pad; no marginal series of suckers in adults (except in *Onychoteuthis meridiopacifica* Rancurel & Okutani, 1990).

*Photophores.* Photophores present in mantle cavity: one small anteriorly on ink sac near anus (sometimes visible through the mantle in fresh specimens), one large posteriorly on intestine over posterior extremity of ink sac, appear as circular cream-coloured pads; ventral periphery of each eyeball with one bilobed or two distinct photophores.

#### HABITAT AND BIOLOGY

Species are epipelagic to mesopelagic, mostly strongly eurythermal. Like the paralarvae of other squid, the paralarvae of *Onychoteuthis* have the ability to withdraw their head and arms into their mantle cavity. Adult females lose their tentacles apparently near the time of spawning.

#### SIZE

Small- to medium-sized squids 60–350 mm ML at maturity.

#### REMARKS

They are commonly seen in surface waters at night and can be collected at night by dip net at nightlight stations. Only young squid are captured in standard midwater trawls. It appears that older squid avoid the trawls. They have been known to fly onto the decks of ships.

Two *Onychoteuthis* species have been reported from Australian waters: *O. meridiopacifica* Rancurel & Okutani, 1990 and *O. banksii* (Leach, 1817). The latter taxon was, until recently, known to comprise a species-complex. The taxonomy of *O. banksii* has been partially resolved by Bolstad (2008). *Onychoteuthis banksii* is now known to occur only in the central and north Atlantic, and the Gulf of Mexico. The Australian representatives of this group have not yet been thoroughly re-examined in the light of this work. Until this is done it is not possible to have a good understanding of which species actually occur in Australian waters and where they are distributed. Australian Museum specimens have been examined by the author in the light of Bolstad's work and specimens previously identified as *O. banksii* appear to be *O. aequimanus* Gabb, 1868.

#### DISTRIBUTION

Mostly in tropical and subtropical waters throughout the world's oceans. Absent from the Arctic, upper boral Atlantic, subantarctic and Antarctic waters.

#### REFERENCES

Bolstad *et al.* (2010b), Bolstad *et al.* (2010e), Roper and Jereb (2010r).

### Key to *Onychoteuthis* species

1a. Fin length ~55% ML. Oro-ventral tentacle stalk ridge without chromatophores. 19–23 hooks on manus of tentacular club. Visceral photophores large, circular (O o) ............................ *O. aequimanus*

1b. Fin length ~40% ML. Oro-ventral tentacle stalk ridge with chromatophores. 15–19 hooks on manus of tentacular club. Visceral photophores minute, circular (O o) .................................*O. meridiopacifica*

## *Onychoteuthis aequimanus* Gabb, 1868

*Onychoteuthis aequimanus* Gabb, W. M. (1868). Description of a new species of cephalopod from the South Pacific. *American Journal of Conchology* 4(1), 23–24 [23, pl. 2, fig 1].

Type data: syntypes (2)ANSP A6444 (previous no. 74028).

Type locality: South Pacific Ocean, Society Islands.

#### COMMON NAME

Clubhook Squid

#### DIAGNOSTIC FEATURES

Mantle conical; gladius visible along anterior ~85% of dorsal midline as dark, raised ridge. Tentacular club ~28% ML; 19–23 long, slender hooks, shallowly recurved in lateral profile. Distal-most ventral hook bases may develop into low spike in specimens > 90 mm ML.

*Photophores.* Intestinal photophores circular, well defined; anterior photophores ~40–60% that of posterior photophores. Ocular photophore spans one-third of ventral circumference of eye, uniformly broad.

Fresh-caught specimen, dorsal view (K. Graham).

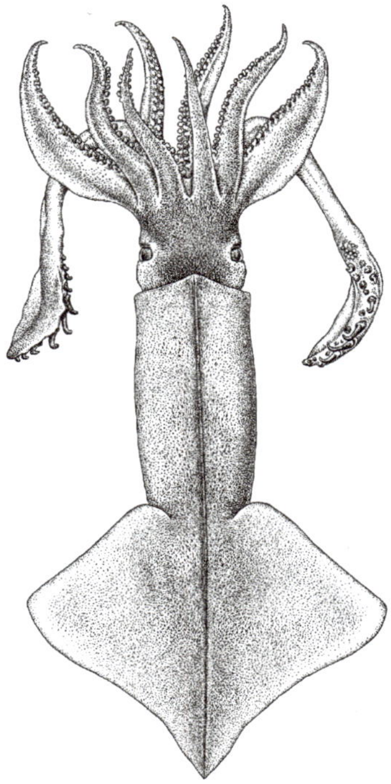

Dorsal view (ABRS, K. Hollis).

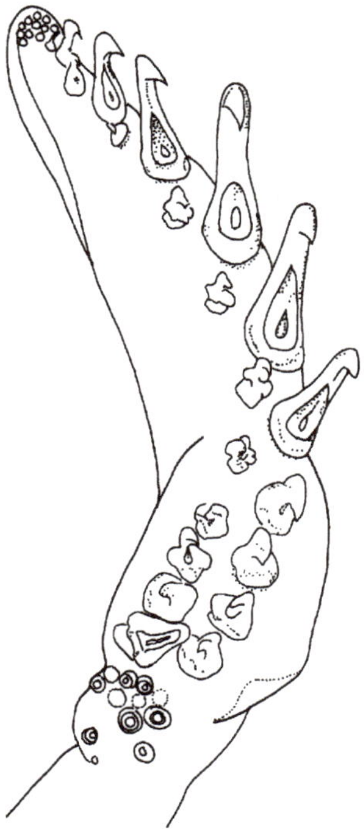

Tentacular club (redrawn from Bolstad 2008).

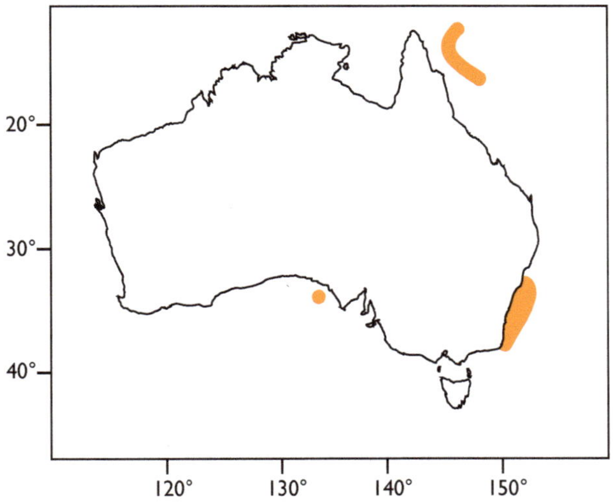


*Onychoteuthis aequimanus* probable distribution. (Australian records are based on few specimens, but the species is likely to occur more widely.)

*Colour.* Chromatophores form reduced band across dorsal half to one-third tentacle stalk width orally, proximal to carpus. Chromatophores absent from oro-ventral tentacle stalk ridge and from small patch on ventral surface of mantle (posterior ~15% ML).

## SIZE

A medium-sized squid; ML up to 300 mm.

## DISTRIBUTION

Pacific Ocean south of 30°N.

## REFERENCES

Bolstad (2008), Bolstad (2010).

# *Onychoteuthis meridiopacifica* Rancurel & Okutani, 1990

*Onychoteuthis meridiopacifica* Rancurel, P. & Okutani, T. (1990). A new species of squid genus *Onychoteuthis* from the Southwest Pacific. *Venus 49*(1), 25–30 [25, figs 1–6].

Type data: holotype NSMT Mo-67008.

Type locality: western South Pacific Ocean (lancetfish stomach), 21°15′S, 155°11.5′E.

## DISTINGUISHING CHARACTERS

The short, broad fins and the presence of a few marginal suckers on the tentacular clubs of adults are diagnostic for this species.

## DIAGNOSTIC FEATURES

Mantle spindle-shaped; bulbous anterior to fin; width 35–40% ML, tapering abruptly posteriorly. Primary folds present; secondary nuchal folds not developed. Fins broadly rhombic, together ovate, short, length ~35% ML;

Preserved specimen (K. Bolstad).

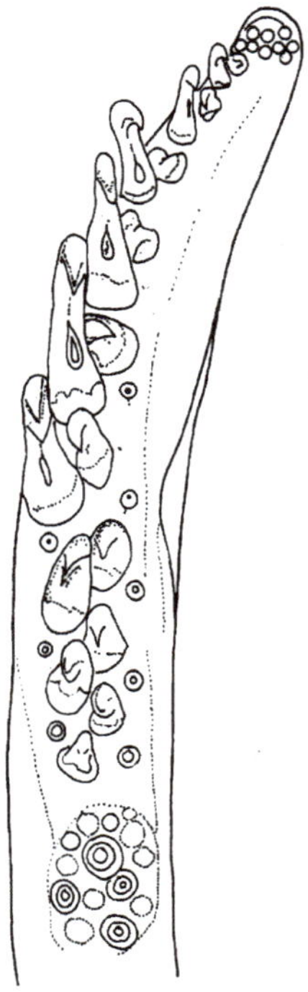

Tentacular club, holotype (M. Shea, redrawn from Bolstad 2010).

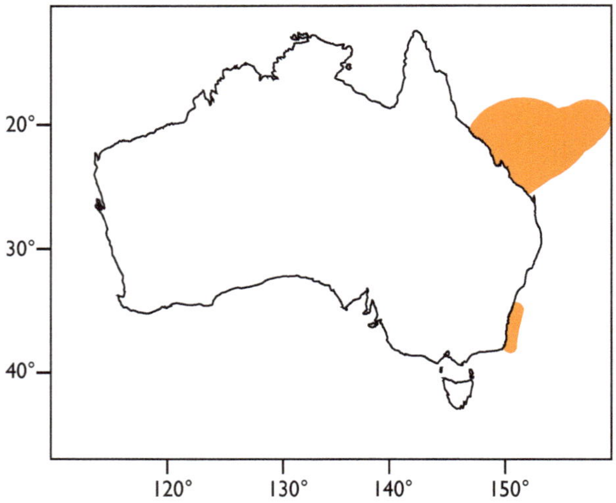


*Onychoteuthis meridiopacifica* distribution. (Australian records are based on few specimens, but the species is likely to occur more widely.)

width 70–80% ML. Funnel groove well defined, V-shaped, broad posteriorly. Arms with 50–60 suckers. Tentacle club length ~20–25% ML, with 16–19 (usually 17 or 18) robust hooks; distal-most ventral hook bases not produced into spike; some proximal marginal suckers remain in adults: three or four in distal series, one or two in ventral series; carpal-locking apparatus with six suckers; terminal pad with few suckers. Mature males with large penis.

*Photophores.* Intestinal photophores present, tiny, may be poorly defined, with anterior only slightly smaller than posterior photophore, does not exceed width of intestine; single ocular photophores uniformly broad, spanning ventral ~25% eye circumference.

*Colour.* Dark reddish purple, darkest medially on dorsal surface of mantle, fins and head; lateral surfaces of mantle and head coppery iridescent. One or two dark 'tail' spots present beneath fins on either side of mantle. Chromatophores absent from tentacular stalk proximal to carpus; chromatophores present along oro-ventral stalk ridge.

### HABITAT AND BIOLOGY

The species is abundant in the central water mass of the South Pacific Ocean. It is epipelagic (possibly to mesopelagic). Males become mature at 40 mm ML.

### SIZE

Small-sized species; maximum recorded ML 90 mm.

### REMARKS

This may be the smallest species in the genus.

### DISTRIBUTION

South-western Pacific Ocean, 16°37′S to 23°56′S, and 162°00′E to 133°15′W. Off the eastern coast of Australia,

0–2000 m (based on three records; nearly all known specimens are ex-stomach contents), in the East Australian current.

#### REFERENCES

Vecchione *et al.* (2008b), Roper and Jereb (2010r).

## *Onykia* Lesueur, 1821

#### DIAGNOSTIC FEATURES

Mantle muscular to moderately soft; dermis of mantle with warty structures or soft wrinkles in adults. Gladius not visible beneath the skin in dorsal midline. Three nuchal folds on lateral side of posterior head; dorsal-most fold with nuchal membrane that curves dorsally and extends to or nearly to nuchal cartilage. Funnel groove with inverted Y-shaped ridge in animals longer than 100 mm ML (except in *Onykia ingens* (Smith, 1881)); anterior margin of funnel groove rounded. Tentacular club in subadults and adults with two medial series of hooks on manus, no suckers; suckers occur only on carpal cluster (with alternating knobs), and minute suckers on distal terminal pad; without suckers on distal portion of club posterior to dorsal pad.

*Photophores.* Absent.

#### SIZE

Medium-sized (275 mm ML) to very large (1.6–2.3 m ML) squids.

#### REMARKS

For many years, several species now recognised as members of *Onykia* were placed in the genus *Moroteuthis*. Tsuchiya and Okutani (1992), demonstrated that specimens historically attributed to *Onykia* Lesueur, 1821, were the young stages of known species of *Moroteuthis* Verrill, 1881, thus making the generic name *Moroteuthis* a junior synonym of *Onykia*.

The paralarvae of some species have a distinctive appearance. They are broad and have a characteristic line of large dorsal mantle chromatophores whose widths increase posteriorly. Small paralarvae in preserved samples have the head mostly withdrawn into the mantle cavity.

#### REFERENCES

Roper and Jereb (2010r), Bolstad *et al.* (2011).

### Key to *Onykia* species

1a. Mantle surface smooth ....................*O. (O.) carriboea*
1b. Mantle surface longitudinally ridged or rugose .....2
2a. Mantle surface longitudinally ridged ........................ ..........................................................*O. (O.) lönnbergii*
2b. Mantle surface rugose ...............................................3
3a. Fin shape saggitate. Y-shaped ridge in funnel groove present ............................................... *O. (O.) 'robsoni'*
3b. Fin shape rhombic. Y-shaped ridge in funnel groove absent................................................... *O. (M.) ingens*

## *Onykia (Onykia) carriboea* Lesueur, 1821

*Onykia carriboea* Lesueur, C. A. (1821). Description of several new species of Cuttle-fish. *Journal of the Academy of Natural Sciences of Philadelphia 2*(1), 86–101, 7 plates [98, pl. 9, figs 1, 2].

Type data: neotype NMNH 40121.

Type locality: USA, North Carolina, Cape Hatteras 36°05.5′N, 69°51.75′W.

#### COMMON NAME

Caribbean Clubhook Squid

#### DIAGNOSTIC FEATURES

Mantle muscular, cross-section nearly ellipsoidal in young, to cylindrical with growth. Fins much wider than long; length nearly 50% ML; width 70–75% ML; fins subrhombic, convex anterior and posterior margins, bluntly rounded lateral angles. Tentacle club with two medial series of ~10–12 hooks each (total 20–24 hooks), two marginal series of small suckers; well-developed carpus with eight or nine suckers, plus their accompanying knobs; dactylus with few very small to minute suckers.

*Colour.* Chromatophores vivid, deep purple, particularly densely packed on dorsal mantle and dorsal head. Skin with reflective, silvery sheen.

*Sculpture.* Skin surface smooth.

#### HABITAT AND BIOLOGY

Paralarvae and juveniles of *Onykia carriboea* live near and at the surface, often in association with *Sargassum* weed. Very little is known about its biology.

*Interest to fisheries.* Currently no interest exists for a fishery.

#### SIZE

The maximum known ML is 100 mm.

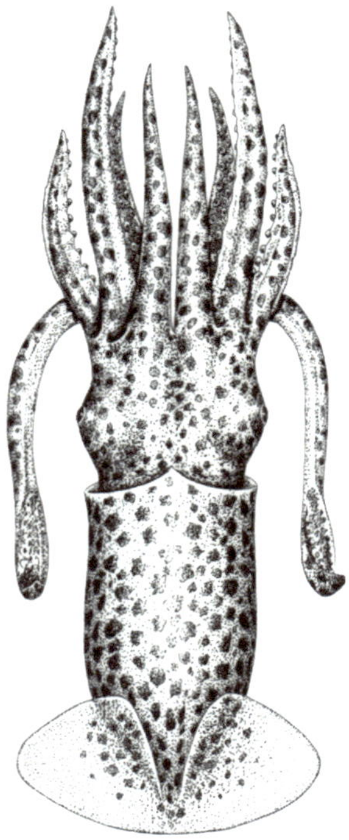

Dorsal view (A. Guerra).

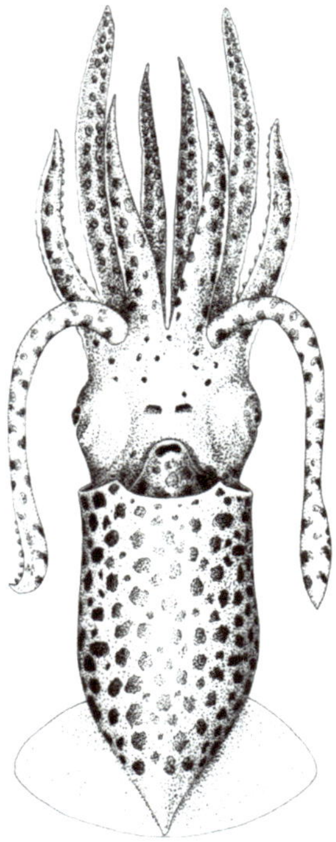

Ventral view (A. Guerra).

Tentacular club (A. Guerra).

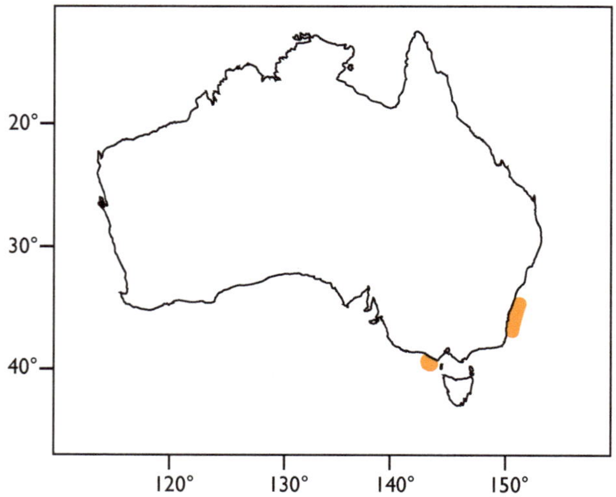


*Onykia carriboea* distribution.

**REMARKS**

*Onykia carriboea* is known, with certainty, only from small juveniles: no adult, mature specimen has been found and-described. This suggests that the name could represent the immature growth stage of other onychoteuthid species (Kubodera *et al.* 1998, Bolstad 2007), perhaps *O. 'robsoni'* (Adam, 1962) (see also Vecchione *et al.* 2003a). However, detailed comparisons of specimens from their respective type localities are needed before this problem can be resolved. In order to stabilise the taxonomy Bolstad (2010) erected a neotype for this species as the type had been lost.

**DISTRIBUTION**

This species is believed to have a cosmopolitan circumglobal distribution, throughout tropical and subtropical/ warm temperate oceans. However, no definitive statement can be made before its specific status is clarified.

**REFERENCES**

Vecchione *et al.* (2003a), Roper and Jereb (2010r).

## *Onykia (Moroteuthopsis) ingens* (Smith, 1881)

*Onychoteuthis ingens* Smith, E. A. (1881). Account of the zoological collection made during the survey of H.M.S. 'Alert,' in the Straits of Magellan, and on the Coast of Patagonia. IV. Mollusca and Molluscoida. *Proceedings of the Zoological Society of London 1881*(1), 22–44 [25, pl. 3 figs 1–1d].

Type data: holotype BMNH 1880.10.8.1 (head only).

Type locality: south-eastern South Pacific Ocean, west coast of Patagonia, Port Riofrio.

**COMMON NAME**

Greater Clubhook Squid

**DIAGNOSTIC FEATURES**

Mantle robust, broad, thick, heavily muscled, not drawn out into sharp 'tail'; skin covered with soft, fleshy warts, as flat, elongate or curved tubercles, in paving-stone

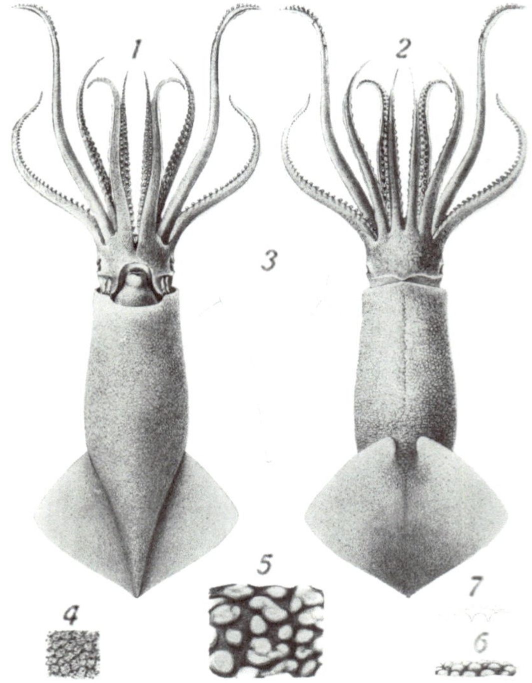


*1*, ventral view; *2*, dorsal view; *3*, ventral view sketch; *4*, detail of mantle; *5*, detail of mantle enlargement; *6*, *7*, mantle lateral view (Pfeffer 1912, Plate 11).

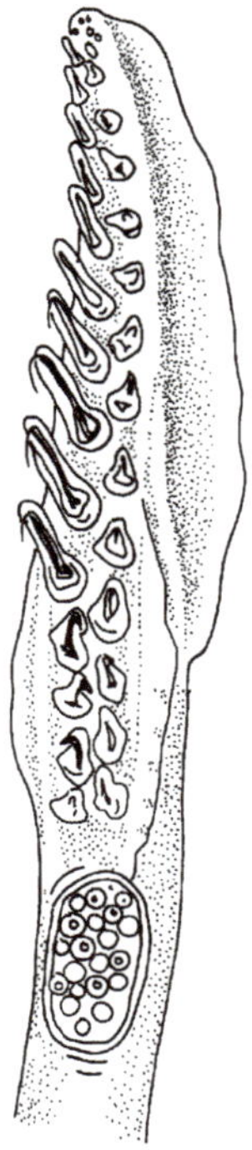

Tentacular club (A. Reid, redrawn from Pfeffer 1912, Plate 12).

pattern. Fins large, broad, with broad fin angles of 50–55° each; fin length ~50% (or slightly larger) of ML; rhomboidal, not attenuate into 'tail'. Longest arms (2 and 3) ~70% ML; arm pair 1 shortest. Tentacular clubs not expanded, with 28–32 hooks in two medial series (13–16 hooks in dorsal series, 13–15 hooks in ventral series); hooks on ventral series much larger than on dorsal series, largest hooks positioned at sixth to eighth pair; carpus with usually 10–11 (range 8–13) suckers plus their associated knobs; distal club pad with usually 14 small suckers (range 13–18). Tentacular club of young *Onykia ingens* has few marginal suckers on proximal dorsal margin and on distal ventral margin. Lower beak in fresh animals with very distinctive broad, thin undarkened region of chitin.

*Colour.* Pale cream or violet flesh-red.

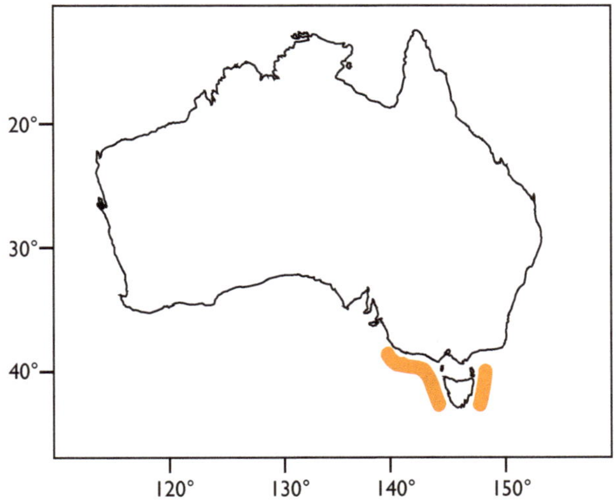


*Onykia (Moroteuthopsis) ingens* distribution. (Australian records are based on few specimens, but the species is likely to occur more widely.)

## HABITAT AND BIOLOGY

Adult *O. (M.) ingens* apparently are primarily benthic/epibenthic in lower sublittoral and bathyal zones, having been caught in demersal trawls at 300–1450 m in continental slope and bathyal waters. Juveniles and subadults are epi- to mesopelagic. Females apparently undergo strong ontogenetic descent, but males do not show this behaviour. The species is therefore important in transporting energy from epipelagic waters to deeper benthic environments due to this migration as paralarvae develop into adults. It is an ammoniacal species.

The cost of reproduction for this squid has been analysed using measures of condition and tissue biochemistry by Jackson *et al.* (2004). The weights of several structures drop dramatically after spawning with the mantle, for example, showing an increase in water indicating remobilisation of energy to fuel reproduction. Males implant spermatophores directly into a variety of internal and external tissues of the female (Jackson and Jackson, 2004). As might be expected, the diet of *O. (M.) ingens*, as determined from stomach contents and fatty acid profiles, varies temporally and ontogenetically as a result of varying availability of prey species (Phillips *et al.* 2003a, b). In the case of *O. (M.) ingens*, the primary prey consists of myctophid fishes and krill.

McArthur *et al.* (2003) have examined mercury concentrations in the muscle tissue of *O. (M.) ingens* and Patagonian toothfish (*Dissostichus eleginoides* (Smitt, 1898)). They showed that (unlike toothfish) *O. (M.) ingens* had a low mercury concentration in the mantle tissue and postulated that squid may have a mechanism that prevents bioaccumulation of mercury in the mantle and presumably other body tissues.

Sands *et al.* (2003) examined RAPD allele frequencies in *O. (M.) ingens* and found high levels of variation between individuals but little, though some, subtle population variation between two widely separated sites (Falkland Islands and Macquarie Island). This study showed that stretches of the Southern Ocean between the two sampling sites may not be an effective barrier to gene flow, and it may be that egg and paralarval dispersal by strong Southern Ocean currents may be the cause of some gene flow. While these results provide preliminary evidence for population structuring more samples and more informative genetic markers are needed to identify processes driving population structure.

*Interest to fisheries.* This species is extremely abundant in sub-Antarctic waters and is believed to have some fishery potential but, as a neutrally buoyant, ammoniacal species, this is currently problematic. However, if the ammoniacal problem can be solved there is potential for a huge fishery. Sands *et al.* (2003) found some evidence for population structure among samples of *M. ingens* from around the Falkland Islands and Macquarie Island, inferred from RAPD analyses. Should a fishery ever develop, the population structure of the species, formerly thought to comprise a single panmictic population in the Southern Ocean, will need to be further investigated.

#### SIZE

Large-sized species; maximum ML 520 mm.

#### DISTRIBUTION

*Onykia ingens* is a notalian circumpolar species in sub-Antarctic waters north of the Antarctic Polar Front, south of the Subtropical Front. It occurs on the Patagonian shelf, in southern Chile and the Prince Edward, Crozet, Kerguelen Islands and southern New Zealand islands (Antipodes, Campbell, etc.) as well as southern Australian and southern South Africa waters. While described as circumpolar, the continuity of the distribution is uncertain due to inadequate sampling over deep ocean areas (Sands *et al.* 2003).

#### REFERENCES

McArthur *et al.* (2003), Vecchione *et al.* (2003b), Roper and Jereb (2010r).

### *Onykia* (*Onykia*) *loennbergii* (Ishikawa & Wakiya, 1914)

*Moroteuthis lönnbergi* Ishikawa, C. and Wakiya, Y. (1914). On a new species of *Moroteuthis* from the Bay of Sagami, *M. lönnbergii. Journal of the College of Agriculture, Imperial University of Tokyo* 4(7), 445–460 [445, pls 45, 46].

Type data: neotype NSMT Mo-76203.

Type locality: north-western North Pacific Ocean, Japan, Sagami Bay, 35°12.8′N, 139°28.2′E.

#### COMMON NAME(S)

Hooked Squid, Japanese Clubhook Squid

#### DISTINGUISHING CHARACTERS

This species is very similar to *Onykia 'robsoni'* (Adam, 1962). They differ in the form of the dermal skin pads (that are short, fine ridges, fusing with one another *v.* tubercles or warts in *O. 'robsoni'*). The fins are more rhombic in shape than those of *O. 'robsoni'*. They differ in the cross-sectional shape of the rostrum (triangular *v.* oval).

#### DIAGNOSTIC FEATURES

Mantle robust, muscular; skin rugose, with longitudinal ridges and warts in skin; posterior tip drawn out into moderately long, pointed 'tail'. Eyes with thick, pigmented lobes on each side of indentation in eyelid. Fins large, rhomboidal, slightly drawn out posteriorly; fin length 50–60% ML; fin width, broad, up to 50–55% ML, roughly equal to fin length; fin angles ~35–40°. Longest arms (4)

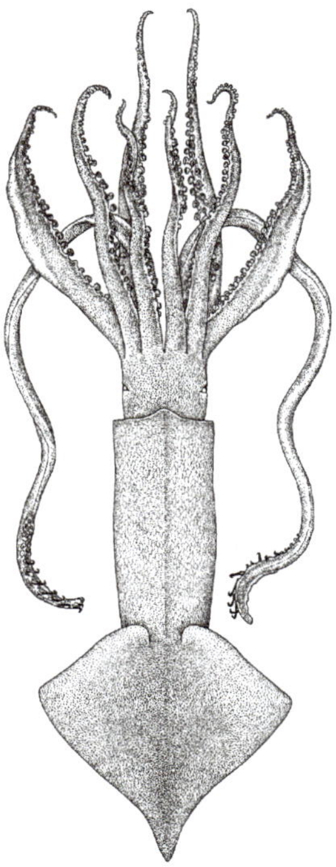

Dorsal view (ABRS, K. Hollis).

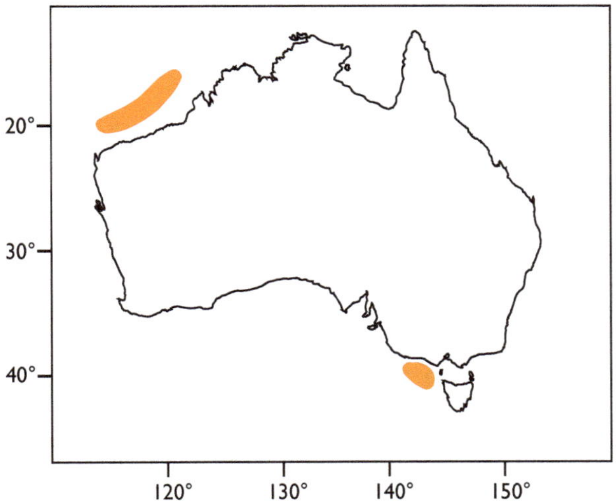


*Onykia (Onykia) loennbergii* distribution. (Australian records are based on few specimens, but the species is likely to occur more widely.)

60% ML. Sucker rings chitinous, smooth. Tentacular club long, unexpanded, with ~25–30 hooks in two median series on manus; fourth to sixth hooks largest in dorsal series, sixth or seventh hook largest in ventral series; seven or eight small suckers plus associated knobs on carpus; 10–13 minute suckers on extreme tip of dactylus. Hectocotylus absent. Rhachidian tooth of radula with two large lateral cusps; first lateral tooth with a single large lateral cusp.

*Colour.* In life, reddish purple with closely spaced, longitudinal ridges creating timber-like grain.

### HABITAT AND BIOLOGY

*Onykia loennbergi* is an oceanic, ammoniacal, neutrally buoyant species (epipelagic, mesopelagic, bathyal) from temperate and subtropical waters. Its exact depth distribution is unknown, but it has been caught in open nets at depths between 730 and 920 m. However, it also occurs in depths of <200 m, as evidenced by it being preyed upon by northern fur seals.

*Interest to fisheries.* Undetermined and not fished commercially, but currently unsuitable at full maturity because of its ammoniacal composition. The species is fished commercially in other parts of the world.

### SIZE

Maximum ML 360 mm.

### DISTRIBUTION

This species is distributed in the western North Pacific Ocean off eastern Honshu, Japan in the north, southward to Tsugaru Strait, the Kuroshiro Current, into the Indian Ocean (Saya-de-Malha Bank), and southward to the North West Shelf, Australia. Records also from southern Australia.

### REFERENCES

Vecchione *et al.* (2007), Roper and Jereb (2010r).

## *Onykia (Onykia) 'robsoni'* (Adam, 1962)

*Moroteuthis robsoni* Adam, W. (1962). Cephalopodes de l'Archipel du Cap-Vert, de l'Angola et du Mozambique. *Memórias da Junta de Investigacões do Ultramar, series 2, 33*, 9–64 [24, text figs 2, 3, pl. 1].

Type data: neotype SAMC S4110.

Type locality: south-eastern South Atlantic Ocean, Angola, 30°03'S, 14°23'E.

### COMMON NAME

Rugose Clubhook Squid

### DIAGNOSTIC FEATURES

Mantle long, slender, tip very long, pointed; skin rugose, covered with rounded, fleshy, irregular tubercles, or warts (not ridges). Fins heart-shaped, very long, attenuate; length 53–62% ML; narrow, width 51–70% ML (in animals of 156–370 mm ML); together form very sharply

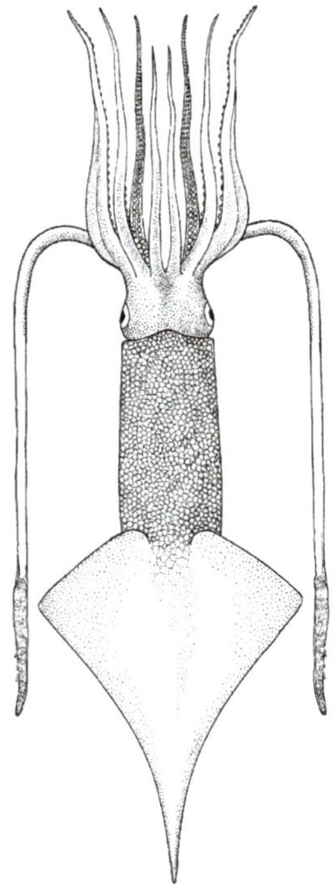

Dorsal view (FAO, E. D'Antoni).

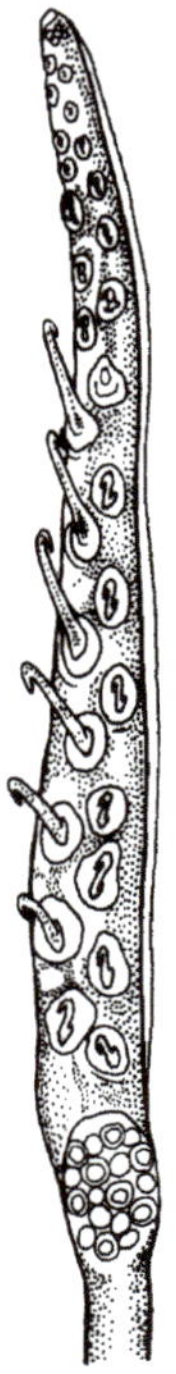

Tentacular club (FAO, E. D'Antoni).

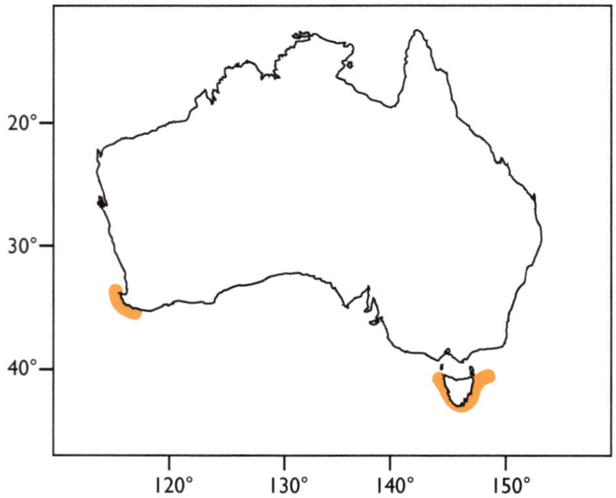


*Onykia* (*Onykia*) *'robsoni'* distribution. (Australian records are based on few specimens, but the species is likely to occur more widely.)

pointed, lanceolate 'tail'; each fin angle 30–40°. Tentacular club very narrow, unexpanded; carpus with 10–12 small suckers plus their associated knobs; manus with 26–32 hooks in two median series; dactylus with 12–17 minute suckers. Arms attenuate, longest (4) ~57–86% ML. Rhachidian and first lateral teeth with accessory cusps. Deep reddish colouration.

#### HABITAT AND BIOLOGY

*Onykia 'robsoni'* is an oceanic species, both pelagic and benthic. It is taken with open midwater trawls fished between 250 and 550 m depths (exact depth distribution unknown). It is very likely a species-complex, so the species from Australian waters is probably not *O. 'robsoni'*. It is a non-ammoniacal species.

*Interest to fisheries.* This species is caught by deep-set bottom trawls; it is believed to have some fishery potential in stages before full maturity. As it is an ammoniacal species, this is currently problematic.

#### SIZE

Large-sized species: maximum ML 630 mm.

#### DISTRIBUTION

The distribution of *Onykia 'robsoni'* is circumglobal in southern subtropical and notalian regions, extending south to the Antarctic Polar Front. *Onykia 'robsoni'* occurs in the southern Atlantic to South Georgia, off southern and south-western South Africa, as well as off southern Australia (although see above) and southern New Zealand.

#### REFERENCES

Bolstad (2010), Roper and Jereb (2010r), Vecchione *et al.* (2011).

### *Walvisteuthis* Nesis & Nikitina, 1986

The genus is currently monotypic; however, the very broad distribution of this species, suggests that '*Walvisteuthis virilis*' probably represents a species-complex.

### *Walvisteuthis virilis* Nesis & Nikitina, 1986

*Walvisteuthis virilis* Nesis, K. N. and Nikitina, I. V. (1986). A new family of abyssal squids (Cephalopoda, Oegopsida) from the south-eastern Atlantic. *Zoologichesky Zhurnal 65*(1), 47–54 [47, figs 1–3].

Type data: holotype ZMMGU Y-264.

Type locality: eastern South Atlantic Ocean, Walvis Ridge, 33°06'S, 02°7'E.

#### COMMON NAME

Whale Squid, Walvis Ridge Squid

#### DIAGNOSTIC FEATURES

Mantle broadly conical, bluntly rounded posteriorly. Tissues of mantle, fins, head, appendages watery,

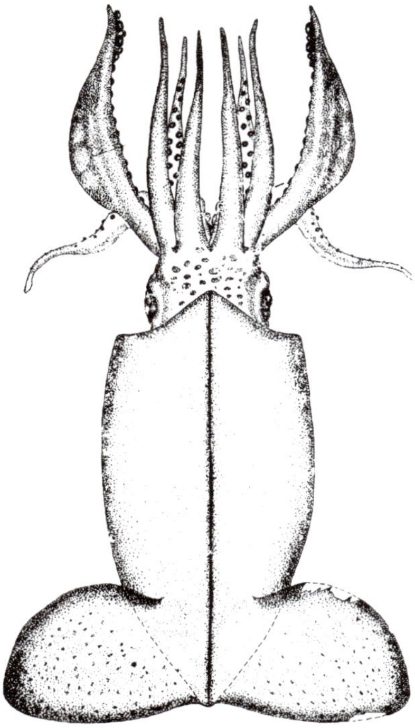

Dorsal view (M. Shea, redrawn from Nesis and Nikitina, 1986).

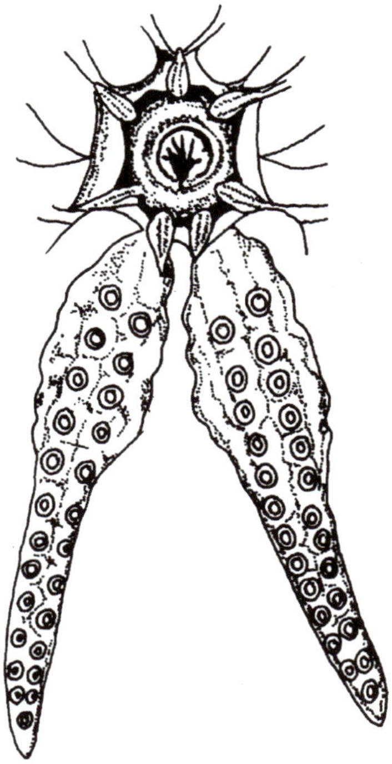

Buccal connectives and arms 4 (M. Shea, redrawn from Nesis and Nikitina, 1986).

semigelatinous. No warts, ridges, wrinkles on surface of mantle. Fins very short (about one-third of ML), broad, transversely oval, width ~90% ML; posterior lobes meet in midline, not separate. Head short, broad, narrower than mantle. Eyes large, sinus present; olfactory papilla tongue-shaped. Nuchal folds absent. Funnel wide and long, reaches level of mid-eye. Funnel-locking cartilage a simple, straight groove, slightly broader distally; mantle component a long, narrow, straight ridge. Buccal membrane with seven lappets; connectives DDVV – that is, arm 4 connectives attach to ventral edge of buccal membrane. Arms short, ~50% ML; formula 4 > 3 = 2 > 1; arms 1 without keels, arms 2 with low keel the entire length, arms 3 with large, broad keel in distal portion; arms 4 with well developed keel, unusually broad along proximal portion. Arm suckers biserial, large, short-stalked, flattened, no hooks; three or four pairs of greatly enlarged suckers on midportion of arms 3. Tentacles, thin, weak, non-muscular, short; length only approximates arm length; club occupies more than half tentacle length; carpus with well developed, dispersed, locking apparatus, eight or nine suckers, less than half diameter of normal arm suckers; five or six pairs of minute, widely spaced, medial suckers in two series on long stalks on manus; a few marginal suckers may occur. Spermatophore storage sac and penis extremely well developed, at least 50% ML, reaches mantle opening; hectocotylus absent.

*Photophores.* No visceral photophores.

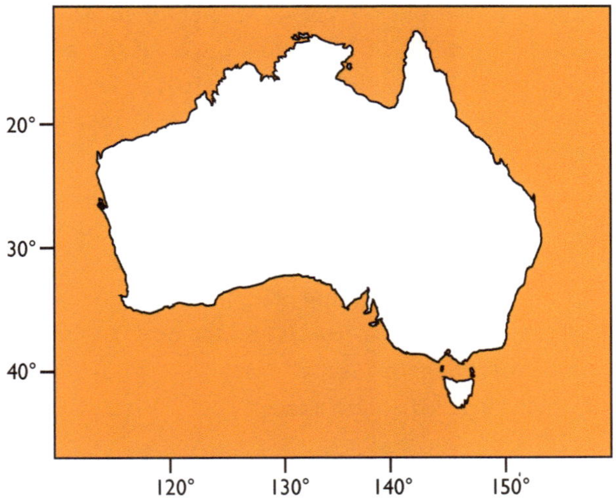


*Walvisteuthis virilis* distribution.

## HABITAT AND BIOLOGY

*Walvisteuthis virilis* is an epipelagic to upper mesopelagic species. Paralarvae are known as small as 1.9 mm ML. The smallest paralarvae are extremely slender with mantle width ~25% ML. Between this size and ~4 mm to

5 mm ML they are easily recognised by their slender appearance and a distinctive elongate patch of large chromatophores along the dorsal midline. Eyes dorsoventrally elongate and strongly bulge from head. Club suckers biserial, large (about same size as arm suckers), prominent. At roughly 4 mm to 5 mm ML paralarvae undergo morphological change. They become relatively broad, eyes become hemispherical and fins become much more prominent. At 7 mm ML squid is very broad for its length. By 12 mm ML the V-shaped funnel groove and distinctive rostrum are present and hooks begin to form in the medial-ventral series on the club. Chromatophores small and scattered but larger on dorsal surfaces than on ventral surfaces.

#### SIZE

Small-sized species; maximum ML to 103 mm.

#### REMARKS

On the ventral surface of the head the chromatophores lie deep within the tissue and are not readily visible in preserved specimens. The status of this taxon has recently has been reviewed by Tsuchiya (unpublished data) who concluded that *W. virilis* is a synonym of the onychoteuthid *Onykia rancureli* Okutani, 1981. The peculiar features noted by Nesis and Nikitina (1986) appear to be modifications that arise at sexual maturity. In fact, *O. rancureli* had been widely recognised as being very different from other members of the genus and several authors have suggested that a new generic name is needed (e.g. Tsuchiya and Okutani (1992)). This generic name is considered to be *Walvisteuthis* and the new combination *Walvisteuthis rancureli* is used here.

*Interest to fisheries.* None.

#### DISTRIBUTION

Worldwide in tropical to temperate seas. A southern subtropical to notalian bathypelagic form, it has been found in the topical Indian Ocean, the temperate and tropical South Atlantic and in the tropical north and south Pacific Ocean.

#### REFERENCE

Roper and Jereb (2010r).

## Pholidoteuthidae Voss, 1956

The family is monotypic, represented by *Pholidoteuthis* Adam, 1950.

### *Pholidoteuthis* Adam, 1950

A single species, *Pholidoteuthis massyae* (Pfeffer, 1912), is represented in Australian waters.

### *Pholidoteuthis massyae* (Pfeffer, 1912)

*Tetronychoteuthis massyae* Pfeffer, G. (1912). Die Cephalopoden der Plankton-Expedition. Zugleich eine monographische ubersicht der Oegopsiden Cephalopoden. *Ergebnisse der Plankton-Expedition der Humboldt-stiftung 2*, 1–815 [102, pl. 14 figs 15–19].

Type data: holotype ZMH (not found 2001).

Type locality: eastern North Atlantic Ocean, 48°N, 15°W.

#### COMMON NAME(S)

Coffee Bean Scaled Squid, Scaled Squid

#### DIAGNOSTIC FEATURES

Mantle elongate, thick-walled, spongy, moderately to almost completely weakly muscled, tapers posteriorly to moderately pointed 'tail'. Funnel-locking cartilage a straight, simple groove. Fins short with distinct, small anterior lobes, rhomboidal in outline, do not form elongate 'tail', fin length 35–45% ML, fin width 50–70% ML. Arms relatively short, 30–60% ML, robust, subequal in length; suckers biserial, without hooks. Tentacles long, clubs virtually unexpanded, long; club suckers tetraserial (no hooks), compressed in late juveniles and adults into ovoid, elongate structures with slit-like openings; long sides of inner rings lined with acute teeth; manus length about twice as long as dactylus length; carpal-locking mechanism absent; small flaps with short membranes occur along lateral edges of club adjacent to stalks of lateral rows of suckers, not attached to protective membranes. Buccal membrane with seven lappets; connectives to arms 4 attach to ventral margins.

*Colour.* Colour dull mauve or purple, or lighter pinkish tone. In large individuals, dark epidermis almost always abraded away from fins, head and arms (leaving them patchy white).

*Photophores.* Absent.

*Sculpture.* In animals from ~20–25 mm ML, mantle spongy, covered with tightly packed, round dermal structures (papillose tubercles) that consist of solid, densely packed material histologically reminiscent of elastic cartilage, evenly distributed throughout dense connective tissue matrix; these tubercles are roughly mushroom-shaped in cross-section, with slightly concave central disc and thick base; periphery of tubercle contains 7–10 conical

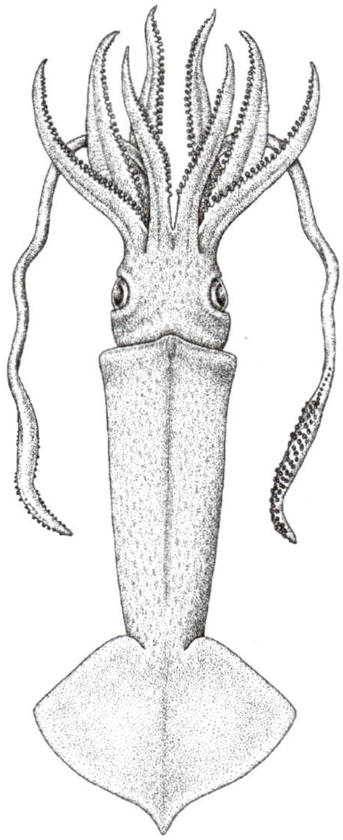

Dorsal view (ABRS, K. Hollis).

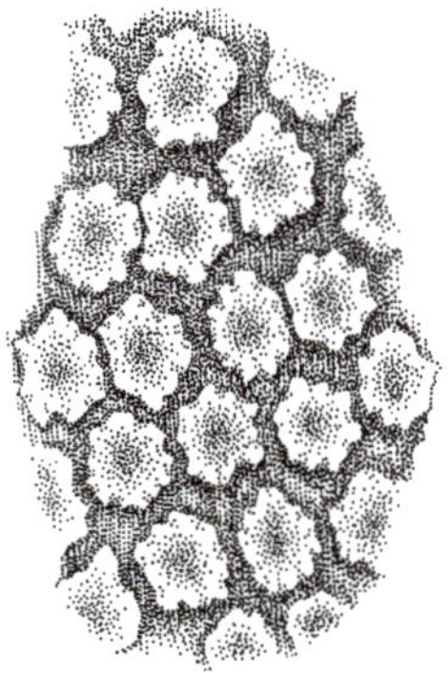

Enlargement of scales (ABRS, K. Hollis).

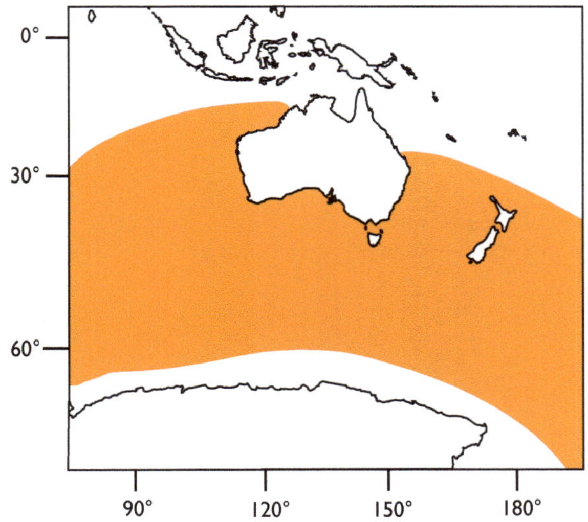


*Pholidoteuthis massyae* distribution.

papillae, each with two to four prongs. Tubercles terminate abruptly on ventral mantle at about midpoint of fins.

## HABITAT AND BIOLOGY

*Pholidioteuthis* have been caught in demersal trawls in continental slope waters during the day but not at night, suggesting they may disperse into midwater at night. The vertical distribution of adults encompasses mesopelagic to bathypelagic, bathybenthic zones. Paralarvae can occur in the upper 200 m, while adults are mesopelagic to bathypelagic, concentrated in the 600–800 m zone; descent to bathyal bottom depths coincides with maturation. Spawning apparently occurs on or near the bottom, to at least 1500 m.

Formerly, *Pholidoteuthis* species were considered relatively rare squids. They are now considered to be geographically widespread and numerous. Net captures indicate that paralarvae and juveniles occur in the upper 500 m, while adults occur in daytime bottom trawls to at least 1500 m. Adults probably leave the bottom at night and disperse into the bathy-mesopelagic zone. Paralarvae and juveniles very widely dispersed by currents; circumglobally in the case of *P. massyae*.

*Interest to fisheries.* Currently this species is of no interest to fisheries because of its inaccessibility in deep water and its spongy consistency and ammoniacal tissue. When a processing technique is developed, this species should be considered a potential resource because of its size, abundance and broad distribution.

## SIZE

Maximum recorded ML 720 mm. Large-sized squids; maximum ML up to 800 mm.

## REMARKS

The papillose tubercules on *P. massyae* are believed to function to reduce hydrodynamic drag over the surface of the mantle, thus increasing the swimming efficiency. For details concerning the tubercles see Roper and Lu (1990).

## DISTRIBUTION

A widely distributed, cosmopolitan species in southern subtropical, southern temperate, and sub-Antarctic waters of the Atlantic, Pacific and Indian Oceans. Oceanic, cosmopolitan in tropical and temperate seas.

## REFERENCES

O'Shea *et al.* (2007), Roper and Jereb (2010s).

## Psychroteuthidae Thiele, 1920

The family is monotypic, represented by *Psychroteuthis glacialis* Thiele, 1920.

## *Psychroteuthis* Thiele, 1920

### *Psychroteuthis glacialis* Thiele, 1920

*Psychroteuthis glacialis* Thiele, J. (1920). Die Cephalopoden der deutschen Südpolar-Expedition 1901–1903. *Deutsche Südpolar-Expediton, 1901 to 1903, Zoologie 16*(8), 433–465 [440, pl. 53, figs 9–16].

Type data: lectotype ZMB Moll.-110014.

Type locality: Southern Ocean, 66°S, 89°E.

#### COMMON NAME

Glacial Squid

#### DIAGNOSTIC FEATURES

Mantle elongate, muscular, tapers to pointed 'tail'. Fins saggitate, rhomboidal, muscular, length 55–60% ML; anterior fin lobes distinct; posterior margins join mantle to form pointed 'tail'. Eyes proportionally very large. Funnel-locking cartilage a very slightly curved, narrow, simple groove. Arms muscular, subequal in length; arm suckers biserial; sucker rings smooth except on distal-most suckers. Tentacles long, muscular; tentacular clubs expanded; carpal-locking region extends proximally along ventral edge of tentacular stalk with row of alternating knobs and suckers (some may be biserial) and dorsally along distal portion; manus with tetraserial suckers proximally, ventromedial series with greatly enlarged suckers, those of dorsomedial series somewhat enlarged; distal manus and dactylus with four to seven series of suckers. Buccal connectives attach to dorsal borders of arms 4 (formula DDVD).

*Photophores?* Large, composite photophore on tips of arms 3 in mature males, consisting of transverse light-producing plates arranged in zigzag pattern and covered with black-pigmented shields. (Herring 2007 states that these may not actually be photophores.) Photophores on mantle and head absent.

#### HABITAT AND BIOLOGY

*Psychroteuthis glacialis* is an oceanic species reported from <200–920 m, taken with benthic and benthopelagic trawls. Paralarvae and juveniles occur at or near the surface. This has been determined by stomach contents findings, indicating they are frequently preyed upon by many species of seabirds, penguins and marine mammals. As adults they undergo ontogenetic descent into the depths where they are a major prey species for deeper-diving forms. It is a non-ammoniacal species. Size-frequency data indicate that the species has a 2 year life cycle; males mature earlier and at a smaller size than females.

*Interest to fisheries.* Currently no specific fishery exists, but the species is of major importance as prey for both exploited and unexploited predatory species. Furthermore, since the musculature is firm, non-ammoniacal, and since the species is large (up to 440 mm ML) and apparently

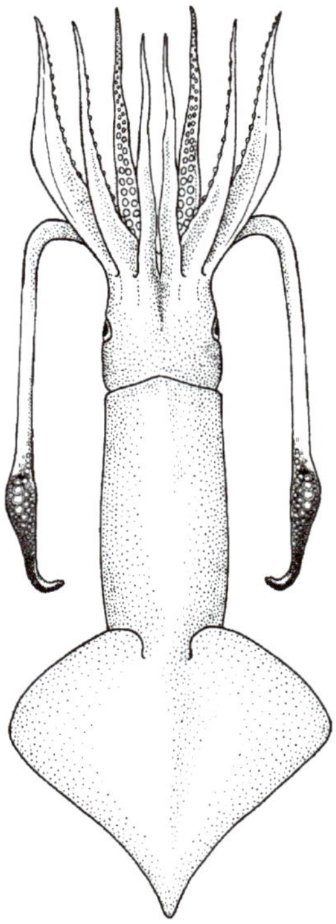

Dorsal view (FAO, E. D'Antoni).

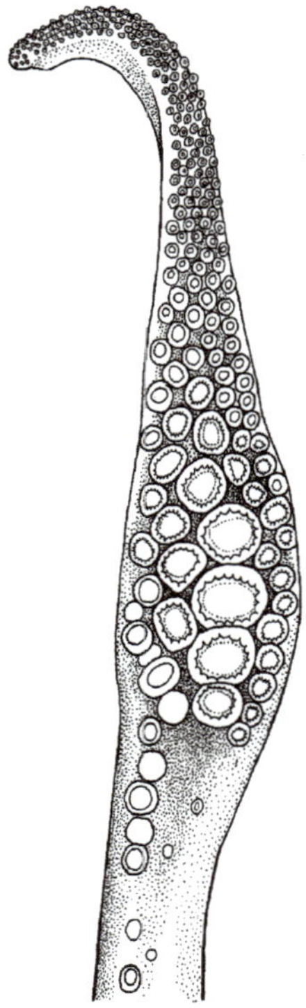

Tentacular club (FAO, E. D'Antoni).

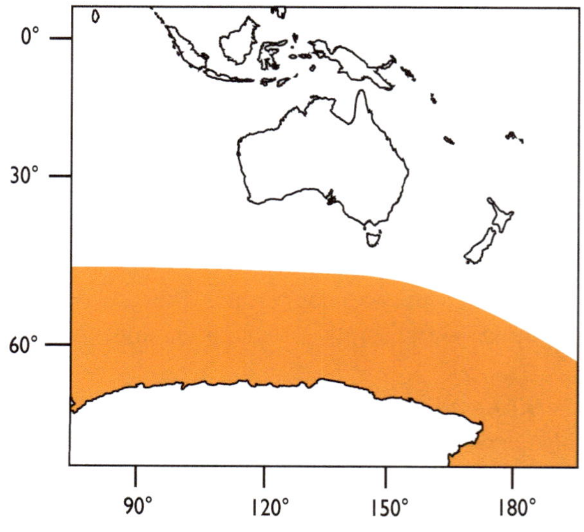


*Psychroteuthis glacialis* distribution.

very abundant, it should be considered a potentially valuable resource. This seems more likely now that deep trawling activities continue to probe mesopelagic and upper bathypelagic depths.

#### SIZE

A medium-sized squid; maximum ML up to 440 mm.

#### REMARKS

*Psychroteuthis glacialis* was originally described from incomplete specimens from the stomachs of penguins and seals. Specimens of *P. glacialis* now are very abundant in research collections as a result of several decades of intensive research and collections in Antarctic waters, as well as on Antarctic marine and terrestrial predators. It is the predominant squid in waters close to the Antarctic continent in the Southern Ocean.

Gröger *et al.* (2000) published a detailed analysis of beak length to body size and mass measurements in a wide size range of this species. They showed that the lower rostral length is a good predictor of the squid's ML and wet weight. This analysis provides essential information for future use in biomass estimates in Southern Ocean top predators, as beaks of *P. glacialis* occur frequently among the stomach contents of Antarctic seabirds, seals and toothed whales.

Paralarvae show a strong similarity to paralarvae of Histioteuthidae, especially with regard to the shape of the fins, and in tentacular club suckers being smaller than the arm suckers and the anterior position in the mantle cavity of the digestive gland. The structure is also very similar between the families.

#### DISTRIBUTION

This species occurs circumglobally in Antarctic waters. It is an Antarctic endemic extending north to the Sub-Antarctic Front.

#### REFERENCE

Roper and Jereb (2010s).

## Pyroteuthidae Pfeffer, 1912

#### COMMON NAME

Fire Squids

#### DIAGNOSTIC FEATURES

Nuchal folds absent. Tail acute, not fleshy, supported by needle-like terminal conus of gladius; vesicles absent. Fins subterminal; posterior lobes distinct. Straight mantle–funnel locking apparatus. Arm suckers biserial. Hooks present on arms 1–3 (at least). Tentacles with permanent constriction and angular bend near base; stalk 'ligament' and vein leave tentacle distal to base and not in membrane. Right or left arm 4 hectocotylised. Tentacular club hook presence or absence varies with genus; armature on manus always tetraserial. Buccal crown with seven or eight supports; connectives attach to dorsal margins of arms 4. Secondary buccal connectives attach to ventral margins of arms 1 and 2. Ink sac embedded in digestive gland. Oviduct single, unpaired. Nidamental glands present; oviducal glands normal on one side, may be reduced or absent on other side.

*Photophores.* Photophores present on viscera, eyeballs, and tentacles; absent from fins, mantle, funnel, head, and arms.

#### HABITAT AND BIOLOGY

Pyroteuthids are among the more common squids found in the midwaters of the open ocean. They are epipelagic as paralarvae and juveniles, and then as adults they descend to upper mesopelagic habitats, from which they undergo diel vertical migrations into the epipelagic zone at night.

#### SIZE

A small-sized squid; maximum ML 60 mm.

#### REMARKS

The two genera currently recognised in the family Pyroteuthidae are *Pyroteuthis* and *Pterygioteuthis*. Together with Ancistrocheiridae and Enoploteuthidae, the Pyroteuthidae form a group known as the 'Enoploteuthid group' (see 'Remarks' under Ancistrocheiridae).

#### DISTRIBUTION

Species in this family are very broadly cosmopolitan, distributed worldwide in tropical to temperate waters.

### Key to genera of the Pyroteuthidae

1a. Hooks present on tentacular clubs and on arms 4; arm hooks biserial over entire length; club hooks in a single row; right ventral arm hectocotylised. Twelve photophores on ventral side of eyeball (none lidded), nine large (of different sizes) and three small; 10 photophores in mantle cavity include three in a row transverse to the longitudinal axis of body at the level of gills; six or seven, sometimes eight, photophores on tentacular stalk of young and adults. Right ventral arm hectocotylised, without toothed plate. Only left oviduct developed; right oviduct may be present but significantly reduced.................................. *Pyroteuthis*

1b. Hooks absent on tentacular club; no hooks, or not more than two hooks, on arms 4; hooks few, only in one or two series in middle of arms 1–3. Fourteen or 15 photophores on ventral side of eyeball (one lidded), 10 large and four or five small; eight photophores in mantle cavity, including only one at level of gills on longitudinal axis of body; four photophores on tentacular stalk. Left ventral arm hectocotylised; toothed plate present. Only right oviduct developed in females .............................................. *Pterygioteuthis*

## *Pyroteuthis* Hoyle, 1904

### COMMON NAME
Firefly Squid

### DISTINGUISHING CHARACTERS
Similar in appearance to *Pterygioteuthis* except adults are larger and have broader heads. See key to genera.

### DIAGNOSTIC FEATURES
Mantle conical, 'tail' sharply pointed. Fins rounded, not joined posteriorly. Arms with more than 13 hooks (arms 4 with > 16 hooks); hooks and suckers on arms 1–3, in ventral series only or biserial. Right arm 4 hectocotylised with hooks modified by development of secondary cusp; toothed plate absent. Tentacles with permanent constriction and bend near base (tentacle elbow). Club manus with one series of hooks and three series of suckers. Eight buccal supports with primary connectives attached to dorsal margins of arms 4; secondary connectives attach ventrally to arms 1 and 2. Both oviducts present, right may be reduced.

*Photophores.* Numerous photophores in body cavity on viscera, around eyes and along tentacle stalks. Eyeball without lidded photophores; six or seven separate photophores in tentacular stalk; three anterior abdominal and small brachial photophores.

*Colour.* Buccal membrane reddish to purplish in colour.

### HABITAT AND BIOLOGY
The quick, brilliant bioluminescent flashes observed in *Pterygioteuthis* have not been seen in *Pyroteuthis* spp. because they lack lidded photophores and the other photophores such as the branchial photophores are less specialised.

### SIZE
Mantle lengths up to 45 mm.

### REMARKS
The major differences between the species are found in the arrangement of tentacular photophores, form of the hectocotylus and shape of the hooks on the hectocotylus. Two species are found in Australian waters: *P. addolux* Young, 1972 and *P. margaritifera* (Rüppell, 1844).

### DISTRIBUTION
Most of the tropical and temperate regions of the world; absent from the tropical eastern Pacific.

### REFERENCES
Young and Mangold (2009), Roper and Jereb (2010t).

### Key to *Pyroteuthis* species

1a. Tentacular club with four large ellipsoid photophores (three are double). Distal-most photophore near base of carpal cluster. Two smaller spherical photophores, one at base of tentacular stalk and one at base of club and one smaller spherical to ovoid photophore near base of tentacle. Hectocotylus with 10 proximal hooks, each small with primary cusp bearing smooth inner edge and short secondary cusp; flap short, almost semi-circular; six or seven hooks opposite flap; 6–15 suckers at tip ............................... *P. addolux*

1b. Tentacular club with three large, ellipsoid photophores (all are double photophores). Distal-most large photophore positioned well proximal to the base of the carpal cluster. Three smaller, spherical photophores, two at base of tentacular stalk and one at base of club. Hectocotylus with 13–19 proximal hooks; each proximal hook large with primary cusp bearing smooth inner edge and large, rounded secondary cusp; fleshy, elongate pad distal to proximal hooks (unlike flap in *P. addolux*); three hooks opposite flap; 0–13 suckers at distal tip ...................... ............................................................... *P. margaritifera*

## *Pyroteuthis addolux* Young, 1972

*Pyroteuthis addolux* Young, R. E. (1972). The systematics and areal distribution of pelagic cephalopods from the seas off southern California. *Smithsonian Contributions to Zoology 97*, 1–159 [22, pls 3b, 4i–q, 4s].

Type data: holotype SBMNH 34962.

Type locality: eastern North Pacific Ocean, 31°39′N, 133°16′W.

Ventral view (R. Young).

### DISTINGUISHING CHARACTERS

Differs from other species in the genus in the arrangement of the photophores on the tentacles and the structure of the hectocotylus. It also has shorter arms relative to the ML than the other species.

### DIAGNOSTIC FEATURES

Hectocotylus with 10 proximal hooks, each small with primary cusp bearing smooth inner edge and short seconday cusp; flap short, almost semicircular; six or seven hooks opposite flap; 6–15 suckers at tip.

*Photophores.* Four large ellipsoid photophores (three are double). Distal-most photophore near base of carpal cluster. Two smaller, spherical photophores, one at base of tentacular stalk and one at base of club and one smaller spherical to ovoid photophore near base of tentacle. Paralarvae with a series of chromatophores on distal portion of each tentacle (however, these are easily lost during capture); small branchial photophores and larger anal photophores; large ocular photophores; few chromatophores on mantle.

### HABITAT AND BIOLOGY

Adults mesopelagic and probably ascending to the epipelagic zone at night. (Off Hawaii they are found at 450–500 m during the day and 150–200 m at night and show a clear pattern of vertical migration.) When disturbed, this species has been observed emitting a long-duration flash where tentacular and other photophores glow in a Y-shaped pattern display. To produce the pattern, the tentacles are held straight and rigid at a 45° angle to the body axis. If touched, the tentacles undulate and are not held in a rigid position. They also utilise the photophores for countershading.

*Interest to fisheries.* No interest in a commercial fishery exists at present.

### SIZE

The maximum ML is 50 mm.

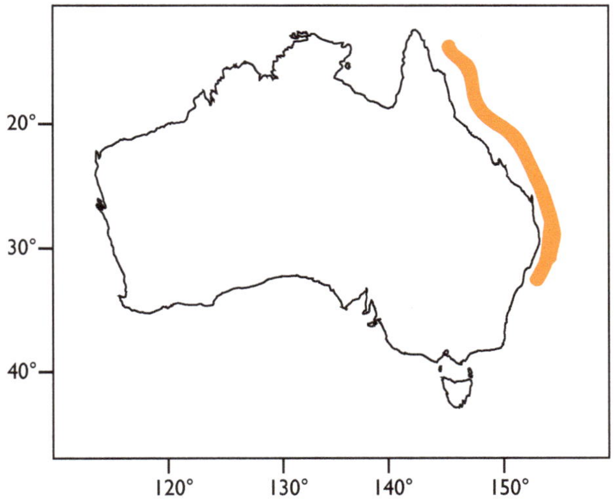


*Pyroteuthis addolux* distribution. (Australian records are based on few specimens, but the species is likely to occur more widely.)

### DISTRIBUTION

Worldwide in tropical and temperate regions. Absent from the tropical eastern Pacific.

### REFERENCES

Roper and Jereb (2010t), Young and Mangold (2012).

## *Pyroteuthis margaritifera* (Rüppell, 1844)

*Enoploteuthis margaritifera*, Rüppell, E. (1844). Intorno ad alcuni cefalopodi del mare di Messina: lettera del Dr. Eduardo Rüppell di Frankfort sul Meno al Prof. Anastasio Cocco. *Giornale del Gabinetto Letterario di Messina* *5*(27–28), 129–135 [130 [2], fig. 1].

Type data: syntypes (2) 1993012 BMNH; lectotype 26930.

Type locality: Sicily, Messina.

### DISTINGUISHING CHARACTERS

Most easily separated from other members of the genus by the arrangement of photophores on the tentacles and the structure of the hectocotylus.

### DIAGNOSTIC FEATURES

Hectocotylus (right arm 4) with 13–19 proximal hooks; each proximal hook large with primary cusp bearing smooth inner edge and large, rounded secondary cusp; fleshy, elongate pad (unlike flap in *Pyroteuthis addolux*) distal to proximal hooks; three hooks opposite flap; 0–13 suckers at distal tip.

*Photophores.* Tentacular club with three large, ellipsoid photophores (all are double photophores). Distalmost large photophore positioned well proximal to base of carpal cluster. Three smaller, spherical photophores: two at base of tentacular stalk and one at base of club. Spherical to ovoid superficial photophore absent.

### HABITAT AND BIOLOGY

*Pyroteuthis margaritifera* undergoes diel vertical migration. Spawning is restricted to a short period in summer in New Zealand waters and the largest mature females were larger than the males.

A study by Brandt (1983) showed this species to be associated with a warm core eddy of the East Australian Current during summer night-trawls. This species was found to be more common at 250 m than at shallower depths.

### SIZE

Mantle length up to 50 mm.

### REMARKS

The systematic status of this taxon over its wide range is unknown and differences have been noted in the number of tentacle photophores over its range. Those from southern and south-western Australian waters have six photophores, including the smaller number two photophore. A review of this genus is needed.

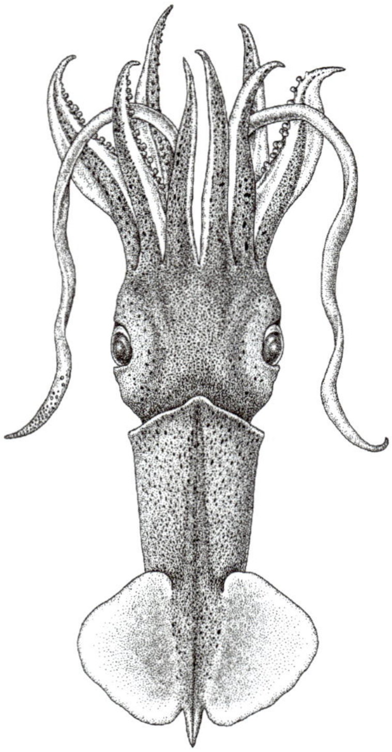

Dorsal view (ABRS, K. Hollis).

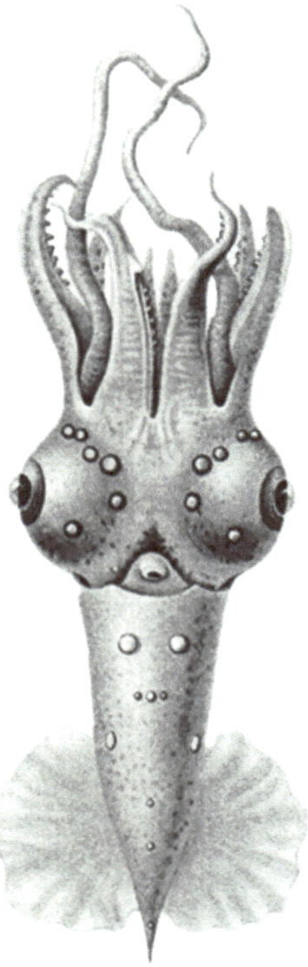

Ventral view (ABRS, K. Hollis).

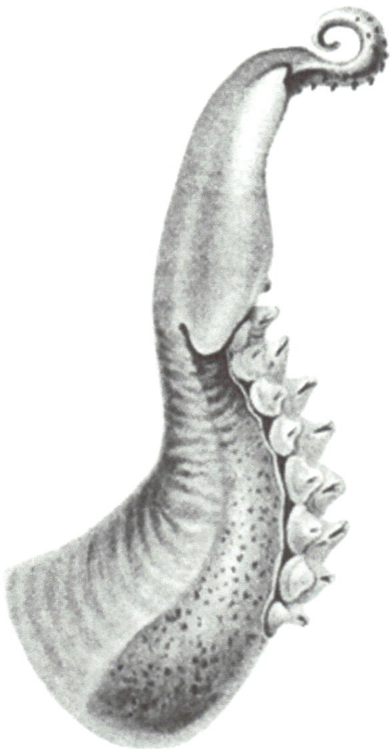

Hectocotylus, lateral view (Plate XI, Chun 1910).

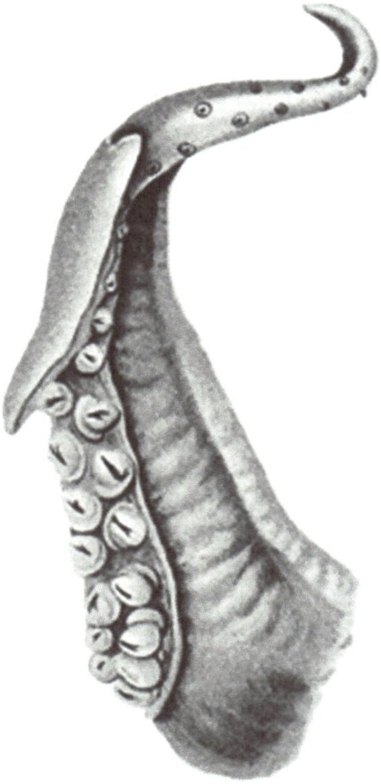

Hectocotylus, oral view (Plate XI, Chun 1910).

Left tentacular club (Plate XI, Chun 1910).

Eye with photophores (Plate XIV, Chun 1910).

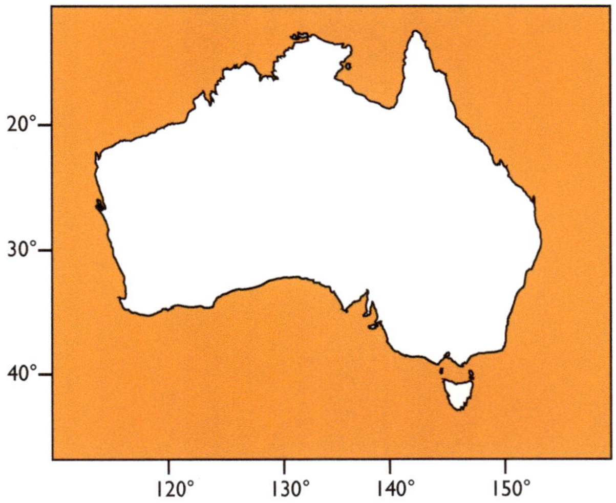


*Pyroteuthis margaritifera* distribution.

#### DISTRIBUTION

Widely distributed throughout the tropical and temperate Atlantic, Indian and South Pacific Oceans. Absent from the easten Pacific.

#### REFERENCES

Young and Mangold (1996), Roper and Jereb (2010t).

### *Pterygioteuthis* Fischer, 1896

#### COMMON NAME

Enope Squid

#### DISTINGUISHING CHARACTERS

Pair of photophores at the base of the gills are much larger than those seen in *Pyroteuthis* and useful for distinguishing paralarvae.

#### DIAGNOSTIC FEATURES

Arms with fewer than eight hooks in one or two transverse series in middle portions of arms 1–3. Males with maximum of two hooks on arms 4 in *P. giardi*. Female arms 4 generally lack hooks. Male left arm 4 hectocotylised, with toothed plate, and large swollen regions containing folded, glandular tubes. Tentacular club suckers tetraserial; hooks absent. Only right oviduct present.

*Photophores.* Eyeball with 14 or 15 ventral lidded photophores; 10 large, four or five small; eyeball photophore number six is lidded. Four photophores embedded in tentacular stalk. A single anterior abdominal (intestinal) photophore; pair of photophores at base of gills, and pair of anal photophores.

#### HABITAT AND BIOLOGY

The lidded eye photophore can produce a brilliant flash. Closing the eyelid presumably conceals the reflector that lies behind the photogenic tissue. This would provide important concealment from predators that use luminescent search lights.

#### SIZE

Species of *Pterygioteuthis* are the smallest of all oceanic squids.

#### REMARKS

*Pterygioteuthis* are among the most abundant and widespread mesopelagic cephalopods in temperate and tropical oceans.

A recent study by Lindgren (2010a) using morphological and molecular characters examined the systematics of this genus in the eastern tropical Pacific Ocean. Two distinct clades were recognised: a *P. giardi* group and a *P. gemmata* group. She concluded that, 'the distribution of *Pterygioteuthis* in the Indo-West Pacific remains unclear due to a lack of available specimens' (Lindgren 2010a: 397).

Three species are currently recognised: *Pterygioteuthis giardi* Fischer, 1895, *P. gemmata* Chun, 1908 and *P. microlampas* Berry, 1913. *Pterygioteuthis giardi* and *P. gemmata* occur in Australian waters. *Pteygioteuthis microlampas* is known from New Zealand and Hawaii.

Each species is very broadly distributed. It is likely that more species will be discovered following detailed review.

#### REFERENCES

Lindgren (2010a), Roper and Jereb (2010t), Lindgren *et al.* (2011).

### Key to *Pterygioteuthis* species

1a. Hooks on arms 1–3 uniserial. Distal suckers on arms 2 and 3. Tentacular club suckers enlarged. Hectocotylus plate with many small teeth. Female arms 4 with biserial suckers. Four small eye photophores. Buccal membrane with papilla.............................. *P. gemmata*

1b. Hooks on arms 1–3 biserial. No distal suckers on arms 2 and 3. Tentacular club suckers not enlarged. Hectocotylus plate with two large teeth. Female arms 4 without suckers. Five small eye photophores. Buccal membrane with ridges........................*P. giardi*

## *Pterygioteuthis gemmata* Chun, 1908

*Pterygioteuthis gemmata* Chun, C. (1908). Über Cephalopoden der Deutschen Tiefsee-Expedition. *Zoologischer Anzeiger 33*(2), 86–89 [87].

Type data: lectotype ZMB Moll.-101929a.

Type locality: Atlantic Ocean, 5°51′N, 21°40′W.

#### DIAGNOSTIC FEATURES

Armature biserial: arms 1–3 with a single series of hooks proximo-ventrally, suckers on distal arm tips; dorsal series with suckers along entire length. Arm 1 distal suckers normal, not globular; hooks gradually decrease in size distally. Arms 3: five to seven hooks in males and females. Female arms 4 with biserial suckers. Left arm 4 of males hectocotylised; plate with many small teeth. Tentacular club slightly expanded; club suckers enlarged. Buccal membrane with papillae.

*Photophores.* Eye with 10 large and four minute photophores. Tentacles with four photophores.

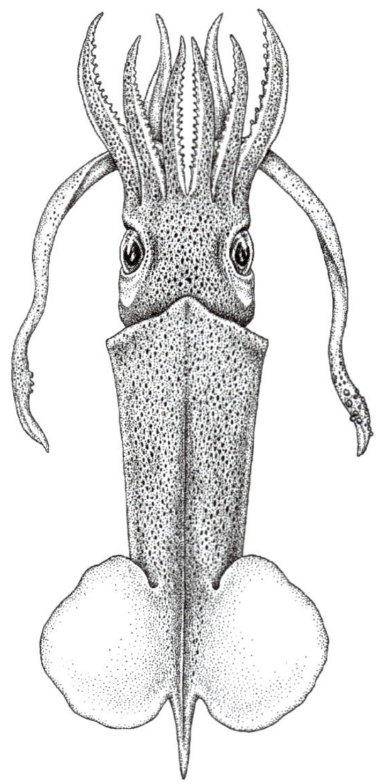

Dorsal view (ABRS, K. Hollis).

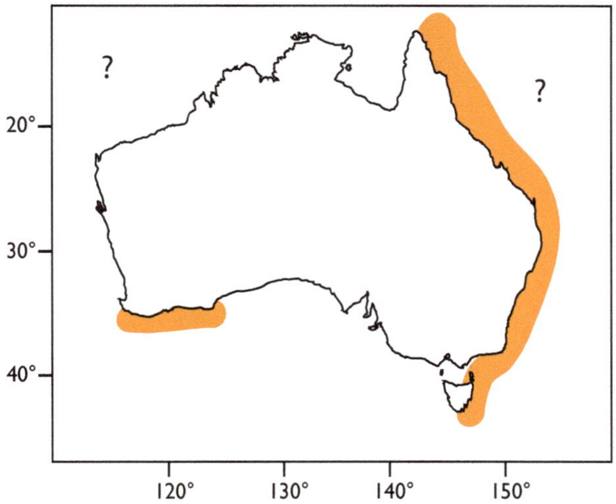


*Pterygioteuthis gemmata* known distribution. (Australian records are based on few specimens, but the species is likely to occur more widely.)

*Colour.* Chromatophores concentrated at elbow, along stalk and at base of tentacle club, extend across entire aboral surface.

### HABITAT AND BIOLOGY

Adults are mesopelagic, found at 300–600 m during the day off southern California, and then they ascend into the upper 200 m of the epipelagic zone during nightly vertical migrations. In the Canary Basin the species occurs at 50–200 m at night and descends to 400–800 m during the day. Little is known about the biology and behaviour of this species; however, spawning is restricted to a short period in summer in New Zealand waters and the largest mature females are larger than the males.

*Pterygioteuthis gemmata* was shown by Brandt (1983) to be associated with a warm core eddy of the East Australian Current. The length-frequency distributions of this species were larger outside than inside the eddy. This may reflect the effect of eddies on growth, spawning times, reproduction and survival.

*Interest to fisheries.* No interest exists at present.

### SIZE

The maximum ML reaches 40 mm.

### DISTRIBUTION

*Pterygioteuthis gemmata* is principally an anti-equatorial species in the Atlantic, Pacific and Indian Oceans.

### REFERENCE

Roper and Jereb (2010t).

## *Pterygioteuthis giardi* Fischer, 1895

*Pterygioteuthis giardi* Fischer, H. (1896). Note preliminaire sur le *Pterygioteuthis Giardi*, cephalopode nouveau recueilli dans le cours de l'Expedition scientifique du Talisman (1883). *Journal de Conchyliologie 43*(4), 205–211 [211, pl 9 figs 1–7].

Type data: holotype MNHN 3-7-727.

Type locality: off Morocco.

### COMMON NAME

Roundear Enope Squid

### DIAGNOSTIC FEATURES

Mantle small, acutely tapered to short, sharp 'tail'. Fins rounded with large anterior and posterior lobes; do not meet posteriorly. Arms short. Arms 1: proximal four to six hooks distinctly larger than distal hooks; distal suckers globular. Hooks biserial on arms 1–3; fewer than eight per arm, only in middle section; tips of arms 2 or 3 lack suckers and hooks; adult males with few hooks (usually no more than two) and no suckers; one or two hooks on right arm 4 in adult males only; hooks and suckers absent from arms 4 in females. Arms 3: four to six suckers proximal to two to four hooks in males. Left arm 4 hectocotylised; hectocotylus plate with two large teeth. Tentacles long, thin; without hooks; four series of suckers on manus; not enlarged. Buccal membrane with ridges. Oviduct developed on right side only; left oviduct absent.

*Photophores.* Eight visceral photophores. Ventroposterior surface of each eye studded with ten large and five small iridescent photophores; photophore six lidded. Tentacles with two spherical photophores.

*Colour.* Tentacular stalks lack chromatophores on distal half; club without chromatophores, not expanded.

### HABITAT AND BIOLOGY

This is a predominantly mesopelagic oceanic species with a depth distribution that ranges from just below the surface to ~500 m. It is known to undertake diel vertical migrations; ascending into the epipelagic zone

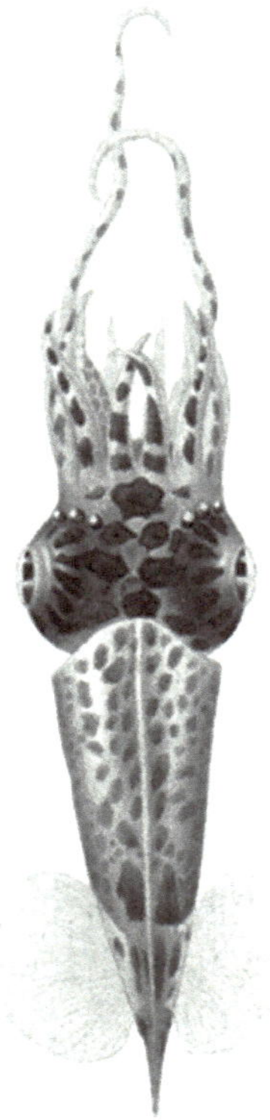

Dorsal view (Plate XII, Chun 1910).

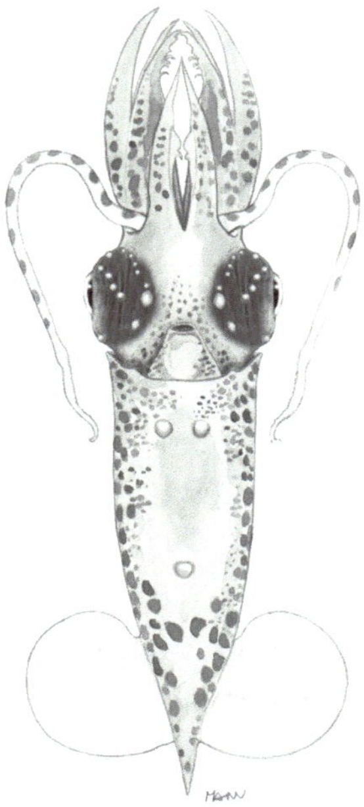

Ventral view (FAO, E. D'Antoni).

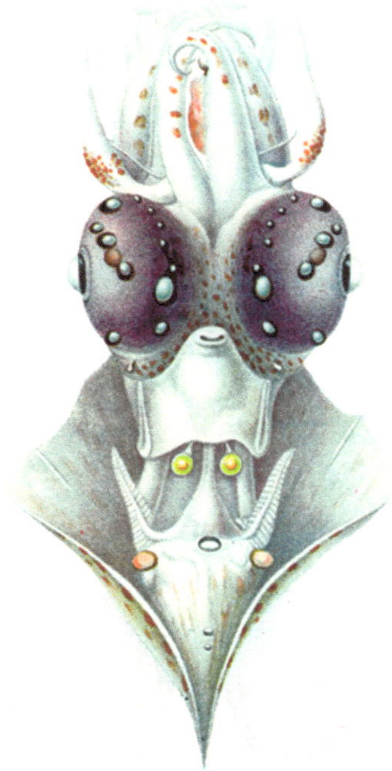

Ventral view, with dissected mantle cavity (Plate XIV, Chun 1910).

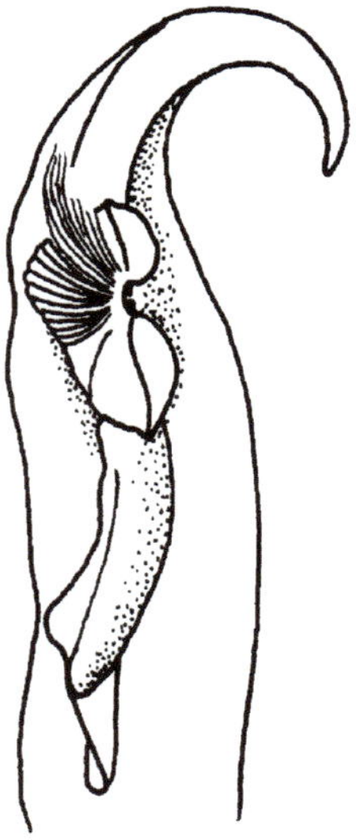

Hectocotylus, oral view (FAO, E. D'Antoni).

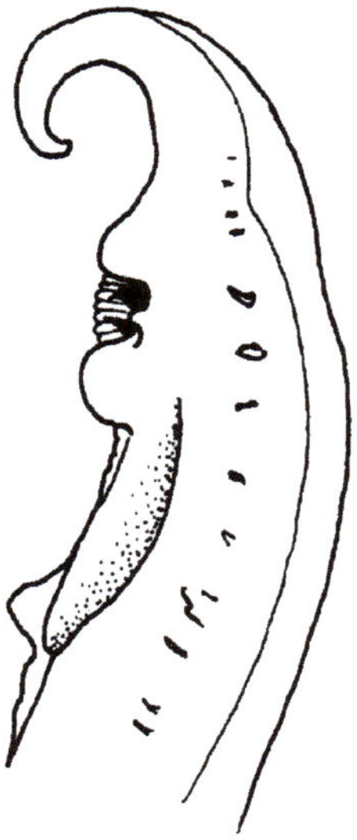

Hectocotylus, lateral view (FAO, E. D'Antoni).

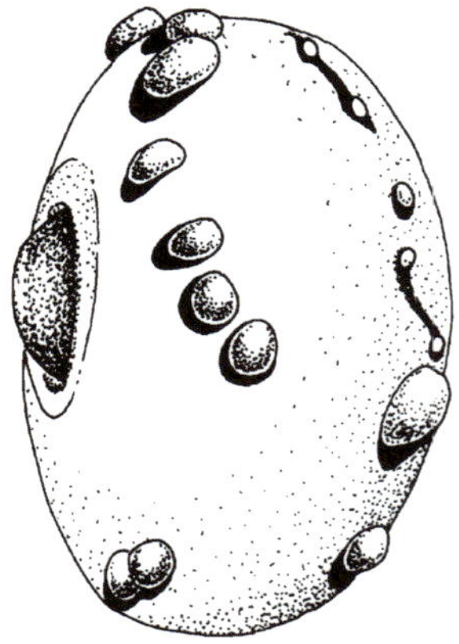

Eye with photophores (Plate XIV, Chun 1910).

at night. It was among the most abundant species caught during summer night-trawls within a warm core eddy of the East Australian Current in the Tasman Sea (Brandt 1983). It is not known whether this species occurs commonly in eddy source water of the Coral Sea.

*Interest to fisheries.* None at present.

## SIZE

The maximum ML attained is 40 mm.

## REMARKS

Lindgren (2010a) found no variation in the morphological traits she examined among 15 specimens from the North Atlantic and the North Pacific.

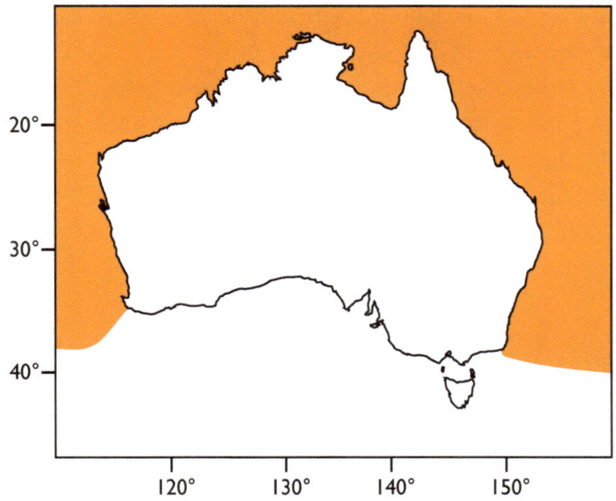


*Pterygioteuthis giardi* known distribution.

Riddell (1985) noted that *Pterygioteuthis giardi* occurs off New Zealand, although he reported the presence of four tentacle photophores on his specimens indicating they may be *P. hoylei* or an undescribed taxon (Lindgren 2010a). Nesis (1982/1987) distinguishes two subspecies: *P. giardi giardi* (Atlantic) and *P. g. hoylei* (Indo-Pacific) based on differences in the size of the hooks on the right arm 4 of males. These forms were first given this taxonomic status by Pfeffer (1912) based on descriptions by Hoyle (1904) of the latter form from the eastern equatorial Pacific. Further study of this taxon is needed to resolve its taxonomy.

#### DISTRIBUTION

This species/subspecies-complex is distributed worldwide as a cosmopolitan tropical and subtropical form in oceanic waters.

#### REFERENCE

Roper and Jereb (2010t).

## Thysanoteuthidae Keferstein, 1866

Thysanoteuthidae is monotypic, represented by *Thysanoteuthis rhombus*.

#### DISTINGUISHING CHARACTERS

The most characteristic traits are the long, broad, rhomboidal fins that occupy the entire length of mantle, the well-developed ⊣-shaped mantle–funnel locking cartilages, the presence of the anterior anal photophore in immature squid and the unique shape and structure of the gladius.

Octopoteuthidae (*Octopoteuthis*, *Taningia*) and Ancistrocheiridae (*Ancistrocheirus lesueuri* (d'Orbigny, 1842)) also have long rhomboid fins, but biserial hooks instead of suckers on the arms and tentacular clubs; in addition, all have obvious photophores on the ventral mantle, arm tips, or internally. Unlike *Thysanoteuthis* these taxa are rarely encountered near the surface.

The loliginid, *Sepioteuthis lessoniana* Lesson, 1830, has fins extending the length of the mantle but these are oval-shaped; in addition the mantle-locking apparatus is simple and straight. Also, *S. lessoniana* is found only in continental shelf waters.

## *Thysanoteuthis* Troschel, 1857

As for family.

## *Thysanoteuthis rhombus* Troschel, 1857

*Thysanoteuthis rhombus* Troschel, F. H. (1857). Bermerkungen über die Cephalopoden von Messina. *Archiv für Naturgeschichte, Berlin 23*(1), 40–76 [70, pl. 4 fig 12, pl. 5 figs 1–4].

Type data: type repository unresolved.

Type locality: Mediterranean Sea, Sicily, Messina.

#### COMMON NAME(S)

Diamondback Squid, Rhomboid Squid

#### DIAGNOSTIC FEATURES

Mantle thick, muscular (in adults up to 60–85% body weight), tapering to blunt, rounded tip posteriorly. Head width similar to mantle width. Funnel with prominent funnel groove. Fins rhomboidal, extending entire length of mantle in adults; not fused together along dorsal midline; three-quarters of fin length. Funnel-locking cartilage subtriangular with rounded angles, ⊣-shaped, with long, narrow longitudinal groove and short, broad transverse groove; knobs at junction of grooves; mantle component with corresponding ridges. Nuchal-locking apparatus well developed, with two large, swollen, hook-like knobs on mantle component that interlock with two conforming depressions and knobs on the nuchal component. Arms relatively short, strongly muscled. Arms with no sexual dimorphism in shape or external morphological traits in adults; differences occur in length of arms 3 in immature squids of 120–450 mm ML (arms 3 length in males 75–110% ML, in females ~50% ML). Arm formula 3 > 2 > 4 > 1 (or 4 = 1). Hooks absent. Arm suckers biserial with sharp teeth; sucker rings with 20–26 sharp conical teeth. Well-developed, wide, protective membranes expanded on long cirri-like trabeculae; especially well developed on arms 3, less so on arms 2; arms 1–3 with distinct aboral keels. Left ventral arm hectocotylised in males; in mature males hectocotylised distal region occupies sucker rows 13–20 and 30–35% of arm length; modified part of arm with tiny untoothed suckers with smooth horny rings and two oval glandular enlargements. Tentacles relatively short, strongly muscled; tentacular clubs widened, relatively long; manus with tetraserial suckers that extend to tip of dactylus; suckers on manus of clubs with 15–29 fine, strongly pointed, conical teeth; carpal-locking apparatus extends proximally along stalk as two series of alternating knobs and suckers. Buccal connectives attached to ventral border of ventral arms 4; formula DDVV. Male reproductive system with very long, narrow spermatophore storage sac (40–60% ML) and long muscular penis; females with hypertrophied oviducal glands (40–52% ML) with very large lancet-like second section (30–40% ML) and relatively small oviducts, with distribution of the main ovarian afferent and efferent blood vessels in mesentery protruding dorsally into the ovary.

*Photophores.* One well-developed anal photophore on ink sac near anal papilla in young squid from 60 to 350 mm ML; photophore gradually reduced, non-functional in adults.

### HABITAT AND BIOLOGY

*Thysanoteuthis rhombus* is an oceanic epipelagic to mesopelagic species that inhabits open oceanic waters in excess of 400 m bottom depth where the surface water temperature is generally > 20°C. They are mainly caught in the upper 50 m of the water column. Juveniles are capable of leaping out of the water but do not 'fly' (glide) in the same manner as ommastrephids and onychoteuthids. Adults are slow swimmers and are often observed in monogamous pairs, although groups of up to 20 have also been observed.

Paralarvae and juveniles live in the epipelagic zone. Subadults and adults (300–850 mm ML) make daily vertical migrations in the upper 600–800 m. At night most of the population migrates towards the surface (0–50 m), and then in daytime descends to depths from 400–800 m, mainly 450–650 m. *Thysanoteuthis rhombus* is a relatively passive migrant that undertakes long seasonal migrations in the warm waters of peripheral oceanic currents. During most of its life cycle it does not swim actively. Rather, it propels itself slowly by gentle undulation of its long, broad, diamond-shaped fins. However, when it encounters danger, *T. rhombus* is capable of a powerful reactive jet of short duration. This low-energy lifestyle enables the rhomboid squid to maintain very high growth rates, among the highest of all squids. By the age of 300 days the squid reaches 800 mm ML and 17.5 kg body weight. Males reach maturity mainly at 400–550 mm ML (age 170–200 days), and females mature at 550–650 mm ML (age 230–250 days). The lifespan is 1 year.

The spermatophore storage sac stores 10–40 spermatophores 60–100 mm long. The potential fecundity of females is 2.5–5 million oocytes, which show asynchronous development. The size of ripe eggs is 1.6–1.8 mm. Egg numbers in the oviducts range up to 140 000. Mating occurs in the 'head to head' position. In females, eight to 30 seminal receptacles are located on the ventral part of the buccal membrane. They are thought to be multiple spawners with spawning lasting throughout the year in tropical waters and during warm seasons in peripheral regions. Spawning is intermittent, with multiple fillings and evacuations of the oviducts. Spawned egg masses contain 32 000–75 000 embryos. The pelagic egg mass is a large cylinder (0.6–2.0 m long, 0.11–0.30 m diameter) of transparent, gelatinous mucus with pink-violet embryos that are distributed in a double spiral in the outer layer. These masses can be found floating in surface waters.

Hatching size is 1.1–1.6 mm ML and development includes a paralarval stage. No pronounced geographical variability occurs in the main morphological features and ecological-population parameters (e.g. size and age at maturity). The same phenomenon is observed for genetic variability: mitochondrial DNA sequences differ very little between conspecific individuals from the western and eastern Pacific Ocean. The population density is low, mainly from 0.01 to 100 specimens per square kilometre. The total minimum estimated biomass within the vast geographical range of the species is only ~1.5–2.5 million tonnes. The greatest abundance occurs in the productive zones of the open ocean and in coastal areas at the periphery of the range, mainly in the south-western part of the Sea of Japan. The social organisation is unique among squids, consisting of pairs of same-sized males and females that occur at an immature stage of < 100 mm ML, and they are probably monogamous, remaining together constantly during their

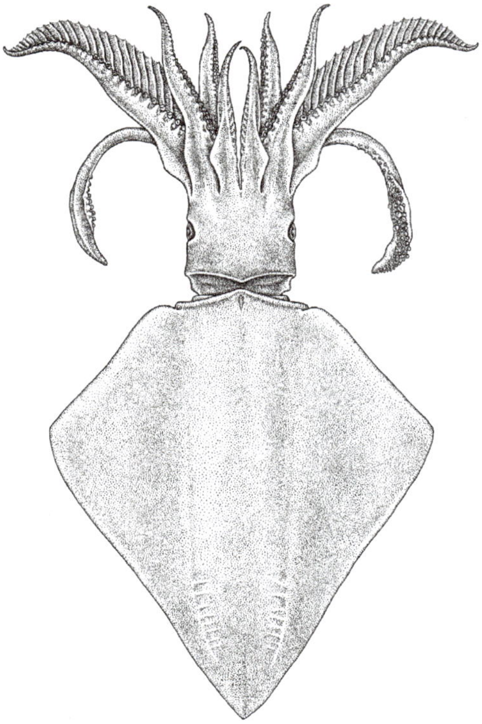

Dorsal view (ABRS, K. Hollis).

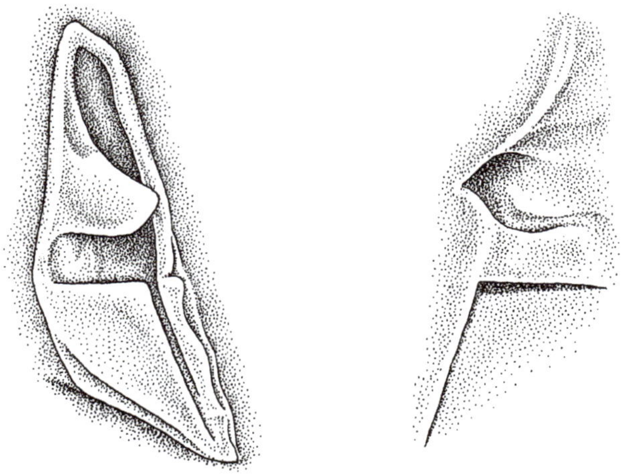

Funnel- (left) and mantle-locking (right) cartilages (ABRS, K. Hollis).

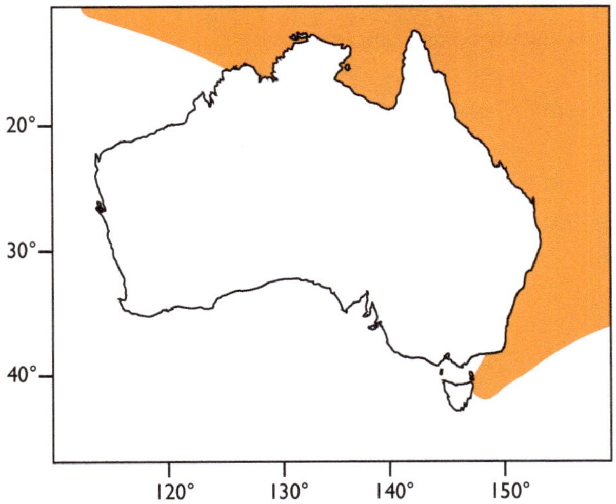


*Thysanoteuthis rhombus* known distribution.

*Thysanoteuthis rhombus* egg mass (Lyle Vail).

lifetimes. Small schools of pairs may be locally common. The reproductive and behavioural strategies are adaptations to their low population density and inactive lifestyle.

*Interest to fisheries.* This species is not a target of fisheries in Australian waters but has been collected using drifting, sometimes baited, jigs and set nets in other parts of its range. It is renowned for its tender flesh that is sometimes eaten as sashimi and forms the basis of a very highly profitable fishery in Japan.

## SIZE

These are large nektonic squids; adult size 100 cm or even 130 cm ML. The maximum body weight known is 24 kg to 30 kg, and some are probably heavier.

#### DISTRIBUTION

Occurs circumglobally in tropical and warm subtropical oceanic waters.

#### REFERENCE

Roper and Jereb (2010u).

### *Order Sepiida*

This clade includes the cuttlefish genera *Metasepia*, *Sepiella* and the currently 'catch-all' genus *Sepia*. The latter genus is currently under review. Some researchers group the Sepiidae, Sepiolidae and the Sepiadariidae into a superfamily (the Sepioidea) but this grouping does not appear to be supported by molecular data. Many generic names exist within Sepiida, but until our understanding of relationships improves it is difficult to apply them (Allcock *et al.* 2014). While morphological data are available, they have yet to be subjected to rigorous phylogenetic analysis. Several monophyletic groups have been recovered in molecular studies (Bonnaud *et al.* 2006, Yoshida *et al.* 2010b, Lindgren *et al.* 2012) and there is some congruence between trees. The subgenera *Acanthosepion* Rochebrune, 1884 and *Doratosepion* Rochebrune, 1884 are well supported. However, to date, only a few of the ~100 nominal species have been included in these analyses and the taxa for inclusion have been selected opportunistically. Several different genes have been sequenced in a range of studies that include various taxa, making it difficult to build on information already available in GenBank. Examination of a large number of species and a consistent set of genes will help resolve these relationships.

Most molecular work on the sepiids to date has focused on populations of particular, usually commercially fished, species often using microsatellite markers to identify stocks for management purposes.

## Sepiidae Keferstein, 1866

#### COMMON NAME

Cuttlefishes

#### DIAGNOSTIC FEATURES

Mantle robust, slightly flattened dorsoventrally, may be broad or slender, oval, oblong or nearly circular in outline; anterior dorsal margin projected anteriorly, not fused with head. Fins narrow, located dorsolaterally on mantle, approximately equal to ML; posterior fin lobes free, not connected to each other. Head slightly narrower than mantle; eyes prominent, covered by transparent membrane and conspicuous secondary fold on eyelid. Mouth surrounded by 10 appendages (eight arms and two tentacles). Arms with two to four suckers in each transverse row. Males of most species with hectocotylised (usually left) ventral arm(s) 4 for holding and transferring spermatophores; when present, hectocotylus usually consists of region of reduced suckers and may be swollen and crenulated by transverse folds. Tentacular clubs with four or more suckers in each transverse row; tentacles retractile into pockets on the ventrolateral sides of the head between arms 3 and 4. Arm and club suckers with chitinous rings. Funnel–mantle locking cartilage curved or angular. One pair of gills with no branchial canal between afferent and efferent blood vessels. Digestive gland bilobed. Buccal membrane present, with or without suckers; each radula tooth unicuspid. Olfactory organ a ciliated pit.

*Cuttlebone.* Internal calcareous cuttlebone located dorsally in mantle underneath skin; length equal to ML except in *Metasepia*; shape ranges from lanceolate to oval or diamond-shaped. Dorsal side a calcareous plate, ventral side finely laminate, porous, and comprised of thin septa separated by transverse calcareous rods.

#### HABITAT AND BIOLOGY

Cuttlefishes inhabit the continental shelf and upper slope to a maximum depth of ~1000 m. They are primarily bottom-dwellers over a range of habitats, including rocky, sandy, and muddy substrates, seagrass, seaweed and coral reefs. They are slower swimmers than the more streamlined squids. Cuttlefishes are able to attain neutral buoyancy by regulating the relative amounts of gas and fluid in the chambers of the cuttlebone, and they are able to hover in midwater, with fins acting as stabilisers. The cuttlebone length, width, septal spacing and structural morphology are correlated with maximum habitat depth: the deepest living species all share the characteristics of cuttlebones with closely packed septa and modified sutures. All large species such as *Sepia apama* Gray, 1849, *S. latimanus* Quoy & Gaimard, 1832, and *S. 'pharaonis'* Ehrenberg, 1831 are restricted to much shallower depths (usually < 200 m) and show very different septal spacing and sutures than the deeper water species. Some species migrate seasonally in response to temperature changes and aggregate, usually in shallow water, at spawning time.

Within a species, individuals may attain sexual maturity at very different sizes, depending upon the combined effects of several factors including temperature, light and diet. Males use their hectocotylus to transfer sperm packets to females. When they are ready to spawn, the eggs are fertilised externally and the eggs (relatively few in number) are individually attached to various substrates in clusters; the time required for development depends upon temperature. The young are well developed when they hatch and immediately adopt an adult lifestyle. The lifespan is between 18 and 24 months, although males of some species may live longer. Post-spawning mortality is high in females. Cuttlefishes feed on a wide range of invertebrates and bony fishes.

Apart from *Sepia apama* in the Upper Spencer Gulf of south Australia, there have been no fisheries stock structure studies of Australian cuttlefishes. Catches are often reported as mixed species.

#### SIZE

Up to 500 mm ML, and 12 kg in weight.

#### REMARKS

In the descriptions that follow, depth ranges and distributions are based on whole animals, not cuttlebones. Cuttlebones may drift well outside the true distributional range of a species, so referring to ranges that include beachwashed cuttlebones is misleading (Reid, 2016). When accessing electronic databases, or literature pertaining to distributions, it is important to determine whether the data refer to whole animals or beachwashed cuttlebones (as is true for *Nautilus* and *Spirula*).

To assist with identification, species tentatively recognised as belonging to doratosepion are treated separately in the keys to *Sepia* below.

### Key to genera of the Sepiidae (see Fig. 14)

1a. Cuttlebone diamond-shaped in outline; cuttlebone much shorter than mantle, located in the anterior 50~60% of mantle; dorsal anterior edge of mantle without tongue-like projection .................................... ................................................*Metasepia* (*M. pfefferi*)

1b. Cuttlebone outline elliptical to lanceolate; cuttlebone length approximately equal to ML; dorsal anterior edge of mantle usually with tongue-like projection ....................................................................2

2a. Gland and gland pore located on the ventral side of the posterior end of the mantle. Mantle-locking apparatus with triangular projection. Cuttlebone inner cone with very short limbs; outer cone a wide, spatulate, chitinised border around posterior end of cuttlebone .......................................... *Sepiella*

2b. Gland and gland pore absent. Mantle-locking apparatus semicircular, without triangular projection. Cuttlebone inner cone with relatively long limbs; outer cone usually calcareous, not obviously spatulate posteriorly......................................... *Sepia*

## *Sepia* Linnaeus, 1758

#### COMMON NAME

Cuttlefish

#### DISTINGUISHING CHARACTERS

Dorsal anterior edge of mantle usually with tongue-like projection. Ventroposterior gland and gland pore absent. Mantle-locking apparatus semicircular, without triangular projection. Cuttlebone inner cone with relatively long limbs; outer cone usually calcareous, not obviously spatulate posteriorly; outline elliptical to lanceolate; cuttlebone length approximately equal to ML.

### Key to *Sepia* species

(*Sepia baxteri* (Iredale, 1940) and *S. reesi* Adam, 1979 are not included in the key as they have been described only from cuttlebones.)

1a. Elongated, narrow body. Cuttlebone long, narrow....2

1b. Body broad. Cuttlebone broad, not narrow ...........3

2a. Some biserial suckers on arms; arms often modified in one or both sexes, sometimes elongate, thread-like or differ in sucker arrangement along length. Posterior end of cuttlebone with two 'wings' at posterior extremity and slightly recurved, forming a 'hood' 'doratosepion-group' ....................................... (see Keys to Australian doratosepion species below)

2b. Arm suckers tetraserial; arms not unusually modified. Posterior end of cuttlebone normal, without 'wings', outer cone long, broad........*S. mira*

3a. Tentacular club suckers differ markedly in size....4

3b. Tentacular club suckers may differ, but not markedly, in size........................................................5

4a. Tentacular club with five or six suckers in each transverse row; protective membranes fused at base of club. Cuttlebone spine without ventral keel............................................................................ *S. chirotrema*

4b. Tentacular club with three or four suckers in each transverse row; protective membranes not fused at base of club. Cuttlebone spine with ventral keel............................................................................ *S. opipara*

5a. Tentacular club suckers all approximately similar sized............................................................................ 6

5b. Tentacular club suckers differ in size........................ 8

6a. Tentacular club with 13 or more tiny suckers in each transverse row; sucker-bearing surface convex........7

6b. Tentacular club with < 13 suckers in each transverse row; sucker-bearing surface flat .................. 14

7a. Cuttlebone inner cone rounded, shiny and thickened; bases of fins with whitish bands; males without longitudinal whitish bars .............. *S. smithi*

7b. Cuttlebone inner cone thin, narrow and not shiny; bases of fins without whitish bands; males with longitudinal whitish bars..................... *S. whitleyana*

8a. Tentacular club with eight suckers in each transverse row..................................................................... 9

8b. Tentacular club with four to six suckers in each transverse row.......................................................... 10

9a. Cuttlebone inner cone limbs broaden posteriorly with distinctive, thick, bulbous swelling............................................................................ *S. 'pharaonis'*

9b. Cuttlebone inner cone limbs narrow anteriorly, broaden posteriorly, narrow, not bulbous.....*S. irvingi*

10a. Cuttlebone spine greatly reduced or absent in adults ........................................................................ 11

10b. Cuttlebone spine present........................................ 12

11a. Head with three flap-like papillae posterior to eyes; dorsal cuttlebone without reticulate sculpture............................................................................ *S. apama*

11b. Head without three flap-like papillae posterior to eyes; dorsal cuttlebone with reticulate sculpture ............................................................................ *S. bandensis*

12a. Cuttlebone inner cone narrow, uniform width throughout length ................................................... 13

12b. Cuttlebone inner cone narrow anteriorly, broadens posteriorly................................................. *S. papuensis*

13a. Males without hectocotylus; cuttlebone spine without keel in both sexes ..................... *S. latimanus*

13b. Males with hectocotylus; cuttlebone spine with keel in both sexes........................................ *S. plangon*

14a. Cuttlebone without spine; cuttlebone dorsal surface rough with irregular calcified projections ..... *S. senta*

14b. Cuttlebone with spine; cuttlebone dorsal surface without calcified projections ................................. 15

15a. Outer margin of cuttlebone inner cone raised to form a flat ledge ....................................................... 16

15b. Cuttlebone inner cone not raised to form a flat ledge.......................................................................... 17

16a. Tentacular club with 10–12 suckers in each transverse row; dorsal and ventral protective membranes the same length ........................................... *S. elliptica*

16b. Tentacular club with five to seven suckers in each transverse row; dorsal and ventral protective membranes differ in length ................................ *S. sulcata*

17a. Tentacular club with four to six suckers in each transverse row .......................................................... 18

17b. Tentacular club with 8–12 suckers in each transverse row .................................................................. 19

18a. Tentacular club dorsal and ventral protective membranes fused at base of club............................20

18b. Tentacular club dorsal and ventral protective membranes not fused at base of club.....................22

19a. Cuttlebone broad, wedge-shaped, flat dorsally and laterally............................................................ *S. plana*

19b. Cuttlebone oval, not flat dorsally and laterally..... 21

20a. Cuttlebone broad, oval; sulcus very indistinct or absent; inner cone whitish .......................... *S. mestus*

20b. Cuttlebone oval, not broad; sulcus deep, wide, very distinct; inner cone pink............................ *S. rozella*

21a. Cuttlebone acute, pointed posteriorly; cuttlebone inner cone narrow throughout; sulcus shallow, narrow ......................................................... *S. hedleyi*

21b. Cuttlebone bluntly rounded posteriorly; cuttlebone inner cone narrow anteriorly, broadens posteriorly; sulcus deep, wide ......................... *S. novaehollandiae*

22a. Hectocotylus absent in males. Cuttlebone rounded anteriorly and posteriorly; cuttlebone inner cone narrow anteriorly, broadens posteriorly, raised to form a thick, rounded ledge..................... *S. grahami*

22b. Hectocotylus present in males. Cuttlebone inverted V-shape anteriorly, acute posteriorly; cuttlebone inner cone limbs narrow throughout, not raised to form a rounded ledge................................ *S. cultrata*

### Key to doratosepion species (males only)

1a. Arms 3 elongate, longer than remaining arms; or arms 3 modified with protective membrane thickened forming transverse ridges distally..................2

1b. Arms 3 not elongate, not longer than remaining arms; arms 3 protective membranes not modified .............3

2a. Arms 3 elongate, longer than remaining arms. Arms 1 with purplish band on each side of dorsal midline, arms 2–4 with median longitudinal purplish band ................................................ *S. braggi*

2b. Arms 3 not elongate, not longer than remaining arms. Arms 1–4 without distinctive markings........ *S. cottoni*

3a. Club with four to six suckers in each transverse row, suckers differ in size, some suckers towards posterior end of club enlarged. Cuttlebone inner cone does not form a distinctly raised, flattened ledge posteriorly ................................................................... 4

3b. Club with eight suckers in each transverse row, sucker all similar sized, small. Cuttlebone inner cone forms a distinctly raised, flattened ledge posteriorly ........................................................... *S. koilados*

4a. Arms 1 not elongate, not longer than remaining arms, distal tips not expanded and leaf-shaped. Arms without distinctive purplish pigmentation ....5

4b. Arms 1 elongate, longer than remaining arms, distal tips expanded, leaf-shaped. All arms with purplish bands and blotches............ *S. tanybracheia*

5a. Cuttlebone anterior striae inverted U-shape or straight; inner cone narrow posteriorly.................. 6

5b. Cuttlebone anterior striae M-shaped; inner cone broad posteriorly ........................................... *S. rhoda*

6a. Suckers present on dorsal and ventral side of modified portion of hectocotylised arm 4. Without orange-pink lustrous band dorsally along base of fins ........... 7

6b. Suckers absent from ventral side of modified portion of hectocotylised arm 4. Narrow orange-pink lustrous band dorsally along base of fins.......*S. vercoi*

7a. Club suckers relatively similar in size, suckers towards posterior end of club not greatly enlarged. Arm suckers biserial proximally, tetraserial distally ...................................................................... *S. bidhaia*

7b. Club suckers differ in size, three or four enlarged suckers towards posterior end of club. Arm suckers tetraserial........................................................ *S. limata*

### Key to doratosepion species (females only)

Female *Sepia tanybracheia* Reid, 2000 are unknown. The few female *S. cottoni* Adam, 1979 available for study are in poor condition and have not been well described. In using the key below, arm and club sucker indices (ASIn and CLSI respectively) are a measure of sucker diameters divided by dorsal ML.

1a. Arms 2 not elongate, not longer than remaining arms; cuttlebone anterior striae inverted U-shape or straight ................................................................... 2

1b. Arms 2 elongate, longer than remaining arms; cuttlebone anterior striae M-shaped ...........*S. rhoda*

2a. Club with four to six suckers in each transverse row, suckers differ in size, some suckers towards posterior end of club enlarged; cuttlebone inner cone does not form a raised, flattened ledge posteriorly.................. 3

2b. Club with eight suckers in each transverse row, suckers all similar sized, small; cuttlebone inner cone forms a raised, flattened ledge posteriorly... ................................................................. *S. koilados*

3a. Sucker diameter index for largest arm sucker (ASIn) < 2.4; sucker diameter for largest club sucker (ClSI) < 2.4 ................................................................... 4

3b. Sucker diameter index for largest arm sucker (ASIn) > 2.4; sucker diameter for largest club sucker (ClSI) > 2.4....................................... *S. cottoni*

4a. Arms 1–4 without distinctive markings................. 5

4b. Arms 1 with purplish band on each side of dorsal midline, arms 2–4 with median longitudinal purplish band................................................ *S. braggi*

5a. Arms 2 slightly longer than remaining arms; arms 1–3 with some biserial suckers. Tentacular club dorsal protective membrane extends beyond carpus along stalk. Dorsoposterior end of cuttlebone covered with a smooth, white, glaze-like substance .... *S. vercoi*

5b. Arms 2 not longer than remaining arms; arms 1–3 arm suckers tetraserial. Tentacular club dorsal protective membrane terminates at posterior end of carpus. Dorsoposterior end of cuttlebone not covered with a smooth, white, glaze-like substance ................ 6

6a. Suckers arms 1–3 tetraserial proximally, biserial distally. Club suckers relatively similar in size, suckers towards posterior end of club not greatly enlarged (ClSI 0.43–0.85).......................... *S. bidhaia*

6b. Suckers arms 1–3 biserial proximally, tetraserial medially, biserial distally. Club suckers differ in size, three or four enlarged suckers towards posterior end of club (ClSI 1.20–1.70) ................. *S. limata*

### *Sepia apama* Gray, 1849

*Sepia apama* Gray, J. E. (1849). Catalogue of the Mollusca in the Collection of the British Museum. 1. Cephalopoda Antepedia. London: British Museum (Natural History) 164 pp. [103].

Type data: holotype BMNH 1846.5.25.2.

Type locality: Australia: South Australia, Port Adelaide.

#### COMMON NAME(S)

Giant Australian Cuttlefish, Giant Cuttlefish

#### DIAGNOSTIC FEATURES

Very large, robust species. Mantle broadly oval. Fins wide. All arms subequal in length in both males and females; protective membranes wide, well developed. Arm suckers tetraserial. Left ventral arm hectocotylised: sucker size slightly reduced proximally (6–10 transverse rows of equal-sized suckers); oral surface of modified region not wide, fleshy, but normal, as on opposite arm. Club crescent-shaped; sucker-bearing surface flattened, with four or five suckers in each transverse row; suckers differ markedly in size, some median suckers greatly enlarged. Swimming keel of club very broad, extends well proximal to carpus; dorsal and ventral protective membranes joined at base of club; separated from stalk by membrane. Buccal membrane without suckers; in females with single median spermathecae in ventral part. Eggs spherical.

Three flat, semicircular, flap-like papillae posterior to eyes.

*Cuttlebone.* Outline oval, or oblong (juveniles broadly oval, wider in anterior half, becoming elongate in adults); bone bluntly rounded anteriorly; acuminate, rounded posteriorly; dorsal surface creamy white, evenly convex posteriorly, flat anteriorly; dorsal surface slightly granulose with irregular longitudinal ridges; indistinct dorsal median rib present in large specimens, broadens anteriorly, bordered laterally by indistinct grooves; lateral ribs indistinct. Spine present in juveniles, lost in adults. Striated zone flat, or slightly concave; last loculus flat in large animals, or convex in smaller ones; shallow, wide, sulcus extends along striated zone only, or sulcus absent. Anterior striae are inverted U-shape in smaller specimens, becoming straight in large specimens. Inner cone limbs are narrow anteriorly, broaden slightly posteriorly; in large specimens thickened into rough callus on inner margin of outer cone.

*Colour.* Reddish brown. All arms have whitish transverse bars and spots bordered by darker pigment. Dorsal mantle with fine, irregular, reticulate, white bands and spots, some fused to form irregular transverse bands; bands joined laterally into longitudinal broken white band adjacent to fins. Fins pale, with narrow white band along outer margin.

*Sculpture.* Three flat, semicircular, flap-like papillae posterior to eyes.

#### HABITAT AND BIOLOGY

*Sepia apama* is a neritic demersal species with cryptic habits, commonly found in relatively shallow waters around rocky reefs where it occupies and defends a crevice or cave (Aitken and O'Dor 2004). It feeds on fishes, crabs and other crustaceans, moving out of its den for brief hunting trips during the day. As with all cuttlefishes, it is highly dependent on excellent camouflage for predator avoidance and sneaking up on prey.

Largely because of a mass spawning event, easily accessible to scuba divers, that occurs annually* in Spencer Gulf, South Australia, the biology of *Sepia apama* (or at least the Upper Spencer Gulf population of the species) has been more extensively studied than any other Australian cuttlefishes.

During a highly ritualised and stereotyped courtship, males display broad bands of dark chromatophores that pulsate across the dorsal mantle as part of their visual display. Males often guard females to ward off other males. Small males have been observed to mimic the colouration and behaviour of females, using this diversion to sneak close to females to mate with them without the awareness of the larger, aggressive males. This change in skin patterning, posture and tactics can occur instantly and repetitively. In addition to mimicry, they operate by mating with females when the consort is occupied warding off other males ('open stealth') and may meet females under rocks ('hidden stealth'). Hanlon *et al.* (2005) has found that, despite the high rate at which females reject copulation attempts, strong mate guarding behaviour by consort males and the high levels of sperm competition, these sneaker males can often meet with fertilisation success. Mating takes place head to head and sperm packages (spermatangia) are placed by

* The numbers of animals participating in the mass spawning event at Whyalla has been decreasing dramatically in recent years. It has been reported that the event did not occur at all in 2012. The reason for this decline is as yet unknown. At the time of writing, however, in May 2014 an ABC news story reported that the cuttlefishes appear to be returning to spawn in at least one site in the Upper Spencer Gulf (www.abc.net.au/news/2014-05-20/cuttlefish-swarm-back-to-sa-breeding-site/5463854).

males onto the buccal area. Sperm are stored in the female's spermathecae ventral to the beak. Naud *et al.* (2005) showed that paternal sperm were more likely to have come from spermatangia than from the sperm receptacles. These results confirm a genetic polyandrous mating system and suggest that the fertilisation pattern with respect to the sperm stores used is not random. Naud and Havenhand (2006) showed that spermatangia and spermatophores can retain and release live sperm for long periods, with sperm resuspended from spermatangia taken from mated females in the field remaining motile for up to 100 hours at 12°C (ambient water temperature during the spawning season). The observed longevity increases the potential for sperm competition in this species. Naud *et al.* (2004), in a behavioural and genetic assessment of reproductive success, showed that females mate with multiple males and females' eggs had multiple paternity. Males of any size or status can obtain successful fertilisations.

Hall and Hanlon (2002) noted an operational sex ratio of males to females as 4:1, resulting in intense male competition. Both sexes sought and accepted multiple mates. Head to head mating had a mean duration of 2.4 minutes. The long interval between individual eggs being laid (minimum 6.5 minutes) allows multiple matings to occur between laying eggs and provides the opportunity for multiple paternity, sperm competition and female choice. Males used multiple characteristic reproductive behaviours with respect to size and status (i.e. lone or paired). Males paired with and defended females, rather than defending egg-laying sites. Paired males guarded females pre- and post-copulation. Lone males either searched for lone females, challenged paired males for the takeover of females through dramatic agonistic displays, or watched for opportunities for extra-pair copulations (EPCs). Males used 'open stealth', 'hidden stealth' and female mimicry (small males only) to achieve 'sneaker' extra-pair copulations. Females showed direct male choice by rejecting 70% of the 122 mating attempts, at times rejecting large males to mate with smaller males. Lone females moved extensively in search of egg-laying sites and interacted with many males, possibly to increase male competition. By aggregating to spawn and spawning asynchronously, females may also increase male–male competition through indirect mate choice.

Schnell *et al.* (2015) examined mating behaviour in *S. apama* in the field in Upper Spencer Gulf, and in a laboratory study of specimens collected around Sydney in New South Wales. In both populations, females displayed a white lateral stripe along the base of the fins when they were not receptive to mating. Mating history predicted the likelihood of white stripe display, with those that had recently mated more likely to display the stripe and less likely to adopt receptive postures. Male mating behaviour was not affected by the females' expression of the stripe, but they were more likely to select unfamiliar females rather than those with which they had previously mated. Familiarity with the male did not affect their mating behaviour response. Clearly, discrimination tactics are used for mate assessment and differences between the sexes exist in their ability to discriminate among potential mates. Schnell *et al.* (2015) suggest that there may be underlying selective pressures that are driving the disparities between the sexes.

Southern Australian animals spawn from April to September. Mass spawning occurs in the Upper Spencer Gulf from April to June, with peak spawning during May to June. Lemon-shaped eggs ~1 cm in length are laid daily in subtidal crevices and hatch after 3–5 months. Low incubation temperatures ~12°C (that mitigate potential problems of diffusive gas exchanges) may be a factor that limits the distribution of the species to cool southern waters. Based on the analysis of regular growth increments in *S. apama* (again from Upper Spencer Gulf), Hall *et al.* (2007) concluded that this population has two alternative life cycles. The first involves rapid juvenile growth during the first summer after hatching, with maturity reached within 7 or 8 months. These individuals return to spawn in their first year as small individuals. The second cycle involves much slower juvenile growth during the first summer with maturity deferred until their second year when they return to spawn as much larger individuals. There was no indication in this study to suggest that either type returns to the spawning aggregation the following year. At the end of the spawning season in August they disappear. Whether they die or move elsewhere is unknown. Other alternative hypotheses were suggested by Hall *et al.* (2007) that might explain what appears to be a two alternate life cycle scenario (mixing of alternate life cycles or stocks, for example) but no evidence for alternatives has been discovered to date.

Zylinski *et al.* (2011) statistically examined the correlation intensity between neighbouring colour pixels in *Sepia apama* exhibiting both camouflage colouration and conspicuous signalling *v.* the background colouration. The measured image statistics of camouflaged animals closely resembled their backgrounds, while signalling animals differ significantly from their backgrounds. While these results are not unexpected, the statistical analysis of image parameters in a quantitative way may provide the basis for a general set of rules about crypsis and signals. *Sepia apama* with its repertoire of colouration and patterns in a single visual habitat proved to be an ideal study organism for this research. Hanlon *et al.* (2007) showed that *Sepia apama* also use their diverse camouflage repertoire at night. Using a remotely operated vehicle (ROV) equipped with a video camera and a red light in the Upper Spencer Gulf spawning ground, they showed that *S. apama* ceased sexual signalling and reproductive behaviour at dusk, settled to the bottom and quickly adopted colour patterns to produce camouflage tailored to different backgrounds. During the day, only 3% of cuttlefishes were camouflaged on the spawning ground, but at night 86% (71 of 83 cuttlefishes) were camouflaged. Nocturnal visual predators (such as bream, mulloway and snapper) may provide the selective pressure for this camouflage patterning.

Aitken and O'Dor (2004) examined respirometry and swimming dynamics in *Sepia apama* and concluded, based on three laboratory-reared specimens, that cuttlefishes cannot swim at high speeds for extended periods. They showed that at 15°C and a swimming speed of 0.06 m $s^{-1}$, the gross cost of transport was calculated at 10.1 $kg^{-1}$ $m^{-1}$, with a net cost of 4.1 $kg^{-1}$ $m^{-1}$. Aitken *et al.* (2005) studied the behaviour and energetics of representatives of a South Australian population (Boston Bay, House Reef) as revealed by radio acoustic positioning and telemetry (RAPT). They showed that the maximum range (the extreme points of the area covered by the cuttlefish) for an individual varied from 90 m (juvenile) to 550 m (adult). The home range is the area of the environment occupied over the monitoring period and was found to be 5300 $m^2$ to a maximum of 27 300 $m^2$. They tend to be diurnal as the average distance travelled by day was greater than that of the night, with their foraging behaviour at 3.7% by day and 2.1% per night. This is not surprising, as they are exclusively visual hunters (Boucher-Rodoni *et al.* 1987). They spend ~95% of the day resting in dens (e.g. overhanging rocks, caves etc.) and appear to be 'sit and wait' predators, taking advantage of prey in and around their den. Through predator avoidance, efficient foraging and a relatively quiescent lifestyle, energy can be channelled towards growth.

Payne *et al.* (2013) estimated that *Sepia apama* from Upper Spencer Gulf have an optimal swimming speed of 6 or 7 cm $s^{-1}$ and most individuals are incapable of maintaining swimming above 15 m $s^{-1}$. They concluded, based on these measures, that dispersal potential was consistent with the observed fine scale population structure in this species (i.e. restricting their dispersal potential to < 140 km over the required 60 days between when juveniles feeding in northern Spencer Gulf migrate to the Pt Lowly breeding grounds). This distance is consistent with the approximate distance between Port Lowly and the barrier of genetic divergence to the south that has been recognised in molecular studies (Donellan unpublished data).

Battam *et al.* (2010) calculated the energy value of eastern Australian cuttlefishes as prey for *Diomedea* albatrosses that rely on seasonally available *S. apama* as a source of nourishment when undergoing moult. The gross energy value of *S. apama* bodies was 20.9 kJ $g^{-1}$ dry mass, with a protein content typically > 75% dry mass and a 1% fat content. In eastern Australia, off Wollongong, New South Wales (39°29′S, 150°50′E) the density of *S. apama* is relatively high through winter and spring, similar to that seen in Spencer Gulf during the breeding season (Hall and Hanlon 2002), though no dense breeding aggregation has yet been discovered.

Kawashima *et al.* (2013) have determined the complete mitochondrial genome for *S. apama*, but it is not stated in their paper where the specimens were actually collected from.

*Interest to fisheries.* This species is taken as by-catch in prawn and mixed species trawl fisheries, by hook-and-lines or speared by divers. It is commonly available in fish markets along the southern coast of Australia, sold for human consumption and as bait. Few details on quantities or origin of catches are available. It may have potential for aquaculture (Nabhitabhata 1995).

## SIZE

Maximum size up to 560 mm ML (Battam *et al.* 2010 unpublished data) and weight in excess of 10.5 kg. This is the world's largest cuttlefish. Those in eastern Australia are much larger than those from elsewhere.

## REMARKS

Recent morphological and molecular analyses have shown that specimens of *Sepia apama* from the east coast (New South Wales), south-west coast (Western Australia) and southern Australia (South Australia, Victoria) comprise three genetic groups. However, these populations are not geographically isolated. The data suggest that populations of *S. apama* may have been geographically isolated in the past and have come into secondary contact, particularly in south-eastern Australia (Kassahn *et al.* 2003).

Donellan (unpublished data) has recognised five distinct populations. Work is continuing to determine

Anterior view (B. Cole).

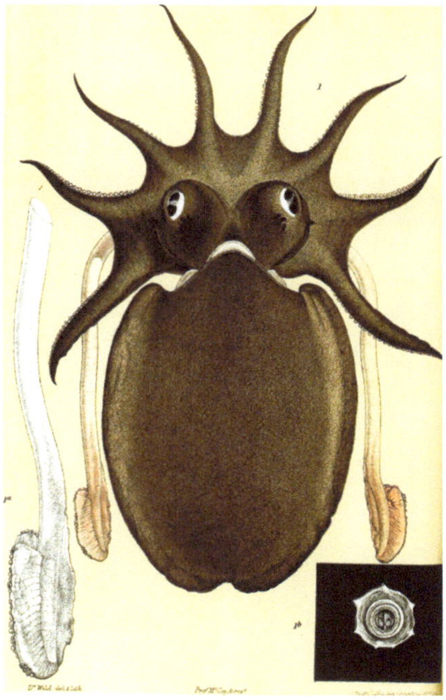

Dorsal view.

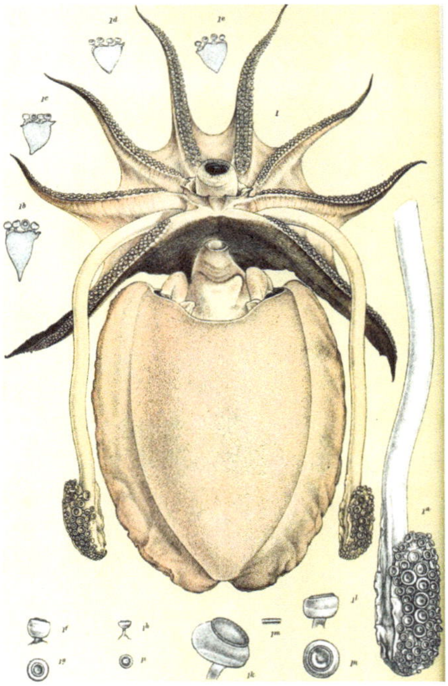

Ventral view.

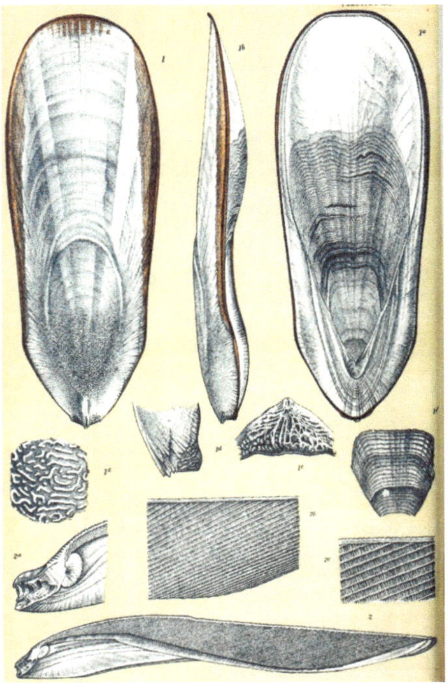

Cuttlebone details. (Bottom left and right figures: Plates 188–190. *Prodromus of the Zoology of Victoria* 1885–1890).

whether unique specific status for any of the five populations that have been identified as genetically distinct is justified. Whether specific status is warranted or not, each of these populations should be treated as distinct units in future studies of the biology of this taxon. Populations should also be considered as distinct for fisheries management purposes.

The examination of dicyemid parasites has shown, to some extent, a similar pattern of disjunct *S. apama* populations. Dicyemids are generally highly host-species specific and can be used as biological 'tags' to assist in the resolution of cephalopod taxonomy and assist in host species' identification. Catalano (2013b) described *Dicyema coffinense* Catalano, 2013 from *S. apama* collected in Coffin Bay, South

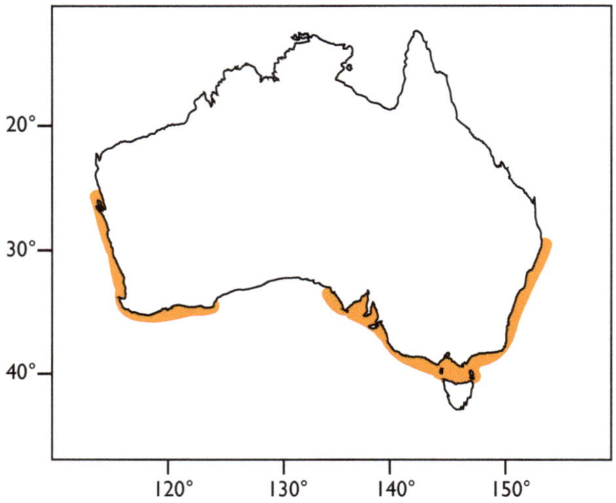


*Sepia apama* known distribution. *Note:* The distribution is based on whole animals only, not beachwashed cuttlebones.

Australia, and *D. koinonum* Catalano, 2013 from Gulf St Vincent (middle and lower Gulf) and Spencer Gulf (upper, middle and lower Gulf) *S. apama* populations. In addition, *Dicyemennea spencerense* Catalano, 2013 (with type host *S. novaehollandiae* Hoyle, 1909) was found to occur in Spencer Gulf *S. apama* (upper, middle and lower Gulf). *Dicyema multimegalum* Catalano, 2013 was described from *S. apama* collected at Cronulla and North Bondi, New South Wales.

Interestingly, among these recently described parasites, *D. koinonum* and *D. spencerense* are found in both *S. apama* and *S. novaehollandiae*. As stated above, in general dicyemid parasites are highly host specific. However, there are previous records of one dicyemid species found in two or more host species. For example *Pseudicyema truncatum* (Whitman, 1883) has previously been recorded from *S. officinalis*, *S. elegans* and *S. orbignyana* (Catalano, personal communication). These results are consistent with an as yet unpublished phylogeny of the Sepiidae, based on molecular characters, that has shown *S. apama* and *S. novaehollandiae* to be sister taxa (Moussalli, personal communication).

The author is currently comparing the morphology of *S. apama* specimens from each of the five distinct populations recognised by Donnellan (unpublished data), and a molecular study is ongoing (A. Lindgren, unpublished data). The resolution of the taxonomy of *S. apama* requires redescription of *S. apama* from the type locality (Gulf St Vincent, South Australia). Unfortunately the type specimen is a cuttlebone only. An added complication is that it appears that the eastern Australian *S. apama* may be genetically mixing with the 'true' *S. apama* population at the southern edge of its range. Further work is needed to determine the status of the Upper Spencer Gulf *v.* lower Spencer Gulf/south-eastern Australian populations, and the eastern and western Australian populations.

Given that distinct populations are recognised, it is important that in future studies the location of any study animals is given. Behaviours and other biological features of *S. apama* may vary according to the population studied and may not be uniform across its entire distribution.

### DISTRIBUTION

Southern Australia from 29°49′12″S, 153°19′48″E to 25°18′35″S, 113°37′47″E. Known depth range 1–84 m.

### REFERENCE

Reid *et al.* (2005b).

## *Sepia bandensis* Adam, 1939

*Sepia bandensis* Adam, W. (1939). Notes sur les Cephalopodes, XI. – Sepia bandensis sp. nov. *Bulletin du Musée royal d'Histoire naturelle de Belgique 15*(18), 1–7 [1, figs 1–4].

Type data: syntypes IRSNB IG-10542 and 10652.

Type locality: Indonesia: Banda Sea, Banda Neira.

### COMMON NAME

Stumpy Spined Cuttlefish

### DIAGNOSTIC FEATURES

Mantle broadly rounded. Head short and broad and arms short. All arms keeled aborally and with narrow protective membranes. Arm suckers tetraserial. Left ventral arm of male hectocotylised: suckers of fifth proximal row to eleventh row smaller than remaining suckers. Suckers in ventral row smaller than those in dorsal row. Tentacular club short (about one-fifth ML), detached from stalk; dorsal and ventral protective membranes wide, fused basally; dorsal membrane separated from swimming membrane by a deep groove that extends distally to level of median suckers; swimming membrane longer than club. Five suckers in each transverse row, suckers in median longitudinal row larger than rest, central three suckers largest. Swimming keel extends beyond base of club; dorsal and ventral protective membranes joined at base of club, separated from stalk by a membrane.

*Cuttlebone.* Cuttlebone outline broad, oval; bluntly rounded anteriorly and slightly acuminate posteriorly; dorsal surface evenly convex; calcified with reticulate sculpture; dorsal median and lateral ribs absent. Spine reduced to tiny, blunt tubercle. Sulcus shallow, narrow,

Dorsal view (M. Norman).

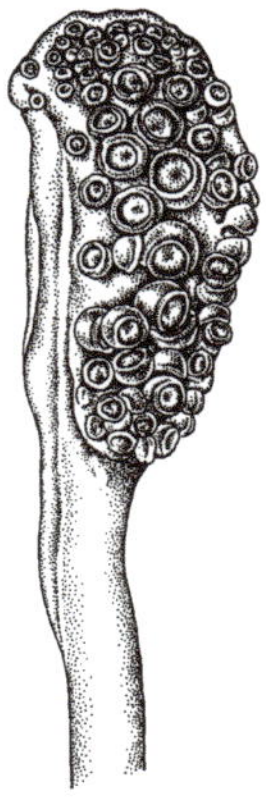

Tentacular club (ABRS, K. Hollis).

Cuttlebone, dorsal view (D. Paul).

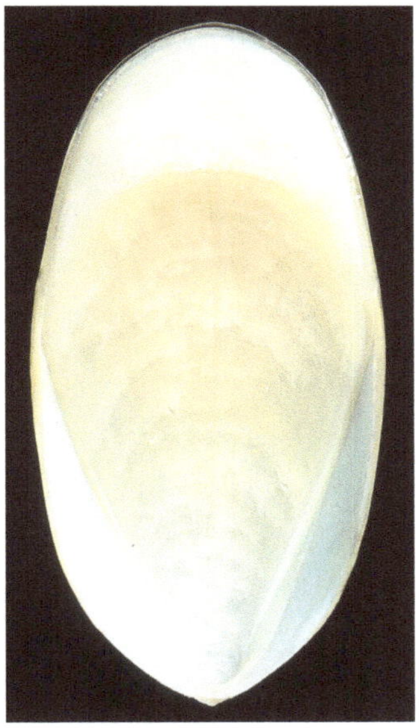

Cuttlebone, ventral view (D. Paul).

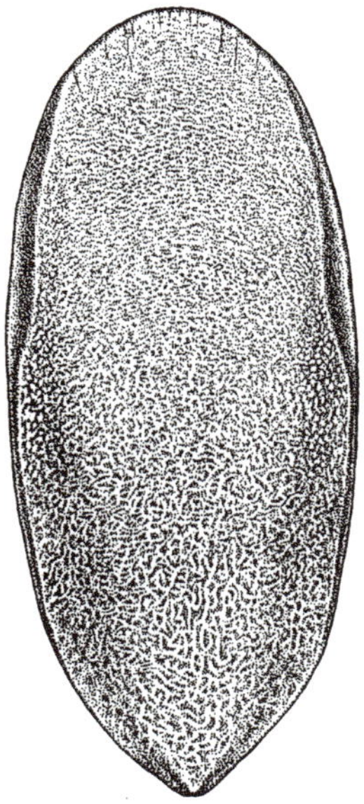

Cuttlebone, dorsal view (ABRS, K. Hollis).

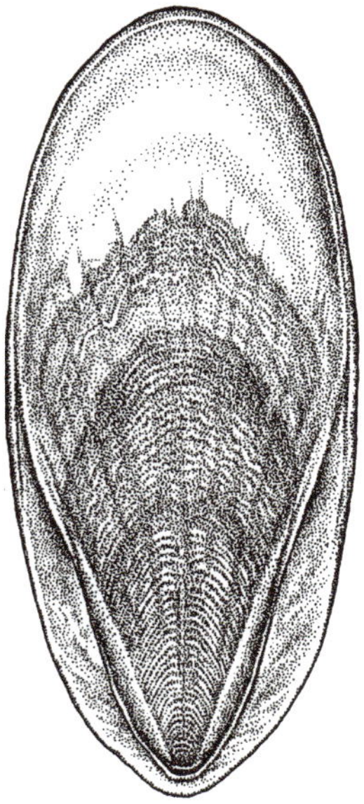

Cuttlebone, ventral view (ABRS, K. Hollis).

extends along striated zone only. Anterior striae are inverted U-shape. Inner cone limbs are narrow anteriorly, broaden posteriorly.

*Colour.* Reddish brown, light brown, or greenish yellow-brown when fresh, and with whitish mottle. Dorsal mantle has white blotches concentrated into short, longitudinal bands joined laterally into longitudinal broken white band adjacent to fins. It often shows a pair of brown patches on the posterior end of the mantle, often accompanied by white patches over the eye orbits. Head with scattered white spots. All arms have whitish transverse bars and spots bordered by darker pigment. Fins pale, with narrow white band along outer margin and small luminescent blue spots basally.

*Sculpture.* Dorsal mantle with longitudinal row of ridge-like papillae along each side, adjacent to base of each fin and scattered short ridges dorsal to cuttlebone (corresponding to whitish bars). Head and arms with short, scattered, bar-like papillae positioned dorsally and laterally.

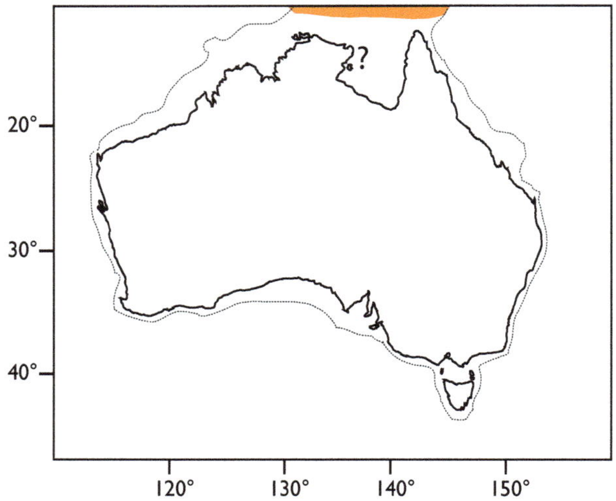


*Sepia bandensis* known distribution. *Note:* The distribution is based on whole animals only, not beachwashed cuttlebones. (Australian records are based on few specimens, but the species is likely to occur more widely.)

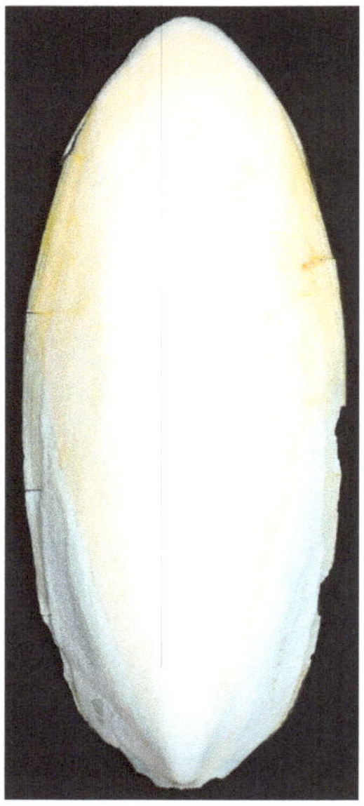

Dorsal view (holotype AM C.133317, L. Mott).

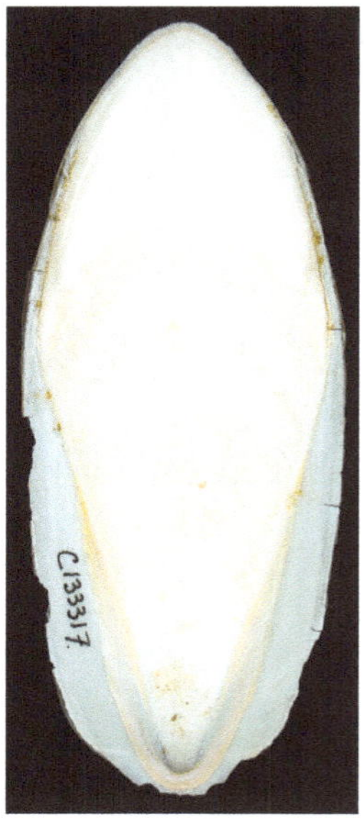


Ventral view (holotype AM C.133317, L. Mott).

## HABITAT AND BIOLOGY

Found in coastal shallow waters over coral reefs and sand. Night-active. This species has been observed 'walking', rather than swimming, using arms 3 (or 4) and pair of raised flaps on the ventral mantle. It is often found in association with sea cucumbers and sea stars.

## SIZE

Up to 70 mm ML. Animals from the Alas Strait, Indonesia reported at ~40 g (males) and 45 g (females).

## DISTRIBUTION

Tropical Indo-Pacific: Philippines, Malaysian Borneo, Java, Sulawesi, Alas Strait, eastern Indonesia, New Guinea and northern Australia.

## REFERENCE

Reid *et al.* (2005b).

# *Sepia baxteri* (Iredale, 1940)

*Blandosepia baxteri* Iredale, T. (1940). Marine molluscs from Lord Howe Island, Norfolk Island, Australia and New Caledonia. *Australian Zoologist 9*, 429–443 [442].

Type data: holotype AM C.133317.

Type locality: Australia: Lord Howe Island.

## DISTINGUISHING CHARACTERS

It is similar to *Sepia bandensis* Adam, 1939, but the bone in *S. baxteri* is slightly acuminate, while in *S. bandensis* it is more evenly rounded.

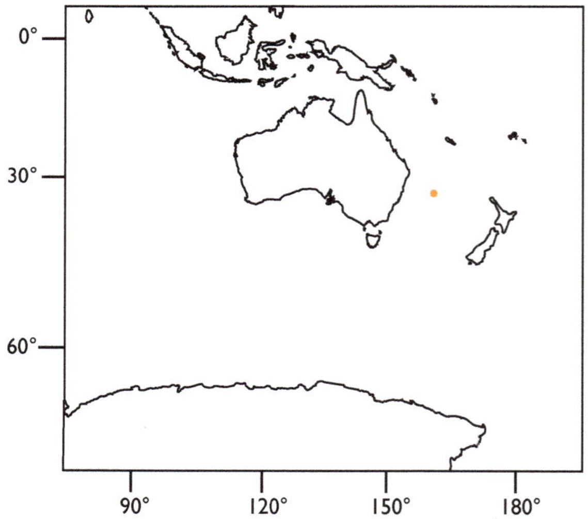


*Sepia baxteri* distribution. *Note:* The distribution is based on beachwashed cuttlebones only.

## DIAGNOSTIC FEATURES

Cuttlebone outline elongate oval, slightly acuminate anteriorly, rounded posteriorly. Dorsal surface brownish pink or yellowish, evenly convex, slightly granulose with faint dorsal and lateral ribs. Inner cone U-shaped, slightly wider and thickened posteriorly, yellowish. Outer cone white, weakly calcified. Spine reduced to blunt tubercle. Sulcus absent but bone with shallow median depression that extends along full length of bone. Striated zone long; anterior striae are broad inverted V-shape. Inner cone limbs are narrow anteriorly, broaden, thicken posteriorly.

## HABITAT AND BIOLOGY

Unknown.

## SIZE

Up to 74 mm ML.

## REMARKS

This species is only known from numerous cuttlebones collected on Lord Howe Island, New South Wales. Whole animals are needed to resolve the status and identity of this species.

## DISTRIBUTION

Lord Howe Island 31°32′58″S, 159°04′57″E.

## REFERENCE

Reid *et al.* (2005b).

### *Sepia bidhaia* Reid, 2000

*Sepia bidhaia* Reid, A. L. (2000). Australian cuttlefishes (Cephalopoda: Sepiidae): the 'doratosepion' species complex. *Invertebrate Taxonomy 14*, 1–76 [4, figs 2–7].

Type data: holotype MV F.82478.

Type locality: Australia: Queensland, 17°55.8′S, 146°58.2′E.

## DISTINGUISHING CHARACTERS

*Sepia bidhaia* is very similar to *S. limata* (Iredale, 1926), though the two species differ in several characters. The club suckers of *S. limata* clearly differ in size; in *S. bidhaia* the club suckers are all relatively similar sized. The arrangement of arm suckers differs between the two species. In males, the suckers are tetraserial in *S. limata*, while those of *S. bidhaia* are biserial proximally and tetraserial distally. The arm suckers of *S. limata* females are biserial proximally, tetraserial medially and biserial distally on all arms. Female *S. bidhaia* have tetraserial suckers proximally on all arms, with biserial suckers on the distal ends of arms 1–3. *Sepia bidhaia* is found at greater depths (200–304 m) than *S. limata* (43–146 m).

## DIAGNOSTIC FEATURES

Small species. All arms subequal in length in both males and females. Arm sucker arrangement differs between sexes. In males, sucker rows on arms 1–4 biserial proximally, tetraserial distally. In females, suckers on arms 1–3 tetraserial proximally, biserial at distal arm tips; arms 4 suckers tetraserial. Female arms 1 and 4 protective membranes wide, enclose distal suckers; arms 2 and 3 distally slightly thickened, fleshy, suckers displaced laterally. Greater proportion of arms 2 and 3 with distal biserial suckers than arms 1 and 4. Left ventral arm hectocotylised: sucker size normal proximally (eight or nine transverse rows), reduced medially, much smaller than normal suckers (eight rows), with suckers in two ventral series of modified portion very slightly smaller than those in two dorsal series. Oral surface of modified region wide, swollen, fleshy, with transversely grooved ridges; with shallow

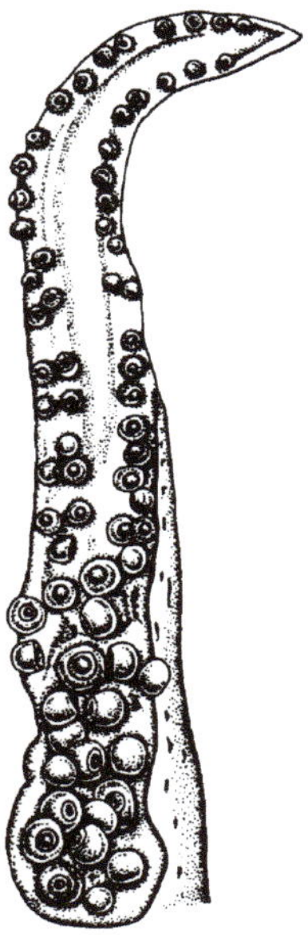

Hectocotylus (A. Reid).

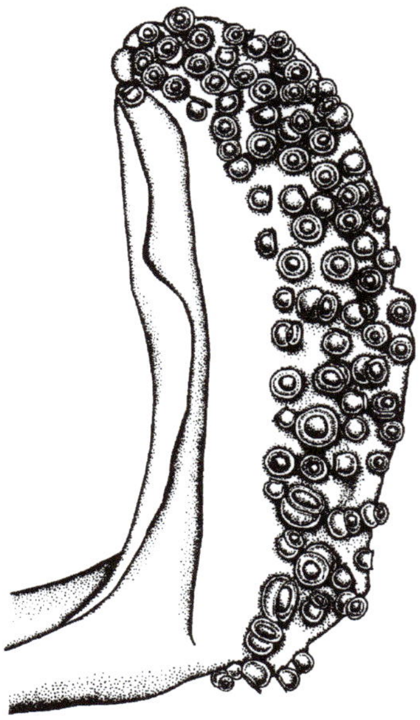

Tentacular club (A. Reid).

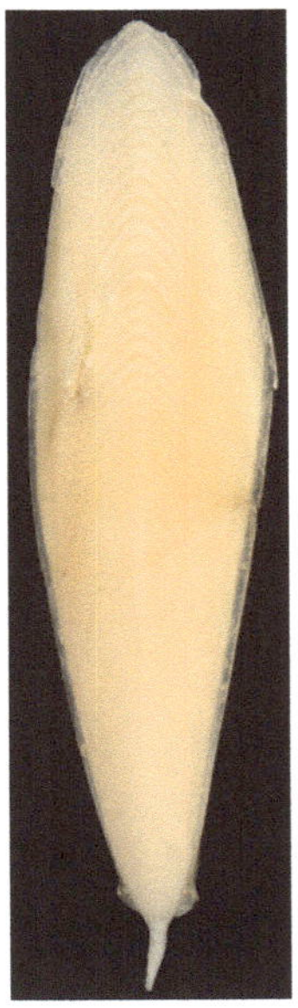

Cuttlebone, dorsal view (D. Paul).

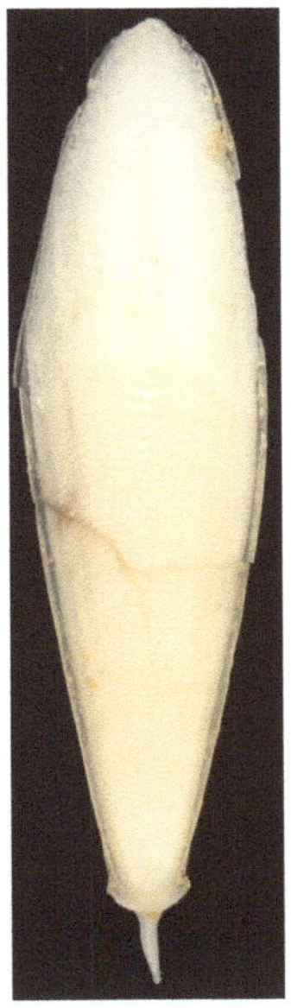

Cuttlebone, ventral view (D. Paul).

Enlargement of cuttlebone posterior end, ventral view (A. Reid).

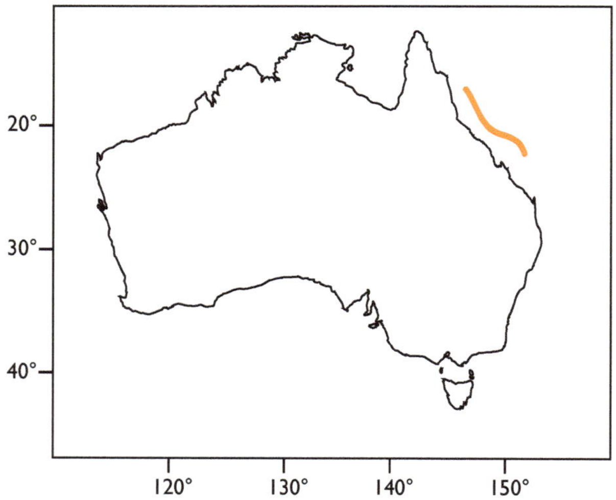


*Sepia bidhaia* distribution. *Note:* The distribution is based on whole animals only, not beachwashed cuttlebones.

median furrow. Suckers in two dorsal and two ventral series displaced laterally. Tentacular club crescent-shaped; short; sucker-bearing surface flattened, with four suckers in each transverse row; suckers differ slightly in size. Swimming keel of club extends slightly beyond carpus; dorsal and ventral protective membranes not fused at base of club; joined to stalk; dorsal and ventral membranes same length, terminate at posterior end of carpus. Buccal membrane without suckers.

*Cuttlebone.* Cuttlebone outline oblong (widest anteriorly); strongly convex in lateral view; bone acuminate, acute anteriorly and posteriorly; recurved ventrally. Dorsal surface creamy white; evenly convex; granulose, calcified dorsally, particularly posteriorly (not as pronounced on flared outer cone); dorsal median rib indistinct, broadens anteriorly; lateral ribs absent. Chitin borders lateral margins of cuttlebone. Spine present, without keel. Striated zone convex; sulcus shallow, narrow, extends entire length of cuttlebone. Anterior striae slightly convex to straight. Inner cone limbs strap-like anteriorly, narrower posteriorly; thickened slightly, not raised to form ledge posteriorly. Outer cone limbs expanded posteriorly forming two short 'wings', directed ventrally, forming recurved cup-like structure.

*Colour* (alcohol preserved specimens). Pale buff pinkish brown, or purplish brown; head with few scattered chromatophores and with scattered orange spots; arms with orange spots. Arm and club sucker rings pale brown, or yellow brown. Dorsal mantle chromatophores concentrated medially over cuttlebone and with orange spots. Ridges (see below) orange-pink in colour.

*Sculpture.* Dorsal mantle with longitudinal row of up to six ridge-like papillae along each side, close to base of each fin and with scattered short ridges dorsal to cuttlebone; ventral mantle with longitudinal row of six narrow ridges along each side close to fins; anterior-most pair and posterior two pairs shorter than rest.

### SIZE

Males up to 37 mm ML; females up to 57 mm ML.

### DISTRIBUTION

South-western Pacific: Australia, Queensland, off the Great Barrier Reef, 17°20′42″S, 146°41′30″E to 22°11′33″S, 153°11′43″. Known depth range 181–304 m.

### REFERENCE

Reid (2000).

## *Sepia braggi* Verco, 1907

*Sepia braggi* Verco, J. C. (1907). Notes on South Australian marine Mollusca with descriptions of new species. 6. *Transactions of the Royal Society of South Australia 31*, 213–214 [213, pl. 27 fig. 6].

Type data: holotype, SAMA D.14130, cuttlebone.

Type locality: Australia: South Australia, Glenelg.

### COMMON NAME

Slender Cuttlefish

### DISTINGUISHING CHARACTERS

*Sepia braggi* males differ from *S. limata* Iredale, 1926 and *S. rhoda* (Iredale, 1954) males in having elongate third arms. *Sepia braggi* does not have a hectocotylus; the left ventral arms are hectocotylised in *S. limata* and *S. rhoda*. The arrangement of the arm suckers differs among these three species. Female *S. rhoda* have elongate second arms. While the second and third arms of *S. braggi* females are slightly longer than the other arms, they do not differ to the same extent as those of mature *S. rhoda* females. The anterior striae on cuttlebones are M-shape in *S. rhoda* and the posterior inner cone is very broad, while in *S. braggi* the anterior striae are an inverted U-shape and the inner cone narrow. The cuttlebone is broader in *S. limata* than the *S. braggi* cuttlebone. *Sepia braggi* of both sexes differ from all other Australian narrow-boned species in its distinctive arm pigmentation.

### DIAGNOSTIC FEATURES

Male and female arms differ in relative lengths; male (mature) arms 3 elongate; female arms 2 and 3 longer than arms 1 and 4 (arms 2 usually longer than arms 3). Distal arm tips strongly attenuate in both sexes, suckers enclosed

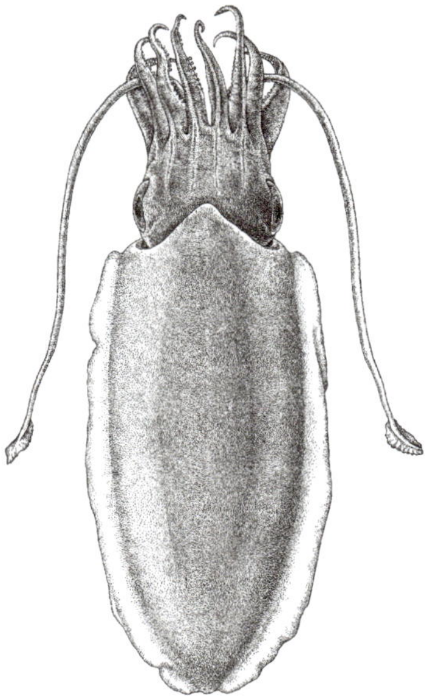

Dorsal view (ABRS, K. Hollis).

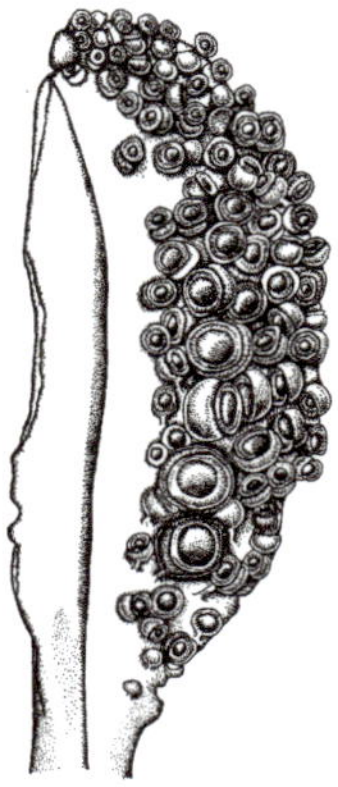

Tentacular club (A. Reid).

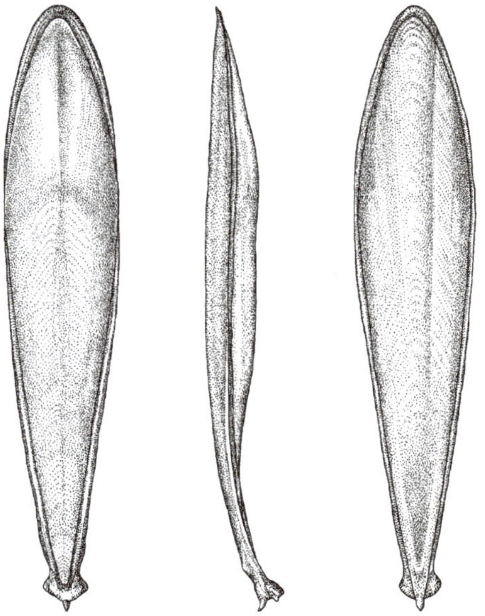

Cuttlebone, ventral view (left), side view (centre) and dorsal view (right) (ABRS, K. Hollis).

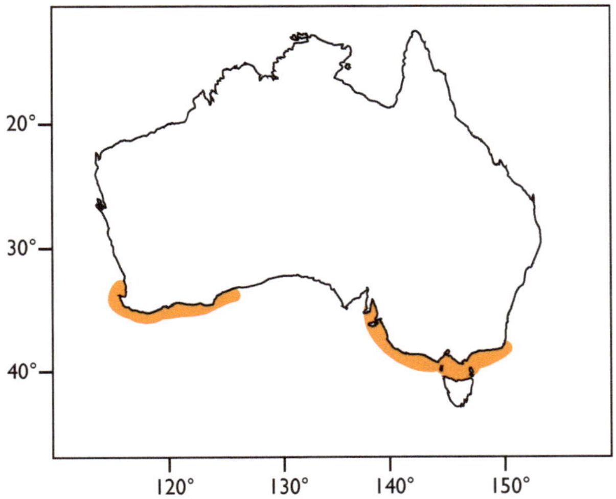


*Sepia braggi* distribution. *Note:* The distribution is based on whole animals only, not beachwashed cuttlebones.

by protective membranes. Arm suckers tetraserial proximally, biserial at distal arm tips (biserial region over greater proportion of arms 2 and 3 than of arms 1 and 4); biserial suckers reduced and displaced laterally, with gap between them on arms 2 of females and arms 3 of males. Hectocotylus absent. Club slightly recurved, short, sucker-bearing surface flattened, with four to six suckers in each transverse row; suckers differ in size, five or six median suckers twice diameter of rest. Swimming keel of club extends proximally slightly beyond carpus; dorsal and ventral protective membranes not joined at base of club but fused to tentacular stalk; dorsal and ventral membranes same length, extend proximal to carpus along stalk. Gills with 26–30 lamellae per demibranch. Buccal membrane without suckers. Buccal membrane in females without spermathecae. Spermatophores 2.4–4.4 mm long; smallest male recorded with well-developed spermatophores in spermatophoric sac is 24.1 mm ML. Eggs oval, up to 9. 4 mm long, 5. 5 mm wide.

*Cuttlebone.* Cuttlebone outline lanceolate, broadest in anterior third at anterior end of striated zone; acuminate, acute, anteriorly and posteriorly; strongly recurved ventrally; dorsal surface pinkish; evenly convex; calcified medially, slightly granulose with irregular longitudinal ridges (particularly posteriorly). Spine and posterior end (approximately one quarter) of bone covered with an ochre-coloured, smooth, glaze-like substance. Dorsal median rib distinct anteriorly, indistinct posteriorly; rib broadens anteriorly; lateral ribs absent. Chitin present as wide bands bordering lateral margins of cuttlebone. Spine short, pointed, curves dorsally; keel(s) absent; fine, radiating ribs between outer cone and spine. Striated zone and last loculus convex; sulcus deep, narrow, flanked by rounded ribs, extends entire length of cuttlebone. Anterior striae are inverted U-shape, incurved medially, slightly V-shaped following sulcus. Inner cone limbs are strap-like anteriorly, narrower posteriorly, U-shape; slightly raised into rounded posterior ridge; thickened, shiny (yellowish); outer cone limbs are expanded posteriorly into two short 'wings', directed ventrally, to form recurved cup-like structure.

*Colour.* In preserved specimens pale buff to pinkish brown. Arms 1 have a darkly pigmented purplish band on each side of dorsal midline. Arms 2–4 with median longitudinal purplish band along their aboral surfaces. All arms have short transverse bars and blotches lateral to these longitudinal bands. Arm and club sucker rings mid-brown. Fins without markings along base.

*Sculpture.* Dorsal mantle has scattered short ridges dorsal to cuttlebone and up to five orange-pink ridges at base of each fin. Head with short, scattered papillose bars posterior to eyes; arm papillae present; same as those on head.

### HABITAT AND BIOLOGY

*Sepia braggi* is a demersal species.

*Interest to fisheries.* Taken as by-catch with other cuttlefishes along southern Australia, then used locally as bait and food.

### SIZE

Males up to 49 mm ML; females up to 80 mm ML.

### REMARKS

The pigmentation on the arms is very distinctive and clearly visible even in small specimens. While arms 3 are obviously elongate in adult males, they are similar in length to the remaining arms in immature males (less than ~30 mm ML).

### DISTRIBUTION

Southern Indo-Pacific: southern Australia from southern New South Wales, 36°22′48″S, 150°7′12″E, to Western Australia, 33°57′0″S, 115°4′11″E. Australian endemic. Known depth range 20–128 m.

### REFERENCE

Reid (2000).

## *Sepia chirotrema* Berry, 1918

*Sepia chirotrema* Berry, S. S. (1918). Report on the Cephalopoda obtained by the F.I.S. Endeavour in the Great Australian Bight and other southern Australian localities. *Biological Results of the Fishing Experiments carried on by the F.I.S. 'Endeavour', 1909–14* 4, 203–298 [268, text figs 56–61, pls 74–79].

Type data: holotype AM E.2459 (not found).

Type locality: Australia: South Australia, Investigator Strait, south of Kangaroo Island.

### DIAGNOSTIC FEATURES

Mantle broad, oval. All arms subequal in length in both males and females; arm suckers tetraserial. Left ventral arm hectocotylised. Oral surface of modified region as for remaining arm, not wide, fleshy. Hectocotylised arm distally attenuate, compressed. Club sucker-bearing surface flattened, with five or six suckers in each transverse row. Suckers differ dramatically in size, some greatly enlarged. Swimming keel of club extends well beyond carpus; dorsal and ventral protective membranes fused at base of club; joined to stalk; dorsal and ventral membranes same length.

*Cuttlebone.* Cuttlebone acuminate, acute anteriorly; bluntly rounded posteriorly. Dorsal surface pinkish; flat medially, flat laterally; texture rough. Dorsal median rib present, distinct, sides approximately parallel; lateral ribs present, distinct. Spine present, short, pointed; curves dorsally; keel(s) absent. Striated zone concave; sulcus extends along striated zone only; deep, wide. Anterior striae are V-shaped, U-shaped posteriorly. Inner cone limbs narrow anteriorly, broaden posteriorly; not raised to form ledge posteriorly; not thickened. Outer cone lateral limbs flared ventrolaterally; limbs continue as narrow ledge ventral to spine.

*Sculpture.* Papillae present on head; positioned between eyes and nuchal cartilage.

### SIZE

Up to ~200 mm ML.

### DISTRIBUTION

Southern Indo-Pacific: Southern Australia from 36°32′59″S, 136°30′E to 19°15′S, 117°43′11″E. Known depth range 120–220 m.

### REFERENCE

Reid *et al.* (2005b).

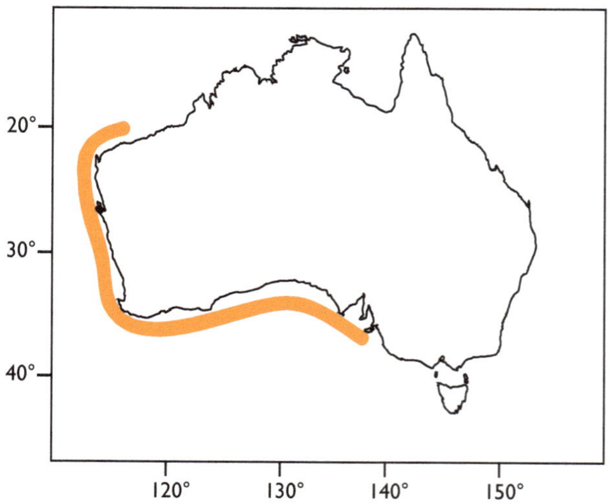


*Sepia chirotrema* distribution. *Note:* The distribution is based on whole animals only, not beachwashed cuttlebones.

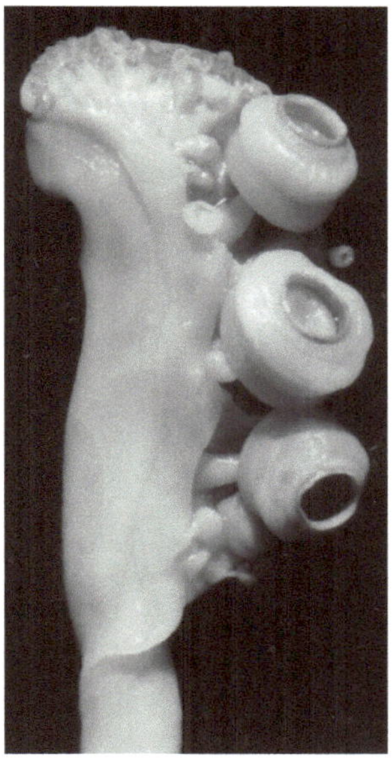

Tentacular club (C. Rowley).

Cuttlebone, dorsal view (D. Paul).

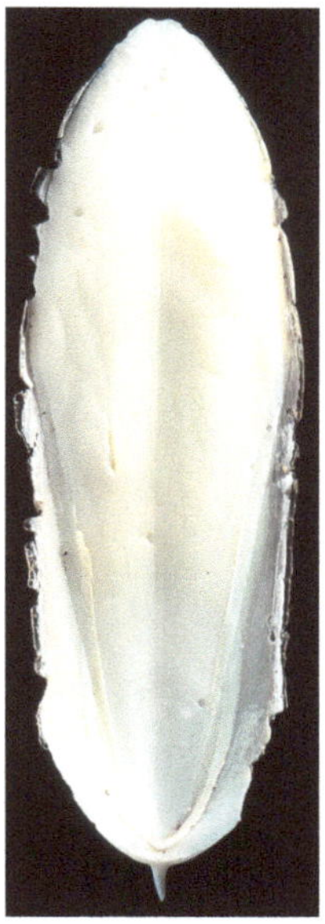

Cuttlebone, ventral view (D. Paul).

*Sepia chirotrema*

## *Sepia cottoni* Adam, 1979

*Sepia cottoni* Adam, W. (1979). The Sepiidae (Cephalopoda, Decapoda) in the collections of the Western Australian Museum. *Records of the Western Australian Museum 7*, 113–212 [193, pl. 11, figs 1–6].

Type data: holotype WAM 435-65.

Type locality: Australia: Western Australia, west of Lancelin, 31°54′S, 114°55′E.

### DISTINGUISHING CHARACTERS

The species occurs sympatrically with at least four other species: *Sepia koilados* Reid, 2000, *S. braggi* Verco, 1907, *S. rhoda* (Iredale, 1954) and *S. vercoi* Adam, 1979. The bone of *S. koilados* is very distinctive and clearly differs from that of *S. cottoni*. The pigmentation on the arms of *S. braggi* can be used to distinguish it from *S. cottoni*. Mature female *S. rhoda* differ from *S. cottoni* in having elongate arms 2, while those of *S. cottoni* are subequal. In addition, the cuttlebone anterior striae are M-shaped in *S. rhoda* and inverted U-shape in *S. cottoni*. In *S. vercoi*, the club dorsal protective membrane extends beyond the carpus, while it is the same length as the carpus in *S. cottoni*. The suckers on arms 3 and 4 of *S. vercoi* females are tetraserial, while in *S. cottoni* the proximal and distal-most suckers are biserial on these arms.

### DIAGNOSTIC FEATURES

Male arms elongate (twice as long as in female); protective membranes in both sexes wide, well developed (though disappearing on distal half of arms 1, 2 and 4 in males). Male arms 3: protective membranes after about the seventh sucker row wide and thick with transverse ridges, ridges alternate, unite inner margin of one membrane and outer margin of opposite membrane, each ridge with two small suckers: an inner one at base of protective membrane and smaller sucker near outer margin of opposite membrane. On ventral membrane, outer suckers on margin; on wider, thicker dorsal membrane, suckers halfway down transverse ridges. Arm sucker arrangement same in both sexes: biserial proximally, tetraserial medially, biserial at distal arm tips. Male arm suckers on each arm biserial basally, suckers normal in size, medial seven or eight rows of tetraserial suckers large, distal biserial suckers reduced, then absent on threadlike tips of arms 1, 2 and 4 and thickened tip of arm 3. Dorsal suckers proximally on arm tips, ventral suckers absent near distal tip of arms (arrangement of suckers on these arms in alternating pairs similar, but not as pronounced as on arm 3). Hectocotylus absent. Club with three or four suckers in each transverse row. Suckers differ markedly in size, some greatly enlarged, twice diameter of rest. Swimming keel of club extends slightly beyond carpus; dorsal and ventral protective membranes not fused at base of club.

*Cuttlebone.* Cuttlebone outline lanceolate (broadest in anterior third); strongly recurved ventrally. Dorsal surface pinkish; granulose (with irregular longitudinal lines). Dorsal median rib present, distinct anteriorly, indistinct posteriorly. Spine present, long, pointed; curves dorsally; keel(s) absent. Sulcus extends entire length of cuttlebone; shallow, narrow; not flanked by rounded ribs. Anterior striae are inverted U-shaped, incurved medially,

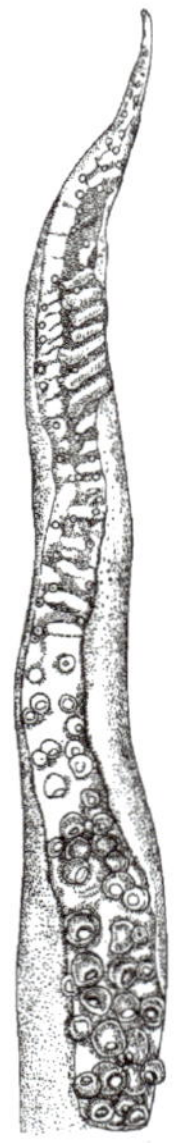

Hectocotylus (FAO, E. D'Antoni)

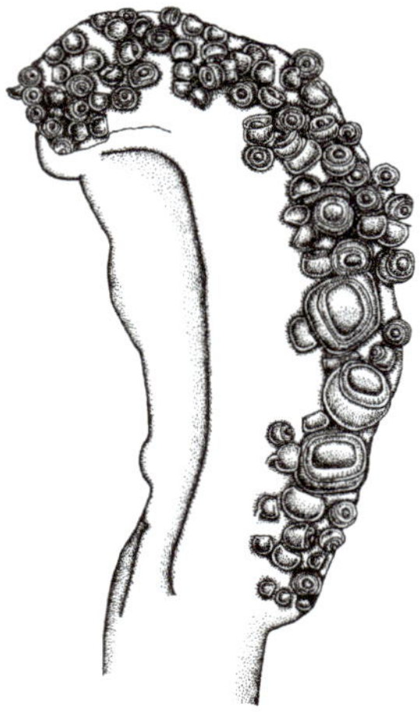

Tentacular club (A. Reid).

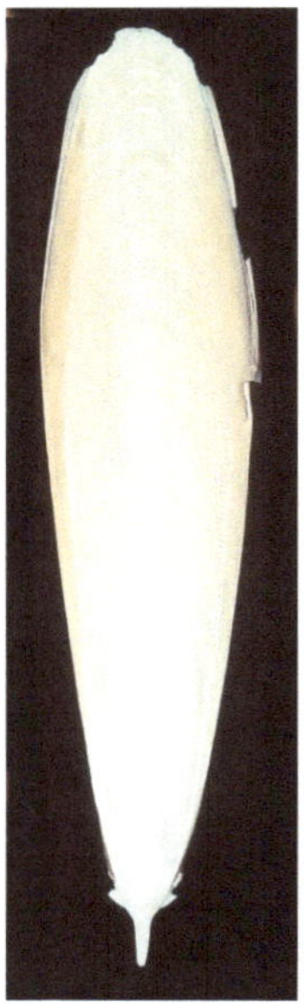

Cuttlebone, dorsal view (D. Paul).

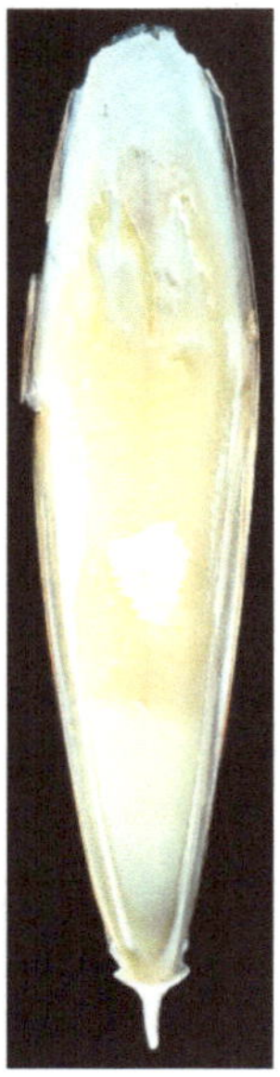

Cuttlebone, ventral view (D. Paul).

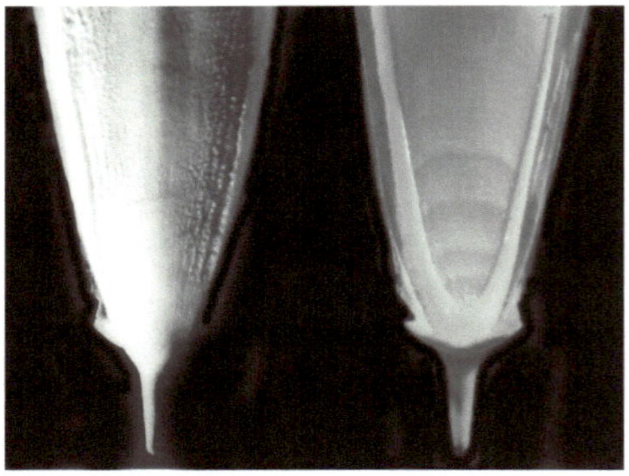

Posterior end of cuttlebone dorsal view (left), ventral view (right) (A. Reid).

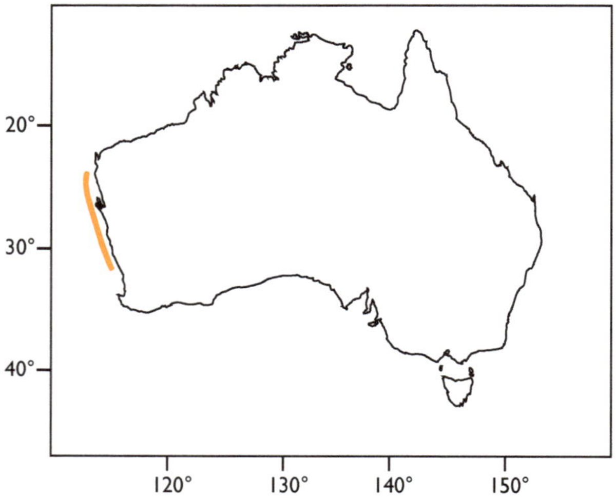


*Sepia cottoni* distribution. *Note:* The distribution is based on whole animals only, not beachwashed cuttlebones.

slightly V-shaped following sulcus. Inner cone limbs narrow, strap-like anteriorly, broaden posteriorly; slightly raised to form rounded posterior ledge; thickened. Outer cone limbs expanded posteriorly forming two short 'wings', directed ventrally, forming recurved cup-like structure.

### SIZE

Up to 55 mm ML.

### REMARKS

Dunning *et al.* (1994) report that *S. cottoni* occurs in the Gulf of Carpentaria, and Adam (1979) mentions a cuttlebone collected from North Stradbroke Island off Queensland, 27°31.5′S, 121°27′E, implying a northern Australian distribution although there is some doubt regarding the correct identification of these specimens and these records should be treated with caution.

### DISTRIBUTION

South-eastern Indian Ocean: Australia, Western Australia, from 19°04′24″S, 118°47′30″E, to 33°45′12″S, 114°28′36″E. Known depth range 83–128 m.

### REFERENCE

Reid (2000).

## *Sepia cultrata* Hoyle, 1885

*Sepia cultrata* Hoyle, W. E. (1885). Diagnoses of new species of Cephalopoda collected during the cruise of H.M.S. Challenger. Part 2. The Decapoda. *Annals and Magazine of Natural History (5)*16, 181–203 [198].

Type data: holotype BMNH 1889.4.24.75.

Type locality: Australia: New South Wales, off Twofold Bay, H.M.S. Challenger Stn 163A, 36°59′S, 150°20′E.

### COMMON NAME

Knifebone Cuttlefish

### DIAGNOSTIC FEATURES

All arms subequal in length in both males and females. Arm suckers tetraserial. Left ventral arm hectocotylised:

Fresh-caught specimen, dorsal view (K. Graham).

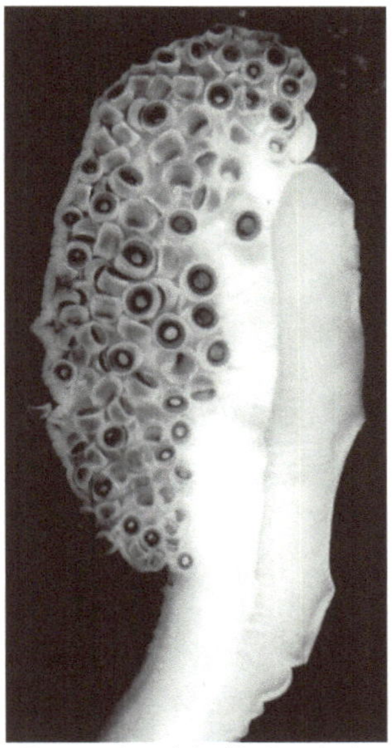

Tentacular club (C. Rowley).

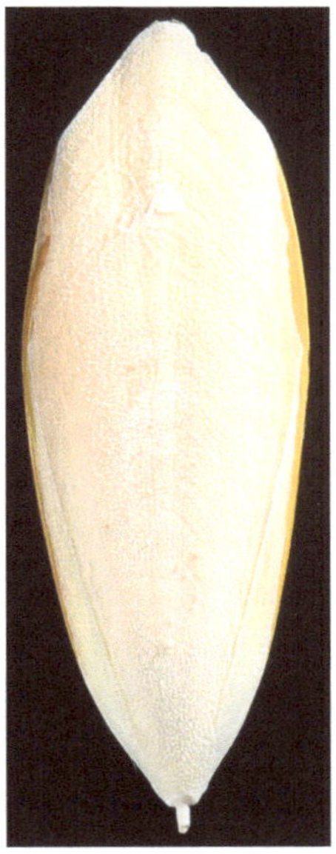

Cuttlebone, dorsal view (D. Paul).

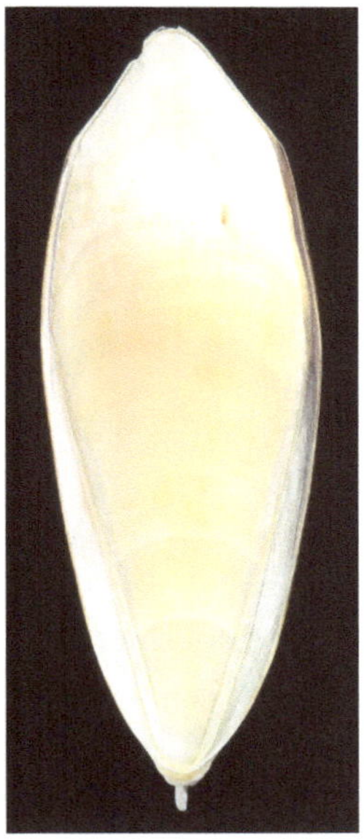

Cuttlebone, ventral view (D. Paul).

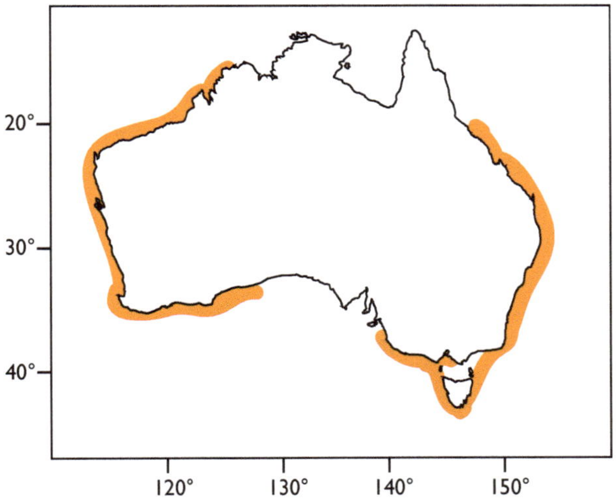


*Sepia cultrata* distribution. *Note:* The distribution is based on whole animals only, not beachwashed cuttlebones.

suckers size normal proximally (approximately seven transverse rows), greatly reduced medially (five or six rows), with suckers in two dorsal series smaller than those in two ventral series (in some specimens, two dorsal rows appear merged as if in a single row); oral surface of modified region swollen; suckers in two dorsal and two ventral series displaced laterally, with gap between them. Club crescent-shaped, sucker-bearing surface flattened, with five or six small suckers in each transverse row. Swimming keel of club extends well proximal to carpus; dorsal and ventral protective membranes not joined at base of club; dorsal and ventral membranes differ in length; dorsal membrane extends proximal to carpus along stalk, ventral membrane terminates at proximal end of carpus. Buccal membrane without suckers.

*Cuttlebone.* Cuttlebone outline oval; obtuse anteriorly; acuminate, acute posteriorly; dorsal surface pinkish, convex posteriorly, flat anteriorly; some granulation on lateroposterior margins bordered by longitudinal ridge; dorsal median rib present; rib broadens anteriorly; lateral ribs indistinct. Spine straight. Striated zone concave in extreme posterior part, slightly convex in anterior part; last loculus slightly concave; sulcus absent. Anterior striae slightly convex to straight. Inner cone limbs are uniform width, very narrow, U-shaped posteriorly; thickened.

*Colour.* Pale buff pinkish brown. Fresh-caught specimens pale cream. Cuttlebone cream or salmon-coloured dorsally.

*Sculpture.* Dorsal mantle with scattered short papillae along each side close to fins and with longitudinal row of ridge-like papillae adjacent to base of each fin; ventral mantle with longitudinal row of four or five narrow ridges along each side close to fins.

## HABITAT AND BIOLOGY

*Interest to fisheries.* Species taken as by-catch of prawn and mixed species trawl fisheries.

## SIZE

Up to 120 mm ML.

### DISTRIBUTION

Southern Australia from southern Queensland, 17°45′S, 146°55′11″E, south-westward to Western Australia, 13°49′48″S, 125°E, including Tasmania. Known depth range 132–803 m (main catches 300–500 m); outer shelf and upper bathyal.

### REFERENCE

Reid *et al.* (2005b).

## *Sepia elliptica* Hoyle, 1885

*Sepia elliptica* Hoyle, W. E. (1885). Diagnoses of new species of Cephalopoda collected during the cruise of H.M.S. Challenger. Part 2. The Decapoda. *Annals and Magazine of Natural History (5)*16, 181–203 [189].

Type data: syntypes BMNH 1889.4.24.71–72, BMNH 1889.4.24.73–74.

Type locality: Arafura Sea, 9°59′S, 139°42′E, 08°56′S, 136°05′E.

### COMMON NAME

Ovalbone Cuttlefish

### DIAGNOSTIC FEATURES

Mantle oval; dorsal anterior margin triangular, acute. All arms subequal in length in both males and females; protective membranes narrow. Arm suckers tetraserial. Left ventral arm hectocotylised: sucker size normal proximally (seven or eight rows), slightly reduced medially (seven rows), with suckers in two dorsal series of modified portion smaller than those in two ventral series; suckers then normal size to arm tip. Oral surface of modified region not wide, fleshy, but normal, as on opposite arm. Club sucker-bearing surface flattened, with 10–12 minute suckers in each transverse row; suckers all similar size. Swimming keel of club extends well proximal to carpus; dorsal and ventral protective membranes joined at base of club; dorsal membrane forms shallow cleft at junction with stalk.

*Cuttlebone.* Cuttlebone outline oval; bone very angular, V-shaped anteriorly; bluntly rounded posteriorly; dorsal surface creamy white; dorsal surface evenly convex; texture smooth, not pustulose; dorsal median rib indistinct, broadens anteriorly; lateral ribs indistinct. Chitin surrounds entire margin of cuttlebone. Spine short, pointed, curves dorsally, keel(s) absent. Striated zone concave; last loculus convex; sulcus deep, wide. Anterior striae are inverted U-shape. Inner cone limbs are narrow anteriorly, broaden posteriorly; outer margin of inner cone raised into flat posterior ledge; ledge whitish (sometimes with a thin rim of chitin on outer margin); ledge not thickened.

*Colour.* Live animals with pale transverse stripes across dorsal surface of mantle and white line along base of fins.

### HABITAT AND BIOLOGY

*Sepia elliptica* occurs mainly in coastal waters. This species was collected between 10 and 62 m in Gulf of Carpentaria surveys. A broad size range of both sexes was found to

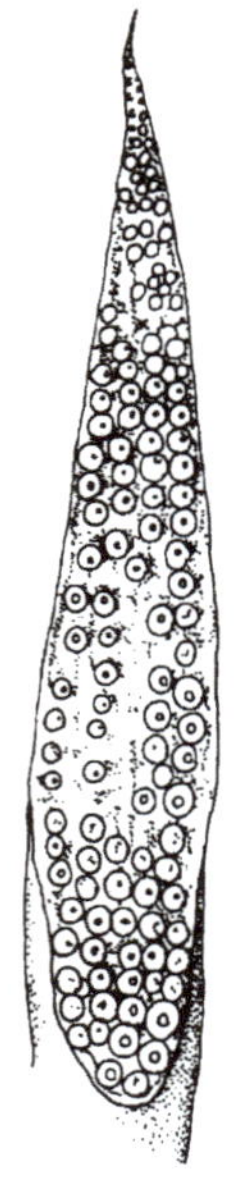

Hectocotylus (FAO, E. D'Antoni).

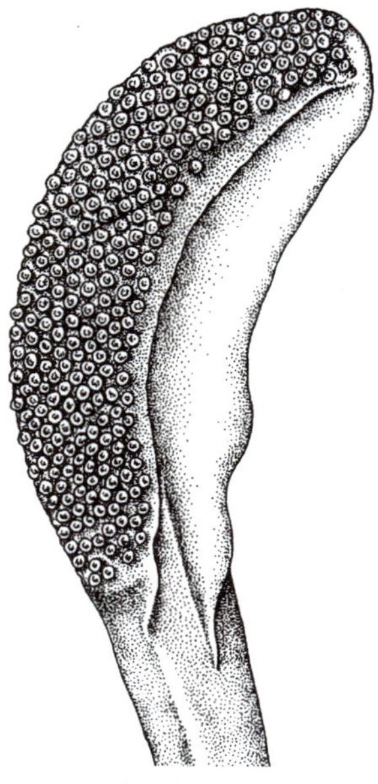

Tentacular club (subadult) (ABRS, K. Hollis).

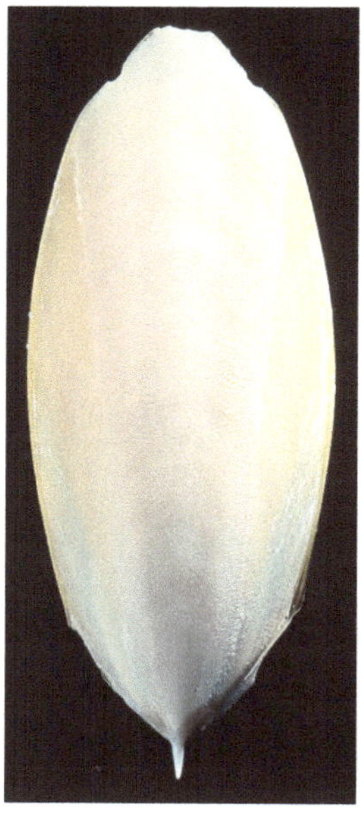

Cuttlebone (ABRS, K. Hollis).

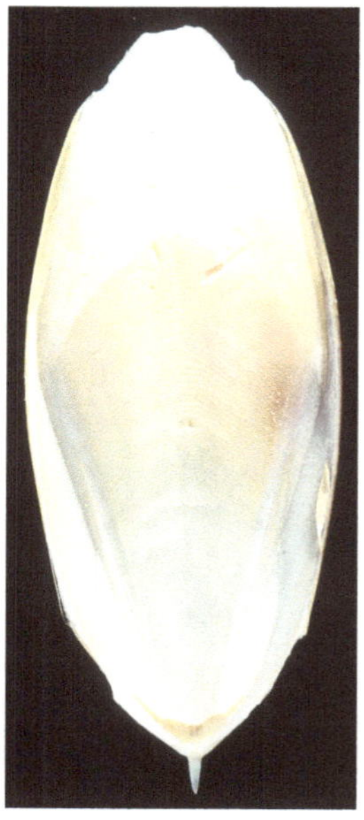

Cuttlebone, dorsal view (D. Paul).

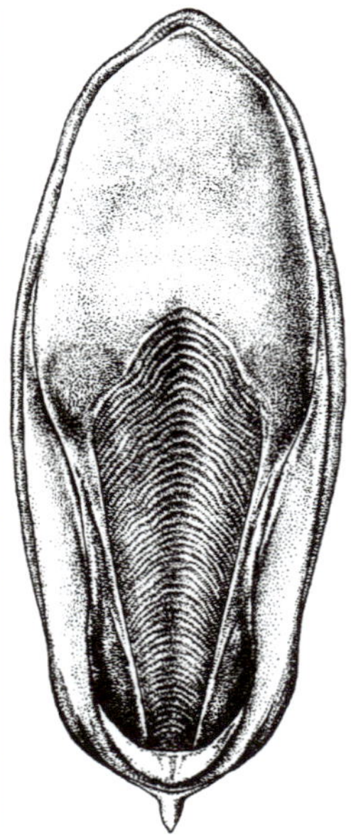

Cuttlebone, ventral view (D. Paul).

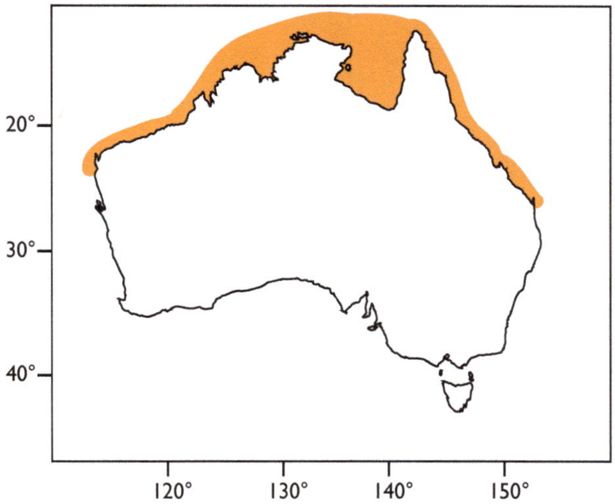

*Sepia elliptica* distribution. *Note:* The distribution is based on whole animals only, not beachwashed cuttlebones.

occur in the Gulf over most of the year, indicative of an extended spawning season. Several experimental studies were carried out on growth of this species in captivity; results showed that both water temperature and feeding levels affect somatic growth but with different effects on the muscular tissue development and protein composition. No direct relationship was found between individual growth rates and food availability.

Inhabits relatively deep water and moves into shallower water for spawning.

*Interest to fisheries.* *Sepia elliptica* is taken as by-catch of prawn and mixed species trawl fisheries throughout its range. The population structure observed in the Gulf of Carpentaria by experimental trawl surveys and the relative high catch rates of trawlers in some areas suggest that *S. elliptica* is a promising fisheries resource.

### SIZE

Up to 175 mm ML.

### REMARKS

Duc (1978, 1993) refers to *S. elliptica* in Vietnamese waters; however, this probably is a misidentification of *S. esculenta* (Hoyle, 1885). The occurrence of *S. elliptica* in Indian waters has yet to be confirmed.

### DISTRIBUTION

Northern Australia from Western Australia, 26°40′12″S, 153°40′48″E, to Queensland, Capricorn Island Group, 22°15′S, 114°15′E, including Gulf of Carpentaria. Vietnam and India (doubtful records). Possibly tropical Indo-Pacific (conspecificity to be confirmed). Known depth range 13–142 m.

### REFERENCE

Reid *et al.* (2005b).

## *Sepia grahami* Reid, 2001

*Sepia grahami* Reid, A. L. (2001). A new cuttlefish, Sepia grahami, sp. nov. (Cephalopoda: Sepiidae) from eastern Australia. *Proceedings of the Linnean Society of New South Wales 123*, 159–172 [160, figs 1–5].

Type data: holotype MV F.87757.

Type locality: Australia: New South Wales, 29°33′S, 153°25′E–29°32′S, 153°25′E.

### COMMON NAME

Ken's Cuttlefish

### DIAGNOSTIC FEATURES

Mantle oval. All arms subequal in length in both males and females; distal arm tips not attenuate. Arm suckers tetraserial. Hectocotylus absent. Club short, slightly recurved, sucker-bearing surface flattened, with four or five suckers in each transverse row (usually four, rarely five); suckers differ only slightly in size, three or four suckers slightly enlarged. Swimming keel of club extends well proximal to carpus; dorsal and ventral protective membranes not joined at base of club; dorsal membrane extends proximal to carpus along stalk, ventral membrane terminates at proximal end of carpus.

*Cuttlebone.* Cuttlebone oval; dorsal surface creamy white, with slight pinkish tinge; dorsal median rib absent, lateral ribs indistinct. Spine with ventral keel. Striated zone concave, sulcus shallow, narrow, indistinct. Anterior striae are inverted U-shape; last loculus flat. Inner cone limbs are narrow, strap-like anteriorly, broaden posteriorly. Inner cone limbs are raised into rounded ledge posteriorly, ledge thickened, yellowish or ochre coloured.

*Colour.* Pinkish brown. Mantle with dorsal 'eyespots'.

*Sculpture.* Dorsal mantle with longitudinal series of up to six ridges along each side, close to fins; prominent ear-shaped lobe dorsal to eye and two lobes ventral to eye, anterior-most lobe largest.

### HABITAT AND BIOLOGY

*Interest to fisheries.* There is no official information on this species but it is commonly confused with *Sepia*

Freshly caught specimen, dorsal view (K. Graham).

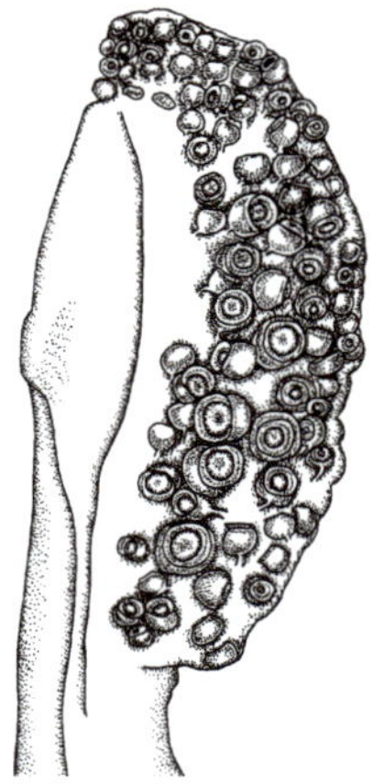

Tentacular club (A. Reid).

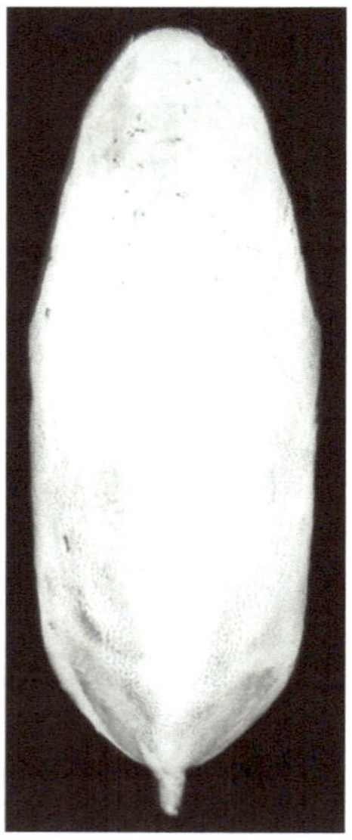

Cuttlebone, dorsal view (L. Mott).

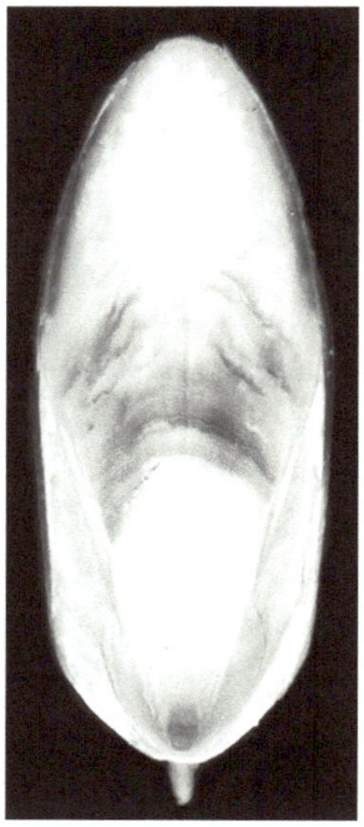

Cuttlebone, ventral view (L. Mott).

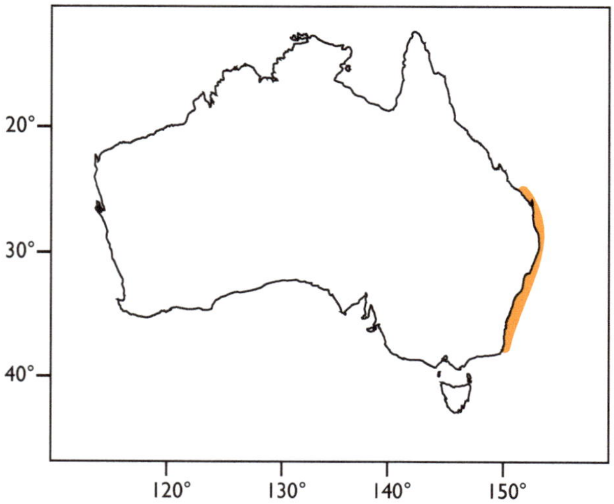


*Sepia grahami* distribution. *Note:* The distribution is based on whole animals only, not beachwashed cuttlebones.

*rozella* (Iredale, 1926) and *S. mestus* Gray, 1849. *Sepia rozella* commonly appears in fish markets in Sydney.

### SIZE

Males up to 66 mm ML; females up to 82 mm ML.

### DISTRIBUTION

South-western Pacific: Australia, New South Wales, from 28°37′47″S, 153°24′E to 36°40′12″S, 150°3′E. Known depth range 2–84 m.

### REFERENCE

Reid *et al.* (2005b).

## *Sepia hedleyi* Berry, 1918

*Sepia hedleyi* Berry, S. S. (1918). Report on the Cephalopoda obtained by the F.I.S. Endeavour in the Great Australian Bight and other southern Australian localities. Biological *Results of the Fishing Experiments carried on by the F.I.S. 'Endeavour', 1909–14 4*, 203–298 [258, text figs 48–50, pls 71–72].

Type data: holotype AM E.2464.

Type locality: Australia: South Australia, Investigator Strait, south of Kangaroo Island.

### COMMON NAME(S)

Hedley's Cuttlefish, King Cuttlefish

### DIAGNOSTIC FEATURES

Mantle broad, oval. All arms subequal in length in both males and females. Arm suckers tetraserial. Left ventral arm hectocotylised: sucker size normal proximally (six to eight transverse rows), reduced medially (9–10 rows), with suckers in two dorsal series of modified portion much smaller than those in two ventral series. Oral surface of modified region wide, swollen, fleshy, with transversely grooved ridges; two dorsal and two ventral series of suckers displaced laterally, with gap between; suckers in two ventral series aligned in a single row. Club crescent-shaped, moderate length, sucker-bearing surface flattened, with 9–12 suckers in each transverse row; all club suckers of similar, small, size. Swimming keel of club extends well proximal to carpus; dorsal and ventral protective membranes not joined at base of club but are fused to tentacular stalk; dorsal and ventral membranes same length; extend proximal to carpus along stalk. Buccal membrane without suckers; in females extends ventrally with two spermathecae.

*Cuttlebone.* Cuttlebone outline oblong to elongate diamond-shaped; acuminate, acute, anteriorly and

Fresh-caught specimen (CSIRO Science Image AF9586).

Cuttlebone, dorsal view (D. Paul).

Cuttlebone, ventral view (D. Paul).

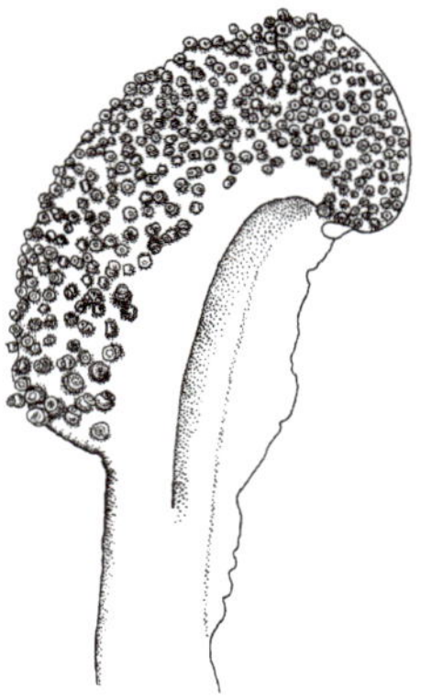

Tentacular club (A. Reid).

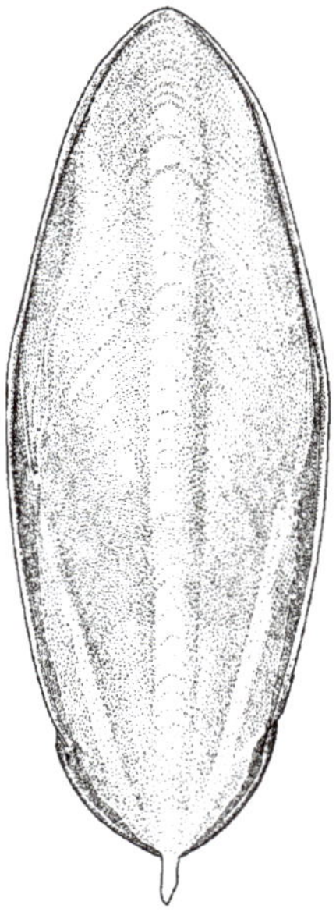

Cuttlebone, dorsal view (A. Reid).

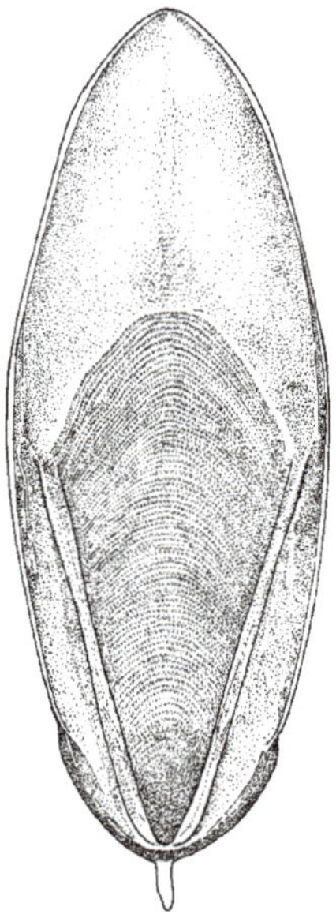

Cuttlebone, ventral view (A. Reid).

posteriorly; dorsal surface creamy white; dorsal surface evenly convex; granulose; spine and posterior tip of bone covered with a smooth glaze-like substance. Dorsal median rib distinct; sides approximately parallel; ribs bordered laterally by distinct grooves; lateral ribs indistinct. Chitin surrounds entire margin of cuttlebone. Spine short, pointed, straight, parallel to bone, keel(s) absent. Striated zone flat; last loculus convex; sulcus shallow, narrow, extends entire length of cuttlebone. Anterior striae are inverted U-shape. Inner cone limbs are uniform width, form narrow V-shape posteriorly, thickened, shiny; outer cone calcified, narrow anteriorly, broadens posteriorly.

*Colour.* Buff pinkish brown. Arms without markings. Dorsal ridges orange-pink in colour.

*Sculpture.* Dorsal mantle has longitudinal row of up to six ridge-like papillae along each side, adjacent to base of each fin.

### HABITAT AND BIOLOGY

*Sepia hedleyi* is taken as by-catch of prawn and mixed species offshore trawl fisheries.

### SIZE

Males up to 83 mm ML; females up to 108 mm ML.

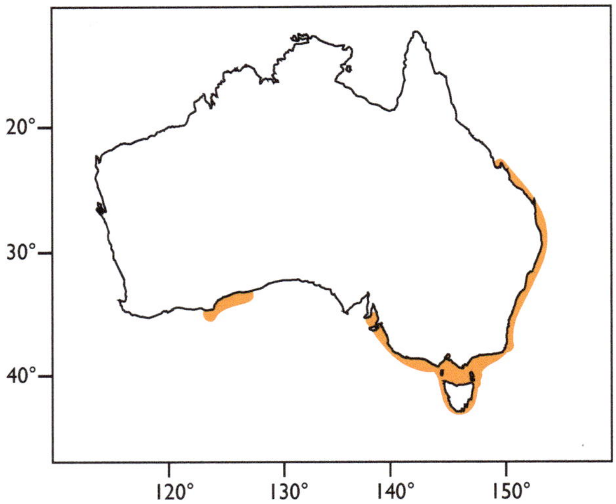


*Sepia hedleyi* distribution. *Note:* The distribution is based on whole animals only, not beachwashed cuttlebones.

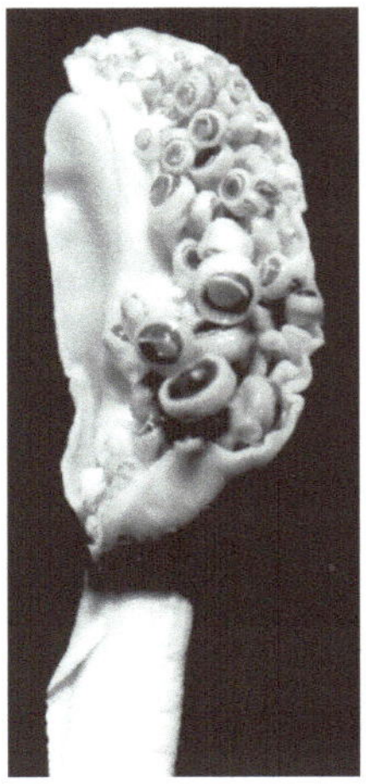

Tentacular club (C. Rowley).

Cuttlebone, dorsal view (D. Paul).

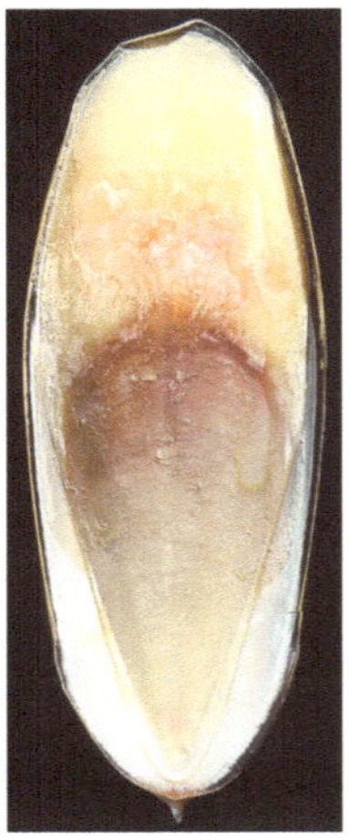

Cuttlebone, ventral view (D. Paul).

## REMARKS

Formerly known as *S. rex*, a synonym of *S. hedleyi*.

## DISTRIBUTION

Southern Indo-Pacific: Australia, Queensland, from off the Great Barrier Reef, 23°26′59″S, 151°55′11″E, southwards around southern Australia to Western Australia, 33°10′12″S, 127°55′12″E. Known depth range 47–1092(?) m. Australian endemic.

## REFERENCE

Reid (2001).

# *Sepia irvingi* Meyer, 1909

*Sepia irvingi* Meyer, W. Th. (1909). Cephalopoda. pp. 329–335 in Michaelsen, W. & Hartmeyer, R. (eds) *Die Fauna Südwest-Australiens.* Jena: G. Fischer Vol. 2 [333, figs 7–10].

Type data: syntypes WAM 4203, ZMH 3842.

Type locality: Australia: Western Australia, Cockburn Sound, Garden I., 32°11′S, 115°43′E; Port Royal (King George Sound), 35°03′S, 117°58′E; Warnbro Sound, 32°20′S, 115°43′E.

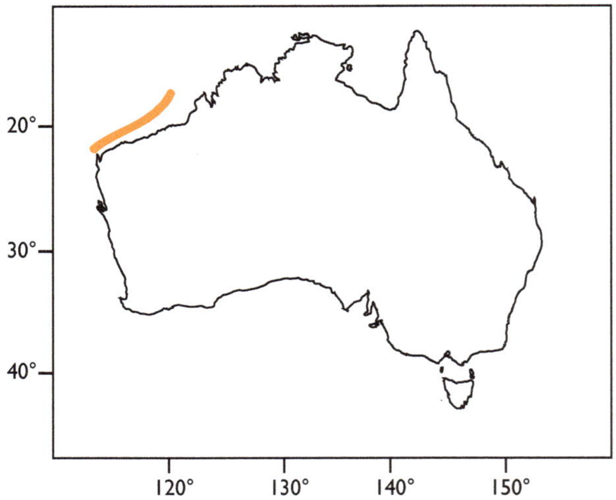


*Sepia irvingi* distribution. *Note:* The distribution is based on whole animals only, not beachwashed cuttlebones.

## DIAGNOSTIC FEATURES

Mantle broad. All arms subequal in length in both males and females. Non-hectocotylised arm suckers tetraserial. Hectocotylus absent. Club sucker-bearing surface flattened; with eight suckers in each transverse row; suckers differ markedly in size, some greatly enlarged. Swimming keel of club extends well beyond carpus; dorsal and ventral protective membranes fused at base of club; separated from stalk by membrane; dorsal and ventral membranes same length; dorsal membrane forms deep cleft at junction with stalk.

*Cuttlebone.* Cuttlebone outline oblong. Bone bluntly rounded anteriorly and posteriorly. Dorsal surface creamy white; evenly convex. Dorsal median rib present, indistinct, broadens anteriorly; lateral ribs present, indistinct. Spine present, short, pointed. Striated zone and last loculus flat. Sulcus shallow, narrow. Anterior striae are inverted U-shaped. Inner cone limbs narrow anteriorly, broaden posteriorly; not raised to form ledge posteriorly; thickened. Outer cone broadens posteriorly.

## SIZE

Up to 100 mm ML.

### DISTRIBUTION

Western Australia, from 18°49′S, 118°29′E to 19°12.06′S, 116°12′E. Known depth range 130–170 m.

### REFERENCE

Reid *et al.* (2005b).

## *Sepia koilados* Reid, 2000

*Sepia koilados* Reid, A. L. (2000). Australian cuttlefishes (Cephalopoda: Sepiidae): the 'doratosepion' species complex. *Invertebrate Taxonomy 14*, 1–76 [24, figs 7, 14–18].

Type data: holotype MV F.82459.

Type locality: Australia: Western Australia, North West Shelf, 18°36′S, 118°2–4′E.

### DIAGNOSTIC FEATURES

Small to moderate-sized species. All arms subequal in length in both males and females. Arm suckers tetraserial in both sexes. Female dorsal and ventral arm sucker pairs widely spaced with distinct gap between on arms 1–3. Left ventral arm hectocotylised; sucker size normal proximally, reduced distally, then normal to arm tip; from proximal to distal end of arm; suckers in two dorsal series smaller than remaining suckers; two dorsal and two ventral series displaced laterally; suckers markedly reduced. Oral surface of modified region wide; protective membranes well developed, with transversely grooved ridges; ventral protective membrane thick, fleshy; with deep median furrow. Club crescent-shaped; sucker-bearing surface flattened, with eight suckers in each transverse row. Suckers all similar size, small. Swimming keel of club extends well beyond carpus; dorsal and ventral protective membranes not fused at base of club; joined to stalk; dorsal and ventral membranes same length, extend beyond carpus along stalk. Buccal membrane without suckers.

*Cuttlebone.* Cuttlebone outline oblong; slightly recurved ventrally; dorsal surface creamy white; evenly convex; calcified, granulose, posteriolateral calcification concentrated in short ridges, bordered laterally by two long calcareous ridges. Dorsal median rib, distinct, rounded, broadens anteriorly; bordered laterally by grooves; lateral ribs distinct. Chitin present, wide bands border lateral margins of cuttlebone. Spine present; keels absent. Sulcus extends entire length of cuttlebone; deep, narrow; flanked by rounded ribs bordered laterally by shallow grooves. Anterior striae are inverted U-shape. Inner cone limbs narrow anteriorly, broaden posteriorly;

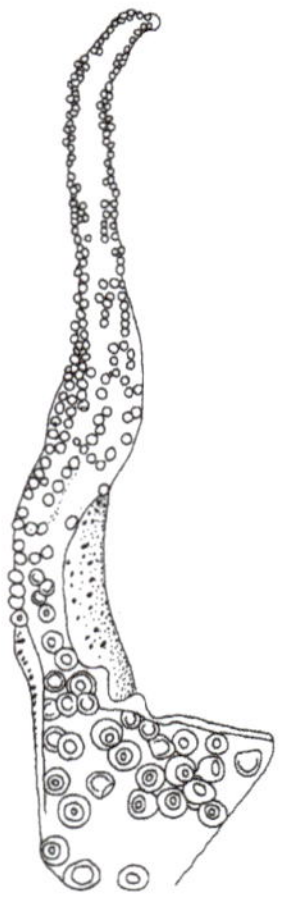

Arm 2, oral view (A. Reid).

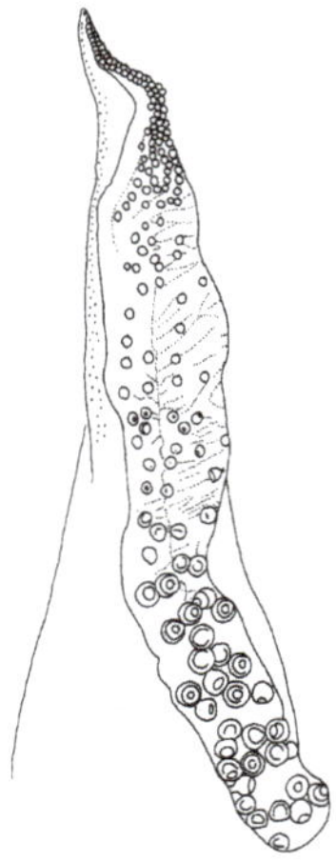

Hectocotylus (A. Reid).

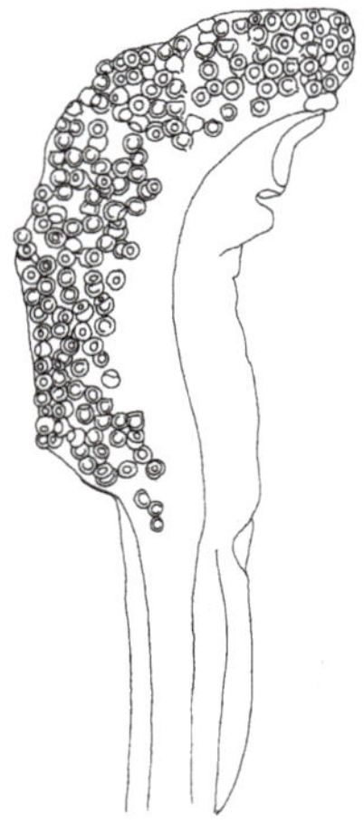

Tentacular club (A. Reid).

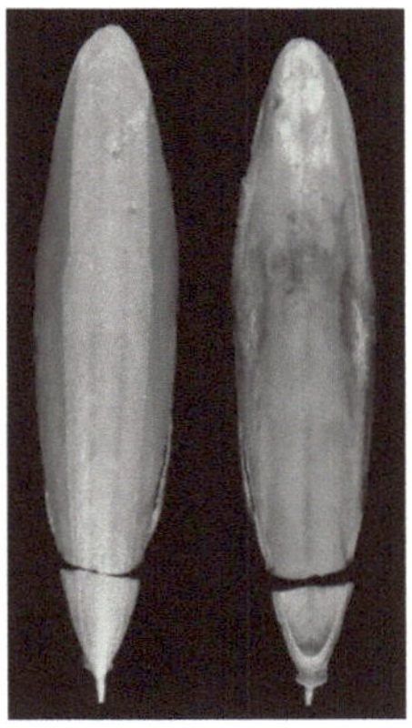

Cuttlebone, dorsal view, left, ventral view, right (A. Reid).

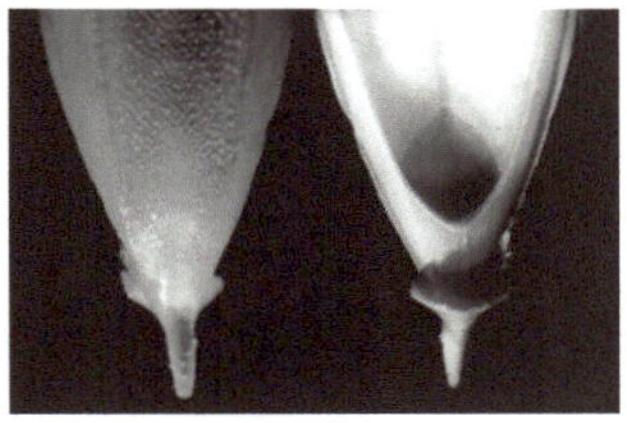

Cuttlebone, posterior end enlargement: dorsal view (right) ventral view (left) enlargement (A. Reid).

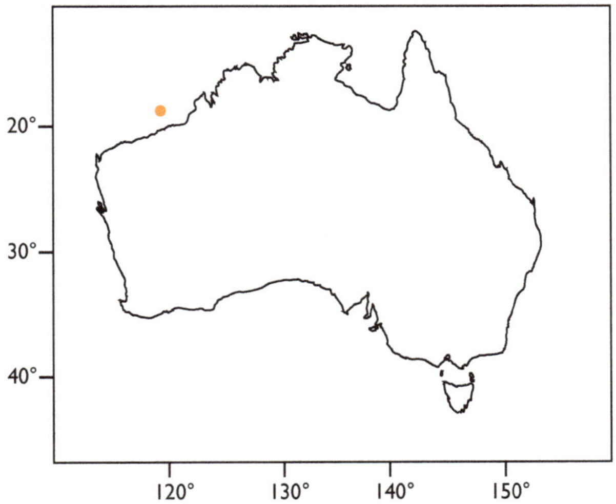


*Sepia koilados* distribution. *Note:* The distribution is based on whole animals only, not beachwashed cuttlebones. (Australian records are based on few specimens, but the species is likely to occur more widely.)

raised to form slightly thickened, flat posterior ledge. Outer limbs expanded posteriorly forming two short 'wings', directed ventrally, forming recurved cup-like structure.

*Colour* (alcohol preserved specimens). Head, arms and dorsal mantle purplish brown, with chromatophores concentrated in broad band medially over cuttlebone; fins pale.

### HABITAT AND BIOLOGY

Habitat silt, mud, rock, limestone.

### SIZE

Males up to 68 mm ML; females up to 58 mm ML.

### REMARKS

*Sepia koilados* sp. nov. occurs sympatrically with *S. rhoda* (Iredale, 1954), *S. vercoi* Adam, 1979 and also possibly with *S. cottoni* Adam, 1979.

### DISTRIBUTION

Type locality: Australia: Western Australia, North West Shelf, 19°12'06"S, 116°25'12"E. Known depth range 116–203 m.

### REFERENCE

Reid (2000).

## *Sepia latimanus* Quoy & Gaimard, 1832

*Sepia latimanus* Quoy, J. R. C. & Gaimard, J. P. (1832). Mollusques. In 'Zoologie du voyage de l'Astrolabe sous les ordres du Capitane Dumont d'Urville, pendant les années 1826–1829. Zoologie'. 2 vols, pp. 1–320 (J. Tastu, Editeur-Imprimeur: Paris) [68, pl. 2, figs 2–11].

Type data: syntypes MNHN 1-6-192.

Type locality: New Guinea, Port Dorey.

### COMMON NAME

Broadclub Cuttlefish

### DIAGNOSTIC FEATURES

Mantle oval. Arm suckers tetraserial. Hectocotylus absent. Club crescent-shaped, sucker-bearing surface flattened, with five or six suckers in each transverse row; suckers differ markedly in size, some greatly enlarged. Swimming keel of club extends well proximal to carpus; dorsal and ventral protective membranes joined at base of club; separated from stalk by membrane; dorsal membrane forms shallow cleft at junction with stalk. Buccal membrane without suckers.

*Cuttlebone.* Cuttlebone bluntly rounded anteriorly and posteriorly; dorsal surface convex posteriorly, flat anteriorly; granulose; dorsal median rib and lateral ribs indistinct, or absent. Spine short, pointed, stout, keels absent. Striated zone concave; last loculus convex; sulcus shallow, narrow, extends entire length of cuttlebone. Anterior striae are inverted V-shape. Inner cone limbs uniform width, narrow, U-shaped posteriorly; thickened into a rounded, shiny ridge; outer cone calcified, narrow anteriorly, broadens posteriorly, deep cup-like.

*Colour.* Light brown, yellowish or dark brown (dark in males during courtship and breeding), with whitish mottle, blotches or spots. Arms have longitudinal white bands on their margins that appear as broad white blotches when the arms are extended, and arms 1–3 have broad, longitudinal brownish bands medially, extending onto head. Dorsal mantle has transverse saddle mark, small white and brown spots, narrow brown transverse bands, and bold, white, transverse stripes and spots dorsally (spots and blotches are present in most colour patterns; saddle-mark shown only occasionally; transverse stripes shown by males during the breeding season). In live animals, the eyes are typically yellow around the ventral margins. Fins pale with white, transverse stripes extending onto mantle and narrow, white band along outer margin.

*Sculpture.* Dorsal mantle covered with numerous large papillae and with series of elongate papillae along each side, adjacent to base of each fin. Head papillose.

### HABITAT AND BIOLOGY

*Sepia latimanus* is day-active. During hunting it appears to mesmerise prey by displaying rhythmic colour bands along the body. It feeds on fishes and crustaceans. During the breeding season, males establish a territory, defending a coral head (typically *Porites*) among which females lay eggs following mating. Courtship is highly ritualised and stereotyped, and it incorporates striking visual displays. Males often guard females to ward off other males. Mating takes place head to head and spermatophores are placed in the females' buccal membrane ventral to the mouth. The eggs harden after laying and are very difficult to extract from the coral. They hatch after 4–6 weeks, and the young immediately hide among the coral and coral

Female preparing for egg laying (M. Norman).

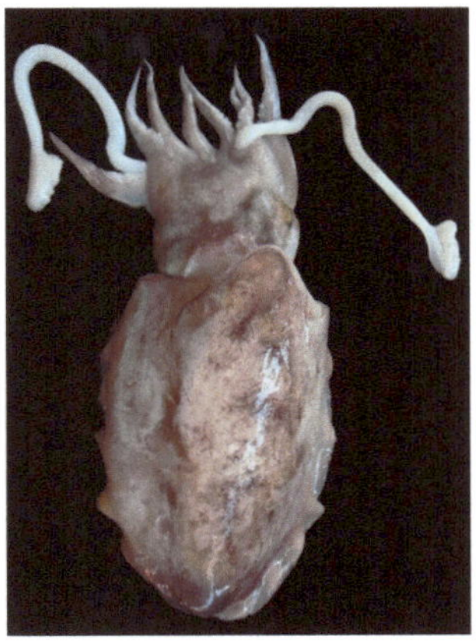

Fresh-caught specimen, dorsal view (A. Reid).

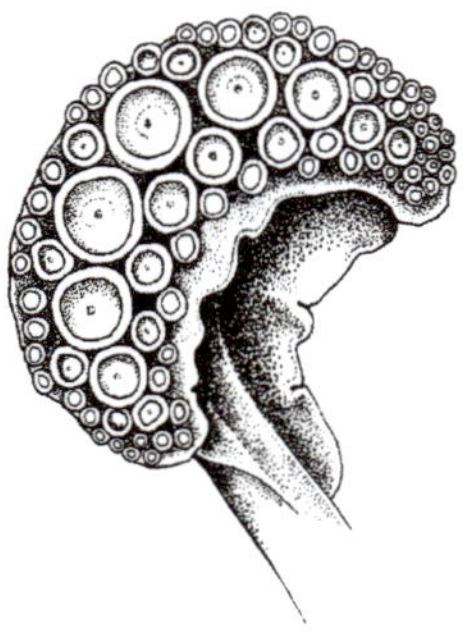

Tentacular club (FAO, E. D'Antoni).

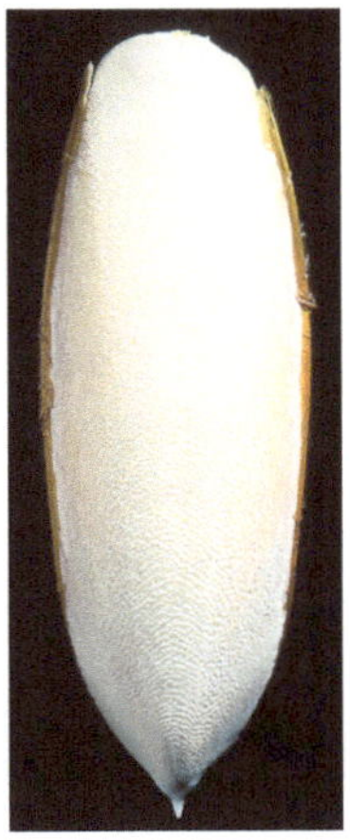

Cuttlebone, dorsal view (D. Paul).

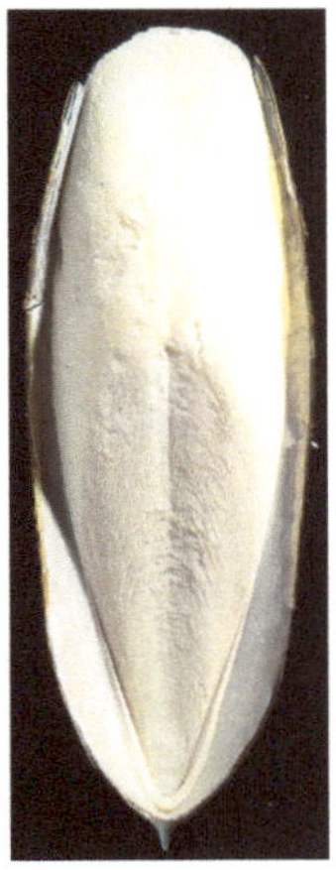

Cuttlebone, ventral view (D. Paul).

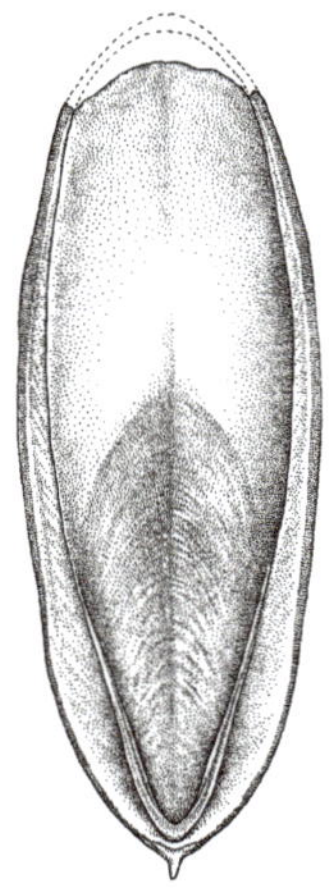

Cuttlebone, ventral view (ABRS, K. Hollis).

rubble. Juveniles often mimic mangrove leaves, their colouration and posture complete with the stem, ribs and scattered black spots.

## SIZE

Up to 500 mm ML and weight to 10 kg.

## REMARKS

Conspecificity over the range of this species needs to be confirmed.

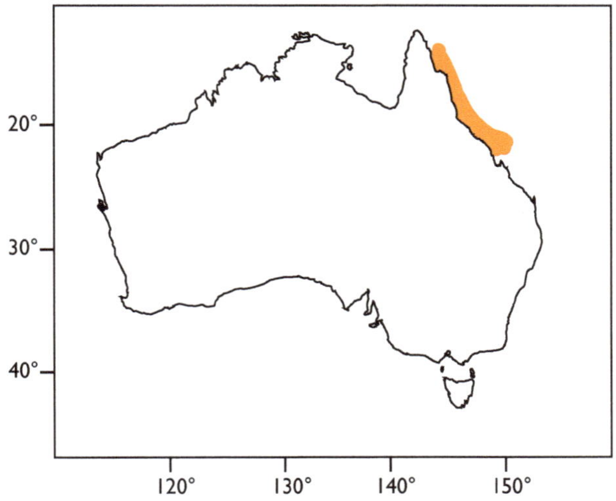


*Sepia latimanus* distribution. *Note:* The distribution is based on whole animals only, not beachwashed cuttlebones.

Kawashima *et al.* (2013) have determined the complete mitochondrial genome for *S. latimanus* purchased at a Japanese fish market.

## DISTRIBUTION

Indo-Pacific: from southern Mozambique, throughout the periphery of the Indian Ocean, Malacca Strait, Melanesian Islands, South China Sea, Philippine Sea and East China Sea, Taiwan Province of China and Japan to southern Kyushu. Indonesia to north-western and north-eastern Australia from 14°35′59″S, 145°37′12″E to 21°26′59″S, 151°40′48″E, the Coral Sea, Palau, Guam, New Caledonia, Fiji and Cocos (Keeling) Islands. Records from Madagascar and south-eastern Australia doubtful. Known depth range 8–45 m.

## REFERENCE

Reid *et al.* (2005b).

# *Sepia limata* (Iredale, 1926)

*Arctosepia limata* Iredale, T. (1926). The cuttlefish 'bones' of the Sydney beaches (Phylum Mollusca – Class Cephalopoda). *Australian Zoologist 4*, 186–196 [193, pl. 23].

Type data: holotype AM C.133316 (cuttlebone).

Type locality: Australia: New South Wales, Manly Beach, 33°48′S, 151°17′E.

## DIAGNOSTIC FEATURES

Small species. All arms subequal in length in both males and females. Distal arm tips in both sexes not markedly attenuate. Arm sucker arrangement differs between sexes.

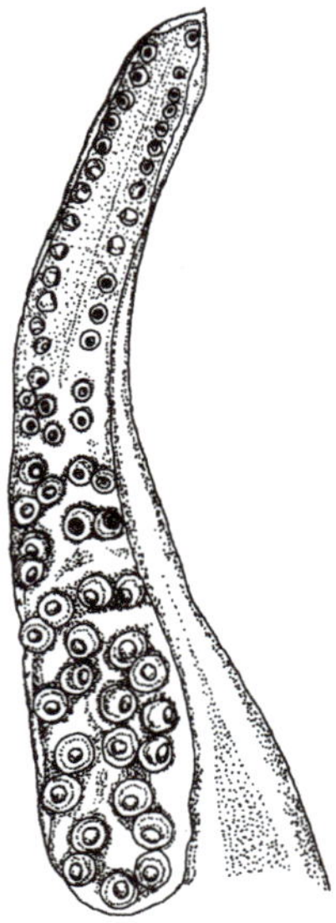

Arm 3, oral view (A. Reid).

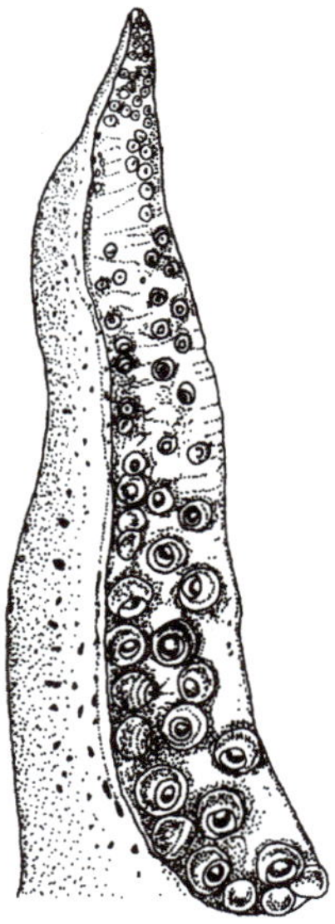

Hectocotylus (A. Reid).

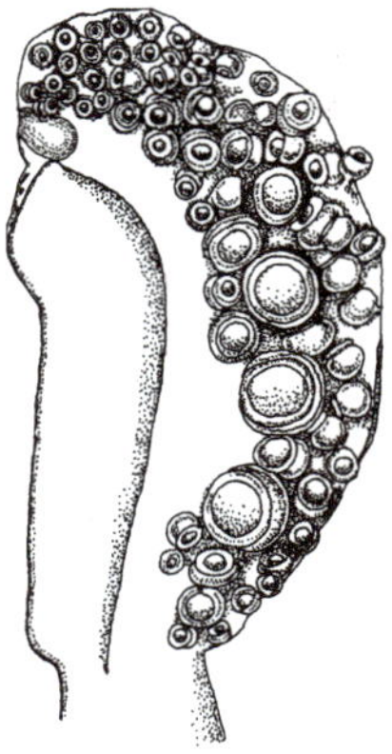

Tentacular club (A. Reid).

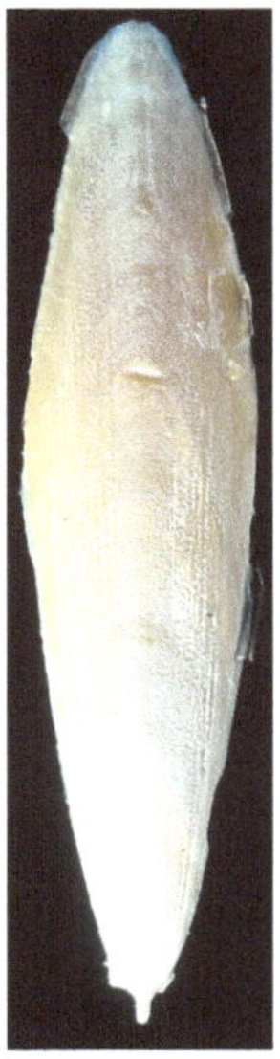

Cuttlebone, dorsal view (D. Paul).

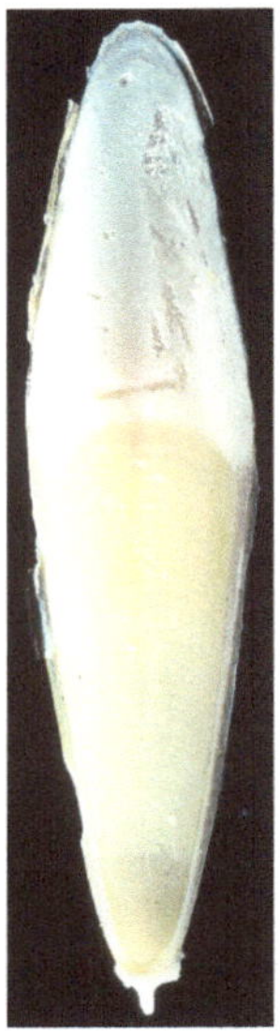

Cuttlebone, ventral view (D. Paul).

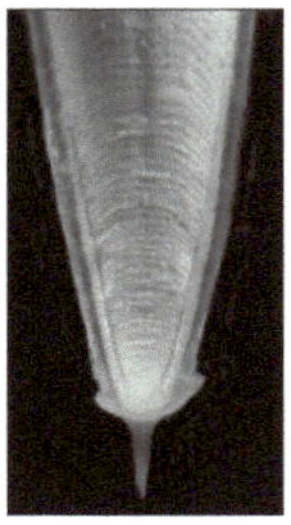

Posterior end of cuttlebone enlargement, ventral view (A. Reid).

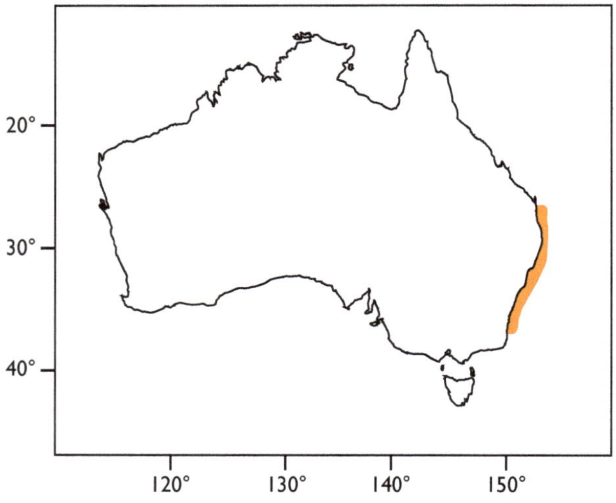


*Sepia limata* distribution. *Note:* The distribution is based on whole animals only, not beachwashed cuttlebones.

In males, arm suckers tetraserial. In females, arm suckers biserial proximally, tetraserial medially, biserial at distal arm tips. Female arms 1 and 4 protective membranes wide, enclose distal suckers; arms 2 and 3 distally slightly thickened, fleshy, suckers displaced laterally. Greater proportion of arms 2 and 3 with distal biserial suckers than arms 1 and 4. Left arm 4 hectocotylised; sucker size reduced distally; from proximal to distal end of arm, seven to nine rows of normal suckers, 8–10 rows of reduced suckers (approximately). Suckers in two medial series smaller than those in dorsal and ventral marginal series; reduced suckers much smaller than normal arm suckers. Oral surface of modified region wide, swollen, fleshy, with transversely grooved ridges; without distinct median furrow. Suckers in two dorsal and two ventral series displaced laterally. Hectocotylised arm not markedly attenuate distally. Club crescent-shaped; sucker-bearing surface flattened, with four suckers in each transverse row. Swimming keel of club extends slightly beyond carpus; dorsal and ventral protective membranes not fused at base of club; joined to stalk; dorsal and ventral membranes same length, terminate at posterior end of carpus.

*Cuttlebone.* Cuttlebone outline lanceolate, not strongly convex in lateral view. Dorsal surface pinkish; evenly convex; entire surface calcified with reticulate granulose sculpture concentrated posteriolaterally and posteriorly in irregular longitudinal ridges. Dorsal median rib present, indistinct, broadens anteriorly; not bordered laterally by distinct grooves; lateral ribs absent. Spine present, long, pointed; dorsoposterior end of cuttlebone without median longitudinal ridge anterior to spine. Sulcus extends entire length of cuttlebone; shallow, narrow; not flanked by rounded ribs. Anterior striae are inverted U-shaped. Inner cone limbs strap-like anteriorly, narrower posteriorly; slightly raised to form rounded posterior ridge; thickened slightly; shiny. Outer cone lateral limbs not flared ventrolaterally; limbs expanded posteriorly forming two short 'wings', directed ventrally, forming recurved cup-like structure.

*Colour* (alcohol preserved specimens). Reddish brown; head with chromatophores concentrated over eye orbits; dorsal mantle chromatophores concentrated medially over cuttlebone. Fins pale.

*Sculpture.* Dorsal mantle with longitudinal row of up to six orange-pink ridge-like papillae along each side, close to base of each fin and with scattered short ridges dorsal to cuttlebone. Ventral mantle with longitudinal row of four or five narrow ridges along each side close to fins.

#### SIZE

Males up to 35 mm ML; females up to 42 mm ML.

#### DISTRIBUTION

South-western Pacific: Australia, southern Queensland to New South Wales, 26°30′S, 153°43′47″E to 33°58′47″S, 151°16′12″E. Known depth range 11–183 m.

#### REFERENCE

Reid (2000).

### *Sepia mestus* Gray, 1849

*Sepia mestus* Gray, J. E. (1849). 'Catalogue of the Mollusca in the Collection of the British Museum. 1. Cephalopoda Antepedia.' (British Museum (Natural History): London.) 164 pp. [108].

Type data: syntypes BMNH 1842.11.9.84, BMNH 1842.11.11.13.

Type locality: Australia.

#### COMMON NAME

Reaper Cuttlefish

#### DIAGNOSTIC FEATURES

Small to moderate-sized species. All arms subequal in length in both males and females. Arm suckers tetraserial in both sexes. Hectocotylus absent. Club slightly recurved; moderate length; sucker-bearing surface flattened, with five or six suckers in each transverse row; suckers differ slightly in size. Swimming keel of club extends well beyond carpus; dorsal and ventral protective membranes fused at base of club; joined to stalk; dorsal and ventral membranes same length. Buccal membrane without suckers.

*Cuttlebone.* Cuttlebone outline broad, oval; dorsal surface creamy white; evenly convex; calcified with reticulate sculpture. Dorsal ribs absent. Spine present, long, pointed, curves dorsally; with ventral keel. Sulcus absent (very faint, irregular furrows, but otherwise absent). Anterior striae inverted U-shaped anteriorly, inner cone broadly U-shaped posteriorly; not raised to form ledge posteriorly. Outer cone moderate width.

*Colour* (alcohol preserved specimens). Reddish brown, or purplish brown. Paired dorsal eye spots present, and posterior to these, a pair of smaller orange-pink coloured spots in most specimens; fins pale without markings at base. Live specimens with whitish mottle, or red. Dorsal mantle sometimes with transverse saddle mark. Eye sockets yellowish.

*Sculpture.* Dorsal mantle with scattered short longitudinal ridges concentrated mid-dorsally (a pair of flaps can be raised on the inner margins of the dorsal eyespots). Ventral mantle with wrinkled patches on each side.

#### HABITAT AND BIOLOGY

*Sepia mestus* is a neritic demersal species found in shallow rocky reefs, often seen by divers, sheltered under overhanging rock ledges. Males are strongly coloured and patterned during breeding; the female is pale and drab in comparison without distinctive markings.

The wrinkled pads on the ventral surface of the mantle are distinctive in this species. These patches may be adhesive glands similar to those seen in *S. papillata* Quoy and Gaimard, 1832 from the south-eastern Atlantic and south-western Indian Ocean, *S. pulchra* Roeleveld and Liltved, 1985 from the south-eastern Atlantic and *S. tuberculata* Lamarck, 1798 (von Byern *et al.* 2011) from Africa. In the latter species these structures are used to attach mechanically to the substratum. In *S. tuberculata* the adhesive area comprises two cell types that simultaneously secrete an adhesive substance. The ultrastructure of the glandular region has not been examined in *S. mestus*. It would be interesting to determine whether the glandular pads perform a similar function to that observed in other sepiids. As *S. mestus* occurs in shallow water and is often found under rock ledges, some form of adhesion would be useful to help immobilise the animal enabling it to resist the shearing forces of moving water.

Boletzky and Roeleveld (2000) have discussed ventral adhesion as a 'settling response' to hard substrates observed in young cuttlefishes, which involves well-defined dermal structures. In some species this behaviour is retained in the adult stage.

#### SIZE

Males up to 77 mm ML; females up to 124 mm ML.

Live animal, frontal view, top (M. Dowse).

Live animal, dorsolateral view (B. Saunders).

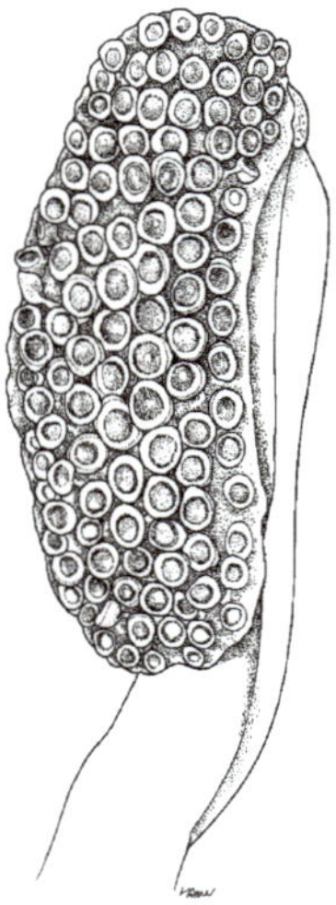

Tentacular club (A. Reid).

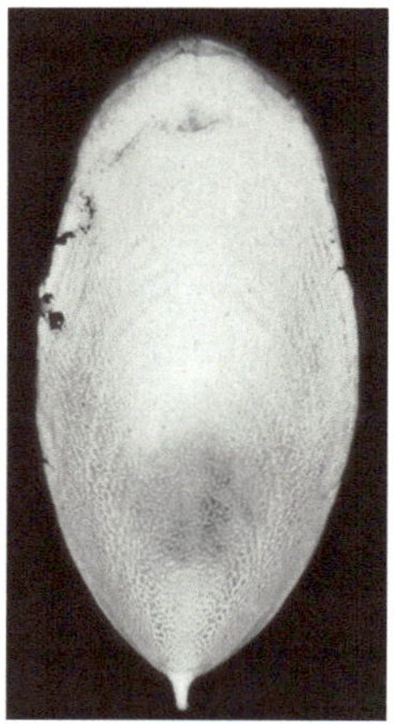

Cuttlebone, dorsal view (A. Reid).

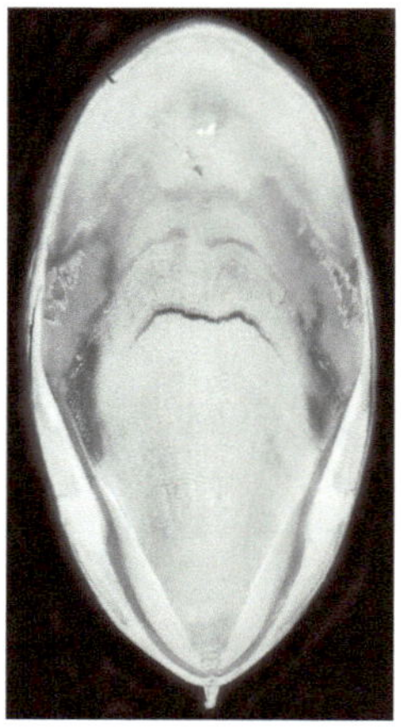

Cuttlebone, ventral view (A. Reid).

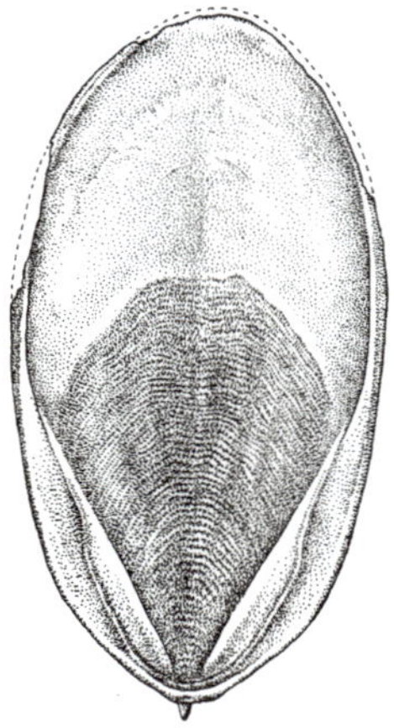

Cuttlebone, ventral view (K. Hollis).

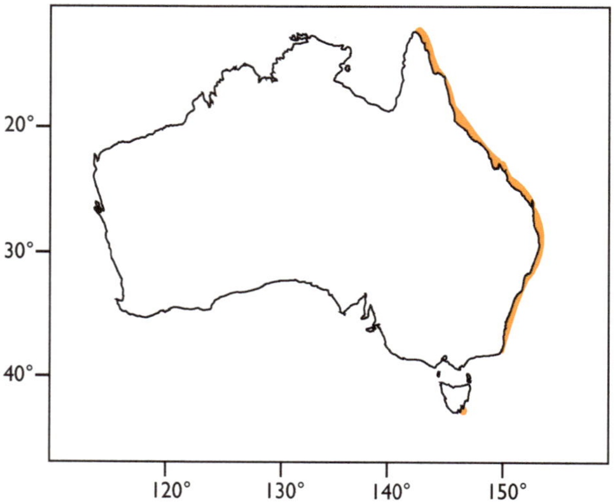


*Sepia mestus* distribution. *Note:* The distribution is based on whole animals only, not beachwashed cuttlebones.

## REMARKS

Work is in progress by the author to resolve the taxonomy of *S. mestus* from north-eastern Australia. It is likely that some specimens in collections have been misidentified and may belong to other species.

## DISTRIBUTION

Eastern Australia from 10°58′48″S, 144°1′12″E to 43°6′S, 147°15″. Records from China and Vietnam are now known to be misidentifications. *Sepia mestus* is an Australian endemic. Known depth range 1–70 m.

## REFERENCE

Reid and Lu (2005).

## *Sepia mira* (Cotton, 1932)

*Tenuisepia mira* Cotton, B. C. (1932). Notes on Australian Mollusca, with descriptions of new genera and new species. *Records of the South Australian Museum 4*, 537–547 [546, figs 7–9].

Type data: holotype SAMA D.10507 (cuttlebone).

Type locality: Australia: Queensland, North West Islet, 23°18′S, 151°42′E.

### DIAGNOSTIC FEATURES

Small species. Male arms 4 slightly longer than other arms, arms 1–3 all similar length; female arm lengths subequal. Arm suckers tetraserial. Left ventral arm hectocotylised: sucker size reduced proximally, only slightly smaller than normal arm suckers, with suckers equal in size across rows. Oral surface of modified region wide, swollen, fleshy, with transversely grooved ridges. Club, with three suckers in each transverse row; suckers differ markedly in size, some greatly enlarged. Swimming keel of club extends well beyond carpus; dorsal and ventral protective membranes not fused at base of club; joined to stalk; dorsal and ventral membranes same length, terminate at posterior end of carpus.

*Cuttlebone.* Cuttlebone bluntly rounded anteriorly and posteriorly; dorsal surface pinkish; evenly convex; granulose, calcified dorsally, particularly posteriorly. Dorsal ribs absent. Spine long, pointed, directed dorsally; with ventral keel. Striated zone and last loculus convex. Indistinct, shallow, narrow sulcus extends entire length of cuttlebone. Anterior striae are broad inverted U-shaped, or slightly convex to straight. Inner cone limbs raised to form rounded, thickened, shiny ledge posteriorly. Outer lateral limbs flared ventrolaterally.

*Colour* (alcohol preserved specimens). Reddish brown. Dorsal mantle with orange spots. Distinctive pigment spots on ventral side of club, spots in radiating rows extending from central region towards ventral margin.

*Sculpture.* Dorsal mantle with longitudinal row of orange-pink ridge-like papillae along each side, close to base of each fin and with scattered short longitudinal ridges concentrated mid-dorsally.

### SIZE

Up to 55 mm ML.

### DISTRIBUTION

Eastern Australia from 14°40′11″S, 145°28′11″E to 30°16′48″S, 156°13′11″E. Known depth range 20–72 m.

### REFERENCE

Reid (1998).

Fresh-caught specimen, cuttlebone removed (K. Graham).

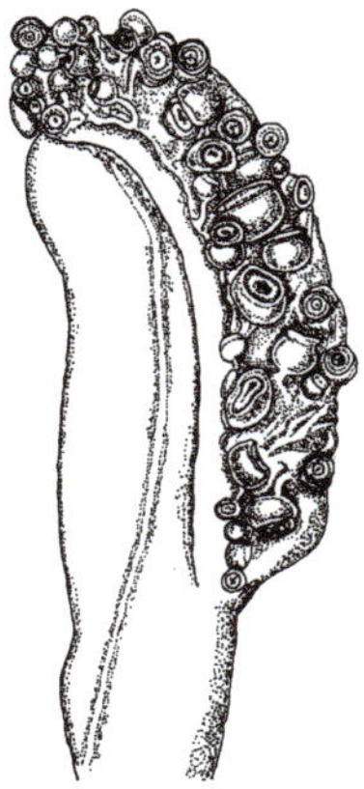

Tentacular club (A. Reid).

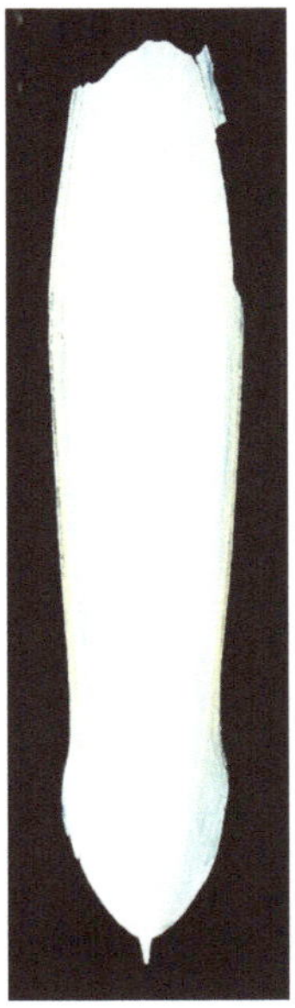

Cuttlebone, dorsal view (D. Paul).

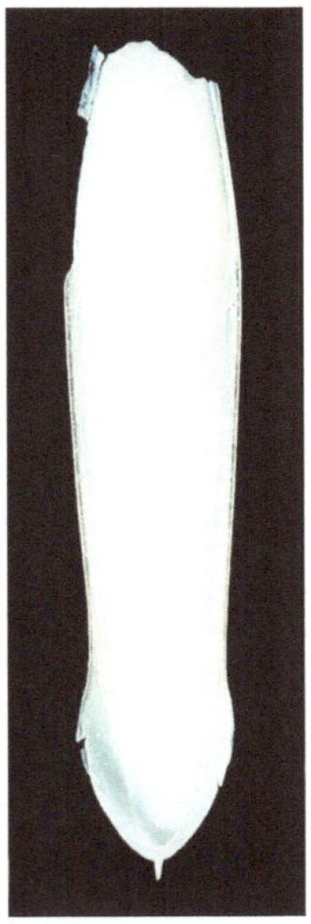

Cuttlebone, ventral view (D. Paul).

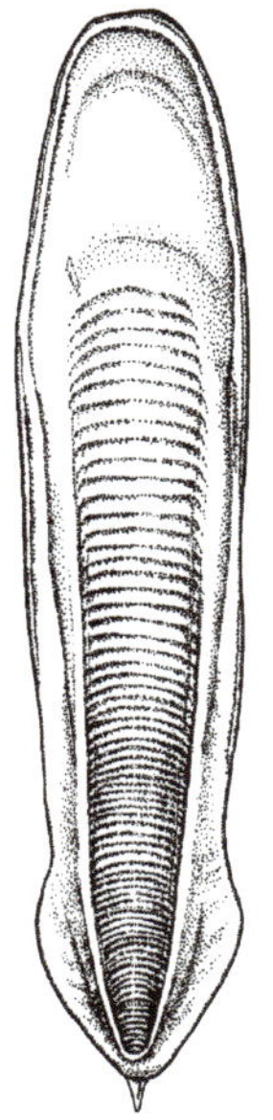

Cuttlebone, ventral view (A. Reid).

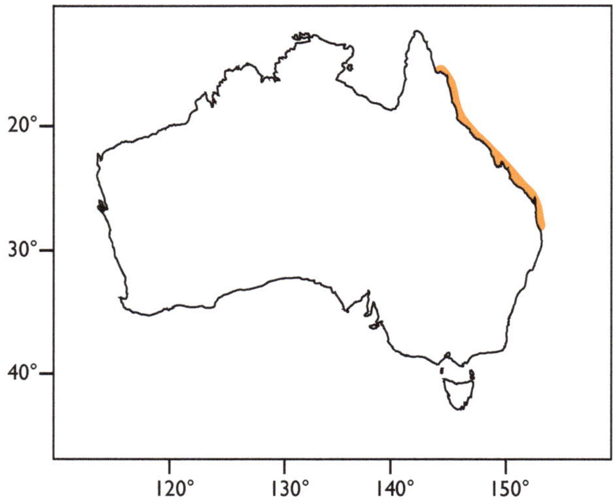


*Sepia mira* distribution. *Note:* The distribution is based on whole animals only, not beachwashed cuttlebones.

## *Sepia novaehollandiae* Hoyle, 1909

*Sepia novaehollandiae* Hoyle, W. E. (1909). A catalogue of recent Cephalopoda. Second Supplement, 1897–1906. *Proceedings of the Royal Physical Society of Edinburgh 17*, 254–299 [266].

Type data: syntypes MNHN 1-6-194 1 (cuttlebone).

Type locality: Australia: South Australia, Kangaroo Island, 35°50′S, 138°03′E.

### DIAGNOSTIC FEATURES

All arms subequal in length in both males and females. Arm suckers tetraserial in both sexes. Left ventral arm hectocotylised: sucker-bearing surface flattened. Club with eight suckers in each transverse row; suckers differ slightly in size, second and third dorsal series enlarged. Swimming keel of club extends well beyond carpus; dorsal and ventral protective membranes not fused in small specimens, fused at base of club in large specimens; dorsal and ventral membranes same length.

*Cuttlebone.* Cuttlebone dorsal surface pinkish; indistinct dorsal median rib present, broadens anteriorly; lateral ribs indistinct. Spine short, pointed; curves dorsally; keel(s) absent. Striated zone convex; last loculus flat; deep wide sulcus extends along striated zone. Anterior striae are V-shaped. Inner cone limbs narrow anteriorly, broaden posteriorly; not raised to form ledge. Outer cone lateral limbs flared ventrolaterally.

*Colour.* Freshly caught specimens cream to orange-brown with short, mainly longitudinal bars on dorsal mantle; bars primarily transverse on head and arms. Base of fins with two distinctive, luminescent yellow-green bars and a lateral row of spots. Side of head and arms with numerous greenish blue or white bands. Cuttlebone dorsal surface pinkish.

### SIZE

Up to 77 mm ML.

### REMARKS

Catalano (2013b) described three dicyemid parasites from the renal appendages of *Sepia novaehollandiae*: *Dicyema koinonum*

Fresh-caught, specimen, dorsal view (WAM).

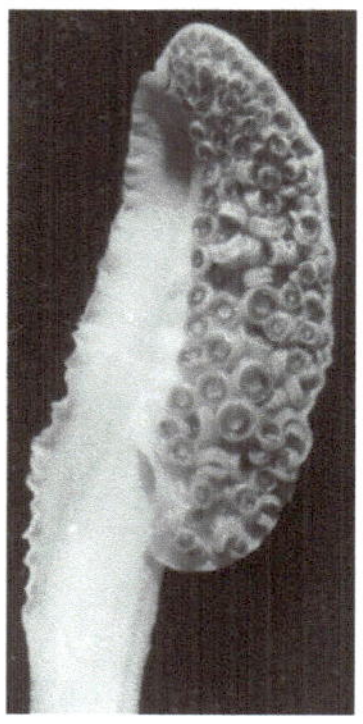

Tentacular club (C. Rowley).

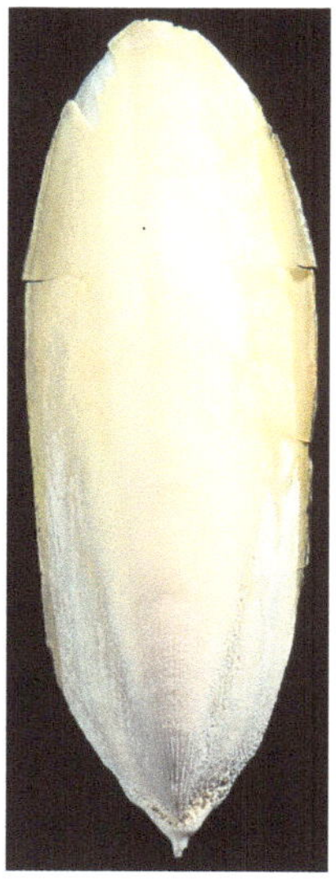

Cuttlebone, dorsal view (D. Paul).

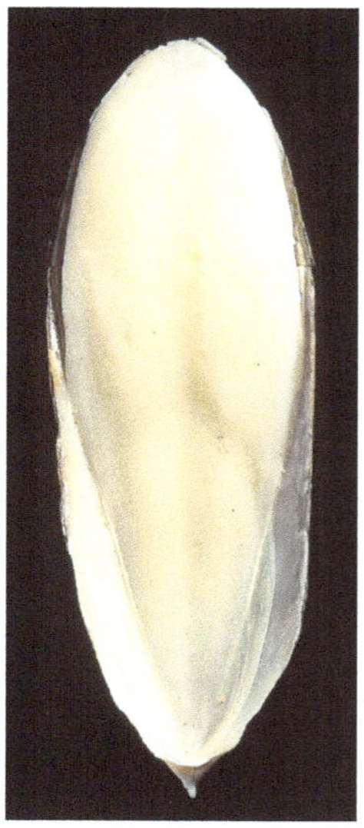

Cuttlebone, ventral view (D. Paul).

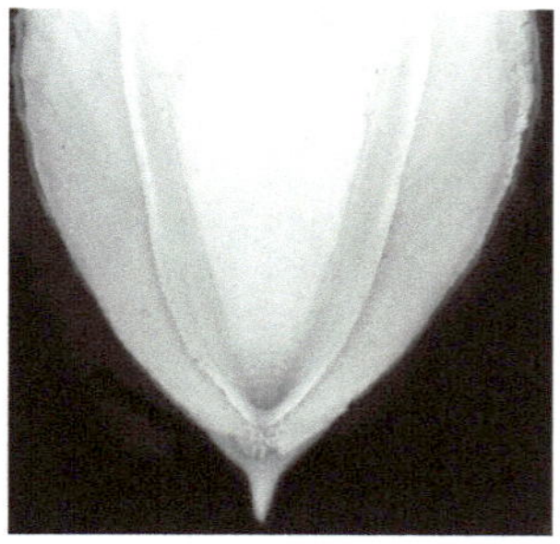

Enlargement of cuttlebone, ventral view (A. Reid).

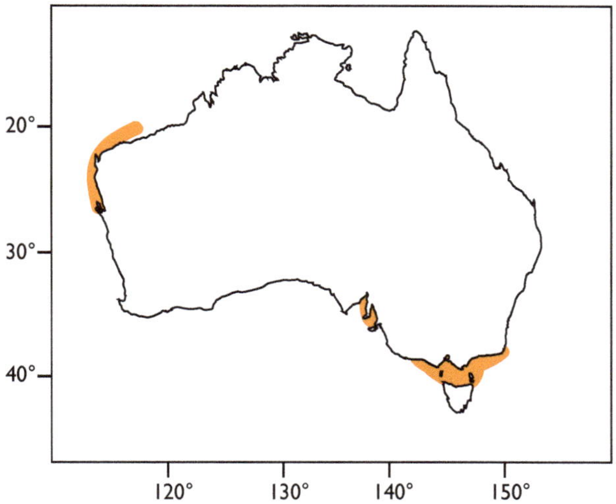


*Sepia novaehollandiae* distribution. *Note:* The distribution is based on whole animals only, not beachwashed cuttlebones.

Catalano, 2013 from *S. novaehollandiae* collected from Gulf St Vincent and Spencer Gulf, South Australia (this species also occurs in *Sepia apama* from the same locations), *D. vincentense* Catalano, 2013 from *S. novaehollandiae* collected from Gulf St Vincent, South Australia, and *Dicyemennea spencerense* Catalano, 2013 from *S. novaehollandiae* and *S. apama* collected from Spencer Gulf, South Australia. *Sepia novaehollandiae* is the type host for *D. spencerense* and *D. vincentense.* Dicyemids are highly host species specific and can be used as biological 'tags' to assist in the resolution of cephalopod taxonomy and assist in host species' identification. The presence of different parasites in *S. novaehollandiae* from different locations reflects some geographical population structuring in *S. novaehollandiae.*

#### DISTRIBUTION

Southern Australia from 37°25′47″S, 149°56′59″E to 18°55′47″S, 118°45′E. Known depth range 15–348 m.

#### REFERENCE

Reid *et al.* (2005b).

### *Sepia opipara* (Iredale, 1926)

*Glyptosepia opipara* Iredale, T. (1926). The cuttlefish 'bones' of the Sydney beaches (Phylum Mollusca–Class Cephalopoda). *Australian Zoologist 4*, 186–196 [191, pl. 22 figs 7, 8].

Type data: holotype AM C.133330 (cuttlebone).

Type locality: Australia: Queensland, Capricorn Island Group, Masthead Island, 23°32′S, 151°44′E.

#### COMMON NAME

Magnificent Cuttlefish

#### DIAGNOSTIC FEATURES

Mantle broad, oval. All arms subequal in length in both males and females; protective membranes narrow. Arm suckers tetraserial. Left ventral arm hectocotylised: sucker size normal proximally (five or six transverse rows), reduced medially (six or seven rows), and then normal size to arm tip. Oral surface of modified region not wide, fleshy, but normal, as on opposite arm. Club sucker-bearing surface flattened, with three or four suckers in each transverse row; suckers differ dramatically in size, middle of club with four or five greatly enlarged suckers, second proximal one the largest, obliquely one small lateral sucker and one minute marginal sucker dorsal to the large sucker, one small lateral and four minute marginal suckers ventral to it; suckers proximal and distal to the large suckers are small. Swimming keel of club extends well proximal to carpus; dorsal and ventral protective membranes not joined at base of club but fused to tentacular stalk; dorsal and ventral membranes of equal length; extend proximal to carpus along stalk; dorsal membrane forms deep cleft at junction with stalk.

*Cuttlebone.* Cuttlebone outline oblong; bone bluntly rounded anteriorly; acuminate, acute posteriorly; dorsal surface pinkish; dorsal surface flat medially and laterally; texture rough, with irregular calcified projections; dorsal median rib distinct; sides approximately parallel, flanked by broad grooves; lateral ribs distinct. Chitin borders lateral and anterior margins of cuttlebone. Spine short, pointed, curves dorsally, with ventral keel. Striated zone flat; last loculus convex; sulcus shallow, narrow. Anterior striae are inverted U-shape; limbs of inner cone extend anteriorly to end of striated zone. Inner cone limbs are uniform width, narrow, U-shape posteriorly, thickened; outer cone calcified, narrow anteriorly, broadens posteriorly.

*Colour.* Freshly caught animals dark tan or brown, dorsally, with darker blotches. Base of fins banded with longitudinal iridescent orange bars. Cuttlebone pinkish dorsally.

#### HABITAT AND BIOLOGY

*Interest to fisheries.* Species taken as by-catch of prawn and mixed species trawl fisheries. The size of this species and its broad distribution indicate that it has potential for exploitation.

#### SIZE

Up to ~150 mm ML.

#### DISTRIBUTION

Northern Australia from 36°57′S, 151°45′E to 26°25′12″S, 112°35′59″E. Known depth range 37–252 m.

Fresh-caught specimen, dorsal view (K. Graham).

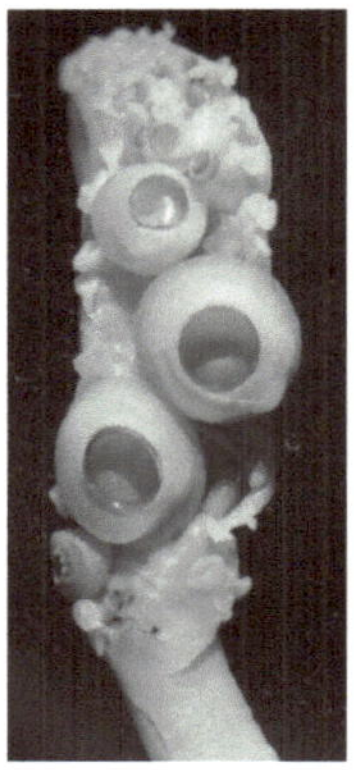

Tentacular club (ABRS, K. Hollis).

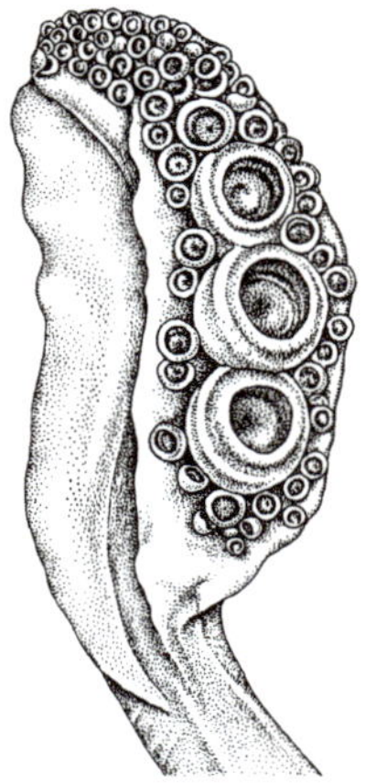

Tentacular club (C. Rowley).

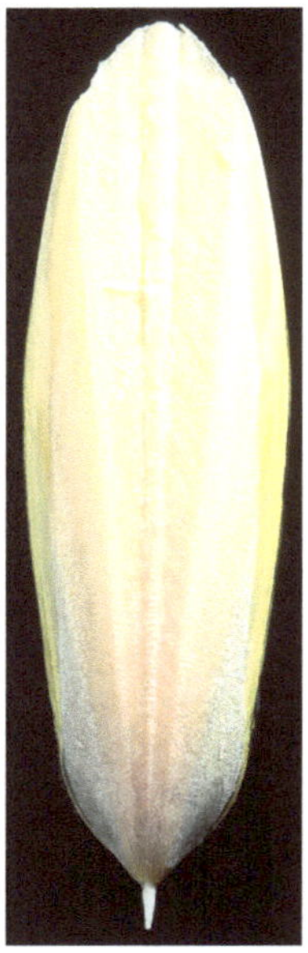

Cuttlebone, dorsal view (D. Paul).

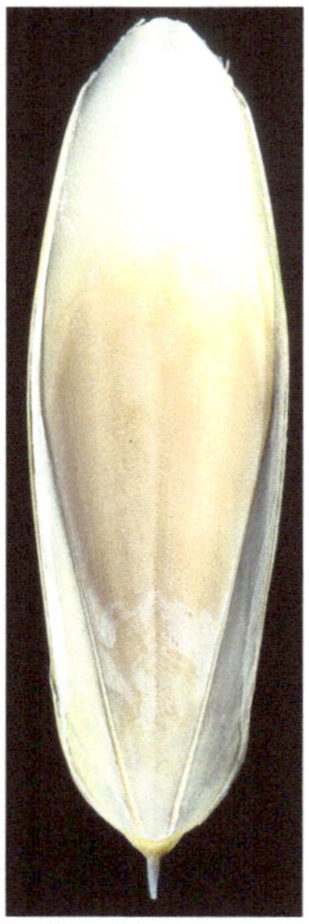

Cuttlebone, ventral view (D. Paul).

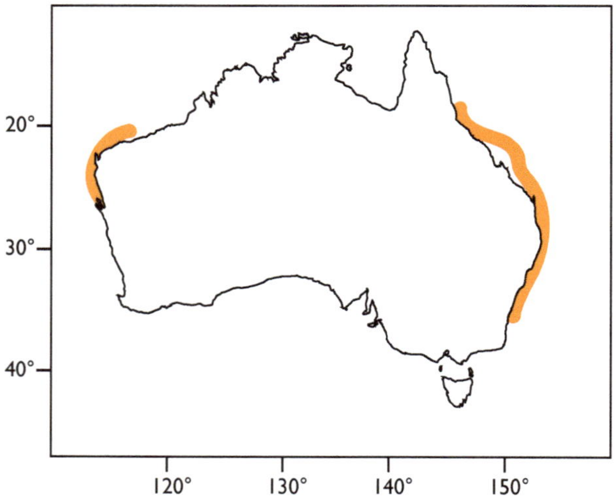


*Sepia opipara* distribution. *Note:* The distribution is based on whole animals only, not beachwashed cuttlebones.

**REFERENCE**

Reid *et al.* (2005b).

## *Sepia papuensis* Hoyle, 1885

*Sepia papuensis* Hoyle, W. E. (1885). Diagnoses of new species of Cephalopoda collected during the cruise of H.M.S. Challenger. Part 2. The Decapoda. *Annals and Magazine of Natural History (5)*16, 181–203 [197].

Type data: syntypes BMNH 1889.4.24.68.

Type locality: Arafura Sea, south of Papua, HMS Challenger Stn 188, 9°59′S, 139°42′E.

**COMMON NAME**

Papuan Cuttlefish

**DIAGNOSTIC FEATURES**

Mantle oval. All arms subequal in length in both males and females; protective membranes wide, well developed. Arm sucker arrangement differs between sexes: in males, sucker rows on arms 1–3 tetraserial proximally, biserial at distal tips, arms 4 suckers tetraserial; in females, arm suckers tetraserial. Hectocotylus absent. Club sucker-bearing surface flattened, with five or six suckers in each transverse row; suckers differ markedly in size. Swimming keel of club extends well proximal to carpus; dorsal and ventral protective membranes not joined in small specimens, joined at base of club in large specimens; dorsal protective membrane much longer than ventral protective membrane, extending proximal to carpus along stalk.

*Cuttlebone.* Cuttlebone outline oblong; bone bluntly rounded anteriorly and posteriorly; dorsal surface creamy white; dorsal surface convex medially, flat laterally; texture smooth; dorsal median and lateral ribs distinct; median rib broadens anteriorly. Chitin borders lateral and anterior margins of cuttlebone. Spine long, pointed, curves dorsally, with ventral keel. Striated zone and last loculus concave; shallow, wide, sulcus, extends entire length of cuttlebone. Anterior striae are inverted U-shaped. Inner cone limbs are narrow anteriorly, broaden posteriorly.

*Colour.* Light brown. Dorsal mantle has white blotches or spots. Paired dorsal eye spots present. Live animals dark greenish brown with white blotches, matching the colour of their habitat in seagrass and algal beds.

Live animal, lateral view (C. Roper).

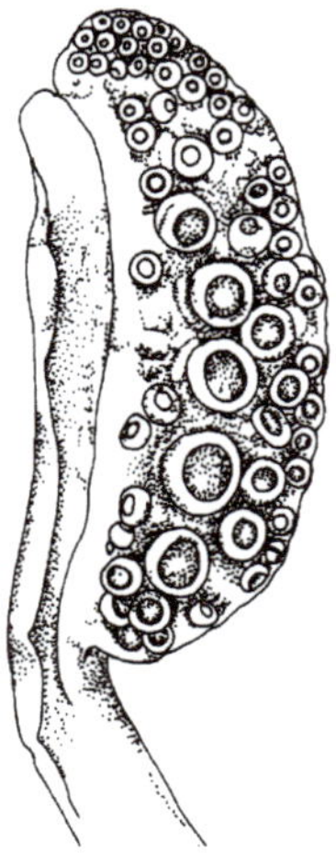
Tentacular club (A. Reid).

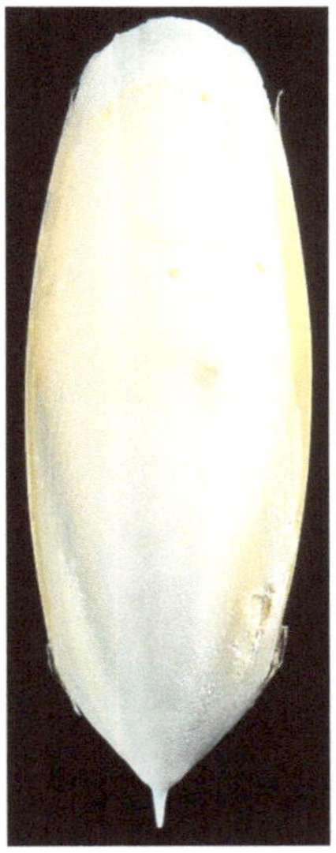
Cuttlebone, dorsal view (D. Paul).

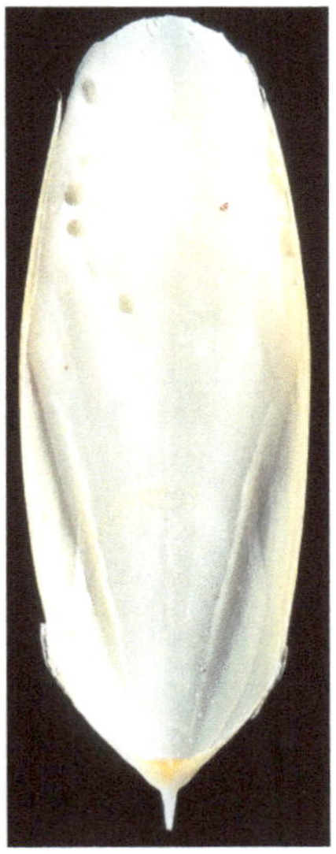
Cuttlebone, ventral view (D. Paul).

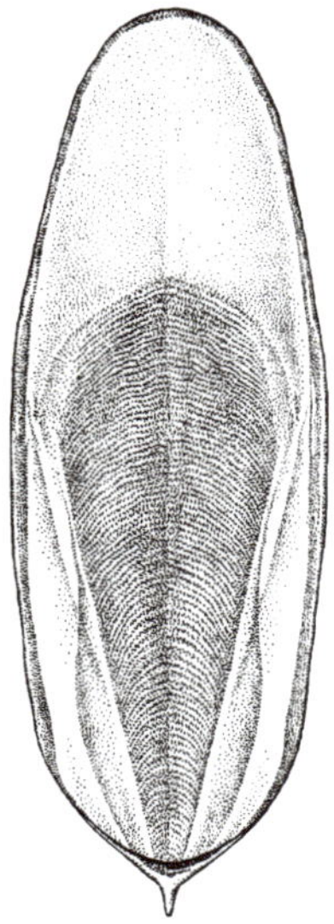
Cuttlebone, ventral view (ABRS, K. Hollis).

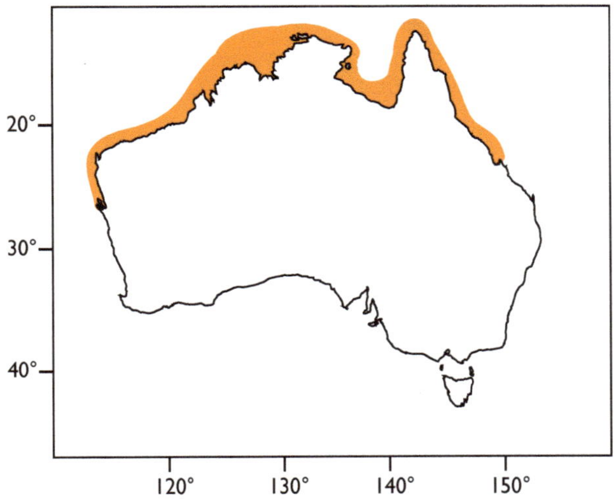

*Sepia papuensis* distribution. *Note:* The distribution is based on whole animals only, not beachwashed cuttlebones.

*Sculpture.* Dorsal mantle with series of elongate papillae along each side, adjacent to base of each fin, rest of mantle covered with numerous small papillae. Head with two pairs large papillae located over anterior and posterior ends of each eye, numerous scattered, small papillae and one on each eyelid. Arms 1–3 with papillae.

## HABITAT AND BIOLOGY

*Sepia papuensis* is a shelf species found on silt, sand, or muddy substrates. This species appears to be night-active and has been observed foraging in beds of seaweed and seagrass. Small individuals have been observed to extend flattened arms to mimic the shape of the seaweed *Halimeda* as camouflage. Behavioural studies have shown that the species exhibits a very wide range of colour patterns associated with substrate colour and texture, and in response to potential attack by predators.

*Interest to fisheries.* Species taken as by-catch of prawn and mixed species trawl fisheries. This species was collected in small numbers in demersal trawl and dredge surveys in the Gulf of Carpentaria, Australia in 1990 and 1991 and probably forms part of the by-catch of Taiwanese trawl fisheries in the region.

## SIZE

Up to 110 mm ML.

## REMARKS

Catalano (2013a) described two new dicyemid parasites, *Dicyema papuceum* Catalano 2013 and *Dicyema furuyi* Catalano 2013 from the renal appendages of *S. papuensis* collected from Shark Bay, Western Australia. Dicyemids are highly host species-specific and can be used as biological 'tags' to assist in the resolution of cephalopod taxonomy and assist in host species' identification.

## DISTRIBUTION

Indo-West Pacific: Philippine Islands, Indonesia, Bali, Ternate, Arafura and Coral Seas, northern, north-eastern and north-western Australia from 23°26′59″S, 151°55′11″E to 25°19′12″S, 113°37′47″E, including Gulf of Carpentaria. Known depth range 4–104 m.

#### REFERENCE

Reid *et al.* (2005b).

### *Sepia 'pharaonis'* Ehrenberg, 1831

*Sepia pharaonis* Ehrenberg, C. G. (1831). Symbolae Physicae Evertebrata. 1. Mollusaca. *Berolini ex Officina Academica.* [4].

Type data: lectotype ZMB Moll.-000083.

Type locality: Red Sea, Massaouah, Gulf of Suez. (There is some confusion on this point with two locations referred to in the original description: one in the Gulf of Suez and the other in modern day Eritrea. See Anderson *et al.* 2007.)

#### COMMON NAME

Pharaoh's Cuttlefish

#### DIAGNOSTIC FEATURES

Mantle oval. All arms subequal in length in both males and females. Arm suckers tetraserial. Left ventral arm hectocotylised: sucker size normal proximally (10–12 rows), reduced medially (six rows), with suckers in two dorsal series of modified portion much smaller than those in two ventral series, and then normal-sized suckers distally to arm tip. Oral surface of modified region wide, swollen, fleshy, with transversely grooved ridges; with shallow median furrow; suckers in two dorsal and two ventral series displaced laterally, with gap between them. Club sucker-bearing surface flattened, with eight suckers in each transverse row; suckers differ markedly in size, five or six median suckers enlarged (three or four of these are greatly enlarged). Swimming keel of club terminates at proximal end of carpus; dorsal and ventral protective membranes not joined at base of club; dorsal and ventral membranes same length; extend proximal to carpus along stalk; dorsal membrane forms shallow cleft at junction with stalk. Buccal membrane with a few, minute suckers (each lappet bearing one or two small suckers).

*Cuttlebone.* Outline oblong; bone bluntly rounded anteriorly; acuminate, acute, posteriorly; dorsal surface creamy white; dorsal surface evenly convex; texture smooth; dorsal median rib distinct, rib broadens anteriorly; lateral ribs indistinct. Chitin borders lateral and anterior margins of cuttlebone. Spine short, pointed, curves dorsally, keel(s) absent. Striated zone concave; last loculus flat; deep, wide, sulcus extends entire length of cuttlebone; sulcus flanked by rounded ribs. Anterior striae are inverted U-shaped. Inner cone limbs are narrow anteriorly, broaden posteriorly with distinctive thick, bulbous swelling.

*Colour.* Pale brownish or reddish purple. Head and arms with transverse zebra-stripe pattern. Dorsal mantle has white blotches or spots, transverse saddle mark, and transverse zebra-stripe pattern (saddle mark in females; stripes especially in males; small specimens may show stripe markings or few markings). Fins with longitudinal white band at base, bordered by narrow band of ground-coloured pigment along each side; white stripe solid on anterior 75% of body, interrupted by blocks of ground-coloured pigment on posterior 25% of mantle.

*Sculpture.* Dorsal mantle with series of elongate papillae along each side, adjacent to base of each fin, or covered with numerous small papillae.

#### HABITAT AND BIOLOGY

*Interest to fisheries.* The species contributes ~90% of the cuttlefishes caught off Australia by Chinese pair trawlers. Off the North West Shelf and Timor Seas, sepiids (mainly *S. 'pharaonis'*) tend to replace squids as the dominant cephalopods caught. Domestic fisheries in these waters take this species as by-catch of prawn and mixed species trawl fisheries.

#### SIZE

Up to 420 mm ML; weight to 5 kg. Males usually smaller than females.

#### REMARKS

Based on morphological studies, molecular research, and comparative reproductive biology, animals currently referred to as *S. 'pharaonis'* are known to represent a species-complex, but relationships within the complex remain unresolved. The most recent molecular study by Anderson *et al.* (2011) compared cytochrome oxidase subunit I and a nuclear gene region from ~50 specimens over the range of the taxon. These data were compared with 16S rRNA sequence data that had been studied previously (Anderson *et al.* 2007). These workers recognised five clades, but relationships among them were mostly poorly supported. All trees from the resulting phylogenetic analyses did reveal strong support for an Australian

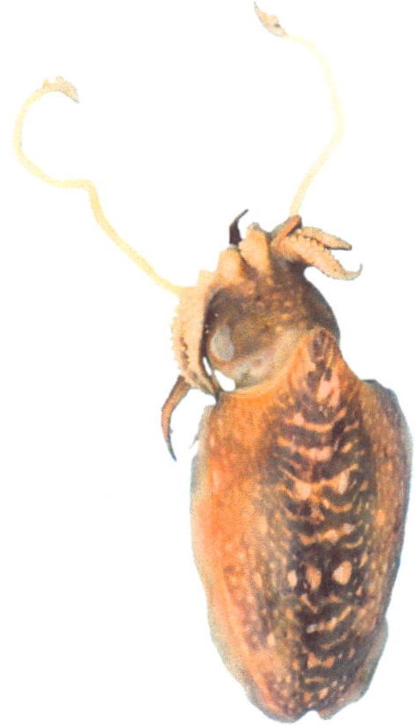

Fresh-caught specimen, dorsal view (WAM).

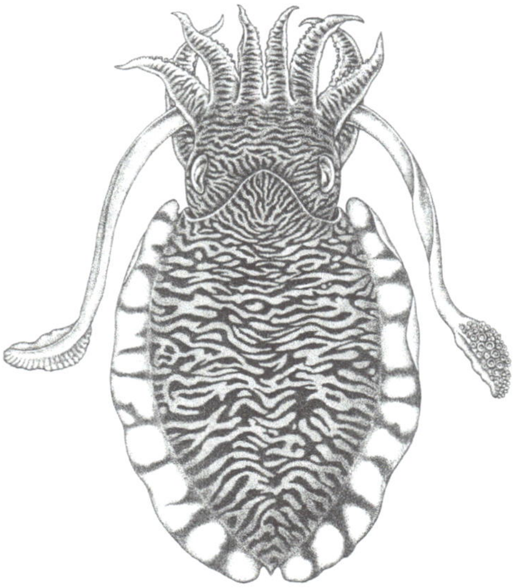

Dorsal view (ABRS, K. Hollis).

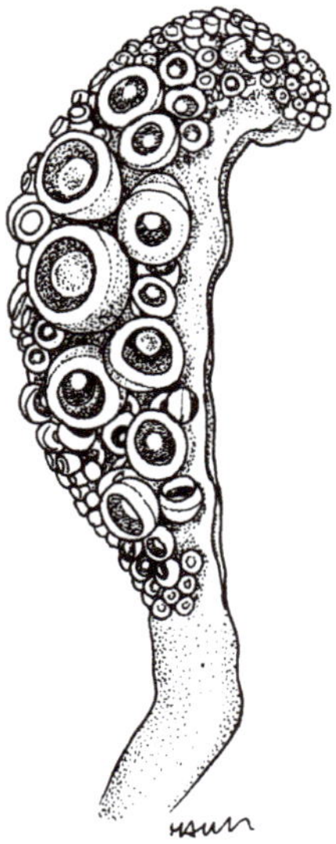

Tentacular club (FAO, E. D'Antoni).

Cuttlebone, dorsal view (D. Paul).

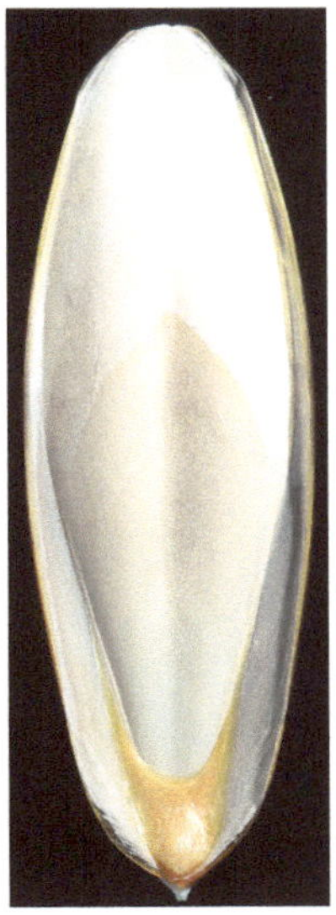

Cuttlebone, ventral view (D. Paul).

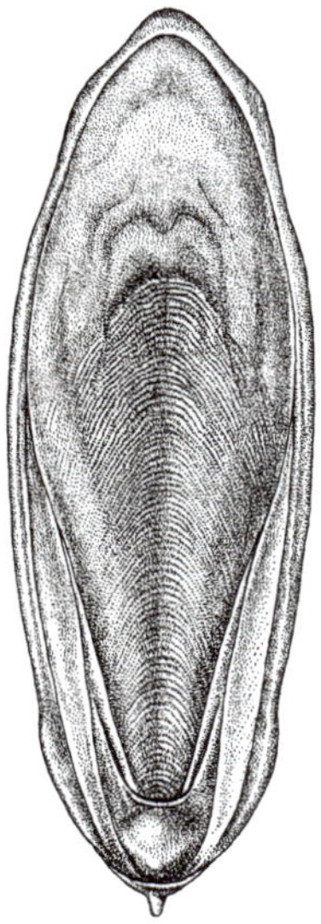

Cuttlebone, ventral view (ABRS, K. Hollis).

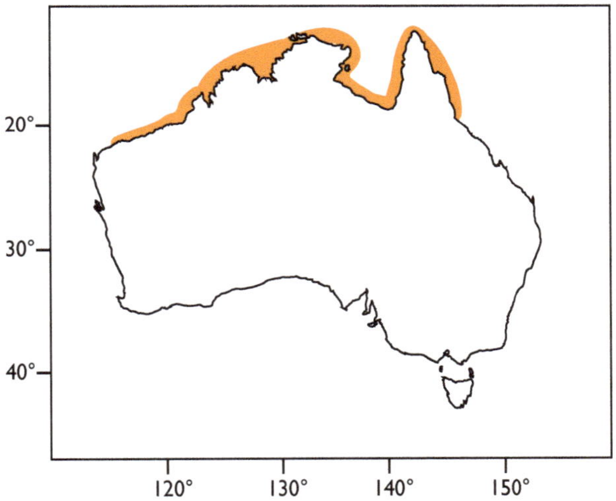


*Sepia 'pharaonis'* distribution. *Note:* The distribution is based on whole animals only, not beachwashed cuttlebones.

clade and a Red Sea/Gulf of Aden clade (corresponding to *S. 'pharaonis' sensu stricto*). The Australian samples constituted a separate, rather distantly related group.

Some morphological differences between these forms have been observed (Norman 2000), but more detailed morphological analyses from across the range of this complex are needed to properly resolve species' boundaries. The hectocotylus differs between *S. 'pharaonis'* from Japan and Australia. Given its significant interest to fisheries, the status of this putative species needs to be examined and specimens from throughout its range compared with those from the type locality to accurately determine its population structure and/or species boundaries.

A detailed morphological description of *S. 'pharaonis'* from Australian waters is needed to compare with representatives of this complex from elsewhere.

Kawashima *et al.* (2013) have determined the complete mitochondrial genome for *S. 'pharaonis'*. It is not stated in the paper where this specimen was collected, but it was obtained from the South Australian Museum. The molecular study by Yoshida *et al.* (2010b) placed *Sepia 'pharaonis'* in the subgenus *Acanthosepion* Rochebrune, 1884.

## DISTRIBUTION

*Sepia 'pharaonis'* species-complex: Indian Ocean and western Pacific, including the Red Sea and Arabian Sea south to Zanzibar and Madagascar, Andaman Sea to South China Sea, East China Sea, Taiwan, Japan (Kyushu and possibly southern Honshu), eastern Indonesia and Australia. Northern Australia from 18°36′S, 146°28′47″E to 20°26′59″S, 115°1′48″E. *Sepia 'pharaonis'* is a neritic demersal species. Known depth range 5–124 m.

### REFERENCE

Reid *et al.* (2005b).

## *Sepia plana* Lu & Reid, 1997

*Sepia plana* Lu, C. C. & Reid, A. L. (1997). Two new species of cuttlefish from northwestern Australian waters with a redescription of *Sepia sulcata* Hoyle, 1885 (Sepiidae: Cephalopoda). *Records of the Western Australian Museum 18*(3), 277–310 [279, figs 2–9].

Type data: holotype MV F.77206.

Type locality: Australia: Western Australia, North West Shelf, 14°15′S, 121°59′E.

### DIAGNOSTIC FEATURES

Large, robust species. All arms subequal in length in both males and females. Arm suckers tetraserial. Left ventral arm hectocotylised: sucker size slightly reduced proximally (approximately eight transverse rows), with suckers in dorsomarginal series of modified portion smaller than remaining suckers. Oral surface of modified region wide, swollen, fleshy, with transversely grooved ridges. Club crescent-shaped; sucker-bearing surface flattened, with 8–10 suckers in each transverse row; swimming keel extends well beyond carpus; dorsal and ventral protective membranes not fused at base of club; joined to stalk; dorsal and ventral membranes same length, extend beyond carpus along stalk.

*Cuttlebone.* Outline wedge-shaped; dorsal surface creamy white; flat medially and laterally; granulose. Indistinct dorsal median rib present, broadens anteriorly; lateral ribs present, distinct. Spine present, short, pointed; Striated zone and last loculus flat; sulcus absent. Anterior striae are broad inverted U-shaped.

*Colour* (alcohol preserved specimens). Pale buff pinkish brown.

*Sculpture.* Body papillae present. Ventral mantle with longitudinal series of four or five orange-pink narrow ridges along each side close to fins.

### SIZE

Males up to 99 mm ML; females up to 151 mm ML.

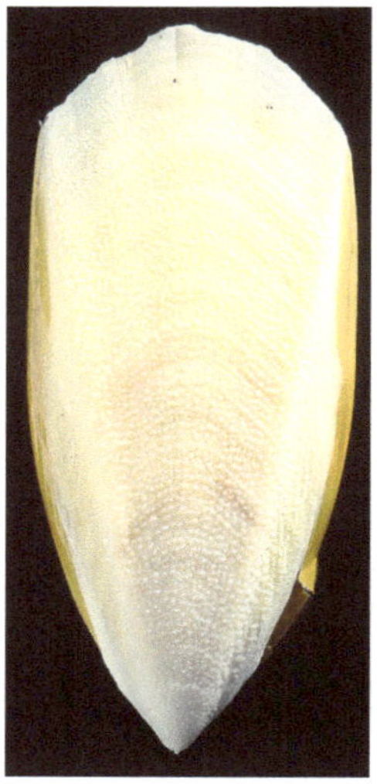

Cuttlebone, dorsal view (D. Paul).

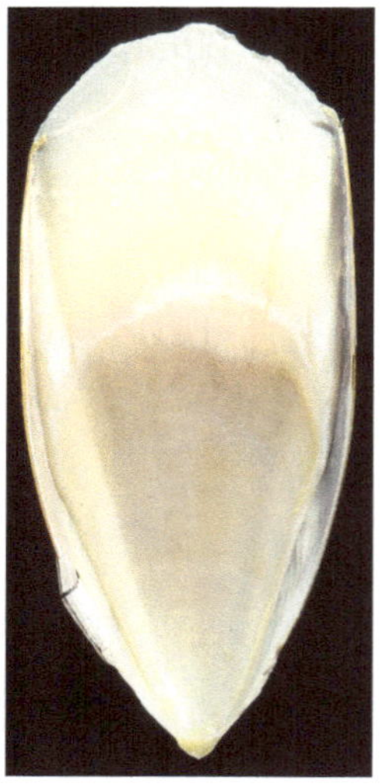

Cuttlebone, ventral view (D. Paul).

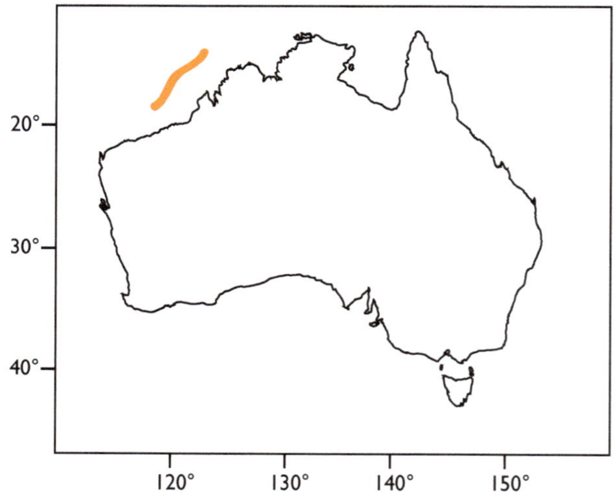


*Sepia plana* distribution. *Note:* The distribution is based on whole animals only, not beachwashed cuttlebones.

### REMARKS

The structure of the cuttlebone with its distinctive flattened dorsal and lateral margins distinguishes *Sepia plana* from all other Australian species.

### DISTRIBUTION

Australia: Western Australia, North West Shelf, 12°4′12″S, 122°49′47″E to 17°55′12″S, 118°15′36″E. Known depth range 430–460 m. Cuttlebones found off eastern Australia possibly belong to the closely related *S. subplana* Lu & Boucher-Rodoni, 2001.

### REFERENCE

Lu and Reid (1997).

## *Sepia plangon* Gray, 1849

*Sepia plangon* Gray, J. E. (1849). 'Catalogue of the Mollusca in the Collection of the British Museum. 1. Cephalopoda Antepedia.' (Spottiswoodes & Shaw: London.) 164 pp. [104].

Type data: syntypes BMNH 1963.I.26.W 1 animal without cuttlebone, BMNH 1842.II.9.85 1 broken cuttlebone.

Type locality: Australia: New South Wales, Port Jackson, 33°51'S, 151°16'E.

### COMMON NAME

Mourning Cuttlefish

Live animal (M. Dowse).

Mating pair with female (left) and male (right), right side of male animal mimicking female body pattern (M. Garwood).

### DIAGNOSTIC FEATURES

All arms subequal in length in both males and females. Arm suckers tetraserial. Left ventral arm hectocotylised: sucker size normal proximally (five rows), slightly reduced medially (five rows), with suckers on modified portion all similar in size, and then normal-sized suckers distally to arm tip. Oral surface of modified region not wide, fleshy. Club sucker-bearing surface flattened, with five suckers in each transverse row; suckers differ markedly in size, some greatly enlarged (median series largest followed by those immediately bordering ventral series). Swimming keel of club extends well proximal to carpus; dorsal and ventral protective membranes joined at base of club but fused to tentacular stalk; dorsal and ventral membranes same length; extend proximal to carpus along stalk.

*Cuttlebone.* Cuttlebone outline oblong; acuminate, acute anteriorly; bluntly rounded posteriorly; dorsal surface pinkish; dorsal surface convex medially, flat laterally; granulose; dorsal median rib distinct; sides approximately parallel; lateral ribs indistinct. Spine long, pointed, straight, parallel to bone; with ventral keel. Striated zone concave; last loculus flat; sulcus deep, narrow, extends entire length of cuttlebone; sulcus flanked by rounded ribs (giving striae a wavy appearance). Anterior striae are inverted V-shape. Inner cone limbs are uniform width, narrow, U-shaped posteriorly.

*Colour.* Reddish purple, light brown, or dark brown, or sometimes with whitish mottle. During the breeding season, the head and dorsal mantle of males have narrow, irregular light-coloured transverse stripes. Eye sockets white. Fins pale with broad light greenish band along base.

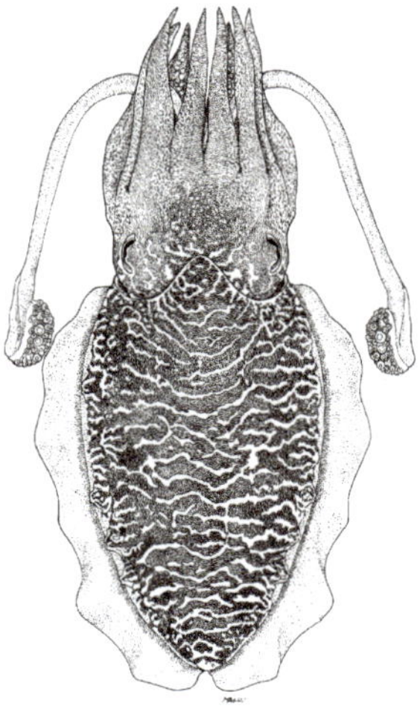

Dorsal view (FAO, E. D'Antoni).

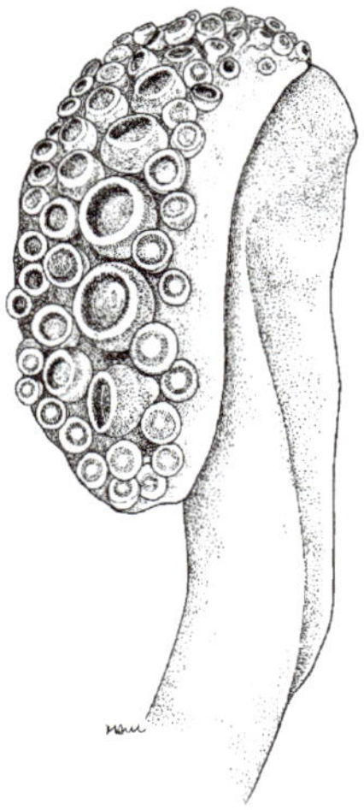

Tentacular club (FAO, E. D'Antoni).

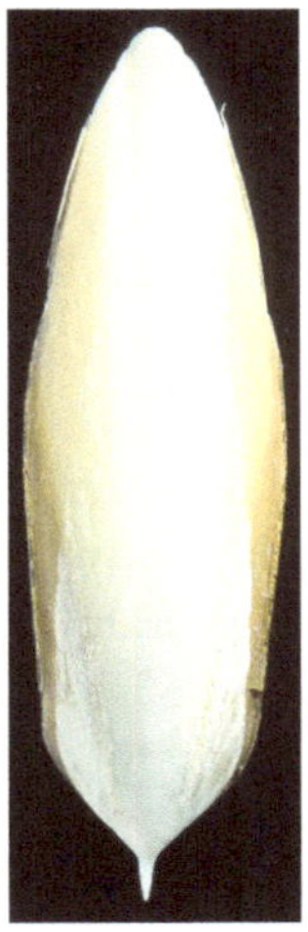

Cuttlebone, dorsal view (D. Paul).

Cuttlebone, ventral view (D. Paul).

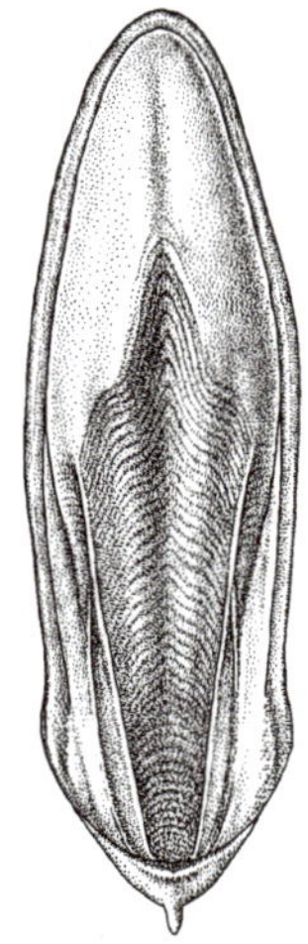

Cuttlebone, ventral view (ABRS, K. Hollis).

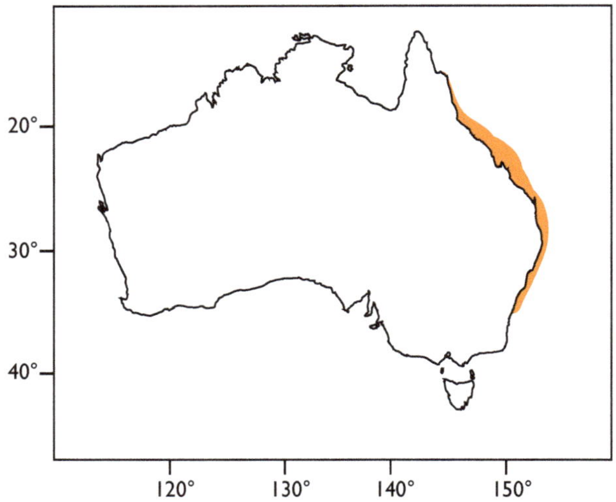


*Sepia plangon* distribution. *Note:* The distribution is based on whole animals only, not beachwashed cuttlebones.

### HABITAT AND BIOLOGY

*Sepia plangon* is frequently found among seaweed and seagrass. It is active during the day, foraging for crustaceans and fishes. It often can be seen resting on the sea floor, raising the anterior end of the body by using its arms as stilts. The breeding season around Sydney is between July to November. Females are plain coloured, with mottled camouflage colouration, while males show a pattern of pulsating stripes on the mantle during interspecific interactions. Males deceive rivals during courtship by displaying female body patterns on one side of the body, while displaying male courtship patterns to receptive females on the other side of the body in a longitudinally split display (Brown *et al.* 2012). This deception, which appears to occur only in the presence of a single rival male, prevents rivals from disrupting courtship and maximises reproductive success for the displaying male.

*Sepia plangon* (and possibly other species) uses high resolution polarisation vision to improve image contrast (Temple *et al.* 2012). It is able to detect differences in electronic field (e-vector) orientations as small as 1°. The use of high resolution polarisation vision significantly enhances visual information. This can aid in accurate background pattern matching and enable this cuttlefish to detect animals, such as predatory fishes, that may be using a less sophisticated form of polarisation background matching. It may also be able to covertly communicate in ways invisible to animals lacking polarisation vision.

*Interest to fisheries.* Species taken as by-catch of prawn and mixed species trawl fisheries.

### SIZE

Up to 135 mm ML.

### DISTRIBUTION

South-western Pacific: eastern Australia from 14°37′48″S, 145°24′E to 34°34′12″S, 151°11′59″E. Known depth range intertidal to 147 m.

### REFERENCE

Reid *et al.* (2005b).

## *Sepia reesi* Adam, 1979

*Sepia reesi* Adam, W. (1979). The Sepiidae (Cephalopoda, Decapoda) in the collections of the Western Australian Museum. *Records of the Western Australian Museum 7*, 113–212 [200, pl. 4 fig. 3].

Type data: holotype WAM 497-76 (cuttlebone).

Type locality: Australia: Western Australia, Salmon Bay, Rottnest Island, 32°00′S, 115°30′E.

### DIAGNOSTIC FEATURES

Cuttlebone elongate, acuminate anteriorly, subrectangular posteriorly. Dorsal surface salmon coloured, with distinct broad median rib bordered by narrow grooves. Cuttlebone nearly smooth anteriorly, weakly granulose posteriorly, granules forming reticulate pattern near posterior end and at each side laterally on outer cone. Bone ventrally with narrow median sulcus along entire length of cuttlebone. Anterior striae inverted V-shaped. Striated zone flanked on each side by broad nearly smooth area, showing weak striations, that is flat or weakly concave, forms rounded angle with nearly flat median part of striated zone. In posterior half of bone, limbs of inner cone separate striated area from lateral smooth zones. Inner cone U-shaped, slightly thickened, shiny, rounded posteriorly. Outer cone slightly enlarged posteriorly, slightly recurved. Posterior end of bone recurved ventrally. Spine rounded and directed dorsally, not curved.

### SIZE

Up to 45 mm ML.

### REMARKS

The type specimen of this species is a cuttlebone collected at Rottnest Island. Only two whole animals have been

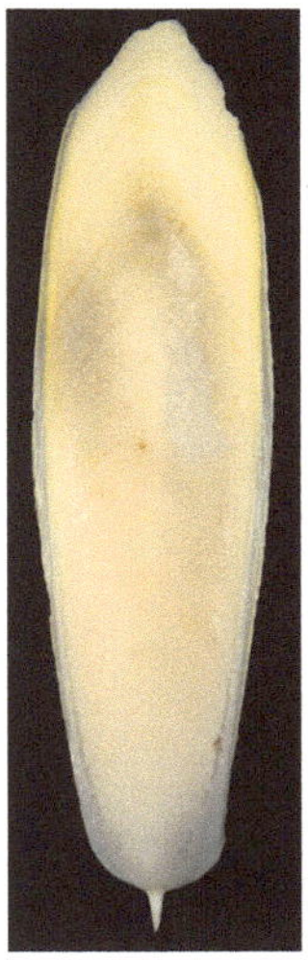

Cuttlebone, dorsal view (L. Mott).

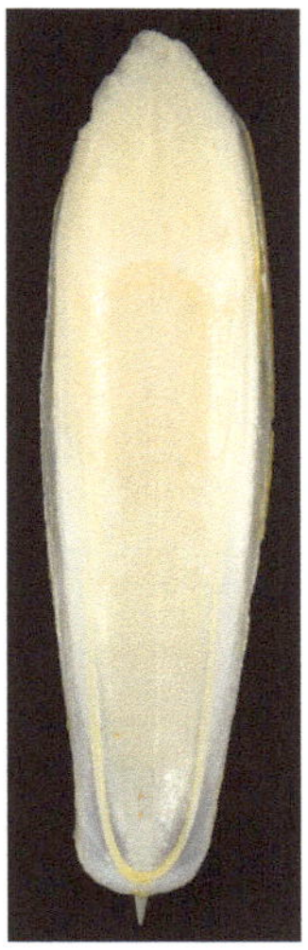

Cuttlebone, ventral view (L. Mott).

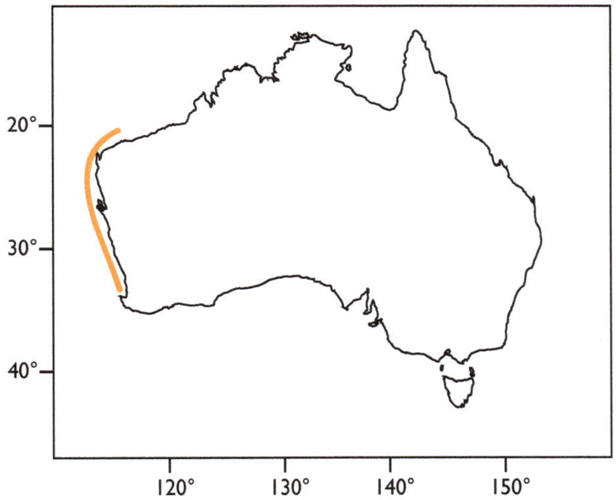


*Sepia reesi* distribution. *Note:* The distribution is based on whole animals only, not beachwashed cuttlebones.

found to date in museum collections: one immature female collected in 146–150 m near Rottnest Island, 32°00′S, 115°15′E (WAM 470-72) and another from off Port Headland, Western Australia, 19°56′39″S, 117°53′51″E, collected in 43 m (MV F.57290). Possibly its range is continuous between these locations.

### DISTRIBUTION

Western Australia from 19°56′39″S, 117°53′51″E to 32°00′S, 115°15′E. Known depth range 43–150 m.

### REFERENCES

Reid *et al.* (2005b), Reid (unpublished data).

## *Sepia rhoda* (Iredale, 1954)

*Arctosepia rhoda* Iredale, T. (1954). Cuttlefish 'bones' again. *Australian Zoologist 12*, 63–82 [75, pl. 4 figs 10–12].

Type data: holotype AM C.133319 (cuttlebone).

Type locality: Australia: Western Australia, Point Cloates, 22°43′S, 113°40′E.

### DIAGNOSTIC FEATURES

Small species. Male and female arms differ in relative lengths. Male arm lengths subequal; female arms 2 greatly elongate. Distal arm tips attenuate in both sexes, suckers enclosed by protective membranes. Non-hectocotylised arm sucker arrangement same in both sexes: arm suckers biserial proximally, tetraserial medially, biserial at distal arm tips (in males only extreme distal arm tips with biserial suckers). Female arms 1–4 protective membranes wide, arms slightly paddle-shaped at tips; biserial suckers displaced laterally with gap between; greater proportion of arms 2 and 3 with distal biserial suckers than arms 1, then 4 (arms 4 only a few proximal rows and distal tip with biserial suckers). Left ventral arm hectocotylised: sucker size normal proximally, reduced medially, with suckers in dorsomedial series of modified portion smaller than dorsomarginal suckers and ventral-most pair in each series usually absent, and then normal-sized suckers to arm tip. Oral surface of modified region wide, fleshy, with transversely grooved ridges; with deep median furrow. Hectocotylised arm distally attenuate, compressed. Club crescent-shaped; sucker-bearing surface flattened, with four or five suckers in each transverse row; suckers differ in size, three or four enlarged suckers towards posterior end of club. Swimming keel of club extends slightly beyond carpus; dorsal and ventral protective membranes not fused at base of club; joined to stalk; dorsal and ventral membranes same length, terminate at posterior end of carpus.

*Cuttlebone.* Cuttlebone outline lanceolate; strongly convex in lateral view. Bone acuminate, acute anteriorly and posteriorly; recurved ventrally. Dorsal surface pinkish, evenly convex; entire surface calcified with reticulate granulose sculpture concentrated medially and on lateral margins in irregular longitudinal ridges. Dorsal median rib distinct, sides approximately parallel; bordered laterally by distinct grooves; lateral ribs indistinct. Spine long, pointed. Striated zone and last loculus convex; deep, narrow sulcus extends entire length of cuttlebone; not

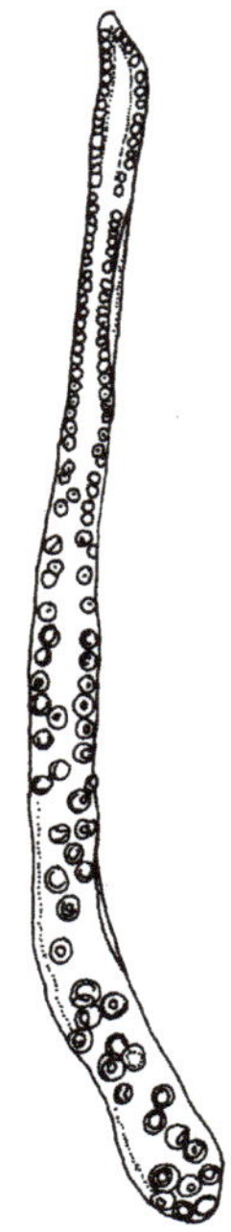

Arm 2, oral view (A. Reid).

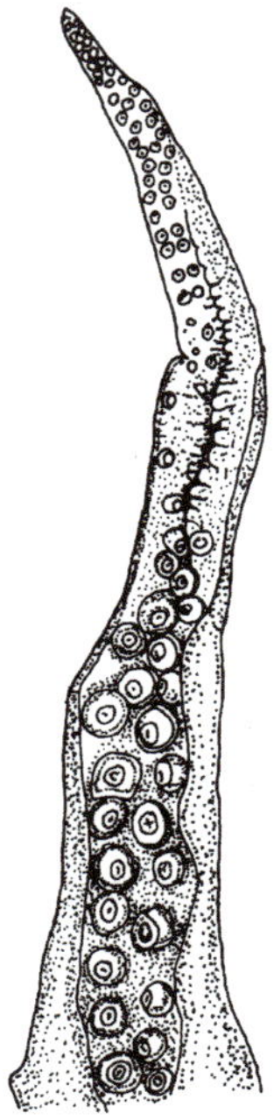

Hectocotylus, oral view (A. Reid).

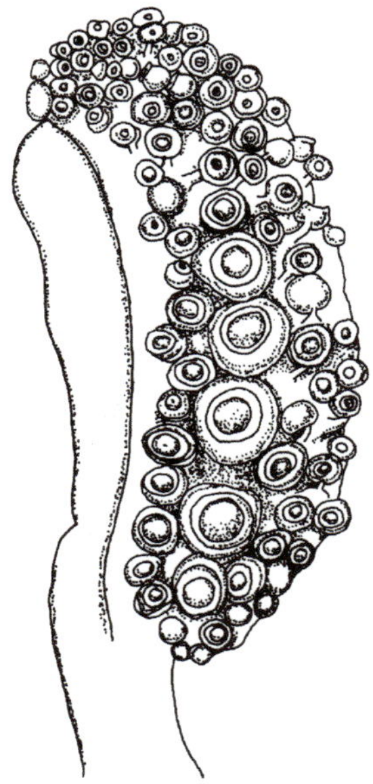

Tentacular club (A. Reid).

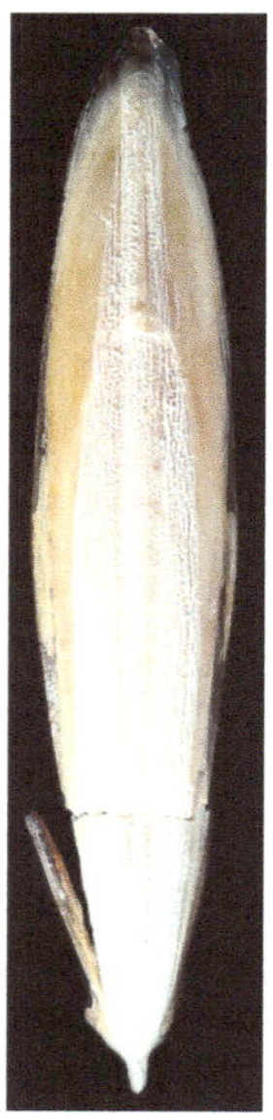

Cuttlebone, dorsal view (D. Paul).

Cuttlebone, ventral view (D. Paul).

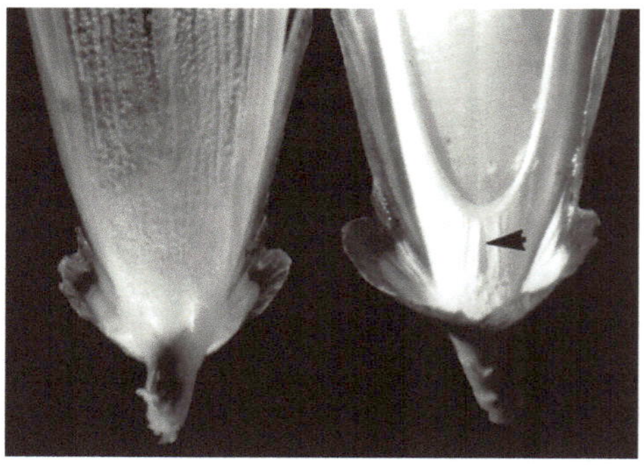

Cuttlebone, ventral view enlargement, left, dorsal view; right, dorsal view, arrow indicates inner cone median longitudial ridge (A. Reid).

flanked by rounded ribs. Last loculus with pronounced median indentation. Anterior striae M-shaped. Inner cone limbs narrow, strap-like anteriorly, broaden posteriorly (sometimes with a median longitudinal ridge); raised slightly to form wide, flat, U-shaped ledge; thickened, dull, not shiny posteriorly. Outer cone limbs expanded posteriorly forming two short 'wings', directed ventrally.

*Colour* (alcohol preserved specimens). Pale buff pinkish brown; head with chromatophores concentrated over eye orbits; dorsal mantle chromatophores concentrated medially over cuttlebone.

*Sculpture.* Dorsal mantle with longitudinal row of up to six orange-pink ridge-like papillae along each side, close to base of each fin (and a series of short bars mediodorsally on mantle in longitudinal rows). Ventral mantle with longitudinal series of four or five narrow ridges along each side close to fins; head papillae present; positioned between eyes. Partially broken transverse bar

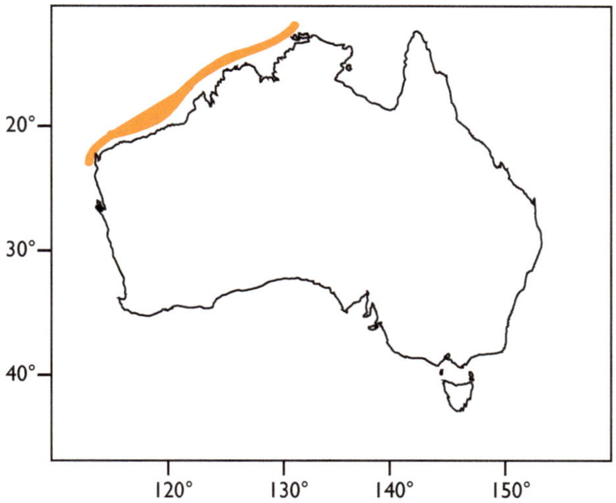


*Sepia rhoda* distribution. *Note:* The distribution is based on whole animals only, not beachwashed cuttlebones.

extending between eyes, and scattered shorter bars over eye orbits. Arm papillae present, same as those on head.

#### SIZE

Males up to 61 mm ML; females up to 58 mm ML.

#### REMARKS

This species occurs sympatrically with *Sepia koilados* Reid, 2000, *S. vercoi* Adam, 1979 and possibly *S. cottoni* Adam, 1979.

#### DISTRIBUTION

Indo-Pacific: northern and north-eastern Australia from the Arafura Sea 10°24′S, 130°23′E to 20°47′S, 114°48′E. Known depth range 63–230 m.

#### REFERENCE

Reid (2000).

### *Sepia rozella* (Iredale, 1926)

*Solitosepia rozella* Iredale, T. (1926). The cuttlefish 'bones' of the Sydney beaches (Phylum Mollusca – Class Cephalopoda). *Australian Zoologist 4*, 186–196 pls 22, 23 [190, pl. 21 figs 6, 7].

Type data: holotype AM C.133336.

Type locality: Australia: New South Wales, Sydney, Manly Beach, 33°48′S, 151°17′E.

#### COMMON NAME

Rosecone Cuttlefish

#### DIAGNOSTIC FEATURES

Mantle oval; dorsal anterior margin triangular, acute. Fins wide, rounded posteriorly, with narrow gap between them. All arms subequal in length in both males and females. Arm suckers tetraserial; female arms 1 protective membranes well developed, enveloping suckers; arm tips attenuate. Male arm suckers normal in size (not greatly enlarged). Hectocotylus absent; only modification being the basal 40% of left ventral arm slightly thicker and oral surface of arm wider than corresponding portion of right ventral arm, with shallow median furrow. Club short, oval; sucker-bearing surface flattened, with four or five suckers in each transverse row; suckers differ slightly in size, four or five enlarged suckers in longitudinal series towards posterior end of club. Swimming keel of club extends well proximal to carpus; dorsal and ventral protective membranes joined at base of club (V-shape); separated from tentacular stalk by membrane; dorsal and ventral membranes same length; dorsal membrane much wider than ventral membrane; dorsal membrane forms deep cleft at junction with stalk. Buccal membrane without suckers. Eggs spherical.

*Cuttlebone.* Cuttlebone oblong, bluntly rounded anteriorly and posteriorly; recurved ventrally; dorsal surface creamy white; dorsal surface evenly convex; entire surface calcified, granulose, sculpture reticulate on posteriolateral half of bone. Spine and posterior tip of bone covered with a smooth glaze-like substance. Dorsal median rib and lateral ribs indistinct. Chitin borders lateral margins of cuttlebone. Spine long, pointed, straight, directed dorsally, with ventral keel and ventral notch at base. Striated zone concave; last loculus flat; sulcus deep, wide, forms distinct V-shape medially, flanked by rounded ribs. Anterior striae inverted V-shape. Inner cone limbs narrow anteriorly, broaden posteriorly, rose coloured; outer margin of inner cone raised into rounded, thickened, posterior ledge; outer cone calcified; narrow anteriorly, broadens posteriorly; outer cone lateral limbs are flared ventrolaterally.

*Colour.* Purplish brown. Head with chromatophores concentrated over eye orbits. Arms without markings. Paired dorsal eye spots absent. Fins pale, without markings at base. Cuttlebone inner cone rose-pink.

*Sculpture.* Dorsal mantle covered with numerous small papillae; ventral mantle with longitudinal row of six narrow reddish purple ridges along each side close to

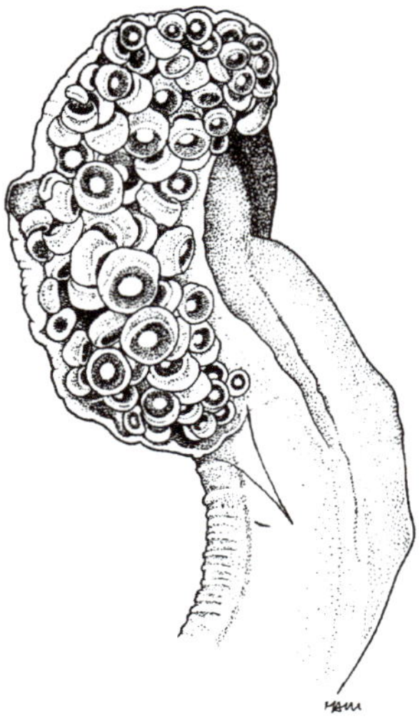

Tentacular club (FAO, E. D'Antoni).

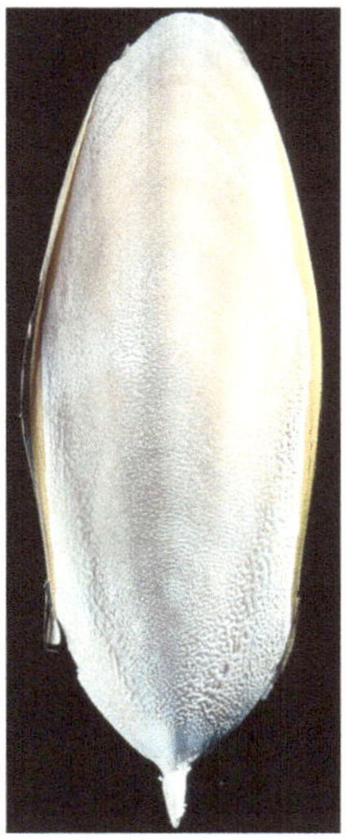

Cuttlebone, dorsal view (D. Paul).

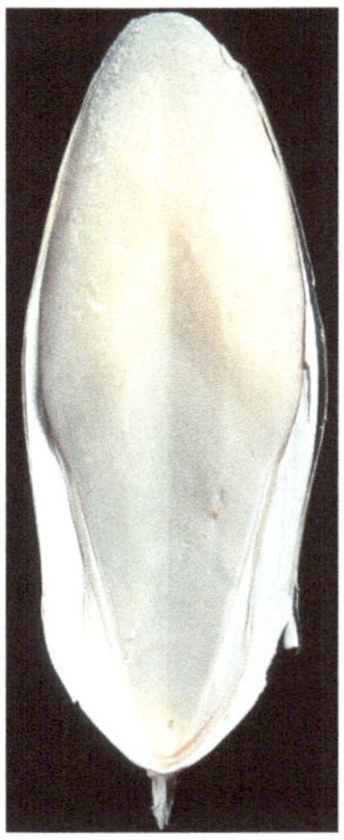

Cuttlebone, ventral view (D. Paul).

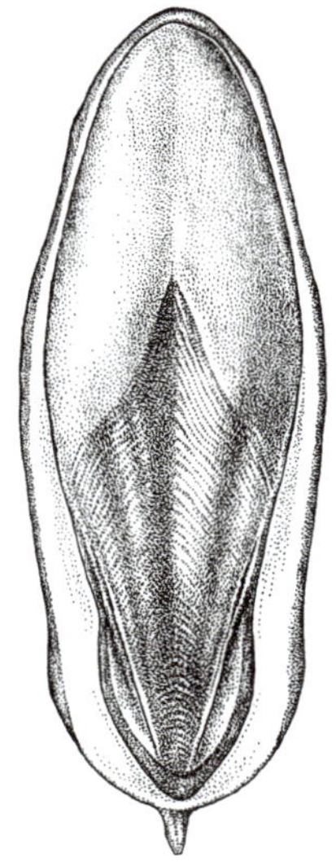

Cuttlebone, ventral view (ABRS, K. Hollis).

*Sepia rozella*

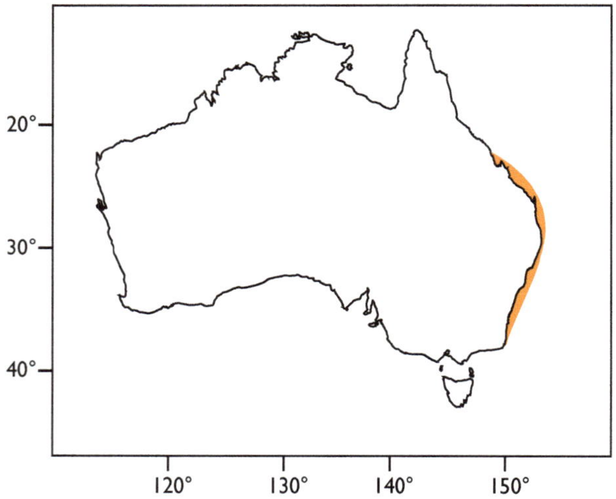


*Sepia rozella* distribution. *Note:* The distribution is based on whole animals only, not beachwashed cuttlebones.

fins; anterior-most pair and posterior two pairs shorter than rest. Head papillose dorsally and laterally.

### HABITAT AND BIOLOGY

*Interest to fisheries.* Species taken as by-catch of prawn and mixed species trawl fisheries. It is the main species caught in ocean trawls.

### SIZE

Up to 140 mm ML.

### DISTRIBUTION

Eastern Australia from 23°26′59″S, 151°55′11″E to 36°49′47″S, 149°58′47″E. Known depth 29–209 m.

### REFERENCE

Reid *et al.* (2005b).

## *Sepia senta* Lu & Reid, 1997

*Sepia senta* Lu, C. C. and Reid, A. L. (1997). Two new species of cuttlefish from northwestern Australian waters with a redescription of *Sepia sulcata* Hoyle, 1885 (Sepiidae: Cephalopoda). *Records of the Western Australian Museum 18*(3), 277–310 [286, figs 9–16].

Type data: holotype MV F.77200.

Type locality: Australia: Western Australia, North West Shelf, 14°45′S, 122°06′E.

### DIAGNOSTIC FEATURES

Small to moderate-sized species. All arms subequal in length in both males and females. Arm suckers tetraserial in both sexes. Left ventral arm hectocotylised: sucker size normal proximally (one transverse row), slightly reduced medially (8–11 transverse rows), with suckers in two dorsal series of modified portion smaller than remaining suckers, then and normal-sized suckers to arm tip. Oral surface of modified region wide, swollen, fleshy, with transversely grooved ridges; without distinct median furrow. Right ventral arm slightly thickened, fleshy, with slightly reduced suckers. Suckers in two dorsal two ventral series displaced laterally. Club crescent-shaped; sucker-bearing surface flattened, with six to eight suckers in each transverse row, Suckers differ slightly in size, small. Swimming keel of club extends slightly beyond carpus; dorsal and ventral protective membranes not fused at base of

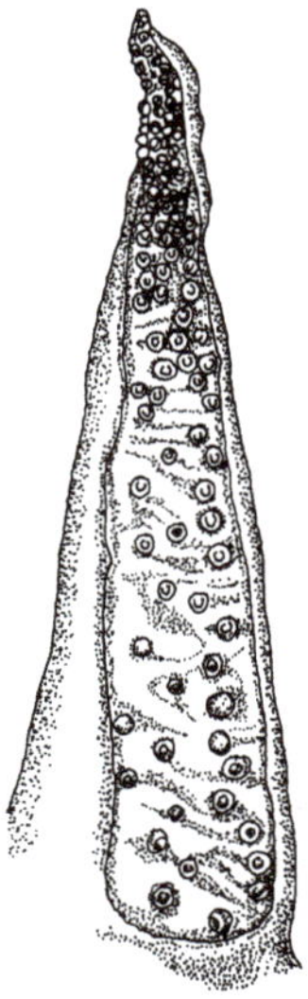

Hectocotylus (A. Reid).

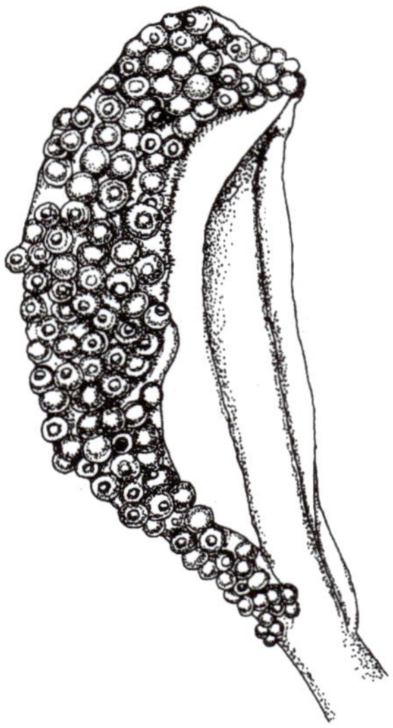

Tentacular club (A. Reid).

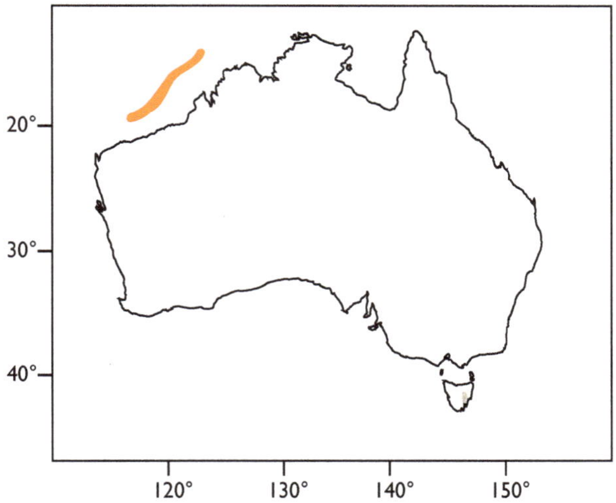


*Sepia senta* distribution. *Note:* The distribution is based on whole animals only, not beachwashed cuttlebones.

Cuttlebone, dorsal view (D. Paul).

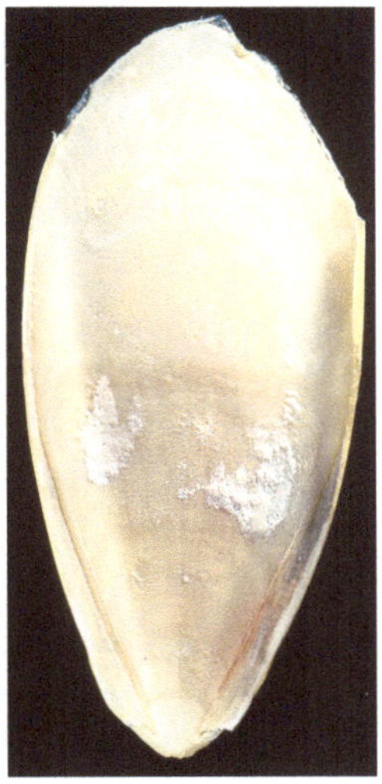

Cuttlebone, ventral views (D. Paul).

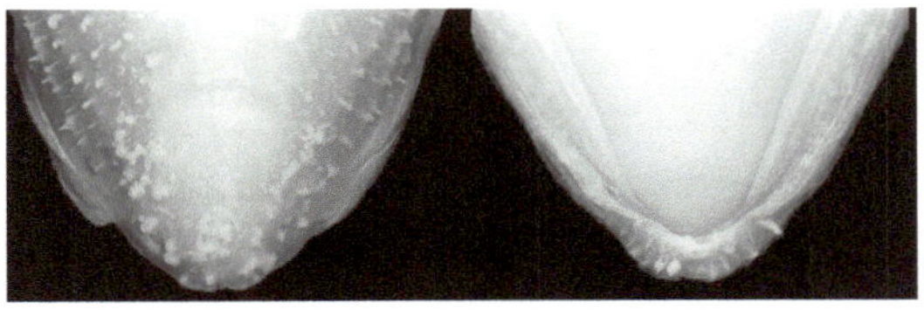

Enlargement of cuttlebone dorsal (left) and ventral view (right) (A. Reid).

club; dorsal and ventral membranes differing in length, dorsal membrane shorter than carpus, ventral membrane terminates at posterior end of carpus. Buccal membrane without suckers.

*Cuttlebone.* Cuttlebone outline oval, triangular, obtuse anteriorly; bluntly rounded posteriorly. Dorsal surface yellowish; flat medially, curved, convex laterally; texture rough, with irregular calcified projections (particularly pronounced towards the distal tip of the sepion). Dorsal median rib present, indistinct, broadens anteriorly; lateral ribs present, indistinct. Chitin present as thin film over entire dorsal surface of cuttlebone. Spine absent. Striated zone flat; sulcus absent. Anterior striae are slightly convex to straight. Inner cone limbs uniform width, narrow, U-shaped posteriorly.

*Colour* (alcohol preserved specimens). Pale buff pinkish brown.

## SIZE

Males up to 62 mm ML; females up to 83 mm ML.

## REMARKS

*Sepia senta* differs from all other sepiids in the structure of the dorsal side of the cuttlebone, with its irregular calcareous projections.

## DISTRIBUTION

South-eastern Indian Ocean: Western Australia, North West Shelf, 12°04′S, 122°51′E to 18°35′S, 117°29′E. Possibly Indonesia. A very similar, probably closely related animal has been found in the Philippines. Known depth range 344–356 m.

## REFERENCE

Lu and Reid (1997).

## *Sepia smithi* Hoyle, 1885

*Sepia smithi* Hoyle, W. E. (1885). Diagnoses of new species of Cephalopoda collected during the cruise of H.M.S. Challenger. Part 2. The Decapoda. *Annals and Magazine of Natural History (5)*16, 181–203 [190].

Type data: syntypes BMNH 1889.4.24.66-67 2♀ (cuttlebone figured by Hoyle (1886 pl. 16: figs 10–12) no longer extant).

Type locality: Arafura Sea, south of Papua, H.M.S. Challenger Stn 188, 9°59′S, 139°42′E.

Compiled from secondary source: Adam, W. & Rees, W. J. (1966). A review of the cephalopod family Sepiidae. Scientific Reports of the John Murray Expedition 1933–1944 11, 1–165.

### COMMON NAME

Smith's Cuttlefish

### DISTINGUISHING CHARACTERS

*Sepia smithi* often is confused with *S. whitleyana* (Iredale, 1926). The two species differ in the nature of the cuttlebone inner cone, which is round, shiny and thickened in *S. smithi* and thin, narrow and not shiny in *S. whitleyana*. The bases of the fins of *S. whitleyana* lack the longitudinal whitish bands that are present in *S. smithi*. Male *S. smithi* lack the distinctive longitudinal white bars that are present in male *S. whitleyana*. The cuttlebone is wider in *S. whitleyana* than in *S. smithi*, particularly in the anterior third of the bone.

### DIAGNOSTIC FEATURES

Mantle broad, oval. Fins wide, anterior origin almost at mantle margin, rounded posteriorly, with narrow gap between them. All arms subequal in length in both males and females. Arm suckers tetraserial. Hectocotylus present; both ventral arms modified. Left ventral arm: eight rows of normal sized suckers (approximately) proximally, five to eight rows of greatly reduced suckers medially, and then normal-sized suckers distally to arm tip. Suckers of hectocotylus in two ventral series are smaller than those in two dorsal series; oral surface of modified region not wide, fleshy, but normal; without distinct median furrow. Male right ventral arm: about six to eight rows of normal-sized suckers proximally, five rows of reduced suckers medially, and then normal-sized suckers to arm tip. Suckers of hectocotylus in dorsal row are smaller than those in two ventral series in modified region; oral surface of modified region very wide, fleshy, sponge-like in texture with pronounced median depression; suckers in two dorsal series aligned in a single row. Both ventral arms of males: suckers in two ventral series are displaced laterally, with gap between, widely spaced (more obviously so on right ventral arm). Club straight, slender, or slightly recurved, long, sucker-bearing surface convex, with 13–22 suckers in each transverse row; all club suckers of similar, minute, size. Swimming keel of club terminates at proximal end of carpus; dorsal and ventral protective membranes not joined at base of club but fused to tentacular stalk; dorsal and ventral membranes same length, terminate at posterior end of carpus; dorsal membrane forms shallow cleft at junction with stalk. Buccal membrane without suckers; in females with single median spermatheca in ventral part. Eggs spherical.

*Cuttlebone.* Cuttlebone outline oblong; bone triangular, obtuse anteriorly; bluntly rounded posteriorly; dorsal surface creamy white, or yellowish; dorsal surface evenly convex; entire surface calcified, granulose with sculpture arranged in series of blunt, transverse, inverted V-shape ridges. Spine and posterior end (approximately one-quarter) of bone covered with an ochre-coloured smooth glaze-like substance. Dorsal median rib and lateral ribs indistinct; median rib broadens slightly anteriorly, bordered laterally by indistinct grooves. Chitin surrounds entire margin of cuttlebone. Spine short, pointed, straight, directed dorsally, keel(s) absent. Striated zone concave; last loculus flat; sulcus shallow, wide, extends along striated zone only. Anterior striae are inverted U-shape. Inner cone limbs are broad, strap-like anteriorly, U-shaped posteriorly; outer margin of inner cone raised into rounded, thickened, shiny, posterior ledge.

*Colour.* Pale buff pinkish brown, light brown, or greyish brown with white blotches and spots, sometimes joined to form irregular transverse bands. Arms 1–3 have a longitudinal band of orange-red pigment along their aboral surfaces. Dorsal mantle has pinkish spots, or blotches and scattered purplish spots and blotches. Paired dorsal eye spots are present (both sexes). Fins in both sexes with longitudinal whitish band lateral to series of orange-pink ridges overlain with regularly spaced transverse purplish blotches.

*Sculpture.* Dorsal mantle covered with numerous small papillae and with series of seven or eight elongate orange-pink papillae along each side, adjacent to base of each fin. Head papillose dorsally and laterally.

### HABITAT AND BIOLOGY

In the Gulf of Carpentaria, *S. smithi* has been caught where bottom depths varied between 7 and 55 m, typically on sand and mud. *Sepia smithi* has been observed at night lying on the sea floor or burying into the sand, its pale colours and skin textures matching the sand surface. This lack of nocturnal activity suggests the species may be day-active.

*Interest to fisheries.* Species taken as by-catch of prawn and mixed species trawl fisheries.

### SIZE

Males up to 140 mm ML; females up to 150 mm ML.

### DISTRIBUTION

Indo-Pacific: northern Australia from 27°13′11″S, 153°45′E to 20°19′47″S, 118°15′E. Known depth range 10–84 m.

### REFERENCE

Reid *et al.* (2005b).

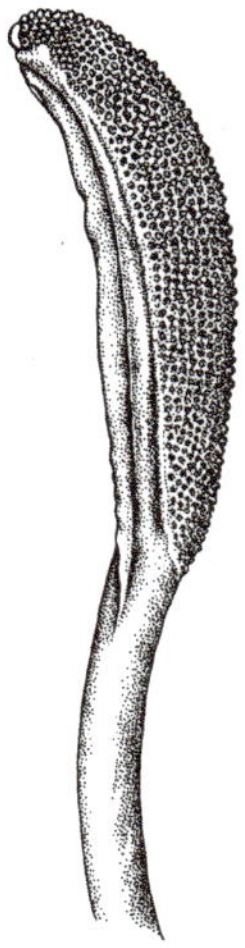

Tentacular club (ABRS, K. Hollis).

Cuttlebone, dorsal view (D. Paul).

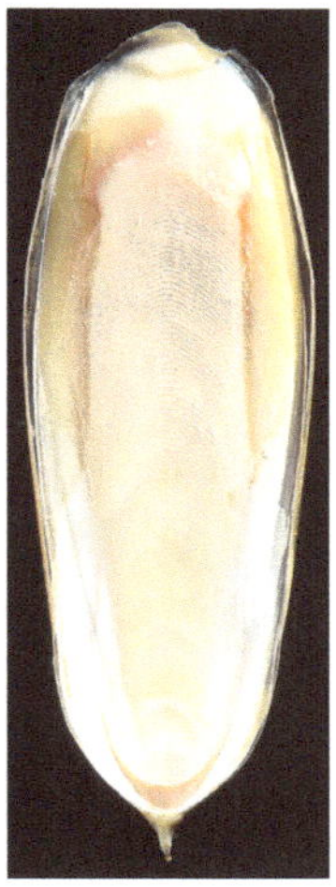

Cuttlebone, ventral views (D. Paul).

Live animal, dorsal view (M. Norman).

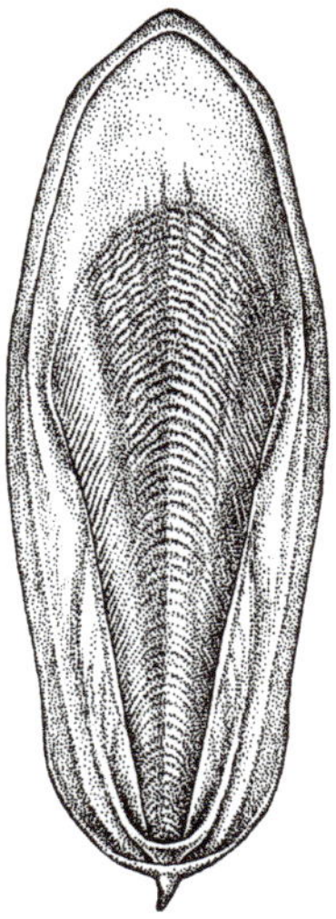

Cuttlebone, ventral view (ABRS, K. Hollis).

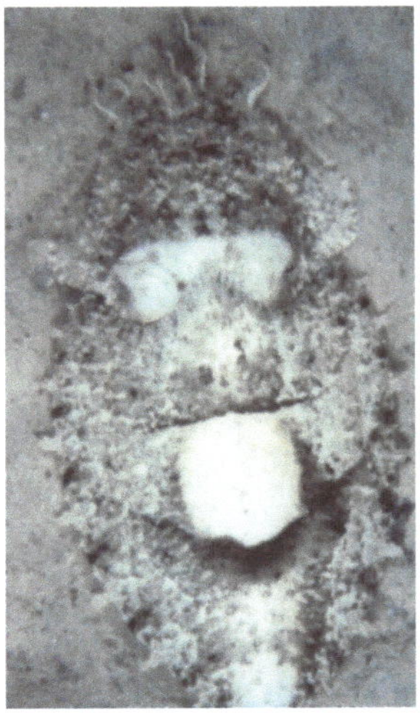

Live animal, dorsal view (M. Norman).

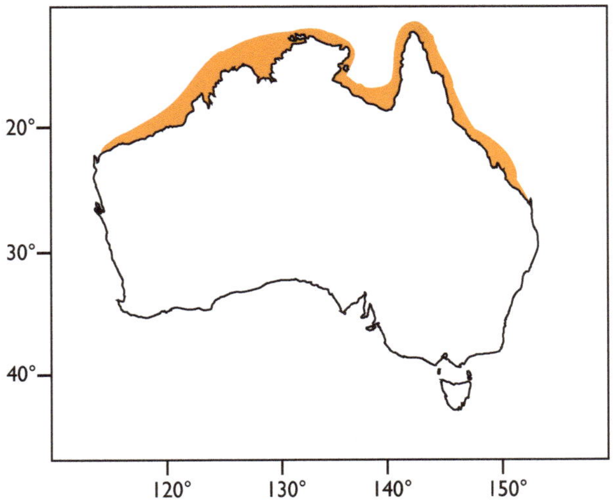


*Sepia smithi* distribution. *Note:* The distribution is based on whole animals only, not beachwashed cuttlebones.

## *Sepia sulcata* Hoyle, 1885

*Sepia sulcata* Hoyle, W. E. (1885). Diagnoses of new species of Cephalopoda collected during the cruise of H.M.S. Challenger. Part 2. The Decapoda. *Annals and Magazine of Natural History (5)16*, 181–203 [192].

Type data: holotype BMNH 89.4.24.77.

Type locality: Arafura Sea, off Kei (as Ki) Island, 5°49′15″S, 132°14′15″E.

### COMMON NAME

Grooved Cuttlefish

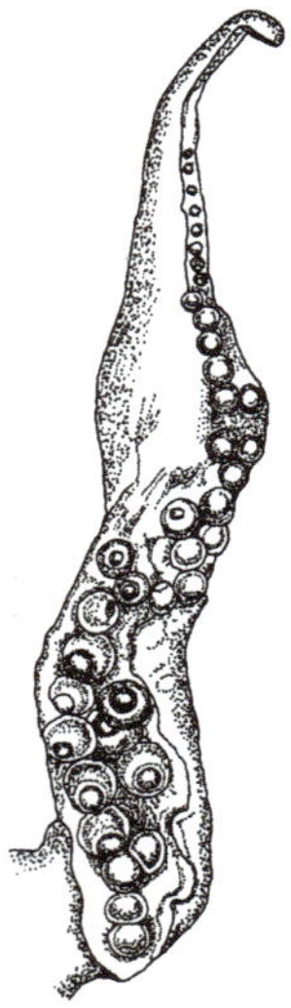

Hectocotylus, left ventral arm (A. Reid).

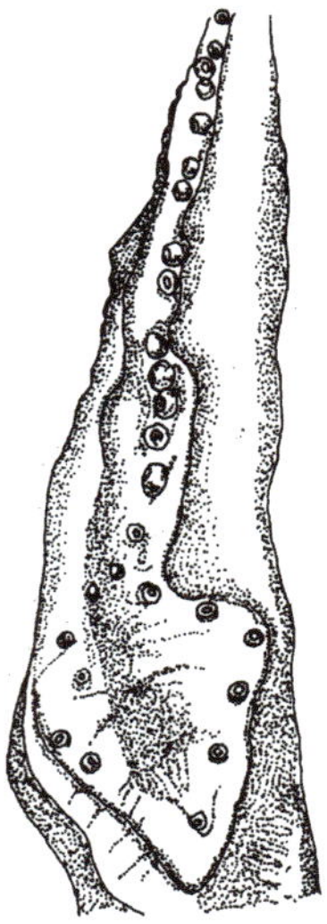

Right ventral arm, male (A. Reid).

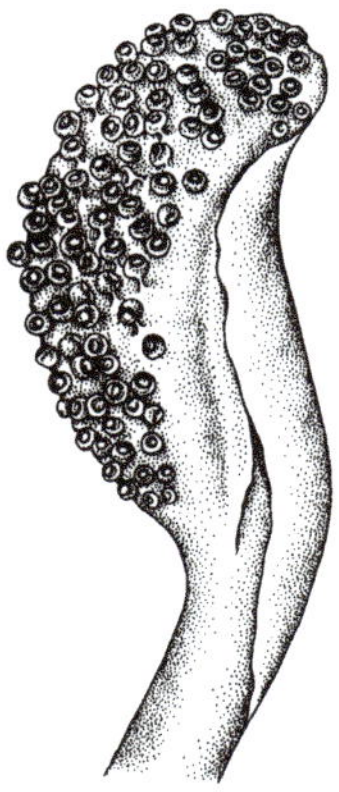

Tentacular club (A. Reid).

### DIAGNOSTIC FEATURES

All arms subequal in length in both males and females. Sucker rows on arms 1 and 3 in males biserial proximally, tetraserial distally (~8–11 rows biserial suckers on arms 1, six to eight rows biserial suckers on arms 2 and 3); suckers on arms 4 biserial at base (approximately two rows), tetraserial distally. Sucker rows on arms 1–3 in females biserial proximally for six or seven rows, tetraserial distally, suckers on arms 4 biserial proximally for two or three rows, rest tetraserial. Distal arm tips strongly attenuate. Left ventral arm hectocotylised: sucker size greatly reduced proximally (~14 transverse rows), two dorsal series of suckers on modified portion smaller than those in two ventral series; two dorsal and two ventral series widely spaced in mature males. Oral surface wide, fleshy with deep longitudinal furrow between swollen protective membranes. Club with five to seven small, subequal suckers in each transverse row; dorsal and ventral protective membranes not joined at base of club.

*Cuttlebone.* Cuttlebone oblong, strongly acuminate, acute, anteriorly and posteriorly. A thick, cartilaginous subdermal layer adheres closely to dorsal side of cuttlebone. Dorsal median and lateral ribs distinct; sulcus shallow, narrow, flanked by prominent rounded ribs bordered laterally by shallow grooves. Inner cone limbs form narrow, raised ledge posteriorly.

*Sculpture.* Dorsal mantle with up to 10 short, longitudinal, orange-pink ridges along each side close to fins; ventral mantle with six longitudinal ridges along each side close to fins.

### HABITAT AND BIOLOGY

Habitat mud, silt, rock.

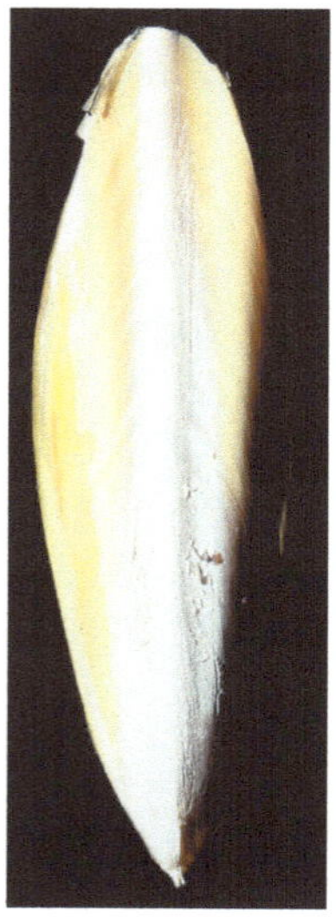

Cuttlebone, dorsal view (A. Reid).

Cuttlebone, ventral view (A. Reid).

Cuttlebone, ventral view, posterior end (A. Reid).

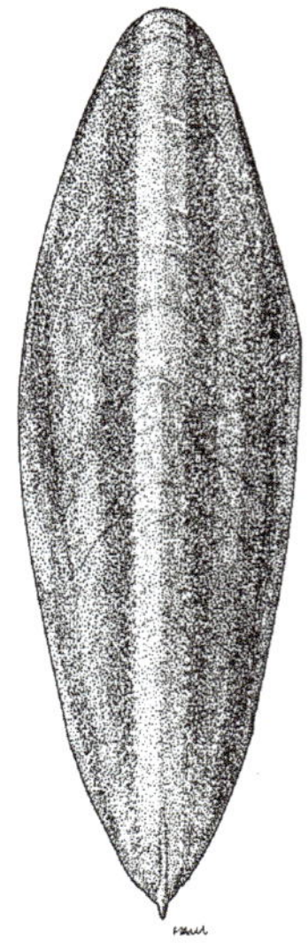

Cuttlebone, dorsal view (FAO, E. D'Antoni).

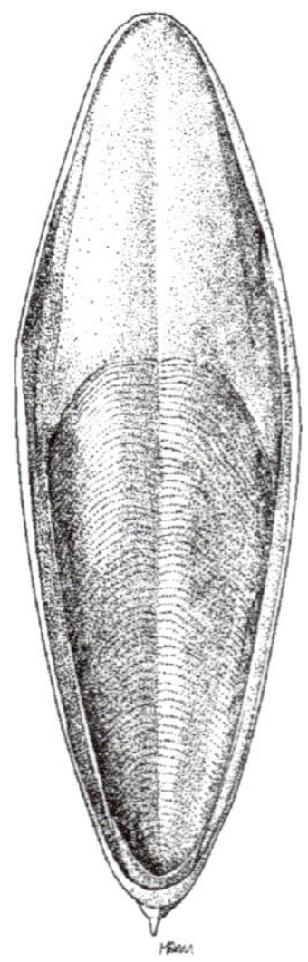

Cuttlebone, ventral view (FAO, E. D'Antoni).

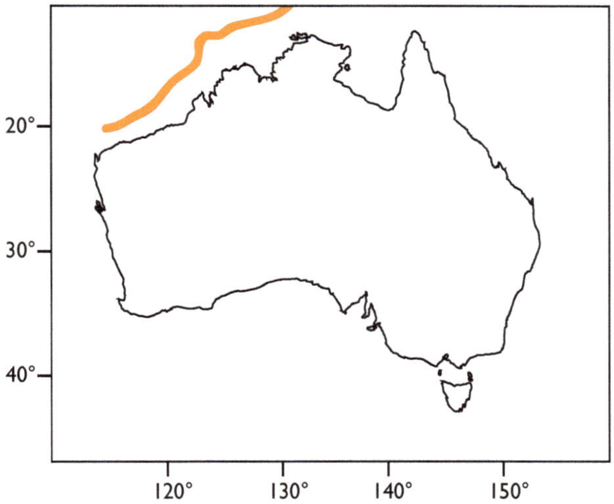


*Sepia sulcata* distribution. *Note:* The distribution is based on whole animals only, not beachwashed cuttlebones.

*Interest to fisheries.* Due to its size and distribution, this species may occur in multispecies trawl catches.

### SIZE

Males up to 68 mm ML; females up to 97 mm ML.

### REMARKS

The known depth range occupied by this species is broad. There is some evidence to suggest that there might be a migration into relatively shallower waters for spawning.

### DISTRIBUTION

Indo-Pacific: Indonesia, Arafura Sea, off Kai Islands, 05°49′15′S, 132°14′15′E, to Australia: Western Australia, North West Shelf, 18°10′S, 118°18′E to 17°17′S, 120°12′E. Known depth range 166–304 m.

### REFERENCE

Lu and Reid (1997).

## *Sepia tanybracheia* Reid, 2000

*Sepia tanybracheia* Reid, A. L. (2000). Australian cuttlefishes (Cephalopoda: Sepiidae): the 'doratosepion' species complex. *Invertebrate Taxonomy 14*, 1–76 [47].

Type data: holotype MV F.82479.

Type locality: Australia: Western Australia, near Ashmore Is, 12°36–37′S, 123°44–46′E.

### DIAGNOSTIC FEATURES

Small to moderate-sized species. Male arms 1 elongate; arms 1 narrow proximally, flared distally, protective membranes broad. Arm sucker arrangement differs between sexes. In males, arm suckers biserial on arms 1, arms 2 and 3 suckers biserial proximally, tetraserial medially, biserial distally, arms 4 tetraserial proximally, biserial distally (arms 2 and 3, suckers biserial on distal third of arms, arms 4 suckers biserial on distal arm tips). Left ventral arm hectocotylised; sucker size normal proximally, reduced medially, and then normal to arm tip; from proximal to distal end of arm, nine rows of normal suckers,

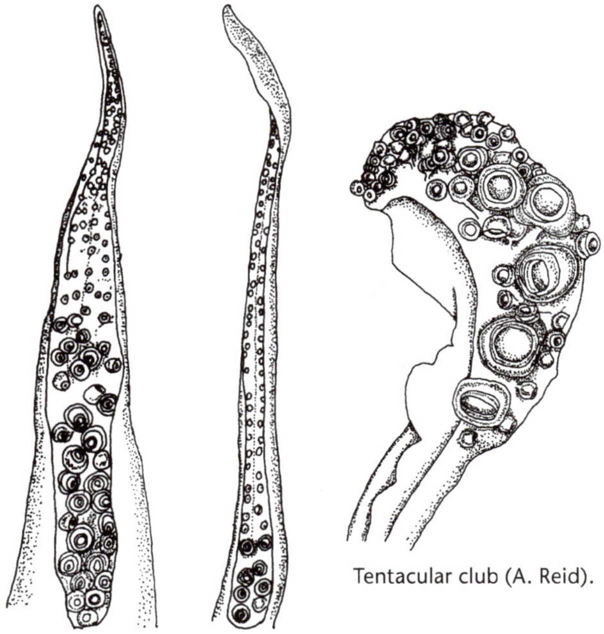

Arm 1, oral view (A. Reid). Hectocotylus (A. Reid). Tentacular club (A. Reid).

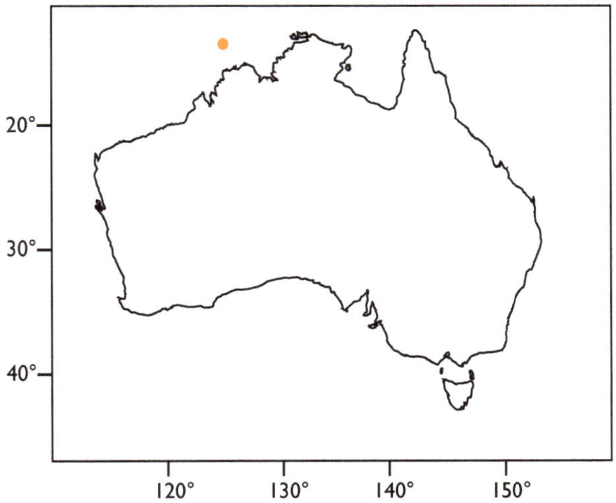


*Sepia tanybracheia* distribution. *Note:* The distribution is based on whole animals only, not beachwashed cuttlebones. (Australian records are based on few specimens, but the species is likely to occur more widely.)

10 rows of reduced suckers (approximately). Suckers equal in size across rows; reduced suckers much smaller than normal arm suckers. Oral surface of modified region wide, swollen, fleshy, with transversely grooved ridges; with deep median furrow. Club crescent-shaped; sucker-bearing face flattened, with four suckers in each transverse row. Suckers differ markedly in size, some greatly enlarged. Swimming keel of club extends well beyond carpus; dorsal and ventral protective membranes not fused at base of club; joined to stalk; dorsal and ventral membranes same length, extend beyond carpus along stalk.

*Colour* (alcohol preserved specimens). Purplish brown; arms 1 aboral tips with median dark maroon band; arms 2–4 distally with partially broken dark band on each side, proximally with dark blotches. Dorsal mantle with scattered reddish spots.

*Sculpture.* Dorsal mantle with longitudinal row of up to six ridge-like papillae along each side, close to base of each fin and with scattered short ridges dorsal to cuttlebone. Ventral mantle with longitudinal row of orange-pink narrow ridges along each side close to fins; papillae present on head and arms: short scattered bars positioned dorsally and laterally.

#### SIZE

Males up to 51 mm ML.

#### REMARKS

Cuttlebone not known.

#### DISTRIBUTION

South-eastern Indian Ocean: known only from type locality 12°37′S, 123°44′E. Known depth range 200–205 m.

#### REFERENCE

Reid (2000).

### *Sepia vercoi* Adam, 1979

*Sepia vercoi* Adam, W. (1979). The Sepiidae (Cephalopoda, Decapoda) in the collections of the Western Australian Museum. *Records of the Western Australian Museum 7*, 113–212 [190, pl. 10 figs 5, 6].

Type data: holotype WAM 441-65.

Type locality: Australia: Western Australia, west of Shark Bay, 25°31′S, 112°29′E.

#### DIAGNOSTIC FEATURES

Small species. Male and female arms differ in relative lengths. Male arm lengths subequal; female arms 2 slightly longer than rest. Arm sucker arrangement differs between sexes: in males, sucker rows on arms 1 and 2 biserial proximally, biserial or tetraserial distally in small specimens, larger specimens with fewer rows of biserial suckers, arrangement predominantly tetraserial; arms 3 and 4 tetraserial (except for hectocotylus as described below). In females, arm suckers tetraserial. Female arms 2 and 3 suckers displaced laterally with gap between (swollen, fleshy). Left ventral arm of male hectocotylised; sucker size normal proximally, reduced medially, and then normal to arm tip; from proximal to distal end of arm, six to eight rows of normal suckers (proximal one to three rows biserial, following six to eight rows tetraserial), five to seven rows of reduced suckers. Suckers in two dorsal series similar in size, ventral pair in each series absent; reduced suckers much smaller than normal arm suckers. Oral surface of modified region ventral side wide, fleshy, with transversely grooved ridges; with deep median furrow. Club crescent-shape, with four suckers in each transverse row; swimming keel of club extends slightly beyond carpus; dorsal and ventral protective membranes not fused at base of club; joined to stalk; dorsal and ventral membranes differing in length, dorsal membrane extends beyond carpus along stalk, ventral membrane terminates at posterior end of carpus.

*Cuttlebone.* Cuttlebone outline lanceolate. Bone bluntly rounded anteriorly; acuminate, acute posteriorly; recurved ventrally. Dorsal surface pinkish; convex posteriorly, flat anteriorly; entire surface calcified with reticulate granulose sculpture concentrated medially and on lateral margins in irregular longitudinal ridges, calcification thickest posteriorly. Dorsal median rib present, indistinct, broadens anteriorly; not bordered laterally by distinct grooves; lateral ribs absent. Spine present, short,

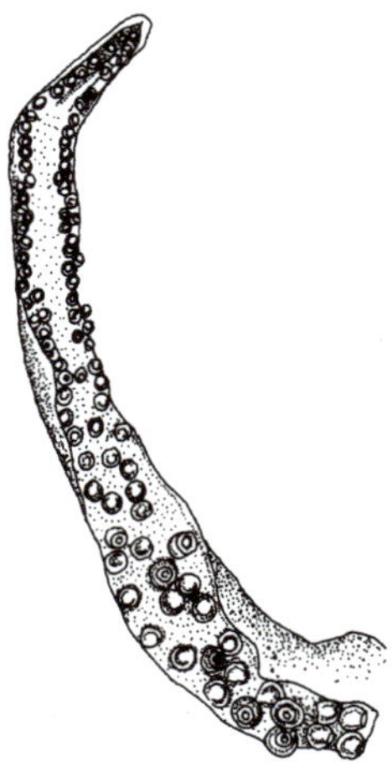

Arm 2 (A. Reid).

Hectocotylus (A. Reid).

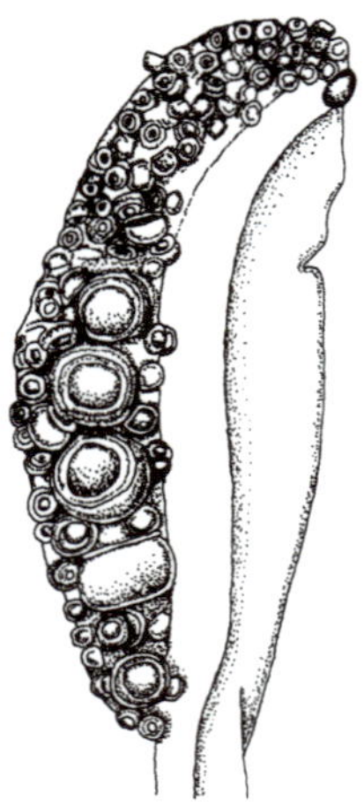

Tentacular club (A. Reid).

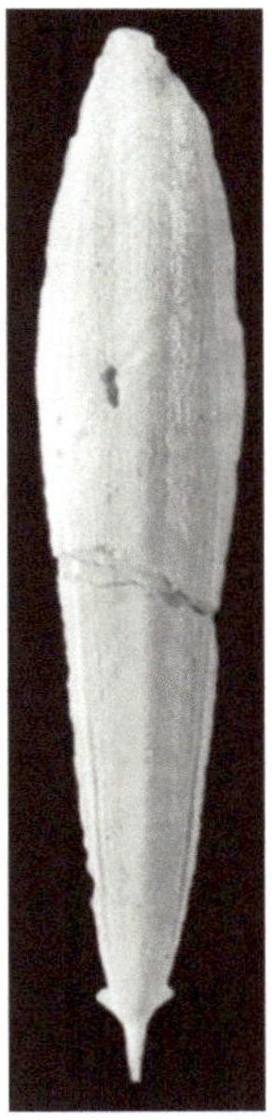

Cuttlebone,
dorsal view (A. Reid).

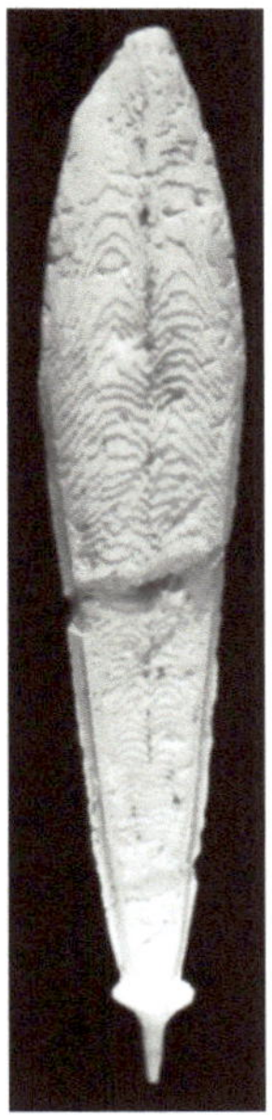

Cuttlebone,
ventral view (A. Reid).

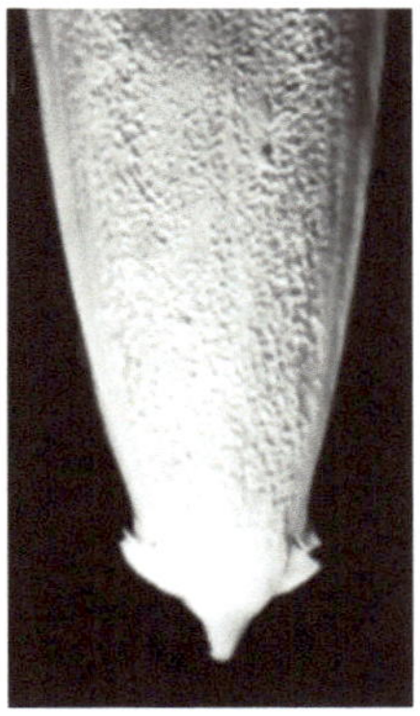

Posterior end of
cuttlebone enlargement,
dorsal view (A. Reid).

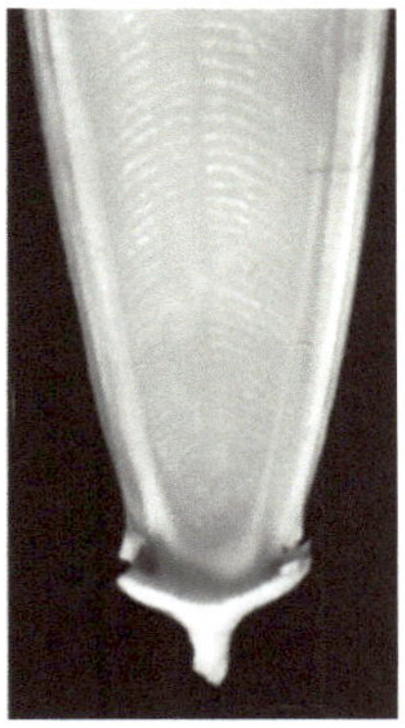

Posterior end of
cuttlebone enlargement,
ventral view (A. Reid).

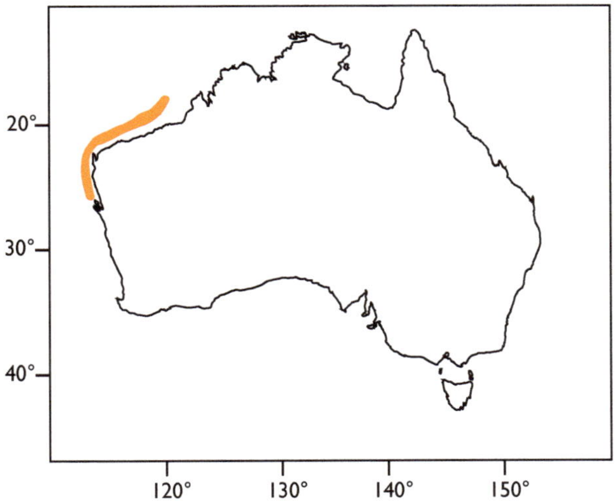


*Sepia vercoi* distribution. *Note:* The distribution is based on whole animals only, not beachwashed cuttlebones.

pointed; straight, directed dorsally; keel(s) absent. Shallow, narrow sulcus extends entire length of cuttlebone; flanked by rounded ribs bordered laterally by shallow grooves. Anterior striae are inverted U-shaped (large specimens), or slightly convex to straight (small specimens). Inner cone limbs narrow, strap-like anteriorly, broaden posteriorly; slightly raised to form rounded posterior ridge; thickened. Outer cone limbs expanded posteriorly forming two short 'wings', directed ventrally.

*Colour* (alcohol preserved specimens). Purplish brown (pale); head with scattered orange spots, with chromatophores concentrated medially and over eye orbits, and luminous silvery-mauve band extending from posterior end of head dorsally over each eye to halfway along arms 3, and four short luminous bars evenly spaced on head between these long bands. Dorsal mantle chromatophores concentrated medially over cuttlebone. Fins with narrow lustrous orange-pink band along base in males.

*Sculpture.* Body papillae present; dorsal mantle with longitudinal row of up to six orange-pink ridge-like papillae along each side, close to base of each fin and with scattered short ridges dorsal to cuttlebone. Short, scattered bars on head and arms.

## SIZE

Up to 46 mm ML (both sexes).

## REMARKS

This species occurs sympatrically with *Sepia koilados* Reid, 2000, *S. cottoni* Adam, 1979 and *S. rhoda* (Iredale, 1954).

## DISTRIBUTION

South-eastern Indian Ocean: Western Australia, 18°31′S, 118°09′E to 25°31′S, 112°29′E. Known depth range 76–201 m.

## REFERENCE

Reid (2000).

## *Sepia whitleyana* (Iredale, 1926)

*Acanthosepion whitleyanum* Iredale, T. (1926). The cuttlefish 'bones' of the Sydney beaches (Phylum Mollusca–Class Cephalopoda). *Australian Zoologist 4*, 186–196 pls 22, 23 [195, pl. 23 figs 9, 10].

Type data: holotype AM C.133331.

Type locality: Australia: New South Wales, Port Macquarie, 31°27′S, 152°55′E.

Distribution: New South Wales (central east coast), Queensland (Gulf of Carpentaria, north-east coast). Known depth range to 128 m.

### COMMON NAME

Whitley's Cuttlefish

### DISTINGUISHING CHARACTERS

*Sepia whitleyana* is very similar and closely related to *S. smithi* Hoyle, 1885. The cuttlebone inner cone is round, shiny and thickened in *S. smithi* and thin, narrow and not shiny in *S. whitleyana*. The bases of the fins of *S. whitleyana* lack the longitudinal whitish bands that are present in *S. smithi*. Male *S. smithi* lack the distinctive longitudinal white bars that are present in male *S. whitleyana*. The cuttlebone is wider in *S. whitleyana* than in *S. smithi*, particularly in the anterior third of the bone.

### DIAGNOSTIC FEATURES

Mantle broad, oval. Fins rounded posteriorly, with narrow gap between them. All arms subequal in length in both males and females. Arm suckers tetraserial. Hectocotylus present; both ventral arms modified. Male left ventral arm: six to eight rows of normal-sized suckers proximally, five or six rows of reduced suckers medially, and then normal-sized suckers to arm tip. Suckers of hectocotylus in two dorsal series are much smaller than those in two ventral series; oral surface of modified region wide, slightly fleshy (but not to same extent as right ventral arm). Male right ventral arm: five to seven rows of normal-sized suckers proximally, five or six rows of reduced suckers medially, and then normal-sized suckers to arm tip; oral surface of modified region wide, fleshy, sponge-like in texture with distinct median furrow; suckers in two dorsal and two ventral series displaced laterally, with wide gap between them; reduced suckers all approximately same size or dorsal and ventral-most rows may be smaller than median two rows on proximal end of modified region. Club long; sucker-bearing surface flattened, with 12–21 small suckers in each transverse row. Swimming keel of club terminates at proximal end of carpus; dorsal and ventral protective membranes not joined at base of club but fused to tentacular stalk; dorsal and ventral membranes same length, terminate at proximal end of carpus; approximately equal width; dorsal membrane forms shallow cleft at junction with stalk. Buccal membrane without suckers; females have a single, median spermatheca in ventral part.

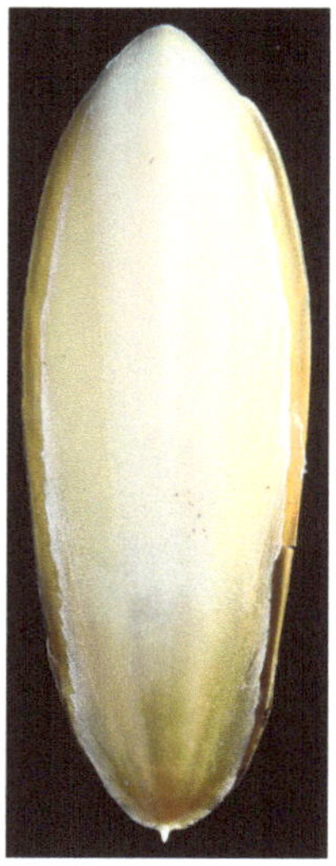

Cuttlebone, dorsal view (D. Paul).

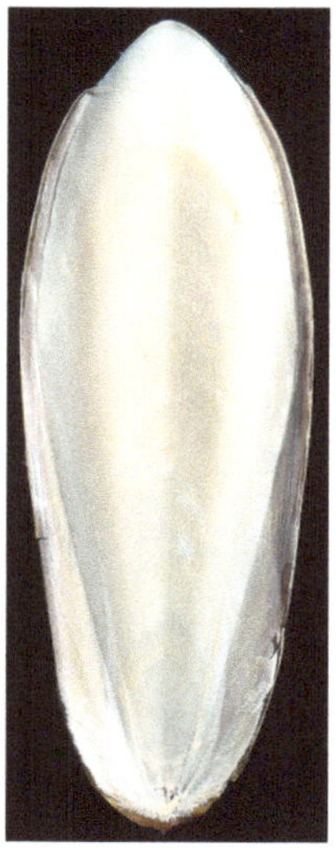

Cuttlebone, ventral view (D. Paul).

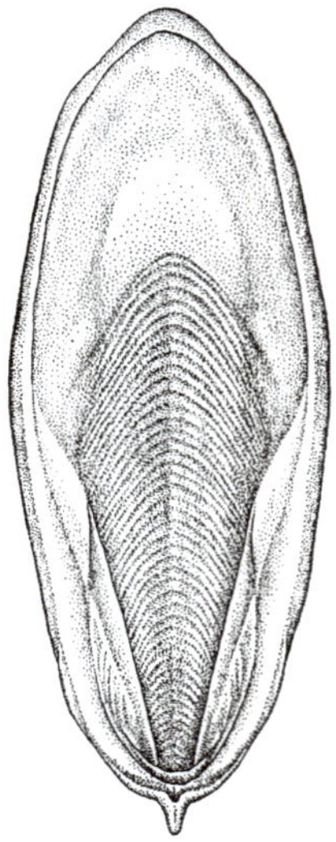

Cuttlebone, ventral view (K. Hollis).

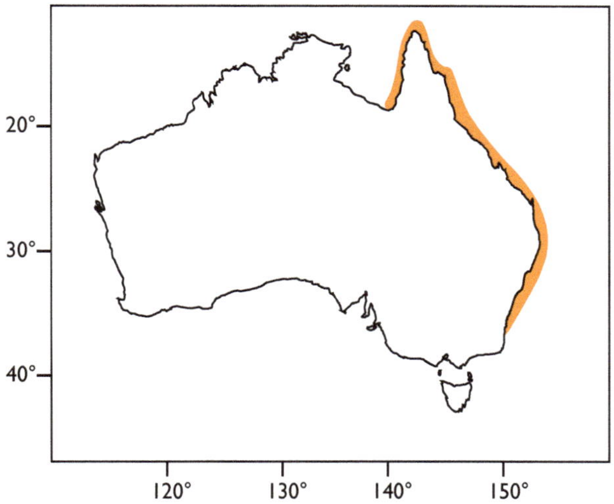


*Sepia whitleyana* distribution. *Note:* The distribution is based on whole animals only, not beachwashed cuttlebones.

*Cuttlebone.* Cuttlebone outline oblong; acuminate, acute, anteriorly; bluntly rounded posteriorly; dorsal surface creamy white; dorsal surface convex medially, flat laterally; granulose, or calcified medially. Spine and posterior tip of bone covered with a smooth, glaze-like substance. Dorsal median rib and lateral ribs absent or indistinct; sides of median rib approximately parallel. Chitin surrounds entire margin of cuttlebone. Spine short, pointed, straight, directed dorsally. Spine keel(s) absent. Striated zone deeply concave; last loculus convex; sulcus deep, wide, extends along striated zone only; sulcus flanked by rounded ribs. Anterior striae are inverted U-shaped; limbs of inner cone short, extend anteriorly to approximately one-third the length of the striated zone. Inner cone limbs are narrow anteriorly, broad U-shaped posteriorly; slightly raised into flat, very narrow, posterior ledge; ledge not thickened, dull, not shiny; outer cone calcified, narrow anteriorly, broadens posteriorly.

*Colour.* Freshly caught animals dark olive or bluish grey with characteristic pattern of wavy longitudinal lines and spots. Head with a few scattered chromatophores. Arms 1–3 have a longitudinal band of pale pigment along their aboral surfaces. Arm and club sucker rings pale brown, or yellow brown. Dorsal mantle has pinkish spots, or blotches and narrow longitudinal whitish bands in males only. Paired dorsal eye spots present. Fins without markings at base, or with approximately six short, dark, transverse bars along each side.

*Sculpture.* Dorsal mantle with longitudinal row of up to nine ridge-like papillae along each side, adjacent to base of each fin.

#### HABITAT AND BIOLOGY

*Interest to fisheries.* Species taken as by-catch of prawn and mixed species trawl fisheries.

#### SIZE

Up to 174 mm ML (both sexes).

#### REMARKS

*Sepia whitleyana* is often confused with *S. smithi* Hoyle, 1885.

#### DISTRIBUTION

South-western Pacific: northern and eastern Australia from 14°37′48″S, 145°24′E to 35°30′S, 150°48′E. Western extent of range unknown. Known depth range 31–53 m.

#### REFERENCE

Reid *et al.* (2005b).

## *Metasepia* Hoyle, 1885

### *Metasepia pfefferi* (Hoyle, 1885)

*Sepia (Metasepia) pfefferi* Hoyle, W. E. (1885). Diagnoses of new species of Cephalopoda collected during the cruise of H.M.S. Challenger. Part 2. The Decapoda. *Annales and Magazine of Natural History (5)*16, 181–203 [199].

Type data: holotype BMNH 10.9.1874.

Type locality: Arafura Sea: south of Papua New Guinea, 9°59′S, 139°42′E.

#### COMMON NAME

Pfeffer's Flamboyant Cuttlefish

#### DIAGNOSTIC FEATURES

Mantle oval, very broad. Arms broad, blade-like; arms 1 shorter than rest. Arm suckers tetraserial. Left ventral arm hectocotylised: oral surface of modified region wide, swollen, fleshy, with transversely grooved ridges and deep median furrow. Club sucker-bearing surface flattened, with five or six suckers in each transverse row; suckers differ markedly in size: three or four median suckers greatly enlarged, occupying most of median portion of club. Swimming keel of club extends well proximal to carpus; dorsal and ventral protective membranes not joined at base of club but fused to tentacular stalk; dorsal and ventral membranes differ in length; extend proximal to carpus along stalk.

Dorsolateral view (B. Saunders).

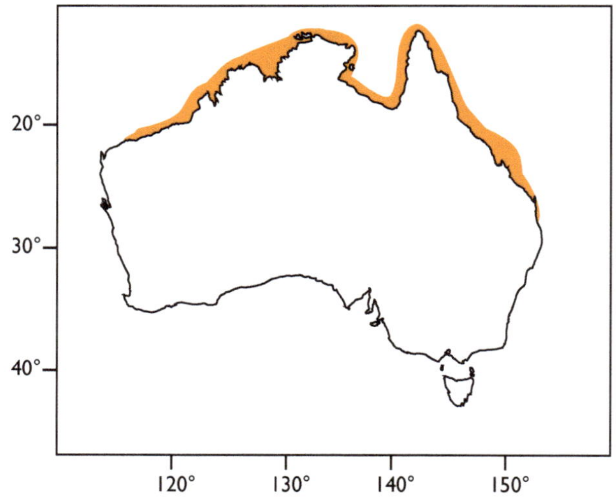


*Metasepia pfefferi* distribution. *Note:* The distribution is based on whole animals only, not beachwashed cuttlebones.

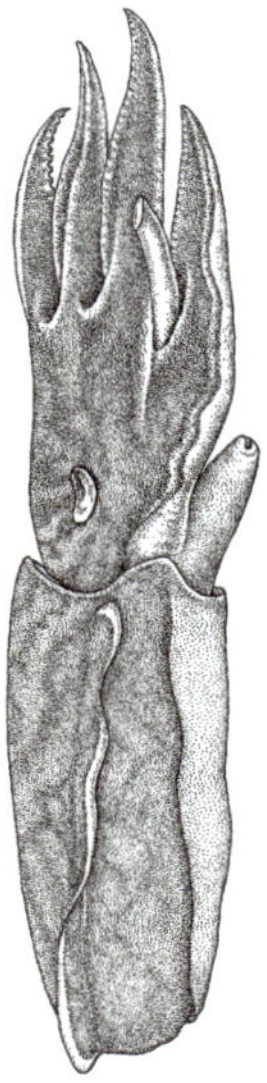

Lateral view (K. Hollis).

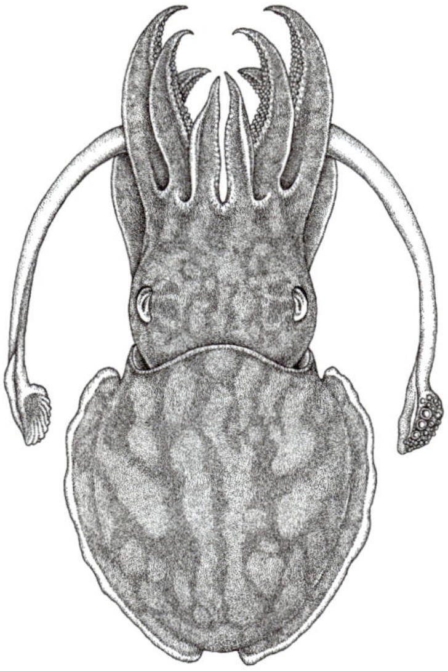

Dorsal view (K. Hollis).

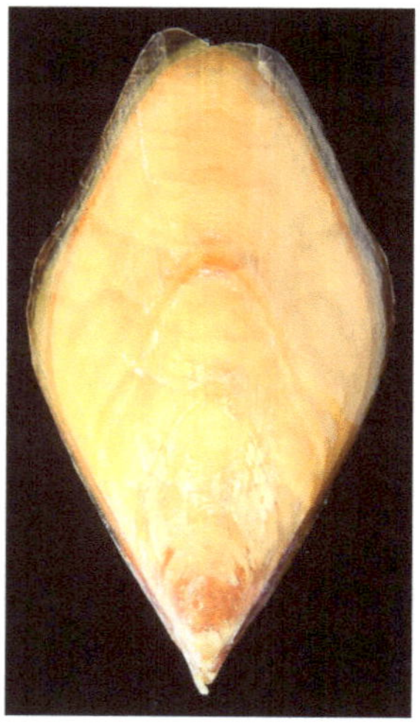

Cuttlebone, dorsal view (D. Paul).

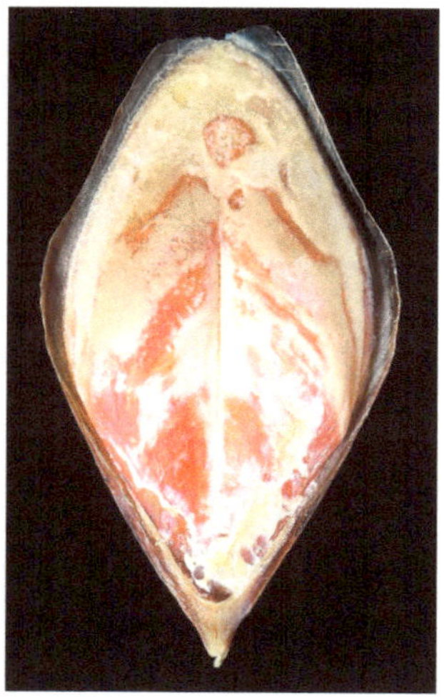

Cuttlebone, ventral view (D. Paul).

*Cuttlebone.* Cuttlebone much shorter than mantle, located in anterior 60–75% of mantle; outline rhomboidal; acuminate, acute, anteriorly and posteriorly; dorsal surface yellowish, evenly convex; texture smooth, not pustulose; dorsal median rib absent. Chitin present as a thin film over entire dorsal surface of cuttlebone. Spine absent or small; chitinous. Striated zone concave; last loculus strongly convex, thick in anterior third; sulcus deep, wide, extends along striated zone only. Anterior striae are inverted V-shaped; limbs of inner cone are very short, uniform width, narrow, U-shaped, thickened slightly posteriorly; outer cone absent.

*Colour.* Base colour dark brown with mobile patterns of white and yellow; purple-pink along arms.

*Sculpture.* Dorsal mantle has three pairs of large, flat, flap-like papillae. Head with papillae over eyes.

### HABITAT AND BIOLOGY

Sand and muddy substrate. Active during the day, hunting fishes and crustaceans. Capable of excellent camouflage while stalking prey. If disturbed or attacked, it quickly changes to a dramatic colour pattern of dark brown, black, white and yellow patches and may show bright red on the arm tips. It has been observed to 'amble' along the sea floor in this colour pattern, rhythmically waving the wide protective membranes on the arms. Eggs are laid singly in crevices or ledges in coral, rock or wood. One female has been observed squeezing her eggs through the hole of a coconut husk. The round, white eggs turn clear as they develop. Juveniles are capable of the same bright colour pattern as adults from birth.

*Interest to fisheries.* None as food. But this species is spectacular in its colour and textural displays and would make an excellent aquarium species. It has been successfully kept (bred through several generations) at the Steinhart Aquarium (Californian Academy of Sciences), MBARI and some home hobbyists.

### SIZE

Up to 60 mm ML.

### REMARKS

Results of recent molecular studies have shown that *Metasepia pfefferi* as currently recognised comprises two species (Moussalli, personal communication).

### DISTRIBUTION

Northern Australia from 26°34′12″S, 153°40′11″E to 20°19′47″S, 118°15′E. New Guinea. Known depth range 4–86 m.

### REFERENCE

Reid *et al.* (2005b).

## *Sepiella* Gray, 1849

### Key to *Sepiella* species

1a. Club with 12–15 suckers in each transverse row; swimming keel shorter than carpus. Cuttlebone strongly convex in lateral view......*S. mangkangunga*

1b. Club with 7–10 suckers in each transverse row; swimming keel terminates at level of carpus. Cuttlebone not strongly convex in lateral view ........ ..........................................................................*S. weberi*

## *Sepiella mangkangunga* Reid & Lu, 1998

*Sepiella mangkangunga* Reid, A. & Lu, C. C. (1998). A new *Sepiella* Gray, 1849 (Cephalopoda: Sepiidae) from northern Australia, with a redescription of *Sepiella weberi* Adam, 1939. *Beagle, Records if the Northern Territory Museum of Arts and Sciences 14*, 71–102 [74, figs 1–11].

Type data: holotype NTM P.10606.

Type locality: Australia: Northern Territory, off Stingray Head, 12°48′S, 130°21′E.

### DISTINGUISHING CHARACTERS

*Sepiella mangkangunga* differs from *S. weberi* Adam, 1939 in several characters. This species has a greater number club suckers in the transverse rows (12–15) than *S. weberi* (7–10). The club swimming keel in *S. mangkangunga* is shorter than the carpus, and is the same length as the carpus in *S. weberi*. The cuttlebone in *S. mangkangunga* is broader, and wider than the *S. weberi* cuttlebone, with a pronounced indentation in the posteriolateral edge of the outer cone. The depression in the last loculus is more pronounced in *S. mangkangunga* than in *S. weberi*. The inner cone is thicker in *S. mangkangunga* than in *S. weberi* (though this difference may be difficult to detect in small *S. mangkangunga* specimens). In addition, the distinctive ovoid spots along the base of the fins in *S. weberi* are not seen in *S. mangkangunga*. The latter species is found in shallow water (3.3–1.1 m), while *S. weberi* occurs at depths of 88–77 m in Australian waters, though it has been found in shallower depths elsewhere.

### DIAGNOSTIC FEATURES

Small to moderate-sized species. All arms subequal in length in both males and females. Left ventral arm hectocotylised: sucker size markedly reduced proximally (eight or nine rows), with suckers in two dorsal series smaller than remaining suckers. Oral surface of modified region wide, swollen, fleshy, with transversely grooved ridges; without distinct median furrow; suckers in two dorsal and two ventral series displaced laterally, two ventral series close together, rows alternate. Club crescent-shaped; sucker-bearing surface convex, with 12–15 suckers in each transverse row. Swimming keel of club shorter than carpus; dorsal and ventral protective membranes not fused at base of club; joined to stalk; dorsal and ventral membranes same length, extend beyond carpus along stalk; dorsal membrane forms shallow cleft at junction with stalk.

*Cuttlebone.* Cuttlebone outline oblong, strongly convex in lateral view, distinctly V-shaped in lateral view, widest at junction of striated zone and last loculus. Dorsal median rib present, distinct, broadens anteriorly; lateral ribs absent. Spine absent. Sulcus extends along striated zone only; shallow, wide; flanked by rounded ribs. Anterior striae are inverted U-shaped (slightly wavy). Limbs of inner cone short, extend anteriorly to junction of striated zone and posterior termination of last loculus. Inner cone limbs thickened; shiny. Outer cone chitinous, spatulate, expanded; posteriolateral wall with pronounced indentation in males and slight indentation in females.

*Colour.* Head with few scattered chromatophores; arms without markings. Dorsal mantle pale, peppered with scattered purple-black chromatophores.

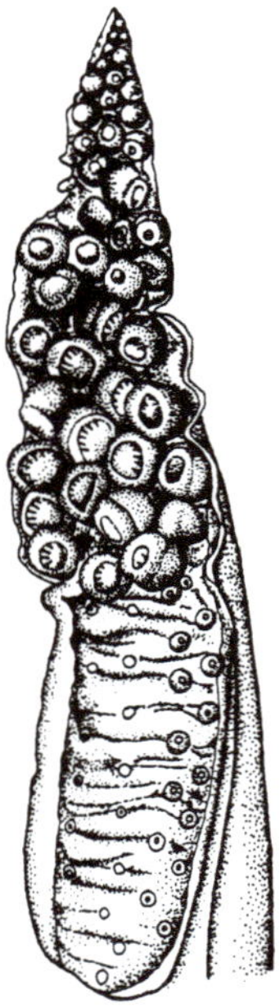

Hectocotylus (A. Reid).

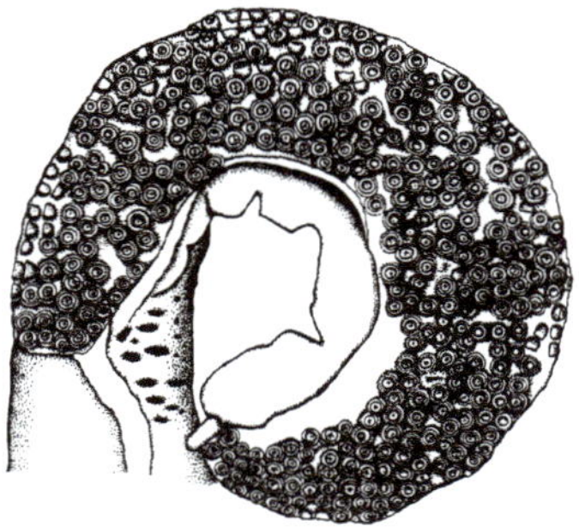

Tentacular club (A. Reid).

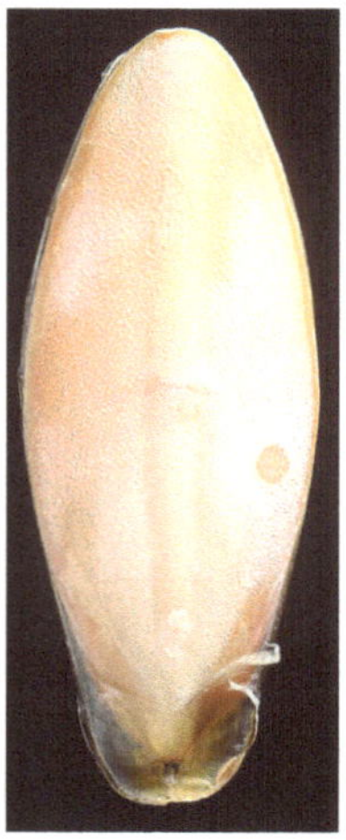

Cuttlebone, dorsal view (D. Paul).

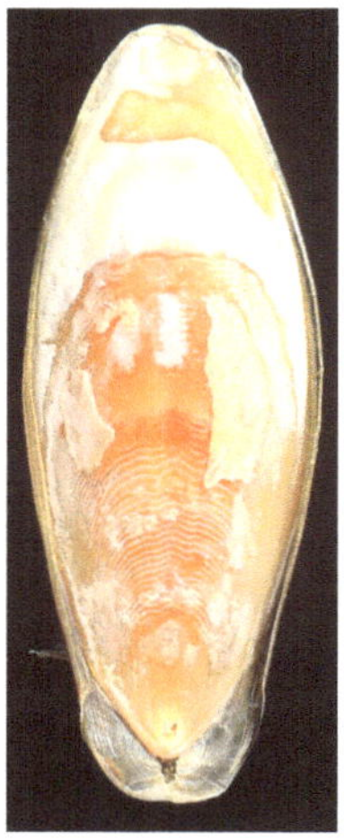

Cuttlebone, ventral views (D. Paul).

### HABITAT AND BIOLOGY

Sand.

### SIZE

Males up to 58 mm ML; females up to 59 mm ML.

### DISTRIBUTION

Indo-Pacific: Australia, northern Australia from 12°52′48″S, 130°16′48″E to 11°57′36″S, 135°5′24″E. Depth from 1.1–3.3 m.

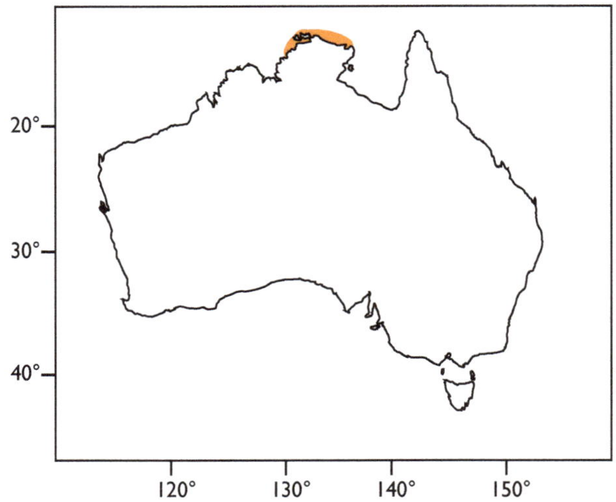


*Sepiella mangkangkunga* distribution. *Note:* The distribution is based on whole animals only, not beachwashed cuttlebones.

### REFERENCE

Reid and Lu (1998).

## *Sepiella weberi* Adam, 1939

*Sepiella weberi* Adam, W. (1939). Céphalopoda. 2e Partie. II Revision des espèces Indo-Malaises du genre Sepia Linné, 1758. III. Revision du genre *Sepiella* (Gray) Steenstrup, 1880. *Siboga-Expeditie 55b*, 1–135 [98, pl 4 figs 1, 2]. 556, 98.

Type data: syntypes ZMA Moll. 2.39.002, Moll. 3.39.001.

Type locality: Timor, off Kupang, surface (attracted to light), 8°35′S, 126°00′E, and Soemba, 10°S 119°56′E, eastern Indonesia, and Soemba, 10°S 119°56′E, eastern Indonesia.

### DIAGNOSTIC FEATURES

All arms subequal in length in both males and females. Arm suckers tetraserial; two median series of suckers in males are larger than dorsal and ventral marginal series, and larger than female arm suckers. Left ventral arm hectocotylised: suckers reduced proximally, much smaller than normal arm suckers (11 or 12 transverse rows), with suckers in two dorsal series smaller than those in two ventral series. Oral surface of modified region wide, swollen, fleshy, with transversely grooved ridges. Suckers of hectocotylus in two dorsal and two ventral series are displaced laterally, with gap between, two ventral series close together, rows alternate. Club crescent-shaped, sucker-bearing surface convex, with 7–10 suckers in each transverse row; dorsal

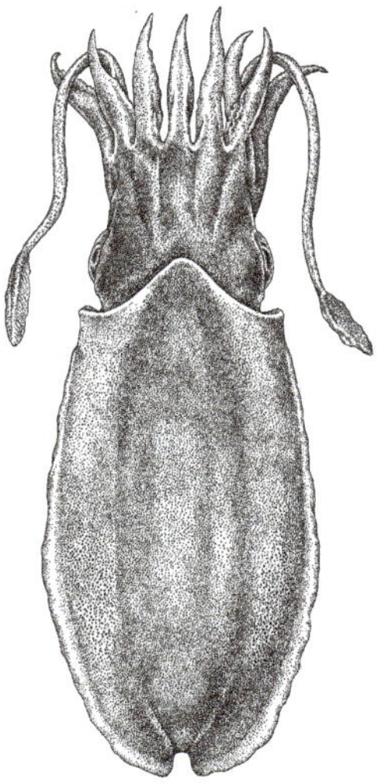

Dorsal view (ABRS, K. Hollis).

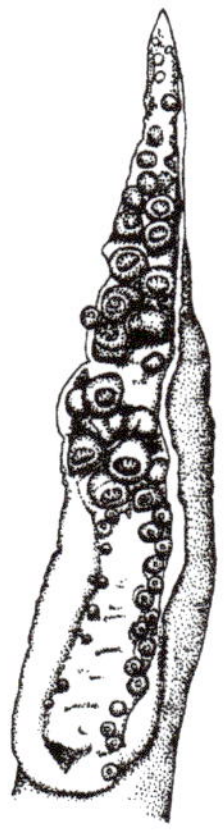

Hectocotylus (A. Reid).

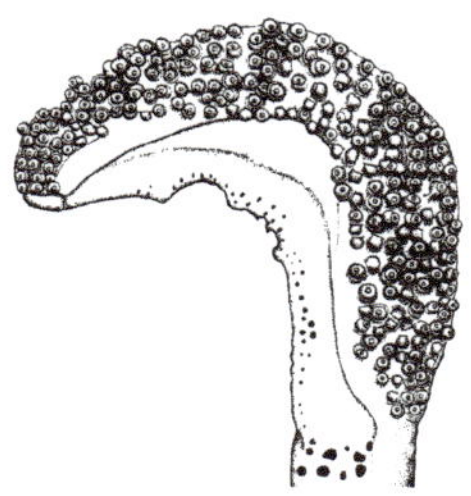

Tentacular club (A. Reid).

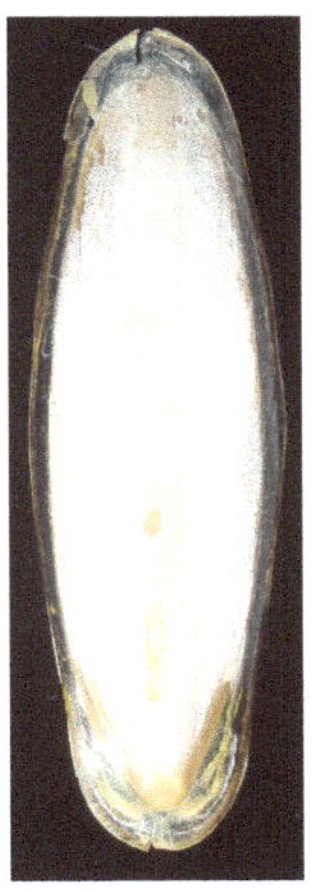

Cuttlebone, dorsal view (D. Paul).

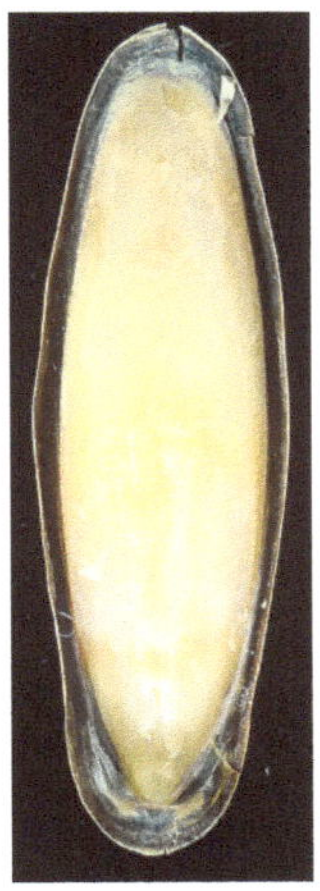

Cuttlebone, ventral view (D. Paul).

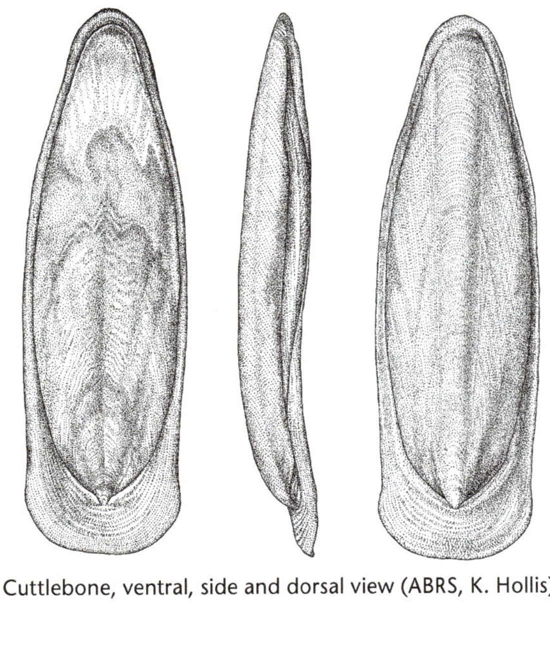

Cuttlebone, ventral, side and dorsal view (ABRS, K. Hollis).

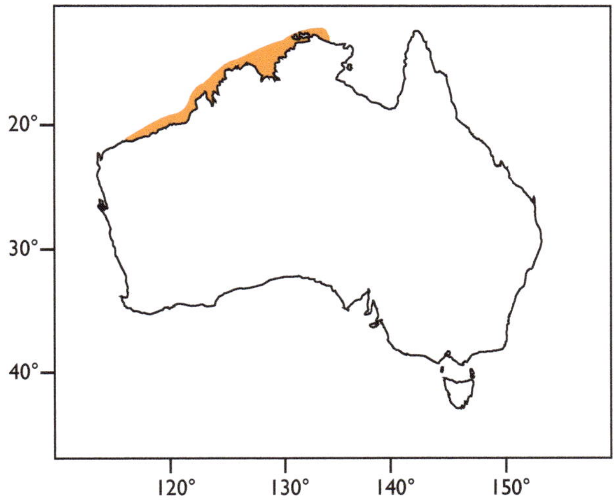


*Sepiella weberi* distribution. *Note:* The distribution is based on whole animals only, not beachwashed cuttlebones.

and ventral protective membranes not joined at base of club. Swimming keel and club protective membranes are equal in length to sucker-bearing surface of club.

*Cuttlebone.* Cuttlebone outline oblong; cuttlebone width 21–31% cuttlebone length in males, 30–33% in females; bone strongly recurved ventrally; dorsal median rib indistinct; rib broadens anteriorly; lateral ribs absent. Striated zone and last loculus convex; sulcus shallow, narrow, flanked by rounded ribs, extends entire length of cuttlebone. Anterior striae are inverted U-shaped, or wavy; limbs of the inner cone are short, thickened slightly; shiny; outer cone chitinous, spatulate, expanded; without indentation on posteriolateral margin.

*Colour.* Purplish brown. Dorsal mantle pale, peppered with scattered purple-black chromatophores. Fins with five or six oval orange-pink spots at base in both sexes, spots slightly larger in males.

## HABITAT AND BIOLOGY

Known depth range shallow waters to 88 m (Australia).

*Interest to fisheries.* Specific reports are not available, but the species may occur in multispecies trawl catches due to its size and distribution.

## SIZE

Males up to 60 mm ML; females up to 70 mm ML.

## DISTRIBUTION

Indo-Pacific: Indonesia (Timor–Soemba), 08°35′S, 126°00′E to 10°S 119°56′E, to north-western Australia, 10°58′12″S, 132°19′11″E to 19°34′47″S, 117°12′E. South

China Sea: Vietnam (unconfirmed records). Known depth range (Australia) 77–88 m.

#### REFERENCE

Reid and Lu (1998).

## Order Sepiolida

This order contains the two families Sepiolidae and Sepiadariidae. Molecular phylogenetic analyses of sepiolids have been limited. Lindgren *et al.* (2012), in a study using four to 10 genes supported the monophyly of Sepiadariidae ( = *Sepioloidea* + *Sepiadarium*) and confirmed Sepiadariidae as a sister taxon to the Sepiolidae, thus supporting a monophyletic Sepiolida.

### Key to sepiolid families

1a. Dorsal mantle fused with head. Internal shell absent. Both ventral arms or left ventral arm hectocotylised ........................................... Sepiadariidae

1b. Dorsal mantle free or fused to head. Internal shell rudimentary, straight and chitinous. One or both dorsal arms, or one dorsolateral arm hectocotylised............................................................ Sepiolidae

## Sepiadariidae Fischer, 1882 in Fisher (1880–87)

#### DIAGNOSTIC FEATURES

Small. Dorsal mantle fused to head. Nuchal cartilage absent. Fins narrow, or short; ear-shaped, with pronounced anterior lobes, or 'earlets'; fin attachment long. Mantle and funnel-locking cartilages absent in *Sepiadarium* (mantle and funnel permanently fused) or with two components in *Sepioloidea*. Non-hectocotylised arm sucker arrangement same in both sexes: biserial, sometimes tetraserial at middle and distal tips of arms. Hectocotylus present on left ventral arm: distal end of hectocotylus without suckers, pedicels modified forming series of transverse lamellae. Tentacle stalk held in deep sheath or web between bases of arms 3 and 4, web encircles base of tentacles on outside and inside forming cutaneous sac. Internal gladius absent. Photophores absent. Dorsal mantle surface covered with large, white leucophores surrounded by smaller red-brown chromatophores.

#### HABITAT AND BIOLOGY

Benthic. All species bury in the sand during the day and emerge at night to feed. All produce mucous from the underside of the body when disturbed.

#### SIZE

Up to 40 mm ML.

#### DISTRIBUTION

Found only in the western central Pacific, including Australia and New Zealand.

### Key to genera of the Sepiadariidae

1a. Ventral mantle permanently connected to funnel by wide muscular band, funnel-locking apparatus absent; fins positioned in posterior half of mantle... .................................................................. *Sepiadarium*

1b. Ventral mantle not permanently connected to funnel, funnel-locking apparatus present; fins extend along most of ML ....................... *Sepioloidea*

## *Sepiadarium* Steenstrup, 1881

### Key to *Sepiadarium* species

1a. Fins 70–80% ML. Arms 4 not shorter than arms 2. Small round red or brown points and black or dark brown spots on dorsal side of mantle and head....... ...................................................................... *S. auritum*

1b. Fin length ~50% ML. Arms 4 shorter or equal to arms 2. Pigment spots light .....................................2

2a. Tentacular club with 6 suckers in each transverse row. Hectocotylised arm distally with a series of conical lamellae, without continuous vertical membrane .............................................. *S. austrinum*

2b. Tentacular club with 8 suckers in each transverse row. Hectocotylised arm distally with a series of transverse paddle-like lamellae bordered by vertical membrane with serrated edge....... *S. kochii*

## *Sepiadarium auritum* Robson, 1914

*Sepiadarium auritum* Robson, G. C. (1914). Cephalopods from the Monte Bello Islands. *Proceedings of the Zoological Society of London 1814*, 677–680 [678, fig. 1].

Type data: holotype BMNH 1914.5.7.1.

Type locality: Australia: Western Australia, Monte Bello Islands, Hermite Island.

#### DIAGNOSTIC FEATURES

Length of fins 75–80% ML. Mantle wider than long. Arms 4 not shorter than arms 2. Arms suckers biserial, becoming smaller and irregular distally. Left ventral arm hectocotylised: proximally with about six pairs of normal suckers;

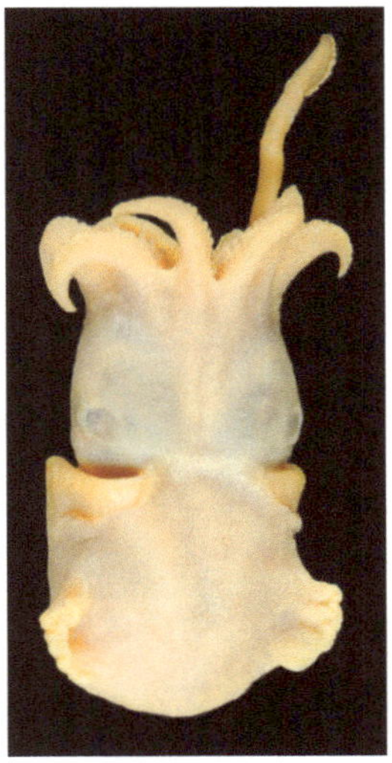

Preserved museum specimen, dorsal view (AM C.456770, L. Mott).

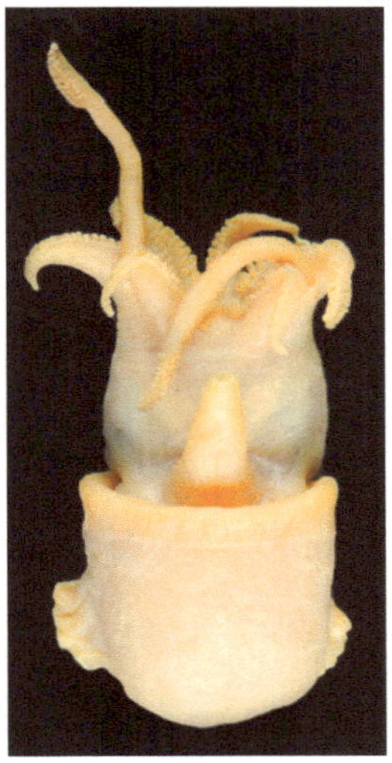

Preserved museum specimen, view (AM C.456770, L. Mott).

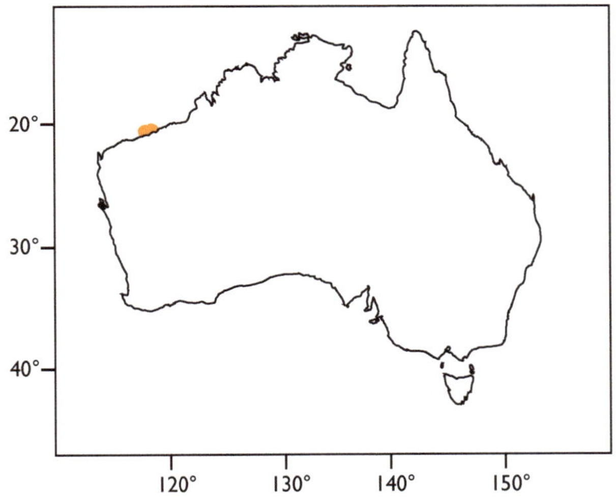


*Sepiadarium auritum* distribution.

suckers replaced by single series of conical lamellae on distal end of hectocotylised arm; arm grooved at distal tip, recurved dorsally, without protective membrane.

*Colour.* Small, round red or brown points and black or dark brown spots on dorsal side of mantle and head.

### SIZE

Type specimen, 11 mm ML.

### DISTRIBUTION

Eastern Indian Ocean: north-western Australia.

### REFERENCE

Reid *et al.* (2005b).

## *Sepiadarium austrinum* Berry, 1921

*Sepiadarium austrinum* Berry, S. S. (1921). A review of the cephalopod genera Sepioloidea, Sepiadarium, and Idiosepius. *Records of the South Australian Museum 1*(4), 347–364 [354].

Type data: holotype SAMA D.17493.

Type locality: Australia: South Australia, St Vincent Gulf.

### COMMON NAME

Southern Bottletail Squid

Live animal (P. Mercurio).

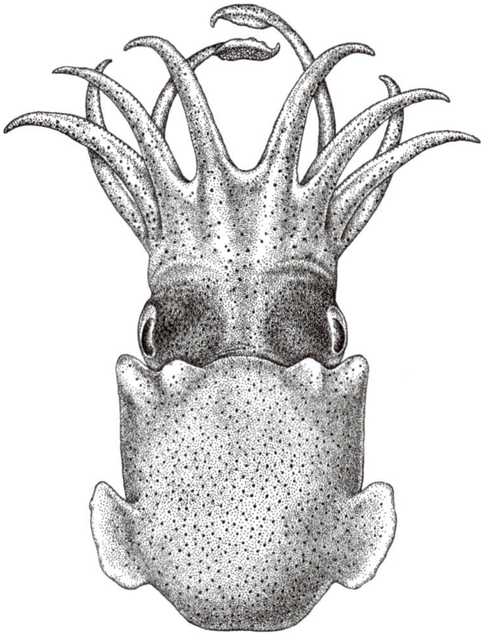

Dorsal view (ABRS, K. Hollis).

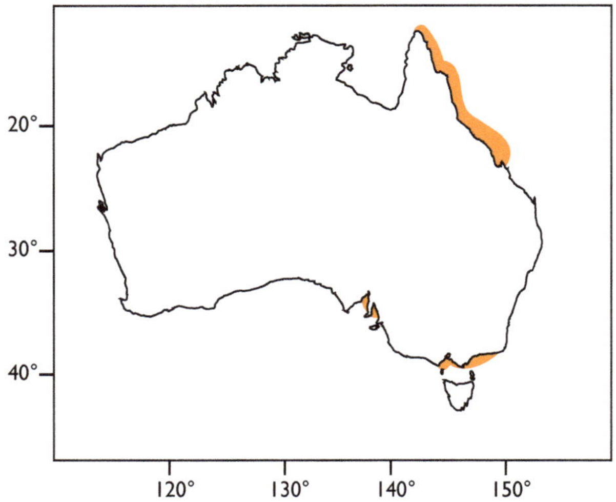


*Sepiadarium austrinum* distribution.

#### DIAGNOSTIC FEATURES

Dorsal mantle fused to head. Mantle cavity divided by thin septum. Ventral mantle fused to each side of proximal end of funnel. Fins short, do not exceed length of mantle anteriorly or posteriorly; attached within posterior half of mantle. Head broad, as wide as mantle. Arms differ in relative lengths in both males and females: arms 1 and 2 longer than 3 and 4. Non-hectocotylised arm sucker arrangement same in both sexes: biserial, sometimes tetraserial at middle and distal tips of arms. Left ventral arm hectocotylised: base of hectocotylus with nine or 10 pairs of normal suckers; suckers replaced by single series of conical lamellae on distal end of hectocotylised arm; arm grooved at distal tip, recurved dorsally, without protective membrane. Club with six suckers in each transverse row; club with wide keel, not bound by distinct folds or continuous membranes.

*Colour.* Body transparent, yellow or orange, with large, ovoid white leucophores.

#### HABITAT AND BIOLOGY

This species is found in sandy habitats in sheltered waters. They bury in the sand during the day, with only the eyes showing; and emerge at night to feed over the sand and near seagrass beds. They can be maintained easily in aquaria, fed on amphipods, isopods and other small crustaceans. They are able to produce slime from glands on the ventral surface of the body. Both sexes mate from an early age, with immature females able to store sperm in a buccal pouch. Sand-coated eggs are attached to the base of seaweed or seagrass. The young settle soon after hatching.

*Interest to fisheries.* Undetermined. This species may have a certain economic interest due to its easy survival in aquaria.

#### SIZE

Up to 30 mm ML.

#### DISTRIBUTION

Southern Indo-Pacific: southern Australia.

#### REFERENCE

Reid *et al.* (2005b).

### *Sepiadarium kochii* Steenstrup, 1881

*Sepiadarium kochii* Steenstrup, J. (1881). *Sepiadarium* og *Idiosepius* to nye Slaegter af Sepiernes Familie. Med Bemaerkninger om de to beslaegtede Former Sepiolidea D'Orb. og Spirula Lmk. *Danske Videnskabernes Selskabs Skrifter, 6 Raekke, Naturvidenskabelig og Mathematisk* 1(3), 211–242 [218, pl. 1, figs 1–10].

Type data: syntype (1) ZMUC, remaining 2 syntypes not extant.

Type locality: Hong Kong: (probably) Deepwater Bay.

#### DIAGNOSTIC FEATURES

Small; mantle dome-shaped. Dorsal mantle fused to head; length of fused portion ~25% ML. Mantle cavity divided by thin septum. Ventral mantle fused to each side of proximal end of funnel. Fins narrow; oblong; short, do not exceed length of mantle anteriorly or posteriorly; posteriorly with wide gap between them; attached along posterior half of mantle. All arms subequal in length in both males and females, arm formula 1 ≥ 2 >3 ≥ 4. Non-hectocotylised arm sucker arrangement same in both sexes: biserial basally, tetraserial over distal 20 transverse rows. Left ventral arm hectocotylised: base of arm suckers normal, not modified; 18–20 fleshy lamellae over distal two-thirds of arm, bordered by fold-like membrane; lamellae thick, longitudinally grooved transverse pads. Club with eight suckers in each transverse row; all suckers of similar minute size; very densely set.

*Colour.* Dorsal mantle surface covered with large, white leucophores surrounded by smaller red- brown chromatophores.

Live animal (R. Caldwell).

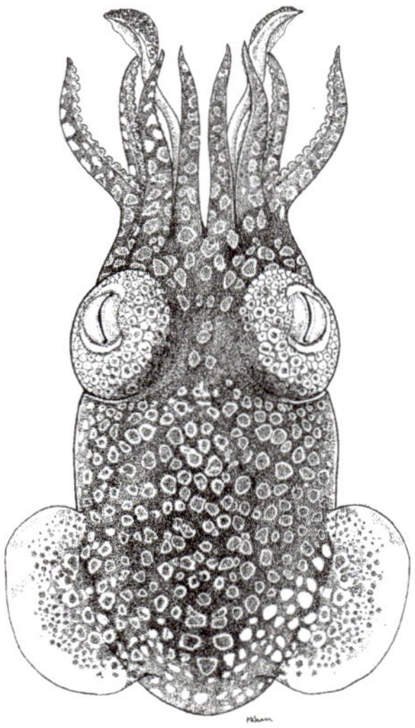

Dorsal view (FAO, E. D'Antoni).

HABITAT AND BIOLOGY

Upper sublittoral; known depth range to 60 m. Inhabits soft sediments. Remains buried during the day, then emerges at night to feed, primarily on small crustacea.

*Interest to fisheries.* None.

SIZE

Up to 30 mm ML.

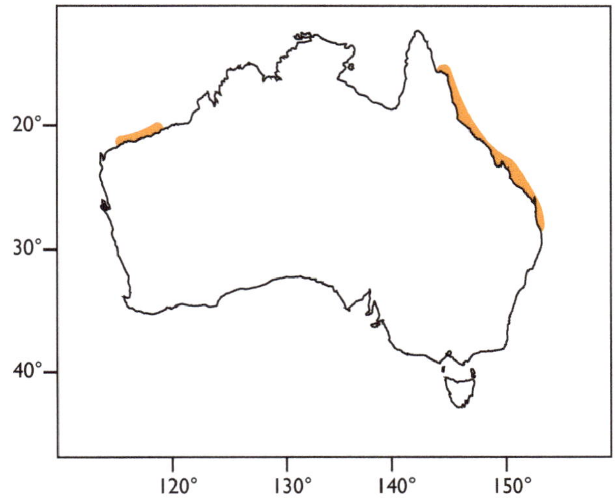


*Sepiadarium kochii* distribution.

DISTRIBUTION

Indo-Pacific: throughout Indo-Malayan waters from Japan (Tokyo Bay, Pacific side and Toyama Bay (Japan Sea side)), Taiwan Province of China and the South China Sea to India; southern Indonesia to New Guinea and northern Australia. The species has been reported from South Australia, though this record is unconfirmed.

REFERENCE

Jereb and Roper (2005).

## *Sepiadarium* sp.

COMMON NAME

Lace Bottletail Squid

DIAGNOSTIC FEATURES

Fins long and narrow, kidney-shaped. Males with modified ventral arm. Mucous glands on ventral mantle. Sand rubble and rocky reef habitat.

*Colour.* Transparent white to uniform dark purple with white lace-like markings.

HABITAT AND BIOLOGY

Sand rubble and rocky reef habitats. It remains buried in sand during the day, emerging at night to feed. Females have a spermatheca ventral to the beak.

SIZE

Up to ~30 mm ML.

### DISTRIBUTION

South Australia and south Western Australia. Shallow water.

### REMARKS

This species has yet to be described.

### REFERENCE

Norman and Reid (2000).

Live animal, dorsal view (P. Mercurio).

## *Sepioloidea* Ferussac & d'Orbigny, 1839

### COMMON NAME

Bottletail Squids

Live animal, dorsolateral view (P. Mercurio).

### DIAGNOSTIC FEATURES

Body short, fins narrow, located medially on each side of body. Mantle fused with head in nuchal area; mantle and funnel-locking cartilage with two components. Arms joined by a wide web; web forms deep sheath around base of tentacles. Left ventral arm hectocotylised; suckers absent distally, pedicels modified forming series of transverse lamellae. Gladius absent. Photophores absent.

### REFERENCE

Reid (2009).

### Key to species of the Sepioloidea

1a. Mantle, head and arms with narrow stripes. Anterior mantle margin fringed. Tentacular club with transverse rows of ~20 suckers....... *S. lineolata*

1b. Mantle, head and arms uniform colour, without stripes. Anterior mantle margin not fringed. Tentacular club transverse rows of ~40 minute suckers ........................................................................ *S. magna*

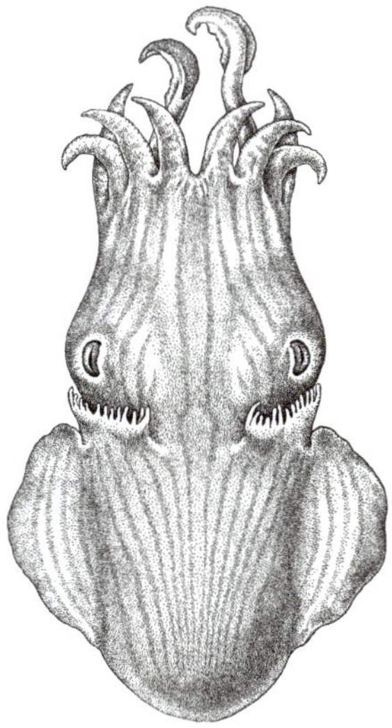

Dorsal view (ABRS, K. Hollis).

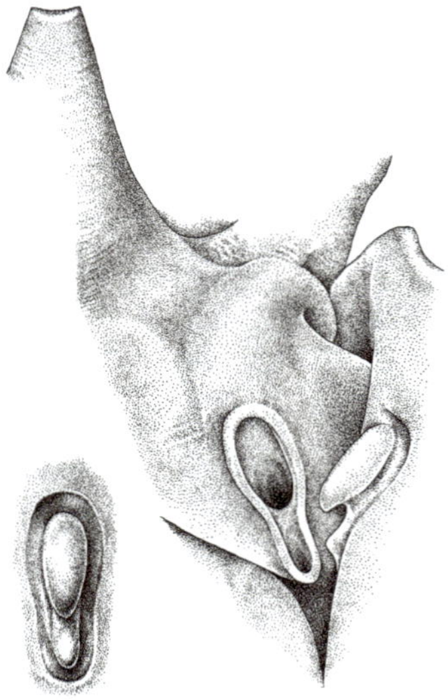

Funnel- and mantle-locking cartilages (ABRS, K. Hollis).

## *Sepioloidea lineolata* (Quoy & Gaimard, 1832)

*Sepiola lineolata* Quoy, J. R. and Gaimard, J. P. (1832). Mollusques. Voyage de decouvertes de l'Astrolabe pendant les annees 1826–1827–1828–1829. *Zoologie* *2*(1), 1–320 [82, pl. 5 figs 8–13].

Type data: holotype MNHN 1-7-23.

Type locality: Australia: New South Wales, Jervis Bay.

### COMMON NAME

Striped Pyjama Squid

### DIAGNOSTIC FEATURES

Ventral mantle connected to funnel by cartilaginous locking apparatus of dash-and-dot projections fitting into corresponding double sockets. Anterior margin of mantle aperture on dorsal mantle fringed with finger-like projections.

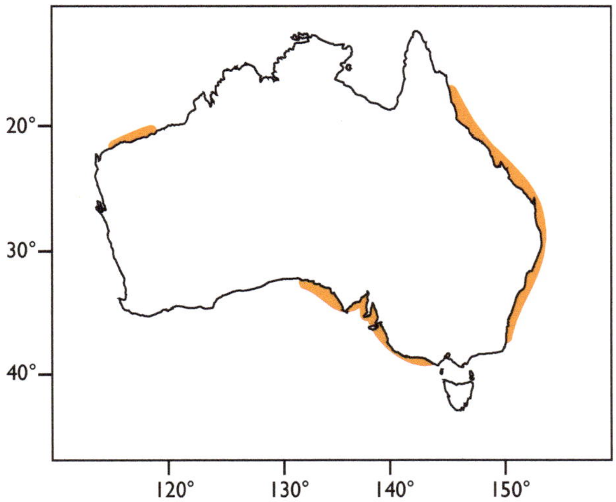


*Sepioloidea lineolata* distribution.

*Colour.* White with many narrow longitudinal stripes of pink to black on dorsal and lateral surfaces.

#### HABITAT AND BIOLOGY

Like many other sepiolids, *S. lineolata* is an ambush predator, hiding under the sand, with only the eyes protruding to search for passing prey. Talbot and Marshall (2010) demonstrated that *S. lineolata* is sensitive to polarised light. While the structure of the photoreceptors in squids, cuttlefishes and octopuses suggest they can use this polarisation sensitivity to enhance levels of contrast in the visual field, Talbot and Marshall's (2010) study was the first to demonstrate this behaviourally. *Sepioloidea lineolata* also possesses a dorsally shifted horizontal pupil slit. Talbot and Marshall (2010) showed that each has a horizontal strip of high density photoreceptors shifted ventrally in the retina, suggesting a change in sensitivity to the dorsal visual field. This implies that this bottom-dwelling species has enhanced visual acuity in the dorsal visual field and relies on eye movements to detect predators and prey while remaining hidden.

*Sepioloidea lineolata* is the host taxon for the dicyemid parasite, *Dicyema pyjamaceum* (Catalano and Furuya 2013).

#### SIZE

Up to 50 mm ML.

#### REMARKS

The distinctive pattern that characterises this small sepiolid has meant that it is readily assigned to a single species. It has yet to be confirmed whether *S. lineolata* is a single species or comprises more than one species grouped under this name. Its disjunct distribution suggests that this requires further investigation.

#### DISTRIBUTION

Southern Indo-Pacific: eastern, southern and western Australia.

#### REFERENCE

Reid (2009).

### *Sepioloidea magna* Reid, 2010

*Sepioloidea magna* Reid, A. (2009). *Sepioloidea magna* sp. nov.: a new bottletail squid (Cephalopoda: Sepiadariidae) from northern Australia. *The Beagle, Records of the Museums and Art Galleries of the Northern Territory.* *25*, 103–109 [103, figs 1–7].

Type data: holotype NTM P.41686.

Type locality: Arafura Sea: due north of Cape Van Diemen, 9°47′S, 130°26′E.

#### COMMON NAME

Great Bottletail Squid

#### DISTINGUISHING CHARACTERS

*Sepioloidea magna* differs from the only other species of this genus found in Australia due to its large size (up to

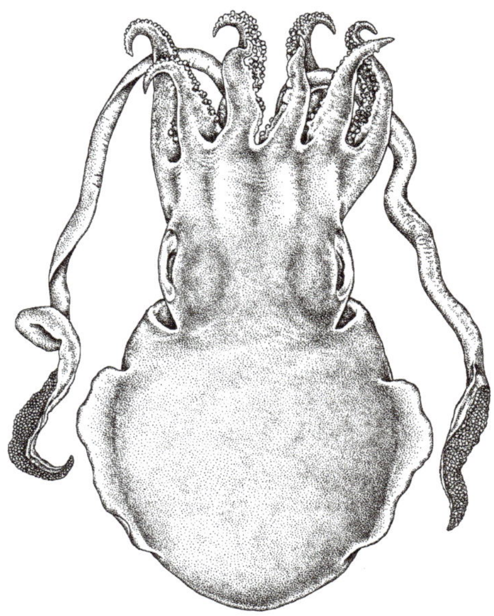

Dorsal view (ABRS, K. Hollis).

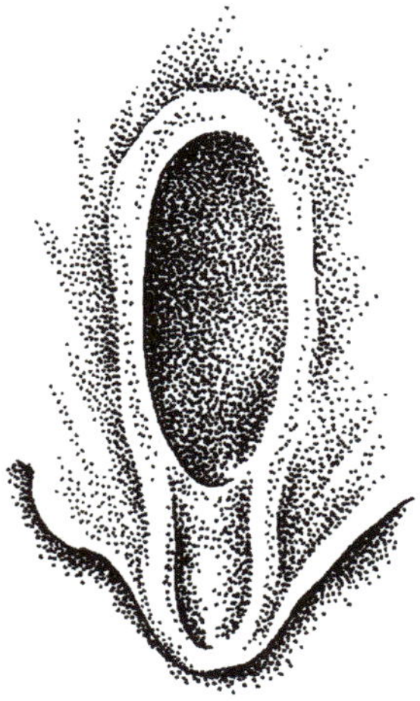

Funnel-locking cartilage (ABRS, K. Hollis).

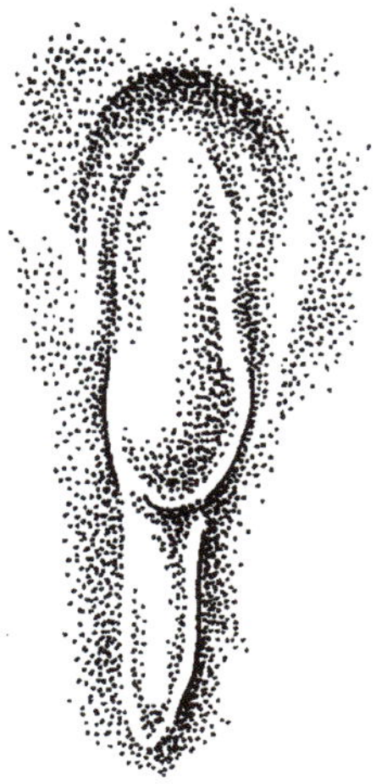

Mantle-locking cartilage (ABRS, K. Hollis).

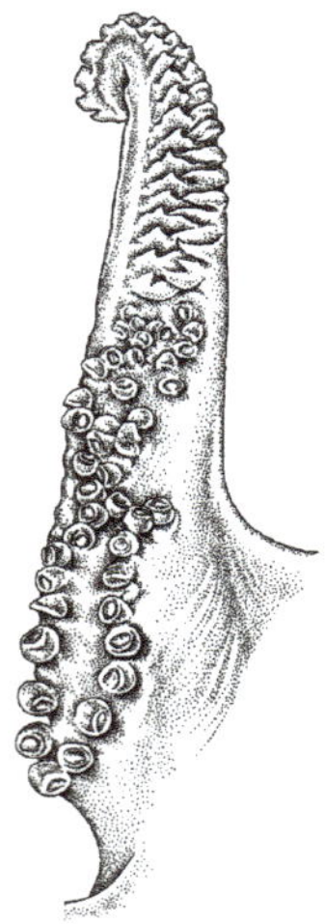

Hectocotylus (ABRS, K. Hollis).

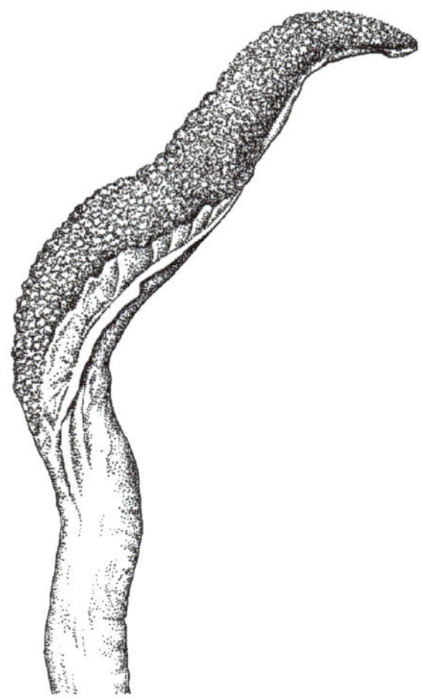

Tentacular club (K. Hollis).

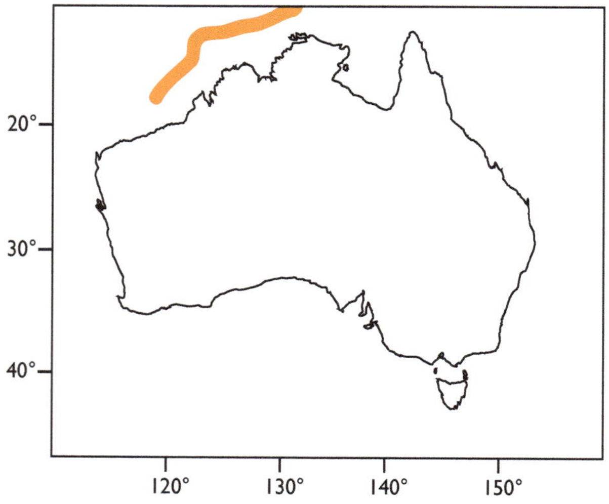



*Sepioloidea magna* distribution.

82 mm ML), absence of body patterning and in the smooth mantle margin. *Sepioloidea lineolata* is distinctly striped, small (up to 50 mm ML) and has a fringed mantle margin.

### DIAGNOSTIC FEATURES

Large sepiadariid. Dorsal arms shorter than other arm pairs. Arm suckers biserial proximally for ~10 rows, tetraserial distally, crowded on distal arm tips. In males, basal 2 or 3 rows of suckers enlarged on arms 2 and 3 and to a lesser extent on arm 4. Suckers similar sized on all arms of females. Distal half of right ventral arm 4 of males swollen, fleshy, suckers widely spaced, with two series on dorsal and ventral sides of arm. Basal half of hectocotylised left ventral arm normal, suckers biserial for 12 rows, tetraserial for next four or five rows; distal half devoid of suckers, with ~22 rows of flap-like lappets that are bilobed dorsally, swollen in proximal 12 rows (approximately); ventral lappets consist of thin flaps joined distally to dorsal lappets; deep furrow between lappets. Arm webs deep. Tentacular club long, slender, with ~40 transverse rows of minute (0.15–0.20 mm diameter) suckers.

*Colour.* Preserved specimens bland, pale purplish grey or cream peppered with tiny chromatophores.

### HABITAT AND BIOLOGY

Nothing is known about the biology of this species. The large eggs indicate that the hatchlings are likely to be benthic as are other members of this family.

### SIZE

Up to 82 mm ML.

### DISTRIBUTION

Timor and Arafura Seas from 8°47′S, 132°21′E to 18°29′S, 117°42′E. Known depth range 225–426 m.

### REFERENCE

Reid (2009).

## Sepiolidae Leach, 1817

### Key to subfamilies and genera of the Sepiolidae

1a. Arms not united by a broad web, or only third and fourth arms united by a broad web; anterior edge of ventral mantle not forming a ventral shield...........2

1b. All arms except the fourth pair united by a broad web; anterior edge of ventral mantle extended into extensive ventral shield anteriorly, covering funnel and sometimes reaching eye level............................. ...................................(subfamily Heteroteuthinae) 3

2a. Dorsal mantle fused to head by cutaneous occipital band; nuchal cartilage absent; left dorsal arm hectocotylised; internal gladius rudimentary, or absent... .............................................(subfamily Sepiolinae) 4

2b. Dorsal mantle not fused to head; nuchal cartilage present; left or both dorsal arms hectocotylised; gladius present.....................(subfamily Rossiinae) 5

3a. Ventral shield covers > 80% of ventral mantle.......6

3b. Ventral shield covers 50% of mantle (anterior half) ................................... *Heteroteuthis (H. dagamensis)*

4a. In male, only distal half of left dorsal arm modified: distal suckers modified into a row of column-like structures with tiny slit-like openings; arm suckers in four or more transverse rows; tentacular club with many tiny suckers in > 10 rows................................................ *Euprymna*

4b. In male, entire hectocotylised arm or basal portion modified with recesses and projections; arm suckers in two transverse rows (sometimes more on tips of ventral arms); tentacular club with < 10 rows of suckers............................................... *Sepiola*

5a. Anal flaps present, well developed; ink sac well developed.........................*Austrorossia (A. australis)*

5b. Anal flaps reduced, or absent; ink sac greatly reduced................................*Neorossia (N. leptodons)*

6a. Funnel-locking cartilage a straight or curved groove with an elongate or deep angular pit. Tentacular organ absent, or extends along most of club sucker-bearing face ..................................................7

6b. Funnel-locking cartilage a straight groove, without a deep pit. Tentacular organ proximal to sucker-bearing face of club, or slightly overlaps.................. ............................................... *Stoloteuthis (S. weberi)*

7a. Head-mantle fusion broad. Posterior fin lobes angled. Male arms without glands. Tentacular organ extends along most of club sucker-bearing face............................................ *Iridoteuthis (I. spp.)*

7b. Head-mantle fusion narrow. Posterior fin lobes rounded. Male arms 1 with glands. Tentacular organ absent.................... *Sepiolina (S. nipponensis)*

## Heteroteuthinae Appellöf, 1898

### DISTINGUISHING CHARACTERS

Small pelagic or benthopelagic sepiolids. Ventral mantle margin projects anteriorly as a ventral shield. Arms 1–3 joined by deep web. Suckers biserial, may be more numerous on tips of arms 4. Tentacular clubs not expanded (except in *Sepiolina*); keel present only at base of club as narrow elongate fold, the 'tentacular organ' (except in *Sepiolina*). Gladius absent.

*Photophores.* A large visceral photophore with large circular lens on ventral surface of ink sac. The ventral shield appears to be involved with diffusing light from the photophore.

*Colour.* Lateral sides of mantle silvery, ventral shield colourful.

### REMARKS

A study by Lindgren *et al.* (2012) that included three species and examined four to 10 genes found the Heteroteuthinae to be polyphyletic.

## *Heteroteuthis* Gray, 1849

### DISTINGUISHING CHARACTERS

Differs from all other heteroteuthins in having a ventral shield confined to the anterior half of the mantle. In other genera, the ventral shield covers 80% or more of the ventral mantle.

### DIAGNOSTIC FEATURES

Dorsal mantle margin free from head. Ventral mantle shield confined to anterior half of mantle. Fins attached posteriorly on mantle, fin length much shorter than ML; posterior lobes rounded. Funnel-locking apparatus a deep, curved groove. Arm suckers biserial. Right arms 1 and 2 hectocotylised in males, asymmetrical, arm bases bare and fused, area of fusion with single large gland; right arm 2 broad proximally and bordered by glands; arm 3 with two or three greatly enlarged suckers. Some arms in females with bare tips. Tentacles as thick as arms with numerous minute suckers; tentacular organ proximal to sucker-bearing face of club.

*Photophores.* Large visceral photophore with two large pores on the surface. Photophores with dense covering of iridophores.

### HABITAT AND BIOLOGY

Capable of producing brilliant luminescence from their bacterial photophore or secreting luminescent clouds via the two large pores on the photophore. The iridophores appear to act as a colour filter and shutter for the bioluminescence.

Hoving *et al.* (2008) reported on reproduction in *Heteroteuthis dispar* (Rüppell, 1844) from the mid-Atlantic Ridge. While this species does not occur in Australian waters it is likely the same adaptations may apply to the Australian species. Unlike other sepiolids, *Heteroteuthis* species are pelagic. Hoving *et al.* (2008) suggest that adaptations to an oceanic lifestyle include having much smaller eggs than other sepiolids, high fecundity, large spermatophores that are few in number, and apparent internal fertilisation (sperm appear to enter into the visceropericardial coelom from the seminal receptacle, indicating that eggs are fertilised internally – a strategy unknown in decapods but present in most octopods). All of these adaptations are typical of taxa that have undergone expansion from inhabiting shelf habitats to the open ocean. They might therefore be expected to experience lower food availability and higher predation rates resulting in evolutionary pressure to increase fecundity. Large spermatophores may be a strategy to minimise the number of mating events needed to reduce the risk of predation during courtship and copulation and ensure the availability of sperm despite possibly limited mating encounters. This suggestion is also supported by the presence of spermatangia in immature females (Hoving *et al.* 2008).

Evidence suggests that, like other sepiolids, *Heteroteuthis* lay their eggs on the ocean floor at moderate depths (< 1000 m). They appear to be common near seamounts and submarine ridges.

### REMARKS

This genus is in need of review.

### DISTRIBUTION

Most oceanic of the subfamily Heteroteuthidinae. Representatives of the genus *Heteroteuthis* are distributed circumglobally in tropical and warm, temperate waters.

### REFERENCES

Young *et al.* (2008), Young *et al.* (2013).

## *Heteroteuthis dagamensis* Robson, 1924

*Heteroteuthis hawaiiensis dagamensis* Robson, G. C. (1924). Preliminary report on the Cephalopoda (Decapoda) procured by the S.S. 'Pickle'. *Report of the Fisheries and Marine Biological Survey of the Union of South Africa 3*, 1–14 [11].

Type data: holotype BMNH Syntypes (2) 1924.9.9.49–50 (one specimen missing).

Type locality: South Africa: Natal Coast.

### DIAGNOSTIC FEATURES

Eyes very large, rounded (~40–45% mantle length); secondary eyelid present. Short, broad nuchal cartilage protrudes from skin. Ventral mantle very broad and swollen, flattening over posterior half. Ventral mantle broadly arched on each side on either side of funnel; ventral shield ~40% of ventral ML. Anterior fin attachment about halfway between mantle margin and posterior tip of mantle; posterior fin lobes rounded. Nuchal cartilage free. Funnel-locking apparatus a deep curved groove; mantle component with strongly curved, slender ridge that terminates anteriorly in a slight depression. Dorsal pad of funnel organ inverted V-shaped; ventral pads large and oval. Arms short, thick, arms 1 shortest, arms increasing in size ventrally, formula 4 > 3 > 2 > 1, or 3 ≥ 4 > 2 > 1. No obvious aboral arm keels or protective membranes in either sex. Moderately deep web joins all arms except arm 4. All arm suckers globular with smooth inner rings; arms 1–3 with biserial suckers, arms 4 with suckers irregularly arranged distally. Right arms 1 and 2 of males hectocotylised: arms longer than left counterparts, with deep joining membrane and swollen marginal membranes on proximal half of both arms with pores and internal white gladular masses. Proximal ventral bilobed projection of right arm 1 overlaps base of right arm 2. Large gland between these arms opens between lobes. Males with some greatly enlarged suckers in dorsal series only of arms 3 from about sucker 10 (counting from proximal end of arm); corresponding suckers in ventral series enlarged slightly; suckers in dorsal series alternate with those in ventral series. Tips of arms 1 and 2 bare in females, devoid of suckers. Tentacular club short, curved, without protective membrane; tentacular organ present; club with numerous minute suckers that diminish in size at proximal and distal ends of club; 18–22 suckers in each transverse row.

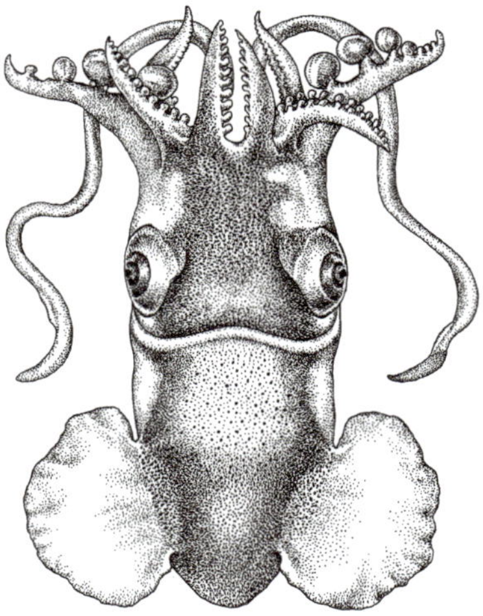

Dorsal view (ABRS. K. Hollis).

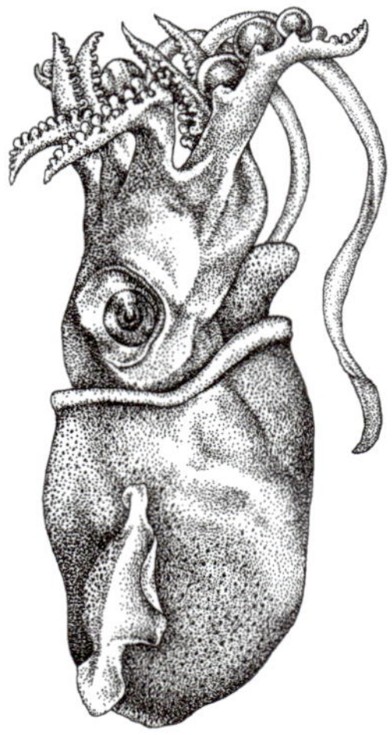

Lateral view (ABRS. K. Hollis).

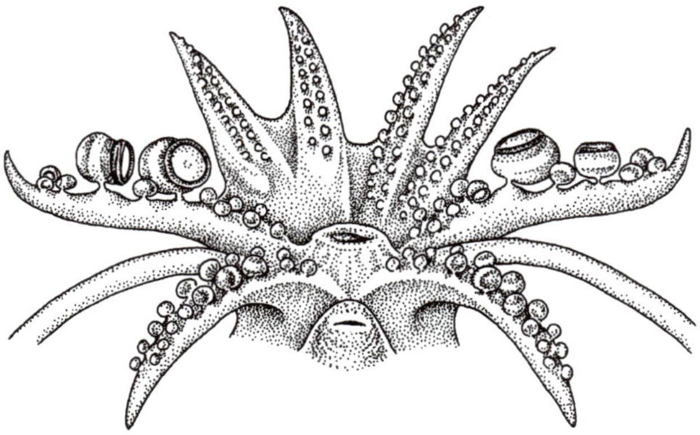

Male arm crown, oral view (ABRS. K. Hollis).

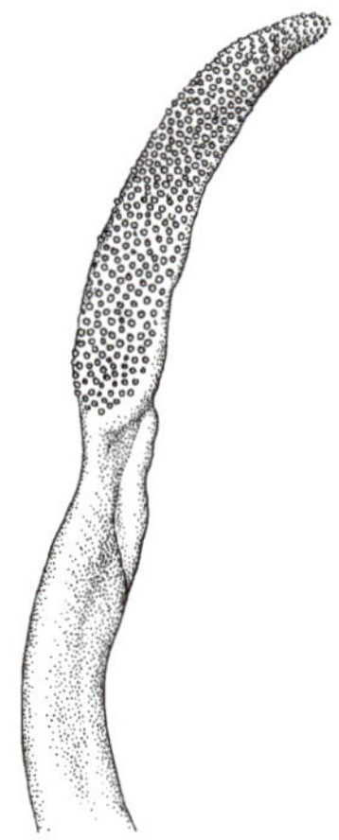

Tentacular club (ABRS. K. Hollis).

*Photophores.* Large green-bronze photophore at base of funnel, covering ink sac. Photophore ~35% ML.

*Colour.* Dark reddish brown with light buff fins, arms and tentacles, iridescent. Luminous glow all over.

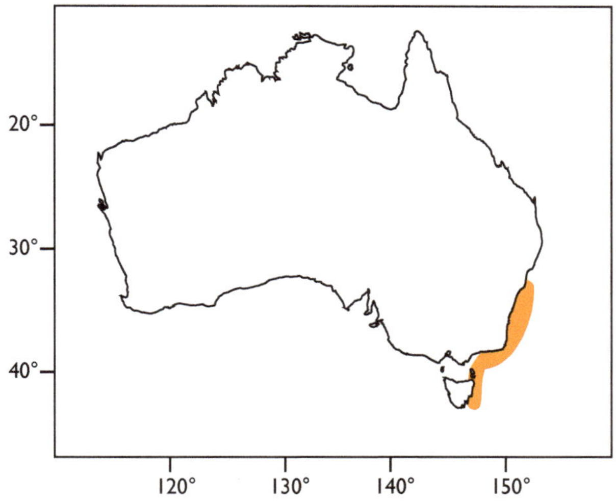


*Heteroteuthis dagamensis* known Australian distribution.

Patch of rich dark purple between eyes. Eyes with golden-bronze hue.

#### SIZE

Up to ~15 mm ML.

#### REMARKS

This genus is in need of a thorough review. The Australian representatives have previously been referred to as *Heteroteuthis serventyi* Allan, 1945, which was described from a female collected at Jervis Bay. There is doubt about the validity of this species. In the original description the only difference cited between *H. serventyi* and *H. hawaiiensis dagamensis* Robson, 1924 is that the ventral mantle does not entirely conceal the funnel and head. Specimens from south-eastern Australia held in Australian Museum collections, including the *H. serventyi* type specimen, were examined by the author and no superficial differences separate this material from *H. dagamensis.* Given *H. dagamensis* has date priority and at this stage it appears that no characters separate the Australian form from *H. dagamensis*, as stated in Vecchione *et al.* (2013) it appears that *H. dagamensis* is the correct species name.

#### DISTRIBUTION

Widespread in the southern hemisphere from South Africa to New Zealand and South America. It is known from south-eastern Australia.

#### REFERENCES

Reid and Jereb (2005), Vecchione *et al.* (2013).

## *Stoloteuthis* Verrill, 1881

### DIAGNOSTIC FEATURES

Mantle broadly or narrowly fused to head dorsally, or free. Ventral shield covers >80% of ventral ML; ventral mantle does not extend beyond middle of eyes. Fin margins rounded posteriorly (except in *Stoloteuthis maoria* from New Zealand). Funnel-locking cartilage a straight groove, or groove with anterior angular pit (unknown in *S. weberi* (Joubin, 1902)). Arm suckers relatively larger in males but no arm asymmetry present. Male arms symmetrical; arms 1 with glands, very broad (but not in *S. weberi*); arms 2 with one to several greatly enlarged suckers (except in undescribed *S.* sp. A from New Caledonia). Arm tips in females not modified (although unknown in *S. weberi*); suckers of arms 4 in females in two identical series. Tentacular organ proximal to sucker-bearing face of club, or barely overlapping

### REMARKS

This genus is in need of review and it is unlikely that the four nominal and one recognised but undescribed species form a monophyletic group.

### REFERENCE

Young *et al.* (2013).

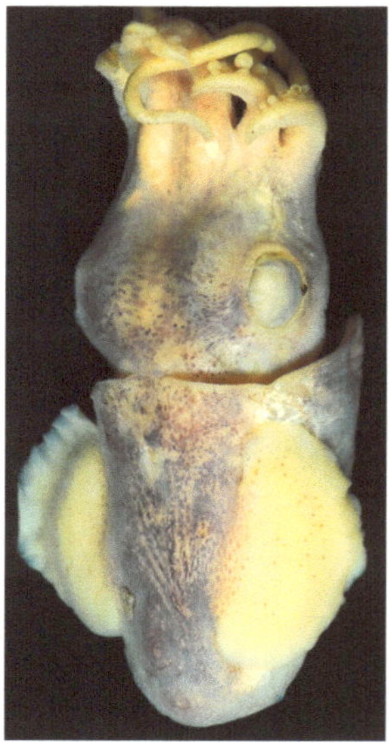
Dorsal view (WAM 34556, A. Reid).

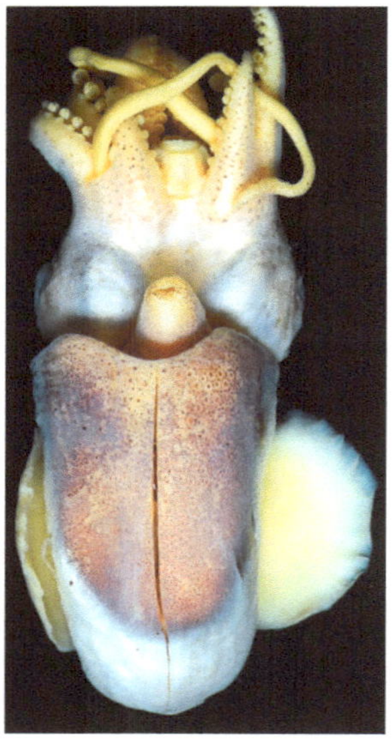
Ventral view (WAM 34556, A. Reid).

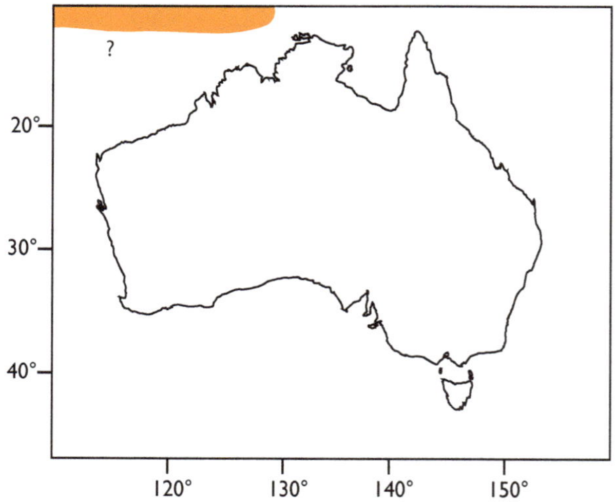

*Stoloteuthis weberi* distribution. (Australian records are based on few specimens, but the species is likely to occur more widely.)

## *Stoloteuthis weberi* (Joubin, 1902)

*Heteroteuthis weberi* Joubin, L. (1902). Observations sur divers Cephalopodes. 7e note: *Heteroteuthis weberi* n. sp. *Bulletin de la Societe scientifique et medicale de l'Ouest, 11*, 401–404 [401].

Type data: holotype MNHN 1-7-281.

Type locality: Indonesia: 7°35′04″S, 117°28′06″E.

### DIAGNOSTIC FEATURES

Dorsal mantle not fused to head. Ventral shield large. Ventral mantle strongly indented medially. Fins attached parallel to long axis of body; fin length 75% of ML; anterior edge of fin extends well beyond anterior edge of mantle; posterior lobe of fin rounded. Arms of mature male subequal in length, arms 2 not longer than arms 1 and 4; symmetrical; suckers reaching to distal tips of arms; three enlarged suckers on arms 2, two suckers twice as large as normal arm suckers, third sucker moderately enlarged. Depth of web between arms 1 in males 33–55% arm length. Club small, with thin keel; club tentacular organ proximal to sucker-bearing face of club, or barely overlapping it; club suckers in transverse rows of 12–15, larger on one side than the other. Ink ejected with a luminous liquid.

### SIZE

Males up to 7 mm ML.

### REMARKS

This species is in need of redescription based on the type and additional material. It is currently only known from the type specimen and was formerly placed in

*Heteroteuthis*. It differs from all other nominal *Stoloteuthis* in that the dorsal mantle is not fused to the head, but it resembles other *Stoloteuthis* species in other characters: the arm sucker arrangement, size of the fins and ventral shield, indentation of the funnel and rounded posterior fin lobe.

#### DISTRIBUTION

Indo-Pacific: central Indonesia. South-western Pacific: south-eastern Australia. Pelagic (epi-mesopelagic) or bathyal, depth 700–900 m.

#### REFERENCES

Allan (1945), Vecchione and Young (2007).

### *Iridoteuthis* Naef, 1912

#### DIAGNOSTIC FEATURES

Dorsal mantle broadly fused to head (fusion approximately reaches posterior midpoints of eyes). Mantle with mid-dorsal arch. Ventral mantle shield large (~80% of ventral ML); extends nearly to anterior margin of eyes; with strong anteriomedial indentation. Posterior fin lobes angled, pointed. Funnel-locking apparatus a straight or curved groove with deep, angular, anterior pit. Fins with pointed posterior lobes. Aboral surfaces of arms in both sexes with distal, expanded, tough membranous keels. Male arms symmetrical, glands absent. Arms 3 in males without enlarged suckers. Females with papillae on oral surface of arms, proximal to suckers. Tentacular club with tentacular organ that extends along the sucker-bearing portion of the club for most of its length.

*Photophores.* Large visceral photophore papillae each with two large, laterally visible tubes. Tubes do not penetrate photophore lens, but enter glandular photophore between lens and ink sac.

#### REMARKS

Three species are currently known. One is *Iridoteuthis iris* (Berry, 1909) and two are undescribed: one from Australia (*Iridoteuthis* sp.) and the other off New Caledonia (*Iridoteuthis* sp. A).

#### REFERENCES

Young and Vecchione (2007b), Young *et al.* (2013), Young and Vecchione (2014).

### *Iridoteuthis* sp.

The Australian species have not yet been properly described. A brief discussion of the three known species is given here as a background for future work.

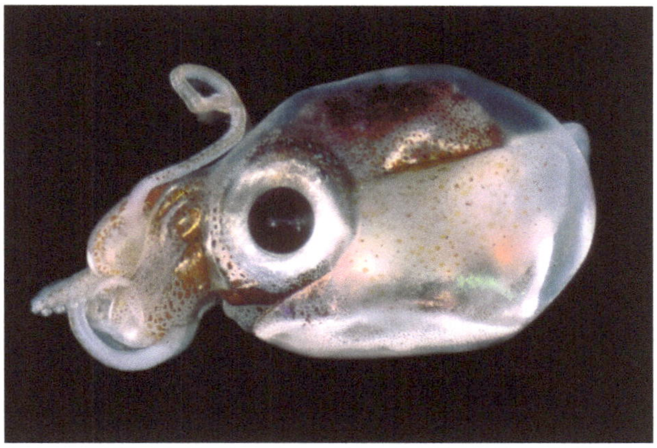

Live animal, lateral view (specimen from Hawaiian waters) (T. Burch).

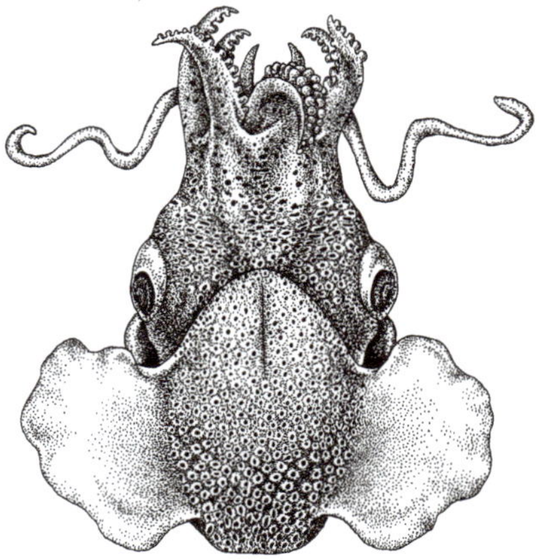

Dorsal view (Australian specimen) (ABRS, K. Hollis).

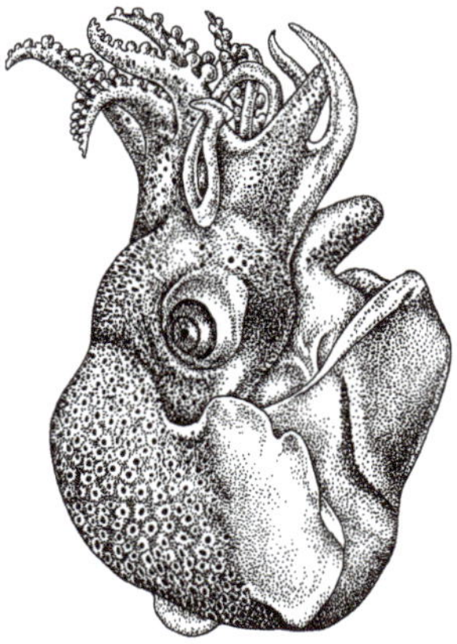

Lateral view (Australian specimen) (ABRS, K. Hollis).

Male arm crown, oral view (Australian specimen) (ABRS, K. Hollis).

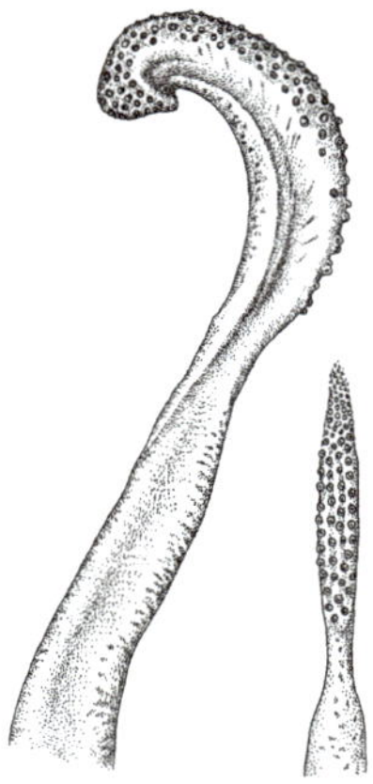

Tentacular club (Australian specimen) (ABRS, K. Hollis).

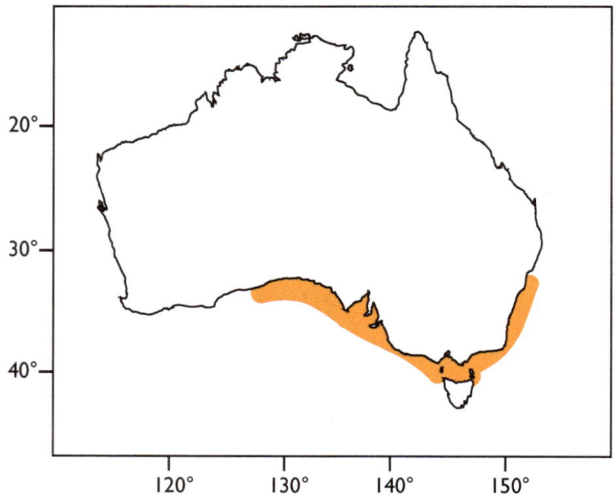


*Iridoteuthis* sp. distribution.

## DIAGNOSTIC FEATURES

*Iridoteuthis iris* (Berry, 1909). Very large eyes. Short mantle projects ventrally anterior to the eye margin; mantle shield covers 80% of ventral ML. Fins unusually large with pointed posterior lobes. *Male arms:* Ventral series of distal arm 1 suckers are larger than dorsal series; protective membranes normal (not elaborate); each arm with 20–28 suckers. Arms 2 with slightly enlarged suckers in middle of arms; 22–31 suckers. Dorsal series of medial arm 3 suckers larger than ventral series; distal to enlarged suckers a short area without suckers and only sucker pedicels; 24–31 suckers. Distal tips of arms 4 with tetraserial suckers; proximal half of each arm with 10–20 relatively large suckers, distal half with 70+ small suckers. Arms 1-3 with aboral flaps. *Female arms*: Distal tips of arms 1 bare; each arm with 18–32 suckers; each arm 2 with 24–45 suckers. Distal region of arms 3 and 4 with elongate stalks but no suckers. Each arm 3 with 18–27 suckers; arm 4 with 6–14 relatively large suckers proximally, followed distally with 10–30 smaller irregularly arranged suckers. Proximal two arm suckers with small, flattened papillae in females. Protective membranes not elaborate or distinctive in either sex. Tentacular club with suckers in 19 transverse rows distally, 11 rows proximally. Known from the Hawaiian archipelago, known depth range 150–300 m. Size: females 28 mm ML, males 24 mm ML.

*Iridoteuthis* sp. This taxon has not been studied in any detail as yet and requires further examination and description. It differs from the above two species (based on the illustration of the male arm crown in Lu and Dunning 1998) in having equal-sized suckers on arms 1 of males; some equal-sized enlarged suckers on arms 2 of males on distal third of arms and some enlarged suckers on proximal end of arms 3. The right dorsal arm 1 of males is extremely swollen on each side of suckers. In addition, the tentacular club appears to only have about four suckers in each transverse row, while those of *I. iris* and *I.* sp. A are much greater in number.

*Iridoteuthis* sp. A. Known from a single mature male and immature female. *Male arms:* Suckers in dorsal and ventral series on arms 1 and 3 of male equal in size; with elaborate ventral protective membranes on arms 1. Suckers on arms 1–3 larger in males than in females. *Female arms:* Distal tips of arms 3 and 4 without elongate distal suckerless stalks. Arms 1 and 3 of females with bare tips. Suckers on arms 4 in two size groups: proximal two-thirds with larger suckers, distal 40% with small suckers. Known from New Caledonia (from stomach of the lancet fish, *Alepisaurus ferox* Lowe, 1833).

## SIZE

~10–15 mm ML.

## REMARKS

The genus needs review before the Australian species is described.

## DISTRIBUTION

New South Wales, Victoria, Great Australian Bight. Depth: upper bathyal.

## REFERENCES

Young and Vecchione (2007b), Young and Vecchione (2014), Young *et al.* (2014).

## *Sepiolina* Naef, 1912

### DIAGNOSTIC FEATURES

Mantle-head fusion narrow. Ventral shield extends for >80% ventral ML. Posterior fin lobes rounded. Funnel-locking cartilage comprised of an elongate anterior pit and short groove. Male arms symmetrical; arms 1 with glands. Enlarged suckers on arms 2 and 3. Tentacular organs absent.

### REFERENCE

Young *et al.* (2013).

Male arm crown, oral view (ABRS, K. Hollis).

Tentacular club (ABRS, K. Hollis).

## *Sepiolina nipponensis* (Berry, 1911)

*Stoloteuthis nipponensis* Berry, S. S. (1911). A new sepiolid from Japan. *Zoologischer Anzeiger, 37*(2), 39–41 [39, fig. 1].

Type data: holotype CASIZ 017976.

Type locality: Japan: Suruga Bay.

### DIAGNOSTIC FEATURES

Dorsal mantle fused to head; junction of mantle and head narrow (~33 mm). Fins wide, about two-thirds ML. Arm suckers biserial in both sexes; female arm suckers numerous, small throughout; male suckers enlarged on arms 2 and 3 and to a lesser extent on arms 4. Hectocotylus present, both dorsal arms modified: oral surface of modified region swollen, fleshy; with transversely grooved ridges. Club with 13–16 suckers in each transverse row; all suckers of similar minute size.

*Photophores.* Rounded photophore inside mantle cavity on ink sac, visible through mantle in fresh specimens.

*Colour.* Body with numerous small chromatophores; ventral mantle margin encircled by silvery iridescent band, ~5 mm wide; chromatophores peppered over band.

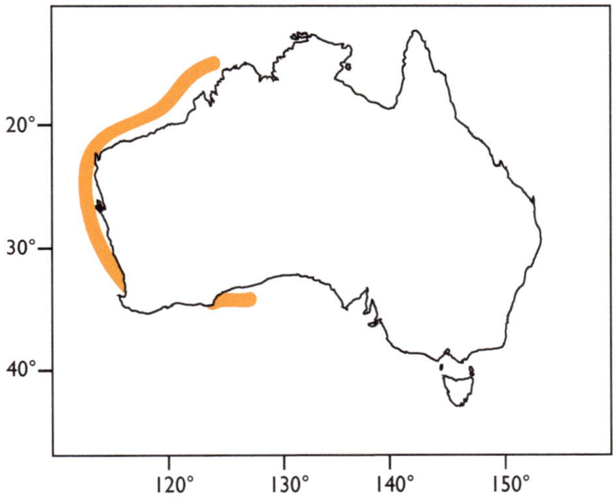



*Sepiolina nipponensis* distribution.

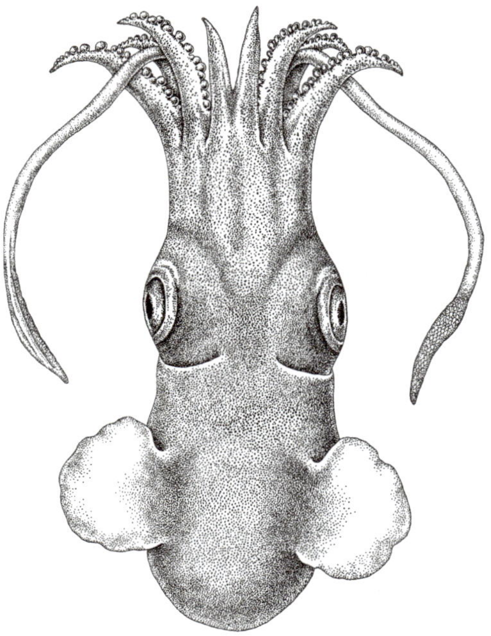

Dorsal view (ABRS, K. Hollis).

### SIZE

Up to 25 mm ML.

### DISTRIBUTION

Western Pacific: southern Japan, Taiwan Province of China, the Philippines and the Great Australian Bight (southern Australia). Neritic.

## Rossiinae Appellöf, 1898

A study by Lindgren *et al.* (2012) that included two species and examined four to 10 genes found the Rossiinae to be monophyletic.

## *Austrorossia* Berry, 1918

This genus is represented by two species in Australian waters: *Austrorossia australis* Berry, 1918 and *A.* sp. 1 Reid (1992).

### *Austrorossia australis* Berry, 1918

*Rossia (Austrorossia) australis* Berry, S. S. (1918). Report on the Cephalopoda obtained by the F.I.S. 'Endeavour' in the Great Australian Bight and other southern Australian localities. *Biological Results of the Fishing Experiments carried on by the F. I. S. 'Endeavour', 1909–14* 4(5), 201–298 [253, text figs 43–47, pls 69–70].

Type data: holotype AM C.148246.

Type locality: Australia: Western Australia, Great Australian Bight southwest of Eucla.

#### DIAGNOSTIC FEATURES

Body smooth, soft. Males mature at smaller sizes and do not grow as large as females. Dorsal mantle free from head (not fused to head). Nuchal cartilage oblong, rounded anteriorly, tapers, slightly narrower posteriorly. Fins wide, ovate, short, do not exceed length of mantle anteriorly or posteriorly; attached within anterior 75% of mantle. Non-hectocotylised arm sucker arrangement same in both sexes: arm suckers biserial; largest suckers on arms 2 and 3 larger than those on arms 1 and 4 in both sexes. Hectocotylus present, both dorsal arms modified: ventrolateral edge of proximal oral surface of hectocotylised arms bordered by swollen glandular crest, inner edge of which forms a deep furrow; glandular crest extends from sucker rows 4–6 to 8–11. Club straight, slender, long; sucker-bearing surface convex; males with 18–26 suckers in each

Dorsal view (K. Graham).

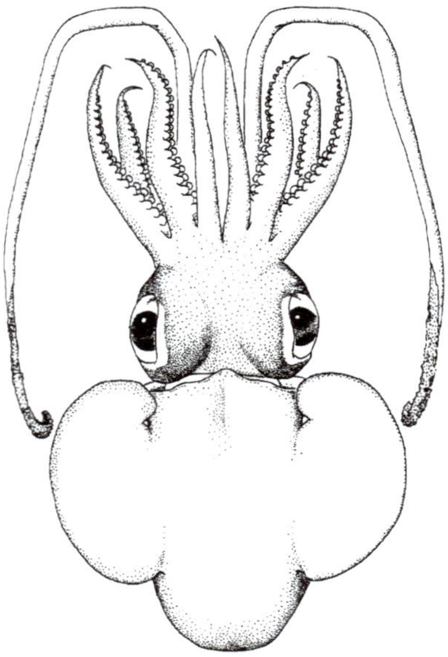

Dorsal view (L. Albertson).

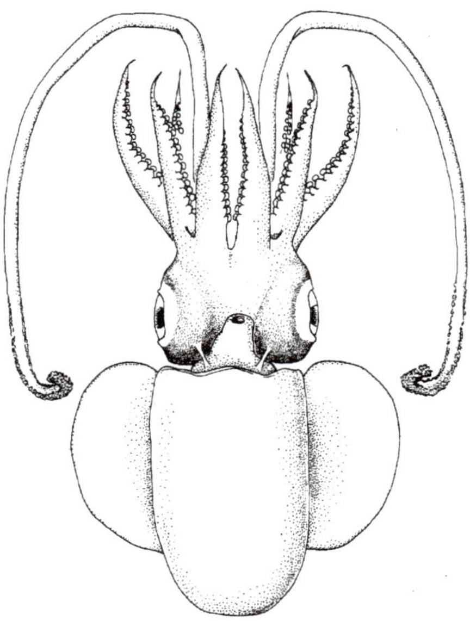

Ventral view (L. Albertson).

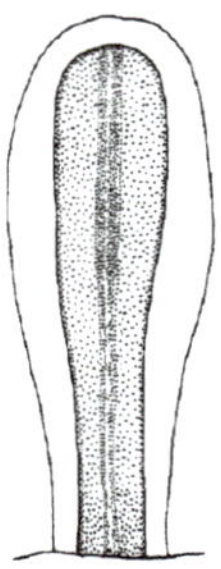

Nuchal-locking cartilage (A. Reid).

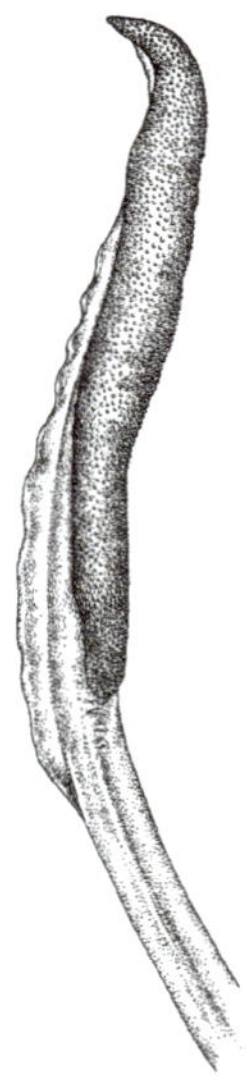

Tentacular club (ABRS, K. Hollis).

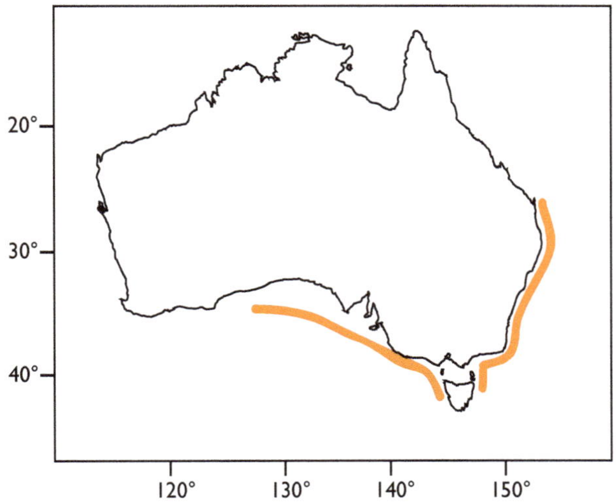

*Austrorossia australis* distribution.

transverse row; females with 25–33 suckers in each transverse row; all suckers of similar minute size. Anal flaps and ink sac well developed. Epirenal bodies present in males only; anal pads present in both sexes. Internal gladius present.

*Colour.* Uniform pinkish to purplish brown.

### HABITAT AND BIOLOGY
Sandy and muddy substrates. Benthic; known depth range 131–665 m. This species is polytelic: ova ripen in small clusters, each individual female spawns several times, probably throughout the year.

### SIZE
Males up to 34 mm ML; females up to 63 mm ML.

### REMARKS
A second *Austrorossia* species is found on the North West Shelf of Western Australia (Reid 1992). While it has been shown to differ from *A. australis*, primarily in quantitative characters, it cannot be morphologically distinguished from the African *A. mastigophora* (Chun, 1915). Molecular work is needed to clarify the status of this species.

### DISTRIBUTION
Indo-Pacific: Australia, Queensland, Raine Island, 11°35′S, 144°04′E, to Western Australia, Great Australian Bight, 34°S 130°50′E.

### REFERENCE
Reid (1991).

## *Neorossia* Boletzky, 1971
There are two species in this family, one of which, *N. leptodons* (Reid, 1992), is found in Australian waters.

## *Neorossia leptodons* Reid, 1992
*Neorossia leptodons* Reid, A. (1992). Taxonomic review of the Australian Rossiinae (Cephalopoda: Sepiolidae), with a description of a new species, *Neorossia leptodons*, and redescription of *N. caroli* (Joubin, 1902). *Bulletin of Marine Science*, *49*(3), 748–831 [797, figs 14, 24–28].

Type data: holotype MV F.57504.

Type locality: Australia: South Australia, Great Australian Bight, 37°18.81′S, 138°36.3′E to 37°17.76′S, 138°35.01′E.

### DIAGNOSTIC FEATURES
Mantle short, broad, very fleshy. Fins attached within posterior 80% of mantle. Nuchal-locking cartlage broad and ovoid, becoming slightly narrower posteriorly. Arm lengths subequal in both sexes, suckers biserial. Club cylindrical, slightly tapered, 8–10 suckers in transverse rows. Ink sac and anal flaps greatly reduced. Anal pads absent.

### HABITAT AND BIOLOGY
Sandy or muddy substrate.

### SIZE
Males up to 42 mm ML; females up to 77. 5 mm ML.

Preserved specimens, male (left) and female (right) (A. Reid).

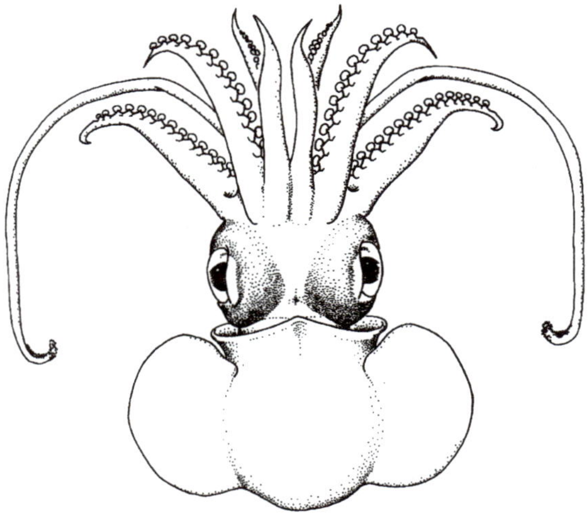

Dorsal view (holotype male, L. Albertson).

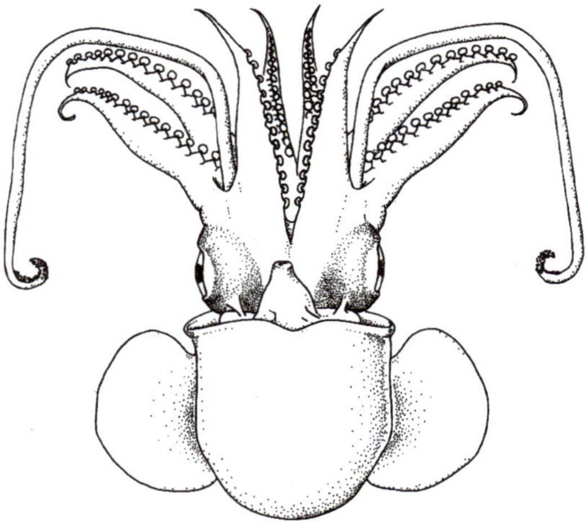

Ventral view (holotype male, L. Albertson).

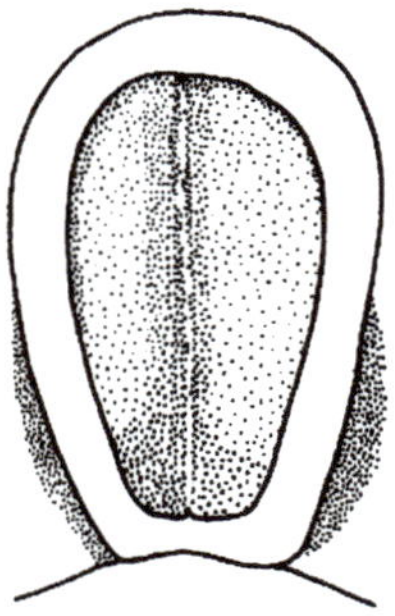

Nuchal-locking cartilage (A. Reid).

**DISTRIBUTION**

South-western Pacific: Australia, New South Wales, 32°08′S, 153°07′E to South Australia, 33°58′S, 131°22′E. Known depth range 130–1110 m.

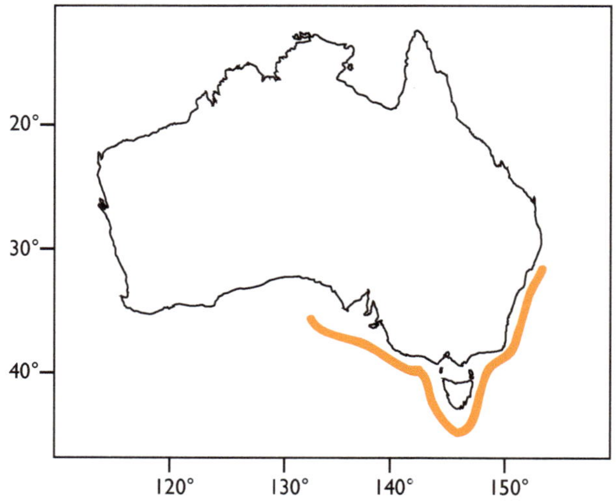


*Neorossia leptodons* distribution.

**REFERENCE**

Reid (1992).

## Sepiolinae Appellöf, 1898

**DIAGNOSTIC FEATURES**

Mantle broad, oval; posterior mantle margin rounded. Dorsal mantle fused to head, or free from head. Mantle cavity divided by thin septum. Fins wide; rounded, semi-circular, or kidney-shaped, with pronounced anterior lobes, or 'earlets'; attached about midway along mantle; fin attachment short, fin length exceeds attachment length. Large eyes covered by corneal membranes. Mantle-locking cartilage simple, linear. Arms short; protective membranes absent in both sexes. Webs between arms 3 and 4 envelop tentacle bases on aboral side only, without forming sac. Hectocotylus present: left dorsal arm, both dorsal arms, or dorsolateral arms modified. Tentacles retractable, club well defined. Only the left oviduct is developed. Eggs large, development direct. Internal gladius present, rudimentary, chitinous or absent.

**SIZE**

Small, up to 80 mm, rarely 100 mm, ML.

**HABITAT AND BIOLOGY**

Benthic or neritic. Represented in tropical, temperate and subpolar waters of all oceans.

**REMARKS**

A study by Lindgren *et al.* (2012) that included 13 species and examined four to 10 genes found the Sepiolinae to be monophyletic.

### Key to Sepiolinae genera

1a. Arm suckers biserial or tetraserial. Male arms 3 not bent inward towards mouth. Distal suckers on male hectocotylised arms greatly modified, with closely packed fleshy papillae formed from enlarged and elongate swollen sucker pedicels........ ..................... *Euprymna*

1b. Proximal arm suckers biserial. Male arms 3 bent inward towards mouth. Male hectocotylised arm modified as a copulatory apparatus with recesses and projections ..................... 2

2a. Occipital band wide (33–50% head width). Light organ on ink sac ..................... *Sepiola*

2b. Occipital band narrow (< 33% head width). Ink sac without light organs..................... *Inioteuthis*

## *Euprymna* Steenstrup, 1887

**DIAGNOSTIC FEATURES**

Broad ligament between head and mantle; commissure greater than one-third of head width. Two or four series of suckers on normal (non-hectocotylised) arms. Tentacular clubs with six or more rows of stalked suckers in each transverse row. Left arm 1 hectocotylised in mature males: distally with two to four series of small suckers on elongate, columnar pedicels; suckers with narrow openings and chitinous rings; basal part of hectocotylised arm with normal suckers and one or two papillae in ventral sucker row, sometimes bearing tiny sucker(s). Enlarged arm suckers in males. Saddle-shaped photophore in mantle cavity, ventral to digestive gland. Gladius absent.

**HABITAT AND BIOLOGY**

*Euprymna* species secrete 'glue' to enable sand to adhere to their bodies so they can effectively hide in sediment. When threatened they release the sand instantly, perhaps to confuse predators. Unlike the 'transitory' adhesion mechanism used by *Idiosepius* (von Byern *et al.* 2008), *Euprymna* use a 'duo-gland system'; one type of secretory cells secretes an adhesive substance, while another type secretes a releasing substance.

**REMARKS**

Norman and Lu (1997) revised this genus and recognised six nominal species, two of which occur in Australia: *E. hoylei* Adam, 1986 and *E. tasmanica* (Pfeffer, 1884). A third species, *E. pardalota*, was described by Reid (2011). A fourth species, *E. stenodactyla* (Grant, 1833), originally described from Mauritius, Indian Ocean, has been reported in various publications to occur in Australia. Norman and Lu (1997), however, state that records from the Indo-west Pacific (Mascarene Islands to Queensland, Australia and Polynesia) are doubtful, so the latter species is not included here. The *E. stenodactyla* type specimen has been lost and the original description is poor, making the resolution of this taxon difficult. A study by Lindgren *et al.* (2012) using four to 10 genes found *Euprymna* to be monophyletic.

As is the case for many sepiolids, female specimens are difficult or impossible to identify based on morphological characters. Most diagnostic features can be ascertained only from examination of the male hectocotylus, specifically the size and arrangement of suckers. Northern Australian *Euprymna* with tetraserial arm suckers require detailed re-examination and, particularly, comparison with nominal species from the Philippines, Indonesia and Thailand (see Aungtonya *et al.* 2011). The group would benefit from a molecular study to perhaps resolve species boundaries among taxa found in the Indo-west Pacific.

**REFERENCE**

Norman and Lu (1997).

### Key to *Euprymna* species

1a. Arms with biserial suckers..................... *E. pardalota*

1b. Arms with tetraserial suckers..................... 2

2a. No enlarged suckers on arms 2 in mature males; three or four enlarged suckers in dorsal and ventral series of ventral arm pair and ventral series only of arm pair 3 ..................... *E. hoylei*

2b. More than 10 moderately enlarged suckers in dorsal and ventral series of arms 2–4; ventral series of arms 2 and 3 with one to three proximal suckers greatly enlarged ..................... *E. tasmanica*

## *Euprymna hoylei* Adam, 1986

*Euprymna hoylei* Adam, W. (1986). Contribution a la connaissance du genre *Euprymna* Steenstrup, 1887 (Mollusca Cephalopoda). *Bulletin de l'Institut royal des Sciences naturelles de Belgique 56*, 131–136 [133].

Type data: holotype WAM 465-65.

Type locality: Sulu Archipelago, 4°31′N, 119°22′E.

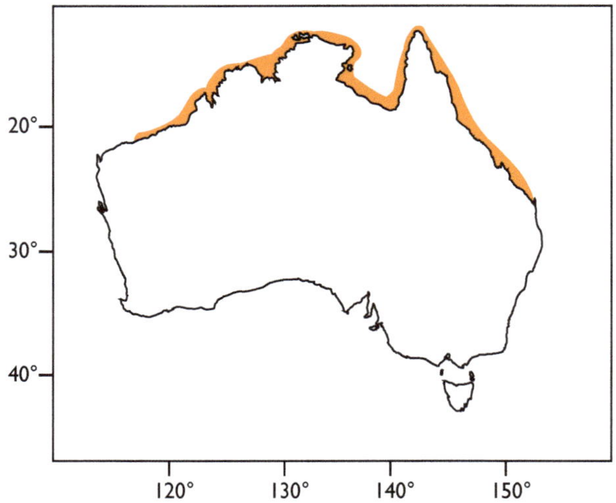


*Euprymna hoylei* distribution.

### DISTINGUISHING CHARACTERS

Similar to *E. hyllebergi* Nateewathana, 1997 but the suckers on arms 2–4 of *E. hyllebergi* do not are not significantly enlarged. *Euprymna hyllebergi* has a large number of stout papillae with slit-like apertures on the distal half of the hectocotylus; these papillae are crowded into rows of four to six suckers medially and in two series distally.

Preserved museum specimen, dorsal view (AM C.303915, L. Mott).

Preserved museum specimen, ventral view (AM C.303915, L. Mott).

### DIAGNOSTIC FEATURES

Males with three or four enlarged suckers in dorsal and ventral series of arm pair 4 and ventral series only of arm pair 3. No enlarged suckers on arms 2.

### REMARKS

See Nateewathana (1997: 472).

### DISTRIBUTION

Tropical Indo-Pacific: western tropical Pacific and northern Australia.

### REFERENCE

Norman and Lu (1997).

## *Euprymna pardalota* Reid, 2011

*Euprymna pardalota* Reid, A. (2011). *Euprymna pardalota* sp. nov. (Cephalopoda: Sepiolidae), a new dumpling squid from northern Australia. *The Beagle, Records of the Museums and Art Galleries of the Northern Territory, 27*, 135–142 [136, figs 1–9].

Type data: holotype NTM P.15796.

Type locality: Australia: Western Australia, Timor Sea, Cartier Reef, 12°32′S, 123°33′E.

### COMMON NAME

Leopard-Spotted Dumpling Squid

## DISTINGUISHING CHARACTERS

Differs from all other Australian *Euprymna* in having biserial arm suckers.

## DIAGNOSTIC FEATURES

Small species, up to 9.5 mm ML. Funnel organ dorsal element spade-shaped. Arm suckers biserial. Left dorsal arm hectocotylised in males: basally, two pairs of normal suckers; third sucker pedicel in ventral row modified into distally directed, enlarged, fleshy papilla, without terminal sucker; following two or three rows suckers normal; distal end of arm bears 9–15 pairs of suckers on enlarged columnar sucker pedicels. Spermatophores with discrete cement body. Spermatophores 46–76% ML.

*Colour.* Preserved specimens cream with large, deep purple, irregularly shaped pigment spots on dorsal and ventral head and mantle. Fins with one to three large spots close to junction with mantle. Aboral surface of arms with mediolongitudinal series of large dark purplish blotches and smaller spots between base of each sucker and extending onto sucker pedicels. Aboral surface of club with series of large blotches close to keel and smaller blotches or bars and spots towards club margin.

Dorsal view (A. Reid).

Ventral view (A. Reid).

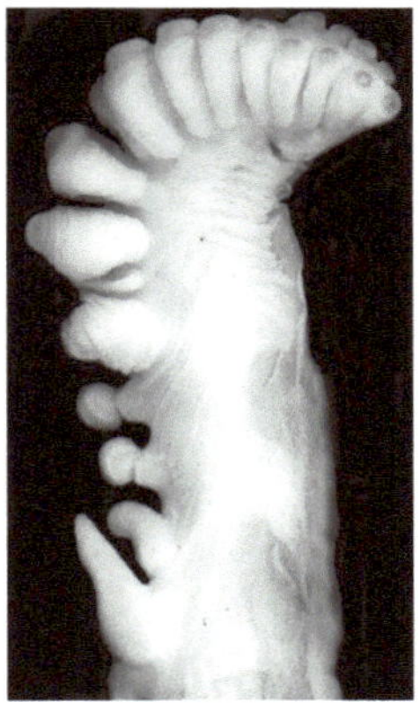

Hectocotylus (A. Reid).

## HABITAT AND BIOLOGY

Coral reef.

## SIZE

Up to 9.5 mm ML. Males and females similar in size.

## REMARKS

*Euprymna pardalota* differs from other *Euprymna* occurring in the region and from the only other species definitely known to occur in Australian waters, *E. tasmanica* (Pfeffer, 1884), in having two series of suckers on the arms rather than four. This trait is known from only one other nominal species of *Euprymna*: *E. phenax* Voss, 1962, which is found in the Philippines.

## DISTRIBUTION

Western Australia from 12°14′S, 122°56′E to Queensland 23°30′S, 152°28′E. Known depth range 0.2–20 m.

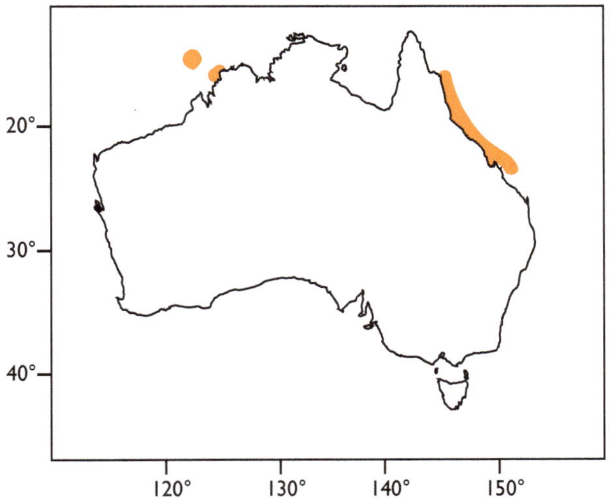


*Euprymna pardalota* distribution.

## REFERENCE

Reid (2011).

## *Euprymna tasmanica* (Pfeffer, 1884)

*Sepiola tasmanica* Pfeffer, G. (1884). Die Cephalopoden des Hamburger Naturhistorischen Museums. *Abhandlungen aus dem Gebiete der Naturwissenschaften, Hamburg* 8(1), 1–30. [6, fig. 7].

Type data: holotype ZMH 3849.

Type locality: Australia: Bass Strait.

### COMMON NAME

Southern Dumpling Squid

### DIAGNOSTIC FEATURES

Dorsal mantle fused to head. Fins wide, rounded, semicircular; short, do not exceed length of mantle anteriorly or posteriorly. Non-hectocotylised arm sucker arrangement same in both sexes: arm suckers tetraserial. Dorsal and ventral series of suckers on arms 2–4 in males enlarged; ventral marginal series of arms 2 and 3 with one to three greatly enlarged suckers basally (8–11% ML); dorsal and ventral marginal series of arms 2–4 with >10 enlarged suckers (diameter 4–7% ML). Hectocotylus present, distal half of left dorsal arm modified: third and/or fourth proximal suckers in ventral series elongated into long papilla(e), each bearing a tiny sucker; base of hectocotylus with 29–38 normal suckers; distal end of hectocotylised arm with sucker pedicels enlarged and tightly packed to form two double series of columnar

Live animals mating (P. Mercurio).

Live animal (C. Bryce).

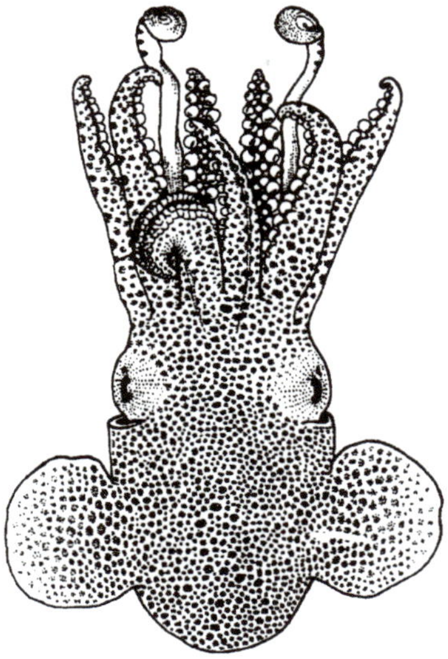

Dorsal view (M. Norman).

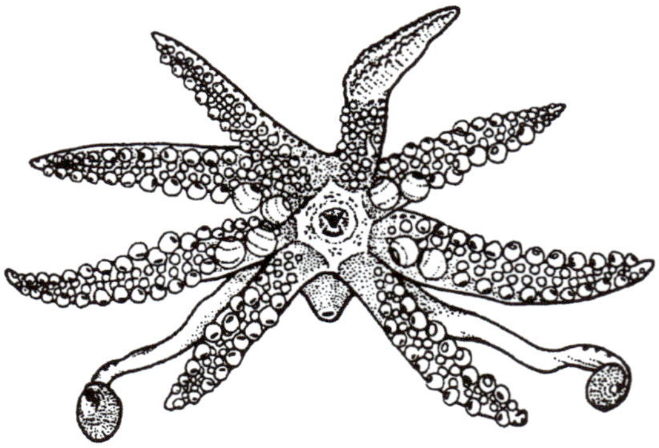

Arm crown, oral view (M. Norman).

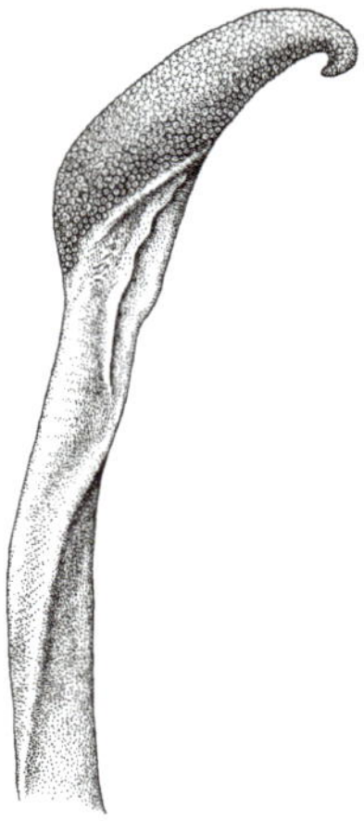

Tentacular club (ABRS, K. Hollis).

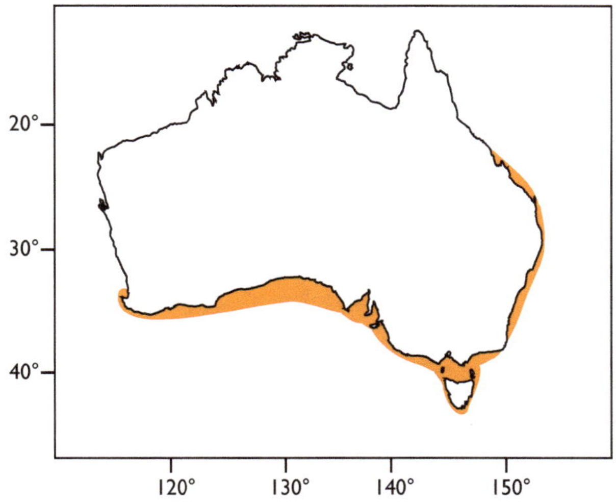


*Euprymna tasmanica* distribution.

structures; suckers reduced with tiny, fleshy, slit-like openings (30–38 in dorsal series, 29–38 in ventral series). Club suckers (many hundreds) all of similar minute size. Saddle-shaped bacterial photophore inside mantle cavity.

#### HABITAT AND BIOLOGY

This species lives in sandy and muddy areas, often in association with seagrass beds. It remains buried in sand during the day, and emerges at night to forage for small crustaceans and fishes. It can 'glue' sand grains to its dorsal body surface to aid in camouflage.

Multiple mating occurs in the laboratory and males increase the number of mantle contractions when mating with recently mated females, suggesting they are able to detect recent mating history. At the conclusion of mating the hectocotylus is enlarged. Spawning occurs in spring and summer, with females laying multiple clutches containing 6–163 eggs per clutch over 1–121 days. Egg number per clutch declines over the spawning period. Females lay pale orange eggs in loose clumps, usually at the base of seaweed or seagrass. In the laboratory, at higher ambient water temperatures, the rate of clutch production increased and egg size decreased (Squires *et al.* 2013).

#### SIZE

Up to 40 mm ML.

#### REMARKS

The taxonomic status of members of this genus is largely unresolved.

#### DISTRIBUTION

Southern Indo-Pacific: eastern and south-eastern Australia, from Brisbane to Shark Bay, Western Australia.

#### REFERENCES

Norman and Lu (1997), Squires *et al.* (2013).

### *Inioteuthis* Verrill, 1881

#### DIAGNOSTIC FEATURES

Sepiolines without internal light organs. Suckers biserial on all arms. Tentacle club suckers in eight transverse rows; males with left arm 1 hectocotylised: a basal part, a broad copulatory apparatus with an apparent deep excavation caused by lengthening of sucker stalks of dorsal and ventral sucker rows, and a distal part. (Bello, personal communication.)

#### REMARKS

Three species belonging to *Inioteuthis* have been described: *I. capensis* Voss, 1962, *I. japonica* (Orbigny, 1845), and *Inioteuthis maculosa* Goodrich, 1896. Two additional species from northern Australia have been found by the author among museum collections. This genus has not previously been recognised from Australia and the status of these two species is currently under review. No descriptions of the female mantle cavity organs are available in the literature.

### *Sepiola* Leach, 1817

#### DIAGNOSTIC FEATURES

Sepiolines with pair of internal kidney-shaped light organs on ventral surface of ink sac. Suckers biserial on arms 1 to 3 and at least proximally on arms 4. Tentacle club suckers in four to eight oblique rows. Females with ear-shaped bursa copulatrix on left ventral side of mantle cavity. Males with left arm 1 hectocotylised: suckers and sucker stalks typically arranged basally, a copulatory apparatus comprised of modified suckerless stalks, and a distal part. (After Bello 2013.)

#### REMARKS

Many species belonging to this genus are found in other parts of the world. The Australian representatives are yet to be formally described.

### *Sepiola* sp.

#### COMMON NAME

Southern Bobtail Squid

#### DIAGNOSTIC FEATURES

Fins rounded and relatively small. Arm suckers biserial. Mature males with greatly enlarged suckers on arms 1 and 3. Males without modifications to left arm 1.

#### HABITAT AND BIOLOGY
Shallow water on sand and mud; often associated with seagrass.

#### SIZE
Up to 20 mm ML.

#### DISTRIBUTION
Southern Western Australia.

#### REMARKS
Only known from a few museum specimens and some live animal photographs and observations. Occurs with *Euprymna tasmanica* (Pfeffer, 1884) and is easily distinguished by the much larger purple-brown chromatophores over the body and arms.

## *Order Spirulida*

### Spirulidae Owen, 1836
The family Spirulidae is monotypic, represented by *Spirula spirula* (Linnaeus, 1758). The phylogenetic position of *Spirula* remains unresolved at this time, perhaps due to the difficulties of obtaining tissue; it has not been included in many molecular studies (Allcock *et al.* 2014). Lindgren *et al.* (2012) in a multi-gene cephalopod phylogeny study found *S. spirula* to be sister to a clade that included *Chtenopteryx* and *Bathyteuthis* (Bathyteuthoidea).

### *Spirula* Lamarck, 1799
As for family.

### *Spirula spirula* (Linnaeus, 1758)
*Nautilus spirula* Linne, C. (1758). Systema Naturae per Regna tria Naturae, Secundum Classes, Ordines, Genera, Species cum Characteribus, Differentiis, Synonymis, Locis. 824 pp. Holmiae. [710].

Type data: repository unresolved.

Type locality: America.

#### COMMON NAME
Ram's Horn Squid

#### DISTINGUISHING CHARACTERS
Coiled internal shell comprised of a series of chambers that functions to regulate buoyancy.

#### DIAGNOSTIC FEATURES
Mantle cylindrical, thin, muscular; dorsal anterior margin triangular, acute (not fused to head). Ventral mantle with two tongue-like projections from between which the funnel protrudes. Fins narrow, ovate, with fringed anterior margins; short, do not exceed length of mantle anteriorly or posteriorly; attached dorsolaterally on posterior end of mantle (almost perpendicular to longitudinal axis of body). Mantle-locking cartilage a simple straight ridge; funnel-locking cartilage a simple, straight depression. Eyes large; ventral eyelids present. Arms increase in length dorsally to ventrally. All arms except fourth pair united by broad webs. Non-hectocotylised arm sucker arrangement same in both sexes: arm suckers tetraserial, or in six series. Hectocotylus present, both ventral arms modified: right hectocotylised arm grooved, concave, with spoon-like expansion, pointed tip and two finger-like outgrowths; left hectocotylised arm round in cross-section with two spoon-like and one finger-like outgrowth with soft papillae at distal tip. Club straight, slender; not expanded, same width as stalk; with 12–16 suckers in each transverse row; all suckers of similar small size. Buccal membrane in females with spermatheca. Radula absent. Spirally coiled internal shell present, located in posterior end of animal; shell comprised of > 30 chambers in adults.

*Photophores.* Large photophore present between fins, surrounded by annular fold of skin.

*Colour.* *Spirula spirula* is normally covered in a red-brown to silvery skin that is often lost in trawled animals.

#### HABITAT AND BIOLOGY
This is a mesopelagic species, inhabiting 600–700 m during the day and found in depths < 300 m at night. Its common name is derived from the coiled shell, large numbers of which are frequently washed ashore. The capture of young at depths between 1000 and 1750 m suggests that females possibly lay eggs on the bottom of continental slopes. This observation is also supported by stable isotope analyses of the shell, which can be used as a biological proxy to interpret ontogenetically related environmental changes (Luckeneder *et al.* 2008). Luckeneder *et al.*'s (2008) study showed that the young hatch in deep, cold seawater < 1000 m depth at temperatures ~4–6°C but move to warmer (12–14°C) mid-water habitats during growth, and then move back into deeper water as adults. Eggs are small and the smallest known paralarva has a 1.5 mm ML with

two shell chambers. The species attains sexual maturity at ~30 mm ML (after 12–15 months of life), and the lifespan is estimated to be ~18–20 months (based on a study of specimens caught by fisheries around Fuerteventura Island (Canary Islands)). Live animals have been seen to retract their head and arms into their mantle and close the opening with the pointed 'flaps' formed by the mantle. They take up a vertical position, head downwards, when at rest. The internal calcified shell provides osmotically regulated buoyancy control.

*Interest to fisheries.* No direct interest to fishery exists, but beachwashed shells are sold in the shell trade.

Fresh-caught specimen, dorsal view (K. Graham).

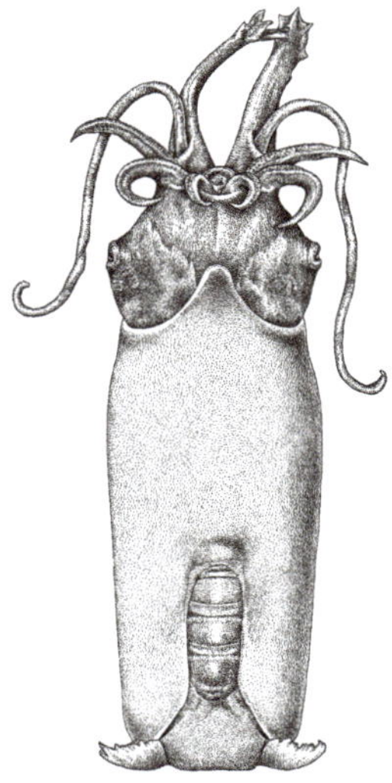

Dorsal view (ABRS, K. Hollis).

Shell (ABRS, K. Hollis).

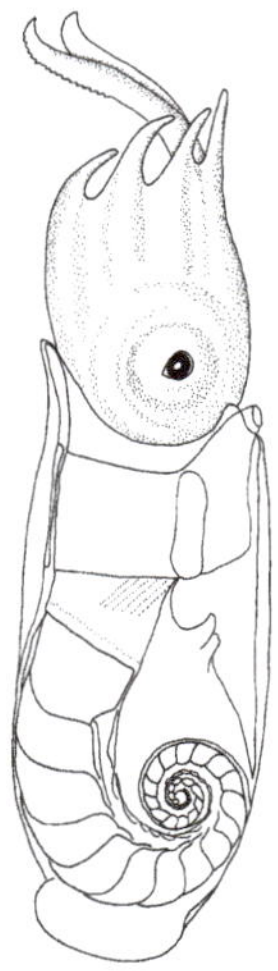

Diagram of whole animal in lateral view to show position of shell (A. Reid).

### SIZE

Rarely exceeds 45 mm ML.

### REMARKS

Only one species, *Spirula spirula*, has been recognised and described. Whether all *Spirula* belong to one species or more than one has yet to be fully examined. Neige and Warnke (2010) applied morphometric analyses to an examination of 110 *Spirula* shells from five geographical areas. Their findings challenge the monospecific status of *Spirula* in showing that shell characteristics varied with geographical origin: specimens from Madagascar, New Zealand and Brazil are larger than those from north-west Africa and Australia. Whether these differences reflect geographic variation or disjunction among populations will require detailed morphological and molecular analyses to unravel.

Warnke (2012) recently compared populations from disjunct ranges in the eastern Atlantic (South Africa and the Canary Islands, Morocco) using the amplified fragment length polymorphism (AFLP) technique. These results (although based on a limited number of samples) revealed a single genetic cluster, indicating there is probably only one population living in eastern Atlantic waters.

In addition, Haring *et al.* (2012) found through the analysis of three mitochondrial genes from a specimen collected off the eastern Australian coast (specifically Avalon Beach, New South Wales) that the molecular distances between populations from the Atlantic and Pacific are extremely low (< 1%). The genetic data so far suggest continuous genetic exchange among *Spirula* populations that may reflect a much wider distribution of *Spirula* than currently indicated by live animal captures.

Based on comparison of haemocyanin sequences that have been used to calibrate a molecular clock for many molluscan species, Warnke *et al.* (2011) estimate a divergence time of Spirulidae and Sepiida at 150 ± 30 million years ago.

Strugnell *et al.* (2006) successfully extracted 12S rDNA from beachwashed *Spirula* shells from Mission Beach, north Queensland. A 310 bp sequence was obtained and was subsequently shown to be identical to that of tissue obtained from a fresh specimen from Fuerteventura Island, Spain.

### DISTRIBUTION

*Spirula* is distributed circumtropically worldwide from ~30°N to ~30°S. As with many other epi- and mesopelagic circumtropical species, *Spirula*'s distribution is

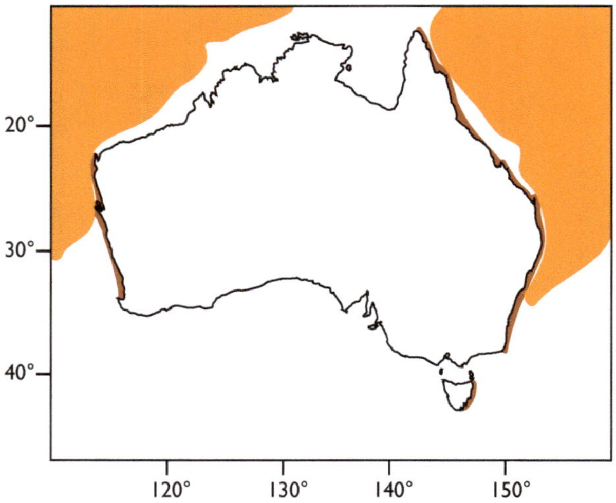


*Spirula spirula* distribution. *Note:* The distribution is based on shell and whole animal data (shells may have drifted). Darker orange on the map refers to beachwashed shells.

patchy. It occurs near oceanic islands and the continental shelf in tropical and subtropical oceanic waters, where the water temperature at 400 m is 10°C or warmer. Current distribution maps are misleading because most include both beachwashed shell and whole animal distributions. A more accurate indication of the currently known distribution of *Spirula* is given in Haring *et al.* (2012, Fig. 1).

#### REFERENCE

Reid (2005c).

## Superorder Octopodiformes

This taxon includes the orders Vampyromorphida and Octopoda.

#### DIAGNOSTIC FEATURES

Members of this group have eight arms and no tentacles (although the second pair of arms in *Vampyroteuthis* are modified as threadlike cirri). Head fused to mantle dorsally. Buccal crown absent. Arms 4 not modified, united by web (except in some argonautoids). Suckers radially symmetrical, without internal chitinous armature. Fins (when present) with cartilaginous axial support (cartilage lost in subadult *Vampyroteuthis*). Shell a gladius, cartilage-like fin support, pair of stylets, or absent.

## *Order Octopoda*

This clade includes all benthic, demersal and pelagic octopods. It includes the suborders Cirrata and Incirrata. The classification shown below follows the family level revision of the order Octopoda proposed in Strugnell *et al.* (2014). Generic assignment follows Norman and Hochberg (2005).

### Key to Octopoda suborders

1a. Pair of paddle-shaped fins. Series of cirri adjacent to arm suckers. Radula usually absent ...................... ....................... Suborder Cirrata (Opisthoteuthidae)

1b. Fins absent. No cirri adjacent to arm suckers. Radula always present................. Suborder Incirrata

## Suborder Cirrata

This clade includes the cirrate (finned) octopods.

#### DIAGNOSTIC FEATURES

Body and arms semi-gelatinous; body with pair of round to elongate fins (two pairs in juvenile vampire squids), supported by cartilaginous structure. Mantle opening narrow, located on the ventral side of the body, tightly fitting around funnel. Arms with single longitudinal series of suckers between digits of thin skin (cirri). Arm webs usually deep and thin (in some groups as an inflatable layer). Small suckers embedded in flesh with soft lined cups (no horny sucker ring). Mature males of some groups with enlarged suckers; no arm tip modifications in males. Radula present, reduced, or absent. Ink sac absent.

*Colour.* Skin white to dark red-brown, uniform in colour.

*Sculpture.* Skin sculpture absent.

#### REMARKS

Cirrate octopods include some of the largest invertebrates of the deep sea and are conspicuous members of benthopelagic and bathypelagic communities (Vecchione *et al.* 2002). Recent trawling, primarily in the northern hemisphere, has shown that the relative abundance of some cirrates, especially opisthoteuthids, may be quite high. Two of the three cirrate families, the Opisthoteuthidae and Cirroctopodidae, are found in Australian waters.

The phylogenetic relationships among cirrate octopods were investigated by Piertney *et al.* (2003) using mitochondrial 16S rDNA sequences. The derived phylogeny supported the traditional separation of cirrate families based on web form. The genera with a single web (*Opisthoteuthis*, *Grimpoteuthis*, *Luteuthis* and *Cirroctopus*) are clearly distinct from those with an intermediate or secondary web (*Cirroteuthis*, *Cirrothauma* and *Stauroteuthis*). The cirrates with a single web could be separated into three groups: (1) *Opisthoteuthis* species (with evidence of two groupings within this broader group); (2) *Grimpoteuthis* and *Luteuthis*; and (3) *Cirroctopus* species. These researchers suggest the following revisions in the classification of the cirrates: (1) *Cirrothauma*, *Cirroteuthis* and *Stauroteuthis* to be united in the Cirroteuthidae; (2) *Grimpoteuthis* and *Luteuthis* be placed in the Grimpoteuthidae; (3) *Opisthoteuthis* placed in Opisthoteuthidae; and (4) *Cirroctopus* placed in a new family, Cirroctopodidae.

This arrangement was reflected in the revised taxonomy of Collins and Villanueva (2006) and is the classification followed here. Only two families, Cirroctopodidae and Opisthoteuthidae, are found in Australian waters.

#### REFERENCE

Jereb *et al.* (2014).

### Key to Cirrata families

1a. Fins large, paddle-like. Shell V-shaped .................... ........................................................Cirroctopodidae (*C. mawsoni*)

1b. Fins small, subterminal. Shell U-shaped ................. ........................................................Opisthoteuthidae

## Cirroctopodidae

#### COMMON NAME

Finned Octopods

#### DIAGNOSTIC FEATURES

Relatively muscular, body not elongate. Eyes tilt dorsally; lenses present. Fins large. Secondary arm web absent. Suckers relatively small; enlarged suckers absent in mature males. Cirri of moderate length, equal to maximum sucker diameter. Gills hemispherical in form. Posterior salivary glands absent. Ink sac absent. Radula absent. Shell V-shaped with spike-like lateral walls.

#### SIZE

Mantle length to 180 mm. Total length >600 mm.

#### DISTRIBUTION

Southern hemisphere.

#### REFERENCE

Hochberg *et al.* (2014).

### *Cirroctopus* Naef, 1923

Cirroctopodidae is monogeneric, represented by *Cirroctopus*, with four species, one of which, *C. mawsoni* (Berry, 1917), is found in Australian waters.

### *Cirroctopus mawsoni* (Berry, 1917)

*Stauroteuthis* (?) *mawsoni* Berry, S. S. (1917). Cephalopoda. *Scientific Reports of the Australasian Antarctic Expedition 1911–1914, (series C) 14*(2), 1–39 [8, text figs 1–4; pl 10, fig. 1].

Type data: holotype AM C.40886.

Type locality: Antarctica, Southern Ocean, off Mertz Glacier, Adelie Land, 66°55′S, 145°21′E.

#### DIAGNOSTIC FEATURES

Body firm, smooth, almost sepioliform, compressed dorsoventrally, rounded below but only slightly arched over fins. Fins large with squared, truncate outer margins; anterior margins straight, posterior margin oblique; join to become continuous with each other posteriorly, forming a broad, flat fold of integument. Slight ridge along mantle margin from anterior junction with fins to anterior mantle margin. Funnel small, short. Arms unequal, formula 1 >2 >3 >4. Oral surface with single series of small, closely packed discoidal suckers, flanked on each side by series of short, blunt, cylindrical papilliform cirri that alternate with suckers almost to tips of arms. Suckers with minute openings and indistinct radial grooves. Arm

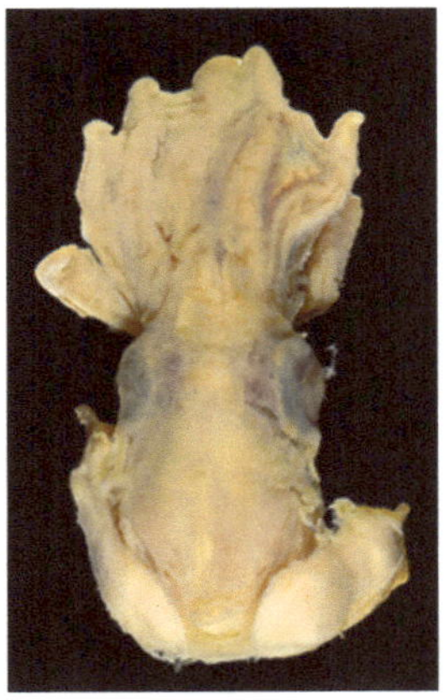

Holotype, dorsal view (AM C.40886, R. Springthorpe).

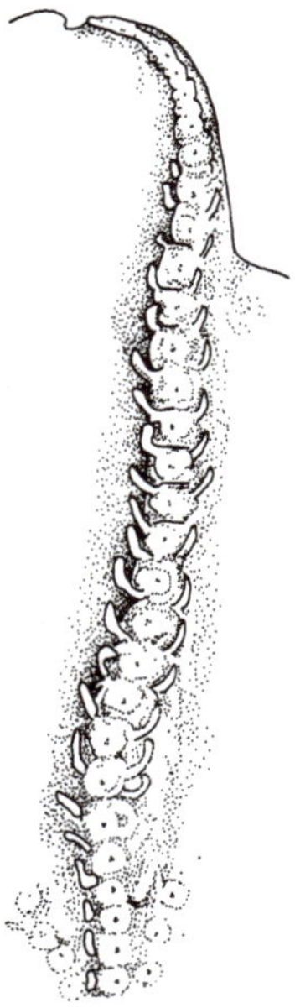

Left dorsal arm (A. Reid, redrawn from original description).

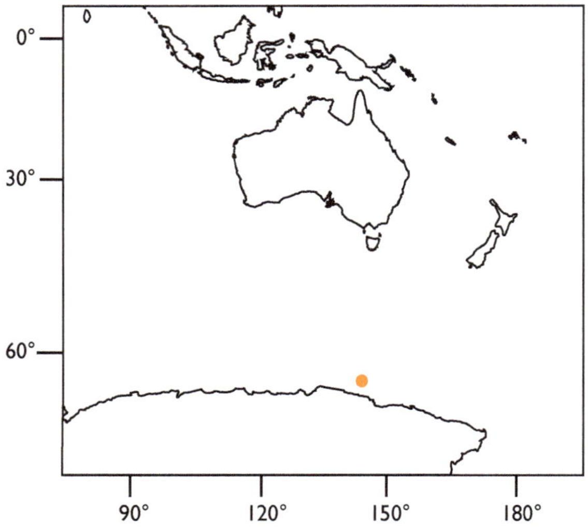


*Cirroctopus mawsoni* distribution.

webs deep, particularly on first pair of arms, which are connected almost to their distal tips. Webs decreasing in depth from dorsal to ventral arms and extend about halfway along ventral arms.

*Colour.* Fresh preserved specimens pale brownish cream; eyes and oral side of web slate coloured; integument reddish.

### SIZE

Mantle length to 12 mm. Total length 32 mm.

### REMARKS

This species is known only from the type specimen, which is immature.

### DISTRIBUTION

Known only from the type locality: Antarctica, Southern Ocean, off Mertz Glacier, Adelie Land, 66°55′S, 145°21′E.

## Opisthoteuthidae

### DISTINGUISHING CHARACTERS

Cirri short (0.4–2.5 × largest sucker diameter) compared to members of the Cirroteuthidae with very long cirri (up to 8 × largest sucker diameter).

### DIAGNOSTIC FEATURES

Body foreshortened along anterioposterior axis. Secondary web absent; web nodules (support) present in some genera. Suckers and cirri extend to arm tips. Suckers not modified. Radula reduced or absent. Simple U-, V-, or W-shaped shell that reaches posterior end of body; lateral wings not expanded or only slightly expanded. Areolar spots present in some genera, including *Opisthoteuthis*.

### HABITAT AND BIOLOGY

Opisthoteuthids alternate between sitting on the ocean floor and swimming just above it. The reduction of the radula is thought to reflect the habit of the prey being swallowed whole.

### REMARKS

One genus, *Opisthoteuthis* Verrill, 1883 is recognised in this family (Collins and Villanueva 2006). It is represented by two species, both of which are found in Australian waters.

### REFERENCES

Vecchione *et al.* (2008a), Hochberg *et al.* (2014).

## *Opisthoteuthis* Verrill, 1883

### COMMON NAME

Flapjack or Pancake Devilfishes

### DISTINGUISHING CHARACTERS

Fins small. One or two regions (proximal and distal) of enlarged suckers in mature males. Sucker apertures without fleshy tooth-like structures. Optic lobe kidney-shaped; white body penetrated by two or more separate optic nerve bundles. Shell broad U-shaped. Areolar spots present.

### DIAGNOSTIC FEATURES

Body flattened in anterioposterior axis (sometimes due to preservation), fresh specimens ovoid, mantle reduced to hump lying on extended arms, connected by thick webs. Tissue semi-gelatinous. Mantle aperture a narrow circular slit. Eyes large, diameter often 50% of head width. Fins small, fin cartilage U-shaped. Single bar-shaped cartilage for support of fins. Kidney-shaped optic lobe; white body penetrated by two or more separate optic nerve bundles. Funnel short, directed posteriorly. Proximal and/or distal region of arms of mature males with enlarged suckers, often with complex alignment. Arm suckers flanked by alternating short cirri; cirri may be retractile into pockets. Arm webs almost reach distal arm tips; web nodules, or web supports present or absent. Longitudinal gill axis perpendicular to body axis. Shell broad, U-shaped, lateral walls of wings not parallel; outer surface of saddle with groove; wing terminates in some species as elongate, simple pointed cone; termination complex in some species.

*Colour.* Areolar spots present, may be difficult to detect in some species.

### HABITAT AND BIOLOGY

Opisthoteuthids are thought to be primarily benthic. They live in the bathyal depths (125–2250 m), on the bottom or above it but rarely ascend to midwater. However, they are capable of swimming by contraction of the arm-web complex. In some species, swimming may be used for prey capture.

### REMARKS

Reliable identification is only possible for mature males. Most species are not fully described and are poorly known.

### REFERENCES

Villanueva *et al.* (2008), Hochberg *et al.* (2014).

## Key to *Opisthoteuthis* species

1a. Arm suckers small, all similar in size. Six gill lamellae. Grey tones dominate colouration.............................................................................*O. persephone*

1b. Proximal five to seven arm suckers larger than rest. Males with three or four enlarged suckers in distal parts of arms 2 and 3 near web edge. Eight gill lamellae. Brown tones dominate.................*O. pluto*

## *Opisthoteuthis persephone* Berry, 1918

*Opisthoteuthis persephone* Berry, S. S. (1918). Report on the Cephalopoda obtained by the F.I.S. 'Endeavour' in the Great Australian Bight and other southern Australian localities. *Biological Results of the Fishing Experiments carried on by the F.I.S. 'Endeavour', 1909–14*, *4*(5), 201–298, 67 figures, 29 plates [290, text figs 66, 67, pls 81, 82, 85–88].

Type data: holotype AM C.148253.

Type locality: Australia: Bass Strait, 42 miles south and east of Genoa Peak.

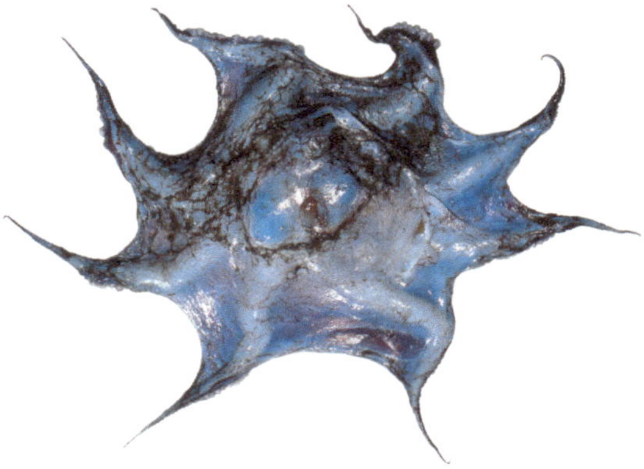

Aboral view (K. Graham).

Oral view (K. Graham).

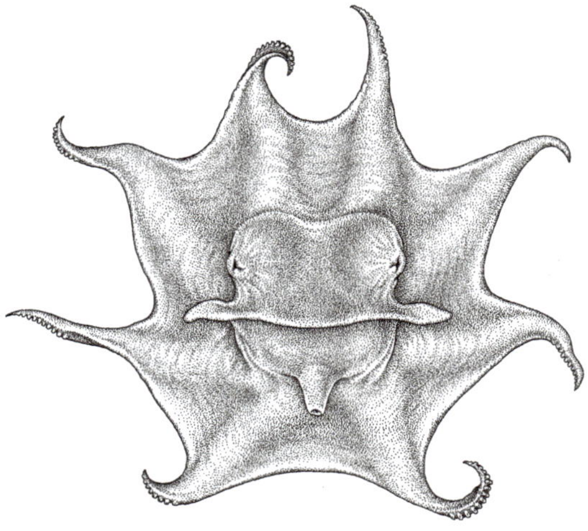

Aboral view (ABRS, K. Hollis).

### DISTINGUISHING CHARACTERS

Funnel organ larger and more slender than that of *Opisthoteuthis pluto* Berry, 1918. Differs from *O. pluto* in its distinctive colouration, particularly over the oral side of the arms and web, and the firm composition of arms.

### DIAGNOSTIC FEATURES

Berry (1918) describes the eyes in this species as 'rather small and inconspicuous', and the fins as 'minute, tenuous'. Arm suckers small, even-sized, in single series of up

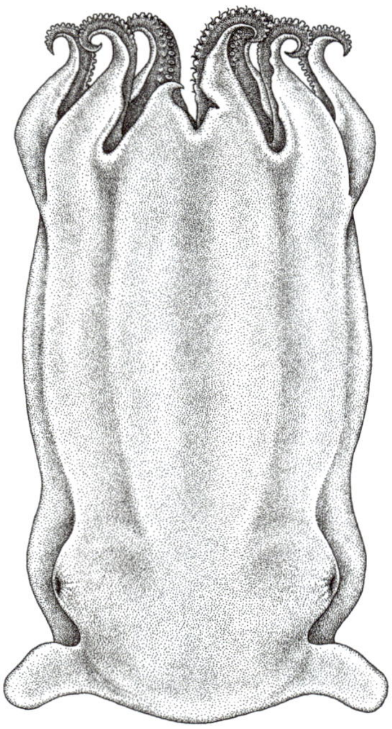

Dorsal view (ABRS, K. Hollis).

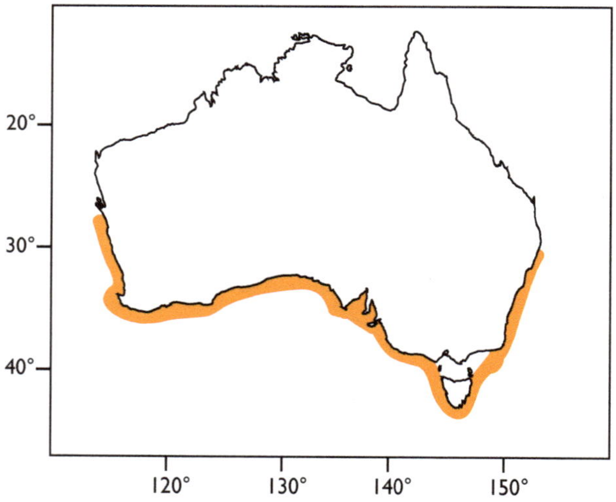


*Opisthoteuthis persephone* distribution.

to 93. Funnel slender; funnel organ a pair of small, oval pads. Cirri with narrow pockets into which they can retract. Holotype specimen with arm suckers numbers: arm 1, 78 suckers; 2, 78; 3, 72; 4, 72. Digestive gland bilobed. Gills with six lamellae per demibranch. Shell broad U-shaped, saddle with posterior groove.

*Colour.* Aboral surfaces of preserved specimens nearly colourless or light grey with traces of superficial membrane lightly pigmented with streaks of brown. Oral surfaces of superficial integument colourless and transparent, interbrachial parts very loose; deeper layer dark bluish slate, paler along arms and over ring-like zone close to mouth. Suckers and cirri yellowish brown. Areolae very small, each with clear central area.

#### HABITAT AND BIOLOGY

Primarily benthic.

#### SIZE

The holotype has an arm span of 330 mm from arm tip to arm tip.

#### REMARKS

*Opisthoteuthis persephone* needs to be redescribed from new material from the type locality. It is sympatric with *O. pluto* Berry, 1918.

#### DISTRIBUTION

Off southern Australia: Great Australian Bight and Bass Strait, depths ~275–550 m.

#### REFERENCES

Young and Vecchione (2003a), Hochberg *et al.* (2014).

### *Opisthoteuthis pluto* Berry, 1918

*Opisthoteuthis (Teuthidiscus) pluto* Berry, S. S. (1918). Report on the Cephalopoda obtained by the F.I.S. 'Endeavour' in the Great Australian Bight and other southern Australian localities. *Biological Results of the Fishing Experiments carried on by the F.I.S. 'Endeavour', 1909–14*, *4*(5), 201–298, 67 figures, 29 plates. [284, text figs 64, 65, pls 81–84].

Type data: AM holotype not extant.

Type locality: Australia: Great Australian Bight, 129°28′E.

#### DIAGNOSTIC FEATURES

Eyes moderately large. Fins near posterior edge of eyes, large, flabby, oblong. Funnel long and narrow. Funnel organ a pair of small, well-separated oblique oval pads. Arms subequal in length; suckers in 80–85 rows on each arm. Female suckers five to seven largest, sucker rows one to three very small on all arms. Male with greatly enlarged suckers proximally and distally; proximal suckers equal in size on all arms; greatly enlarged suckers distally on arms 2 and 3 (or 2–4) near web edge. Both suckers and cirri appear to be completely retractile. Gills with eight lamellae per demibranch. Digestive gland bilobed. Shell broadly U-shaped, saddle with groove along posterior surface; wings with vertical thickening,

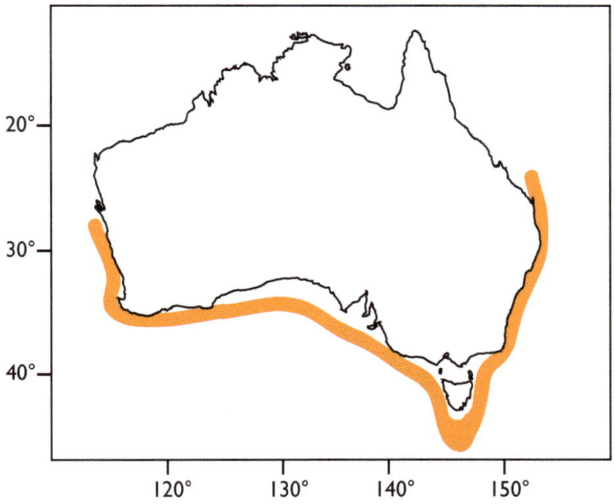


*Opisthoteuthis pluto* distribution.

and then rapidly tapering, outer surface slightly excavated at point of expansion.

*Colour.* Aboral surfaces of preserved specimens dark brown or slate grey. Oral surface chocolate to purple, paler towards margins of webs and around mouth; dorsal surface with areolar spots with clear, sharply defined nucleus surrounded by narrower zone of dark chocolate, and then wider, paler region surrounded by paler slate zone, merging into colour of general integument.

#### HABITAT AND BIOLOGY

Primarily benthic.

#### SIZE

The largest specimen originally described is the holotype with an arm span of 540 mm.

#### REMARKS

*Opisthoteuthis pluto* is sympatric with *O. persephone*.

#### DISTRIBUTION

Off south-western Australia. Great Australian Bight, depth 275–1100 m.

#### REFERENCES

Nesis (1982/1987); Young and Vecchione (2003b), Hochberg *et al.* (2014).

## Suborder Incirrata

This clade, which comprises all the non-cirrate octopods, includes the pelagic families Bolitaenidae, Vitreledonellidae, Amphitretidae, the argonautoids and the benthic cirrate octopods.

### Key to incirrate superfamiles

1. Funnel–mantle locking apparatus present ................ ........................................................ Argonautoidea
2. Funnel–mantle locking apparatus absent ................. ........................................................ Octopodoidea

## Superfamily Argonautoidea Naef, 1912

The Argonautoidea comprises a small group of pelagic incirrate octopods that inhabit most of the world's tropical and temperate oceans. The group is characterised by having marked sexual dimorphism with small or dwarf males and an unusual method of copulating that involves the transfer of a detached hectocotylus, which bears the spermatophores from the male into the female's mantle cavity. The group includes four monogeneric families – Argonautidae, Tremoctopodidae, Ocythoidae and Alloposidae – and appears to be the sister group to all other incirrate octopods (Allcock *et al.* 2014).

Strugnell and Allcock (2010) in the first phylogenetic study to include representatives of each of the four families showed a sister-taxon relationship between *Argonauta* and *Ocythoe* and between *Tremoctopus* and *Haliphron* (Alloposidae). Information from the fossil record was used to calibrate divergence times. Divergence times varied depending on which of two evolutionary scenarios are hypothesised: either that the early ocythoid has a shell and its absence is a secondary loss or that it never had a shell. The most recent common ancestor is estimated to date from the late Tertiary (~45.6 million years ago) if the absence of a shell in this lineage is a secondary loss. Ancestral Argonautoidea are assumed to have been benthic in habit. A transition to a pelagic existence may have been a response to global anoxic events that occurred in the Cenozoic (~60.5 million years ago) and also in the Cretaceous (124.5–97 million years ago).

### Key to families of the Argonautoidea (after Lu and Dunning 1998)

1a. Cephalic water pores absent .................................. 2
1b. Cephalic water pores present.................................. 3
2a. Body gelatinous. Funnel embedded in head tissue. Web deep between all arms. Third right arm of male hectocotylised, developed in inconspicuous sac in front of right eye ............................................ ........................... Alloposidae (*Haliphron atlanticus*)
2b. Body muscular. Funnel not embedded in head tissue. Web shallow between all arms; dorsal arms

of female with distal expansion for secretion of shell. Third left arm of male hectocotylised, developed in sac under left eye....................Argonautidae

3a. Two pairs of cephalic water pores at base of dorsal and ventral arms. Dorsal and dorsolateral arms much longer than ventrolateral and ventral arms (tips of dorsal arms may be autotomised in females); right ventrolateral arm of male hectocotylised. Dorsal and dorsolateral arms connected by wide web. Females do not construct a brood shell for carrying eggs......... ......................Tremoctopodidae (*Tremoctopus gracilis*)

3b. One pair of cephalic water pores at base of ventral arms. Dorsal and ventral arms much longer than dorsolateral and ventrolateral arms; right ventrolateral arm of male hectocotylised. Web absent. Females do not construct a brood shell for carrying eggs.....................Ocythoidae (*Ocythoe tuberculata*)

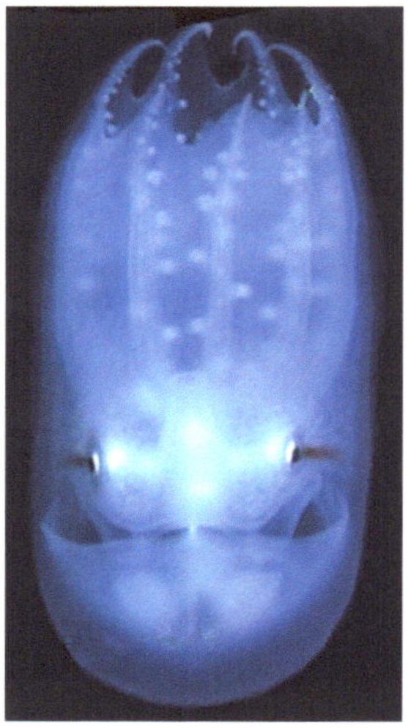

Live animal photographed in Monterey Bay (© 1995 MBARI).

Dorsal view, female (Plate L, Fig. 2a, from orginal description).

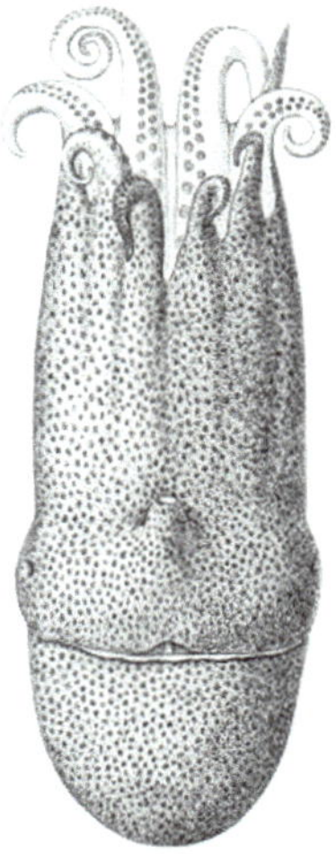

Ventral view, female (Plate L, Fig. 2, from original description).

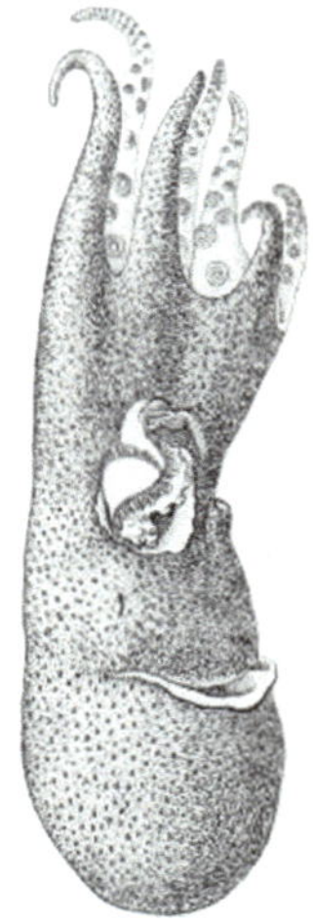

Lateral view, male (Plate L, Fig. 1, from orginal description).

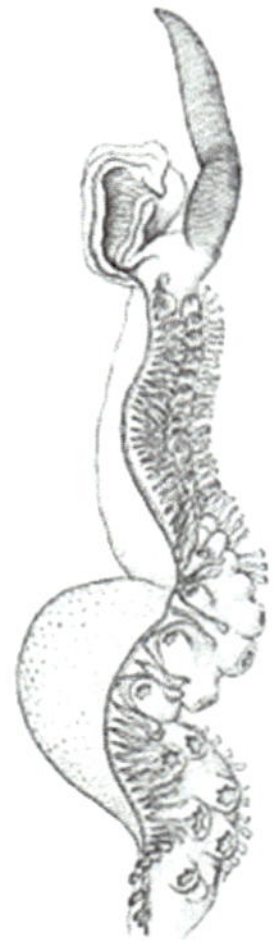

Hectocotylus (Plate L, Fig. 1a, from orginal description).

## Alloposidae Verrill, 1881

The family Alloposidae is currently monotypic, represented by *Haliphron atlanticus* Steenstrup, 1861.

### *Haliphron* Steenstrup, 1861

#### DIAGNOSTIC FEATURES

Body tissue gelatinous, smooth. Cephalic water pores absent. Funnel embedded in head tissue. Funnel-locking apparatus simple; fused in adults. Third right arm of males hectocotylised, developed in sac anterior to right eye. Webs deep between all arms. Suckers biserial, can form single series basally. Ink sac present. Remnant of true shell, a short, thick gelatinous stylet present.

#### *Haliphron atlanticus* Steenstrup, 1861

*Haliphron atlanticus* Steenstrup, J. (1861). Foreviste derpaa en Afdeling af Museets seneste Forogelser af Blaeksprutteklassen, saerligen Octopodformer fra Middelhavet, og gav Bemaerkninger om de enkelte Arter, fornemlig efter det Veranyske Cephalopodvaerk. *Videnskabelige Meddekekser fra deb Naturhistoriske Forening i Kjobenhavn, 1860*, 332–333 [332].

Type data: holotype ZMUC.

Type locality: Atlantic Ocean.

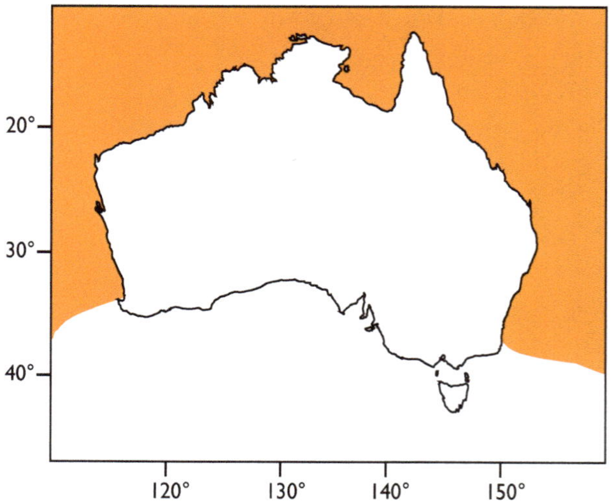



*Haliphron atlanticus* distribution.

#### DISTINGUISHING CHARACTERS

*Haliphron* has a remnant of a true shell, unlike other argonautouids: it is a short, thick, almost gelatinous stylet. It differs from *Argonauta* in having a hydrostatic organ. In addition *Haliphron* attains giant sizes, reaching 400 mm mantle length and a total length of 2 m.

### DIAGNOSTIC FEATURES

Mantle short and broad, head wide; eyes large, diameter up to one-third ML. Arms short with deep webs; formula 1 >2 >3 >4. Funnel long; opens anterioventrally to eyes. Funnel-locking apparatus complex, comprised of hook-like muscular folds on funnel corners with corresponding ridge/grooves on mantle, fused in adults. Funnel organ W-shaped. Arms short; suckers biserial distally, uniserial proximally; enlarged suckers absent. Hectocotylus develops in an inconspicuous pocket in front of right eye (males appear to have only seven arms); when mature, hectocotylus protrudes from pouch, it detaches at mating; hectocotylus with papillate lateral fringe from base to spermatophore reservoir; open spermatophore groove leads to terminal spermatophore reservoir branched off arm at base of penile filament; penile filament hangs free, not enclosed within sac; two rows of large suckers present. First lateral tooth of radula much smaller than rhachidian or second lateral tooth. Hydrostatic organ (swim bladder dorsal to the digestive system) present. Water pore absent. Stylets relatively small, wide, thick, drop-like bodies embedded in lateral walls of mantle.

*Colour.* Densely pigmented reddish purple.

### HABITAT AND BIOLOGY

Females brood their eggs, which are attached to the oral side of the arm bases near the mouth. It has been suggested that *H. atlanticus* may not be entirely pelagic, spending more of its life at the bottom, especially on continental slopes. It may undergo diel migration, moving to surface water at night. Observations of live animals made from a submersible show that they use the deep arm webs to swim with a medusoid motion.

### SIZE

Females very large, up to 690 mm ML, or total length up to 2 m; males up to 100 mm ML, 300 mm total length (i.e. smaller than females, but not dwarf).

### DISTRIBUTION

Circumglobal in the Atlantic, Indian and Pacific Oceans between 43°N and 45°S. Depth range surface to at least 1260 m and up to 6787 m.

### REFERENCE

Finn (2014a).

## Argonautidae Cantraine, 1841

## *Argonauta* Linnaeus, 1758

### COMMON NAME

Paper Nautilus

### DISTINGUISHING CHARACTERS

Third left arm of males hectocotylised. Males of other Argonautoidea (Alloposidae, Ocythoidae and Tremoctopodidae) possess modified third right arms. Within the family, species are distinguished based on the following traits (as determined by Finn 2013): males – relative arm sucker counts, hectocotylised arm sucker counts, gill lamellae counts; females – arm formulae, relative arm sucker counts, gill lamellae counts, presence or absence of shell rib tuberculation, inter-keel tuberculation (paired or paired or alternating), keel width uniform or widening during growth.

### DIAGNOSTIC FEATURES

Body firm, mantle thin and muscular. Mantle-locking apparatus well developed with pit and nodule lock; funnel organ \ Λ / shape. Suckers biserial on long stalks, connected to marginal membrane; interbrachial web narrow, cephalic water pores absent. Distal tips of dorsal arms of female with broad, expanded, membranous, glandular flaps that secrete and hold shell; sexual dimorphism pronounced; males dwarf, with entire left third arm hectocotylised, developed in pouch beneath eye, and with further growth hangs free in the arm crown and reaches around same size as body; autotomised when mature; female produces large, fragile external shell in which she broods her eggs. Stylets absent.

### HABITAT AND BIOLOGY

This pelagic octopus spends its entire life in the open ocean. Little is known about its biology, but it probably feeds on pelagic molluscs, small fishes and crustaceans. During mating, males rupture the pouch containing the modified arm. This is then loaded with sperm, detaches, and is presented to the female. This detached arm manoeuvres to the gill cavity of the female where it attaches to her gills until required for fertilisation. Single females have been found with multiple male arms attached to the gills. The shell is secreted using enlarged webs on the end of the first arm pair and functions as a brood chamber for the eggs and a flotation device for the female argonaut. The shell is not homologous with the true molluscan shell. Young are brooded first in the oviducts (as is the case in *Ocythoe*), and then eggs are laid in strings and attached to the shell until the planktonic larvae hatch. Members of this genus are continuous spawners.

Mass strandings of female argonauts occur at various times, particularly along the southern Australian coastline, perhaps due to currents, winds, or perhaps following massing due to a favourable and concentrated food source. The shells are popular for sale in the shell trade. In the northern hemisphere (where more studies have been undertaken), argonauts form a significant component of the diets of pelagic fishes.

## REMARKS

This genus has been recently revised by Finn (2013), who showed that there are only four species of *Argonauta* worldwide: *A. argo* Linnaeus, 1758; *A. hians* Lightfoot, 1786; *A. nodosus* Lightfoot, 1786 and *A. nouryi* Lorois, 1852, despite > 50 species names occurring in the literature. The plethora of names has resulted from new species being coined on the basis of variation in shell morphology. Finn's (2013) revision has shown that this variation can often be explained by changes in the shape of the shell during growth. All but *A. nouryi* are found in Australian waters.

Female argonauts continue to grow when mature and indices using dorsal ML as a size standard tend to decrease with increasing mantle lengths. This makes comparison among descriptions difficult.

*Argonauta*, unlike other argonautids, does not possess a swim bladder. Females appear to compensate by trapping air in the shell and using it as a flotation device.

## DISTRIBUTION

Worldwide in tropical and temperate oceans.

## REFERENCES

Finn and Norman (2011a), Finn (2013), Finn (2014b).

### Key to *Argonauta* species

(After Finn 2013, with information on male *A. argo* from Naef 1923 and Voss 1956 *fide* Finn 2013.)

*Males*

1a. Normal arms with 17–20 suckers; hectocotylised arm with 58–64 suckers. Eight to 10 gill lamellae per demibranch ........ *A. nodosus*

1b. Normal arms with 10–11 suckers; hectocotylised arm with 37–44 suckers. Six or seven gill lamellae per demibranch ........ *A. hians*

1c. Normal arms with 12 or 13 suckers; hectocotylised arms with 95 suckers, gill lamellae per demibranch unknown ........ *A. argo*

*Females*

1a. Arms 4 longer than arms 2 ........ *A. argo*

1b. Arms 2 longer than arms 4 ........ 2

2a. Arms 1 with 280–360 suckers, arms 2 and 3 with 140–220 suckers, arms 4 with 80–135 suckers. 14–21 gill lamellae per demibranch ........ *A. nodosus*

2b. Arms 1 with 135–160 suckers, arms 2 and 3 with 70–115 suckers, arms 4 with 30–50 suckers. 9–13 gill lamellae per demibranch ........ *A. hians*

*Shells*

1a. Lateral ribs composed of rows of separate tubercles ........ *A. nodosus*

1a. Lateral ribs smooth, continuous ........ 2

2a. Keel narrow, consistent width; keel tubercles consistent size and shape ........ *A. argo*

2b. Keel wide, width and keel tubercle size increases with shell growth ........ *A. hians*

## *Argonauta argo* Linnaeus, 1758

*Argonauta argo* Linnaeus, C. (1758). Systema naturae per regna tria naturae, secundum classes, ordines, genera, species cum characteribus, differentiis, synonymis, locis. 824 pages. Holmiae [708].

Type data: syntypes.

Type locality: 'Pelago, M. Indico, Mediterrane'.

## COMMON NAME

Greater Argonaut

## DIAGNOSTIC FEATURES

Males with 12 or 13 suckers on normal arms, 95 suckers on hectocotylised arm. Female arms 4 longer than arms 2 (arm formula 4 > 2 > 3), typically 2.1–3.6 × ML, reflexed posteriorly; arms 4 longer than arms 2; 350–380 suckers on arms 1, 190–270 suckers on arms 2 and arms 3, 180–240 suckers on arms 4. Gills with 16 or 17 lamellae per demibranch. Radula with seven teeth and two marginal plates in each transverse row, rhachidian teeth without lateral cusps (acuspid). Shell with smooth, continuous lateral ribs; keel narrow and of uniform width; keel

Live female with egg case, side view (V. Taylor).

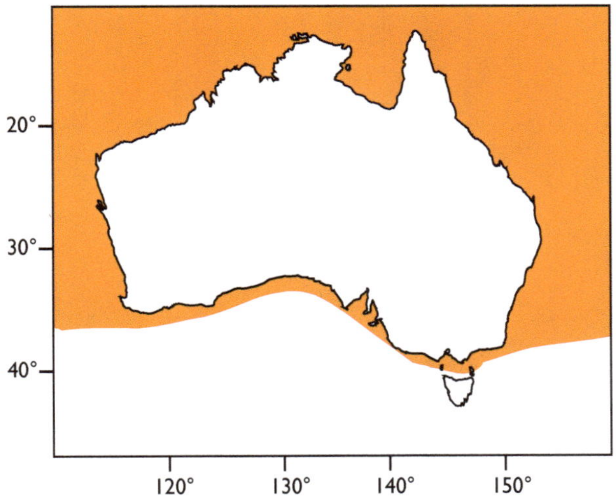


*Argonauta argo* distribution.

tubercles consistent size and shape; no inter-keel tuberculation.

*Colour.* Deep maroon to silver.

### HABITAT AND BIOLOGY

Little is known of the biology of *Argonauta argo* in Australian waters. Some Japanese specimens were observed using surface-acquired air trapped in their shells to attain neutral buoyancy (Finn and Norman 2010; Finn 2010).

### SIZE

Females up to 97 mm ML, shells of Australian females up to 206 mm long. Largest shell examined by Finn (2013) measured 268 mm.

### REMARKS

According to Finn (2013), *A. argo* can display considerable variation the shape of the aperture edge near the axis. In the past this has led to confusion in defining the species.

### DISTRIBUTION

*Argonauta argo* is found circumglobally in both hemispheres between ~40°N and 40°S. In Australian waters, this species is most commonly encountered in southern Western Australia from Rottnest Island to Albany. It is occasionally found over the North West Shelf, Western Australia and southern Australia, including South Australia, Bass Strait, northern Tasmania and eastern Victoria; and shells have been found on beaches from Frazer Island, Queensland to Sydney and Lord Howe Island. While relatively common in the Indian Ocean and southern Western Australia it is rare in South Australia and Victoria, probably drifting from western Australia on the easterly flowing Leeuwin current.

### REFERENCES

Finn (2013), Finn (2014b), Grove and Finn (2014).

## *Argonauta hians* [Lightfoot], 1786

*Argonauta hians* Lightfoot, J. (1786). A Catalogue of the Portland Museum. 194 pp. [44, 139, 174].

Type data: repository unresolved.

Type locality: not designated.

### COMMON NAME

Brown Argonaut

### DIAGNOSTIC FEATURES

Male hectocotysed third left arm extremely long, ~4 × ML and 8 × opposite arm length (arm formula 1 ≥ 4 = 2 > 3). Males with 10 or 11 suckers on all normal arms; 37–44 suckers on hectocotylised arm. Female arms moderate length, ~1 or 2 × ML, reflexed posteriorly over mantle. Female arms 2 longer than arms 4, arm formula 2 ≥ 3 > 4. 135–160 suckers on arms 1, 70–115 suckers on arms 2 and 3; 30–50 suckers on arms 4. Males with six or seven gill lamellae per demibranch; females with 9–13 lamellae per demibranch. Radula with seven teeth and two marginal plates in each row. Rhachidian tooth with well-developed single lateral cusp (tricuspid).

Fresh-caught female specimen with egg case (K. Graham).

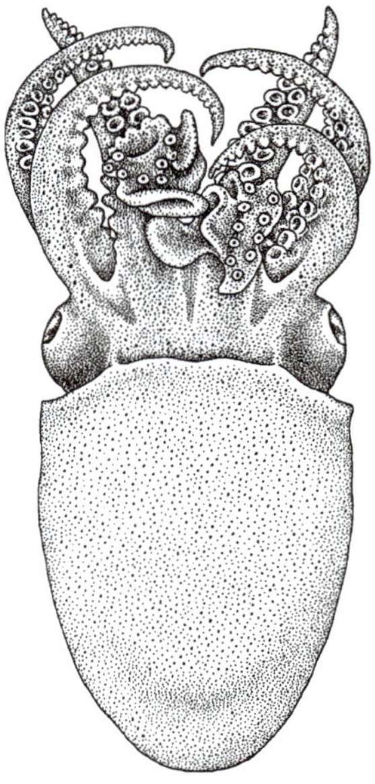

Female, dorsal view (K. Hollis, ABRS).

Male, dorsal view (male and female drawn to scale) (K. Hollis, ABRS).

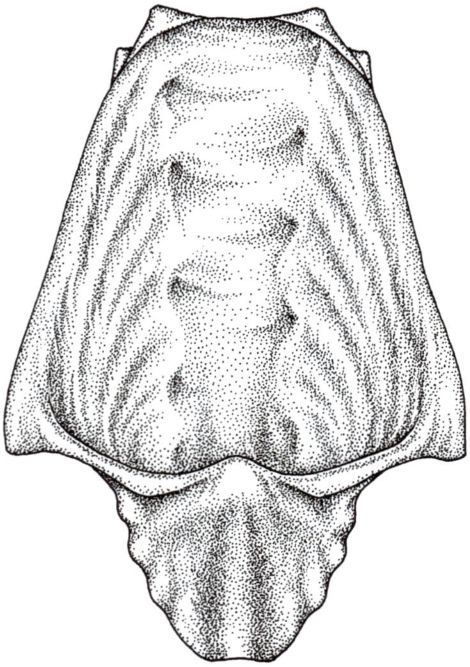

Egg case, anterior view (K. Hollis, ABRS).

Egg case, side view (K. Hollis, ABRS).

*Shells.* Shells (found around Australia) 75 mm length; ribs smooth, continuous, radiating from central axis towards keel; keel width and tubercle size increase with shell growth. Rib count up to 41 in large shells with ~1.5 ribs per keel tubercle; up to 29 tubercles in larger shells; ears often present. Stain pigmentation >67% of keel tubercles.

*Colour.* In life deep maroon to silver. Dorsal mantle with large chromatophores grading to small chromatophores ventrally. Dorsal surface of anterior funnel covered with dense chromatophores.

## HABITAT AND BIOLOGY

Detached male hectocotyli found within the mantle cavity and funnel of females. According to Finn (2013) Australian females are mature between 17–19 mm ML and egg laying begins between 19–21 mm ML. There have

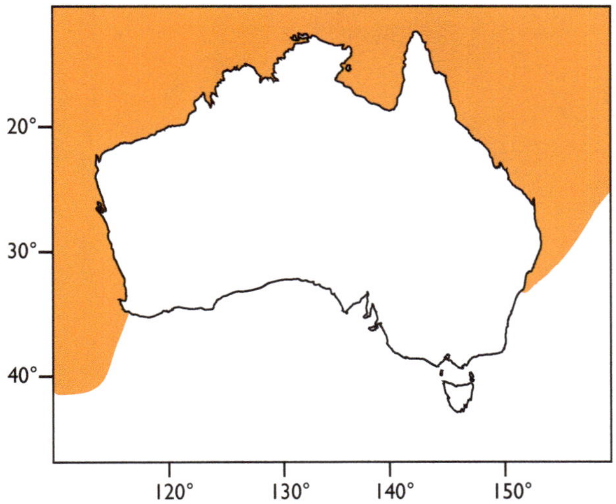


*Argonauta hians* distribution.

been no live observations of these animals in Australian waters.

#### SIZE

Small. Males up to 7 mm ML, 40 mm total length. Females up to 40 mm ML, 118 mm TL. Largest shell found in Australian waters 75 mm long.

#### REMARKS

The shell of this species is highly variable. *Argonauta boettgeri* Maltzan, 1881 is considered by Finn (2013) to be a synonym of *A. hians*.

#### DISTRIBUTION

In Australia, found from the North West Shelf, between Exmouth Gulf and Rowley Shoals, known from Cape Peron (south of Fremantle), Western Australia to Groote Eylandt, Gulf of Carpentaria, Northern Territory and from Cape York, Queensland, south to Botany Bay, New South Wales and Lord Howe Island.

#### REFERENCES

Finn (2013), Finn (2014b).

### *Argonauta nodosus* [Lightfoot], 1786

*Argonauta nodosa* Lightfoot, J. (1786). A Catalogue of the Portland Museum. 194 pp. [96, 172].

Type data: repository unresolved.

Type locality: South Africa: Cape of Good Hope.

#### COMMON NAME(S)

Southern Argonaut, Tuberculated Argonaut

#### DIAGNOSTIC FEATURES

Female mantle conical, posterior mantle with upward torsion following shape of shell. Head small, embedded within mantle. Eyes large, protruding, constricted basally. Funnel-locking cartilage comprised of rounded, posteriorly directed nodules on inside of mantle wall corresponding to depressions on outer edges of funnel. Arm lengths unequal, 1.7–2.4 × ML (arm formula $2 > 3 \geq 4$ or $2 > 4 \geq 3$); dorsal arms of females with wing-like webs; arms 2 longer than arms 4 in females. Web deepest between dorsal arms. Arms with two series of suckers; without enlarged suckers; females with 280–360 suckers on arms 1, 140–220 suckers on arms 2 and arms 3, 80–135 suckers on arms 4. Radula with seven teeth and two marginal plates in each row, without lateral teeth. Males with 17–20 suckers on all normal arms; 58–64 suckers on hectocotylised arm. Male arm formula $4 \geq 1 = 2 > 3$. Detached hectocotylus with 58–64 suckers in two rows. Gills with 14–21 lamellae per demibranch (females); 8–11 lamellae per demibranch (males).

*Shell.* Shell with concentric series of knobbled ridges, laterally flattened. Ribs numerous, close together, sometimes branching as they radiate towards keel. Lateral ribs composed of rows of separate tubercles; keel width and keel tubercle sizes increase with shell growth; no inter-keel tuberculation. Flared shell 'ears' or 'horns' may be present in small specimens, becoming less conspicuous in larger animals. Dark staining on keel tubercles on first third of shell.

*Colour.* In life orange to deep maroon and purple dorsal webs covering external shell surfaces maroon to reflective silver.

#### HABITAT AND BIOLOGY

Female *Argonauta nodosus* strand in large numbers in particular years in southern Australia, New Zealand and elsewhere. Based on museum collections, this appears generally to occur from April to October and December to January. There is no regular periodicity in this occurrence. There may be a correlation between the occurrence of *A. nodosus* in southern Australian waters and schools of euphausiids (small shrimp-like crustaceans, or krill) on which they feed.

Eggs of different stages have been recovered from shells, suggesting there is a periodicity in egg laying. The eggs are tiny, numbering many thousands and the young hatch as planktonic larvae.

Male, side view (B. Reid).

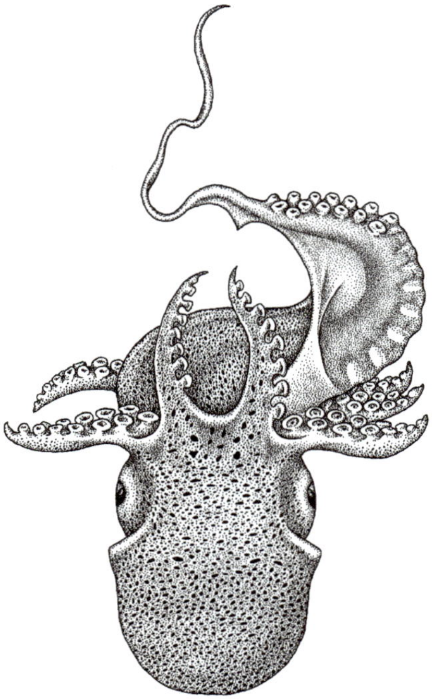

Male, dorsal view (ABRS, K. Hollis).

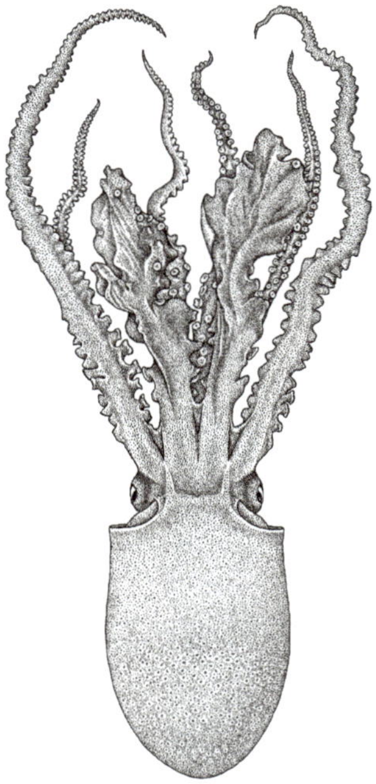

Female (right) (not to scale), dorsal view (ABRS, K. Hollis).

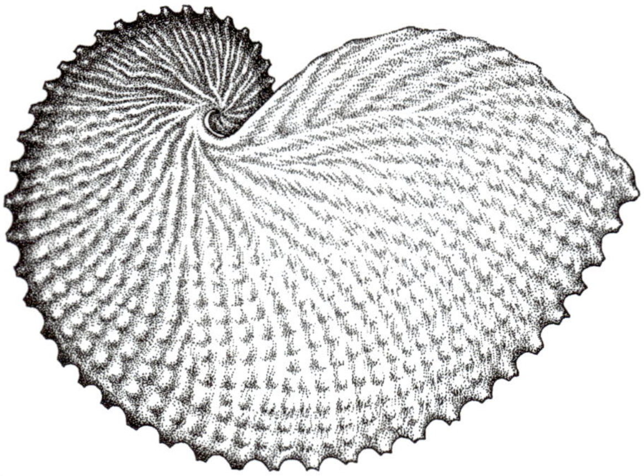

Egg case, lateral view (ABRS, K. Hollis).

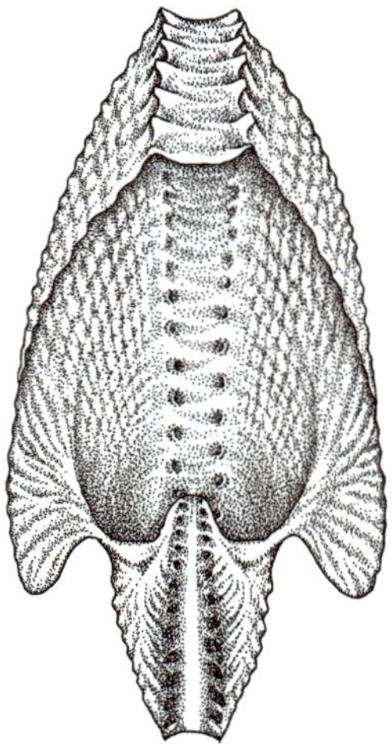

Egg case, anterior view (ABRS, K. Hollis).

## SIZE

Females up to 13 cm ML (shell length up to 24 cm); males up to 11 mm ML and 61 mm total length.

## REMARKS

Immature females may be confused with *A. hians.*

Wolfe *et al.* (2012, 2013) and Smith *et al.* 2012) have examined the shell, or brood chamber, of *A. nodosus* shells using electron backscatter diffraction to characterise the crystallography of biomineralisation and changes that occur after exposure of the shell to reduced pH. The mineralogy of the shell was shown to be 100% calcite with an intermediate Mg-calcite concentration (5.1 Wt%). Magnesium calcite is more susceptible to dissolution at low pH than other skeletal mineral forms, making argonaut brood chambers particularly susceptible to loss of structural integrity and strength as a result of ocean acidification. These studies, however, were conducted only on the shells in the absence of the animal. It is not known whether the females may be able to compensate for this loss of mineralisation at low pH.

## DISTRIBUTION

Southern hemisphere between 10°S and 44°S. Known within Australia from Queensland, Great Barrier Reef, Capricorn Bunker Group, south to Bruny

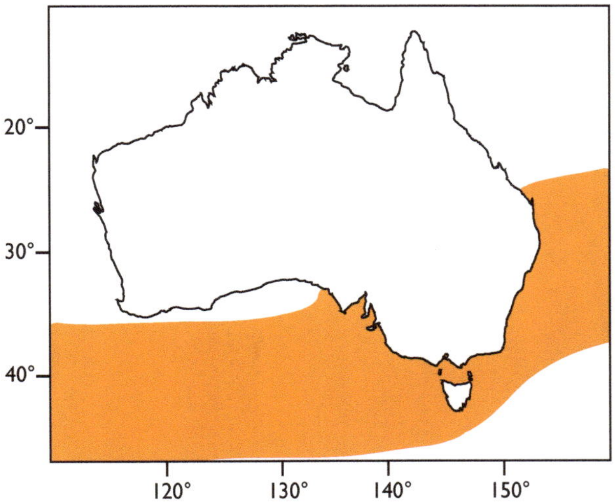


*Argonauta nodosus* distribution.

Island, Tasmania and west to Port Lincoln, South Australia.

#### REFERENCES

Finn and Norman (2011a), Finn (2013), Finn (2014b).

## Ocythoidae Gray, 1849

The family Ocythoidae is monotypic, represented by *Ocythoe tuberculata* Rafinesque, 1814.

#### DISTINGUISHING CHARACTERS

Arms 2 and 3 much shorter than arms 1 and 4. Reticulate ridge pattern on ventral surface of mantle in females.

## *Ocythoe* Rafinesque, 1814

As for family.

### *Ocythoe tuberculata* Rafinesque, 1814

*Ocythoe tuberculata* Rafinesque, C. S. (1814). Precis des Decouvertes et Travaux Somiologiques de Mr. C. S. Rafinesque-Schmaltz entre 1800 et 1814 ou choix raisonne de ses principales Decouvertes en Zoologie et en Botanique. 56 pp. Palermo. [29].

Type data: repository unresolved (type eaten by Rafinesque).

Type locality: not designated [Sicily?].

#### COMMON NAME

Football Octopus

#### DISTINGUISHING CHARACTERS

Arms 2 and 3 much shorter than arms 1 and 4. Reticulate pattern on ventral surface of mantle in females. Hectocotylus carried within a sac that extends from the base of the third right arm (hectocotylus not carried within a sac on the head).

#### DIAGNOSTIC FEATURES

Mantle muscular. Ventral surface of mantle in adults and subadults has permanent ridges forming reticulate pattern. Funnel-locking apparatus formed by permanent, turgid and strongly recurved corners of funnel. Pair of water pores at base of ventral arms leads to extensive water-filled spaces between eyes and arm bases. Dorsal and ventral arms much longer than lateral arms. Females with swimbladder that connects to mantle cavity via muscular duct on right side. Detachable hectocotylus contained in sac extending from base of right arm 3 in males. Hectocotylus without papillate lateral fringes. First lateral teeth of radula greatly reduced.

#### HABITAT AND BIOLOGY

This is a pelagic species that occupies near surface waters (at least at night) in the temperate regions of the world's oceans. The muscular mantle suggests that this is an excellent swimmer; the gas-filled swimbladder in females providing neutral buoyancy. The pygmy males and small females are often found occupying the tests of oceanic salps, holding on with arms curved backwards in the manner used by *Argonauta* to hold its shell. *Ocythoe tuberculata* is a continuous spawner (eggs in various stages of development present in the oviducts in any one time), and the only known ovoviviparous cephalopod (i.e. they give birth to live young that hatch internally). Eggs are ~2 mm in length and number at least 100 000 in large females. Hatchlings are recognised by retarded development of the lateral arms.

#### SIZE

Females large with mantle lengths to 310 mm. Males ~30 mm ML.

#### REMARKS

*Ocythoe* are one of few known cephalopods with a true swimbladder. The population structure of *O. tuberculata* has not been investigated; the distribution suggests that separate populations may exist.

#### DISTRIBUTION

Found throughout the world's oceans in temperate latitudes.

#### REFERENCES

Mangold *et al.* (2010a), Finn (2014c).

Male, dorsal view (ABRS, K. Hollis).

Female, dorsal view (ABRS, K. Hollis).

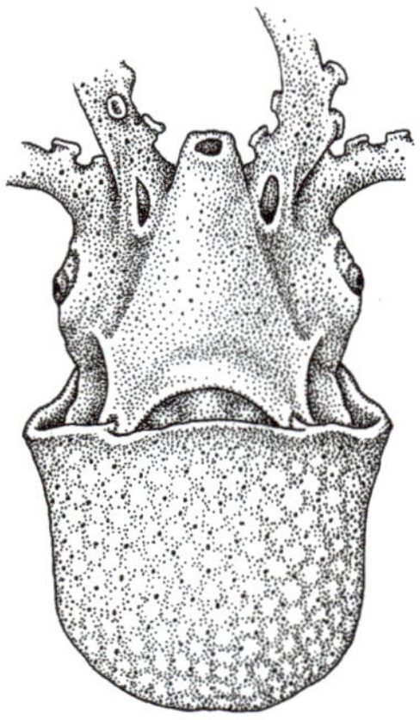

Female funnel (ABRS, K. Hollis).

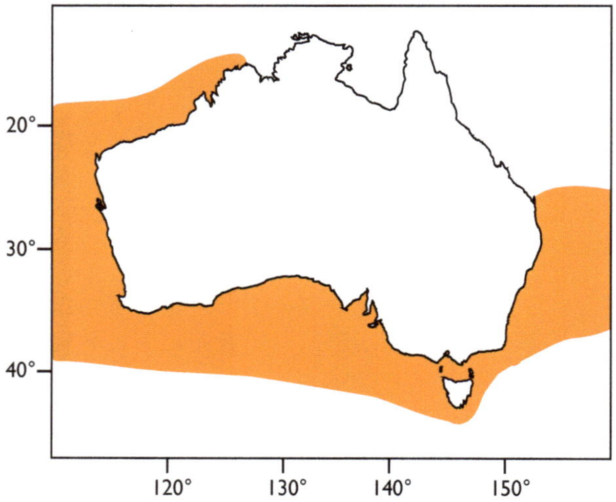


*Ocythoe tuberculata* distribution.

## Tremoctopodidae Tryon, 1879

### *Tremoctopus* Delle Chiaje, 1830

#### DIAGNOSTIC FEATURES

Females large, muscular, ~1 m or more in total length. Arms 1 and 2 of females much longer than arms 3 and 4; deep, extensive web between arms 1 and 2, virtually absent between other arms; two pairs of conspicuous pores in web at base of arms: one pair dorsally and one pair ventrally adjacent to funnel. Large ocelli on dorsal web. Hectocotylus develops in sac beneath right eye; proximal half of hectocotylus with papillate lateral fringes. Swim bladder present, dorsal to digestive system. First lateral teeth of radula much smaller than rhachidian or second lateral teeth.

#### HABITAT AND BIOLOGY

Web and slender arm tips can be autotomised along visible fracture lines. The arm web is normally rolled up and held close to the mouth. It is extended only when the octopod is threatened and is presumably used to distract predators. Eggs are attached to rod-like structures that are carried in the arms of females. Young individuals carry broken tentacles of the Portuguese man o' war (a siphonophore) on the suckers of arms pairs 1 and 2. These borrowed tentacles presumably have a defensive function.

#### SIZE

Males dwarf (~17 mm ML), females large, up to 50 cm ML.

#### REMARKS

There are four species in this genus, only one of which is found in Australian waters. The dorsal arms of *Tremoctopus* have glandular stripes (parallel to the arms) that correspond in position to the shell secreting glands of *Argonauta* and may be homologous.

#### DISTRIBUTION

Surface waters of tropical and subtropical oceans.

#### REFERENCES

Thomas (1977) for review, Mangold *et al.* (2010b), Finn (2014d).

### *Tremoctopus gracilis* (Eydoux & Souleyet, 1852)

*Octopus gracilis* Souleyet, F. L. (1852). Mollusques, volume 2, 664 pages. In: Voyage autour du monde execute pendant les annees 1836 et 1837 sur la corvette La Bonite. Zoologie par M. M. Eydoux et Souleyet. 2 volumes and atlas of 101 plates. Paris [13, pl 1, figs 8, 9].

Type data: syntype MNHN 5-8-1106, other syntype not extant.

Type locality: Pacific Ocean, west of Paris, 8°N, 106°W.

#### COMMON NAME

Blanket Octopus

#### DISTINGUISHING CHARACTERS

*Tremoctopus gracilis* males have 19–22 pairs of transverse suckers on the distal portion and 27–29 pairs on the proximal portion of the hectocotylised arm. *Tremoctopus robsoni* Kirk, 1884 from New Zealand has 27 or 28 pairs and 9–15 pairs respectively.

#### DIAGNOSTIC FEATURES

Two pairs of conspicuous pores in web at base of arms: one pair dorsally and one pair ventrally adjacent to funnel. Hectocotylised arm of males with 19–22 pairs of transverse suckers on the distal portion and 27–29 pairs on the proximal portion. Hectocotylus with fringe of papillae extending full length of the proximal section. Males with 9–11 gill lamellae, females with 13–16 gill lamellae.

Fresh-caught female, dorsal view (S. Leporati).

Fresh-caught female, ventral view (S. Leporati).

*Colour.* Typically with silvery sides and very dark purple/blue dorsal surface.

#### HABITAT AND BIOLOGY

This pelagic species is an intermittent terminal spawner.

#### SIZE

Female ML up to 320 mm ML, total length to 1 m. Male ML up to 15.5 mm.

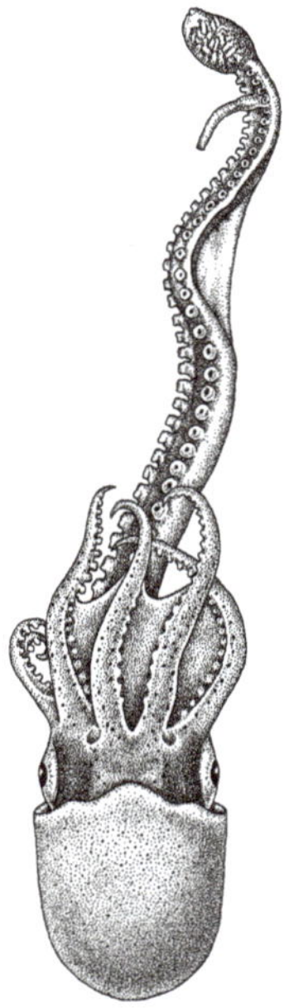

Male, dorsal view with hectocotylus extended (ABRS, K. Hollis).

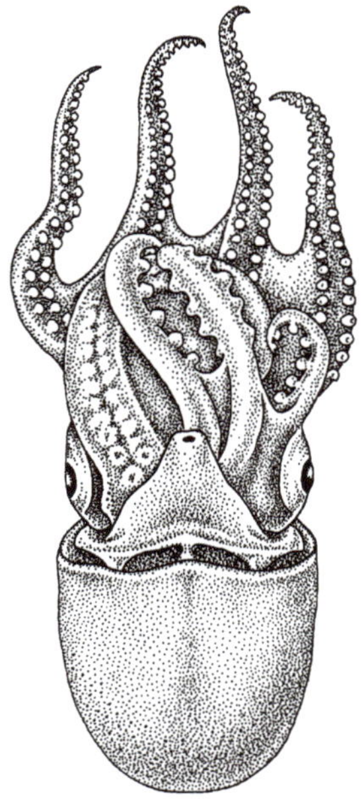

Male, ventral view with hectocotylus rolled up (ABRS, K. Hollis).

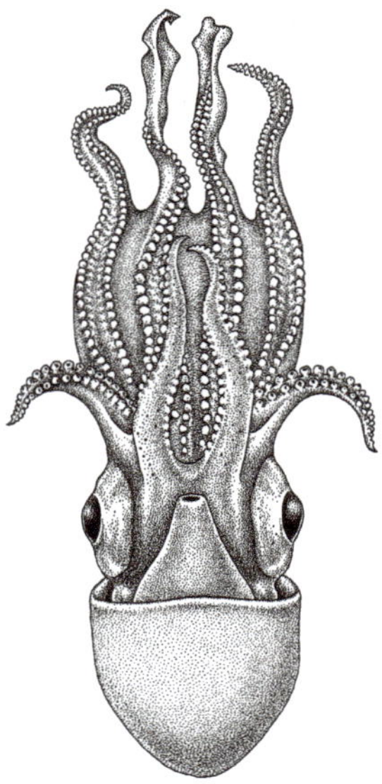

Female, ventral view (ABRS, K. Hollis).

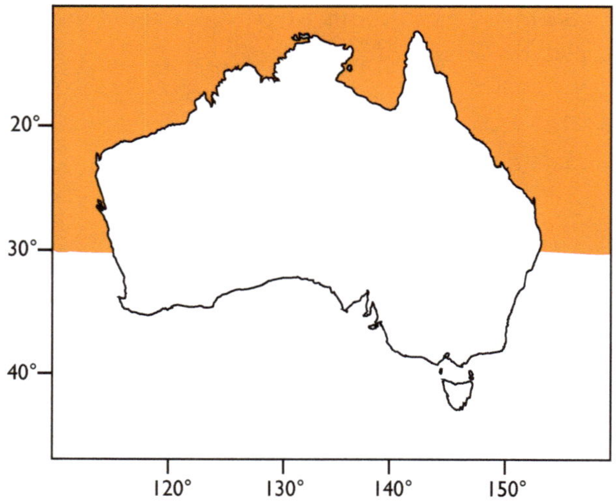


*Tremoctopus gracilis* distribution.

#### REMARKS

Zeidler (1989) reported the occurrence of this species off New South Wales, Western Australia and South Australia. It was formerly thought that the species was circumtropical. While the east Australian current regularly carries tropical species southwards, these warm core eddies generally disintegrate before approaching Bass Strait. Zeidler (1989) postulates that the winter-flowing Leeuwin current is responsible for occasionally carrying *T. gracilis* to southern Australian waters. However, the species is unlikely to survive for long periods in cooler waters. Some museum specimens were obtained from strandings and a specimen apparently surviving in the warmer waters of Gulf St Vincent had clearly not eaten for a considerable time.

#### DISTRIBUTION

Epipelagic in tropical and subtropical open oceans.

#### REFERENCES

Zeidler (1989), Finn (2014d).

## Superfamily Octopodoidea d'Orbigny, 1839 in Ferrusac and d'Orbigny, 1834–1848

Males lack hectocotylus autotomy. At present we are unable to determine morphological traits that unambiguously diagnose octopodoid families; however, there is strong support based on the analysis of molecular characters for the monophyly of each family recognised in Strugnell *et al.* (2014), so that classification is followed here. Past classifications have heavily relied on the presence/absence of an ink sac to define higher-level octopodoid relationships. Strugnell *et al.*'s (2014) study has shown that the presence or absence of an ink sac has arisen multiple times during the evolutionary history of the group and is unlikely to be informative in resolving high-level phylogenetic relationships.

### Key to families and genera of the Octopodoidea (after Lu and Dunning 1998, Norman *et al.* 2013, Norman *et al.* 2014, Strugnell *et al.* 2014)

1a. Body and arms soft, semi-gelatinous. Eyes telescopic, positioned close together on dorsal surface of head (except in Bolitaeninae). Funnel fused to ventral mantle forming two openings to mantle cavity .................................................. Amphitretidae

1b. Body and arms muscular or semi-gelatinous. Eyes lateral, round to oblong, not telescopic. Funnel free from ventral mantle, single opening to mantle cavity ............. 2

2a. Arm suckers uniserial ................................................ ................ Eledonidae (*Eledone*), Megaleledonidae 3

2b. Arm suckers biserial ................................................ ......................... Enteroctopodidae, Octopodidae 7*

3a. Ink sac present ........................................................ 4

3b. Ink sac absent........................................................ 5

4a. Funnel organ UUP-shaped. Five gill lamellae per demibranch. Mature male arm tips with pads of spongy tissue. Dorsal mantle and arm crown with four rows of dark red-brown papillae ....................... ........................................................ *Eledone* (*E. palari*)

4b. Funnel organ V V or W-shaped. Six or more gill lamellae per demibranch. Mature males without pads of spongy tissue on arm pads. Dorsal mantle and arm crown without four series of dark red-brown papillae ........................................................ 6

5a. Skin with raised conical or composite papillae hardened with cartilaginous wart-like inclusions (less obvious in frozen specimens). Funnel organ V V-shaped. Arms 2–4 × ML............................. *Graneledone* (*G. antarctica*)

5b. Skin without hardened papillae, sculpture soft or skin completely smooth. Funnel organ UU-shaped. Arms 1.3 × ML........... *Microeledone* (*M. mangoldae*)

6a. Large muscular species (up to 280 mm ML) with loose skin. Gills with 10–13 lamellae per demibranch. Oesophagus with swelling but no distinct crop.................................... *Megaleledone* (*M. setebos*)

* These two families cannot be unambiguously separated on the basis of morphological traits alone. Strugnell *et al.* (2014) diagnose Enteroctopodidae on the basis of molecular and morphological characters (see below). Members of both families are included in the key to Octopodidae below.

6b. Small muscular species (up to 100 mm ML), skin not markedly loose. Gills with six to no lamellae per demibranch. Distinct crop present as side branch of oesophagus ............7

7a. Two short, longitudinal ridges on mid-dorsal posterior mantle. Ligula groove of hectocotylus spoon-like with transverse ridges ....*Adelieledone adelieana*

7b. No short, longitudinal ridges on mid-dorsal posterior mantle. Ligula groove with transverse ridges.... ............*Pareledone*

## Amphitretidae Hoyle, 1886

Mantle, arms and webs transparent. Eyes telescopic, tubular in shape, oriented vertically (except in Bolitaeninae). Ventral mantle fused to funnel, restricting mantle aperture to two lateral openings. Internal shell absent. Arm suckers uniserial for most of arm length, biserial on distal tips beyond web margins. Third right arm of male modified, consisting of distinct calamus and whip-like distal tip. Digestive gland elongate and narrow, oriented vertically in live animal to minimise silhouette seen by predators from below. Digestive viscera dorsal to digestive gland. Radula comb-like with multiple cusps on central and first two lateral teeth. Ink sac present.

#### HABITAT AND BIOLOGY

Amphitretids are inhabitants of midwater in the open ocean.

### Key to subfamilies of the Amphitretidae

1a. Eyes normal, lateral and laterally compressed, not telescopic and tubular ............Bolitaeninae

1b. Eyes telescopic, oriented vertically, tubular or rectangular in shape ............2

2a. Eyes rectangular in shape with lateral lens; not joined basally. Lateral teeth not comb-like with multiple cusps ............*Vitreledonellinae* (*V. richardi*)

2b. Eyes tubular, oriented vertically; joined basally. Radula comb-like with multiple cusps on lateral teeth ............*Amphitretinae* (*Amphitretus*)

## Subfamily Amphitretinae Hoyle, 1886

## *Amphitretus* Hoyle, 1885

### Key to *Amphitretus* species

1a. Gills with 10 lamellae per demibranch. Non-hectocotylised arms with 22–32 suckers (usually <30). Ligula of hectocotylus without suckers but has two series of papillae. Hectocotylised arm sucker count 27 or 28............*A. pelagicus*

1b. Gills with eight lamellae per demibranch. Non-hectocotylised arms with 24–41 suckers (usually >30). Ligula of hectocotylus without suckers and papillae. Hectocotylised arm sucker count 21–24.... ............*A. thielei*

## *Amphitretus pelagicus* Hoyle, 1885

*Amphitretus pelagicus* Hoyle, W. E. (1885). Diagnosis of new species of Cephalopoda collected during the cruise of H.M.S. 'Challenger' – I. The Octopoda. *Annals and Magazine of Natural History, (series 5) 15,* 222–236 [235].

Type data: holotype BMNH 1890.1.24.5.

Type locality: off Kermadec Islands, 29°55′S, 178°14′W.

#### COMMON NAME

Telescope Octopus

#### DIAGNOSTIC FEATURES

Eye bases close together in V-shaped arrangement. Normal arms with 22–32 suckers. Webs deep, greater than two-thirds of arm length. Males with modified third right arm consisting of whip-like tip bearing two rows of low papillae and distinct calamus. Hectocotylised arm with 27 or 28 suckers. Gills with 10 lamellae per demibranch. Cigar-shaped, elongate liver held in vertical orientation in live animal.

#### HABITAT AND BIOLOGY

This midwater species lives between the epipelagic and bathypelagic zones. Some evidence suggests that smaller animals (<30 mm long) live at depths <150 m while adults have been collected to 2000 m. The telescopic eyes are

Arm crown (AM C.115340, J. Field).

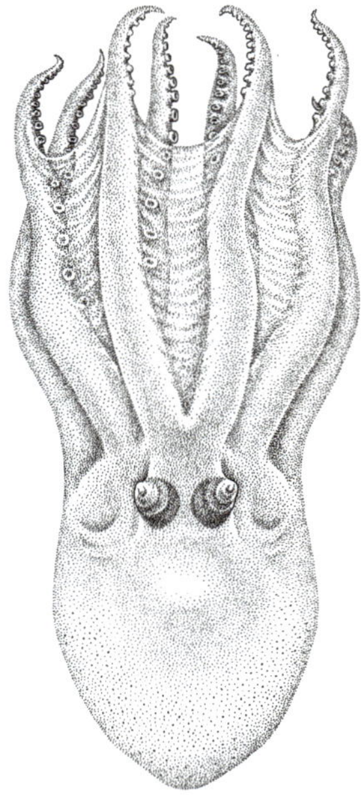

Dorsal view (ABRS, K. Hollis).

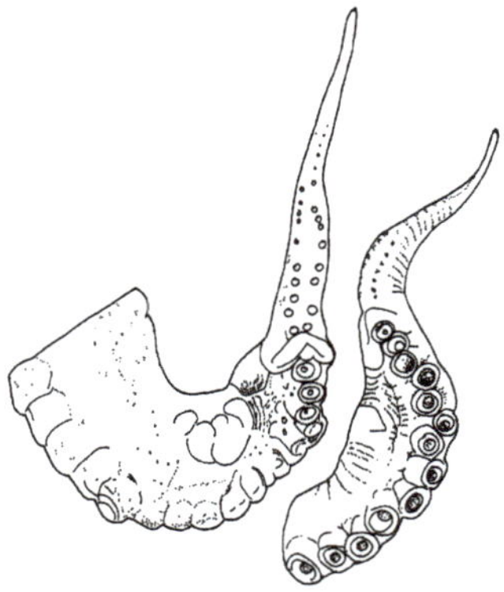

Distal portion of hectocotylus (A. Reid, redrawn from O'Shea 1999).

Enlargement of arm crown (AM C.115340, J. Field).

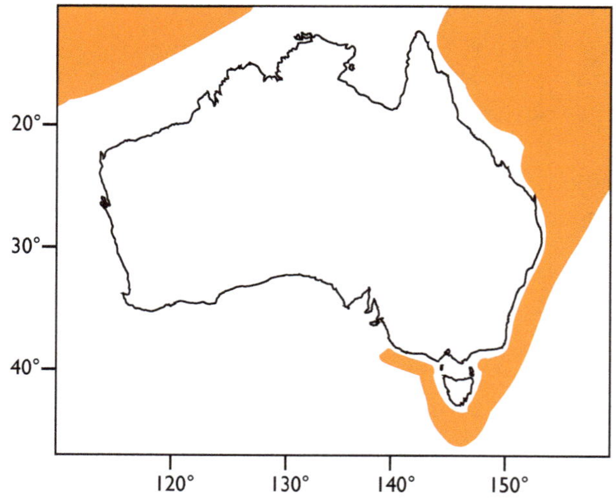


*Amphitretus pelagicus* distribution.

always oriented upwards. It minimises its own silhouette by being mostly transparent and maintaining both eyes and the digestive gland in a vertical orientation.

## SIZE

Mantle length to 100 mm, total length ~200–300 mm.

## REMARKS

Appears to be closely related to *Vitreledonella richardi* Joubin, 1918. South-eastern Australian *Amphitretus* have been identified in museum collections as *A. pelagicus*. The identity of these specimens needs to be checked.

## DISTRIBUTION

Open waters of the Indian, Pacific and Atlantic Oceans. Known depth range 100–2000 m.

**REFERENCE**

Norman and Finn (2014a).

## *Amphitretus thielei* Robson, 1930

*Amphitretus thielei* Robson, G. C. (1930). Cephalopoda, I. Octopoda. *Discovery Report 2*, 371–401 [383, figs 11–13, pl. 3].

Type data: holotype BMNH 1993013.

Type locality: coast of South Africa, 34°13′S, 15°49′E.

**DIAGNOSTIC FEATURES**

Gills with eight lamellae per demibranch. Non-hectocotylised arms with 24–41 suckers (usually >30). Ligula of hectocotylus without suckers and papillae. Hectocotylised arm sucker count 21–24.

**SIZE**

Mantle length to 135 mm, total length >200 mm.

**REMARKS**

This species is poorly known.

**DISTRIBUTION**

Temperate waters of southern Australia, New Zealand and off South Africa. Known depth ranges to 1145 m.

**REFERENCE**

Norman and Finn (2014a).

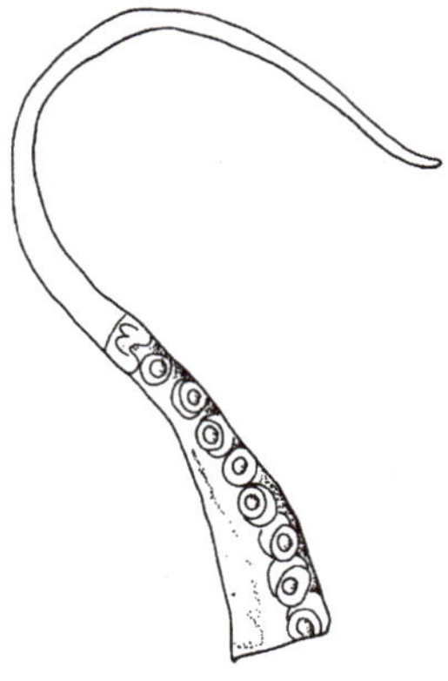

Distal portion of hectocotylus (A. Reid, redrawn from O'Shea 1999).

## Subfamily Bolitaeninae Chun, 1911

Small, gelatinous, mostly transparent octopuses inhabiting tropical and temperate waters worldwide. Body oval. Internal shell absent. Eyes laterally compressed. Arms short, shorter than ML, with single row of suckers. Radula with comb-like teeth bearing multiple sharp cusps on central (rhachidian) tooth and first two lateral teeth rows. Long axis of digestive gland parallel to body axis. Digestive viscera posterior to digestive gland. Large chromatophores over the body. Maturing females with oval light organ surrounding mouth.

**HABITAT AND BIOLOGY**

Members of this subfamily are meso- to bathypelagic and are found in the open ocean to depths of 1400 m. Little is known of their biology or behaviour. Females brood eggs by holding them in their arm crown. The mass is held together by the attachment of stalks to the capsules of other eggs. Off Hawaii the young are released at ~800 m and move up into shallower waters (150–250 m). As they grow they descend gradually to adult depths of ~800–1400 m.

**REMARKS**

Two genera, each with a single species, are currrently recognised: *Bolitaena* and *Japetella*. (*Bolitaena* may, in fact, contain more than one species.)

### Key to genera of the Bolitaeninae*

1a. Eyes small and extend away from brain on long optic stalks. Mature males with third left arm hectocotylised with elongate ligula................ *Bolitaena*

1b. Eyes relatively large, diameter approximately equal with optic stalk length. Mature males without hectocotylised arm............................................ *Japetella*

## *Bolitaena* Verrill, 1884

Norman and Finn (2014b) list a single widespread species, *Bolitaena pygmaea* (Verrill, 1884), in this genus. *Bolitaena sheardi* (Allan, 1945) is mentioned here because the type locality is Cape Byron in New South Wales and, as far as can be ascertained, the status of the species has not formally been determined. Norman and Finn (2014b) do not mention *B. sheardi* in their comprehensive review of the world's octopods.

* Immature specimens difficult to distinguish.

## *Bolitaena pygmaea* (Verrill, 1884)

*Eledonella pygmaea* Verrill, A. E. (1884). Second catalogue of the Mollusca, recently added to the fauna of the New England coast and the adjacent parts of the Atlantic, consisting mostly of deep sea species, with notes on others previously recorded. *Transactions of the Connecticut Academy of Sciences* 6(1), 139—294 [145, pl 32, fig. 2].

Type data: holotype NMNH 35268.

Type locality: Atlantic Ocean, 37°12′20″N, 69°39′W.

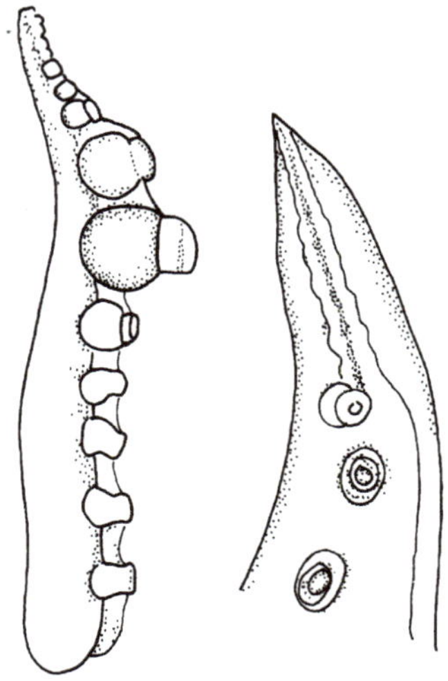

Arm tip and ligula (A. Reid, redrawn from Norman and Finn 2014b).

### COMMON NAME

Pygmy Pelagic Octopod

### DIAGNOSTIC FEATURES

Eyes small on long optic stalks. Male third arm hectocotylised with elongate ligula and one to three greatly enlarged suckers. Posterior salivary glands of mature males are greatly enlarged. Young animals with iridescent digestive gland and eyes.

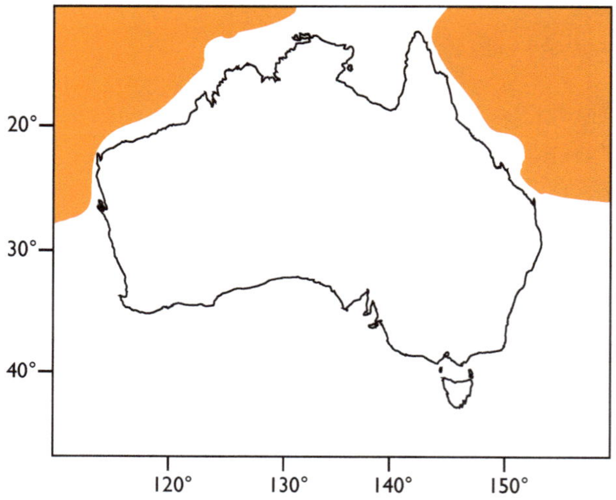


*Bolitaena pygmaea* distribution.

### HABITAT AND BIOLOGY

Young animals tend to occur in the shallower end of the range. As they reach sexual maturity the iridescence of the digestive gland is lost and they migrate to deeper waters. Pigmentation increases with maturity and the arms become relatively longer.

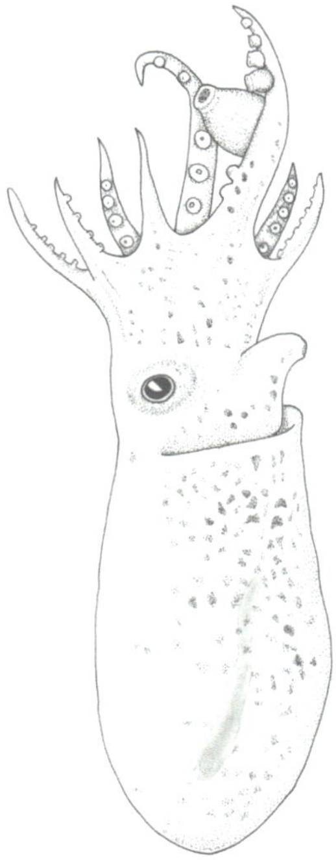

Lateral view (A. Reid, redrawn from Chun (1910) Plate LXXXI Fig. 3 male from Michael Sars expedition).

### SIZE

Mantle length to ~60 mm.

### DISTRIBUTION

Tropical and subtropical waters worldwide. Known depth range 100–1400 m, typically over deeper water.

### REFERENCE

Norman and Finn (2014b).

## *Bolitaena sheardi* (Allan, 1945)

*Eledonella sheardi* Allan, J. (1945). Planktonic cephalopod larvae from the eastern Australian coast. *Records of the Australian Museum 21*(6), 317–350, 4 plates [345, pl. 26, figs 22–27].

Type data: holotype AM C.126032.

Type locality: Australia: New South Wales, Cape Byron, 28°44′S, 153°57′E.

### DIAGNOSTIC FEATURES

Body broad, rounded, continuous with head dorsally, free ventrally. Head short, wide. Mantle extends ventrally to midpoint of eye. Ventral mantle joined to viscera by muscular band. Eyes very large, positioned laterally but swollen dorsally and ventrally into prominent prolongations; covered with skin, luminescent; with mucous pore on ventral side. Funnel slipper-shaped, fused to ventral surface of head along entire length, rounded anteriorly, pointed posteriorly; opening halfway down funnel, conspicuous, oblong, wide, hidden by mantle. Arms short, arms 3 longest, rest subequal. Five or six suckers on each arm in single series, perhaps with tiny suckers at arm tips; webs shallow. No visible hectocotylus. Gills with five to seven lamellae per demibranch.

*Colour.* Yellowish with minute, dark chromatophores over body; darker area over eyeballs. Arms with two parallel darker bands.

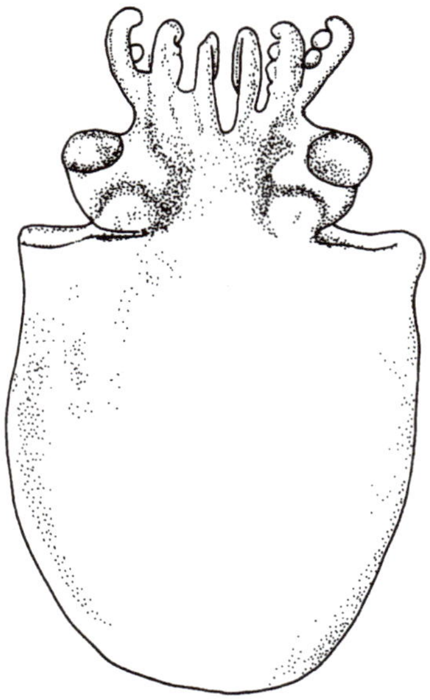

Dorsal view (A. Reid, redrawn from original description).

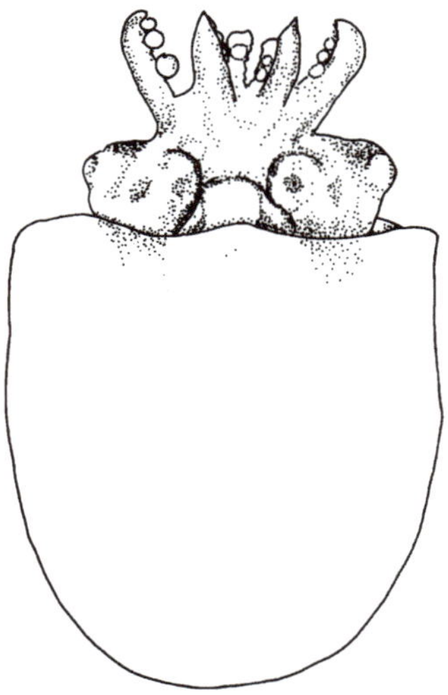

Ventral view (A. Reid, redrawn from original description).

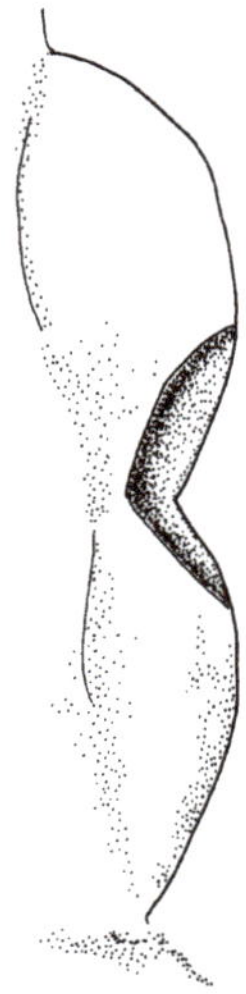

Side view of funnel (A. Reid, redrawn from original description).

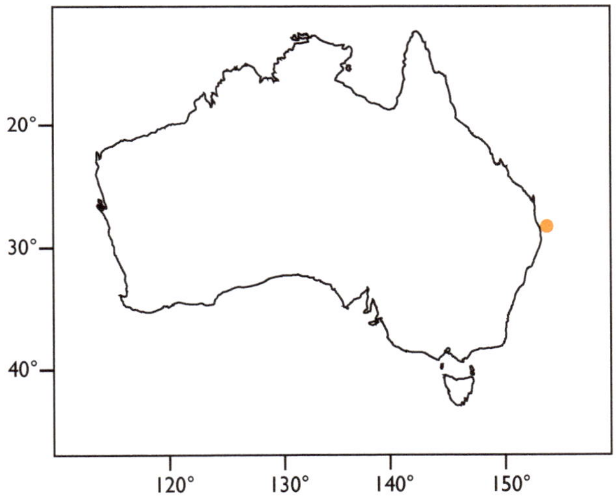


*Bolitaena sheardi* distribution.

### SIZE

Type specimen 7 mm total length.

### REMARKS

The status of this species is uncertain. It was described from a single immature specimen.

### DISTRIBUTION

Only known from the type locality, Cape Byron, New South Wales.

## *Japetella* Hoyle, 1885

*Japetella* is currently monotypic, represented by *Japetella diaphana* Hoyle, 1885.

## *Japetella diaphana* Hoyle, 1885

*Japetella diaphana* Hoyle, W. E. (1885). Diagnosis of new species of Cephalopoda collected during the cruise of H.M.S. 'Challenger' – I. The Octopoda. *Annals and Magazine of Natural History, (series 5) 15*, 222–236. [232].

Type data: holotype BMNH 1890.1.24.8.

### COMMON NAME

Diaphanous Pelagic Octopod

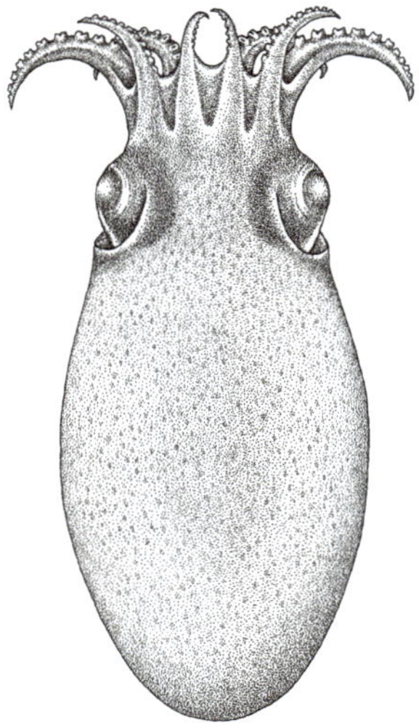

Dorsal view (ABRS, K. Hollis).

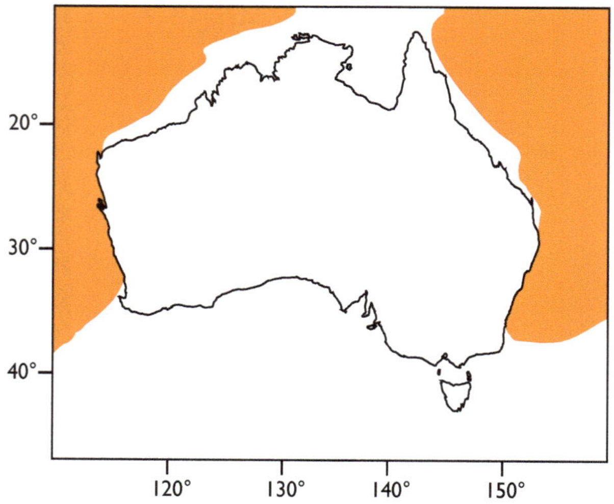


*Japetella diaphana* distribution.

### DISTINGUISHING CHARACTERS

Reaches about twice the length of *Bolitaena pygmaea* (Verrill, 1884). Relatively large eyes on short optic stalks (*Bolitaena* has longer optic stalks.)

### DIAGNOSTIC FEATURES

Eyes relatively large, diameter equal to optic stalk length. Suckers slightly enlarged on distal two-thirds of arms 3 in males. Near sexual maturity, suckers on right arms 3 of males abruptly enlarged, otherwise no other modification (but no fully mature male has ever been captured). Male salivary glands about twice the size of those of females. Digestive gland and eyes iridescent in subadults.

*Colour.* Iridophores in skin, overlying dorsal arms and head; iridophores oriented in plane transverse, rather than parallel to body axis.

*Photophores.* Females with circumoral light organ.

### HABITAT AND BIOLOGY

Gravid females have been found at ~1000 m depth. Young are found at 200–300 m and greater depths. Hatchlings are 3 mm ML. The life history may be similar to that of *B. pygmaea*. Spawned females have been seen, using remotely operated vehicles, floating at ~800 m holding clusters of interconnected eggs between their arms. At this depth the light is too low for them to be visible. They don't feed while brooding and probably die after the young hatch after several months. Hatchlings either swim or are carried by the female to the upper 200–300 m where food is presumably more plentiful. Young *Japetella* descend to mesopelagic depths as they grow.

### SIZE

Mantle length up to 85 mm, total length up to 160 mm.

### REMARKS

Females develop a large circumoral photophore that degenerates after spawning (Herring 2007).

### DISTRIBUTION

Tropical and subtropical regions worldwide, extending into cooler boreal waters in the north Pacific. Known depth range 200–1000 m.

**REFERENCES**

Young (2013), Norman and Finn (2014b).

## Subfamily Vitreledonellidae Robson, 1932

Gelatinous with transparent mantle, arms and webs. Eye a vertical rectangle with lateral lens; tapered rostrum on ventral surface of eye. Optic lobes of brain widely spaced with long optic nerve stalks. Ventral mantle edge free from funnel (i.e. not fused). Single, wide mantle opening. Internal shell absent. Arm suckers uniserial. Third left arm of mature male modified as bulbous oval tip bearing a slender papilla. Gills with outer demibranch only. Lateral teeth of radula are not comb-like, without numerous cusps. First lateral tooth with extra cusp compared to the Octopodidae. Digestive gland elongate and narrow, held vertically in live animal to minimise silhouette from below. Digestive viscera dorsal to digestive gland. Ink sac present. Chromatophores simple.

**HABITAT AND BIOLOGY**

These pelagic octopuses are found from the surface to at least 1000 m, typically over deep water beyond the continental shelf. Video sequences obtained from submersible vehicles suggest that the male envelops the female within his webs during mating. Little else is known about the biology and behaviour of these octopuses.

### *Vitreledonella* Joubin, 1918

*Vitreledonella* is monotypic, represented by *Vitreledonella richardi* Joubin, 1918.

#### *Vitreledonella richardi* Joubin, 1918

*Vitreledonella richardi* Joubin, L. (1918). Estudes preliminaires sur les Cephalopodes recueillis au cours des croisieres de S.A.S. Le Prince de Monaco. 6e Note: *Vitreledonella richardi* Joubin. *Bulletin de l'Institut Oceanographique 340*, 1–40, 29 figures [1, figs 1–29].

Type data: holotype MOM, station 3223.

Type locality: Atlantic Ocean, 30°50′N, 25°43′W.

**COMMON NAME(S)**

Glass Octopus, Glass Octopod

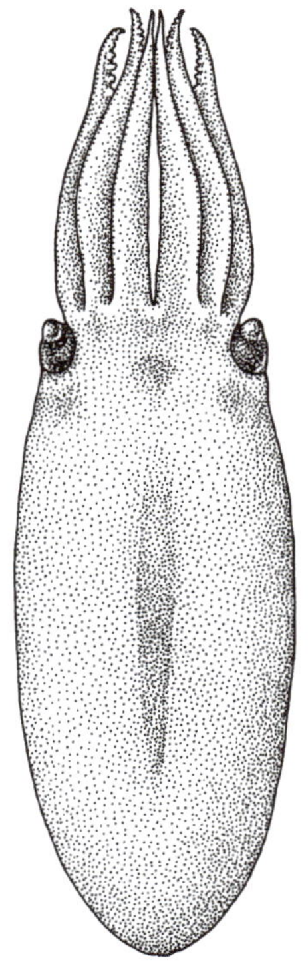

Dorsal view (ABRS, K. Hollis).

*Vitreledonella richardi* distribution.

**DIAGNOSTIC FEATURES**

Largest arm suckers distal to webs. Webs deep, about two-thirds of arm length. Gills with seven lamellae per outer demibranch.

*Colour.* Simple orange chromatophores over mantle, arms and webs.

#### SIZE
Mantle length up to 110 mm; total length to ~450 mm.

#### DISTRIBUTION
*Vitreledonella richardi* is found in tropical and temperate waters worldwide.

#### REFERENCE
Norman and Finn (2014d).

## Eledonidae Grimpe, 1921

### *Eledone* Leach, 1817

#### DIAGNOSTIC FEATURES
Medium to large; mantle ovoid, broad. Funnel organ shape varies among species. Arms short, stout, subequal, 2–3 × ML. No arm autotomy. Webs moderate to deep, deepest ~30–40% arm length; dorsal and ventral webs shorter than lateral webs. Arm suckers uniserial. Males of several species with enlarged suckers. Suckers near arm tips of mature males modified into plate-like ridges or elongate filaments. Third right arm of males hectocotylised, shorter than opposite arm. Copulatory organ small 3–4% arm length, in form of wrinkled rounded tip or simple ligula without calamus. Gills with 8–13 lamellae per demibranch. Ink sac and anal flaps present. Spermatophores medium-sized, eggs variable among species. Cartilaginous stylets present.

*Colour.* Pattern variable.

*Sculpture.* Body covered with fine, closely set warts. Skin ridge around lateral mantle sometimes present.

#### HABITAT AND BIOLOGY
Known depth range 5–500 m. Typically found on soft sediment substrates, but also found on rubble and rocky reefs.

#### SIZE
Up to 250 mm mantle length, 550 mm total length.

#### REMARKS
This genus is in need of revision. The Australian species differs morphologicaly from other species in the genus, so is only tentatively assigned to this genus.

#### DISTRIBUTION
Generally restricted to cooler waters of the Atlantic Ocean and southern Africa. (Australian *Eledone* an anomaly).

#### REFERENCE
Norman *et al.* (2014).

### *Eledone palari* Lu & Stranks, 1992

*Eledone palari* Lu, C. C. and Stranks, T. N. (1992). *Eledone palari*, a new species of Octopus (Cephalopoda: Octopodidae) from Australia. *Bulletin of Marine Science 49*(1–2) (1991), 73–87 [figs 1–6].

Type data: holotype MV F.57849.

Type locality: Australia: Queensland, east of North Stradbroke Island, 27°35′S, 153°50′E.

#### COMMON NAME
Spongetip Octopus

#### DIAGNOSTIC FEATURES
Soft bodied, gelatinous, rounded species. Body dorsoventrally flattened with obvious lateral skin ridge. Funnel organ UU-shaped. Arms short, 1.5–2.5 × ML, approximately equal in length. Webs deep, 38–67% of arm length. No sucker enlargement in either sex. Male hectocotylus (third right arm) ~70–100% length of opposite arm with 31–50 suckers. Ligula well developed, 5–9% of hectocotylised arm length, groove open with weak transverse ridges; calamus well developed, 50–75% of ligula length. All other arm tips in mature males bear pad of spongy tissue of unknown function. Gills with five lamellae per demibranch. Ink sac opens directly to exterior, anterior to anus. Anal flaps absent. Spermatophores robust, two-thirds of

Fresh-caught specimen (K. Graham).

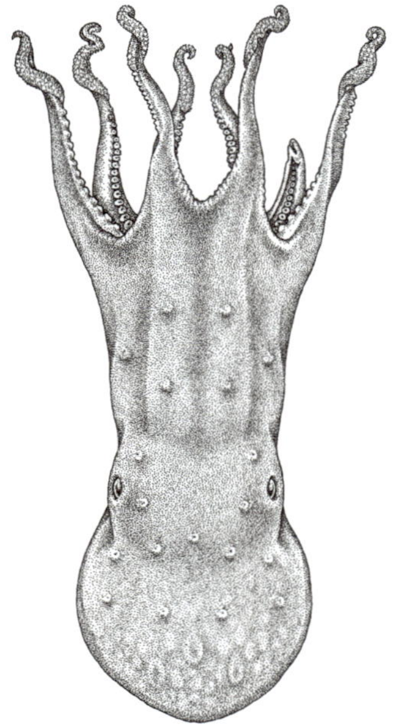

Dorsal view (K. Graham).

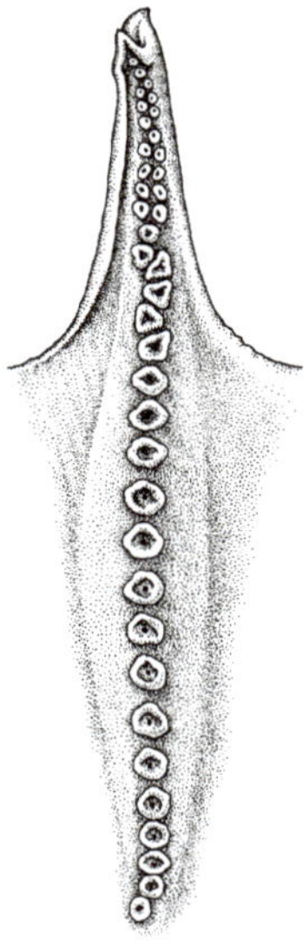

Hectocotylus (ABRS, K. Hollis).

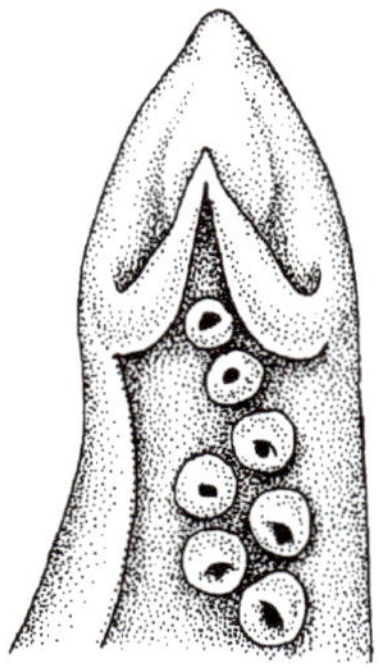

Enlargement of hectocotylus, ventral view (ABRS, K. Hollis).

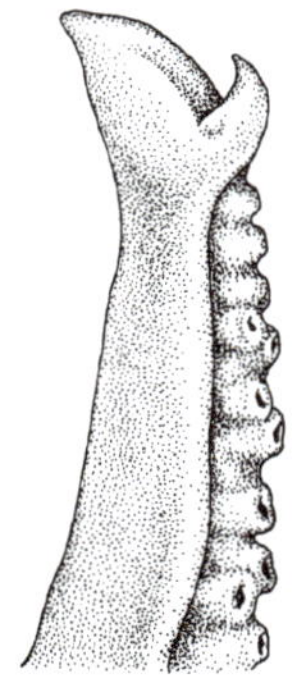

Enlargement of hectocotylus, side view (ABRS, K. Hollis).

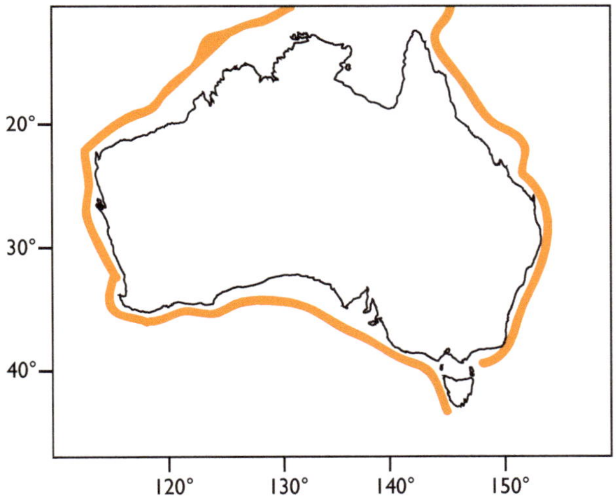


*Eledone palari* distribution.

ML. Eggs large, up to 16 mm long. Stylets large and robust (one-third of ML).

*Colour.* Four rows of dark red-brown papillae on dorsal mantle and arm crown. Cream to brown on upper surfaces, slightly lighter ventrally. Live animals yellow, with iridescent green sheen.

*Sculpture.* Small, low, rounded papillae over upper surfaces. Around four rows of larger dark red-brown papillae on dorsal mantle and arms. Small, low, rounded papillae over upper surfaces. Around four rows of larger dark red-brown papillae on dorsal mantle and arms. Thick skin ridge around lateral margin of mantle.

### HABITAT AND BIOLOGY

Unknown. Species only known from trawl specimens.

*Interest to fisheries.* None.

### SIZE

Up to 76 mm ML, total length 180 mm.

### REMARKS

While currently treated under the genus *Eledone*, the generic placement of this taxon awaits a major revision of the family (Norman and Hochberg 2005).

### DISTRIBUTION

Australia and Indonesia. Known depth range 200–600 m.

### REFERENCES

Lu and Stranks (1992), Norman *et al.* (2014).

## Enteroctopodidae Strugnell et al. 2014

Octopod with biserial suckers and hectocotylus with non-laminate ligula and well developed calamus. Using primer pair 5′-GGTGGAAATGGAGCCCACGT-3′ and 5′-GTGATGKACCAGTC-CATTGTAAGC-3′-, basepairs 182–200 of *octopine dehydrogenase* (ODH) gene as follows: CGATTCTGTGGTTCATCCA (after Strugnell *et al.* 2014).

## *Benthoctopus* Grimpe, 1921

### DISTINGUISHING CHARACTERS

Ligula moderate to large (LLI between 6 and 16), elongate, typically with closed ligula groove. Raised laminae absent.

### DIAGNOSTIC FEATURES

Deepwater incirrate octopodids. Mantle globose to ovoid. Funnel organ UU- or W-shaped. Arms 1.5–4 × ML, subequal, or arms pairs 1 and 2 longest; arm autotomy absent. Suckers biserial. Ligula of hectocotylus (third right arm) slightly to moderately excavated with indistinct midrib, smooth or bearing low, often indistinct rugae, never laminate. Gills with 12 lamellae per demibranch. Crop present, usually with diverticulum. Ink sac absent, anal flaps absent. Radula with multicuspid rachidian teeth. (After Strugnell *et al.* 2009.)

*Colour.* Uniform cream, grey, or purple. Some species have reverse countershading and some have dark oral web.

*Sculpture.* Smooth, or with low, rounded papillae. No large ocular or mantle papillae. Lateral skin ridge present or absent.

### REMARKS

A study based on mitochondrial and nuclear genes by Strugnell *et al.* (2009) that included 11 species of deep-sea octopodoids showed that the nine representatives of *Benthoctopus* examined are a monophyletic group. The two species *B. levis* (Hoyle, 1885) and *B. thielei* Robson, 1932 were not included in this analysis. A later study (Strugnell *et al.* 2011) showed that *Benthoctopus* species with a high latitude distribution in the southern hemisphere actually represent a paraphyletic group comprising three independent clades. The results suggest that the *Benthoctopus* clade originated in relatively shallow northern hemisphere waters. *Benthoctopus* species distributed in the Southern Ocean are representatives of polar emergence from the deep sea. Submerging Antarctic bottom water and emerging circumpolar deep water along the Antarctic continental margin provide unique connections between the deep sea and coastal Antarctic waters. These connections in conjunction with the 500–900 m deep Antarctic continental shelf have possibly allowed deep-sea fauna, in this case, to emerge across the Antarctic continental shelf.

Two species are known to occur in Australian waters: *B. levis* and *B. thielei*. Two species – one from off Victoria, recognised in Strugnell *et al.* (2011) as *Benthoctopus* sp. E, and another from off the coast of Adèlie Land (*Benthoctopus* cf. *rigbyae*) – are yet to be described. The former three species belong to a single clade (together a sister taxon to *B. johnsonianus* Allcock *et al.* (2006) from the north-east Atlantic), while the latter species belongs to a clade that includes *B. rigbyae* Vecchione *et al.* (2009) and several other undescribed species found close to the Antarctic coast with a probable circum-Antarctic distribution. Strugnell *et al.* (2011) suggest that these clades with non-overlapping distributions appear to represent independent invasions of this region from the deep sea in the northern hemisphere. The estimated divergence time of the two clades found in Australian waters is 14 million years ago. During this time there was an increased production and northward spreading of Antarctic bottom water, facilitating direct connections between the deep sea and coastal Antarctic waters.

### REFERENCES

Strugnell *et al.* (2009), Strugnell *et al.* (2011), Norman *et al.* (2014).

### Key to *Benthoctopus* species

1a. Web depth 33–40% of longest arm length. Sucker diameter 6–8% ML. Ligula length 7–9% of hectocotylised arm length ...................................... *B. levis*

1b. Web depth 25% of longest arm length, short between dorsal arms. Suckers 9% ML. Ligula length 13% of hectocotylised arm length .............. *B. thielei*

## *Benthoctopus levis* (Hoyle, 1885)

*Octopus levis* Hoyle, W. E. (1885). Diagnosis of new species of Cephalopoda collected during the cruise of H.M.S. 'Challenger' - I. The Octopoda. *Annals and Magazine of Natural History, (series 5) 15*, 222–236 [229].

Type data: holotype and paratype BMNH 1889.4.24. 43–44.

Type locality: Southern Ocean, off Heard Island, 52°59′S, 73°33′E.

### DIAGNOSTIC FEATURES

Web depth 33–40% of longest arm length. Sucker diameter 6–8% ML. Ligula length 7–9% of hectocotylised arm length.

### HABITAT AND BIOLOGY

Known depth range 13–137 m.

*Benthoctopus levis* from Joubin, L. (1891–1897). Results of the scientific expeditions undertaken on his yacht by Albert I Crown Prince of Monaco. Fascicule XVII Cephalopods collected on the expeditions of the Princesse-Alice (1891–1987), Plate 3, Fig. 7.

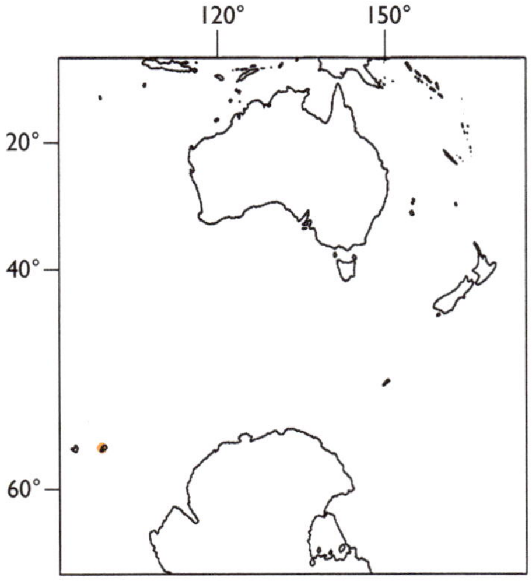

*Benthoctopus levis* distribution.

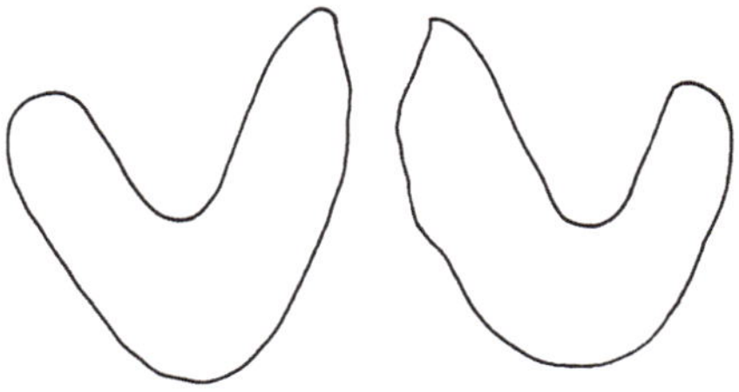
Funnel organ (A. Reid, redrawn from original description).

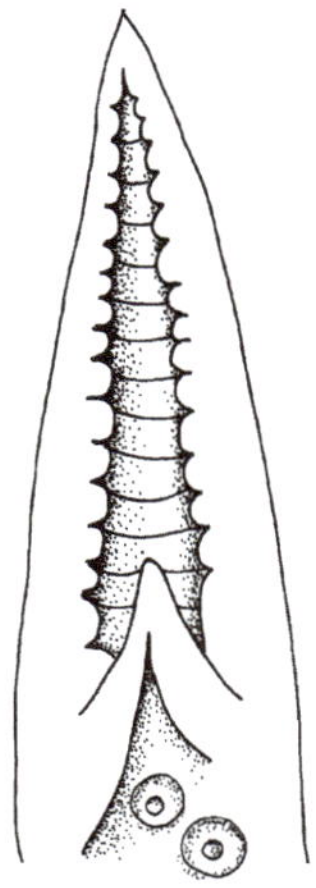
Hectocotylus (A. Reid, redrawn from original description).

**SIZE**

ML to 50 mm; total length to 180 mm.

**REMARKS**

*Benthoctopus levis* is an example of a radiation into Southern Ocean waters associated with polar emergence. *Benthoctopus levis* and *Benthoctopus thielei* Robson, 1932 are sister taxa and are estimated to have diverged 6 million years ago. These species form a monophyletic group with an undescribed species from off the south of Australia (Strugnell *et al.* 2011).

**DISTRIBUTION**

Only known from Heard Island. Known depth range 13–404 m.

**REFERENCES**

Nesis (1982/1987), Strugnell *et al.* (2011), Norman *et al.* (2014).

## *Benthoctopus thielei* Robson, 1932

*Benthoctopus thielei* Robson, G. C. (1932). A Monograph of the Recent Cephalopoda. Part II. The Octopoda. 359 pages, 79 figures, 6 plates. London: British Museum [233, text-figs 37–39].

Type data: holotype ZMB (not found 2000).

Type locality: Kerguelen Is., Gazelle Harbor.

**DIAGNOSTIC FEATURES**

Skin smooth. Funnel organ W-shaped. Web depth 25% of longest arm length, short between dorsal arms. Suckers 9% ML. Ligula 13% of hectocotylised arm length.

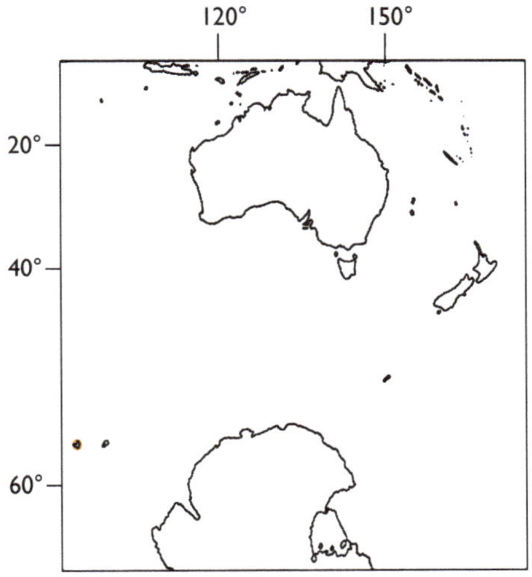

*Benthoctopus thielei* distribution.

**SIZE**

Up to 65 mm ML.

**REMARKS**

Bustamante *et al.* (1998) examined the concentrations of cadmium, copper and zinc in octopuses caught in Kerguelen shelf waters. They found that, compared with levels found in European cephalopods, cadmium concentrations were very high in *Benthoctopus thielei* (27.3–54.4 μg/g dry weight). The digestive gland accumulated 90% of the total cadmium in the whole animal (215 μg/g). This is despite the area experiencing no direct anthropogenic inputs of this metal; the area is presumably unpolluted. These researchers hypothesise that

*B. thielei* plays an important role in cadmium transfer in the food chain from their prey up to top vertebrate predators in the region.

#### DISTRIBUTION

Kerguelen Plateau. Known depth range 126–507 m (just north of the Southern Ocean in the Polar Frontal Zone).

#### REFERENCES

Lu and Stranks (1994), Strugnell *et al.* (2011).

## Megaleledonidae Taki, 1961

#### REMARKS

A landmark study by Strugnell *et al.* (2008b) that included several southern endemic and deep-sea taxa from this family postulates that the Southern Ocean is a centre of origin for this group of deep-sea octopuses. In contrast to the polar emergence of the deep-sea genus *Benthoctopus* discussed above (Strugnell *et al.* 2011) it appears that this clade of octopods arose through polar submergence, originating in Antarctica and moving into the deep sea. It appears that this clade of Southern Ocean octopodoids radiated into the deep sea, eventually reaching the northern hemisphere. Strugnell *et al.* (2008b) suggest that thermohaline circulation acted as an evolutionary driver with the Southern Ocean its centre of origin. The estimated divergence times suggest that the polar emergence of the *Benthoctopus* species occurred within a similar timeframe as the polar submergence of this clade, suggesting that the 'thermohaline expressway' is bi-directional. This lineage diverged ~33 million years ago and radiated at 15 million years ago. Both of these dates are critical in the development of thermohaline circulation.

### Key to genera of the Megaledonidae

1a. Dorsal mantle, head, eyes and bases of arms 1 and 2 covered with rosette-like clusters of wart-like tubercles ................................................ *Graneledone* (*G. antarctica*)

1b. Dorsal mantle, head, eyes and bases of arms 1 and 2 not covered with rosette-like clusters of wart-like tubercles ...................................................................... 2

2a. ML <20 mm in adult. No ridge present around lateral mantle margin. Funnel organ UU-shaped. Arms very short, ~1.3 × ML ........................................................................ *Microeledone* (*M. mangoldae*)

2b. ML >20 mm in adults. Ridge, or trace of ridge around lateral mantle margin. Funnel organ V V- or W-shaped. Arms usually > 1.3 × ML................. 3

3a. Ink sac greatly reduced, anal flaps absent ............... ......................................... *Megaleledone* (*M. setebos*)

3b. Ink sac and anal flaps present ................................ 4

4a. Mid-dorsal mantle with two long integumentary ridges. Hectocotylus ligula groove with transverse ridges ............................ *Adelieledone* (*A. adelieana*)

4b. No integumentary ridges on mid-dorsal mantle. Hectocotylus ligula groove without transverse ridges ......................................................... *Pareledone*

## *Adelieledone* Allcock, Hochberg, Rodhouse & Thorpe, 2003

#### DISTINGUISHING CHARACTERS

Differs from the genus *Pareledone* by the transverse ridges in the ligula groove of the hectocotylus, the sharp tip of the lower beak, the enlarged posterior salivary glands, the absence of stylets, and skin sculpture, especially the presence of two longitudinal integumentary ridges on the dorsal mantle.

#### DIAGNOSTIC FEATURES

Mantle saccular. Funnel organ W-shaped. Web depth medium (WDI 20–40). Arms with small (ASI 4–10) uniserial suckers. Third right arm of male hectocotylised: ligula large (LLI 10–16), ligula groove long, well marked and deep with transverse ridges. Gills with six to eight lamellae per demibranch. Ink sac present or reduced, crop well developed, posterior salivary glands large (up to twice the length of the buccal mass). Diverticulum of terminal organ not coiled. Ink sac present. Spermatophore length index 60–80. Rostral tip of lower beak sharp. Radula with nine elements. Stylets absent.

*Colour.* Chromatophores in connective tissue covering the digestive gland.

*Sculpture.* Two short, longitudinal integumentary ridges on mid-dorsal posterior mantle.

## REMARKS

Three species have been placed in this genus, all Antarctic endemics, one of which, *Adelieledone adelieana* (Berry, 1917), is found within Australian Antarctic Territory.

## REFERENCE

Allcock *et al.* (2003a).

### *Adelieledone adelieana* (Berry, 1917)

*Moschites adelieana* Berry, S. S. (1917). Cephalopoda. *Scientific Reports of the Australasian Antarctic Expedition 1911–1914, (series C) 14*(2), 1–39 [17, text figs 10–13, pl. 11, fig. 5, pl. 12, figs 6–8].

Type data: holotype AM C.40889.

Type locality: Southern Ocean, Adélie Land, off Mertz Glacier, 66°55′S, 145°21′E.

## DIAGNOSTIC FEATURES

Mantle elongate ovoid, demarked from mantle by moderate constriction. Head wider than mantle (88–118% of mantle width); eyes large, project above surface of head. Funnel large, stout; funnel organ W-shaped, outer limbs as long as median limbs. Arms short (1.9–2.2 × ML in mature animals) and stout, arm formula 4 > 3 > 2 > 1; arms 65–69% of total length. Arm suckers uniserial, raised from arm suface, small, without enlargement. Male hectocotylus (third right arm) shorter than opposite arm with 22–28 suckers. Ligula long, 9–15% of third right arm length in mature animals, with well-marked groove crossed by approximately eight transverse ridges. Gills with six or seven lamellae per demibranch. Eggs 8–9 mm long.

*Colour.* Purple-pink to purple-grey dorsally, widely scattered papillae on dorsal surface; paler ventrally, ventral surface smooth and creamy white. When stimulated, animals become darker purple-red to purple-brown dorsally, cream ventrally. One white spot on mid-dorsal brachial crown; one broad spot on mid-dorsal mantle. White transverse bar between eyes. Ocelli absent.

*Sculpture.* Some minor papillation; fine, rounded and widely scattered papillae. One large supraocular papilla. Two longitudinal ridges on mid-dorsal posterior mantle; traces of ventrolateral ridge.

## HABITAT AND BIOLOGY

Males mature at 40 mm ML, females at 45 mm ML. The species has been found on mud and sand bottoms with pebbles and rocks and among sponges, bryozoans and gorgonaceans.

## SIZE

Up to 45 mm ML, 160 mm total length.

## REMARKS

*Adelieledone polymorpha* (Robson, 1930), also from the Antarctic Peninsula, differs from *A. adelieana* in having relatively longer arms (mean ALI (arm 1) 190.2 in *A. polymorpha v.* 154.6 in *A. adelieana*). The normal arm sucker number is greater in *A. polymorpha* (63 *v.* 46), as is the hectocotylised arm sucker number (30–34 *v.* 22–28). There may be more gill lamellae (seven or eight) in *A. polymorpha* than in *A. adelieana* (six or seven).

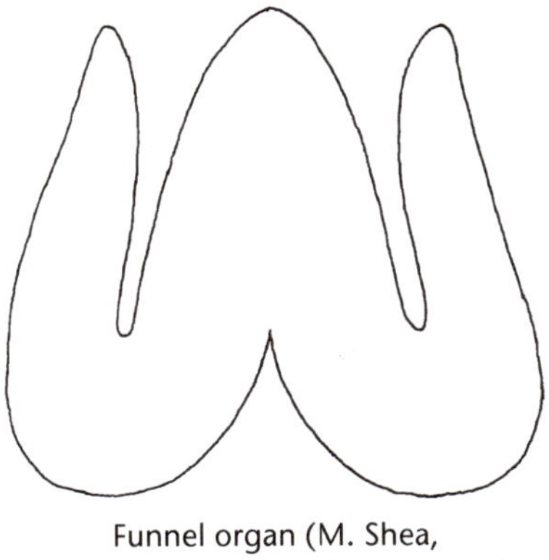

Funnel organ (M. Shea, redrawn from original description).

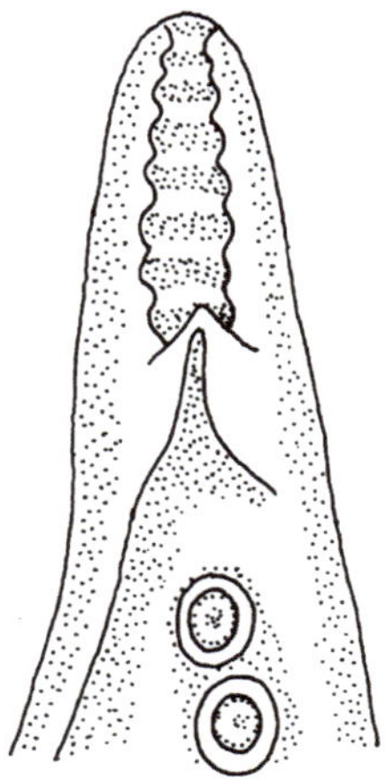

Hectocotylus, oral view (M. Shea, redrawn from original description).

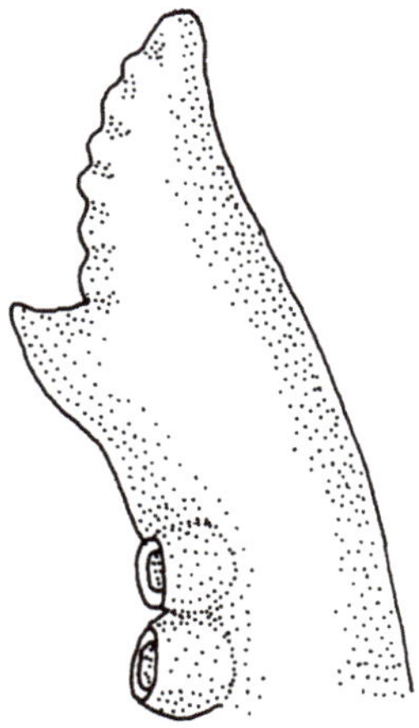

Hectocotylus, lateral view (M. Shea, redrawn from original description).

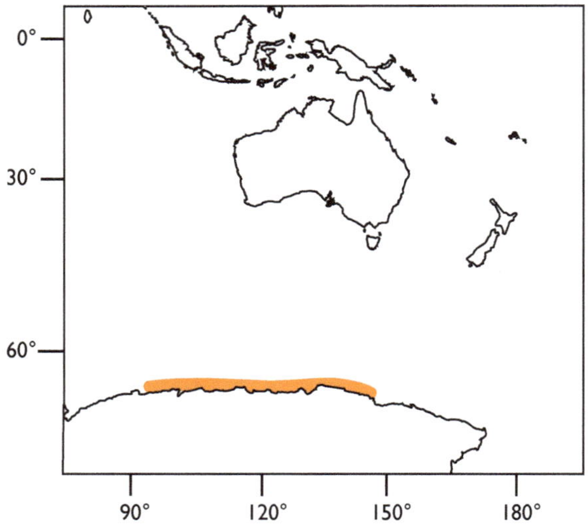


*Adelieledone adelieana* distribution.

#### DISTRIBUTION

On the continental shelf and slope of Antarctica in depths of 139–680 m (–2.2–1.7°C).

#### REFERENCES

Lu and Stranks (1994), Allcock *et al.* (2003a).

### *Graneledone* Joubin, 1918

#### DIAGNOSTIC FEATURES

Robust, deep-water species. Mantle spherical to ovoid. Funnel organ V V-shaped. Arms 2–4 × ML; subequal. No arm autotomy. Web depth variable among species, 15–40% longest arm. Arm suckers uniserial, non-enlarged. Third right arm of males hectocotylised with tip clearly differentiated into ligula and calamus; calamus long in most species. Gills with five to eight lamellae per demibranch. Ink sac and anal flaps absent. Spermatophores 1.2–1.8 × ML. Eggs large. Stylets present, non-mineralised.

*Colour.* Uniform grey to pink-purple, often darker around eye rim and on oral surface of webs and arms.

*Sculpture.* Dorsal mantle, head, eyes and bases of arms 1 and 2 covered with rosette-like clusters of raised rugose, cartilaginous (wart-like) tubercles. No skin ridge around edge of mantle margin.

#### HABITAT AND BIOLOGY

*Graneledone* are deep-water species attaining depths of at least 3000 m. Observations of *Graneledone* as observed at ~2600 m depth in the north Pacific indicate that they attach their eggs to hard substrate and brood them through development (Voight and Grehan, 2000). Based on remotely operated vehicle (ROV) observations of a single animal, it has been shown that *G. boreopacifica* Nesis, 1982 in the Monterey Submarine Canyon, off central California, conducts the longest known egg-brooding period of any animal species: 53 months (Robison *et al.* 2014). If this observation can be extrapolated to other deep-sea species, it appears that these may be among the longest-lived of all cephalopods.

#### SIZE

ML up to 170 mm, total length to 683 mm.

#### REMARKS

The genus contains at least 10 species. Frozen specimens often appear smooth as freezing masks the cartilaginous inclusions. The masking of this trait has led to some taxonomic confusion. Strugnell *et al.* (2008b) showed that members of this genus arose from shallow-water Antarctic ancestors and moved into deeper habitats along an isothermic gradient from southern polar waters.

Bustamante *et al.* (1998) examined the concentrations of cadmium, copper and zinc in octopuses caught in Kerguelen shelf waters (just outside Australian Territory). They found that, compared with levels found in European cephalopods, cadmium concentrations were very high (30.7–47.1 µg/g dry weight) in *Graneledone* sp. (probably *G. gonzalezi* Guerra, González and Cherel, 2000). The digestive gland accumulated 90% of the total cadmium in the whole animal (369 µg/g). These researchers hypothesise that *Graneledone* plays an important role in cadmium transfer in the food chain up to top vertebrate predators in the region.

#### DISTRIBUTION

Deep waters of the southern hemisphere and northern Pacific Ocean.

#### REFERENCE

Norman *et al.* (2014).

### *Graneledone antarctica* Voss, 1976

*Graneledone antarctica* Voss, G. L. (1976). Two new species of octopods of the genus *Graneledone* (Mollusca: Cephalopoda) from the Southern Ocean. *Proceedings of the Biological Society of Washington 88*(42), 447–458, 3 figures [448, figs 1, 2].

Type data: holotype NMNH 729679.

Type locality: Southern Ocean, 74°05.6′S, 175°05.2′W.

#### DIAGNOSTIC FEATURES

Mantle short, broad and distinctly flattened dorsoventrally. Body muscular with overlying thick layer of gelatinous

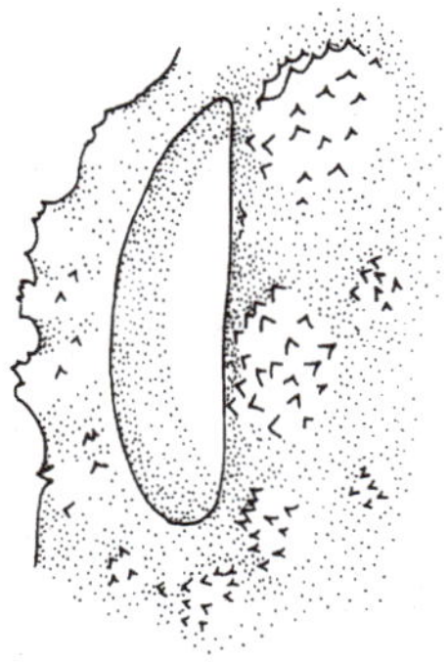

Enlargement of head to show wart-like tubercles (A. Reid, redrawn from original description).

Holotype, dorsal view (A. Reid, modified from original description).

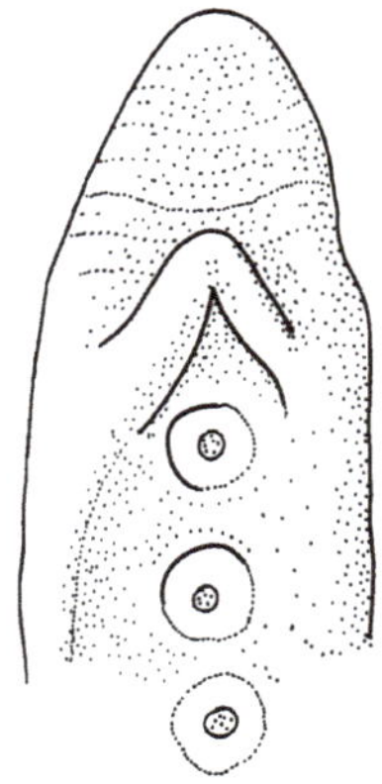

Hectocotylus (A. Reid, redrawn from original description).

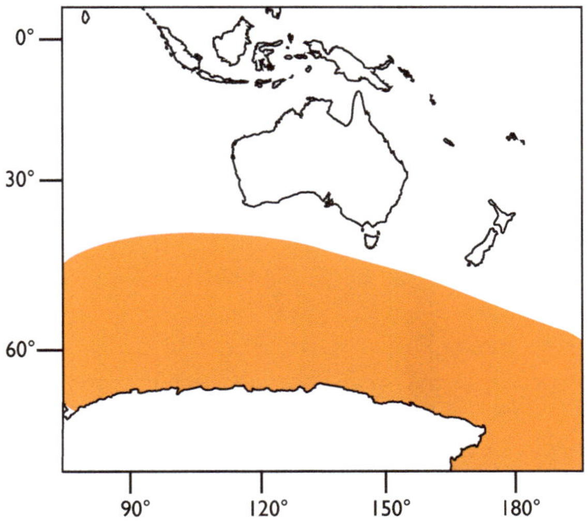


*Graneledone antarctica* distribution.

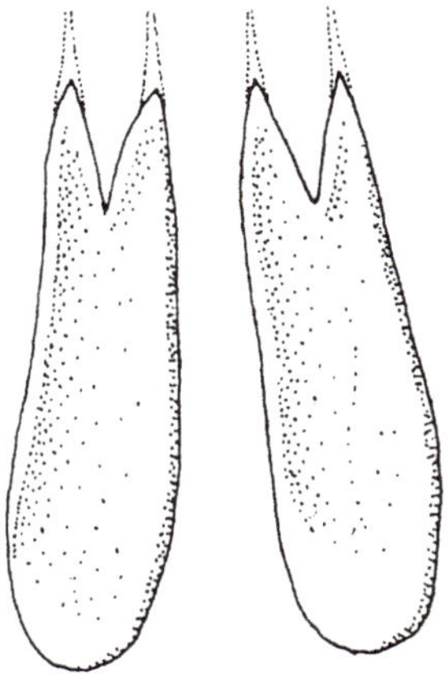

Funnel organ (A. Reid, redrawn from original description).

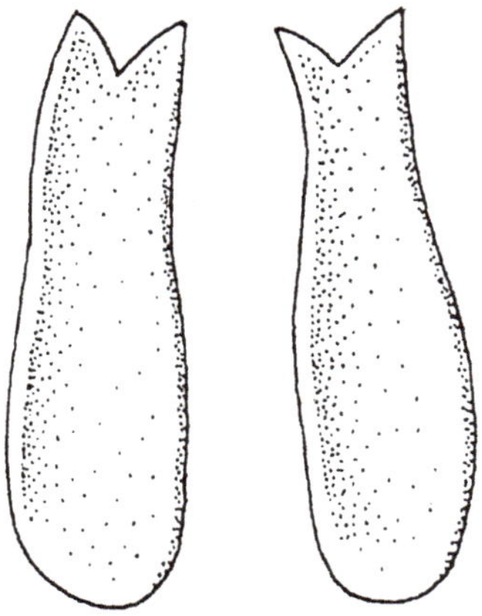

Funnel organ (A. Reid, redrawn from original description).

material, particularly posteriorly; surface covered by thin, tough skin. Head width almost equal to mantle width, eyes large, conspicuous. Funnel organ V V-shaped (but variable; see figures), composed of two elongate oval pads slightly to moderately split anteriorly, rounded posteriorly. Arms long, stout, formula 1 >2 >3 >4 (arms 1 or 2 always longest). Web moderately deep, extends only a short distance along arms dorsally and to distal tips of arms 1–3 ventrally. Arm suckers small. Third right arm of male hectocotylised: ligula small, spoon-shaped, pointed distally and with thickened margins; oral surface with 12–13 low, fleshy folds; calamus low and blunt. Gills with six lamellae per demibranch.

*Colour.* Ethanol-preserved specimens pale yellowish brown dorsally, with purplish hues at base of brachial crown and on and between arm bases. On this background colour, mantle, head, funnel and base of arms reddish brown suffused with purple. Spermatophore groove yellowish.

*Sculpture.* Mantle, head and arms covered with numerous, evenly spaced, close-set warts, consisting of a raised mound bearing from one to >12 small, cone-shaped papillae. Papillae absent from arm pair 4 but liberally scattered on dorsal side of arm pairs 2 and 3. Distinct line of fine warts on lateral mantle arranged like a keel around entire lateral margin. No warts ventrally except for a few around eyes. Distinct circlet of warts around eye that are larger than those elsewhere on the body in the form of two to three supraocular papillae.

## HABITAT AND BIOLOGY

Little is known about the biology of this species, although polychaete setae have been found among stomach contents.

## SIZE

Up to 45 mm ML.

## DISTRIBUTION

Southern Ocean, Antarctica, Ross Sea. Depth to 2341 m.

## REFERENCE

Norman *et al.* (2014).

## *Megaleledone* Taki, 1961

*Megaleledone* is currently monotypic, represented by *Megaleledone setebos* (Robson, 1932).

## *Megaleledone setebos* (Robson, 1932)

*Graneledone setebos* Robson, G. C. (1932). A Monograph of the Recent Cephalopoda based on the collections in the British Museum (Natural History). Part II. The Octopoda. 359 pages, 79 figures, 6 plates. London: British Museum. [313, text fig. 72.]

Type data: holotype. BMNH 1919.12.30.27.

Type locality: Antarctica: Cape Evans, McMurdo Sound.

Juvenile (M. Rauschert).

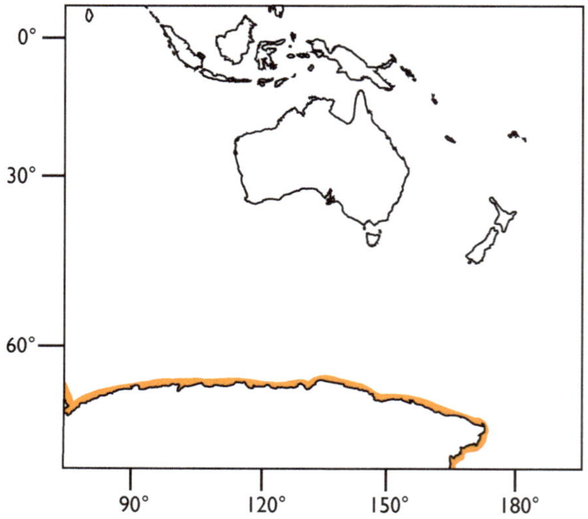


*Megaleledone setebos* distribution.

## COMMON NAME

Giant Antarctic Octopus

## DISTINGUISHING CHARACTERS

Lacks the epidermal warts characteristic of *Graneledone*.

## DIAGNOSTIC FEATURES

Body plump, soft gelatinous consistency, surface smooth, skin loose, muscular tissue firm. Mantle transversely ellipsoid, very broad, mantle aperture wide, head narrower than mantle. Eyes small, do not project far above surface of head. Funnel organ V V-shaped, limbs thick, outer limbs as long as median limbs. Arms plump, formula 4 = 2 = 3 > 1, short, arms ~2–3 × ML in mature animals. No arm autotomy. Web well developed, deepest between lateral arms, deepest web > 40% of arm length; web margins extend to arm tips. Arm suckers uniserial, relatively large (ASI 5.9–*8.7*–18.6), ~40–69 suckers on each normal arm. No enlarged suckers. Male hectocotylus (third right arm) slightly shorter, 90–95% length of opposite arm with 35–40 suckers. Ligula small, 3–4% of third right arm length in mature animals; ligula groove long, well marked and shallow with approximately 10 transverse ridges; calamus long, pointed, ~40% of ligula length. Oesophagus with swelling only but no distinct crop. Ink sac greatly reduced, anal flaps absent. Ink duct opens to exterior anterior to anus. Gills with 10–13 lamellae. Spermatophores large, ~80–90% ML. Ovarian eggs large, up to 42 mm long. Stylets present, non-mineralised.

*Colour.* In life, resting animals cream-pink to grey-pink mottled with red-pink patches, and cream-white to pink-white ventrally. When stimulated becomes darker in colour (dark purple-pink dorsally and light purple-pink ventrally). White spots or bars absent. Ocellae absent.

*Sculpture.* Fine, rounded and widely scattered papillae on dorsal surface; no larger papillae in ocular region. Ventrolateral integumentary ridge present.

### HABITAT AND BIOLOGY

Taki (1961) reported a range of prey items in the stomach of this species, including Foraminifera, Polychaeta, Crustacea, Mollusca, fishes and seaweed. He also noted that this species seems to show more adaptations to the littoral zone rather than the deep sea.

This species has been found on mud and sand bottoms with pebbles and rocks, among sponges and bryozoans at temperatures of –1.9–1.4°C. Known depth range 120–761 m.

### SIZE

Up to 235 mm ML, total length to 900 mm.

### DISTRIBUTION

Common in Antarctic waters. The species appears to have a circumpolar distribution.

### REFERENCES

Taki (1961), Allcock *et al.* (2003b), Norman and Hochberg 2005, Allcock *et al.* (2007).

## *Microeledone* Norman, Hochberg & Boucher-Rodoni, 2004

*Microeledone* is currently monotypic, represented by *Microeledone mangoldae* Norman, Hochberg & Boucher-Rodoni, 2004.

## *Microeledone mangoldae* Norman, Hochberg & Boucher-Rodoni, 2004

*Microeledone mangoldi* Norman, M. D., Hochberg, F. G., and Boucher-Rodoni, R. (2004). *Microeledone mangoldi* n. gen. and n. sp., a deep-water pygmy octopus from the Norfolk ridge, New Caledonia (Cephalopoda: Octopodidae). *Molluscan Research 24*, 193–209 [194, figs 1–4].

Type data: holotype MNHN 2015.

Type locality: New Caledonia: Norfolk Ridge, 23°23′S, 167°52′E.

### COMMON NAME

Sickletooth Pygmy Octopus

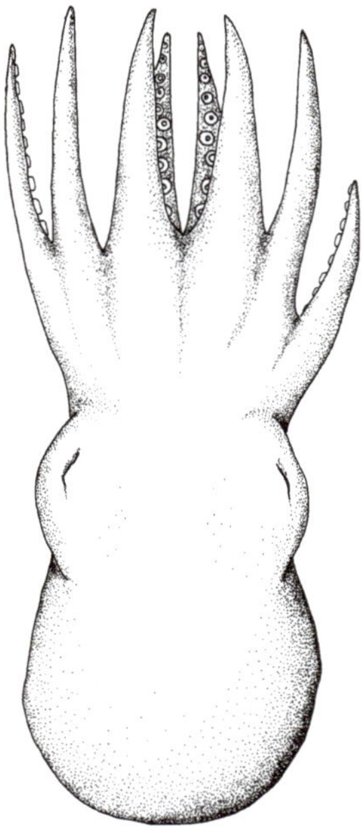

Dorsal view (A. Reid, redrawn from original description).

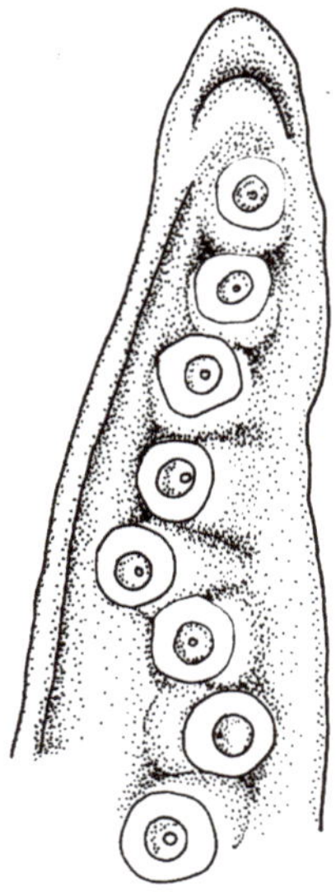

Hectocotylus (A. Reid, redrawn from original description).

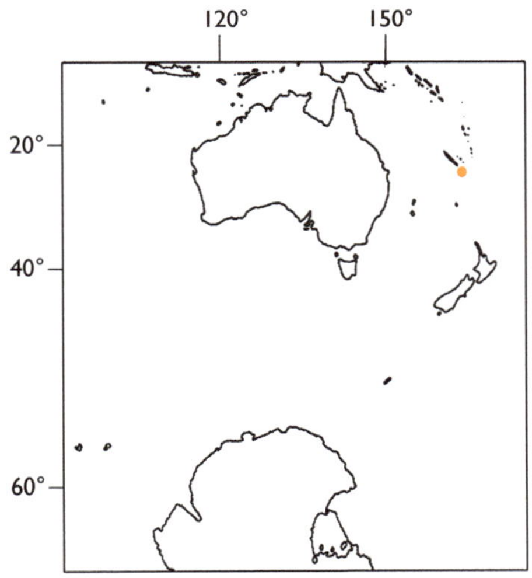



*Microeledone setebos* distribution.

### DIAGNOSTIC FEATURES

Muscular species. Funnel organ UU-shaped, limbs approximately equal length. Arms short, ~1.3 × ML. Arm autotomy absent. Webs deep, approximately equal on all arms, > 40% of arm length (ventral web slightly shallower); web margins absent. Arm suckers uniserial, 31–32 suckers on normal arms; no enlarged suckers. Male hectocotylus (third right arm) approximately ~95% length of opposite arm with 22 suckers (based on single male). Ligula moderate in size, 7.6% of arm length in single male; calamus large, ~45% of ligula length. Radula with seven series of teeth, marginal plates absent. Central rhachidian tooth wide with elongate, flattened and curved central cone, lateral cusps absent. Remaining teeth flattened and blade-like. Oesophagus without distinct crop (swelling only). Gills with four or five lamellae per demibranch.

*Colour.* Pink-cream. Chromatophores absent. Oral surface of arm webs dark maroon for proximal third of arms. Ocelli absent.

*Sculpture.* Skin smooth. No ridge around lateral mantle margin.

#### HABITAT AND BIOLOGY

Bathyal plain, depth 980–1000 m.

#### SIZE

ML to 18 mm, total length to ~46 mm.

#### REMARKS

The description of this animal is based on a single male animal.

#### DISTRIBUTION

Known only from the type locality, New Caledonia, Norfolk Ridge, 23°23′S, 167°52′E.

#### REFERENCE

Norman *et al.* (2014).

### *Pareledone* Robson, 1932

#### DIAGNOSTIC FEATURES

Mantle saccular. Funnel organ W or VV-shaped. Arms with small uniserial suckers. Third right arm of male hectocotylised with distal tip clearly differentiated into ligula and calamus; ligula medium to large (LLI 6–13), ligula groove long, well marked and shallow without transverse ridges. Gills with 6–11 lamellae per demibranch. Crop well developed, posterior salivary glands medium-sized, length approximately equal to length of buccal mass. Rostral tip of lower beak rounded. Radula with nine elements; rhachidian tooth with large central cusp (other cusps may be present). Diverticulum of penis coiled. Spermatophores long and slender. (Diagnosis after Allcock *et al.* 2003a.) Very delicate cartilaginous stylets present.

#### HABITAT AND BIOLOGY

*Pareledone* is the most diverse and abundant octopod genus in the Antarctic and surrounding islands of the southern ocean, and is endemic to that region with species only found south of the Polar Front. It is restricted to the continental shelf and slope margins in depths of < 1000 m. Because of the inaccessibility of their habitat there is still only limited information available about their behaviour and life history. However, studies of gonad maturation suggest that the breeding season extends over a considerable period. Eggs are large (up to 20 mm) so are likely to be incubated for a considerable time, probably over winter with hatchings probably benthic.

#### SIZE

ML to 120 mm.

#### REMARKS

Allcock and Piertney (2002) investigated the phylogenetic relationships of eight species of incirrate octopodids from western Antarctica using molecular sequence data from the mitochondrial 16s rRNA gene. They found *Pareledone* to be polyphyletic. As a result, they removed *P. adeliana* (Berry, 1917) and *P. polymorpha* (Robson, 1930) from the group, placing them in a new genus, later named *Adelieledone* Allcock *et al.* (2003a). The presence or absence of an ink sac was previously used as a trait to define subfamilies. Allcock and Piertney (2002) determined that the loss of an ink sac is an adaption to depth and has no evolutionary significance, resulting in a spurious classification. This view has recently been vindicated in a revisionary study of the polyphyletic Octopodidae by Strugnell *et al.* (2014) resulting in a more meaningful, higher-level classification of the group.

The genus *Pareledone* contains two clades, one containing smooth-skinned species, and the other comprising approximately a dozen papillated species, most of which occur sympatrically on the continental shelf at Elephant Island, South Shetland Islands (Allcock *et al.* 2007). Allcock *et al.* (2010) sequenced 350 specimens of *Pareledone* for the barcoding gene COI. Eleven species were represented. Results suggested a barrier to gene flow to the west of the Antarctic Peninsula, with comparatively warm bottom water acting as a physical barrier. The fast evolving gene, COI, revealed a network of closely related species, suggesting recent, rapid cladogenesis in the group. While COI was shown to be a useful identification tool when a reference set of sequences was available, it was limited in its use to help determine species boundary definitions. Inter- and intra-specific variation in the group was very small, the former possibly due to incomplete lineage sorting and introgressive hybridisation, and the latter suggesting a relatively recent divergence among species within the genus. Glacial cycles and the fragmentation of available habitats were suggested to act in favour of speciation in this genus.

**DISTRIBUTION**

Endemic to the Antarctic. Species only found south of the Polar Front.

**REFERENCES**

Lu and Stranks (1994), Allcock (2005); Allcock *et al.* (2010).

### Key to *Pareledone* species

1a. Integument smooth, without papillae .... *P. harrissoni*
1b. Integument with papillae .......................................... 2
2a. Papillae on ventral mantle cover the entire surface ........................................................ *P. framensis*
2b. No papillae on ventral mantle ................................ 3
3a. Papillae not simple and round, may be large and compound. Supraocular papillae present, irregularly arranged ............................................ *P. charcoti*
3b. Papillae simple and round. Supraocular papillae absent ..................................................... *P. prydzensis*

Live animal, dorsal view (Photo: L. Allcock, U. Piatkowski and M. Vecchione).

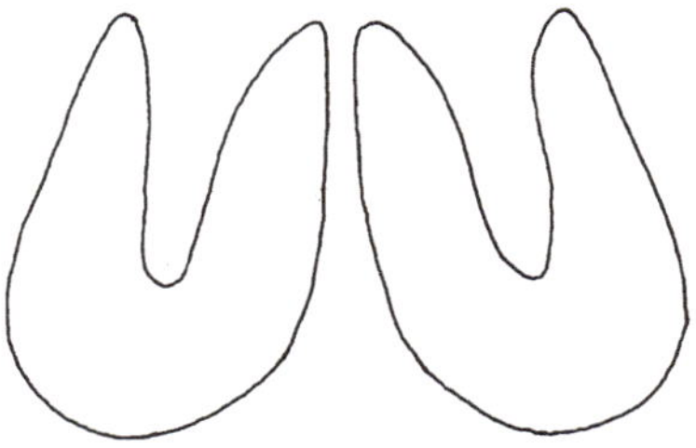

Funnel organ (M. Shea, redrawn from original description).

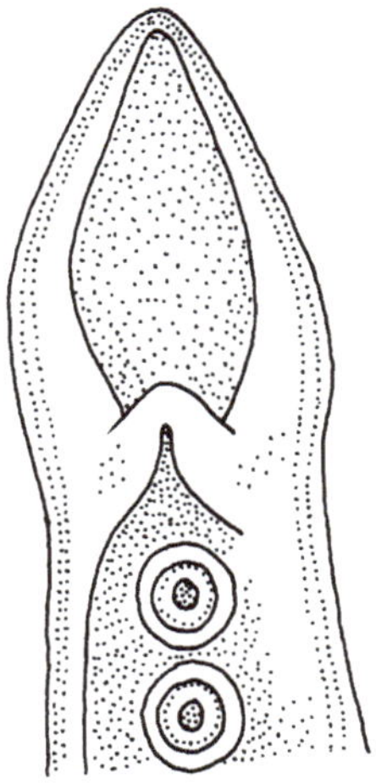

Hectocotylus, oral view (M. Shea, redrawn from original description).

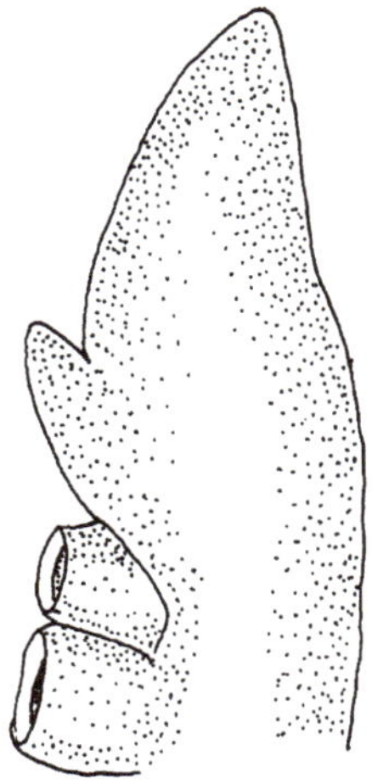

Hectocotylus, lateral view (M. Shea, redrawn from original description).

## *Pareledone charcoti* (Joubin, 1905)

*Eledone charcoti* Joubin, L. (1905). Description de deux Élédones provenant de l'Expédition du Dr Charcot dans l'Antarctique. *Mémoires de la Societé Zoologique de France 18*, 22–31 [22, pl. 3, figs 1, 2].

Type data: holotype MNHN 5-7-1095.

Type locality: Antarctic Peninsula, Booth-Wandel Island [= Booth Island], 65°05′S, 63°55′W.

**DIAGNOSTIC FEATURES**

Mantle spherical, web deep. Funnel organ W- or VV-shaped. Gills with seven to no lamellae per demibranch. Arms short, order usually 3 = 4 >2 >1. Suckers not enlarged. Male hectocotylus (third right arm) shorter than opposite arm, with 26–28 suckers. Opposite arm with 42 suckers. Ligula moderate to large; ligula groove long, well defined and shallow, without marked transverse ridges; calamus distinct, large to very large. Spermatophore storage sac long, storing up to five long (~150 mm) spermatophores. Penis large, diverticulum coiled. Female ovary

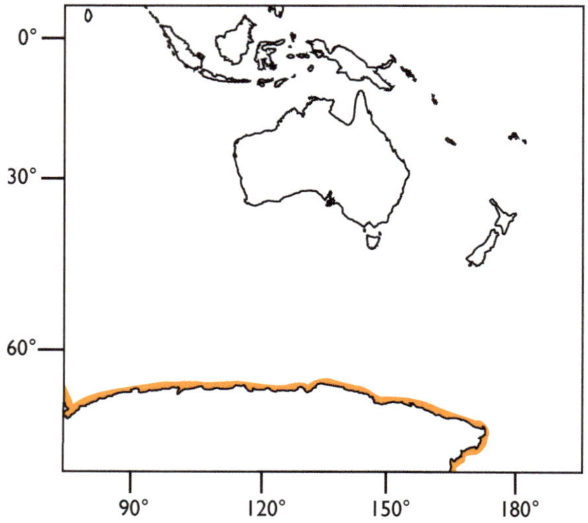


*Pareledone charcoti* distribution.

with up to 80 large (>10 mm) eggs, paired oviducts with large, dark oviducal glands. Ink sac and anal flaps present. Radula with nine elements, rhachidian multicuspid.

*Colour.* Live animals variable, may flush red when disturbed. Ventral mantle creamy white. Often rhomboid of paler marking on dorsal mantle extending from between eyes to posterior mantle. Leucophores may be present between eyes. Freshly dead specimens with dark pink/pale brown hue. Chromatophores sparser ventrally than dorsally, median ventral surface without chromatophores.

*Sculpture.* Papillae simple, irregularly shaped with raised, flat tops; cover entire dorsal surface to, but not beyond, the lateral fold. Papillae on dorsal mantle not arranged in pattern. Lateral fold prominent anteriorly (towards pallial aperture) by slightly raised ridge, sometimes marked by presence of leucophores. Posteriorly no ridge, but dorsal and ventral surfaces delimited by abrupt absence of papillae ventrally. Single, enlarged supraocular papilla over each eye. No other enlarged papillae.

### HABITAT AND BIOLOGY

Off Graham Land 65°05′S, 63°55′W in rock pool, South Shetland Islands 60°50′–63°00′S, 54°44′–61°49′W, 48–392 m; South Orkney Islands 60°43′S, 44°38′W, 16–18 m.

### SIZE

Small, ML to 43 mm.

### REMARKS

Many of the putative *Pareledone* specimens recently examined by Allcock (2005) were found to be misidentified. When she reviewed the papillated *Pareledone* that were all largely being ascribed to *P. charcoti*, Allcock (2007) identified seven new species from the Antarctic Peninsula. Most of the specimens she examined were found in waters shallower than 120 m, while many of the misidentified specimens were from much deeper waters. Her work suggests that the specimens of *P. charcoti* described in Lu and Stranks (1994) do not conform to the redefined species boundaries of *P. charcoti* as their *P. charcoti* has 'fine, rounded and closely set papillae'. The specimens they examined may include more than a single species.

### DISTRIBUTION

Circum-Antarctic. Off Graham Land, 65°05′S, 63°55′W; South Shetland Islands, 60°50′–63°00′S, 54°44′–61°49′W; South Orkney Islands, 60°43′S, 44°38′W. Known depth range intertidal to 392 m.

### REFERENCE

Allcock (2005).

## *Pareledone framensis* Lu & Stranks, 1994

*Pareledone framensis* Lu, C. C.and Stranks, T. (1994). Synopsis of *Pareledone* and *Megaleledone* species, with description of two new species from East Antarctica (Cephalopoda: Octopodidae). *Memoirs of the Museum of Victoria 54*, 221–242 [227, figs 3, 4, 9i–l].

Type data: holotype MV F.65665.

Type locality: East Antarctica, Fram Bank off Mac. Robertson Land 67°11′–67°29′S, 68°50′–70°20′E.

### DISTINGUISHING CHARACTERS

This species is readily distinguished from all other papillated *Pareledone* species by the complete coverage of the ventral mantle with papillae (Allcock *et al.* 2007).

### DIAGNOSTIC FEATURES

Medium-sized *Pareledone.* Mantle spherical, broadly oval; eyes large, prominent. Funnel organ outer limbs three-quarters as long as median limbs. Arms 2.7–3.4 × ML in mature animals, order usually 4 > 3 > 2 > 1. Web shallow, 12–23% of arm length. Arm suckers small. Male hectocotylus (third right arm) shorter than opposite arm, with 44–51 suckers; opposite arm with 61–80 suckers. Ligula medium, 5–8% of third right arm length in mature animals; ligula groove long, well marked and shallow, without transverse ridges; calamus long, pointed. Gills with seven or eight lamellae per demibranch.

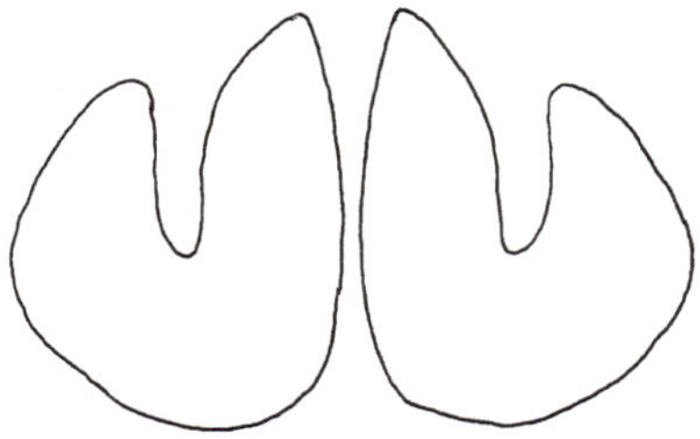

Funnel organ (M. Shea, redrawn from original description).

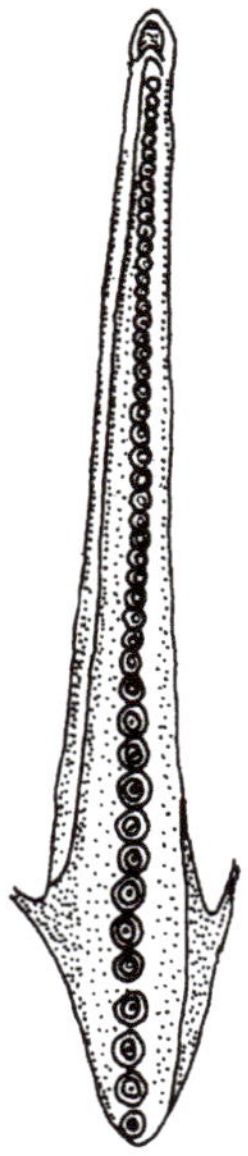

Hectocotylised arm, oral view (M. Shea, redrawn from original description).

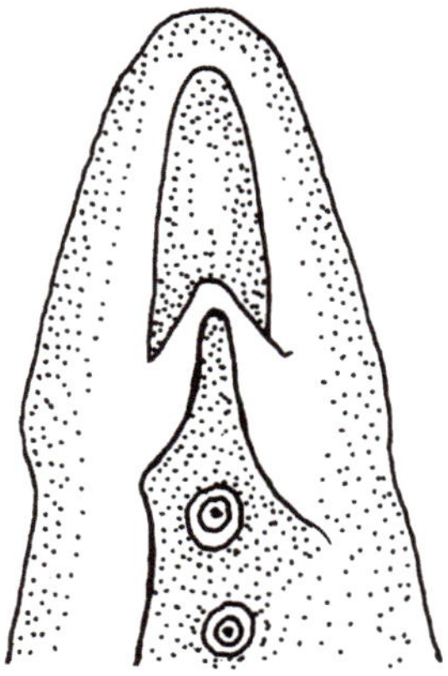

Hectocotylus, oral view (M. Shea, redrawn from original description).

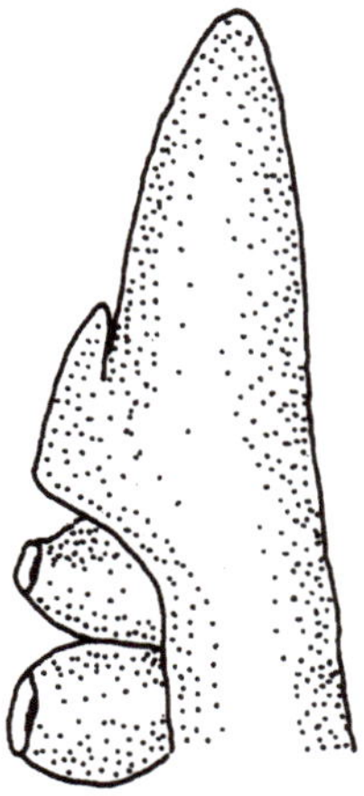

Hectocotylus, lateral view (M. Shea, redrawn from original description).

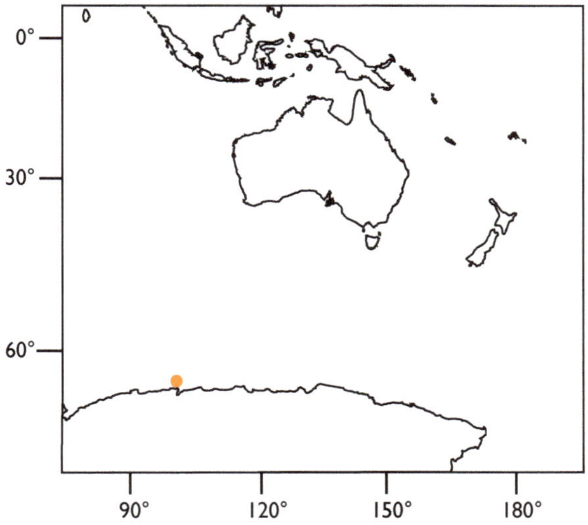



*Pareledone framensis* distribution.

*Colour.* In life, resting animals uniformly yellow-brown to gold dorsally, white to cream-white ventrally. When stimulated, animals become darker in colour, uniformly golden-brown dorsally and yellow-brown ventrally. Large white spots and stripes on mantle: white spots consist of one spot on lateral head, just ventral to each eye; and streak on lateral mantle, posterior to each eye. White transverse bar present between eyes. Dorsal mantle and frontal white spots absent. Lateral head white spots and head bar conspicuous in live animals, often faded and inconspicuous in preserved animals. Ocelli absent.

*Sculpture.* Coarse, rounded and closely set papillae evenly cover dorsum, ventrum and oral surface of web and lateral surfaces of arms. Large unbranched papillae in ocular region, with one very large supraocular papilla, and on dorsal mantle four papillae in diamond pattern. Ventrolateral integumentary ridge absent.

### HABITAT AND BIOLOGY

Inshore species found on mud bottom with pebbles and rocks and among sponges and ascidians. Temperatures -2.2–2.1°C.

### SIZE

Mature males ~50–55 mm ML.

### DISTRIBUTION

Restricted to eastern Antarctica, Fram Bank off Mac. Robertson Land, 67°11′–67°29′S, 68°50′–70°20′E. Known depth range 145–319 m.

### REFERENCES

Lu and Stranks (1994), Allcock (2005).

## *Pareledone harrissoni* (Berry, 1917)

*Moschites harrissoni* Berry, S. S. (1917). Cephalopoda. *Scientific Reports of the Australasian Antarctic Expedition 1911–1914, (series C) 14*(2), 1–39 [24, text figs 21–25, pl. 13 fig. 13, pl. 14 figs 14–16].

Type data: holotype, AM C.40892.

Type locality: Southern Ocean, 65°06′S, 96°13′E.

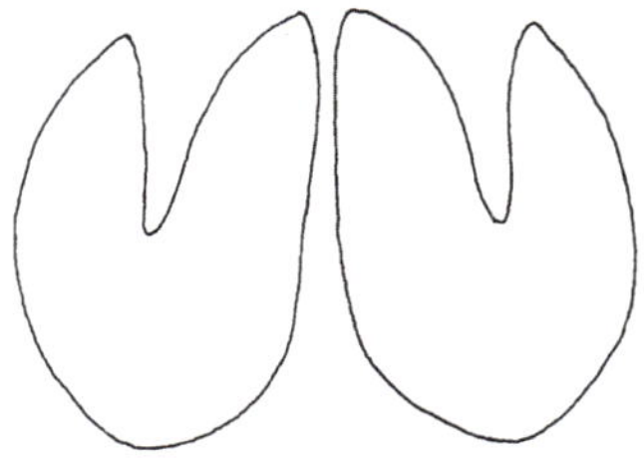

Funnel organ (M. Shea, redrawn from original description).

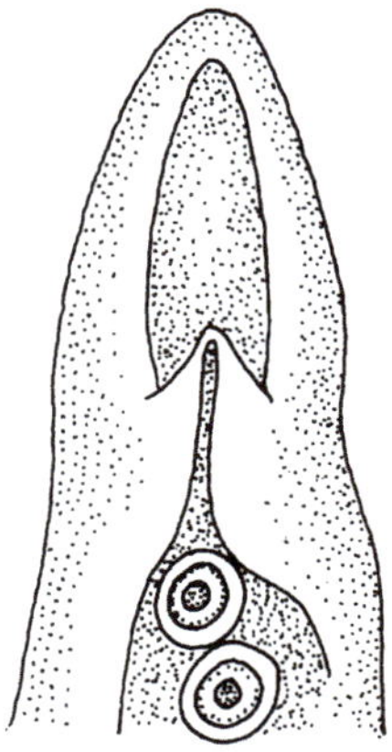

Hectocotylus, oral view (M. Shea, redrawn from original description).

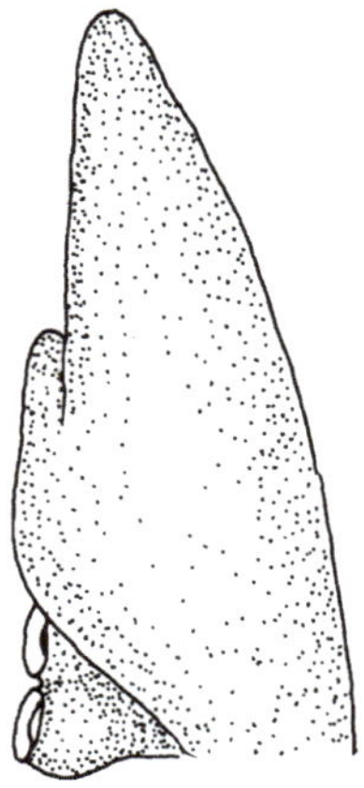

Hectocotylus, lateral view (M. Shea, redrawn from original description).

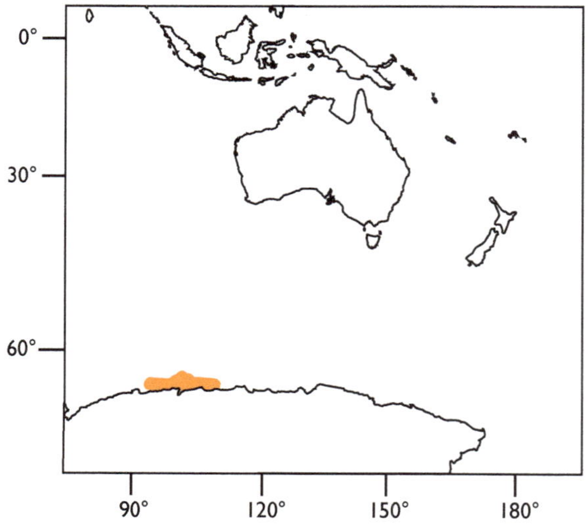


*Pareledone harrissoni* distribution.

### DIAGNOSTIC FEATURES

Mantle spherical to ovoid; skin smooth, loose and wrinkled. Eyes large, project above surface of head. Funnel organ VV-shaped, outer limbs as long as median limbs. Arms long, 1.8–3.6 × ML in mature specimens, order usually 4 > 3 > 2 > 1. Web moderately deep. Arm suckers raised from arm surface, small, without enlargement. Male hectocotylus (third right arm) shorter than opposite arm with 36–50 suckers, opposite arm with 37–79 suckers. Ligula small, 4–11% third arm length, groove long, well defined and shallow, without transverse ridges; calamus long, pointed. Gills with eight or nine lamellae per demibranch. Ink sac present.

Eggs large, 12–14 mm long, 6–7 mm wide.

*Colour.* In life, uniformly pink to purple-pink dorsally, cream-white to pale pink ventrally. When stimulated, animals become darker in colour, uniformly brown-pink to brown-purple dorsally, purple-pink ventrally. Single white spot on mid-dorsal brachial crown; white bar between eyes. Ocelli absent.

*Sculpture.* Fine, rounded and widely scattered papillae on dorsal surface, absent from ventral surface. Large unbranched primary papillae in ocular region, with one large supraocular papilla. Ventrolateral ridge absent.

### HABITAT AND BIOLOGY

Males mature at ~50 mm ML; females mature at ~90 mm ML. Found on mud and sand bottoms with pebbles and rocks among sponges, gorgonaceans and bryozoans at temperatures -2.1–0.6°C.

### SIZE

Up to 100 mm ML.

### REMARKS

This is the largest *Pareledone* found in east Antarctic waters.

### DISTRIBUTION

Restricted to eastern Antarctica from off Mac. Robertson Land (66°59′–67°29′S, 62°49′–68°56′E) off the Amery Ice Shelf, Prydz Bay (66°46′–68°30′S, 71°18′E–78°16′E) and off the Shackleton Ice Shelf (65°06′S–66°44′S, 95°00′E–97°28′E). Known depth range 25–743 m.

### REFERENCE

Lu and Stranks (1994).

## *Pareledone prydzensis* Lu & Stranks, 1994

*Pareledone prydzensis* Lu, C. C. and Stranks, T. (1994). Synopsis of *Pareledone* and *Megaleledone* species, with description of two new species from East Antarctica (Cephalopoda: Octopodidae). *Memoirs of the Museum of Victoria 54*, 221–242 [232, figs 6, 7, 9q–t].

Type data: holotype MV F.65666.

Type locality: Off Amery Ice Shelf, Prydz Bay, 66°48′S, 72°33′E.

### DISTINGUISHING CHARACTERS

*Pareledone prydzensis* is the only species of papillated *Pareledone* with simple round papillae and without enlarged supraocular papillae.

### DIAGNOSTIC FEATURES

Small animals with spherical mantle. Eyes large, do not project far above surface of head. Funnel large, stout, bluntly tapered. Funnel organ VV-shaped, outer limbs three quarters as long as median limbs. Mantle aperture very wide. Arms short, stout, 1.9–2.3 × ML in mature animals, length subequal. Web deep. Arm suckers raised from arm surface, small, without enlargement. Male hectocotylus (third right arm) slightly shorter than opposite arm with 26–29 suckers; opposite arm with 29–36 suckers. Ligula 6–9% of third right arm length in mature animals; groove long, well defined and shallow, without transverse ridges; calamus long, pointed. Gills with six or seven lamellae per demibranch. Ink sac present.

*Colour.* Resting animals grey-purple to pink-purple on dorsal mantle and head; darker pink-purple on dorsal brachial crown, web and arms; cream-white to pink-white ventrally. When stimulated, animals become darker in colour, dark purple brown to black on dorsal arm crown, web and arms and light pink-purple

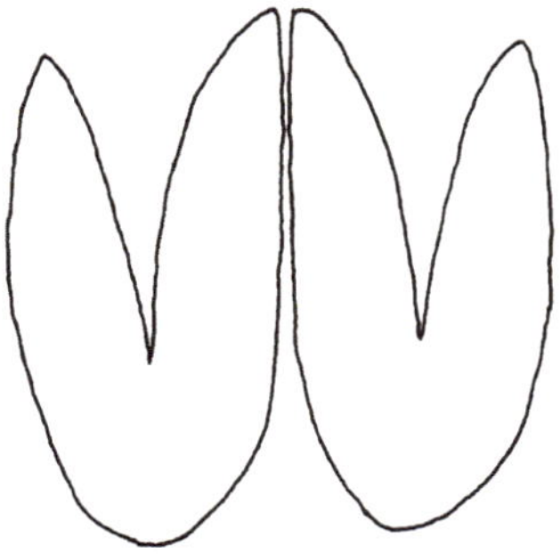

Funnel organ (M. Shea, redrawn from original description).

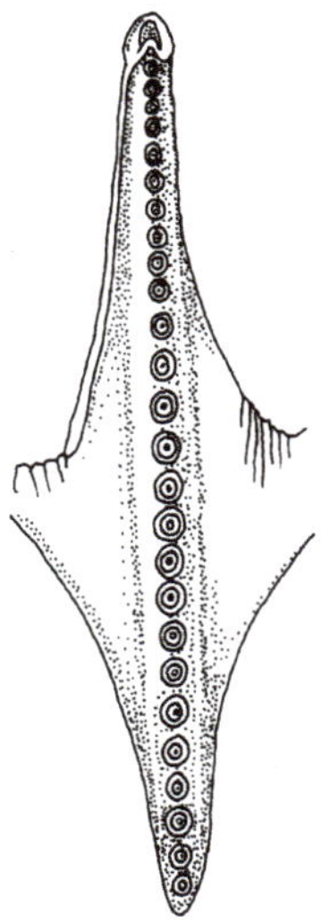

Hectocotylised arm, oral view (M. Shea, redrawn from original description).

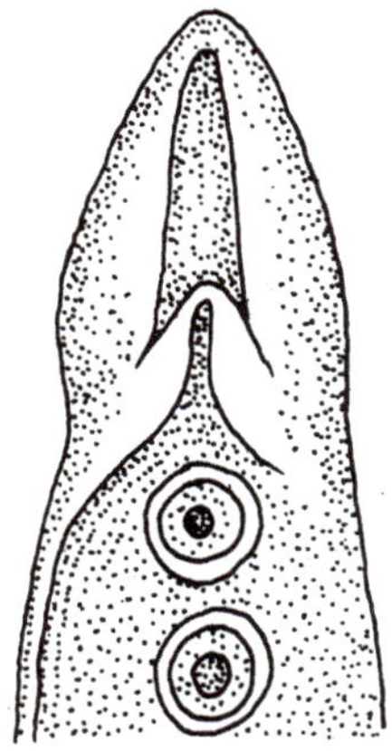

Hectocotylus, oral view (M. Shea, redrawn from original description).

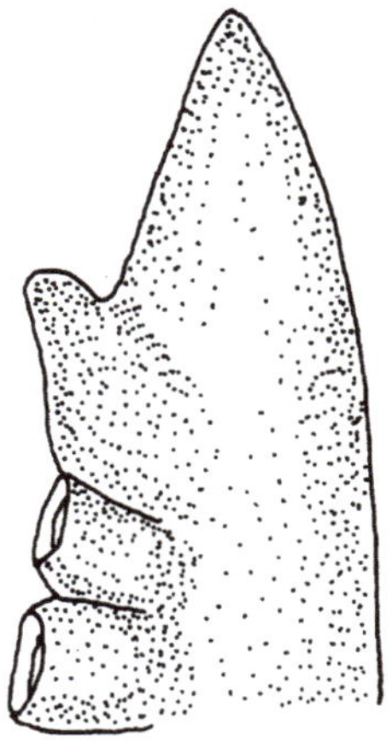

Hectocotylus, lateral view (M. Shea, redrawn from original description).

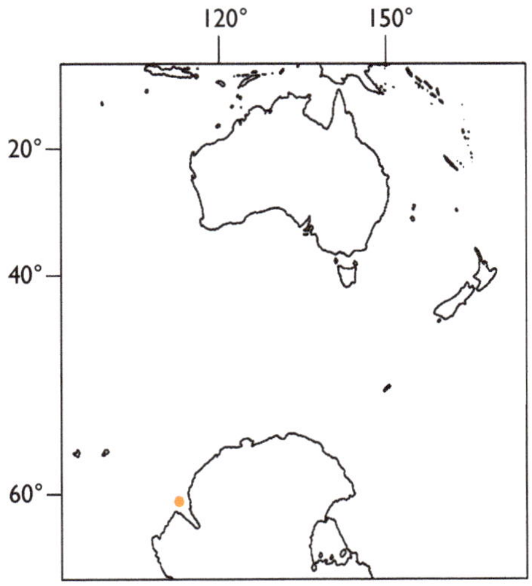


*Pareledone prydzensis* distribution.

ventrally. One white spot on mid-dorsal arm crown, and one broad spot on mid-dorsal posterior mantle. White transverse bar present between eyes. Ocelli absent.

*Sculpture.* Integument with fine, rounded and closely set papillae on dorsal surface (when viewed microscopically); ventral surface smooth; no larger papillae present in ocular region or on dorsal mantle. Ventrolateral integumentary ridge present.

#### HABITAT AND BIOLOGY

Inshore species found on mud and sandy bottoms with rocks, and among sponges and gorgonaceans.

#### SIZE

Males mature at ~25 mm ML.

#### REMARKS

It is likely that this taxon is a species-complex (Allcock *et al.* 2010).

#### DISTRIBUTION

Restricted to eastern Antarctica. Only know from Prydz Bay, off the Amery Ice Shelf, East Antarctica (66°42′–66°48′S, 71°56′–72°37′E). Known depth range 526–676 m.

#### REFERENCE

Allcock (2005).

## Octopodidae d'Orbigny, 1839

#### COMMON NAME

Benthic Octopuses

#### DIAGNOSTIC FEATURES

Muscular octopuses with eight arms. Each arm with biserial suckers. Internal shell absent or reduced to small rod-like stylets. Right arm three in males usually hectocotylised, consisting of a gutter along margin (spermatophoric groove) and modified tip (ligula) used to grip and pass spermatophores into the oviducts of female. Funnel-locking apparatus absent.

#### HABITAT AND BIOLOGY

Bottom-living. Found from intertidal waters down to abyssal depths (> 5000 m). Most (but not all) octopuses hide in lairs or bury in the seabed during the day, emerging at night to feed, mainly on molluscs and crustaceans. Octopuses are solitary predators that typically use their arms to grope into crevices to find food. Captured prey are first injected with poison, which causes paralysis, before being consumed. Some of the larger species are well known to enter lobster traps and consume or remove rock lobsters.

They detect odours on contact using chemosensory cells around the mouth and suckers and can also detect waterborne odours (including those of conspecifics) using receptors in their olfactory organs (Walderon *et al.* 2011).

Males inseminate females using a modified third arm. Sperm packets are deposited directly into the females' mantle cavity and are ready to fertilise the eggs when they are released. Females lay eggs in rock crevices or attach them to sheltered hard surfaces. The young of some species are planktonic and are carried by ocean currents before settling on the bottom. The young of other species settle on the bottom immediately after hatching.

Commercial harvests of octopuses occur throughout the year but vary seasonally and are usually greatest from January to May, with a peak in March. While stock assessment has been studied in Western Australia, where the largest octopus fishery exists, there have been no, or very few, biological studies or stock assessments in other parts of Australia and the long-term effects of harvesting are unknown. The Ocean Trawl Fishery makes the largest commercial harvests, with smaller quantities landed by Prawn Trawl and Ocean Trap and Line. Traps are used extensively in Western Australia.

#### REMARKS

Knowledge of the Australian representatives of this group has increased profoundly over the last two decades thanks to the work of Julian Finn, Mark Norman and Tim Stranks from the Museum Victoria. The former two researchers are excellent and persistent field workers and photographers. Through their work much has been learned about the octopod fauna in this country, with live animal colouration, sculpture and behaviour proving to be important taxonomic tools not previously looked at in such detail.

The classification adopted here follows that in part proposed by Norman and Finn (2001), Norman and Hochberg (2005), and updated and revised in Strugnell *et al.* (2014). These works review the current state of octopus taxonomy, resolve the status of many historical names, and provide revised designations for nominal species in the family. All the nominal species listed below are currently considered to be valid. Some species in Norman and Hochberg (2005) were not assigned to genera; their placement awaits a major revision of the family. If this is the case, a note is made in the 'Remarks' for particular species. Many new species from Australian waters have been recognised in recent years and await description.

Some species, while previously reported from Australian waters, are cited as *nomen dubium* in Norman and Hochberg (2005), so are not treated here. These taxa include *Octopus caerulescens* Blainville, 1826; *O. cornutus* (Owen, 1881); *O. duplex* Hoyle, 1885; *O. pustulosus* Blainville, 1826; and *O. variolatus* Blainville, 1826.

Octopods can be difficult to distinguish, making the construction of a reliable key to representatives of both sexes difficult. To assist in narrowing down probable species it may be necessary to refer to their distributions or depth information. The full list of diagnostic features listed under each species account needs to be consulted to confirm tentative identifications based on keys to genera and species.

### Key to genera of the Octopodidae

1a. Small octopuses with repeated colour pattern of iridescent blue lines or rings over body, arms and webs (markings fade to white in preserved material) .......... *Hapalochlaena*

1b. Small to large octopuses without repeated iridescent markings over body, arms and webs (some species possess a pair of iridescent rings, one on each side of the web between the bases of arms 2 and 3) .......... 2

2a. Ink sac and anal flaps absent .......... 3

2b. Ink sac and anal flaps present .......... 4

3a. Mantle very thin-walled; dark branchial hearts visible through mantle. Arms 6–10 × ML, capable of autotomy (evident as multiple arms severed or regenerating from set plane near arm base). Gills with five or six lamellae per demibranch .......... *Ameloctopus* (*A. litoralis*)

3b. Mantle robust, not thin-walled; branchial hearts not visible through mantle. Arms 4–4.5 × ML, do not autotomise. Gills with seven to nine lamellae per demibranch .......... *Grimpella* (*G. thaumastocheir*)

4a. Arms long, > 4 × ML; capable of autotomy (evident as multiple arms severed or regenerating from set plane near arm base) .......... 5

4b. Arms usually < 4 × ML (except in *Callistoctopus*); not capable of autotomy .......... 6

5a. Arms 4–8 × ML. Two or three enlarged suckers on arms 2 and 3 (slightly enlarged on arms 4) in both sexes. Hectocotylus with calamus. Gills with five to seven lamellae per demibranch. No white U-shaped patch on posterior-dorsal mantle .......... *Abdopus*

5b. Arms 7–10 × ML; arm suckers not enlarged in either sex. Calamus absent. Gills with nine lamellae per demibranch. White U-shaped patch on dorso-posterior mantle .......... *Thaumoctopus* (*T. mimicus*)

6a. Arms long, 5–8 × ML (but without autotomy). Colour pattern of white spots or bars on a red-brown background .......... *Callistoctopus*

6b. Arms usually < 5 × ML. Colour pattern does not include white spots or bars on a red background .......... 7

7a. Web margins along ventral faces of all arms very wide, extend as flared margins along entire length of arms forming loose transparent membranes in preserved specimens .......... *Histoctopus* (*H. zipkasae*)

7b. Web margins along ventral faces of all arms not markedly wide .......... 8

8a. Lateral mantle skin ridge present .......... 9

8b. Lateral mantle skin ridge absent .......... 10

9b. Ligula of hectocotylus with deep transverse ligula groove containing small tooth-like papillae. Mantle without two pairs of darks spots. Three irregular transverse bands over arm crown .......... *Galeoctopus* (*G. lateralis*)

9a. Ligula of hectocotylus without deep transverse groove, tooth-like papillae absent. Two pairs of dark spots on mantle of live and well-preserved animals. Arm crown without irregular transverse bands .......... *Scaeurgus*

10a. Arms typically 2–3 × ML. Distinctive patch and groove skin sculpture extends onto oral surface of web. Pigment extends to oral surface of web. Stylets absent .......... *Amphioctopus*

10b. Arms typically 3–5 × ML. Sculpture and pattern does not extend to oral surface of web. Stylets present .......... *Octopus*

## *Abdopus* Norman & Finn, 2001

### DIAGNOSTIC FEATURES

Small- to moderate-sized octopuses (< 70 mm ML). Eyes moderate-sized, slightly raised. Funnel organ W-shaped with short lateral limbs. Arms long (~4–8 × ML) capable of autotomy (within 3–10 suckers from the mouth) and regeneration. Arm lengths subequal, lateral pairs longer than dorsal pair; formula 3 and/or 4 > 2 > 1; web margins along arms present. Web sectors unequal in length, lateral webs deeper than dorsal webs; 9–20% longest arm; web margins extend to arm tips; formula C/D/E > B > A. Normal arms with 125–260 suckers; 14–30% ML. Mature males with 2–12 enlarged suckers on arms 2 and 3 (slightly enlarged on arm 4). Maximum sucker diameter 14–30% ML. (Suckers slightly enlarged on same arms in females.) Hectocotylised third right arm long, ~80% length of opposite arm, with 90–175 suckers. Ligula small (< 6.5% AL) and narrow; calamus 12–33% ligula length. Gills with five to seven lamellae per demibranch. Ink sac and anal flaps present; salivary glands same length as buccal mass. Terminal organ (penis) linear with simple, small diverticulum; ~< 30% ML. Spermatophores small (~29–65% ML) and narrow (1.3–3.3% spermatophore length) with highly coiled ejaculatory apparatus (40–80 cord whorls) occupying ~30% of spermatophores' length. Small, chitinous stylets present.

*Colour.* Colouration extremely cryptic: cream to dark chocolate brown with white spots present on mantle, arm crown and down arms; complex, mottled. Dorsal white spots present, distinct.

*Sculpture.* Skin highly sculptured in fine, irregular patch-and-groove texture with large, often branched, papillae over dorsal surfaces and arms; four primary papillae in diamond shape on dorsal mantle; one or more supraocular papillae; one large and robust.

### HABITAT AND BIOLOGY

Typically intertidal and shallow subtidal, day-active; found on coral reefs and rubble. Hatchlings probably large and benthic with limited dispersal.

### SIZE

Up to 70 mm ML.

### REMARKS

Norman and Finn (2001) defined the subgenus and subsequently elevated it to generic status.

### REFERENCES

Norman and Finn (2001), Norman and Hochberg (2005), Norman *et al.* (2014).

### Key to *Abdopus* species

Due to insufficient information, *A. tenebricus*, known from only two type specimens, from Bowen, Queensland is not included in this key.

1a. Web depth 5% of longest arm. Normal arms with ~211 suckers. Hectocotylus with 121–137 suckers. Mature males with two or three enlarged suckers on arms 2 and 3 .............................. *A. capricornicus*

1b. Web depth ~15% of longest arm. Normal arms with 220–260 suckers. Hectocotylus with 150–170 suckers. Mature males with three to six enlarged suckers on arms 2 and 3 ........................ *A. aculeatus*

## *Abdopus aculeatus* (d'Orbigny, 1835)

*Octopus aculeatus* d'Orbigny (1835). Voyage dans l'Amerique méridionale, t. V. Mollusques. In Ferussac and D'Orbigny, 1834–1848 Histoire naturelle générale et particulière Cephalopodes Acetabulifères vivants et fossils (Eds M. M. Ferrussac and M. A. d'Orbigny) pp. ix + 361: 53, Poulpe pls 7, 8, 23. (J. B. Bailliere: Paris.) [53].

Type data: holotype MNHN Syntype 4-7-927.

Type locality: Philipines: Manila.

### COMMON NAME

Prickly Octopus

### DIAGNOSTIC FEATURES

Mantle ovoid, length up to ~70 mm. Eyes slightly stalked. Funnel organ W-shaped, with broad limbs; limbs approximately equal in length. Arms long, 5–7 × ML, lateral arms longest, formula typically 3 > 4 = > 2 > 1; arm autonomy and regeneration at level of third to seventh proximal sucker. Web depths ~15% of longest arm length; deepest on lateral arms, shallowest between dorsal arms. Sucker diameter up to 16% ML; muscular, flared, with thick rings and distinct radial cushions; biserial; 220–260 suckers on normal arms in larger animals and 150–170 on male hectocotylised third right arm; three to six enlarged suckers on arms 2 and 3 of mature males; largest at level of ninth to twelfth proximal sucker. Ligula small, 2–3% of arm length in mature males, conical with shallow groove and fine transverse creases; calamus blunt and slightly raised (22–36% of ligula length). Spermatophore groove

Live animal, various postures colour pattern and sculpture (R. Caldwell).

Cluster of eggs bring brooded by the female (R. Caldwell).

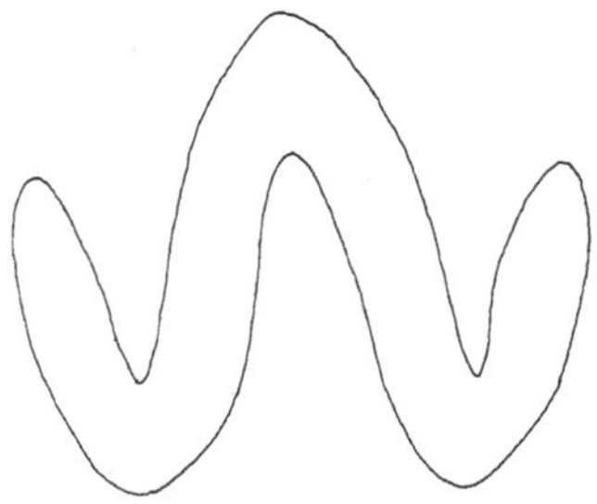

Funnel organ (M. Shea, redrawn from original description).

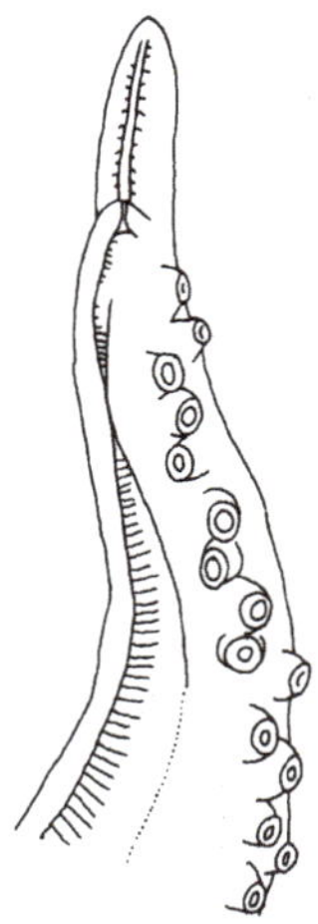

Hectocotylised arm, oral view (M. Shea, redrawn from original description).

well developed, wide, with fine transverse ridges. Gills with six or seven lamellae per demibranch. Eggs small and numerous, capsule length to 3.0 mm.

*Colour.* Live animals mottled brown, green and grey dorsally and laterally. Alarm display: pale green-grey base colour with regular narrow transverse black bars along entire arm length. Pair of transverse white spots visible on dorsal mantle in most colour patterns. Small white spots on dorsal arm crown; white spots between regular dark bars on arms. Preserved specimens pale grey-green with arm bars sometimes visible.

*Sculpture.* Live animals with reticulations forming an irregular patch and groove arrangement. Primary and secondary branched papillae on arm crown and mantle. Four largest primary papillae in crucifix arrangement on dorsal surface of mantle. Single large papilla above and

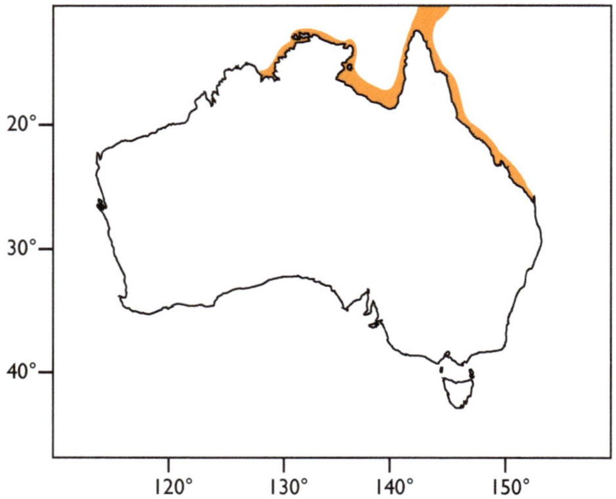


*Abdopus aculeateus* distribution.

slightly posterior to each eye; it can be erected as a horn. Ventral arm crown bordered by broken skin ridge.

### HABITAT AND BIOLOGY

Intertidal, forages on exposed rock and coral rubble reefs and seagrass beds during low tides primarily during the day and has been observed foraging on moonlit low tides at night. This species uses excellent camouflage. Flatfish mimicry occurs as has been reported for other octopuses. Huffard (2006) studied locomotion in this species and found that speed and variation in body form differed during the three dominant modes of movement: crawling, bipedal walking and swimming. Swimming was faster when *A. aculeatus* was dorsoventrally compressed than when the head is raised (which would be expected), but bipedal locomotion can be just as fast as swimming and can also incorporate crypsis and polyphenism. Based on observations of captive animals, copulation occurs 'remotely' between partners, with males extending the hectocotylised arm to deliver spermatophores but otherwise not mounting or positioning close to the female. The size of the eggs suggests that the young are likely to be planktonic.

*Interest to fisheries.* This is one of the most common octopuses sold in the home aquarium trade worldwide.

### SIZE

ML up to 70 mm.

### REMARKS

This species treatment follows Norman and Finn (2001) and is based on the examination of Australian specimens. The species was originally described from the Philippines but only minor morphological differences were noted between the Australian and non-Australian specimens examined by these authors.

### DISTRIBUTION

Taiwan and the Philippines, Vietnam to Australia (tropical Queensland). Known depth range 0–17 m.

### REFERENCES

Norman and Finn (2001), Norman *et al.* (2014).

## *Abdopus capricornicus* Norman & Finn, 2001

*Abdopus capricornicus* Norman, M. D. and Finn, J. (2001). Revision of the *Octopus horridus* species-group, including erection of a new subgenus and description of two member species from the Great Barrier Reef, Australia. *Invertebrate Taxonomy 15*, 13–35 [24, figs 5–8].

Type data: holotype MV F.67020.

Type locality: Australia: Queensland, Capricorn Bunker Group, Tryon Island, 23°15′S, 151°47′E.

### DISTINGUISHING CHARACTERS

*Abdopus capricornicus* differs from *A. aculeatus* (d'Orbigny, 1835) in the following traits (with those for *A. capricornicus* listed first): web depth (~15% *v.* 8.7–11.8% of longest arm); typically lower hectocotylised arm sucker counts (121–137 *v.* 150–170); proportionally shorter ligula (1.4–1.6% *v.* 3.8–6.4% of arm length), longer hectocotylised arm relative to opposite arm (82.4% *v.* 58.8–75% of opposite arm) and longer spermatophores (11.5–13.5 mm *v.* 8.3–9.2 mm) that are smaller relative to ML (28.8–31.9% *v.* 34.6–37.9%) with more sperm cord whorls (63–76 *v.* 40–52).

### DIAGNOSTIC FEATURES

Eyes slightly stalked. Funnel organ W-shaped, lateral limbs shorter than median limbs. Arms ~4.5–5 × ML. Arm autotomy at level of fourth to fifth proximal sucker. Webs ~15% of arm length. Suckers 19–25% ML, biserial. Mature animals with 211 suckers on normal arms (based on single specimen) and 121–137 on male hectocotylised arm. Suckers flared with scalloped rim and distinct radial cushions. Two or three enlarged suckers in mature males at level of eighth to twelfth sucker row on arms 2 and 3. Third right arm in males hectocotylised, slightly shorter than opposite arm. Ligula small, (1.4–1.6% arm length), narrow with shallow groove and many fine transverse creases; calamus

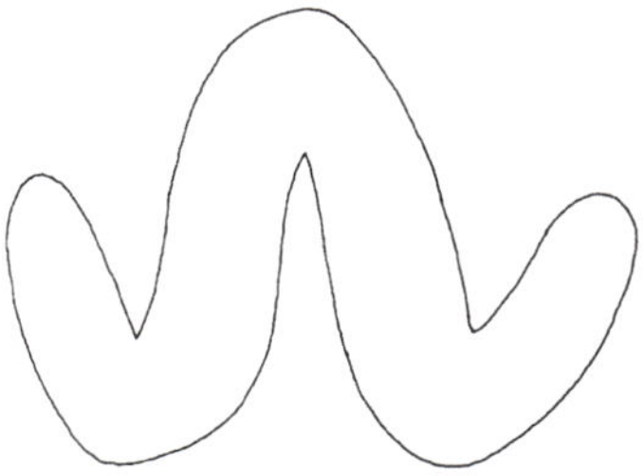

Funnel organ (M. Shea, redrawn from original description).

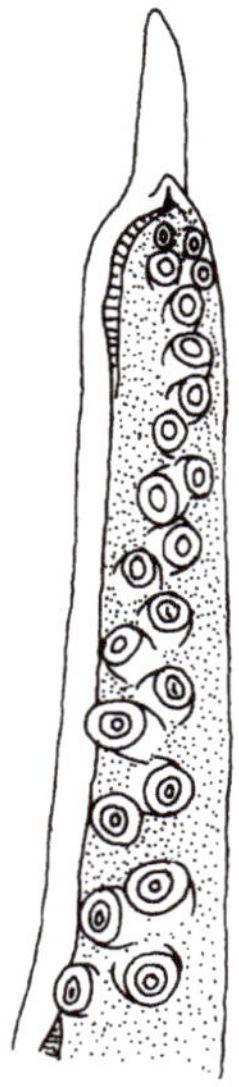

Hectocotylised arm, oral view (M. Shea, redrawn from original description).

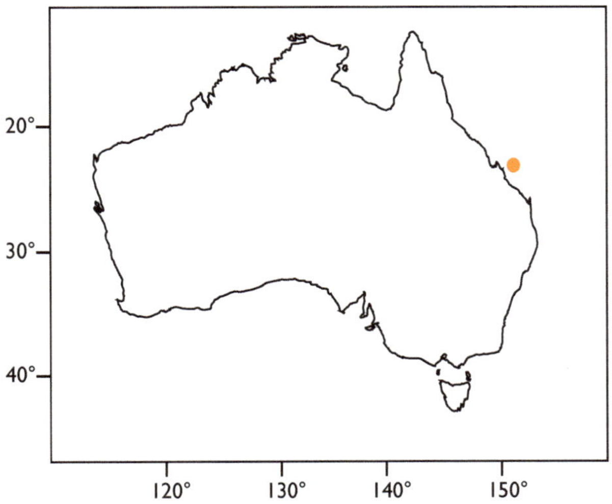


*Abdopus capricornicus* distribution.

short and blunt; spermatophore groove well developed, wide with fine transverse ridges; spermatophore guide poorly developed with shallow notch. Sperm cord in 63–76 regular whorls. Gills with six or seven lamellae per demibranch.

*Colour.* Resting colour patterns pale green, brown or grey. Dorsal white spots present on mantle and frontal white spots on mantle and arm crown. Sometimes with transverse black bars along length of arms with white spots between these. Distinctive alarm display of large pale spots bound by dark borders; sucker rings often darkened on at least dorsal arms. Preserved specimens pale grey-green, with arm bars sometimes visible.

*Sculpture.* Reticulations form fine and irregular patch and groove system over body and arms. Four large primary papillae on dorsal mantle in crucifix arrangement. Large papilla dorsal and slightly posterior to eye.

#### HABITAT AND BIOLOGY

This species has been observed foraging over intertidal coral bedrock and coral rubble flats during daylight low tides.

#### SIZE

ML up to 42 mm.

#### DISTRIBUTION

To date only known from intertidal reefs surrounding Tryon Island in the Capricorn Bunker island group, southern Great Barrier Reef, Australia.

#### REFERENCES

Norman and Finn (2001), Norman *et al.* (2014).

### *Abdopus tenebricus* (E. A. Smith, 1884)

*Octopus tenebricus* Smith, E. A. (1884). Mollusca, pages 34–116, plates 4–7. Report on the Zoological Collections made in the Indo-Pacific Ocean during the Voyage of H.M.S. 'Alert' 1881–2. 684 pp. London: British Museum [35, pl. 4 figs *B*–*B*$^3$].

Type data: lectotype and paralectotype BMNH 1881.11.10.13–14.

Type locality: Australia: Queensland, Port Denison, 20°02′S, 148°15′E.

#### DISTINGUISHING CHARACTERS

According to the original species description, *Abdopus tenbricus* is most similar to *A. aculeatus* (d'Orbigny, 1835) but can be distinguished by the colour, in having different relative arm lengths, 'the narrowness of their lower [oral] surface' (Smith, 1884: 35), the greater prominence of the

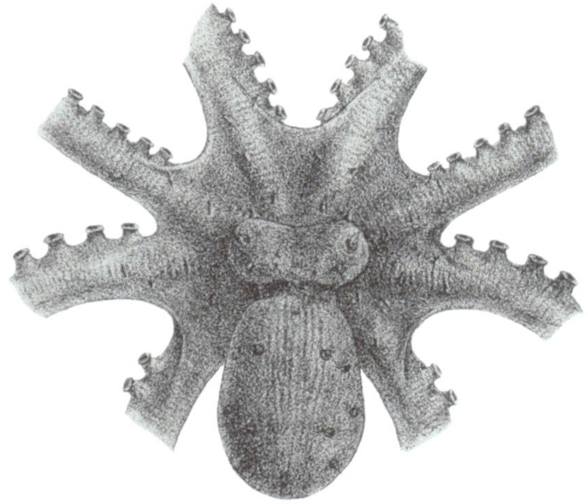

Dorsal view (image *B* from original description).

Oral view of arm crown (image $B^1$ from original description).

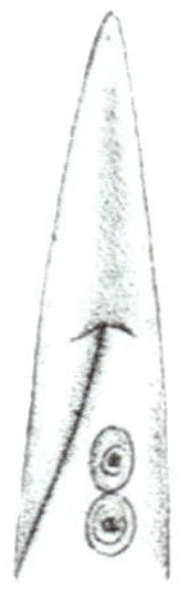

Hectocotylus, oral view (image $B^2$ from original description).

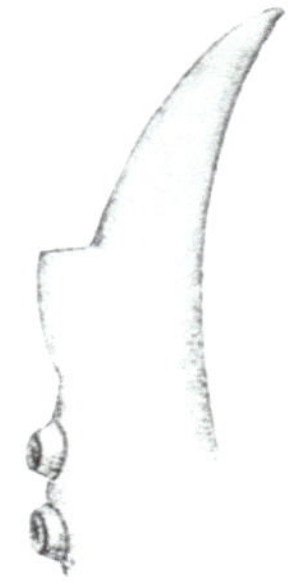

Hectocotylus, lateral view (image $B^3$ from original description).

suckers, the regularity of the uniserial row of four proximal suckers and in having fewer papillae.

## DIAGNOSTIC FEATURES

Mantle ovoid. Head as wide as body; eyes prominent, bulbous. Dorsal arm pair shortest, arms 2 and 3 equal in

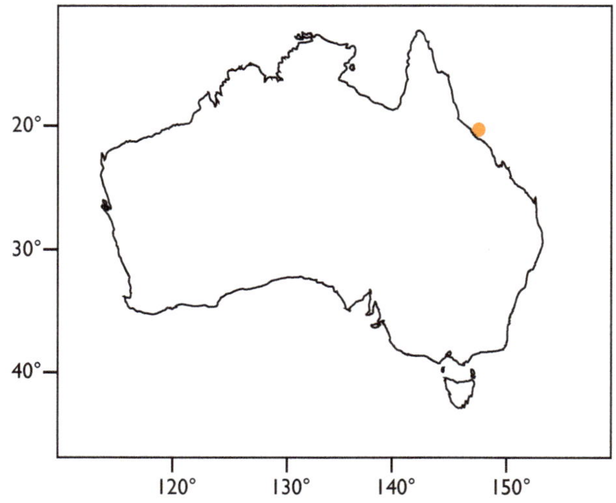


*Abdopus tenebricus* distribution.

length and ventral arm pair slightly shorter. Arm web shallow, extends along arms as narrow ridge. Oral arm surfaces narrow. Arms with 4 uniserial suckers proximally, followed distally by biserial suckers; ~120 suckers on longest arms. Suckers gradually enlarge to about the tenth largest sucker row (diameter ~1.3 mm); suckers of this size present for short distance towards distal arm tip after which they diminish gradually in size.

*Colour.* Uniform dark purplish brown.

*Sculpture.* Dorsal mantle, lateral mantle and head with scattered papillae. Prominent supraocular papilla, and 'a small one in front of each eye and a similar one just behind them appear to be constant in position' (Smith 1884: 35). Largest papillae branched. Arm webs with papillae.

## HABITAT AND BIOLOGY

Sand and rock.

## SIZE

ML up to ~20 mm (measurement from base of funnel to posterior mantle according to the type description is 13 mm).

## DISTRIBUTION

Known only from the two type specimens from Port Denison (Bowen), Queensland. Depth range 5–8 m.

## REFERENCE

Norman *et al.* (2014).

## *Ameloctopus* Norman, 1992

### DIAGNOSTIC FEATURES

Mantle thin-walled; dark branchial hearts visible through mantle. Small head; tiny eyes. Funnel organ greatly reduced to two to four small oval pads. Arms extremely long, > × 6 (up to × 10) ML; capable of autotomy at set plane at level of fourth or fifth sucker pair (counting from proximal end of arm). Gills with five or six lamellae per demibranch. Ink sac and anal flaps absent. Terminal organ without diverticulum. Spermatophores with bulbous, inflated sperm reservoir. Stylets absent.

*Colour.* Arms with banded brown and while colour pattern.

*Sculpture.* Low skin sculpture without papillae.

## *Ameloctopus litoralis* Norman, 1992

*Ameloctopus litoralis* Norman, M. D. (1992). *Ameloctopus litoralis*, gen. et sp. nov. (Cephalopoda: Octopodidae), a new shallow-water octopus from tropical Australian waters. *Invertebrate Taxonomy 6*, 567–582 [569, figs 1–5].

Type data: holotype MV F.57917.

Type locality: Australia: Queensland, Orpheus Island, Pioneer Bay, 18°36′S, 146°29′E.

### COMMON NAME(S)

Banded Drop-arm Octopus, Banded String-arm Octopus

### DIAGNOSTIC FEATURES

Small octopus (up to 30 mm ML); mantle wall thin and semi-translucent; mantle ovoid to greatly elongated. Eyes small, not protruding. Pallial aperture positioned midway along ventral surface of mantle. Funnel organ rudimentary, consisting of two to four small pads: a pair of pads anteriomedially and pair of posterior-lateral pads (.'. arrangement). Arms approximately equal length (in rare intact specimens), greatly elongated (5–9 × ML). Arms sever at base as a decoy for predators, around tenth to twelfth proximal sucker row, close to web margins. Webs very shallow. More than 180 suckers on intact normal arms of larger animals; no enlarged suckers. Suckers biserial and small relative to arm length, ~16% ML; elevated; radial cushions distinct. Male hectocotylus (third right arm) ~20–30% length of opposite arm with 21–38 suckers; no enlarged suckers; arm long in submature males and lacking ligula, but spermatophore groove visible as narrow band of unpigmented tissue extending

Live animal (M. Norman).

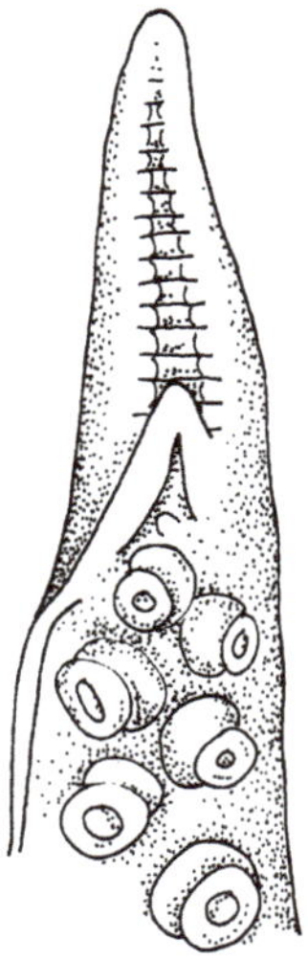

Hectocotylus (A. Reid, redrawn from Norman *et al.* 2014).

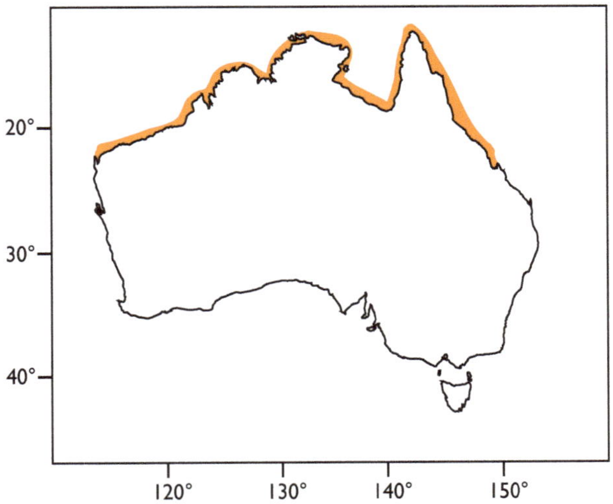


*Ameloctopus litoralis* distribution.

along arm for about one-fifth of its length from web margin. Mature males sever arm at around the fortieth sucker (~20% from proximal end of arm along length) and develop hectocotylus from the arm stump. Ligula small and robust; 12–20% arm length; tapering to dull point with distinct calamus (22–38% ligula length); margins fleshy, copulatory groove shallow with 10–12 transverse ribs. Terminal organ linear, 13–17% ML. Gills with five or six lamellae per demibranch. Posterior salivary glands greatly enlarged. Salivary glands ~2–3 × length of buccal mass. Spermatophores ~38–45% ML, wide (4.8–7.8% spermatophore length), and produced in low numbers (up to 70). Spermatophores with swollen sperm reservoir; ejaculatory apparatus a few simple coils. Sperm cord irregularly packed, not forming regular whorls. Eggs large, 10 mm long (to ~37% ML) and produced in low numbers.

*Colour.* Grey to purple-brown with hearts visible through thin-walled mantle. All arms with regular purple-brown and cream bands along entire length. Blue-black subdermal pigmentation around eyes gives a 'bruised' effect. Dorsal white spots absent.

*Sculpture.* Skin smooth, without papillae or obvious sculpturing. No primary papillae on dorsal mantle, lateral skin ridge absent. Supraocular papillae absent.

#### HABITAT AND BIOLOGY

Eggs have been observed in several stages of development in the mature ovary, suggesting egg laying occurs over an extended period. The size of the eggs and development of the arms in hatchlings suggests that they are benthic. The species is found on intertidal sand, mud and rubble reef flats. Most specimens have been collected from under rocks and rubble where they have excavated lairs. They have also been observed foraging on very low tides at night, sometimes in very shallow water over intertidal sand flats and rubble. They often feed by extending one or more arms to forage among crevices while remaining in the lair. Captive animals have been shown to remain within their shelter and shoot out the long arms to capture passing prey. Arm autonomy is probably a defensive behaviour, with a writhing arm serving as a decoy to predators.

*Interest to fisheries.* None.

#### SIZE

Maximum ML 30 mm. Sexes approximately equal in size.

#### REMARKS

Norman (1992) has postulated that the reduction of the funnel organ in this species is possibly related to the loss of the ink sac as it is thought that the funnel organ secretes mucus to bind expelled ink to form a pseudomorph as a defensive decoy. Perhaps arm autonomy has taken over this role in *Ameloctopus litoralis.*

Young males have a full length, undifferentiated third right arm. The development of the hectocotylus following the autotomy of the third right arm is an unusual trait in this species, giving these octopods full functional use of eight arms for foraging and other functions until sexual maturity.

The Western Australian populations differ in the form of the funnel organ, with the central segments elongate rather than ovoid. Further investigation is needed to fully determine the status of this population.

#### DISTRIBUTION

Tropical coast waters of northern Australia from the southern end of the Great Barrier Reef (One Tree Island, 23°30'S, 116°32'E) to north-western Western Australia (20°28'30"S, 116°32'E). Appears to be restricted to the intertidal zone.

#### REFERENCES

Norman (1992), Norman and Hochberg (2005), Norman *et al.* (2014).

### *Amphioctopus* Fischer, 1882

#### DIAGNOSTIC FEATURES

Funnel organ W-shaped. Arms relatively short, 2–3 × ML. Lateral arm webs deep (~20% arm length); dorsal web very shallow. Hectocotylised arm slightly shorter than opposite arm. Ligula 5–9% of arm length. Gills with 6–11 lamellae per demibranch. Ink sac and anal flaps present. Skin sculptured with regular low, round papillae or regular raised patches defined by distinct grooves. Oral surface of shallow dorsal web with same sculpture and colour pattern as the dorsal surfaces of the body and arms (i.e. fully sculptured and pigmented to the mouth). May be ocellate or non-ocellate and may have dark lines along dorsal edges of arms 1–3. With diamond of longitudinal ridges on dorsal mantle. (Armed spermatophores may be a generic character.)

### HABITAT AND BIOLOGY

Appear to be active primarily at dusk and dawn, when they forage for bivalves, gastropods and hermit crabs.

### SIZE

Small- to medium-sized octopuses (typically < 80 mm ML).

### REMARKS

In the past, this genus was called the *Octopus aegina* species-group by Robson, 1929.

### DISTRIBUTION

Tropical and warm temperate regions.

### REFERENCES

Norman and Sweeney (1997), Norman *et al.* (2014).

## Key to *Amphioctopus* species

1a. Ocelli present, distinct............................................2
1b. Ocelli absent, or present only as an irregular dark spot in same position as ocelli................................3
2a. Ocelli plain, without iridescent ring....*A. exannulatus*
2b. Ocelli contain iridescent ring.................................4
3a. No dark stripe along leading edge of arms 1–3. No enlarged suckers on arms of mature males. Gills with eight or nine lamellae per demibranch ........... .....................................................*A.* cf. *kagoshimensis*
3b. Dark stripe along leading edge of arms 1–3 contrasts with pale suckers. Mature males with four or five slightly enlarged suckers on arms 2 and 3. Gills with 9–11 lamellae per demibranch ........... *A. marginatus*
4a. Ocelli with small, simple pink/purple ring within dark spot. Short, longitudinal brownish bar through eye .......................................................*A. rex*
4b. Ocelli with obvious iridescent ring within dark spot. No short, longitudinal brownish bar through eyes ...............................................................................5
5a. Dark longitudinal lines along leading edge of arms 1–3. Arms without widely spaced, dark transverse bars. Cluster of dark spots dorsal to each eye in flower pattern. Gills with 9–11 lamellae per demibranch .........................................................*A. mototi*
5b. No dark, longitudinal lines along leading edge of arms 1–3. Arms with widely spaced dark, transverse bars. No cluster of dark spots dorsal to each eye in flower pattern ............................ *A. polyzenia*

## *Amphioctopus exannulatus* (Norman, 1993)

*Octopus exannulatus* Norman (1993). *Ocellate octopuses* (Cephalopoda: Octopodidae) of the Great Barrier Reef, Australia: Descriptions of two new species and redescription of *Octopus polyzenia* Gray 1849. *Memoirs of the Museum of Victoria. 53*(2), 309–344 [321, text figs 1e, 7–10, 15c].

Type data: holotype MV F.60143.

Type locality: Australia: Queensland, Moreton Bay, 5 km north of east corner of Mud I, ~27°23′S, 153°15′E.

### COMMON NAME

Plain-spot Ocellate Octopus

### DIAGNOSTIC FEATURES

Moderate-sized robust octopus (to ~50 mm). Funnel organ limbs broad, outer limbs slightly shorter than medial limbs. Arms short (2–3 × ML); dorsal arms shortest, formula 4 > 3 > 2 > 1 or 3 > 4 > 2 > 1. No arm autotomy. Lateral webs deepest, dorsal web shallow, pigmented and sculptured on oral surface, web margins extend to arm tips. Suckers moderate in size, slightly elevated with distinct radial cushions, rim scalloped and often incurved. Normal arms with 120–190 biserial suckers in submature and mature animals (~150 in females; slightly fewer in males). Mature males with two to four enlarged suckers on arms 2 and 3 starting at level of sixth proximal sucker. Gills with seven or eight lamellae per demibranch. Male hectocotylus (third right arm) slightly shorter than opposite arm with 62–77 suckers; spermatophore guide deep with elevated square papillae. Copulatory organ small, ligula conical, 3–6% of arm length, with blunt tip; groove wide and shallow with raised longitudinal midrib; calamus medium-sized, 20–35% of ligula length, slightly raised. Terminal organ large and recurved. Spermatophores very long, up to ~1.5 × ML and produced in low numbers. Eggs small, up to 4 mm long and produced in large numbers (> 5000).

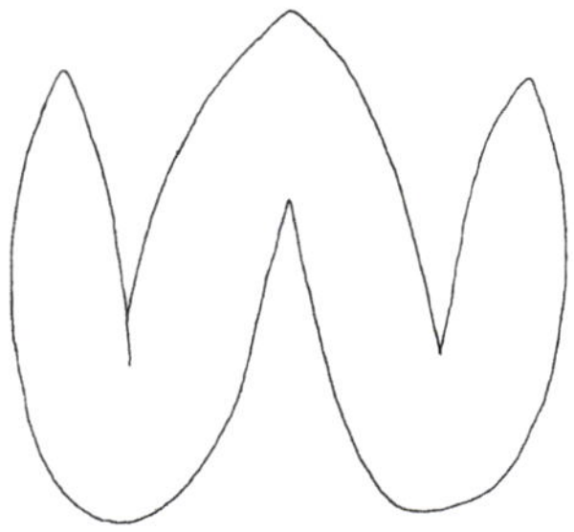

Funnel organ (M. Shea, redrawn from original description).

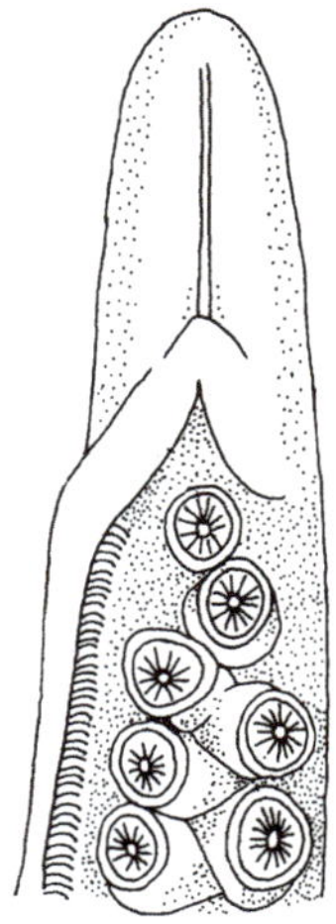

Hectocotylus (M. Shea, redrawn from original description).

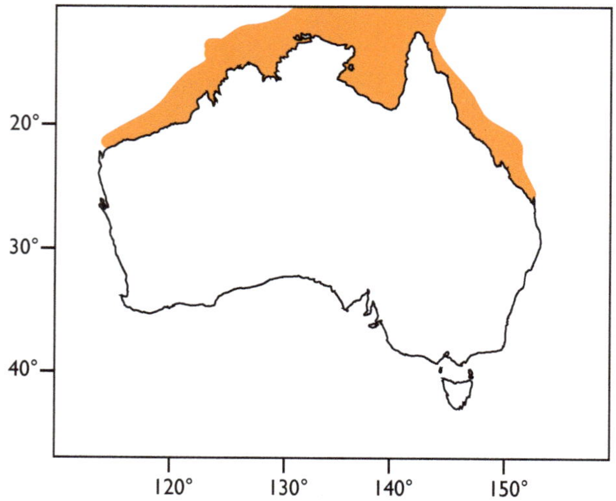


*Amphioctopus exannulatus* distribution.

*Colour.* Base colour variable from grey-white, orange-brown or dark chocolate brown. Alarm pattern consists of four dark stripes over pale base colour. Stripes extend along dorsal mantle and arm crown; inner pair of stripes extend along length of body, joining at midpoint above eyes, continuing as a single medial stripe to margin of dorsal web. Lateral stripes commence from midway along lateral mantle, pass through eye and extend on arm crown to margin of dorsolateral web. All arms with dark leading edges. Plain, dark, well defined ocellus between bases of arms 2 and 3 without iridescent ring. Transverse pair of dorsal and frontal white spots.

*Sculpture.* Skin covered dorsally and ventrally (excluding arm crown) with regular oval to round papillae; ventral papillae almost twice the diameter of dorsal papillae; four longitudinal raised skin ridges on dorsal body in diamond arrangement. Single large branching papilla behind each eye, surrounded by additional smaller papillae.

#### HABITAT AND BIOLOGY

Shallow sand and muddy substrates to depths of at least 84 m. Inhabits open bottom inter-reef substrata and seagrass beds.

*Fisheries importance.* Small harvest in Australia as bait. Caught as by-catch in prawn trawl fisheries.

#### SIZE

Up to 50 mm ML.

#### DISTRIBUTION

Indo-Malayan area from the Philippines to Thailand and northern Australia from Shark Bay, Western Australia 24°49'S, 113°33'E to the Gulf of Carpentaria and south to Moreton Bay, southern Queensland 27°15'S, 153°15'E.

#### REFERENCE

Norman *et al.* (2014).

### *Amphioctopus* cf. *kagoshimensis* Norman & Kubodera, 2006

#### COMMON NAME

Southern Star-eye Octopus

#### DIAGNOSTIC FEATURES

Small, robust species. Gills with eight or nine lamellae per demibranch.

*Colour.* Grooves on body dark, forming mesh-like pattern on arm crown, dorsal arm surfaces and oral surface of dorsal web. Dark longitudinal eye bar

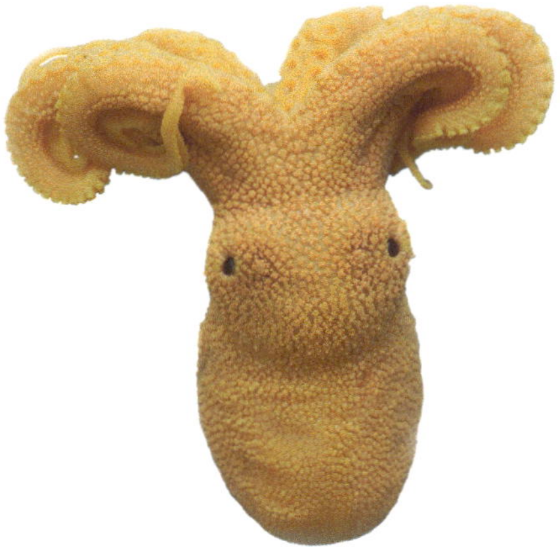

Preserved museum specimen, dorsal view (AM C.159276, A. Reid).

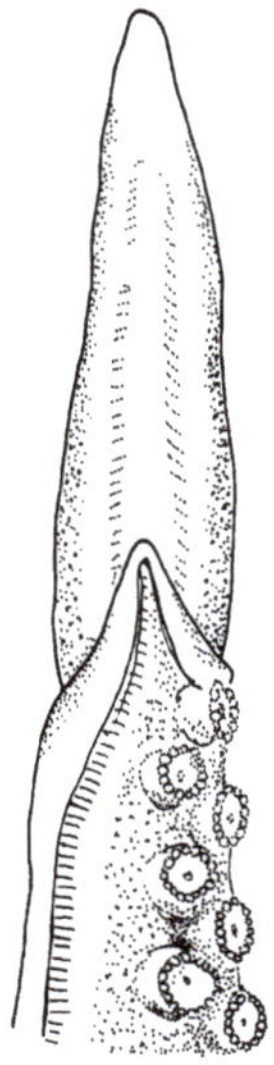

Hectocotylus (A. Reid, redrawn from original description).

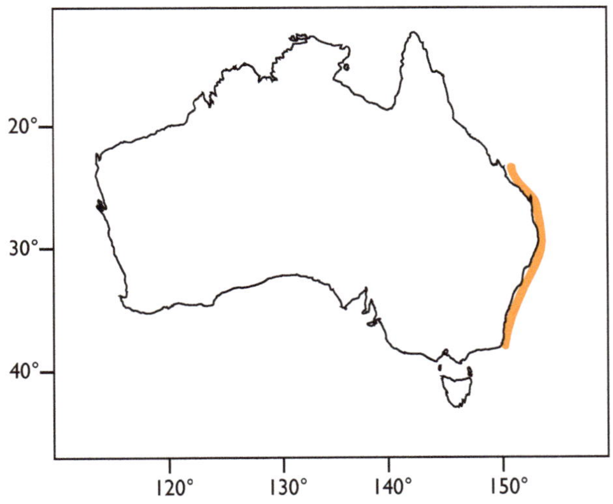


*Amphioctopus* cf. *kagoshimensis* distribution.

bordered by narrow white lines, visible in fresh dead and preserved specimens as four short white lines radiating from eye.

*Sculpture.* Distinct, regular patch and groove arrangement, patches typically pentagonal or hexagonal, largest on ventral arm crown.

### HABITAT AND BIOLOGY

Little is known about this octopus, which has been collected in trawls on soft sediments. The size of the eggs suggest that juveniles are planktonic.

*Fisheries importance.* This species is harvested as bycatch in commercial fisheries in New South Wales.

### SIZE

ML to ~90 mm; total length to 300 mm.

### REMARKS

Very similar, and undoubtedly closely related to Japanese species *A. kagoshimensis* (Ortmann, 1888).

### DISTRIBUTION

Australia: southern Queensland (One Tree Island) south to Tathra, New South Wales; northern New Zealand (as *Octopus* cf. *rugosus* Bosc, 1792). Depth to 115 m.

### REFERENCES

Norman and Kubodera (2006), Norman *et al.* (2014).

## *Amphioctopus marginatus* (Taki, 1964)

*Octopus marginatus* Taki, I. (1964). On eleven new species of the Cephalopoda from Japan, including two new genera of the Octopodinae. *Journal of the Faculty of Fisheries and Animal Husbandry, Hiroshima University* 5(2), 297–343 [304, text figs 47, 48, pl. 5].

Type data: holotype YTKC.

Type locality: Japan: Oita Prefecture, Minami-Amabe County, near Kamae.

### COMMON NAME

Veined Octopus

### DIAGNOSTIC FEATURES

Moderate sized octopus. Funnel organ limbs approximately equal length. Arms short, 2–3 × ML. Dorsal arms slightly shorter than other arms (arm formula 4 = 3 = 2 > 1 or 3 > 4 = 2 > 1). Arm webs deep, up to one-third of arm length, deepest on lateral arms. Web margins extend halfway along arms. No arm autotomy. Up to 150 biserial suckers on each normal arm. Mature males with four or five slightly enlarged suckers on seventh to eleventh row on arms 2 and 3. Male hectocotylus (third right arm) ~80% length of opposite arm with 60–80 suckers. Ligula small (1.5–3.5% of arm length) and well developed calamus, ~50% ligula length. Gills with 9–11 lamellae per demibranch. Ink sac and anal flaps present. Spermatophores long, approximately equal to mantle in length and produced in low numbers. Eggs small, ~3 mm long.

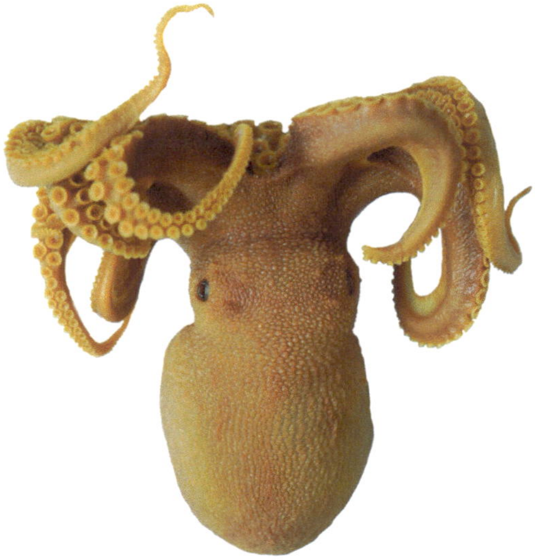

Preserved museum specimen, dorsal view (AM C.164171, A. Reid).

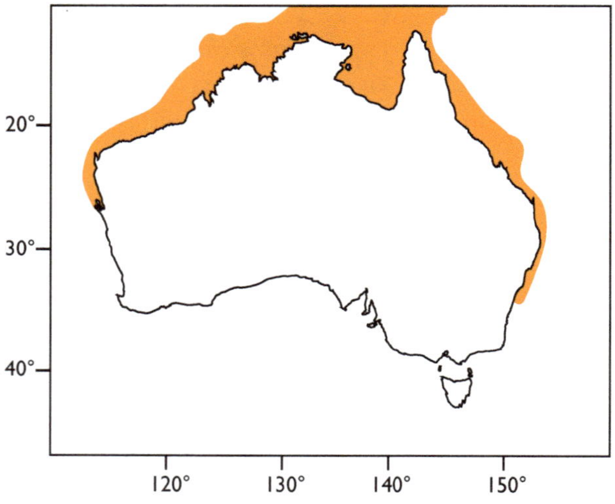


*Amphioctopus marginatus* distribution.

*Colour.* Pattern typically orange-brown to purple with acutely branched dark reticulations defining distinct patches in irregular longitudinal rows; colour pattern extends to mouth on oral surface of dorsal web. Suckers white to pink contrasting against dark brown to black along leading edge of arms 1–3; narrow, transverse 'head bar' visible in live animals; white triangle below each eye; dark reticulations distinctive on lateral arm crown in same position as false eye-spots in ocellate species.

*Sculpture.* Skin sculpture of regular patches separated by distinct grooves; patches arranged in irregular longitudinal lines on anteriolateral edges of mantle where they fuse to form distinctive longitudinal patches. Large papilla over each eye. Sculpture extends to mouth on oral surface of web.

## HABITAT AND BIOLOGY

Known from coastal muddy waters on mud and sand substrates, subtidal to depths of at least 190 m. It makes dens in clamshells as well as discarded human objects such as coconut shells and bottles. It has been observed carrying half a bivalve or coconut shell while looking for another half with which to enclose itself. Females lay up to 100 000 small eggs up to 3 mm long.

A behavioural study of *Amphioctopus marginatus* from India (Sreeja and Bijukumar 2013) showed that primary and secondary defence mechanisms in this species included crypsis (sometimes in motion with an animal's colour pattern, behaviour or other features helping it blend with the background), and hiding and escape behaviour (which included swimming or burrowing into the substrate and jet propulsion for fast escape). Deimatic behaviour was used by captive animals when camouflage failed and they were threatened. Stilt walking and bipedal locomotion (using two arms for locomotion) were observed. Females brooded and aerated the eggs, which took 17 days to hatch.

## SIZE

Maximum ML to ~100 mm, total length to ~300 mm; weight to 400 g.

## REMARKS

This taxon needs to be investigated over it range to determine the presence of cryptic species. May be confused with *A.* cf. *kagoshimensis* Norman & Kubodera, 2006. The two species can be distinguished by the features listed in the key to *Amphioctopus* species.

## DISTRIBUTION

Found in tropical continental waters of the Indian Ocean from the Red Sea and East Africa to South-east Asia and eastern Australia.

## REFERENCES

Norman and Sweeney (1997), Finn *et al.* (2009), Norman *et al.* (2014).

## *Amphioctopus mototi* (Norman, 1993)

*Octopus mototi* Norman (1993). Ocellate octopuses (Cephalopoda: Octopodidae) of the Great Barrier Reef, Australia: Descriptions of two new species and redescription of *Octopus polyzenia* Gray 1849. *Memoirs of the Museum of Victoria 53*(2), 309–344 [329, text figs 1f, 11–14, 15d].

Type data: holotype MV F.60101.

Type locality: Australia: Queensland, Capricorn Bunker Group, Heron Island, 23°50′S, 152°25′E.

### COMMON NAME

Poison Ocellate Octopus

### DIAGNOSTIC FEATURES

Muscular octopus. Funnel organ W-shaped, outer limbs ~90% of length of medial limbs. Arms moderate length, 2.5–3.0 × ML. Lateral arms longest (typically 4 = 3 > 2 > 1). No arm autotomy. Webs 25–40% of arm length;

Live animal, inverse colour displayed when the animal is disturbed (M. Norman).

Live animal, resting colouration (Photos: M. Norman).

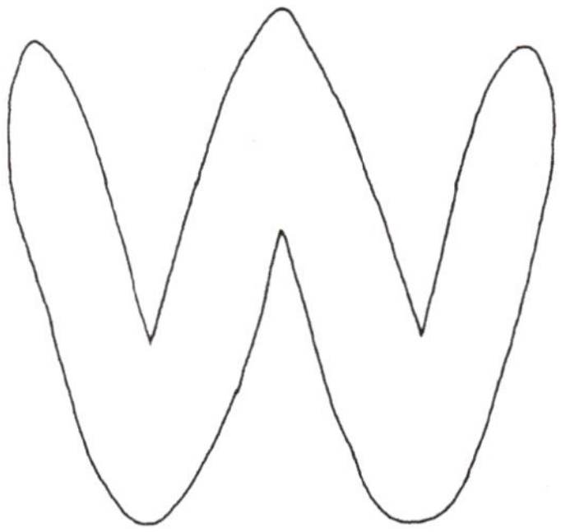

Funnel organ (M. Shea, redrawn from original description).

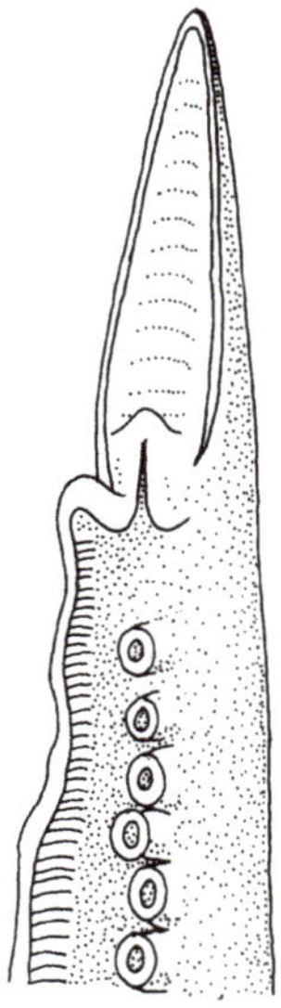

Hectocotylus (M. Shea, redrawn from original description).

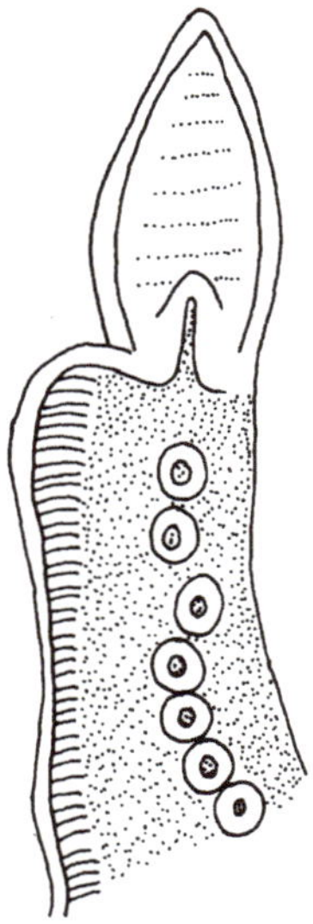

Hectocotylus (M. Shea, redrawn from original description).

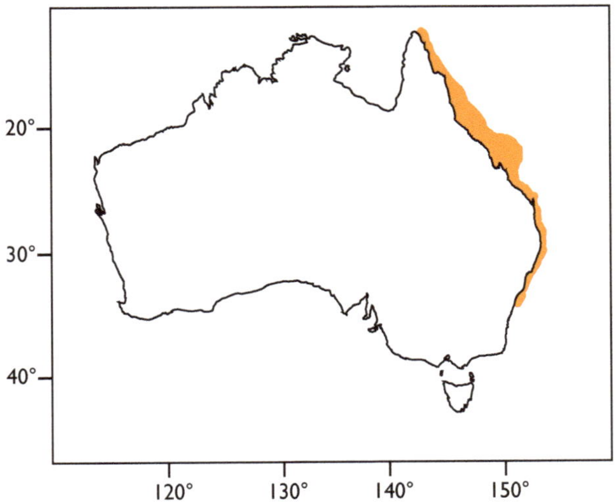



*Amphioctopus mototi* distribution.

deepest on lateral arms, distinctly shallower between dorsal arms; web margins extending to arm tips. In larger animals, 143–176 suckers on each normal arm. Suckers slightly elevated with distinct radial cushions. Slightly enlarged suckers present in mature males on arms 2 and 3, starting at around the fifth proximal sucker. Male hectocotylus (third right arm) ~80–100% length of opposite arm with 95–106 suckers. Spermatophore guide distinct but without papillae. Ligula short and broad with wide groove, 3–7% of arm length; calamus of moderate size, ~40–50% of ligula length; slightly raised. Gills with 9–11 lamellae per demibranch (usually 11). Ink sac and anal flaps present. Spermatophores of moderate size, ~40–50 mm (~60–70% ML), produced in low numbers < 10. Eggs ~6 mm diameter (~8% ML) and produced in large numbers.

*Colour.* Typically orange-brown when resting with scattered dark spots in six longitudinal series. Flower-like arrangement of five dark spots over each eye. Alarm colour pattern of white base colour and six dark maroon longitudinal stripes on mantle, continuing through eye and down leading edges of arms 1–3. Ocellus a dark spot containing narrow iridescent blue ring in live animals; appears as white or superficial ring over dark ocellus in preserved specimens. Well-developed frontal white spot with prominent elongate papilla, and small pair of dorsal white spots midway along dorsal mantle sometimes visible.

*Sculpture.* Skin texture of regular low, rounded papillae. Four longitudinal ridges in diamond arrangement on dorsal mantle. Single small papilla above and slightly behind each eye in centre of 'flower' pattern of spots. Mantle skin ridge absent.

### HABITAT AND BIOLOGY

Occurs on sandy substrates, often associated with coral heads or rubble. Deep lairs are excavated under coral heads or coral rubble on sandy substrate. Appears (from limited observations) to have crepuscular activity patterns. Lairs are surrounded by gastropod shells, which may have contained hermit crabs; stomach contents contain a high proportion of crustacean exoskeletal fragments. Eggs are laid in large numbers in festoons. The small eggs are indicative of planktonic hatchlings.

### SIZE

ML to 70 mm (males); 100 mm (females); total length up to 320 mm. Weight to at least 300 g.

### REMARKS

This octopus may prove to be venomous. On Rapa Island in French Polynesia this species is known as *fe'e mototi*, the 'poison octopus'. Live animals bite objects such as aquarium nets – a behaviour not generally observed in other octopus species.

### DISTRIBUTION

Western and south Pacific from the southern end of the Great Barrier Reef (islands in the Capricorn-Bunker Group), Swains Reefs and Wreck Reef, Coral Sea to New Caledonia and Rapa Island, French Polynesia. Similar form (possibly conspecific) in Okinawa, Japan. Known depth range 1–54 m.

### REFERENCES

Norman (1993b), Norman (2000), Norman *et al.* (2014).

## *Amphioctopus polyzenia* (Gray, 1849)

*Octopus polyzenia* Gray, J. E. (1849). Catalogue of the Mollusca in the Collection of the British Museum. *1. Cephalopoda Antepedia. London: British Museum (Natural History)* 164 pp. [13].

Type data: holotype BMNH 1845.5.14.78.

Type locality: Australia: Northern Territory, Port Essington, Coral Bay, 11°16′S, 132°09′E.

### COMMON NAME

Arm-band Ocellate Octopus

### DIAGNOSTIC FEATURES

Small, robust species. Head narrower than mantle. Funnel organ with moderate limbs. Small octopus with short to moderate arms (2–3 × ML). Dorsal arms shortest, arm formula 4 = 3 >2 >1 or 4 >3 >2 >1. Suckers moderate sized, slightly elevated with distinct radial cushions and scalloped at level of seventh sucker pair. Males with one to three enlarged suckers on all arms. Normal arms with up to ~135 suckers per arm in mature females, ~85–110 in mature males (most on fourth arm). Male hectocotylus (third right arm) ~80–100% length of opposite arm with ~45–55 suckers. Spermatophore guide distinct with small papillae. Copulatory organ small and robust, ligula conical with blunt tip; groove shallow with raised, longitudinal midrib. Terminal organ short and robust. Gills with six or seven gill lamellae per demibranch. Mature males with one to three enlarged suckers on twelfth to fourteenth row on all arms. Spermatophores short, about one-third of ML,

Preserved museum specimen, dorsal view (AM C.164168, A. Reid).

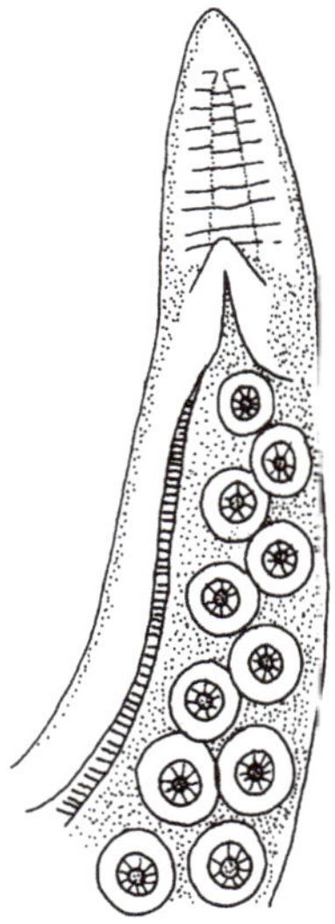

Hectocotylus (M. Shea, redrawn from original description).

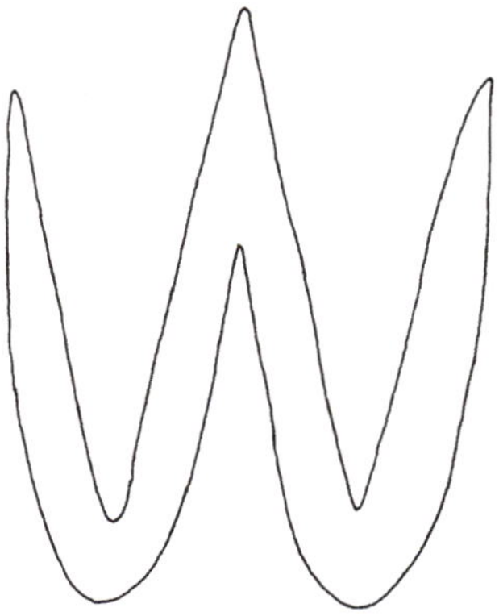

Funnel organ (M. Shea, redrawn from original description).

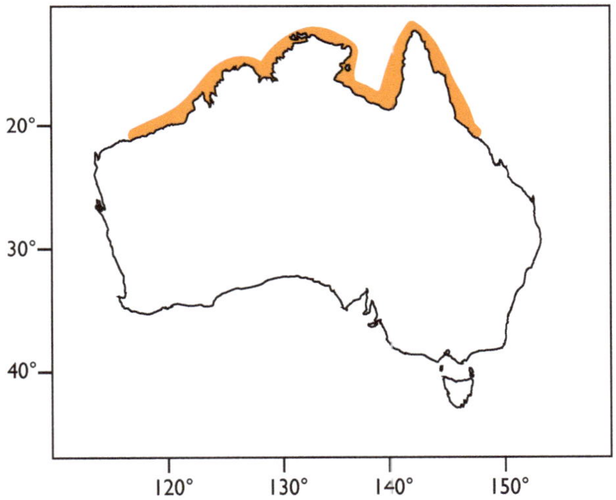


*Amphioctopus polyzenia* distribution.

and produced in low numbers. Sperm cord iridescent gold in colour and coiled in ~45 whorls. Eggs large (to 4.9 mm long), spawned in capsules up to 7.5 mm long and produced in low numbers.

*Colour.* Preserved specimens mottled cream to dark brown; light patches surrounding raised skin ridges form crucifix pattern on dorsal mantle. Ocellus between bases of arms 2 and 3, black oval (may fade on preserved specimens), containing small, iridescent, blue ring in live animals, visible as white or pink superficial ring over dark ocellus in preserved specimens. Widely spaced dark transverse bars on arms, separated by three or four sucker pairs. Two light spots (anterior cream, posterior pinkish) on dorsal arm crown below midpoint of eyes. Faint crucifix pattern of light patches on dorsal body. No dark longitudinal line along dorsal face of arms.

*Sculpture.* Skin with small, low papillae extending onto arm webs. Longitudinal raised skin ridges on dorsal body, largest four in diamond-shaped arrangement. One large and two smaller branching papillae above each eye.

## HABITAT AND BIOLOGY

*Amphioctopus polyzenia* is found in shallow coastal waters on open substrate of rubble, sand or mud to depths of at least 20 m. Eggs in one female at two different stages of development suggest spawning over an extended period. Eggs are laid in short festoons attached by short stalks to a fine central thread. The large eggs and well-developed arms observed in juveniles suggest that the hatchlings are benthic.

*Interest to fisheries.* None, but may be harvested with other northern Australian ocellate octopuses.

## SIZE

Up to 40 mm ML.

### DISTRIBUTION

Coastal waters of northern Australia from Bowen, Queensland 20°01'S, 148°15'E, west to Rosemary Island, Dampier Archipelago, Western Australia 20°29'S, 116°35'E.

### REFERENCE

Norman (1993b).

## *Amphioctopus rex* (Nateewathana and Norman, 1999)

*Octopus rex* Nateewathana, A. and Norman, M. D. (1999). On three new species of ocellate octopuses (Cephalopoda: Octopoda) from Thai waters. *Phuket Marine Biological Center Special Publication 19*(2), 445–462 [142].

Type data: holotype PMBC 17463.

Type locality: Gulf of Thailand, Trat Province, Ko Kood.

### COMMON NAME

King Ocellate Octopus

### DIAGNOSTIC FEATURES

Funnel organ limbs approximately equal length. Arms 2–3 × ML. Lateral and ventral arms longest. (Arm formula typically 4 >3 >2 >1). No arm autotomy. Webs one-fifth to one-third of arm length, deepest on lateral and ventral arms. Larger animals with 134–184 biserial suckers in each normal arm. Mature males with two pairs of enlarged suckers from about the fifth proximal sucker row. Male hectocotylus (third right arm) ~80% length of opposite arm with 63–82 suckers. Ligula 5–9% of arm length; calamus small, ~15% of ligula length. Gills with eight or nine lamellae per demibranch. Ink sac and anal flaps present. Spermatophores ~25 mm long. Eggs small, ~2–3 mm in length.

*Colour.* Preserved animals reddish brown dorsally on head, arms and mantle, white and cream on ventral surfaces. Short brownish black bar through each eye. Narrow dark stripe on dorsolateral surface of arms 1–3. False eye-spots present, containing small, simple pinkish purple iridescent ring. Transverse pair of white spots on dorsal mantle.

*Sculpture.* Pattern of fine, rounded and closely set epidermal tubercles over dorsal and ventral surfaces of arms, head and mantle. Single small papilla over each eye. Mantle ridge absent.

### HABITAT AND BIOLOGY

Found to ~80 m depth in coastal waters on mud and sandy mud substrates. The hatchlings are likely to be planktonic.

*Interest to fisheries.* This species, fished in Thai waters, is marketed in Australia as 'baby octopus'.

### SIZE

Mantle up to 76 mm in length.

### DISTRIBUTION

Tropical continental waters of south and east Asia from India, Thailand and through Indonesia to northern Australia.

### REFERENCE

Norman *et al.* (2014).

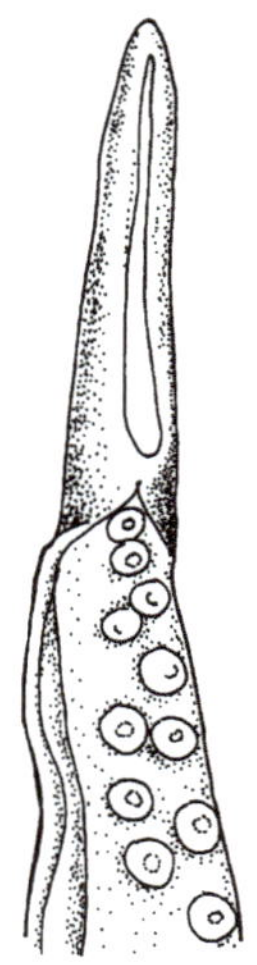

Hectocotylus (A. Reid, redrawn from Norman *et al.* 2014).

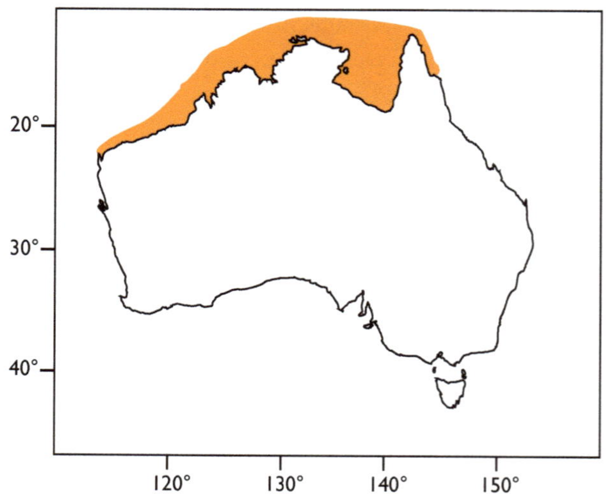


*Amphioctopus rex* distribution.

## *Callistoctopus* Taki, 1964

### DIAGNOSTIC FEATURES

Mantle ovoid to elongate cylindrical. Funnel organ W or UU-shaped. Eyes moderately large, slightly raised. Elongate arms (~5–8 × ML) with dorsal pair longest (arm formula 1 > 2 > 3 > 4); arm autotomy absent. Webs ~5–20% of longest arm (dorsal webs always deepest); margins extend to 70% of arm length. Arms with 192–382 biserial suckers on normal arms. Enlarged suckers absent; suckers 6.4–20.2% ML. Male hectocotylus (third right arm) approximately ~55–77% length of opposite arm with 86–172 suckers. Robust, conical copulatory organ, with deep grooved ligula; ligula large and robust, 4.3–9.1% arm length; calamus 13.6–29.3% ligula length. Terminal organ curled, 10–34% ML; diverticulum long, recurved. Moderate to high gill lamellae counts: 10–15 per demibranch. Ink sac and anal flaps present. Salivary glands ~equal in length to buccal mass. Spermatophore length 38–95% ML; width ~1–3% spermatophore length; 30–55 sperm cord whorls; ejaculatory apparatus with two or three coils. Stylets reduced or absent.

*Colour.* White spots or bars on red background. Dorsal white spots present. Ocelli absent.

*Sculpture.* Scattered low papillae. One large and some small supraocular papillae over each eye.

### HABITAT AND BIOLOGY

Representatives of this genus are found in shallow coastal waters on the continental shelf to ~200 m. Habitats include reef to seagrass beds and soft sediments. It is nocturnally active.

### SIZE

Up to 190 mm ML; total length to 1.3 m.

### REMARKS

*Callistoctopus* have been referred to in earlier publications as the '*Octopus macropus* group' (Norman 1993a, 2000).

### REFERENCES

Norman (1993a), Norman and Hochberg (2005), Norman *et al.* (2014).

### Key to *Callistoctopus* species

1a. Pale cream to pink body with irregular short 'scribbles' on dorsal mantle and arm crown (can be seen in preserved specimens) .................. *C. graptus*

1b. Numerous white or red spots; dark scribbles absent .......................................................................... 2

2a. Dorsal mantle with distinctive arrangement of cream to white longitudinal stripes over pink to red base colour......................................... *C. ornatus*

2b. Dorsal mantle plain or with spots, longitudinal stripes absent.............................................................. 3

3a. Distinctly elongate species with almost tubular arms and short webs; mantle often elongate; deepest web always < 15% of length of longest arm, typically 10%.............................................................. 4

3b. Moderately elongate species with robust arms and deep webs; mantle never greatly elongated; deepest web always > 15% of length of longest arm, typically 20%.............................................................. 5

4a. Colour pattern on dorsal mantle plain red or white, lacking distinctive spots................. *C. aspilosomatis*

4b. Colour pattern on dorsal mantle includes distinctive spots, white on red or red on white ..... 5

5a. More than 230 suckers on normal arms of submature and mature animals; > 100 suckers on hectocotylised arm of males; alarm display in live animals of large red spots over white base colour on dorsal mantle, arm crown and arms .................. .......................................................... *C. dierythraeus*

5b. Typically < 230 suckers on normal arms of submature and mature animals; < 100 suckers on hectocotylised arm of males; alarm display of live animals of white spots over red base colour on dorsal mantle, arm crown, and arms ....... *C. alpheus*

## *Callistoctopus alpheus* (Norman, 1993)

*Octopus alpheus* Norman, M. D. (1993). Four new octopus species of the *Octopus macropus* group (Cephalopoda: Octopodidae) from the Great Barrier Reef, Australia. *Memoirs of the Museum of Victoria 53*(2), 267–308 [270, figs 1–4, 17b].

Type data: holotype MV F.57930.

Type locality: Australia: Queensland, Capricorn Bunker Group, Tryon Island, 23°15′S, 151°47′E.

### COMMON NAME(S)

Capricorn Night Octopus, Capricorn Octopus

### DIAGNOSTIC FEATURES

Moderate to large, robust octopus (up to 80 mm ML). Funnel organ typically W-shaped; outer limbs slightly shorter than medial limbs. Arms moderately long

Live animal (M. Norman).

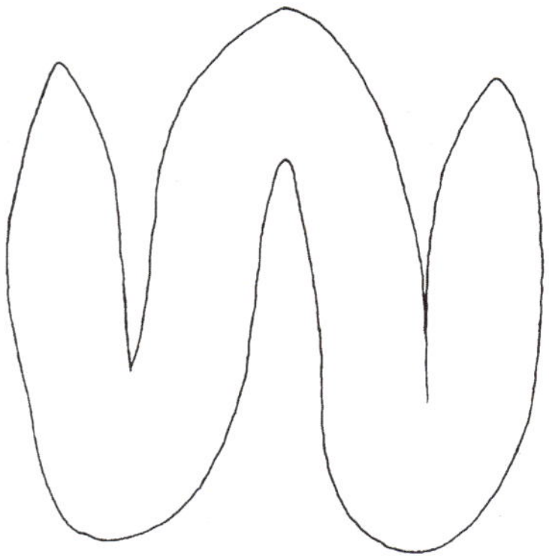

Funnel organ (M. Shea, redrawn from original description).

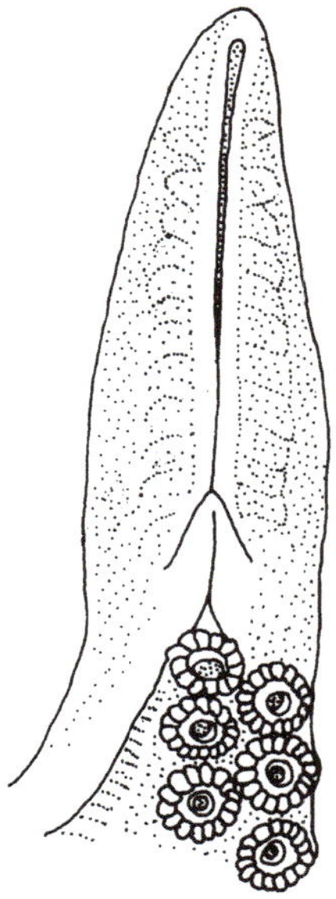

Hectocotylus (M. Shea, redrawn from original description).

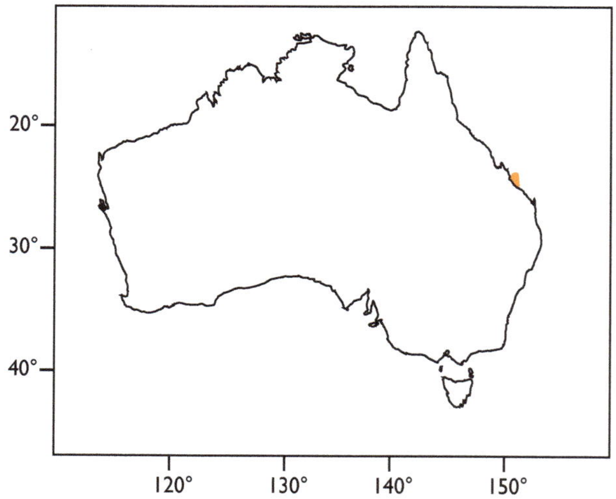


*Callistoctopus alpheus* distribution.

(3–5 × ML). Dorsal arms longer than lateral and ventral arms (arm formula usually 1 > 2 > 3 > 4). Webs moderately deep, deepest on dorsal arm, ventral arms shallowest, deepest 16–25% of longest arm. Web margins extend to arm tips. Suckers medium to large with distinct radial cushions and scalloped outer rim; scalloping exaggerated on small distal suckers forming ring of small digits. Suckers largest on dorsal arms, otherwise none markedly enlarged. In adults 190–230 biserial suckers on normal arms. Male hectocotylus (third right arm) approximately ~70% length of opposite arm with 82–97 suckers. Copulatory organ cylindrical with deep groove and moderate-sized calamus; ligula 5–9% of arm length; ligula groove and spermatophore guide with fine, transverse ridges; calamus moderate-sized, 20–30% of ligula length. Gills with 11–12 lamellae per demibranch. Ink red when released by live animals. Spermatophores large, almost as long as ML; sperm cord coiled in ~40 whorls. Eggs are large, ~8 mm in length and produced in relatively low numbers (< 500).

*Colour.* Orange-brown to red base colour with large evenly spaced white spots over dorsal mantle, arm crown, webs and arms. Spots paired on arms with additional spots close to base of each sucker. Some darker markings on dorsal mantle. Iridescent layer gives animals a green sheen under torchlight at night. Pattern largely lost in preserved specimens. Alarm display of red base colour with large white spots on dorsal mantle, arm crown and arms. Ocelli absent.

*Sculpture.* Scattered low papillae over smooth skin; papillae in centre of white spots on dorsal and lateral mantle. Largest white spots with transverse pair of large papillae. Single larger supraocular papilla directly above each eye, surrounded by low papillae. No skin ridge around lateral mantle.

Live animal (AM slide collection).

### HABITAT AND BIOLOGY

This species is nocturnally active, emerging to forage for crabs and other crustaceans during the night on low tides, probing crevices with long arms to locate prey. They sometimes traverse exposed reef and can be heard 'squeaking' as air is drawn into and out of the mantle. They occupy lairs under coral during the day, concealing the entrance with pieces of dead coral. The large eggs suggest that the hatchlings adopt a benthic habit.

*Interest to fisheries.* None.

### SIZE

Up to 80 mm ML; total length to ~430 mm.

### DISTRIBUTION

Tropical species found on shallow intertidal reef flats in the Capricorn Bunker islands, southern Great Barrier Reef from 23°18'S, 151°47'E to 23°32'S, 151°45'E.

### REFERENCE

Norman *at al.* (2014).

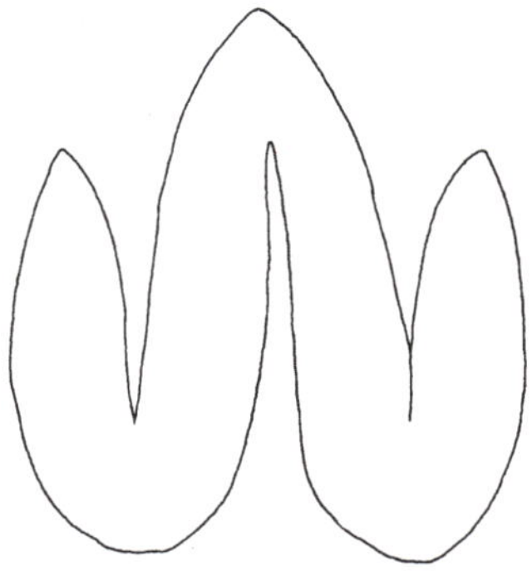

Funnel organ (M. Shea, redrawn from original description).

## *Callistoctopus aspilosomatis* (Norman, 1993)

*Octopus aspilosomatis* Norman, M. D. (1993). Four new octopus species of the *Octopus macropus* group (Cephalopoda: Octopodidae) from the Great Barrier Reef, Australia. *Memoirs of the Museum of Victoria* *53*(2), 267–308 [279, figs 5–8, 17c].

Type data: holotype MV F.67001 (♂ 42.1 mm ML).

Type locality: Australia: Queensland, Frankland Group, Russell Island, 17°14'S, 146°06'E.

### COMMON NAME(S)

Plain-body Night Octopus, Plain Body Octopus

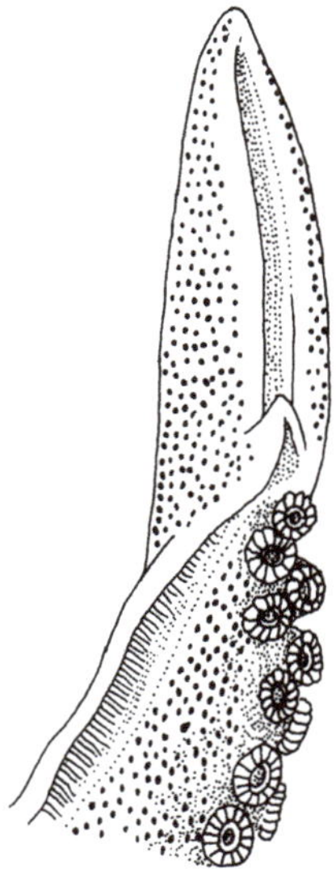

Hectocotylus (M. Shea, redrawn from original description).

### DIAGNOSTIC FEATURES

Moderate-sized octopus. Mantle highly variable in form, often elongate, cylindrical. Funnel organ W-shaped; outer limbs shorter (~70%) than median limbs. Arms 4.5–6.0 × ML. Arms almost cylindrical. Dorsal arms longer than ventral and lateral arms, formula 1 > 2 > 3 > 4; webs shallow, ~10–15% of longest arm; web margins extend to arm tips. In adults 200–260 biserial suckers on normal arms. Male hectocotylus (third right arm) about two-thirds length of opposite arm with 78–95 suckers. No enlarged suckers. Copulatory organ cylindrical with

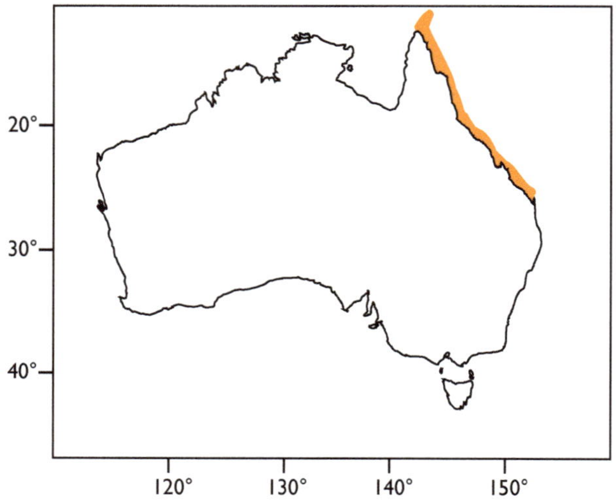

*Callistoctopus aspilosomatis* distribution.

deep, typically closed ligula groove and moderate calamus (~20% ligula length); ligula 6–9% of arm length. Spermatophore guide distinct with moderately deep notch and flattened ridge with square papillae. Gills with 10 or 11 gill lamellae per demibranch. Ink sac and anal flaps present. Spermatophores ~50–80% ML, produced in low numbers (~1–3). Small eggs (< 3 mm diameter) produced in large numbers (> 10 000).

*Colour.* Orange-brown to deep maroon base colour with paired white spots on arm crown, webs and arms. No white spots on mantle. Ocelli absent. Alarm display of red base colour with large white spots on dorsal mantle, arm crown and arms. Arm crown and dorsal arms covered with regular paired white spots over red base colour.

*Sculpture.* Scattered low papillae over smooth skin. Single larger papilla over each eye. Skin ridge absent from lateral mantle.

#### HABITAT AND BIOLOGY

*Callistoctopus aspilosomatis* is found on intertidal to subtidal coral reef flats, emerging to forage during night low tides. It occupies lairs within coal bedrock and under living coral, blocking the entrance during the day with pieces of dead coral. The hatchlings are planktonic.

*Interest to fisheries.* None.

#### SIZE

Up to 80 mm ML, total length to ~440 mm.

#### DISTRIBUTION

This tropical species is known only from coastal reefs and offshore islands of the northern Great Barrier Reef. Possibly also from Okinawa and Moorea in the Society Islands.

#### REFERENCES

Norman (1993a), Norman *et al.* (2014).

### *Callistoctopus dierythraeus* (Norman, 1993)

*Octopus dierythraeus* Norman, M. D. (1993). Four new octopus species of the *Octopus macropus* group (Cephalopoda: Octopodidae) from the Great Barrier Reef, Australia. *Memoirs of the Museum of Victoria* *53*(2), 267–308 [284, figs 9–12, 17d].

Type data: holotype MV F.67007.

Type locality: Australia: Queensland, Gulf of Carpentaria, off west coast of Cape Yorke, 10°51′S, 140°27′E.

#### COMMON NAME(S)

Red-spot Octopus, Red-spot Night Octopus

#### DIAGNOSTIC FEATURES

Large robust octopus, moderately elongate. Funnel organ UU-shaped, outer limbs slightly shorter than medial limbs (75–90%). Arms moderately long (4–5 × ML). Dorsal arms longer than lateral and ventral arms (formula 1 >2 >3 >4). Webs moderately deep 18–28% of longest arm. Web margins extend to arm tips. In adults 200–280 biserial suckers on normal arms. Male hectocotylus (third right arm) ~75% length of opposite arm with 103–125 suckers; no enlarged suckers. Ligula cylindrical and muscular, ~6% of arm length; calamus small, ~15% of ligula length. Gills with 12–14 lamellae per demibranch. Spermatophores large, ~75% ML, produced in low numbers. Eggs large (> 14 mm in length) and produced in low numbers (< 500).

*Colour.* Red base colour with large white spots (resting), or white base colour with red spots (alarm colouration). Foraging animals are orange-brown to red with white spots over dorsal surfaces; alarm display of white base colour with red spots over mantle, arm crown, webs and arms. False eye-spots absent.

*Sculpture.* Scattered rounded papillae over smooth skin. Papillae located in centres of spots. Single slightly larger papilla over each eye surrounded by low punctae. No skin ridge around lateral mantle.

#### HABITAT AND BIOLOGY

This species is found in shallow coastal waters on rubble, sand or mud from intertidal depths to 78 m. It is

Live animal swimming (J. Gaskell).

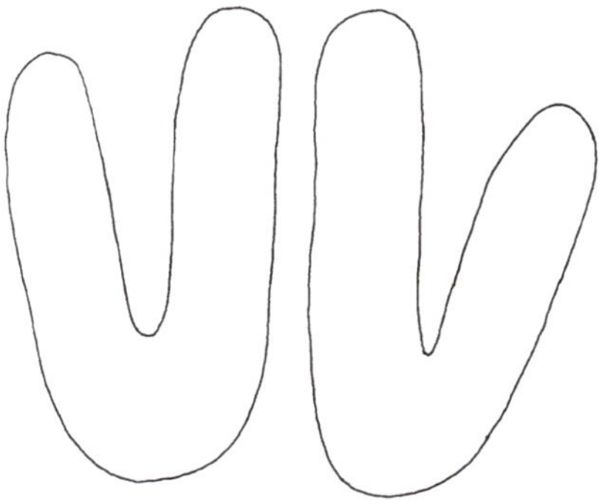

Funnel organ (M. Shea, redrawn from original description).

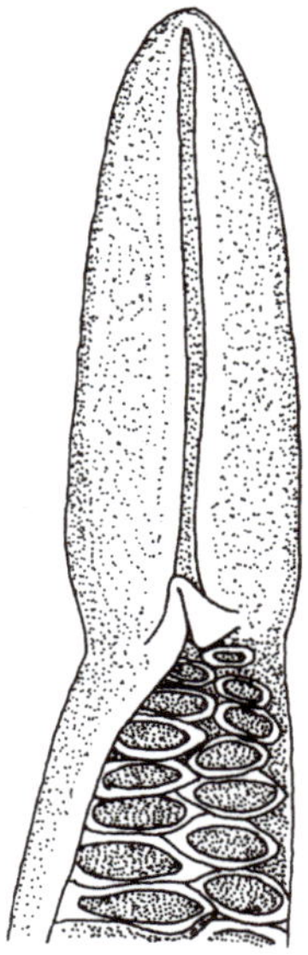

Hectocotylus (M. Shea, redrawn from original description).

Live animal, resting colouration (J. Gaskell).

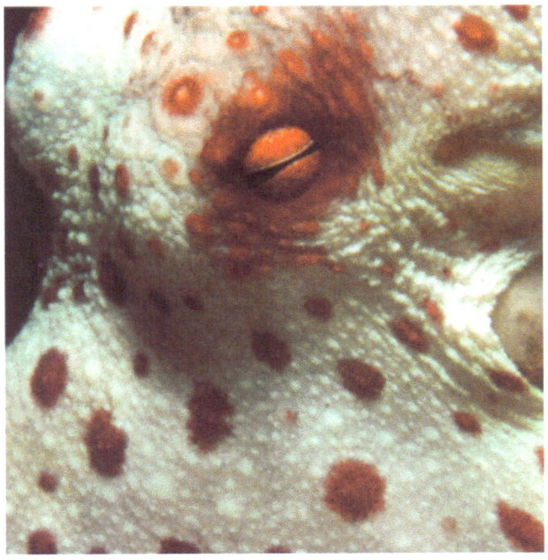

Inverse colour displayed when the animal is disturbed (M. Norman).

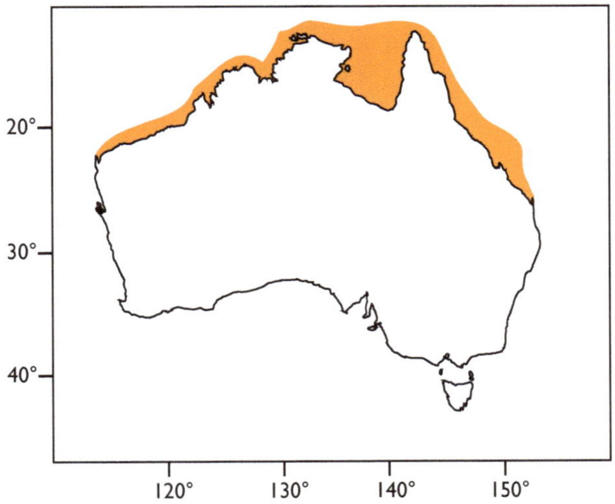


*Callistoctopus dierythraeus* distribution.

nocturnally active. Active lairs are surrounded by clean bivalve shells and crab carapaces. It is also known to eat gastropods, polychaetes and other octopuses.

Steer and Semmens (2003) examined prey handling in this species, and showed that they always initially attempt to pull bivalve shells apart before drilling.

Gastropod shells are always drilled. The size of the prey species did not affect whether shells were pulled or drilled, and the drill or penetration site varied according to the morphology of the prey species. For bivalves, the thinner periphery of the shell was targeted. Gastropods were not drilled in the thinnest part of the shell; elongate species with pronounced aperture lips were drilled in the apical region, close to the columella on the side of the opercula, whereas non-elongate species were drilled immediately above the aperture. It was suggested that the drill site might represent a trade-off between targeting the most effective places to inject paralysing secretions and the mechanically simplest places to drill.

*Interest to fisheries.* Sometimes collected as by-catch.

## SIZE

Up to 135 mm ML, total length to ~800 mm.

## DISTRIBUTION

Northern Australia from north-western Western Australia to the Great Barrier Reef, Queensland.

## REFERENCE

Norman *et al.* (2014).

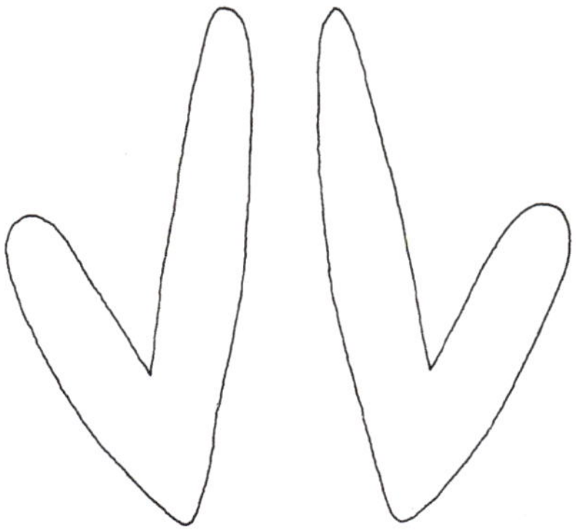

Funnel organ (M. Shea, redrawn from original description).

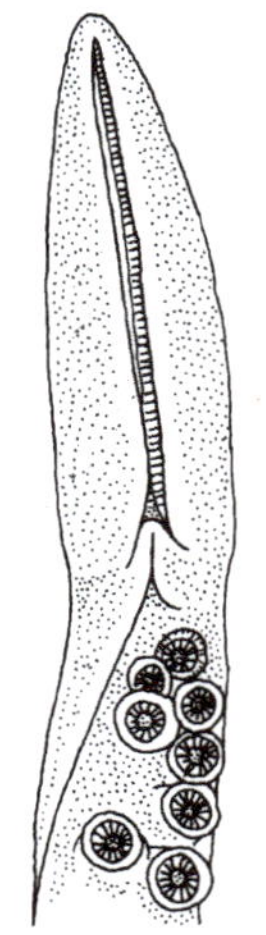

Hectocotylus (M. Shea, redrawn from original description).

# *Callistoctopus graptus* (Norman, 1993)

*Octopus graptus* Norman, M. D. (1993). Four new octopus species of the *Octopus macropus* group (Cephalopoda: Octopodidae) from the Great Barrier Reef, Australia. *Memoirs of the Museum of Victoria 53*(2), 267–308 [296, figs 13–16, 17e].

Type data: holotype MV F.67006.

Type locality: Australia: Queensland, Townsville, Cleveland Bay, 19°11′S, 147°01′E.

## COMMON NAME(S)

Scribbled Octopus, Scribbled Night Octopus

## DIAGNOSTIC FEATURES

Large, robust and muscular species. Funnel organ V V-shaped, outer limbs 2/3 of median limbs. Arms long, 4.5–7.0 × ML. Dorsal arms longer than ventral arms and more robust (arm formula 1 > 2 > 3 > 4). Suckers on longer dorsal arms proportionally larger than other arms, but none markedly enlarged. In larger animals, ~240 biserial suckers on each normal arm. Webs moderately deep,

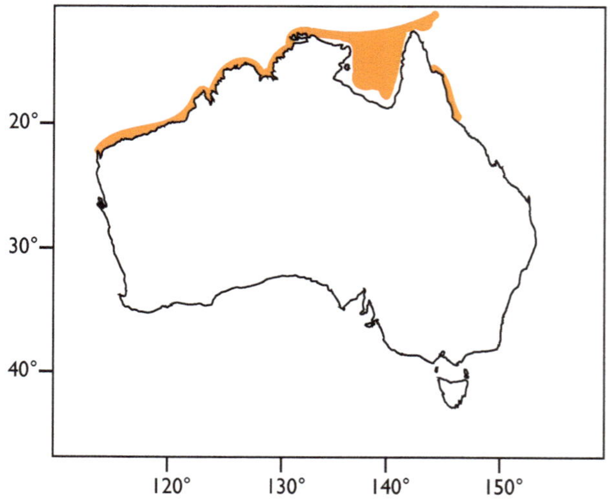



*Callistoctopus graptus* distribution.

deepest ~20% of arm length. Web margins poorly developed. Male hectocotylus (third right arm) ~60–70% length of opposite arm with 80–90 suckers. Ligula large, cylindrical (~6% of arm length), with deep groove;

calamus moderate size, 20% of ligula length. Gills with 13 or 14 lamellae per demibranch. Spermatophores ~90% ML and produced in low numbers. Eggs large (to 28 mm) and produced in low numbers.

*Colour.* Grey, white or orange-brown base colour with dark, irregular spots and short lines ('scribbles') over the dorsal mantle and arm crown. Distal tips of all arms grading to crimson purple.

*Sculpture.* Skin sculpture simple, consisting of scattered low papillae evenly distributed over dorsal surfaces. Single moderate-sized papilla over each eye. No skin ridge around lateral mantle.

#### HABITAT AND BIOLOGY

Known from muddy, coastal waters, living subtidally on soft substrates to depths of 11–36 m. This species has so far been caught only at night, presumably when it is foraging away from its lair. Prey includes crustaceans, bivalves and fishes. Females lay large eggs that are attached singly to the roof of a lair, not in festoons. The egg size suggests that the hatchlings are benthic.

*Interest to fisheries.* Commercially harvested as bycatch in prawn trawl fisheries (caught during trawls over sandy substrates) and sold for human consumption and as bait. It is likely that this species makes up the bulk of the annual catch of octopuses in Queensland waters.

#### SIZE

Maximum ML 190 mm, total length to > 1.3 m.

#### DISTRIBUTION

Found in tropical, muddy coastal waters of northern Australia from the Great Barrier Reef to northern Western Australia.

#### REFERENCES

Norman (1993a), Norman *et al.* (2014).

### *Callistoctopus ornatus* (Gould, 1852)

*Octopus ornatus* Gould, A. A. (1852). Mollusca and Shells. United States Exploring Expedition, during the years 1838, 1839, 1840, 1841, 1842, 12, 1–510 [476, figs 590–590a].

Type data: neotype NMNH 730020.

Type locality: Hawaiian Islands: Oahu Island.

#### COMMON NAME

White-striped Octopus

Live animal (M. Norman).

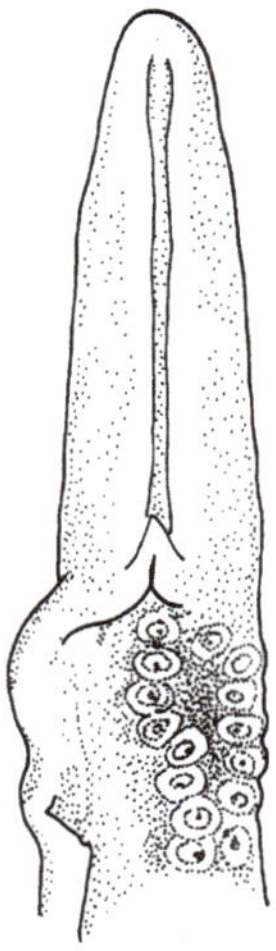

Hectocotylus (A. Reid, redrawn from Norman *et al.* 2014).

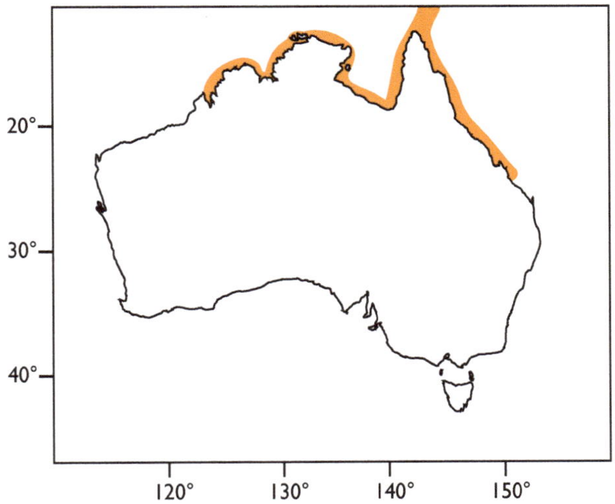


*Callistoctopus ornatus* distribution.

#### DIAGNOSTIC FEATURES

Large and elongate octopus. Funnel organ W-shaped, lateral limbs 75–85% of length of medial limbs. Arms long, 6–8 × ML. Dorsal arms much longer than ventral arms (arm formula 1 > 2 > 3 > 4). Webs shallow, deepest between dorsal arms ~10% arm length; web margins extend as shallow membranes for at least 70% of arm length. In larger animals, 300–400 biserial suckers on each normal arm. Suckers on longer dorsal arms proportionally larger than other arms, but none markedly enlarged. Male hectocotylus (third right arm) ~50–70% length of opposite arm with 150–170 suckers. Ligula cylindrical (up to 6% of arm length), with deep groove; calamus small, ~15% ligula length. Gills with 13 or 14 lamellae per demibranch. Spermatophores small and produced in low numbers. Eggs ~3 mm to 4 mm, ~3% ML.

*Colour.* Brown to deep red base with wide, short, longitudinal white markings; distinctive pattern of longitudinal white bars on dorsal mantle (visible in live, fresh dead, and preserved specimens). Large paired white spots on dorsal arm crown and along arms. Ocelli absent.

*Sculpture.* Skin with small, low papillae over dorsal surfaces; two moderate-sized erectile papillae over each eye. Elongate skin flaps can be raised within longitudinal white bars on mantle.

#### HABITAT AND BIOLOGY

Inhabits clear, tropical waters, from intertidal shallows to a depth of at least 10 m. Found on coral reefs among live and dead coral and coral rubble. Forages exclusively at night, preying primarily on crustaceans and other octopus species. Typically encountered foraging along edges of intertidal reefs adjacent to deeper water during night low tides. Foraging typically consists of moving over rubble at night, investigating holes and crevices with the long dorsal arms. Sometimes encountered swimming at the surface in or near lagoons. Occupies deep lair during the day, the entrance of which is blocked at various levels. Females lay up to 35 000 eggs in festoons, each egg 3–4 mm long. The egg size suggests the young are planktonic. This octopus is harvested on a small scale throughout its range, primarily in local subsistence fisheries. No fishery exists in Australia.

#### REMARKS

Taylor *et al.* (2012) have developed a molecular method of identifying *Octopus vulgaris* Cuvier, 1797, *O. cyanea* Gray, 1849 and *Callistoctopus ornatus* using PCR-RFLP (see 'Remarks' for *O. cyanea*).

#### SIZE

Maximum ML 130 mm, total length to > 1 m; weight to at least 1 kg.

#### DISTRIBUTION

Found widely throughout shallow waters of the tropical Indo-West Pacific region, from Hawaii in the east to the east African coast in the west. Reported as far north as southern Japan (as *Callistoctopus arakawai*) and as far south as New South Wales.

#### REFERENCE

Norman *et al.* (2014).

## *Galeoctopus* Norman, Boucher-Rodoni & Hochberg, 2004

#### DIAGNOSTIC FEATURES

Small muscular octopus, mantle round to squarish. Funnel organ W-shaped. Arms 2.7–4.4 × ML; all arms approximately equal length, dorsal pair slightly shorter. No arm autotomy. Webs deep, deepest around one-third of longest arm. Male hectocotylus (third right arm) ~40–60% length of opposite arm with 150–170 suckers. Ligula 11–20% of hectocotylised arm length with large calamus, > 55% of ligula length; ligula groove enclosed deep within tip of ligula, opening reduced to a transverse crescent-shaped mouth. Floor of groove with paired, raised papillae. Spermatophores large (about equal to ML). Gills with 10 or 11 lamellae per demibranch. Ink sac and anal flaps present. Stylets long, chitinous.

*Colour.* Mottled orange to red-brown on dorsal surfaces of body and arms. Ocelli absent.

*Sculpture.* Numerous raised, rounded papillae on dorsal surfaces, largest papillae stellate. Skin ridge around lateral mantle margin.

#### HABITAT AND BIOLOGY

Known depth range 200–400 m.

#### SIZE

Up to 37 mm ML, total length to 165 mm.

#### DISTRIBUTION

Central and south-west Pacific Ocean.

### REFERENCE

Norman *et al.* (2014).

## *Galeoctopus lateralis* Norman, Boucher-Rodoni & Hochberg, 2004

*Galeoctopus lateralis* Norman, M. D., Boucher-Rodoni, R., and Hochberg, F. G. (2004). The sharkclub octopus, *Galeoctopus lateralis*, a new genus and species of deep-water octopus from the western Pacific Ocean (Cephalopoda: Octotopdidae). *Journal of Molluscan Studies 70*, 247–256 [248, figs 1–5].

Type data: holotype MNHN Holotype 3787.

Type locality: Pacific Ocean: Tonga Tonga, North Kotu group, 19°52′S, 174°42′W.

### COMMON NAME

Sharkclub Octopus

### DIAGNOSTIC FEATURES

Moderately muscular species. Arms 2.7–4.4 × ML; approximately equal in length, dorsal pair slightly shorter. Webs deepest ~20–30% of arm length; dorsal sector slightly shorter, web margins poorly developed. Normal arms with 130–150 biserial suckers. Male hectocotylus (third right arm) ~40–60% length of opposite arm with 35–143 suckers. Enlarged suckers absent. Ligula distinctive, large (11–20% of hectocotylised arm length), fleshy, with deep transverse groove containing raised tooth-like lugs; calamus 57–76% of ligula length. Gills with 10 to 11 lamellae per demibranch. Spermatophores large and produced in low numbers

*Colour.* Mottled orange-brown with three irregular transverse bands across arm crown. False eye-spots (ocelli) absent.

*Sculpture.* Skin with numerous small, rounded papillae interspersed with larger, stellate-shaped papillae. Single large and several smaller papillae over each eye. Skin ridge around lateral margin of mantle.

### HABITAT AND BIOLOGY

Known depth range 200–400 m. Fishes and crustaceans have been found in the stomach.

### SIZE

Up to 37 mm ML, total length to 165 mm.

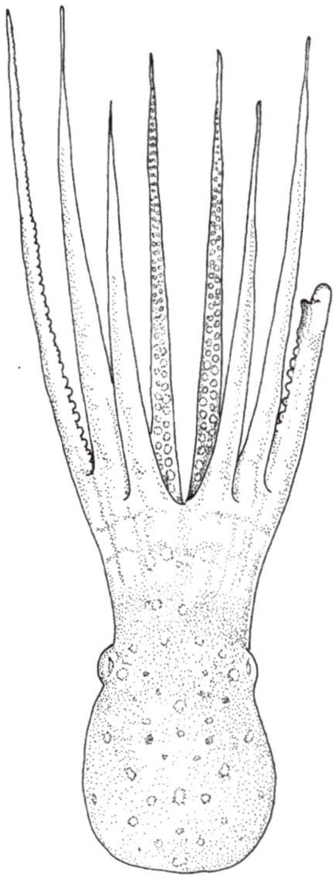

Dorsal view (A. Reid, redrawn from Norman *et al.* 2014).

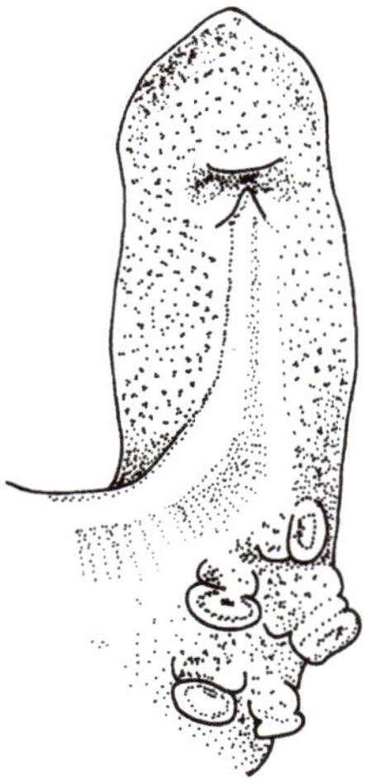

Hectocotylus, oral view (A. Reid, redrawn from Norman *et al.* 2014).

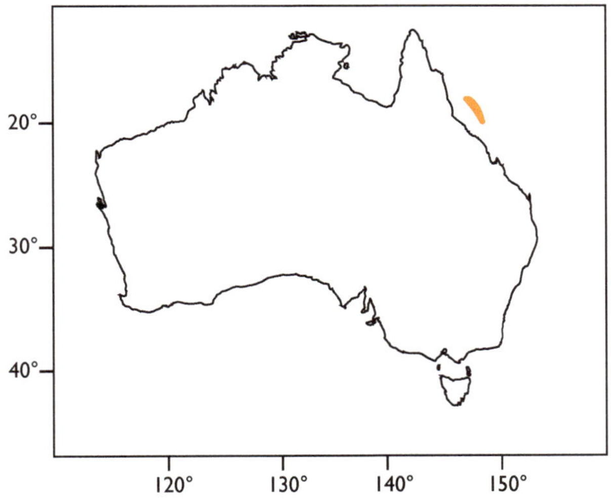


*Galeoctopus lateralis* distribution.

### DISTRIBUTION

Central and south-west Pacific Ocean, including off north-eastern Australia.

### REFERENCE

Norman *et al.* (2014).

## *Grimpella* Robson, 1928

*Grimpella* is currently monotypic, represented *Grimpella thaumastocheir* Robson, 1928.

### *Grimpella thaumastocheir* Robson, 1928

*Grimpella thaumastocheir* Robson, G. C. (1928). Notes on the Cephalopoda. VI. On *Grimpella*, a new genus of Octopoda, with remarks on the classification of the Octopodidae. *Annals and Magazine of Natural History, (series 10)* 2(7), 108–114 [110, figs 1–4].

Type data: holotype BMNH 1928.4.3.5.

Type locality: Australia: South Australia, Port Lincoln.

#### COMMON NAME

Velvet Octopus

#### DIAGNOSTIC FEATURES

Muscular species. Funnel organ UU-shaped, limbs approximately equal in length. Arms 4–4.5 × ML, all similar in length. No arm autotomy. Deepest web up to 32% of arm length; web sectors approximately equal in depth. Approximately 130 biserial suckers on each normal arm. Males with two or three enlarged suckers on each arm, from about the eighth proximal sucker. Male hectocotylus (third right arm) ~70% length of opposite arm with ~52 suckers. Ligula large and robust, 5–7% of arm length; calamus of mature males very large and pointed, ~80% of ligula length. Gills with seven to nine lamellae per demibranch. Ink sac and anal flaps absent. Spermatophores ~80% ML and produced in low numbers. Eggs large, ~15 mm long (45–60% ML). Stylets absent.

*Colour.* Uniform grey-purple to maroon red with iridescent sheen on lateral surfaces. False eye-spots absent.

Live animal (M. Norman).

*Sculpture.* Skin with scattered low, rounded papillae. No large papillae over eyes. No skin ridge around lateral margin of mantle.

#### HABITAT AND BIOLOGY

This is a night-active species found in rubble areas near shore, adjacent to deep water. Known depth range 0–420 m. Eggs are large and attached singly. The young are benthic.

#### SIZE

ML to 50 mm; total length to ~250 mm.

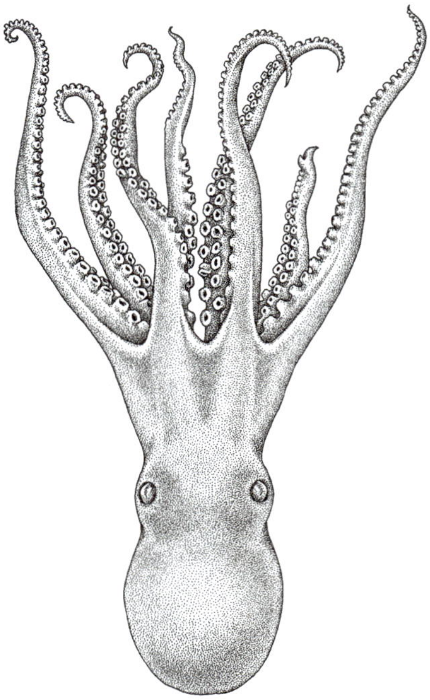

Dorsal view (ABRS, K. Hollis).

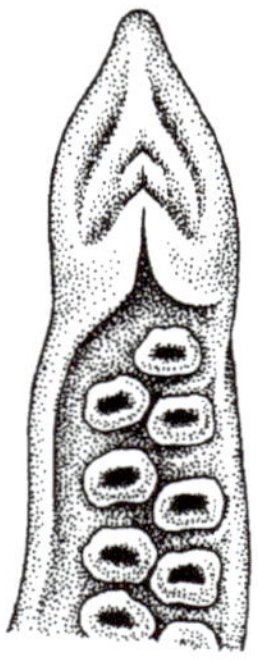

Hectocotylised arm (ABRS, K. Hollis).

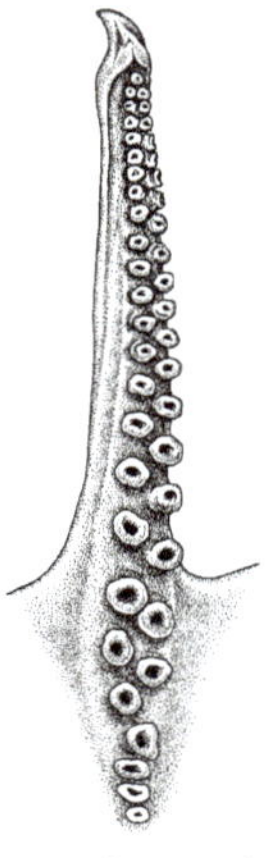

Hectocotylus, oral view (ABRS, K. Hollis).

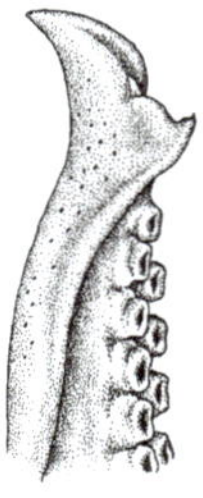

Hectocotylus, side view (ABRS, K. Hollis).

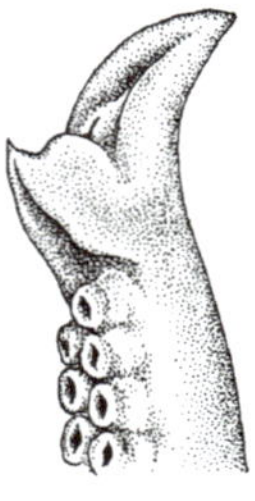

Hectocotylus, ventrolateral view (ABRS, K. Hollis).

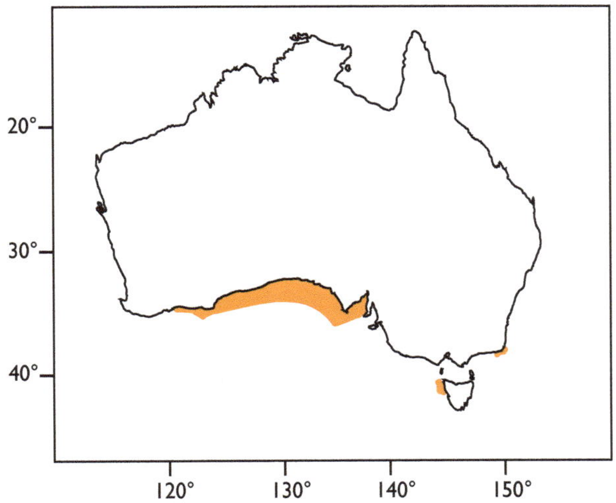


*Grimpella thaumastocheir* distribution.

**DISTRIBUTION**
Southern Australia.

**REFERENCE**
Norman *et al.* (2014).

## *Hapalochlaena* Robson, 1929

**COMMON NAME**
Blue-ringed Octopuses

**DISTINGUISHING CHARACTERS**
Distinctive pattern of iridescent blue rings and/or lines over dorsal surfaces. Ink sac reduced.

**DIAGNOSTIC FEATURES**
Muscular species. Mantle ovoid; posterior tip pronounced in some postures. Funnel organ W-shaped. Arms 1.5–2.0 × ML; lateral arms longest, dorsal arms always shortest. No arm autotomy. Deepest webs (on lateral arms) ~20–35% longest arm length. Suckers biserial. No enlarged suckers. Gills with 5–7 lamellae per demibranch. Ink sac present (reduced in adults of some species); anal flaps present. Third right arm of male hectocotylised, slightly shorter than opposite arm. Ligula and calamus present. Spermatophores short, narrow. Eggs small or large (depending on species), laid in strings and brooded in web by females. Stylets absent.

*Colour.* Distinctive rings and/or lines of iridocytes producing brilliant iridescent blue in life. Ring size varies between species. Transverse pair of white spots on dorsal mantle, slightly anterior to midpoint of mantle.

*Sculpture.* Skin with small, low papillae. Diamond arrangement of longitudinal flap-like papillae on mid-dorsal mantle. Large papilla on posterior tip of mantle. Skin ridge not present around mantle margin.

**REMARKS**
Only three species are currently recognised in Australian waters: *Hapalochlaena maculosa* (Hoyle, 1883) (southern Australia), *H. fasciata* (Hoyle, 1886) (eastern Australia), and *H.* cf. *lunulata* (Quoy and Gaimard, 1832) (northern Australia). The genus includes several undescribed species and is currently being revised by Julian Finn from Museum Victoria.

Blue-ringed Octopuses are extremely venomous and all three species have been reported as responsible for human fatalities. *Direct handing of any Blue-ringed Octopus should be avoided.*

The tetrodotoxin (TTX) venom is produced by symbiotic bacteria. Williams *et al.* (2011) have found that female *H. fasciata* from Sydney, New South Wales and *H. lunulata* from Bali, Indonesia invest TTX into their offspring during the earliest stages of egg formation (TTX was found in the ovaries of females before the eggs had differentiated) and may face an evolutionary trade-off between provision of TTX and retaining TTX for their own defence. Female TTX levels were shown to correlate with those of their offspring. Tetrodotoxin levels increased throughout egg and paralarval development, even after oviposition.

Tests of the defensive use of TTX in *H. lunulata* paralarvae showed that TTX alone is not a deterrent to a range of predators tested. While individual paralarvae were unpalatable to most predators, they did not find TTX-spiked food distasteful, so some other substance may signal to predators the onset of more detrimental symptoms of TTX (Williams *et al.* 2011).

**HABITAT AND BIOLOGY**
Shallow coastal waters on rocky reefs and coral reefs, seagrass and algal beds. Collected at 0–50 m.

*Hapalochlaena* flash iridescent aposomatic body patterns to warn of a tetrodotoxin-laced bite; these rings can also be visible when the animal is resting.

**REMARKS**
According to Julian Finn (personal communication), there have been several reports of Blue-ringed Octopuses swimming near the surface offshore over the years. When

a Blue-ringed Octopus is disturbed it may leave the reef and swim up into the water column. It would seem that they do this more so than other octopus species. There have also been numerous reports of Blue-ringed Octopuses washing up on beaches, often after periods of extreme hot weather. During the life cycle, when most females are brooding eggs, males may move over large areas looking for females.

#### SIZE

ML to 57 mm, total length to 220 mm.

#### DISTRIBUTION

Temperate to tropical waters throughout the Indo-West Pacific, north to Japan and south to southern Australia.

#### REFERENCES

Cheng and Caldwell (2000), Norman *et al.* (2014).

### Key to *Hapalochlaena* species

1a. Blue lines or bars (not rings) on dorsal mantle........ ........................................................................ *H. fasciata*
1b. Blue rings on dorsal mantle....................................2
2a. Cream to yellow base colour with large iridescent blue rings covering dorsal surface of mantle and arms.....................................................*H.* cf. *lunulata*
2b. Grey-green to cream base colour covered with ~50–60 small dark spots containing small iridescent blue rings (0.5–2.0 mm in diameter) ............... ..................................................................... *H. maculosa*

## *Hapalochlaena fasciata* (Hoyle, 1886)

*Octopus pictus fasciata* Hoyle, W. E. (1886). Report on the Cephalopoda. *Report on the Scientific Results of the Voyage of H.M.S. Challenger During the Years 1873–76, Zoology 16*(44), 1–245 [94, pl. 8 fig. 3].

Type data: holotype BMNH 1889.4.24.37.

Type locality: Australia: New South Wales, Port Jackson, 33°50′S, 151°17′E.

#### COMMON NAME

Blue-lined Octopus

#### DIAGNOSTIC FEATURES

Webs of moderate depth, deepest around one-third of arm length. Web deepest on lateral arms, webs between dorsal arms shallowest. Web margins extending to arm tips. In larger animals, ~100 biserial suckers on each normal arm. Male hectocotylus (third right arm) shorter than opposite arm with 32–42 suckers. Ligula of moderate size, 7–12% of hectocotylised arm length; calamus large. Gills with five to seven lamellae per demibranch. Eggs large, ~6–9 mm diameter.

*Colour.* Cream to orange base colour with iridescent blue lines (not rings) on dorsal mantle and single or linked blue rings on arm crown and arms.

#### HABITAT AND BIOLOGY

Occurs on intertidal and shallow rocky reefs. Primarily active at night feeding on small crustaceans. Females carry their eggs within the web, brooding them until hatching. The hatchlings are benthic. This species is extremely venomous. Tetrodotoxin venom is produced in

Live animal (M. Dowse).

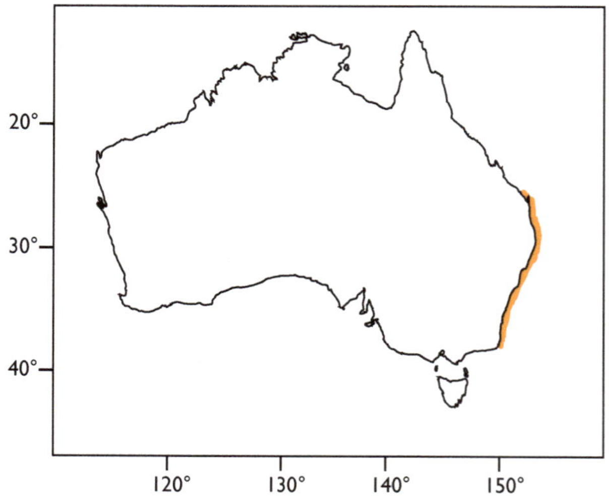


*Hapalochlaena fasciata* distribution.

the salivary glands and distributed throughout the body organs and skin of this species. It has been responsible for at least one human death.

*Interest to fisheries.* No fisheries value.

### SIZE

ML to ~45 mm, total length to ~150 mm.

### DISTRIBUTION

Subtropical waters of eastern Australia from southern Queensland to southern New South Wales. Known depth range 0–20 m.

### REFERENCES

Stranks (1998), Norman *et al.* (2014).

Live animal, dorsal view (specimen from East Timor) (G. Rouse).

## *Hapalochlaena* cf. *lunulata* (Quoy and Gaimard, 1832)

*Octopus lunulatus* Quoy, J. R. and Gaimard, J. P. (1832). Mollusques. *Voyage de decouvertes de l'Astrolabe pendant les annees 1826-1827-1828-1829. Zoologie, 2*(1), 1–320 [86, pl. 6, figs 1, 2].

Type data: syntype MNHN 4-12-973.

Type locality: Papua New Guinea: New Ireland (Bismark Archipelago), Havre Carteret.

### COMMON NAME

Greater Blue-ringed Octopus

Live animal (Western Australia, NW Cape) (C. Bryce).

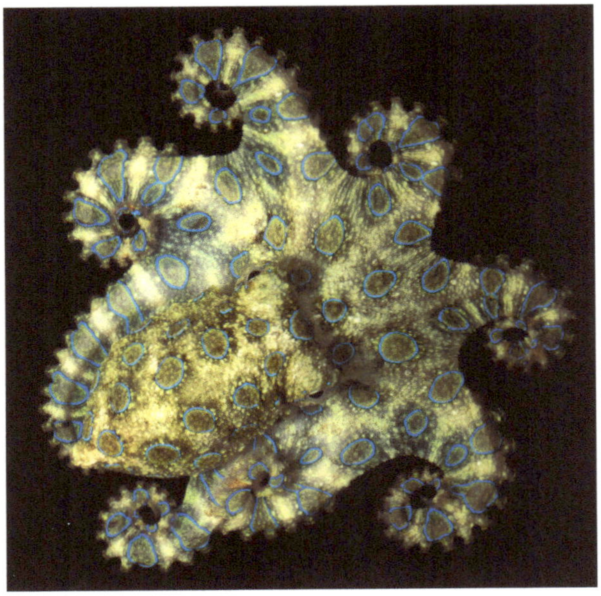

Live animal, dorsal view (specimen from East Timor) (G. Rouse).

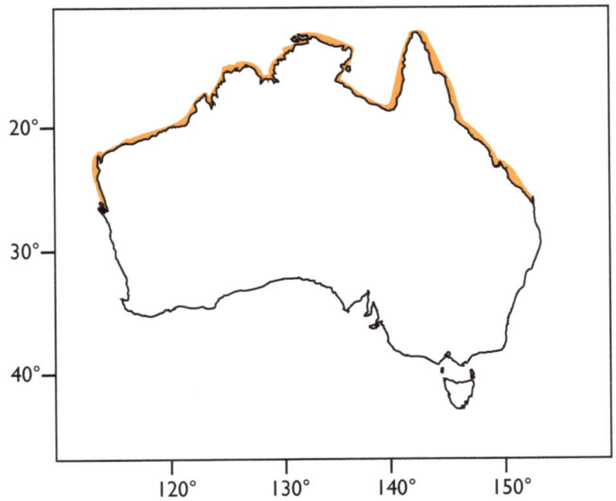


*Hapalochlaena lunulata* distribution.

**DISTINGUISHING CHARACTERS**

*Hapalochlaena* cf. *lunulata* differs from *H. maculosa* (Hoyle, 1883) in the size of the iridescent blue rings, which are ~12 mm diameter in this species and only 1–2 mm diameter in *H. maculosa*.

**DIAGNOSTIC FEATURES**

Tan to orange base colour with large iridescent blue rings, up to ~12 mm diameter, on dorsal mantle, arm crown, and arms.

**REMARKS**

Numerous species, including large-ringed species from Australia, have been treated under this species name. The genus is currently under review. *Hapalochlaena lunulata* brood their eggs until hatching.

**SIZE**

ML up to 50 mm, total length to ~120 mm.

**DISTRIBUTION**

Northern Australian waters.

**REFERENCE**

Norman *et al.* (2014).

Live animal (P. Mercurio).

Live animal (P. Mercurio).

*Hapalochlaena maculosa* with captured stomatopod *Gonodactylaceus falcatus* (R. Caldwell).

## *Hapalochlaena maculosa* (Hoyle, 1883)

*Octopus maculosus* Hoyle, W.E. (1883). On a new species of Octopus (*O. maculosus*). *Proceedings of the Royal Physical Society of Edinburgh 7*, 319–322 [319, pl. 6].

Type data: holotype NMGW Z.78.14.421.

Type locality: Australia.

**COMMON NAME(S)**

Lesser Blue-ringed Octopus, Southern Blue-ringed Octopus

**DIAGNOSTIC FEATURES**

Small muscular octopus. Funnel organ outer limbs ~75% length of medial limbs. Short arms up to 3 × ML. Lateral and ventral arms longest (typically 4 = 3 = 2 > 1). Deepest web ~20–30% of arm length, deepest on lateral arms, shallowest between dorsal arms. Web margins extend to arm tips. Male hectocotylus (third right arm) ~75–80% length of opposite arm with 45–56 suckers. Ligula of moderate size, 7–13% of arm length; calamus large, 33–62% of ligula length. Spermatophores of moderate size, ~43–118% ML. Eggs large, ~7–9 mm.

*Colour.* Cream to grey-green base colour covered with ~50–60 small dark spots containing small iridescent blue rings (~0.5–2.0 mm in diameter) on dorsal mantle

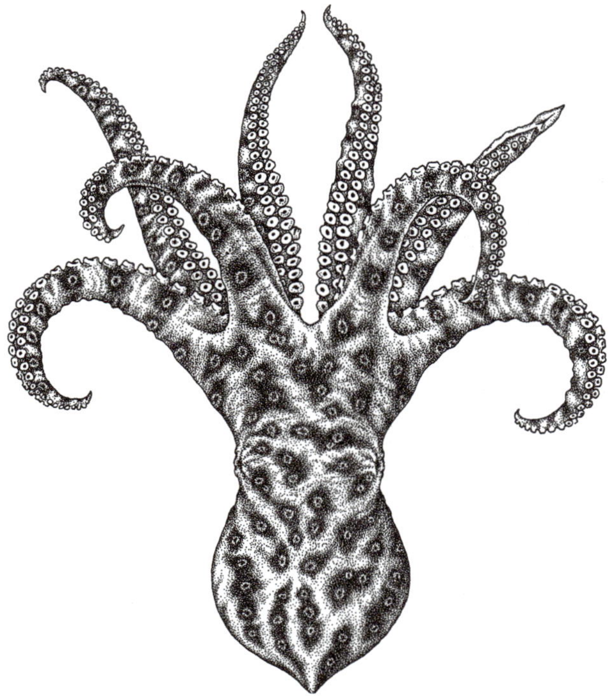

Dorsal view (ABRS, K. Hollis).

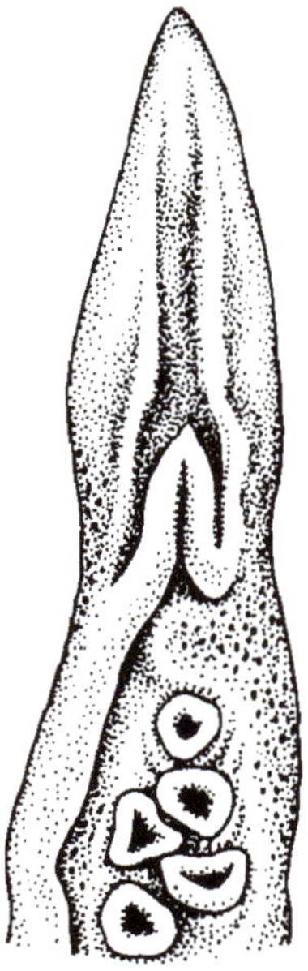

Hectocotylus oral view (ABRS, K. Hollis).

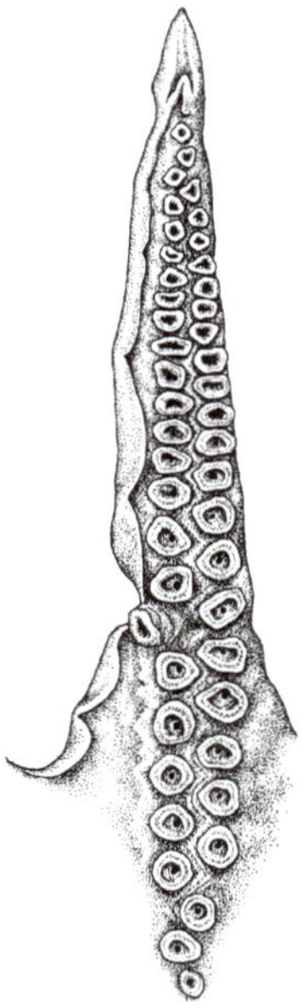

Hectocotylised arm (ABRS, K. Hollis).

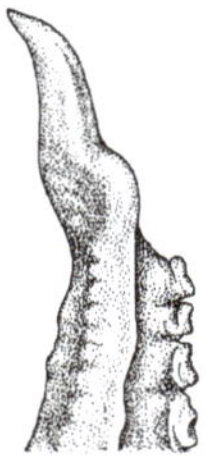

Hectocotylus dorsolateral view (ABRS, K. Hollis).

Hectocotylus ventrolateral view (ABRS, K. Hollis).

and single linked blue rings on arm crowns and arms. Transverse pair of white spots present on dorsal mantle, slightly anterior to mantle midpoint.

*Sculpture.* Skin with wrinkled texture containing small, closely set, low papillae. Diamond of four short longitudinal ridges on dorsal mantle. No skin ridge around mantle edge.

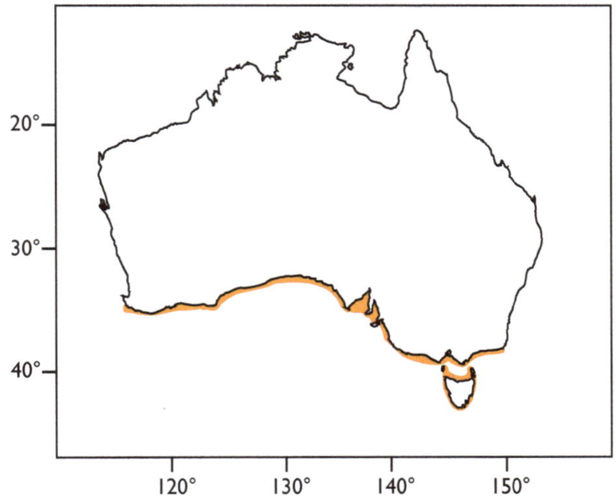


*Hapalochlaena maculosa* distribution.

### HABITAT AND BIOLOGY

Found on intertidal and shallow subtidal reefs, seagrass meadows and rubble areas to depths of at least 55 m. It emerges mainly at night to forage. Females carry eggs within the web. Hatchlings are benthic. Tetrodotoxin has been detected in the eggs and body of this species.

### REMARKS

Extremely venomous with the potential to cause human deaths.

### SIZE

Up to 57 mm ML, total length to ~220 mm.

### DISTRIBUTION

Southern Australia from eastern Victoria to south-western Western Australia.

### REFERENCE

Norman *et al.* (2014).

## *Histoctopus* Norman, Boucher-Rodoni & Hochberg, 2009

### DIAGNOSTIC FEATURES

Mantle muscular, globose. Funnel organ W-shaped. Arms 2.5–3.8 × ML, approximately equal in length (lateral arms slightly shorter). No arm autotomy. Web moderate to deep, deepest 20–30% longest arms, slightly deeper on lateral arms.

Web margins on ventral faces of all arms very wide, form loose semi-transparent membranes in preserved material. Suckers biserial. Males with enlarged suckers on all arms. Gills with eight or nine lamellae per demibranch. Radula with seven series of teeth and marginal plates. Rhachidian tooth with two to three lateral cusps in asymmetrical seriation. Posterior salivary glands moderate to large, approximately equal in length to buccal mass. Distinct crop present as side branch off oesophagus. Ink sac and anal flaps present. Third right arm of male hectocotylised, ~75–85% length of opposite arm. Ligula 4.5–8% of hectocotylised arm length, triangular to spoon-shaped with open groove and raised transverse creases; calamus large and pronounced, ~40–60% of ligula length. Terminal organ (penis) large and T-shaped with diverticulum longer than distal portion. Spermatophores large, equal to, or longer than, ML, bulbous with swollen, short sperm reservoir containing < 10 coils of sperm in a double strand. Distal tip of spermatophore storage sac with reticulate glandular tissue. Eggs moderate size, 6.5–8.0 mm in diameter (7–9% ML). Long chitinous stylets present.

*Colour.* Orange-brown dorsally, cream ventrally. Transverse rows of small, white papillae (leucophores) on dorsal mantle in live animal and some preserved material. Ocelli absent.

*Sculpture.* Evenly spaced, small, rounded papillae. No skin ridge around lateral margin of mantle.

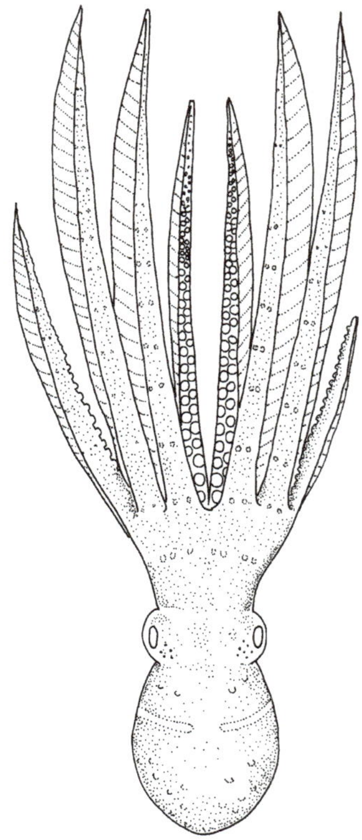

Dorsal view (A. Reid, redrawn from original description).

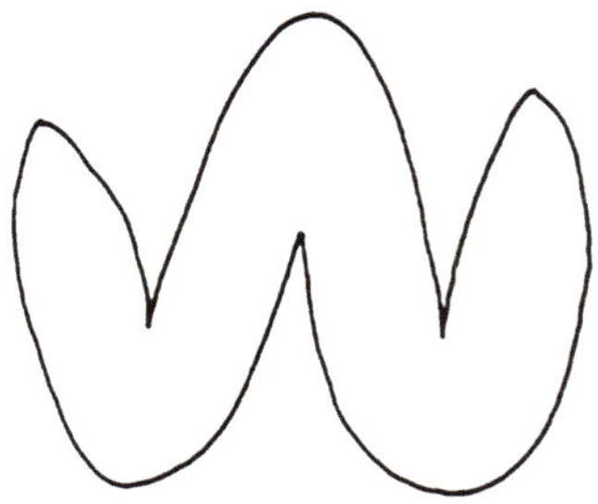

Funnel organ (A. Reid, redrawn from original description).

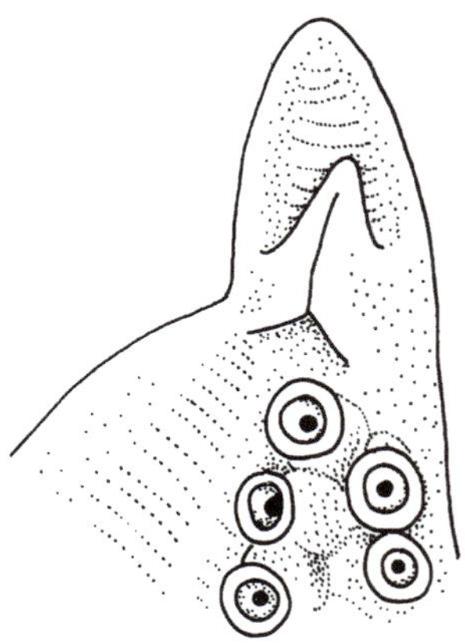

Hectocotylus (A. Reid, redrawn from original description).

### HABITAT AND BIOLOGY

Found from 350 m to 550 m.

### SIZE

Up to 107 mm ML, total length to 380 mm.

### DISTRIBUTION

South-west Pacific Ocean (New Caledonia) and south-eastern Indian Ocean (Western Australia).

## *Histoctopus zipkasae* Norman, Boucher-Rodoni & Hochberg, 2009

*Histoctopus zipkasae* Norman, M., Boucher-Rodini, R., and Hochberg, F. G. (2009). A new genus and two new species of octopuses from Australia and New Caledonia. *Journal of Molluscan Studies 75*, 323–336 [329, figs 5–8].

Type data: holotype MV F.164025.

Type locality: Australia: Western Australia, North West Shelf.

### COMMON NAME(S)

Zippy's Octopus

### DIAGNOSTIC FEATURES

Medium-sized species, ML to at least 107 mm. Funnel organ large. Arm lengths ~3 × ML (longest 2.4–3.7 × ML).

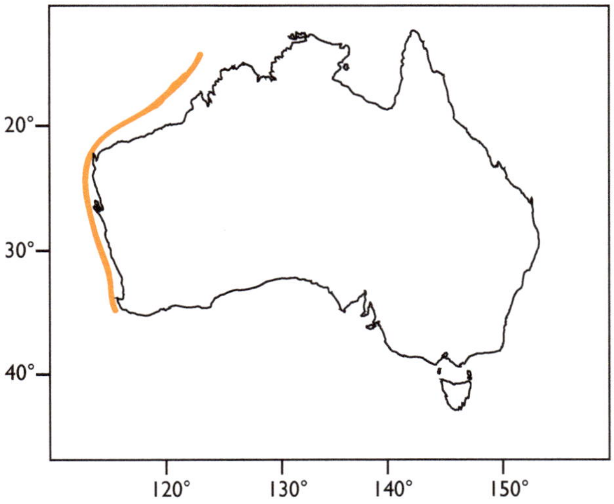


*Histoctopus zipkasae* distribution.

Arms approximately equal in length, dorsal arms slightly shorter. Web sectors very thin and membranous, deep, up to one-third of longest arm length. Webs approximately equal in length, lateral webs slightly deeper. Webs extend as membranous flared margins along entire length of arms, very well developed towards arm tips. Approximately 168 biserial suckers on normal arms. Two or three slightly to moderately enlarged and flattened suckers present on all arms of mature males, diameter up to 15% ML at level of 7th to 13th proximal sucker. Male hectocotylus (third right arm) ~50–70% length of opposite arm with 66–68 suckers. Ligula small to moderate size (4.5–6.2% of arm length), spoon-shaped with robust rings and wide, open groove, floor with ~10–12 simple transverse creases; calamus large and prominent, 34.7–59.7% of ligula length. Gills with nine lamellae per demibranch. Ink sac and anal flaps present. Spermatophores large (0.9–1.1 × ML) and few (3–8). Distal oviducts thickened along length. Eggs of moderate size (6.5–7.8 mm, 7.2–9.1% ML).

*Colour.* Pink-orange dorsally, cream ventrally. Orange-brown basal colour on first three arm pairs. Two transverse rows of small, white spots on dorsal mantle. Regularly spaced white spots present on aboral midline of first three arm pairs.

*Sculpture.* Small, round papillae uniformly scattered on all dorsal surfaces; larger supraocular papillae absent. Lateral mantle ridge absent.

#### HABITAT AND BIOLOGY

Video and still night photography suggests nocturnal to crepuscular activity patterns; however, it is not known whether the captured animals were reacting to the remotely operated camera and lights. They were found on fine sand/silt and low rocky rubble. Known depth range 350–450 m.

#### SIZE

Mantle length to 107 mm; total length to 380 mm.

#### DISTRIBUTION

Indian Ocean, off Western Australia from Perth Canyon to North West Shelf. Known depth range 384–449 m.

#### REFERENCE

Norman *et al.* (2014).

### *Octopus* s. s. Cuvier, 1797

#### DIAGNOSTIC FEATURES

Muscular mantle and arms. Arms typically 3–5 × ML. Water pouches absent. Two series of suckers on all arms, hectocotylus on right side. Male terminal organ ('penis') with diverticulum. Ink sac and anal flaps well developed. Iridescent blue rings or lines absent. Arms without dark leading arm edges. Sculpture on oral surface of dorsal web not a continuation of mantle and aboral web sculpture. Four large primary papillae in diamond arrangement on dorsal mantle. Stylets present.

#### REMARKS

This genus has been shown in several recent molecular studies to be polyphyletic and a major generic revision is required. At this time only a differential diagnosis can be provided for this genus. Members of the genus are artificially grouped by the absence of traits that define other genera in the Octopodidae.

#### REFERENCE

Norman *et al.* (2014).

### Key to *Octopus* species

1a. Raised keel (lateral ridge) around lateral mantle, may be complete or incomplete ............................... 2
1b. Lateral ridge absent ................................................... 6
2a. Eyes large. Arms long, usually > 5 × ML ................ 3
2b. Eyes small, prominent. Arms < 5 × ML .................. 4
3a. Lateral mantle ridge complete, unbroken. Funnel organ W-shaped. Gills with seven or eight lamellae per demibranch. Dorsal mantle and arm crown with scattered pink-red raised patches....... *O. bulbus*

3b. Lateral mantle ridge incomplete, broken. Funnel organ VV-shaped. Gills with nine or 10 lamellae per demibranch. Orange-red with numerous small white spots on dorsal mantle; regular narrow transverse rows of small white spots along arms .... ........ *O. bunurong*

4a. Funnel organ V V-shaped. Mature males with four or five slightly enlarged suckers on lateral arms. Ligula stout, bulbous ........ *O. australis*

4b. Funnel organ VV or W-shaped. Mature males without enlarged suckers on lateral arms. Ligula pointed or conical ........ 5

5a. Mantle lateral ridge may be incomplete posteriorly. Normal arms with ~220 suckers. Ligula elongate, conical, tapering; length 11–16% hectocotylised arm length. Gills with seven or eight lamellae per demibranch. Dark brown bar extends from posterior to each eye, through eyes and onto arms. Maximum ML 106 mm ........ *O. berrima*

5b. Mantle lateral ridge complete. Normal arms with ~150 suckers. Ligula leaf-shaped, pointed; length ~6% hectocotylised arm length. Gills with six lamellae per demibranch. No dark bar extending from posterior to eye onto arms. Maximum ML 25 mm ........ *O. micros*

6a. Ocelli present as dark oval spot bordered by additional thin dark ring ........ *O. cyanea*

6b. Ocelli absent ........ 8

7a. Mature animals < 25 mm ML. Distinct pair of dark chromatophores on ventroposterior mantle and longitudinal row of dark pigment spots along ventral arms ........ *O. bocki*

7b. Mature animals > 25 mm ML. Without pair of dark spots on posterioventral mantle and in row along ventral arms ........ 8

8a. Blue-black dorsal to eyes. Band of chromatophores across medioventral mantle ........ *O. pyrum*

8b. Not obviously blue-black dorsal to eyes. No band of chromatophores across medioventral mantle ... 9

9a. Funnel organ VV- or V V-shaped ........ 10

9b. Funnel organ W-shaped ........ 12

10a. Arms long, up to 7 × ML; dorsal arm pair longest. Hectocotylus < 70% opposite arm length. Supraocular papillae not prominent ........ *O. kaurna*

10b. Arms < 3.6 × ML; arms subequal in length. Hectocotylus > 80% opposite arm length. Prominent supraocular papillae ........ 11

11a. Hectocotylised arm with > 70 suckers. Ligula groove deep. Skin sculpture tile-like. No single papilla forming posterior point in mantle ........ *O. pallidus*

11b. Hectocotylised arm with < 70 suckers. Ligula groove long, shallow with ridges. Skin sculpture with rows of simple papillae but texture not tile-like. Single papilla forms posterior point on mantle ........ *O. superciliosus*

12a. Anal flaps greatly reduced or absent ........ 13

12b. Anal flaps present, not reduced ........ 14

13a. Mature males without enlarged suckers. Hectocotylised arm with ~50 suckers; very short, 20% opposite arm length. Gills with 10 or 11 lamellae per demibranch. Anal flaps absent. Skin smooth ... ........ *O. harpedon*

13b. Mature males with enlarged suckers on arms 2 and 3. Hectocotylised arm with > 80 suckers; not short, 66–84% opposite arm length. Gills with five or six lamellae per demibranch. Anal flaps reduced, tiny. Large papillae can be erected over body ... *O. laqueus*

14a. Dorsal arms longest. Gills with 13–15 lamellae per demibranch. Maximum size 300 mm ML ........ *O. maorum*

14b. Arms subequal in length or lateral arms longest. Gills with six to nine lamellae per demibranch. Maximum size 140 mm ML ........ 15

15a. Lateral arms longest. Mature males with three to five enlarged suckers on arms 2 and 3. Ligula tiny, 1–2% hectocotylised arm length. Body sculptured with rosette- shaped patches and grooves with erect central knobs separated by distinct grooves ........ *O. tetricus* (eastern Australia); *O.* cf. *tetricus* (western Australia)

15b. Arm lengths subequal. Both sexes with 6–10 enlarged suckers on all arms. Ligula 6–10% hectocotylised arm length. Sculptured with tubercles and branched and unbranched papillae and fine, rounded closely set tubercles ........ *O. warringa*

## *Octopus australis* Hoyle, 1885

*Octopus australis* Hoyle, W. E. (1885). Diagnoses of new species of Cephalopoda collected during the cruise of H.M.S. 'Challenger.' Part 1. The Octopoda. *Annals and Magazine of Natural History* 5(15), 222–236 [224].

Type data: lectotype BMNH 1889.4.24.28–29 (♀ 23.3 mm ML); paralectotype BMNH 1889.4.24.28–29 (♂ 12.0 mm ML).

Type locality: Australia: New South Wales, Port Jackson, 33°50′S, 151°17′E.

### COMMON NAME(S)

Hammer Octopus, Southern Octopus

Fresh-caught specimen, dorsal view (K. Graham).

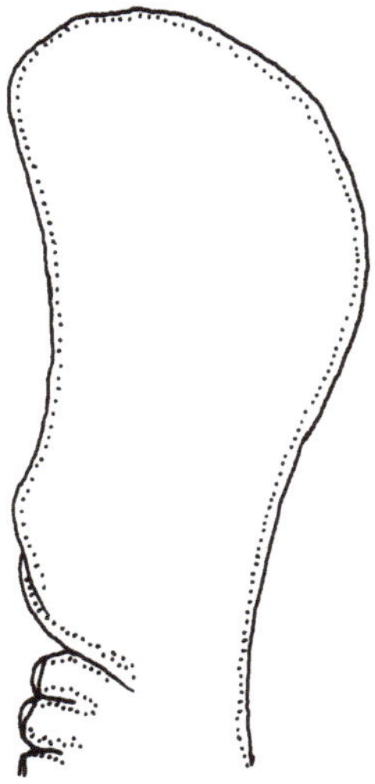

Hectocotylus, lateral view (M. Shea, redrawn from original description).

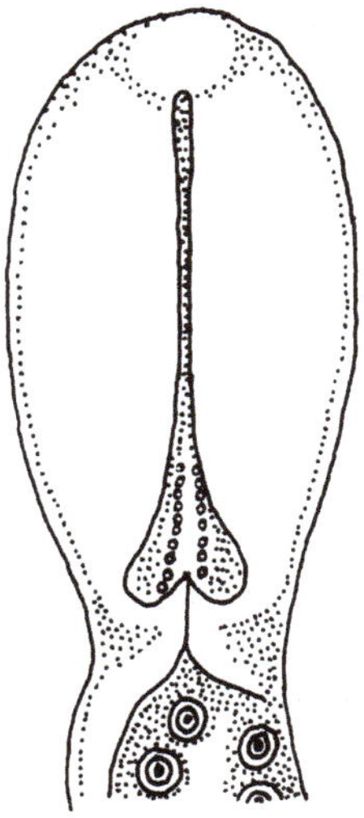

Hectocotylus, oral view (M. Shea, redrawn from original description).

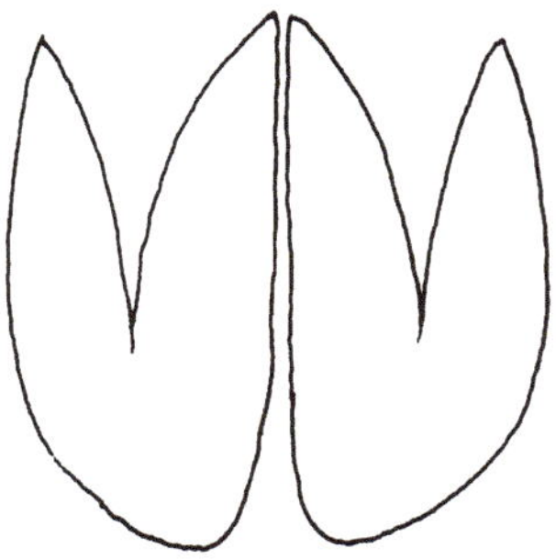

Funnel organ (M. Shea, redrawn from original description).

Hectocotylised arm (M. Shea, redrawn from original description).

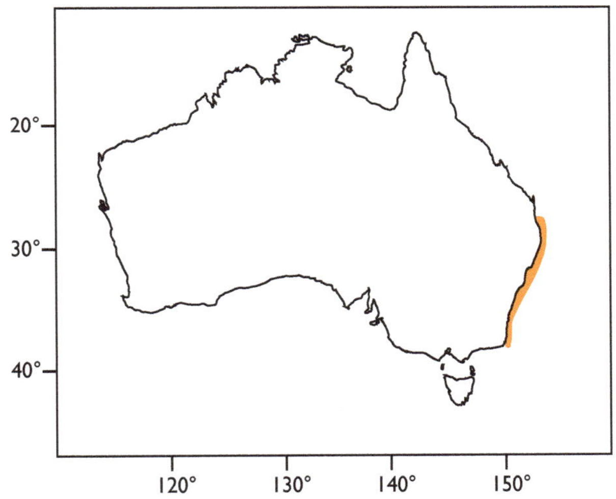


*Octopus australis* distribution (ALA, 12 Aug. 2013).

## DIAGNOSTIC FEATURES

Moderate-sized, robust species. Eyes small, prominent. Arms slender, 2.7–4.5 × ML with slender tips. Lateral arms longest; arm formula typically 3 >2 >4 >1. Webs deep, deepest between lateral arms, up to one-third of arm length. Largest animals with up to 220 biserial suckers on each normal arm; 60–80 on hectocotylised arm of male. Arm suckers without obvious radial grooves. Mature males with four or five slightly enlarged suckers on arms 2 and 3 starting around the sixteenthth proximal sucker. Gills with seven to nine lamellae per demibranch. Funnel organ closely opposed V V-shaped, limbs of approximately equal length. Male hectocotylus (third right arm) ~70–90% length of opposite arm with 60–80 suckers. Ligula large, swollen and club-like, 8–17% of arm length, with a well-marked and deep groove. Two rows of minute papillae along groove; calamus 15–30% of ligula length. Spermatophores 30–40 mm long, 70–90% ML; ejaculatory apparatus tightly coiled, with two or three coils close to the oral end. Eggs large, up to 12 mm long, ~10–20% ML.

*Colour.* Fresh-caught specimens uniform mottled yellow-tan dorsally; white ventrally. Preserved specimens cream to purple-brown dorsally, cream on ventral surfaces. White dumb-bell-shaped marking often visible between eyes. Transverse pair of small white spots on dorsal mantle. False eye-spots (ocelli) absent.

*Sculpture.* Skin texture with small regular rounded patches. Scattered small papillae over dorsal surface and single supraocular papilla over each eye. Mantle with distinct ridge around entire lateral margin, forms a sharply angled peak posteriorly.

### HABITAT AND BIOLOGY

Males mature at ~20–25 mm ML; females mature at ~50–60 mm ML. Common inshore species in eastern Australia. Lives on sand and mud substrates, or among sponges, emerging at night to forage. During the day hides in shells and human refuse, or buries in the sand.

*Interest to fisheries.* Small-scale trawl harvest for human consumption and as bait. Frequently taken as bycatch by prawn trawlers. Constitutes a high proportion of the total octopus fisheries yield from eastern Australia and is the most common species harvested (~70–80% of the landings) in New South Wales.

### SIZE

Up to 72 mm ML (females); males 20–25 mm ML; total length to ~300 mm.

### REMARKS

While currently treated under the genus *Octopus*, the generic placement of this taxon awaits a major revision of the genus (Norman and Hochberg 2005, Norman *et al.* 2014). Some reports of this species from southern Australia probably refer to *Octopus berrima* Stranks & Norman, 1993.

### DISTRIBUTION

Southern Queensland, from Hervey Bay (~25°S) to Jervis Bay, New South Wales (~35°S). Known depth range: intertidal to 134 m.

### REFERENCES

Tait (1982); Stranks and Norman (1993); Norman (2000).

## *Octopus berrima* Stranks & Norman, 1993

*Octopus berrima* Stranks. T. N. & Norman, M. D. (1993). Review of the *Octopus australis* complex from Australia and New Zealand, with description of a new species (Mollusca: Cephalopoda). *Memoirs of the Museum of Victoria 53*(2), 345–373 [355, text figs 3, 6–11].

Type data: holotype MV F.67132 (♂ 58.9 mm ML).

Type locality: Australia: Victoria, Port Phillip Bay off Mordialloc 38°02′S, 145°05′E.

### COMMON NAME

Southern Keeled Octopus

### DIAGNOSTIC FEATURES

Mantle broadly oval with firm consistency, mantle walls thin, muscular. Eyes small, prominent. Funnel organ two closely opposed V-shaped units that may be joined medially; outer limbs three-quarters length of median limbs, or approximately equal length. Arms long, subequal 2–4 × ML in mature animals. Arm formula $2 \geq 3 > 4 > 1$ or, typically $3 = 2 > 4 > 1$. Webs deep; deepest to one-third of arm length, shallowest between dorsal arms. Suckers, none enlarged; ~220 suckers on each normal arm. Male hectocotylus (third right arm) ~70–100% length of opposite arm with 66–78 suckers. Well-developed spermatophore groove with thickened web membrane. Ligula 11–16% of total arm length, recurved orally, conical, elongate with two rows minute papillae along groove; calamus short (15–20% ligula length), acutely pointed. Gills with seven or eight lamellae per demibranch. Ink sac and anal flaps present. Spermatophores relatively long, 35 mm to 50 mm, slender; ejaculatory apparatus tightly coiled without larger coils near the oral end. Eggs large, ~10–14 mm long.

*Colour.* Live, resting animals cream to mottled orange-brown to dark brown in colour with light brown mottling dorsally, white to pale cream ventrally. Dark brown bar runs from behind each eye, through eyes and onto arms. When stimulated, animals become uniform dark brown to purple-brown dorsally, white to pale cream ventrally. Two white spots on dorsal mantle, posterior to eyes. White bar between eyes. Two thin white stripes basally along dorsal arms. Ocelli absent.

*Sculpture.* Fine, rounded and closely set low tubercles, dense on dorsum, ventral tubercles smaller and less prominent. Scattered small, unbranched papillae over dorsum and four large papillae on dorsum in a triangular

Live animal (P. Mercurio).

Live animal (P. Mercurio).

Live animal (P. Mercurio).

arrangement. Single small papilla over each eye and three or four small unbranched papillae around eye region. Ventrolateral ridge around circumference of mantle, more pronounced anteriorly, may be incomplete posteriorly.

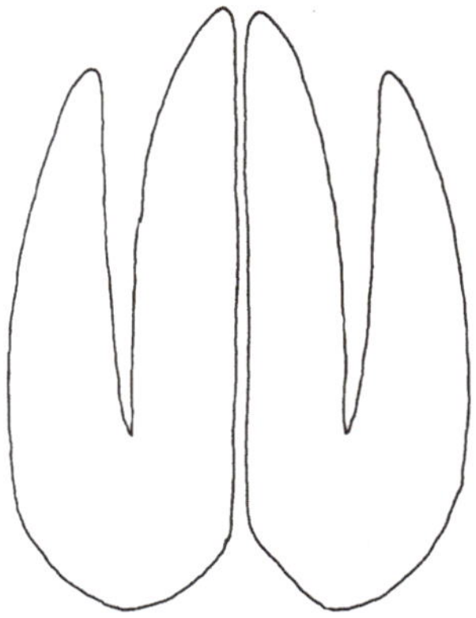

Funnel organ (M. Shea, redrawn from original description).

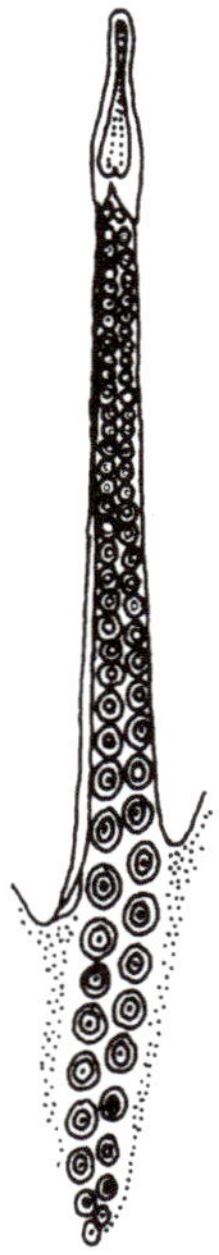

Hectocotylised arm (M. Shea, redrawn from original description).

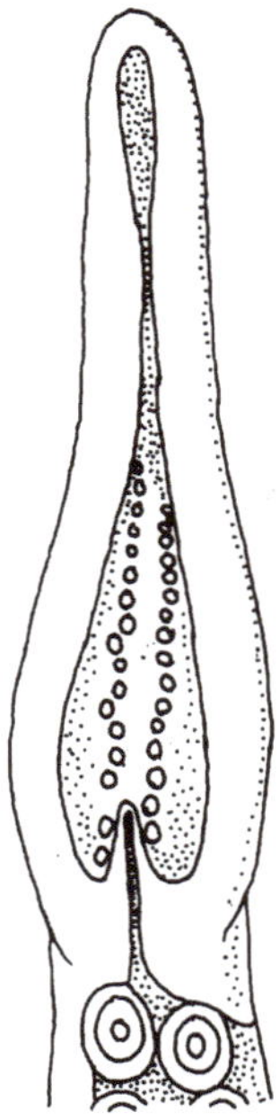

Hectocotylus, oral view (M. Shea, redrawn from original description).

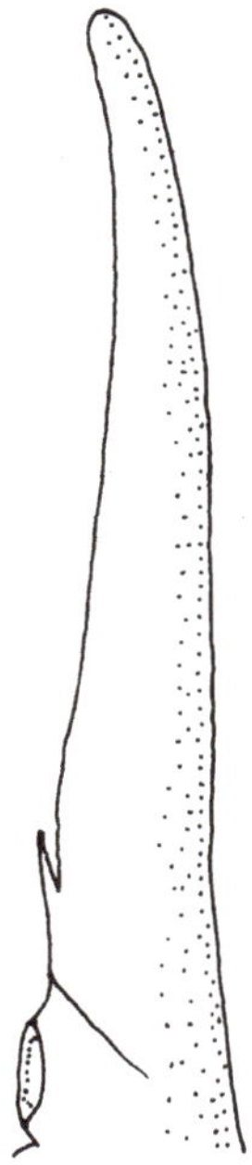

Hectocotylus, lateral view (M. Shea, redrawn from original description).

## HABITAT AND BIOLOGY

Common in inshore waters of eastern Australia, found in areas with sandy or muddy substrates. When mating, males stand on their arm tips, flare their arm webs and display a colour pattern of violet purple arm tips and a cream-coloured mantle when approaching a female before quickly grabbing her and passing spermatophores via the hectocotylus.

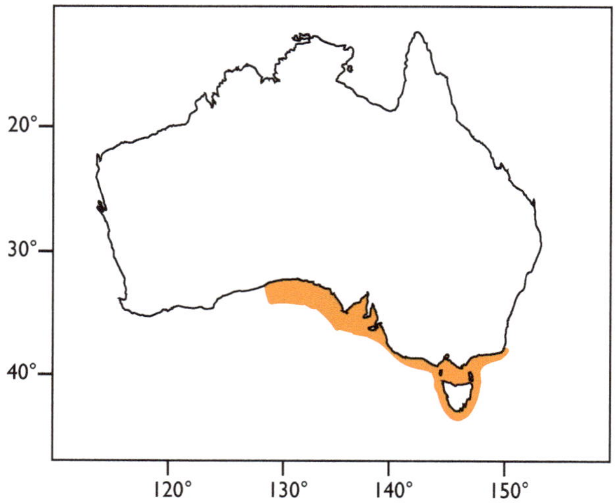


*Octopus berrima* distribution.

Eggs are large and produced in low numbers, with 4–5 mm long hatchlings immediately adopting a benthic existence. Eggs (10–14 mm in capsule length) are attached singly to the substrate by long, thin stalks. Females brood 50–130 eggs for at least 100 days in summer. Hatchlings grow to maximum size in 18–20 months. The species is an opportunistic predator feeding mainly on crustaceans, polychaetes and molluscs, including other octopuses. It forages at night and hides in shells or human refuse (such as bottles) during the day, or buries in the sand, entering arms-first. Before emerging it raises one eye like a periscope to check the surroundings before venturing out to forage.

*Interest to fisheries.* Caught incidentally during scallop and mussel dredging or in low quantities in bait-less pot fisheries. There are intermittent, small-scale fisheries for this species in Victoria, Tasmania and South Australia, with winter catch rates synchronised with mating activity resulting in high catch rates. Animals are mainly used as bait in the longline fishing industry, though several operators service the restaurant trade.

## SIZE

Up to 106 mm ML; total length ~360 mm.

## REMARKS

This species has been formerly identified and described as *Octopus. australis* Hoyle, 1885. While currently treated under the genus *Octopus*, the generic placement of this taxon awaits a major revision of the genus (Norman and Hochberg 2005, Norman *et al.* 2014).

Catalano (2013a) described a new dicyemid parasite, *Dicyemennea floscephalum*, from the renal appendages of *O. berrima* from the Upper Spencer Gulf and Gulf St Vincent, South Australia. Dicyemids are highly host species specific and can be used as biological 'tags' to assist in the resolution of cephalopod taxonomy and assist in host species' identification.

## DISTRIBUTION

South-eastern Australia from the central Great Australian Bight (~132°E) to Twofold Bay, New South Wales (~37°S) including Bass Strait and Tasmania. Depth to at least 250 m.

## REFERENCES

Tait (1982), Stranks and Norman (1993), Stranks (1996), Finn and Norman (2011c), Norman *et al.* (2014).

### *Octopus* cf. *bocki* Adam, 1941

*Octopus bocki* Adam, W. (1941). Notes sur les Cephalopodes, XIV. – Sur une nouvelle espece d'Octopode (*Octopus bocki* sp. nov.) des Iles Fidji. *Bulletin du Musee royal d'Histoire naturelle de Belgique, 17*(35), 1–3. [1].

Type data: holotype NRM.

Type locality: Fiji: Viti Levu, Namuka.

## COMMON NAME

Bock's Octopus

## DIAGNOSTIC FEATURES

Pygmy species. Funnel organ W-shaped, outer limbs shorter than median limbs. Arms 2.5–4 × ML, approximately subequal, arms 1 slightly shorter than rest. Autotomy absent. Arm webs deep, deepest 19–30% of longest arm. Suckers large, tubular in shape with wide acetabulum. Mature males with enlarged, tubular suckers on arms 2 and 3; slightly enlarged suckers on arms 4. Normal arms with 100–116 suckers. Male hectocotylus (third right arm) with ~64 suckers; ligula small and pointed with shallow, wide groove. Gills with five lamellae per demibranch. Ink sac deeply embedded in digestive gland. Spermatophores 38–40% ML; sperm cord with 55–60 regular whorls. Mature eggs 1–2 mm in size.

*Colour.* Preserved specimens with numerous, small, dark red-brown chromatophores over dorsal surface and lateral mantle. Patches of faint iridescent purple on arm crown at base of each arm and anterior to eyes. Similar

Live animal (R. Caldwell).

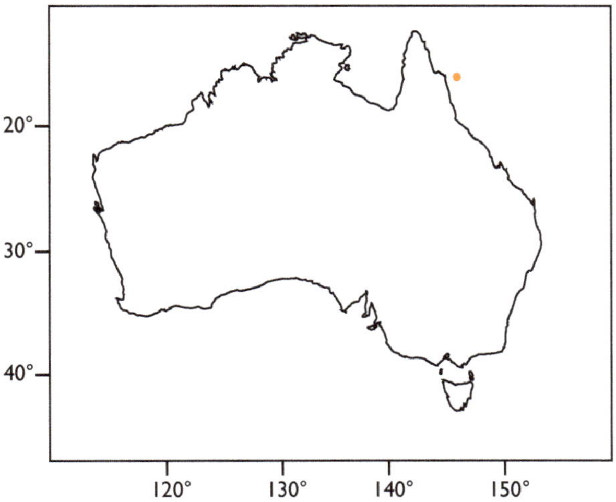


*Octopus* cf. *bocki* distribution.

purple colour in transverse bars on head and as irregular longitudinal lines along dorsal mantle. Distinct pair of dark chromatophores towards posterior end of ventral mantle margin and in single line along aboral midline of each ventral arm. Blue-black pigmentation under skin on eyes produces a 'bruised' appearance. Dorsal white spots absent.

*Sculpture.* Smooth except for small, low granular bumps over eyes. Primary papillae not visible in preserved specimens.

#### HABITAT AND BIOLOGY

Coral reef, rubble, sand and sandy mud.

#### SIZE

Up to 25 mm ML.

#### REMARKS

The diagnostic features listed above are based on the original description and specimens from the Philippines as described in Norman and Sweeney (1997). The photographed specimen from Lizard Island is recognised as similar to *Octopus bocki* Adam, 1941, but the Australian taxon needs to be examined in more detail to determine whether it does belong to this species.

#### DISTRIBUTION

Fiji, Philippines, Australia: Lizard Island.

#### REFERENCES

Norman and Sweeney (1997), Norman (2000).

### *Octopus bulbus* Norman, 2001

*Octopus bulbus* Norman, M. D. (2001). New octopus species from Queensland. *Memoirs of the Queensland Museum 46*(2), 677–690 [680, figs 3–5, 9C. 10D–F].

Type data: holotype MV F.87067 (♂ 41.1 mm ML).

Type locality: Australia: Queensland, east of Mooloolaba, 16–20 miles north of Cape Moreton.

#### COMMON NAME

Bulb-tip Octopus

#### DISTINGUISHING CHARACTERS

*Octopus bulbus* shares some traits with *O. australis* Stranks and Norman, 1993, but the dorsal arms are proportionally longer (4.9–5.6 × ML *v.* 2.7–4.3 × ML) and differ in their relative lengths (arm formula 1 > 2 > 3 > 4 in *O. bulbus v.* 3 > 2 > 4 > 1 in *O. australis*), and webs are deeper in this species (web formula A = B = C > D > E *v.* D > C > B > E > A in *O. australis*). Mature males lack enlarged suckers. The hectocotylus is proportionally shorter in this species than in *O. australis* (56% *v.* 66–86% length of opposite arm). The spermatophores have 24 sperm cord whorls compared with > 60 in *O. australis.*

#### DIAGNOSTIC FEATURES

Mantle elongate to spindle-shaped. Eyes large, pronounced. Funnel organ W-shaped with broad limbs; outer limbs slightly shorter than median limbs. Arms ~5 × ML; dorsal arms longest; arm formula 1 > 2 > 3 > 4. Arm webs shallow, thin. Approximately 200 suckers on intact arms; ~90 suckers on male hectocotylised arm. No enlarged suckers in either sex. Third right arm of males hectocotylised, ~50% length of opposite arm. Ligula large, 8% of hectocotylised arm length; elongate oval with deep, closed groove; ligula groove with very swollen lips; calamus small and sharp (< 20% of ligula length). Seven or eight gill lamellae in each

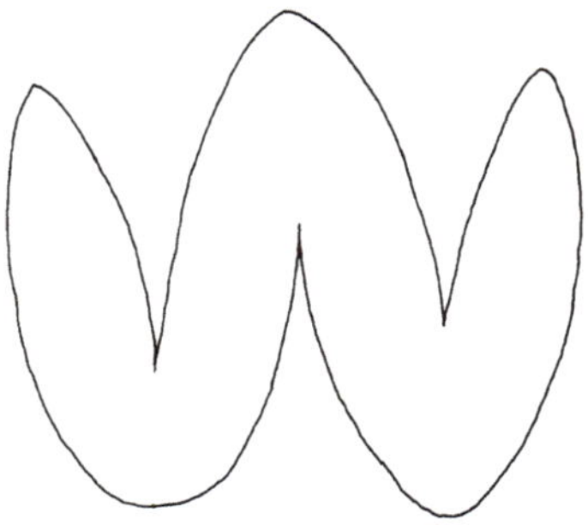

Funnel organ (M. Shea, redrawn from original description).

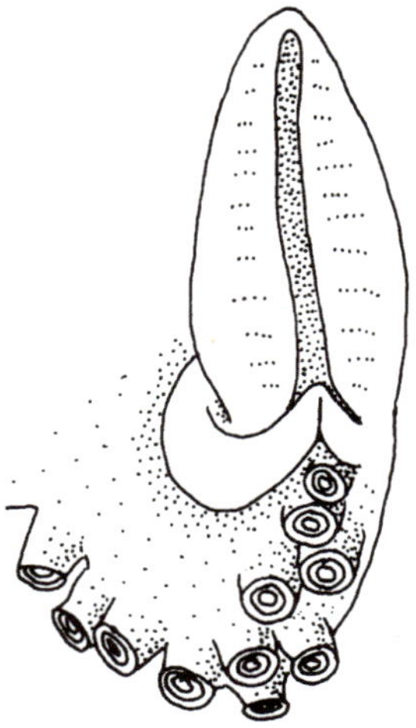

Hectocotylus, oral view (M. Shea, redrawn from original description).

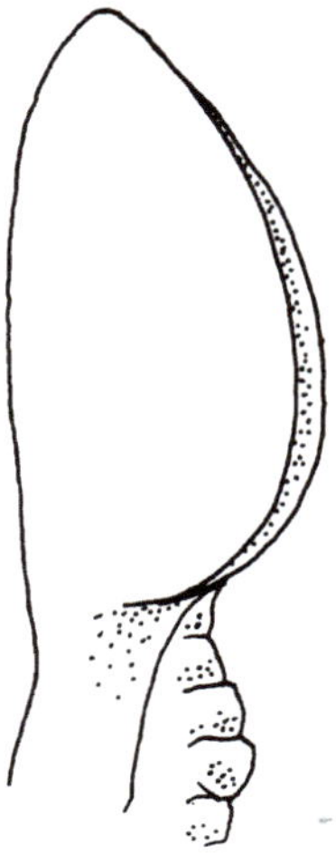

Hectocotylus, lateral view (M. Shea, redrawn from original description).

demibranch. Posterior salivary glands very large and elongate (~2 × length of buccal mass); crop diverticulum present. Anal flaps present. Rhachidian tooth with two or three lateral cusps on each side of median cone. Terminal organ ('penis') linear with diverticulum of similar length to free distal portion. Spermatophores thick and short, ~50% ML; sperm cord with ~24 whorls. Eggs large.

Lateral ridge present; small papillae on ventral mantle within lateral ridges.

*Colour.* Dorsal mantle and arm crown with scattered pink-red raised patches.

*Sculpture.* Lateral ridge present; small papillae on ventral mantle within lateral ridges.

## HABITAT AND BIOLOGY

The large eggs suggest hatchlings are benthic.

## SIZE

Up to 53 mm ML, 340 mm total length.

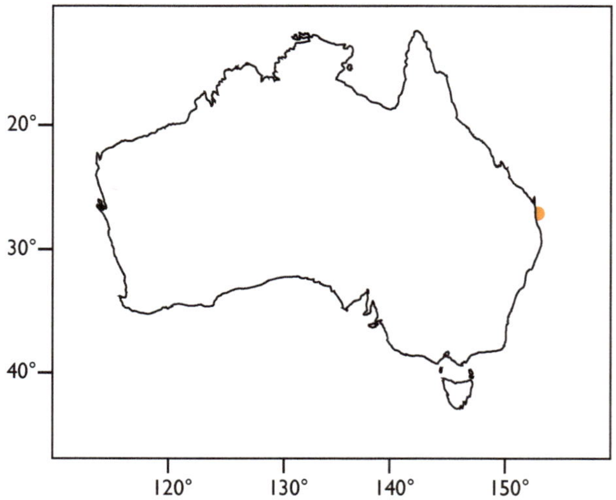


*Octopus bulbus* distribution.

## REMARKS

This species is known from only three specimens. While currently treated under the genus *Octopus*, the generic placement of this taxon awaits a major revision of the genus (Norman and Hochberg 2005, Norman *et al.* 2014).

## DISTRIBUTION

Known only from three specimens collected from southern Queensland, from east of Mooloolaba (~26°24'S 153°36'E) and off Noosa (~26°23'S, 153°41'E). Known depth range 18–195 m.

## REFERENCE

Norman (2001).

# *Octopus bunurong* (Stranks, 1990)

*Octopus bunurong* Stranks, T. N. (1990). Three new species of *Octopus* (Mollusca: Cephalopoda) from south-eastern Australia. *Memoirs of the Museum of Victoria 50*(2), 457–465 [462].

Type data: holotype MV F.53223.

Type locality: Australia: Victoria, Wilsons Promontory, Townsend Point, 38°49'S, 140°16'E.

## COMMON NAME

Southern White-spot Octopus

## DISTINGUISHING CHARACTERS

Can be distinguished from *Octopus kaurna* Stranks, 1990 by the characteristic skin patterning, the larger and more prominent eyes, larger suckers on all arms (ASIn

Fresh-caught specimen, dorsal view (K. Graham).

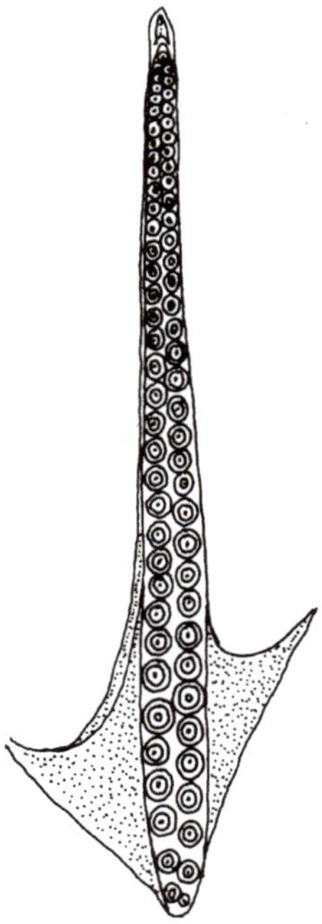

Hectocotylised arm (M. Shea, redrawn from original description).

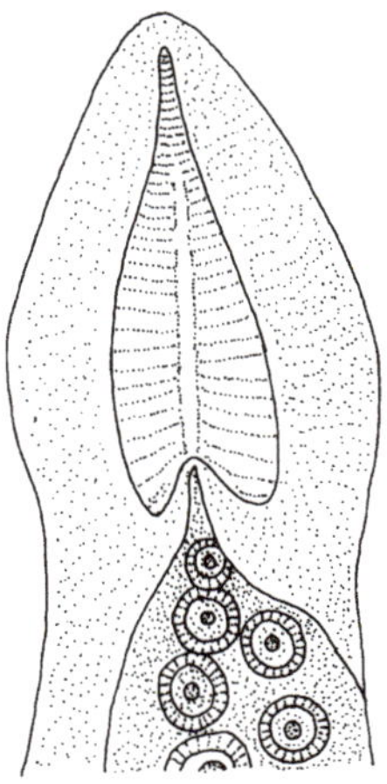

Hectocotylus, oral view (M. Shea, redrawn from original description).

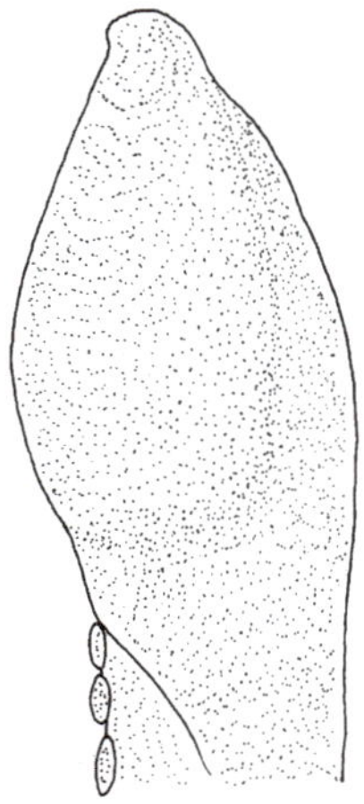

Hectocotylus, lateral view (M. Shea, redrawn from original description).

3.4–6.9–11.9 (*O. bunurong*) *v.* 3.0–5.3–8.7 (*O. kaurna*)), and the longer ligula (LLI 9.0–9.8–11.8 (*O. bunurong*) *v.* 4.7–6.0–8.0 (*O. kaurna*)).

## DIAGNOSTIC FEATURES

Mantle elongate, ovoid. Funnel organ VV-shaped; outer limbs 75% as long as median limbs. Eyes large. Arms very long, 4.1–7.3 × ML, stout; formula 1 >2 >3 >4. No arm autotomy. Webs shallow, deepest ~10–15% arm length.

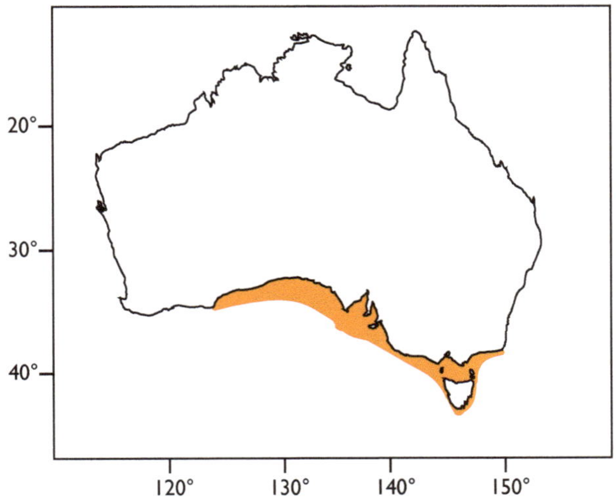



*Octopus bunurong* distribution.

Suckers raised from arm surface; not enlarged. Suckers 15–25 largest (but not markedly enlarged compared to rest). Male hectocotylus (third right arm) ~60% length of opposite arm, with 70–96 suckers. Ligula 9–12% third right arm length; ligula groove long, well marked, deep, with incomplete transverse ridges; calamus short, acutely pointed, 13–22% of ligula length. Each gill demibranch with nine or 10 lamellae. Spermatophores short, slender, ~65% ML. Eggs large 8–10 mm long.

*Colour.* Preserved specimens light brown to red-brown dorsally; paler ventrally. Dorsum mottled with numerous small white spots. Regular narrow transverse rows of small white spots along arms. Able to produce pale colour displays of pink-grey base colour with iridescent green sheen. Surface of tubercles darker than background giving a speckled appearance. Ocelli absent.

*Sculpture.* Integument with fine, rounded tubercles over smooth skin; some irregularly spaced tubercles are larger and more elongate; tubercles largest on dorsum, without large papillae over eyes. Broken ridge of skin around lateral margin of mantle.

## HABITAT AND BIOLOGY

Inshore species living on reefs or rocky areas of sand and among seagrass. Eggs are large and produced in low numbers, with hatchlings immediately adopting a benthic existence.

## SIZE

ML up to 95 mm; total length 475 mm.

### REMARKS

While currently treated under the genus *Octopus*, the generic placement of this taxon awaits a major revision of the genus (Norman and Hochberg 2005, Norman *et al.* 2014).

### DISTRIBUTION

South-eastern Australia from the Great Australian Bight to southern New South Wales including Bass Strait and northern Tasmania. Known depth range 1–130 m.

### REFERENCES

Stranks (1990), Stranks (1996), Norman *et al.* (2014).

Live animal showing eye-spots (M. Norman).

## *Octopus cyanea* Gray, 1849

*Octopus cyanea* Gray, J.E. (1849). Catalogue of the Mollusca in the Collection of the British Museum. *1. Cephalopoda Antepedia. London: British Museum (Natural History)* 164 pp. [15].

Type data: lectotype BMNH 1928.2.4.1. (♂ 112 mm ML).

Type locality: Coast of New Holland.

### COMMON NAME(S)

Day Octopus, Big Blue Octopus

Live animal (L. Vail).

### DIAGNOSTIC FEATURES

Large, robust and muscular species. Funnel organ W-shaped, with broad limbs of approximately equal length. Arms moderate to long, 4–6 × ML. Lateral arms longest, typical arm formula 4 = 3 = 2 > 1. No arm autotomy. Webs of moderate depth, deepest between lateral arms at ~15–23% of arm length. Web margins extend along ventral border of arms to arm tips. Suckers biserial, except first four closest to mouth uniserial. Large animals (mature and submature) with 450–500 biserial suckers on each normal arm. Mature males with 2–4 enlarged suckers on arms 2 and 3, from about the sixth or seventh proximal sucker. Gills with 9–11 lamellae per demibranch. Male hectocotylus (third right arm) ~70–90% length of opposite arm with 180–230 suckers. Ligula tiny, triangular, 1–2% of arm length, conical and bluntly pointed; ligula groove with transverse ridges; calamus small and blunt, 35–40% of ligula length, slightly raised. Spermatophores 30–50 mm long, 30–50% ML, produced in large numbers (hundreds to at least 300). Eggs, 2–3 mm diameter, ~2% ML.

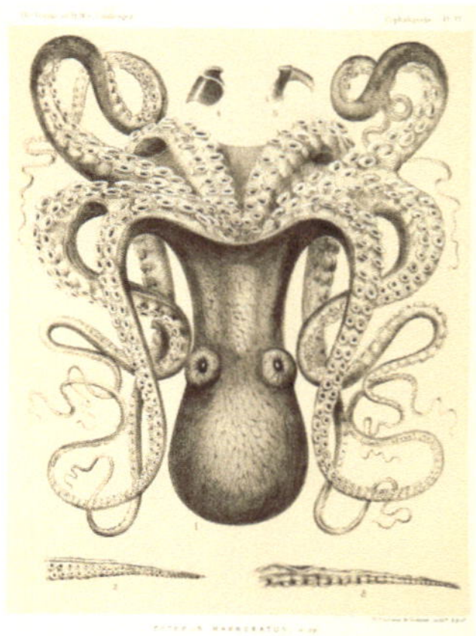

Plate VI from Hoyle 1886. *1*, dorsal view of female; *2*, oral view of hectocoylus; *3*, lateral view of distal tip of hectocotylus; *4*, upper beak; *5*, lower beak.

*Colour.* Colour variable from uniform white to various mottled patterns to uniform dark chocolate brown.

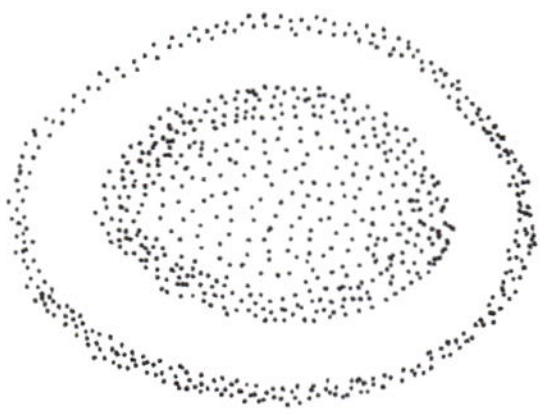

Ocellus (M. Shea, redrawn from Norman 1991).

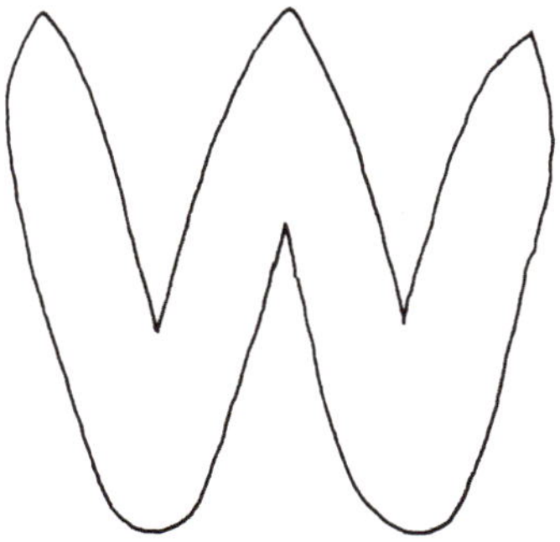

Funnel organ (M. Shea, redrawn from Norman 1991).

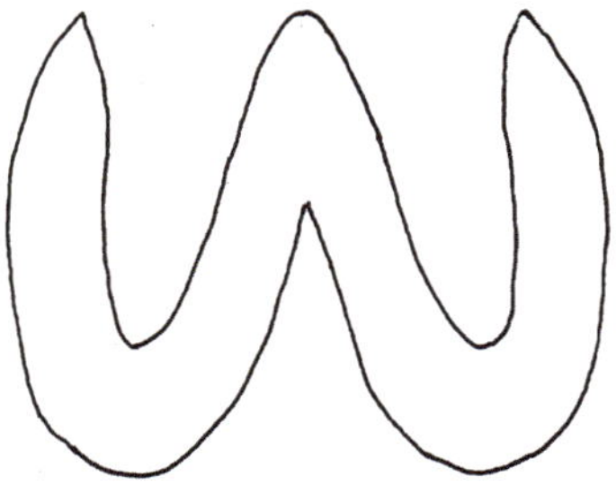

Funnel organ (M. Shea, redrawn from Norman 1991).

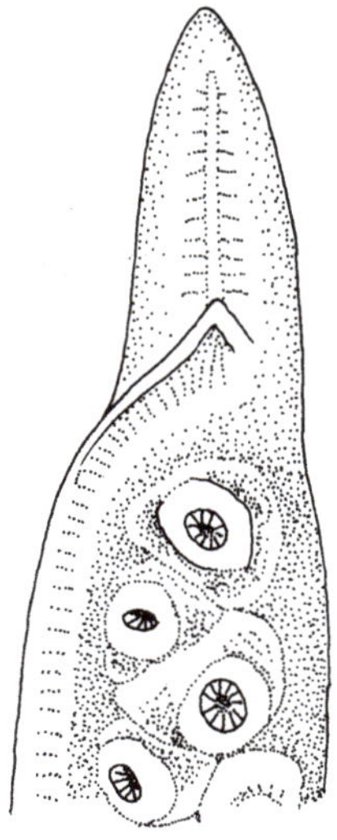

Hectocotylus, oral view (M. Shea, redrawn from Norman 1991).

False eye-spots (ocelli) on arm crown at base of third arms. Ocellus a dark oval patch within pale ring and dark, narrow, outer ring. No iridescent ring. Regular

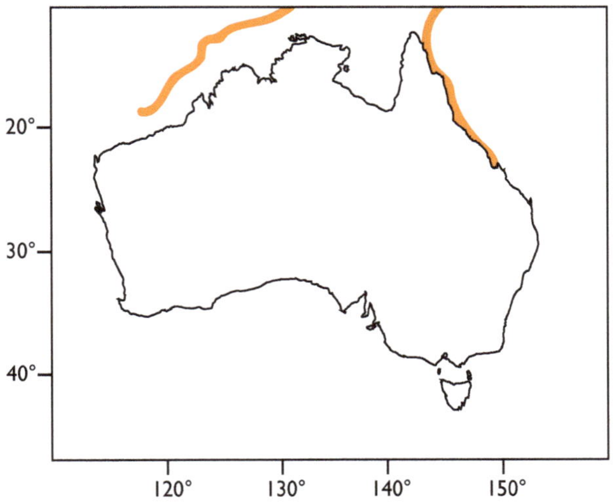


*Octopus cyanea* distribution.

short, dark bars adjacent to suckers along ventral surfaces of all arms in specimens > ~60 mm ML alternating with base of suckers. Arm tips with three to seven longitudinal series of small, cream spots, often pronounced against dark base colour. Transverse pair of white spots on dorsal mantle. Cream spots contain erectile skin ridges.

*Sculpture.* Skin with patch and groove texture forming large circles on dorsal arm crown. Four large circles in diamond arrangement on dorsal mantle, single, large supra-ocular papilla over each eye. Lateral skin ridge absent from around lateral margin of mantle.

## HABITAT AND BIOLOGY

This species occurs on tropical coral reefs from intertidal flats to at least 22 m. It is day-active with higher activity peaks at dusk and dawn. It occupies lairs in coral bedrock, live and dead coral head, and excavations in sand and rubble. Established lairs can often be located due to scattering of empty crab carapaces surrounding opening. This would appear to be this animal's primary or preferred food source, based on the examination of debris around lairs and stomach contents. Mather and Mather (2004) have examined the aptly named 'passing cloud' display of this species as used when hunting crabs. This complex visual display involves changes in colour pattern resulting from selective sequential chromatophore expansion that results in a dark 'cloud' passing forward along the mantle, over the head and down the outstretched arm webs. This display is thought to startle the prey, producing apparent movement and inducing a crab to move while the octopus remains motionless.

*Octopus cyanea* in French Polynesia are known to mimic non-cryptic parrotfishes when swimming above complex reef structures (Hanlon *et al.* 1999).

Females lay up to 600 000 eggs in festoons, each egg ~1 mm long. The small size of the eggs indicates that the hatchlings are planktonic.

Herwig *et al.* (2012) used stylet increment analysis to age a wild population of *O. cyanea* from Ningaloo, Western Australia. They found the population at Ningaloo had rapid, non-asymptotic growth, a short lifespan (< 1 year), coupled with early maturation and high levels of mortality. Males matured at much younger ages and sizes than females. Reproductive activity was concentrated in spring and summer. Size and age at maturity contrasted vastly with populations from other, more equatorial latitudes subjected to higher temperatures, with the Ningaloo females maturing at much smaller sizes. It appears that maturation in this species may be age- rather than size-related. In other parts of the world, this information on population dynamics has been used to influence management practices, with increased mean sizes and catch per unit effort statistics corresponding to fishery closures in peak breeding seasons, thus shifting the average size frequency distributions of populations.

*Interest to fisheries.* Taken on a subsistence basis throughout its range for human consumption. Primarily collected using spears and lures. Commercial harvests occur in Indonesia, Philippines and Hawaii. In some areas, there is evidence of overfishing (Guard and Mgaya 2002). It is taken as an unregulated, recreationally fished species from the intertidal reefs of Ningaloo, Western Australia. Herwig *et al.* (2012) suggested that the female-dominated sex ratios in association with female brooding behaviours indicates that large conspicuous females may be more prone to capture, and the intertidal octopus population has the potential to be negatively affected in an unregulated fishery. Differences in life history traits on a global scale suggests that management of *O. cyanea* populations should be tailored to each unique set of life history characteristics, such as size and age at maturity.

#### SIZE

Up to 160 mm ML; total length > 1 m. Weight to at least 6 kg.

#### REMARKS

Many references to this species mistakenly refer to the Common Sydney Octopus, *O. tetricus* Gould, 1852. The range of *O. tetricus* does not extend further north than Moreton Bay in south-east Queensland (~27°30'S), it lacks ocelli and the zebra stripes on the ventral surface of the arms, and it has much finer reticulations over the skin than *O. cyanea*.

Taylor *et al.* (2012) have developed a molecular method for identifying *O. vulgaris* Cuvier, 1797, *O. cyanea* and *C. ornatus* (Gould, 1852) using PCR-RFLP. They suggest that this cost-effective and unambiguous method of species determination may assist in the management of these three conspecific stocks across the south-western Indian Ocean where they are fished commercially.

While currently treated under the genus *Octopus*, the generic placement of this taxon awaits a major revision of the genus (Norman and Hochberg 2005, Norman *et al.* 2014).

#### DISTRIBUTION

Tropical Indo-West Pacific from the east coast of Africa to Hawaii, southern Japan to northern Australia.

#### REFERENCES

Norman (1991, 1993b), Norman *et al.* (2014).

### *Octopus harpedon* Norman, 2001

*Octopus harpedon* Norman, M. D. (2001). New octopus species from Queensland. *Memoirs of the Queensland Museum 46*(2), 677–690 [677, figs 3–5, 9C, 10D–F].

Type data: holotype AM C.30411 (♂ 56.9 mm ML).

Type locality: Australia: Gulf of Carpentaria, Albatross Bay near Weipa.

#### COMMON NAME

Spaghetti Octopus

#### DISTINGUISHING CHARACTERS

Differs from *Ameloctopus litoralis* Norman, 1992 (another very long-armed species) in gill lamellae count (10 or 11 in *O. harpedon v.* five or six in *A. litoralis*); having a T-shaped terminal organ with diverticulum *v.* linear terminal organ without diverticulum; little pigmentation *v.* arm bands; presence *v.* absence of an ink sac.

#### DIAGNOSTIC FEATURES

Mantle spindle-shaped. Funnel organ unknown. Eyes bulbous. Arms up to 9 × ML; arm formula ~2 > 1 = 3 > 4. Webs extremely shallow. Arm suckers not enlarged. Approximately 270 suckers on non-hectocotylised arms. Male hectocotylus (third right arm) very

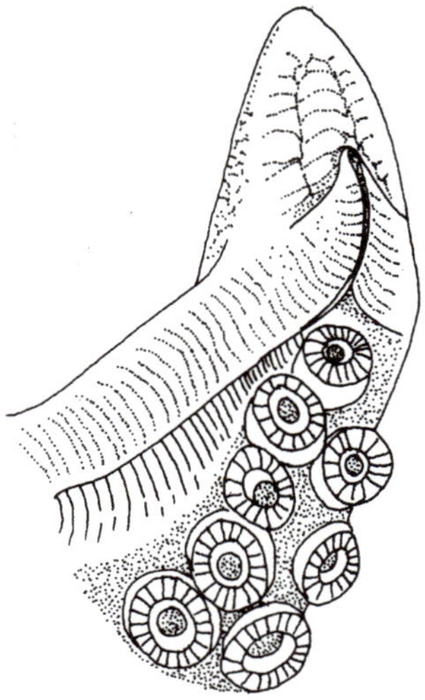

Hectocotylus, oral view (M. Shea, redrawn from original description).

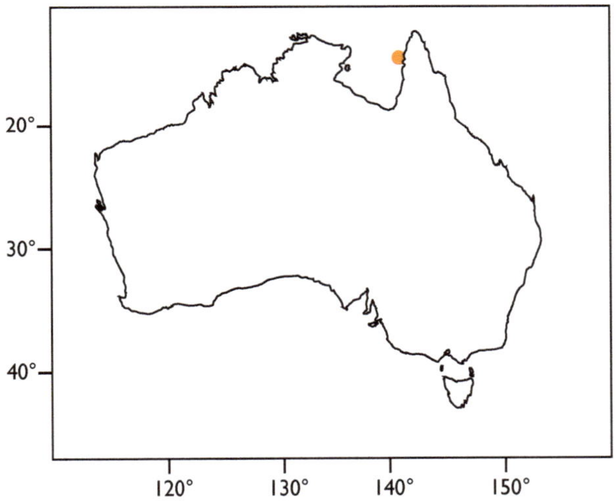


*Octopus harpedon* distribution.

short, ~20% length of opposite arm with 49 suckers. Ligula a small thick-lipped spoon with transverse creases across groove; calamus distinct and pointed, ~50% ligula length; spermatophore groove wide with fine transverse creases. Gills with 10 or 11 lamellae per demibranch. Anal flaps absent (differing from generic diagnosis). Terminal organ ('penis') T-shaped with diverticulum slightly longer than distal portion. Eggs large. Stylets absent.

*Colour.* Unknown in life. Preserved specimens cream (♀) or pink (♂). Dark blue subdermal pigment around eyes.

*Sculpture.* Skin smooth. Lateral ridge absent.

### HABITAT AND BIOLOGY

Shallow coastal waters typically clouded by suspended silt. Possibly uses long arms to probe subterranean fish and crustacean burrows as has been observed in *Ameloctopus litoralis* Norman, 1992. The short, hectocotylised arm suggests mounted copulation rather than copulation from a distance as in other long-armed species. The lack of colour pattern suggests the species may be nocturnally active.

### SIZE

Up to 86 mm ML, total length to 1 m.

### REMARKS

This species was described from one submature specimen of each sex. More detail may be apparent when additional material is found and examined. It shares some affinities with two species currently placed in *Euaxoctopus* Voss, 1971, but further investigation is required to resolve the generic placement of this taxon (Norman 2001).

While currently treated under the genus *Octopus*, the generic placement of this taxon awaits a major revision of the genus (Norman and Hochberg 2005, Norman *et al.* 2014).

### DISTRIBUTION

Known only from the type locality, eastern Gulf of Carpentaria, Albatross Bay near Weipa. Depth to 2 m.

### REFERENCE

Norman (2001).

## *Octopus kaurna* Stranks, 1990

*Octopus kaurna* Stranks, T. N. (1990). Three new species of Octopus (Mollusca: Cephalpoda) from south-eastern Australia. *Memoirs of the Museum of Victoria 50*(2), 457–465 [460, fig. 2].

Type data: holotype MV F.24494.

Type locality: Australia: Victoria, Hobsons Bay, 37°52′S, 144°56′E.

### COMMON NAME

Southern Sand Octopus

### DIAGNOSTIC FEATURES

Mantle elongate, ovoid. Eyes not prominent. Funnel organ V V-shaped. Outer limbs three-quarters length of median

Live animal (P. Mercurio).

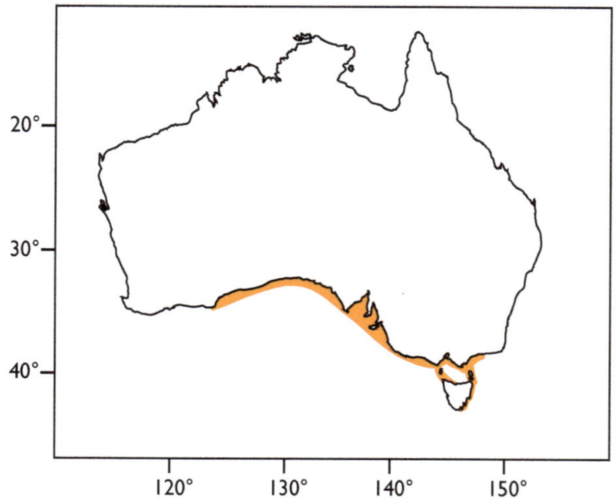


*Octopus kaurna* distribution.

Hectocotylised arm (M. Shea, redrawn from original description).

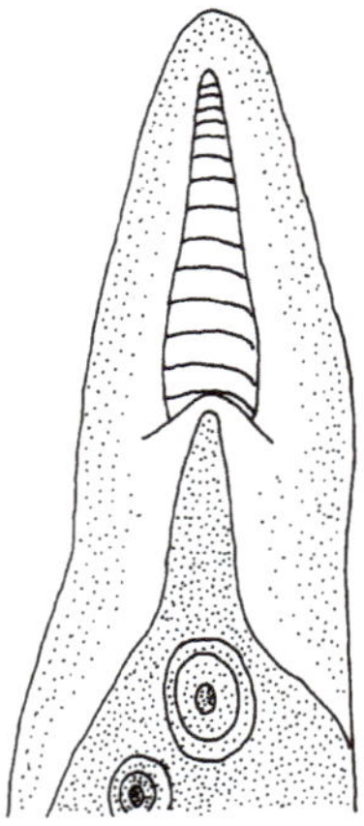

Hectocotylus, oral view (M. Shea, redrawn from original description).

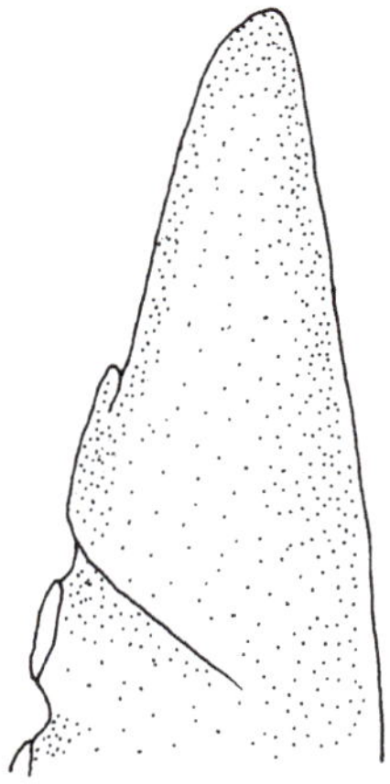

Hectocotylus, lateral view (M. Shea, redrawn from original description).

limbs. Arms very long, slender: up to 7 × ML. Dorsal arms longest, arm formula usually 1 > 2 > 3 > 4. No arm autotomy. Webs shallow, deepest ~10–20% of arm length; web margins extend ~one-third along arms. Arm suckers biserial, deeply set; all similar-sized, small, not enlarged. Male hectocotylus (third right arm) ~55–70% length of opposite arm with 66–129 suckers. Ligula small, 5–8% of third right arm length; ligula groove long, well marked and moderately deep with incomplete transverse ridges; calamus short, 33–48% of ligula length, pointed. Gills with 9–11 lamellae. Spermatophores relatively short, slender. Eggs large 9–11 mm long.

*Colour.* Preserved animal light brown to dark purple dorsally; creamy red to light brown ventrally. Live animals uniform orange-cream to maroon red. Sometimes, while foraging, with pink base colour and dark maroon stripe on each side from mantle to arm tips, running through the eye. Ocelli absent.

*Sculpture.* Smooth, or with fine, rounded and densely set tubercles, largest on lateral mantle surfaces. Supraocular papillae absent. Skin ridge around lateral margin of mantle absent.

## HABITAT AND BIOLOGY

*Octopus kaurna* is a shallow water species found on sand, mud, and seagrass beds. It remains buried during the day in a burrow in the sand formed using mucus to bind the sand grains together. It emerges at night to forage for small crustaceans. Eggs are large and produced in low numbers, with hatchlings immediately adopting a benthic existence.

*Octopus kaurna* is the host taxon for the dicyemid parasite *Dicyema* sp. 1 (Catalano *et al.* 2014).

### SIZE

Up to 85 mm ML; total length to 420 mm.

### REMARKS

While currently treated under the genus *Octopus*, the generic placement of this taxon awaits a major revision of the genus (Norman and Hochberg 2005, Norman *et al.* 2014).

### DISTRIBUTION

South-eastern Australia, from the Great Australian Bight to eastern Victoria, including Bass Strait and northern Tasmania. Known depth range to 50 m.

### REFERENCES

Stranks (1990), Stranks (1996), Finn and Norman (2011b), Norman *et al.* (2014).

## *Octopus laqueus* Kaneko & Kubodera, 2005

*Octopus laqueus* Kaneko, N. and Kobodera, T. (2005). A new species of shallow water *Octopus*, *Octopus laqueus*, (Cephalopoda: Octopodidae) from Okinawa, Japan. *Bulletin of the National Science Museum, Tokyo Series A 31*(1), 7–20 [8, figs 2–7].

Type data: holotype NSMT Mo-75242.

Type locality: Japan: Okinawa Island, Cape Maeda 26°23′N, 127°44′E.

### COMMON NAME

*Sodefuridako* (Pretend Sleeves Octopus)

### DIAGNOSTIC FEATURES

Mantle ovoid. Head narrower than mantle; eyes prominent. Funnel organ W-shaped. Arms medium to long (3.0–4.9 × ML); arm formula 4 = 3 > 2 > 1. Arm webs shallow, thin, shortest between dorsal arm pair, partly transparent, extend to arm tips. No arm autotomy. Arm suckers biserial. Normal arms with 86–160 suckers. Matures at a small size. Males with enlarged suckers on arms 2 and 3 at level of eighth to sixteenth sucker row. Male hectocotylus (third right arm) approximately ~66–84% length of opposite arm with 81–97 suckers. Ligula small and narrow 0.7–2.6% arm length; calamus 16.7–37.5% of ligula length. Gills with five or six lamellae per demibranch. Anal flaps present, tiny, ink sac well developed. Radula with seven transverse series of teeth and two series of marginal plates. Terminal organ straight, not coiled. Spermatophores short to medium length, ejaculatory apparatus with 32–53 coils. Eggs numerous, small, eggs 6.4–12.2% ML.

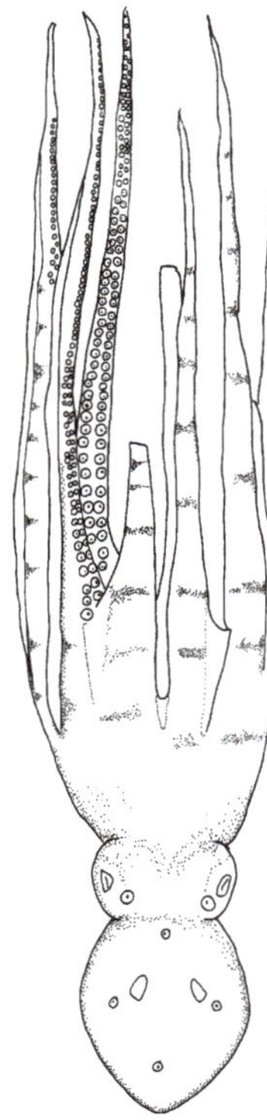

Dorsal view (A. Reid, redrawn from original description).

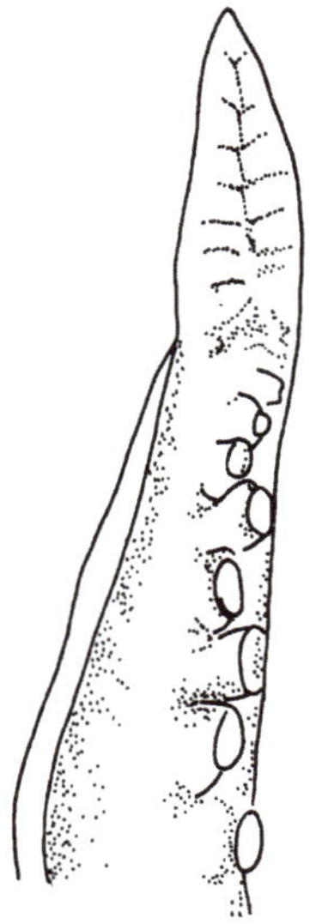

Hectocotylus, oral view (A. Reid, redrawn from original description).

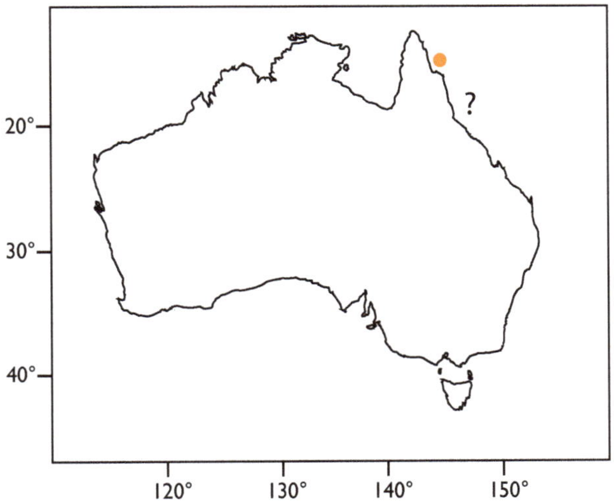


*Octopus laqueus* distribution.

*Colour.* Live animals pale purple or pink, changing to red-brown when alarmed. Preserved animals dark brown. Pair of dorsal white spots present, size varies, not visible in preserved animals. Arms with dark brown transverse bands, rarely observed in live animals but visible after fixation. Arm webs translucent with white lace-like pattern.

*Sculpture.* Dorsal surface slightly textured with irregular, rounded, low warts. Large fingers of skin can be erected on the body, the largest coming off the rear tip like a pointed tail. Primary papillae on dorsal mantle form diamond pattern. Single large primary papilla above each eye.

### HABITAT AND BIOLOGY

This species matures at a small size. It is found on intertidal rocky reefs, hiding during the day and active at night. The small size of the eggs suggests that the juveniles are planktonic.

### SIZE

ML ~24–48 mm.

### REMARKS

Norman *et al.* (2014) reported this species as occurring in Australian waters. The description above is based on the original description of the species. The Australian species are known from only two specimens. Further work is needed to verify conspecificity of the Australian animals with the Japanese form. Norman (2000: 287) stated that, 'it is recognised by the white lace-like markings and the large fingers of skin which can be erected on the body' Kaneko and Kubodera (2005: 14) described the animals as typically white when alive in the diagnosis of the species, but in the description the live animals were described as 'pale purple or pink, changing to red or brown when alarmed'. The lace-like pattern (hence the species name) is reported to occur on the arm webs in the original description.

While currently treated under the genus *Octopus*, the generic placement of this taxon awaits a major revision of the genus (Norman and Hochberg 2005, Norman *et al.* 2014).

### DISTRIBUTION

Australia, northern Great Barrier Reef and Japan. Known depth range 0–18 m.

### REFERENCES

Norman (2000; as *Octopus* sp. 15), Kaneko *et al.* (2006), Norman *et al.* (2014).

## *Octopus maorum* (Hutton, 1880)

*Macroctopus maorum* Hutton, F. W. (1880). *Manual of the New Zealand Mollusca.* Colonial Museum and Geological Survey Department: Wellington. 224 pp. [1].

Type data: Lectotype A, OM'89.03.

Type locality: New Zealand: Blind Bay.

### COMMON NAME

Maori Octopus

### DIAGNOSTIC FEATURES

Mantle broady oval; distinct constriction between head and mantle. Eyes not prominent. Funnel organ W-shaped; outer limbs shorter than median limbs. Arms long, robust with slender tips and of subequal length. Arms up to 5.5 × ML, arm formula 1 >2 >3 >4, dorsal arms longest and thickest. No arm autotomy. Dorsal arm webs deepest, ~20% of arm length. Web margins wide and fleshy, extend to the arm tips. Suckers biserial; enlarged on all arms of mature males and females; approximately six enlarged suckers on all arms starting from around the fifteenth proximal sucker. Male hectocotylus (third right arm) ~60–80% length of opposite arm with 95–135 suckers. Ligula small and narrow (3–6% of third right arm length); calamus moderate in size and pointed, ~25% ligula length. Eggs medium-sized, 6–7 mm long. Gills with 13–15 lamellae on each demibranch.

*Colour.* Colour orange-brown to dark brick red base colour or purple-grey with numerous small iridescent white spots on the mantle, arms, web and dorsal arm crown. Arm faces orange or dark red/ maroon. Ocelli absent.

*Sculpture.* Skin textured with small irregular patches and papillae over all dorsal and lateral surfaces. Able to raise >20 erectile papillae on mantle and upper arm crown. No ridge around mantle margin.

### HABITAT AND BIOLOGY

Found on the continental shelf and upper slope on reefs or rocky areas and among sponges. Smaller animals are generally night-active, while larger adults can be seen foraging during the day. Behavioural observations of a single individual (Vafiadis 1998) showed that the animal foraged during the day by enveloping large rocks with its web, then used its arms to probe cracks and crevices to search for prey. It was also seen to excavate its own hollows, shifting and excavating large (sometimes fist-sized) rocks to make a hollow. Eggs are small in size and produced in large numbers with juveniles temporarily planktonic before setting out to the benthos. Eggs are attached singly by stalks to suitable substrate.

*Octopus maorum* appears to be a spring–summer spawner, with females laying thousands of eggs (6.5–7.5 mm in size) that are attached individually to the substratum. Male and female *O. maorum* collected in south-east Tasmania between March and May 1996 were all mature or close to sexual maturity, with females containing 56 000 and 232 000 eggs in their ovaries (Grubert and Wadley 2000). Ovary weight and fecundity

Live animal (P. Mercurio).

*Octopus maorum* enveloping rock with its arms and arm webs (P. Mercurio).

Female aerating eggs in an aquarum (S. Leporati).

were not correlated with body weight. In males, however, spermatophore length and spermatophore storage sac weight were highly correlated with body weight.

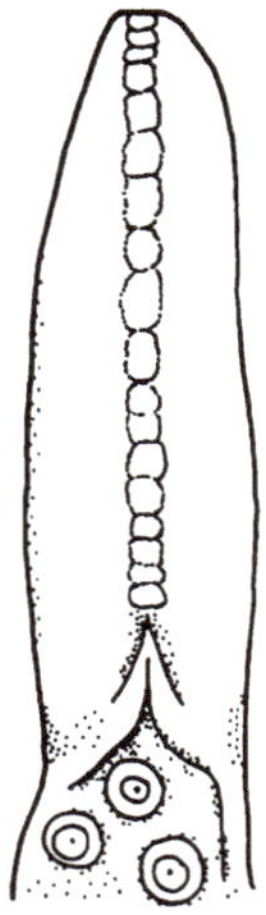

Hectocotylus, oral view (A. Reid, redrawn from Norman *et al.* 2014).

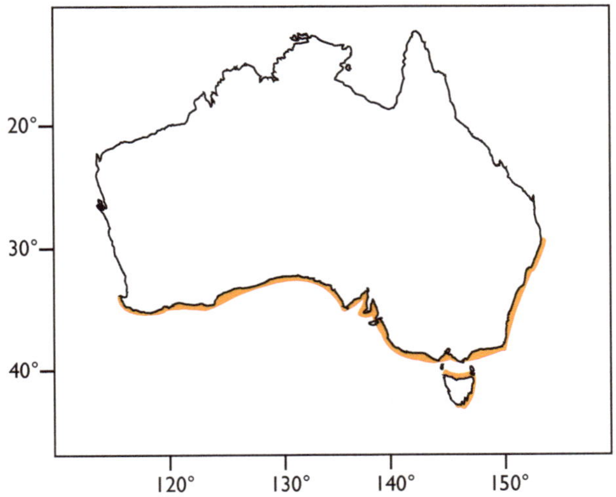


*Octopus maorum* distribution.

Specimens from New Zealand took ~2 months to hatch at 18°C (Anderson 1999). Hatchlings are small to medium in size (total length 5.06 mm) and planktonic with a unique chromatophore pattern. They are reportedly aggressive, readily cannibalising smaller individuals, eggs and other octopus species regardless of size. Grubert *et al.* (1999) have found that *O. maorum* in Eaglehawk Bay, south-east Tasmania, consumes at least four crustacean and six fish species. The remainder of the diet consists of small gastropods and other *O. maorum*.

*Fisheries importance.* This species was formerly a major by-catch of the South Australian rock lobster fishery. It is a pest to lobster fishers, entering pots and consuming lobsters. The introduction of escape hatches in lobster pots in some Australian states has prevented the capture of these octopuses as by-catch. It is sold as food (favoured for Greek-style pickled octopus) and as bait. It is targeted in small-scale fisheries in southern Australia and is collected by hand and seine nets in Tasmania.

### SIZE

Up to 300 mm ML, total length to ~1 m.

### REMARKS

Lobster fishers in northern Tasmania have recorded increases in *O. maorum* in lobster pots, perhaps due to increases in abundance, or as a result of increased activity due to recent water temperature increases attributable to climate change (Pecl 2011).

Doubleday *et al.* (2009) developed five polymorphic microsatellite loci to assess the population genetic structure of *O. maorum* in Australian and New Zealand waters. They found that genetic variation was high and noted divergence among the Australian populations and also between Australian and New Zealand populations. Similarities were found between specimens from South Australia and southern Tasmania, and divergence was found between populations from South Australia and north-east Tasmania, and between two proximate Tasmanian sites. This genetic structuring did not fit an isolation-by-distance model of population differentiation but could be explained largely in relation to regional oceanographic features. In southern Australia the rate and direction of larval dispersal are influenced by passive transport in the Leeuwin current, which moves from west to east around South Australia and the west coast of Tasmania (where it is called the Zeehan Current), and which facilitates connectivity between regions and the East Australian current that flows from north to south along the east coast of Australia, effectively restricting connectivity between regions. Genetic divergence between the Australian and New Zealand populations of *O. maorum* suggests that there is no ongoing population connectivity across the Tasman Sea. It appears that the dispersal potential of *O. maorum* is too limited to maintain panmixia with 2000 km and 4500 m depth between these two continents.

These genetic results are similar to those found in an earlier study, Doubleday *et al.* (2008), in which the elemental composition of the stylets of *O. maorum* were used as spatial discriminators of population structure. They examined five populations: South Australia, New Zealand, north-east Tasmania, south-west Tasmania and Eaglehawk Bay (Tasmania). Their results showed that individuals from Eaglehawk Bay differed from the remaining populations in their elemental stylet composition and should be managed as a separate fishery stock. The elemental signatures of the New Zealand stock also differed considerably from the Australian stock, but Doubleday *et al.* (2008) didn't think that the possibility of trans-Tasman dispersal should be ruled out.

Lalas (2009) attempted to derive size estimates based on *O. maorum* beaks found in prey remains, which he felt could be identified from other octopods for animals weighing > 100 g. Attempts to derive MLs, total length and body mass based on upper and lower beak measurements were unsuccessful due to imprecise regressions thought to reflect the innate variability in octopus morphometrics.

This species has sometimes been incorrectly called *Pinnoctopus cordiformis* Quoy & Gaimard, 1832.

Norman *et al.* (2014) treat this species as an unplaced '*Octopus*' pending a thorough phylogenetic study.

### DISTRIBUTION

Southern and south-eastern Australia and temperate to sub-Antarctic regions of New Zealand. In Australia found from the central Great Australian Bight (~132°E) to central New South Wales, including Tasmania. Known depth range 0–549 m.

### REFERENCES

Stranks (1996), Vafiadis (1998), Anderson (1999), Doubleday *et al.* (2011), Norman *et al.* (2014).

## *Octopus micros* Norman, 2001

*Octopus micros* Norman, M. D. (2001). New octopus species from Queensland. *Memoirs of the Queensland Museum 46*(2), 677–690 [683]. Plates.

Type data: holotype MV F.87070.

Type locality: Australia: Queensland, east of Mooloolaba, 16–20 miles north of Cape Moreton.

### COMMON NAME

Pygmy Keeled Octopus

### DISTINGUISHING CHARACTERS

This species was described from two mature male specimens only and one submature female. It differs from *Octopus bulbus* Norman, 2001 in arm formula (4 = 3 = 2 >1 in *O. micros v.* 1 >2 >3 >4 in *O. bulbus*), shorter arms (2.3–2.7 × ML in *O. micros v.* 4.9–5.6 × ML in *O. bulbous*), lower gill lamella count (6 *v.* 7 or 8), presence of dorsal white spots, and the diamond of papillae on

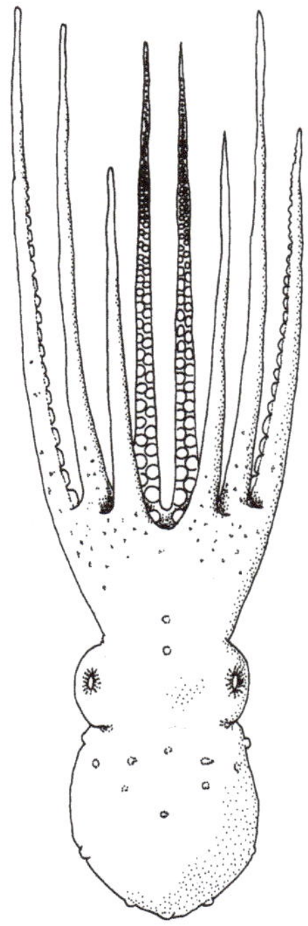

Dorsal view (A. Reid, redrawn from original description).

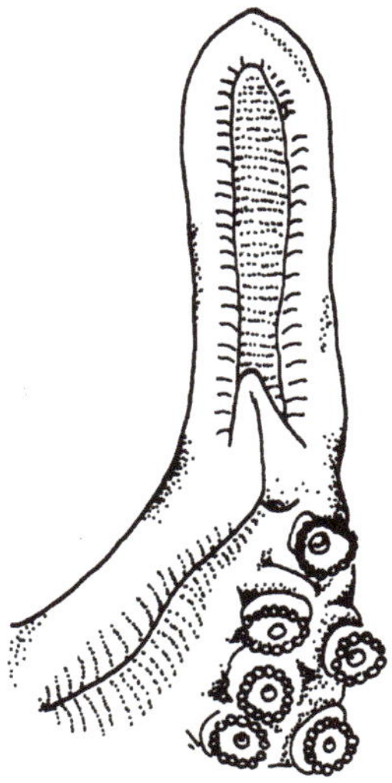

Distal tip of hectocotylus (A. Reid, redrawn from original description).

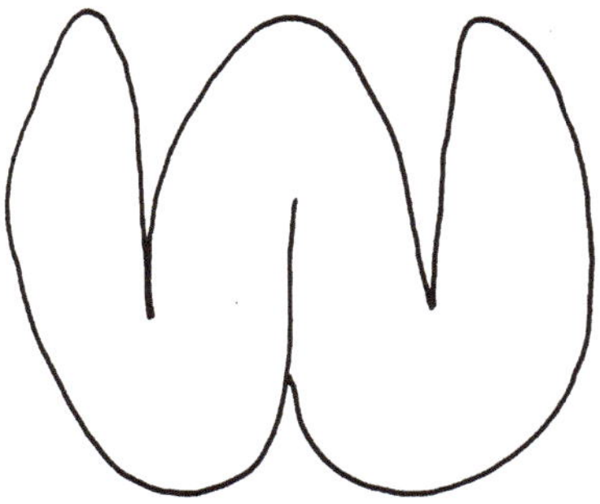

Funnel organ (A. Reid, redrawn from original description).

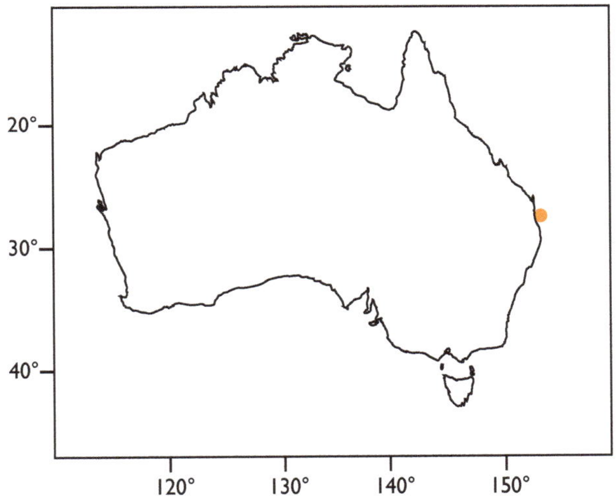


*Octopus micros* distribution.

the dorsal mantle. *Octopus micros* differs from *O. australis* Hoyle, 1885 (also found in the area) in having a higher hectocotylised arm sucker count (83–93 *v.* 62–77), shorter arms (2.3–2.7 *v.* 2.7–4.3 × ML) and a lower gill count (6 *v.* 7–9). Both the latter species share a lateral mantle ridge, similar arm formulae, dorsal white spots and a diamond of primary papillae on the dorsal mantle.

Two other pygmy species have been described so far from Australian waters: *O. warringa* Stranks, 1990 and *O. superciliosus* Quoy and Gaimard, 1832. Both lack the lateral ridge on the mantle.

### DIAGNOSTIC FEATURES

Robust species. Mantle ovoid to spherical. Funnel organ W-shaped with broad limbs; outer limbs similar in length to median limbs. Arms short, 2–3 × ML, approximately equal length, dorsal pair slightly shorter. Lateral webs slightly deeper than dorsal web. No enlarged suckers. Approximately 150 suckers on intact arms of mature animals. Hectocotylised third right arm similar in length to non-hectocotylised arm; with 85–93 suckers. Hectocotylised arm ~80% of length of opposite arm. Ligula leaf-shaped, pointed, ~6% of arm length, with open groove with fine transverse ridges; calamus distinct, sharp, half ligula length. Gills with six lamellae per demibranch. Spermatophore groove well developed, wide with transverse creases; without papillae. Rhachidian tooth with one or two lateral cusps. Posterior salivary glands large; crop diverticulum present, small;

caecum coiled in single whorl. Terminal organ (penis) robust and linear with simple rounded diverticulum. Spermatophore length approximately equal to ML. Sperm cord with regular whorls. Eggs large. Stylets present, non-mineralised.

*Colour.* Preserved specimens red-brown with fine, uniform chromatophores on dorsal surface and lateral arm crown. Dorsal white spots present. Darker purple-brown pigmentation around eyes.

*Sculpture.* Lateral mantle ridge present. Distinct papilla over each eye and four papillae in diamond pattern on dorsal mantle. Large papilla on posterior tip of mantle.

### HABITAT AND BIOLOGY

The size of the eggs suggests that hatchlings are benthic.

### SIZE

Up to 25 mm ML.

### REMARKS

While currently treated under the genus *Octopus*, the generic placement of this taxon awaits a major revision of the genus (Norman and Hochberg 2005, Norman *et al.* 2014).

### DISTRIBUTION

Southern Queensland east of Mooloolaba. Known only from the type locality. Collection depth 166–195 m.

### REFERENCE

Norman (2001).

## *Octopus pallidus* Hoyle, 1885

*Octopus boscii* var. *pallida* Hoyle, W. E. (1885). Diagnoses of new species of Cephalopoda collected during the cruise of the H.M.S. 'Challenger.' Part 1. The Octopoda. *Annals and Magazine of Natural History ser. 5, 15,* 222–236 [223].

Type data: lectotype BMNH 1889.4.24.19.

Type locality: Australia: Victoria, Bass Strait, off East Moncoeur Island, 39°10′S, 146°37′E.

### COMMON NAME

Pale Octopus

### DISTINGUISHING CHARACTERS

The species is superficially similar in integumental sculpture to *Octopus tetricus* Gould, 1852 but clearly differs in the following traits (with *O. pallidus* listed first): arm length 2.5% *v.* 3–4 × ML, arm formula 4 > 3 > 2 > 1 *v.* 3 = 2 > 4 > 1, funnel organ V V-shaped *v.* W-shaped, ligula 10–16% *v.* 1–2% of total third right arm length, eggs 11–13 mm long attached singly to substrate *v.* 2–3 mm long attached in egg strings.

### DIAGNOSTIC FEATURES

Squat, muscular species. Funnel organ closely opposed V V-shape, outer limbs ~75% length of medial limbs. Arms short, stout, ~2.5 × ML; arm lengths subequal, ventral or lateral arms longest, order 4 > 3 > 2 > 1. Arm autotomy absent. Webs deep, deepest on lateral arms, ~25–40% of arm length; web margins extend to arm tips. Normal arms with 200 biserial suckers. Enlarged suckers absent. Male hectocotylus (third right arm) approximately ~80–90% length of opposite arm with 72–86 suckers; spermatophore groove with thickening of web membrane. Ligula 10–16% of arm length; groove deep with incomplete transverse ridges; calamus 30–50% of ligula length, acutely pointed. Gills with seven to nine lamellae per demibranch. Ink sac present. Spermatophores relatively long, ~80% ML. Eggs large (11–13 mm long).

*Colour.* Resting animals brown to mottled cream dorsally, paler ventrally, or uniformly dark brown. When disturbed, a faint orange stripe often present along dorsal arms. Ocelli absent.

*Sculpture.* Integument with closely set rosette-shaped tubercles; largest on dorsum near base of arms, forming tile-like skin sculpture. More than 20 large branched papillae can be raised on all dorsal surfaces, including several large papillae over each eye. Lateral ridge absent.

### HABITAT AND BIOLOGY

*Octopus pallidus* is common in bays and coastal waters of south-eastern Australia, living on sand and among bryozoans, sponges and ascidians on sandy substrates to ~600 m depth. Eggs are large and produced in low numbers, with hatchlings immediately adopting a benthic existence. Leporati *et al.* (2008b) showed, using stylet increment analysis, that *O. pallidus* can reach an age of ~1.6 years and spawning occurs throughout the year, with an optimal spawning period in Bass Strait between late summer and early autumn. In a study involving 503 individuals from Bass Strait waters, males on average were significantly larger than females (550 g *v.* 482 g) and overall growth rates were positively correlated with temperature

Live animal (S. Leporati).

Fresh-caught specimen (J. Keane).

Egg mass (J. Keane).

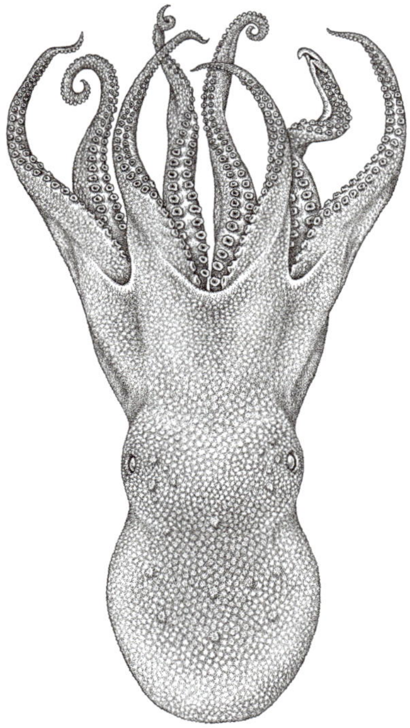

Dorsal view (ABRS, K. Hollis).

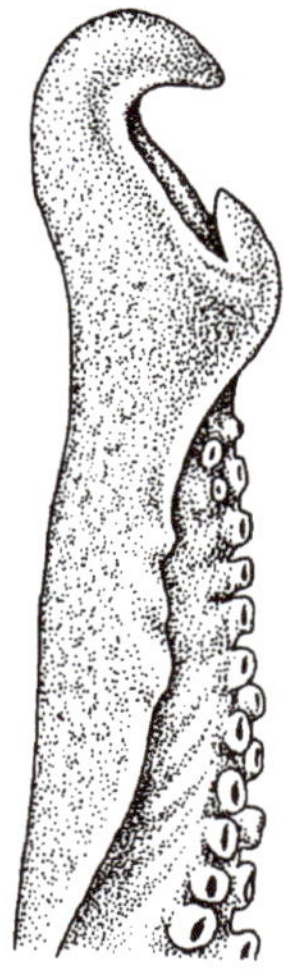

Hectocotylised arm (ABRS, K. Hollis).

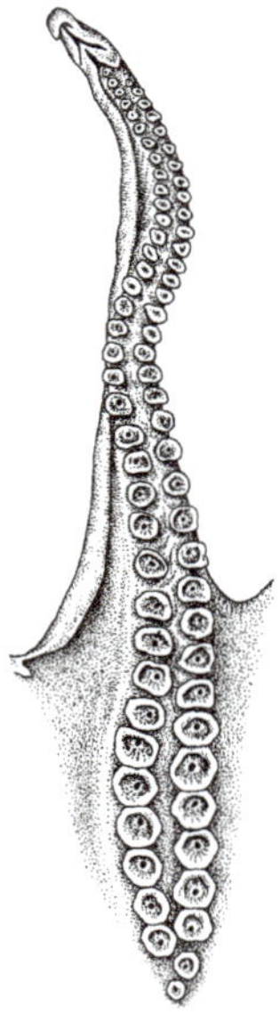

Hectocotylus oral view (ABRS, K. Hollis).

Hectocotylus, dorsolateral view (ABRS, K. Hollis).

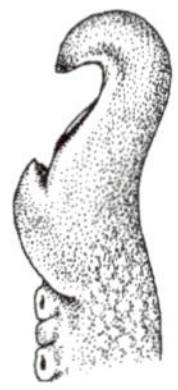

Hectocotylus, ventrolateral view (ABRS, K. Hollis).

at hatching. Males mature at a ML of 50 mm; females mature at a ML of ~60 mm. Leporati *et al.* (2008b) showed no age-size relationship, indicating high levels of growth plasticity. In another study, Leporati *et al.* (2008a) showed that maturation primarily depends on size with little relationship to age and is highly variable between sexes, with females >350 days still maturing in comparison to all males >142 days being mature.

Stylet increment analysis is currently the key method to age octopus; however, currently it is not reliable for all species. Doubleday and Semmens (2011) have found

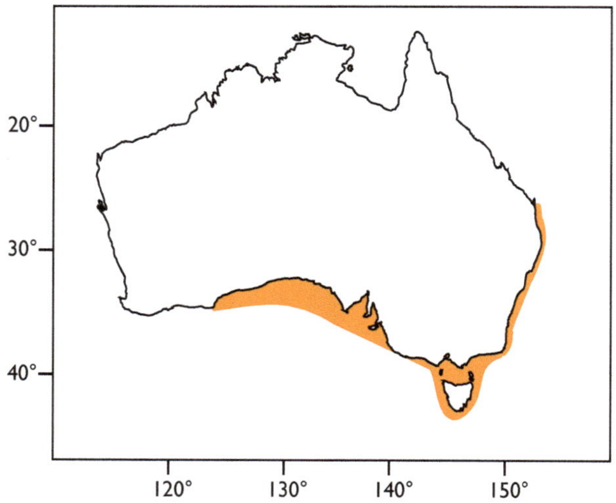


*Octopus pallidus* distribution.

in *O. pallidus* that the age pigment lipofuscin, visible in brain tissue using standard histological techniques, is highly exponentially correlated with age in known-age laboratory-reared animals, and the concentration of lipofuscin was not related to sex, temperature or body weight in same-age individuals.

*Interest to fisheries.* Taken as by-catch by inshore demersal otter trawlers and Danish seiners in Victoria. A small scale fishery occurs in north-western Tasmania using longlines and plastic unbaited pots. Small numbers are taken incidentally from southern rock lobster pots. It is sometimes caught in the waters of southern New South Wales.

### SIZE

Up to 150 mm ML, total length to 540 mm.

### REMARKS

While currently treated under the genus *Octopus*, the generic placement of this taxon awaits a major revision of the genus (Norman and Hochberg 2005, Norman *et al.* 2014).

### DISTRIBUTION

Australian endemic: south-eastern Australia from southern New South Wales to the Great Australian Bight, including Bass Strait and the northern and eastern coasts of Tasmania. Known depth range 7–275 m.

### REFERENCES

Stranks (1988), Stranks (1996), Leporati *et al.* (2008a, 2008b).

## *Octopus pyrum* Norman, Hochberg & Lu, 1997

*Octopus pyrum* (1997). Norman, M. D., Hochberg, F. F. and Lu, C. C. (1997). Mollusca Cephalopoda: Mid-depth octopuses (200–1000 m) of the Banda and Arafura Seas (Octopodidae and Alloposidae). In: A. Crosnier and P. Bouchet (eds.), Resultats des Campagnes MUSORSTOM, volume 16. *Memoirs de Museum National d'Histoire Naturelle, Paris, 172,* 357–383 [365, figs 5–7, 11b].

Type data: holotype MNHN 2029.

Type locality: Banda Sea: Kai Islands, 05°21′S, 132°30′E.

### DIAGNOSTIC FEATURES

Eyes slightly pronounced. Funnel organ W-shaped, outer limbs shorter than median ones. Arms 3–4 × ML, lateral arms slightly longer than other arms. No arm autotomy. Deepest arm webs 20–30% length of longest arms. Suckers small, biserial; 120 suckers on normal arms. No enlarged suckers in either sex. Male hectocotylus (third right arm) ~65% length of opposite arm with 46–55 suckers. Ligula 8–10% of arm length, robust, pear-shaped, thickened at base; groove with median rib and fine transverse ridges; calamus large and pointed (~50% of ligula length). Spermatophore groove wide and thin with transverse ridges. Spermatophore guide distinct, bordered by flattened papillae or digits of skin. Gills with seven or eight lamellae per demibranch. Spermatophores 67–75% ML and produced in low numbers. Stylets clear, non-mineralised.

*Colour.* Orange to crimson base colour dorsally, cream ventrally. Dorsal surfaces scattered with numerous raised leucophores, cream to gold in colour. Blue-black subdermal colour often visible over each eye. Distinctive transverse band of chromatophores across medioventral mantle. Large chromatophores on ventral digestive gland, visible within mantle cavity, adjacent to gills.

*Sculpture.* Numerous raised papillae scattered over dorsal surfaces, including eyes. Lateral mantle ridge absent.

### SIZE

Up to 35 mm ML, total length to 170 mm.

### REMARKS

While currently treated under the genus *Octopus*, the generic placement of this taxon awaits a major revision

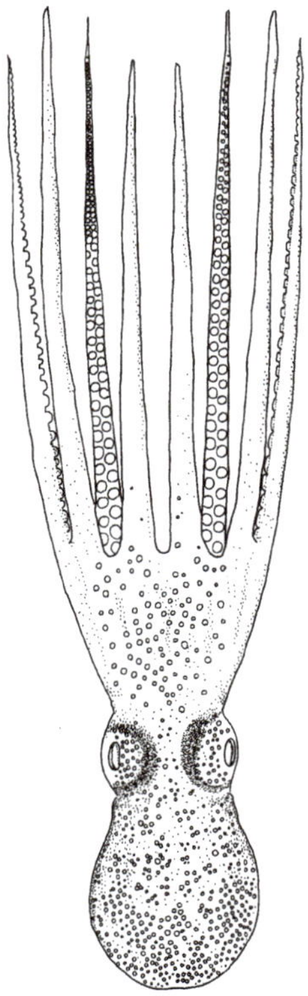

Dorsal view (A. Reid, redrawn from original description).

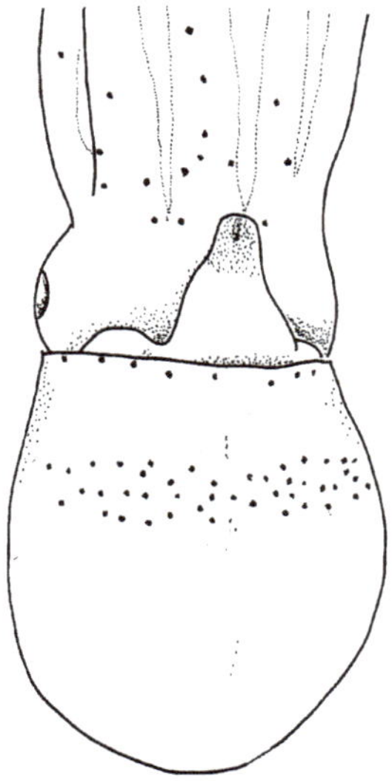

Ventral mantle (A. Reid, redrawn from original description).

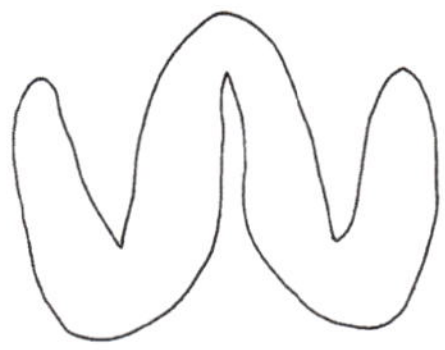

Funnel organ (A. Reid, redrawn from original description).

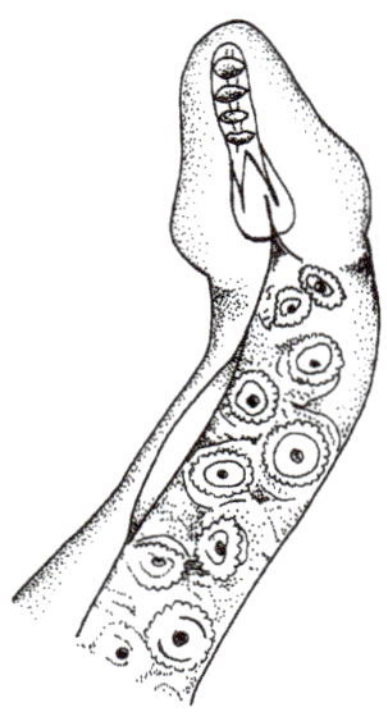

Distal tip of hectocotylus (A. Reid, redrawn from original description).

of the genus (Norman and Hochberg 2005, Norman *et al.* 2014).

## DISTRIBUTION

North-eastern Australia, off Brisbane, Indonesia, Banda and Arafura Seas. Known depth range 329–511 m.

## REFERENCE

Norman *et al.* (2014).

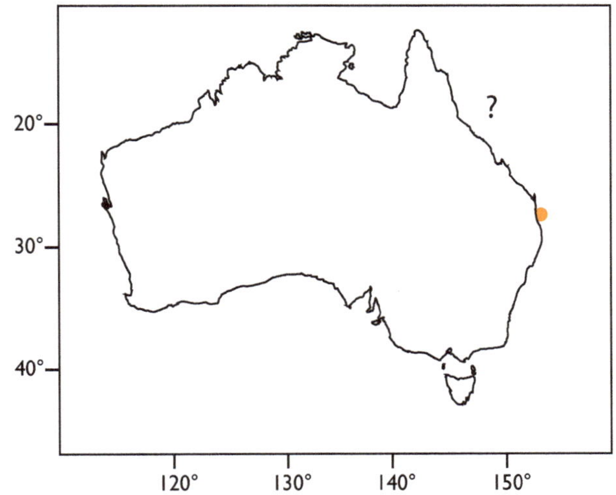


*Octopus pyrum* distribution.

# *Octopus superciliosus* Quoy & Gaimard, 1832

*Octopus superciliosus* Quoy, J. R. and Gaimard, J. P. (1832). Mollusques. *Voyage de decouvertes de l'Astrolabe pendant les annees 1826–1827–1828–1829, Zoologie 2*(1), 1–320 [88, pl. 6, fig. 4].

Type data: lectotype MNHN 5-4-1047.

Type locality: Australia: Bass Strait, Westernport Bay.

## COMMON NAME

Frilled Pygmy Octopus

## DISTINGUISHING CHARACTERS

*Octopus superciliosus* can be distinguished from *O. australis* Hoyle, 1885, *O. berrima* Stranks & Norman, 1993 and *O. warringa* Stranks, 1990 by the characteristic skin patterning (particularly the unbranched papillae on the head and mantle dorsum and traces of a ventrolateral ridge on the mantle but otherwise smooth skin surface), the wide, flattened ligula with many transverse ridges, and in having large eggs that are attached singly to the substrate.

## DIAGNOSTIC FEATURES

Head narrower than mantle and demarked by moderate constriction. Eyes large, project above surface of head. Funnel organ VV-shaped, outer limbs 75% length of median limbs, Arms long, 1.4–3.6 × ML; subequal. Web shallow. Suckers raised above arm surface, no enlarged suckers. Male hectocotylus (third right arm) slightly shorter than opposite arm in length with 46–66 suckers. Ligula wide, medium-sized, 8–13% total arm length, ligula groove long, well marked, shallow with transverse ridges; calamus long, acutely pointed. Gills with six to eight lamellae per demibranch. Eggs 8–11 mm long.

*Colour.* In life, resting animals cream to light brown dorsally, white to cream ventrally; when stimulated, animals change to mottled light and dark brown or uniform dark brown to purple. Preserved specimens dark brown to purple dorsally, cream to light brown ventrally.

*Sculpture.* Unbranched papillae on dorsum and ventrum. Dorsal mantle with about seven subparallel, longitudinal rows of four or five simple, usually elongate papillae along ML. Single large papilla forms posterior point on mantle. One large supraocular papilla and numerous smaller papillae around each eye. Lateral mantle ridge absent but two to four rows of elongate papillae resemble such ridges. Mantle surface otherwise smooth.

### HABITAT AND BIOLOGY

This is an inshore species living on sand or muddy substrate or among sponges and polyzoans. Eggs are large and produced in low numbers, with hatchlings immediately adopting a benthic lifestyle.

Live animal (M. Norman).

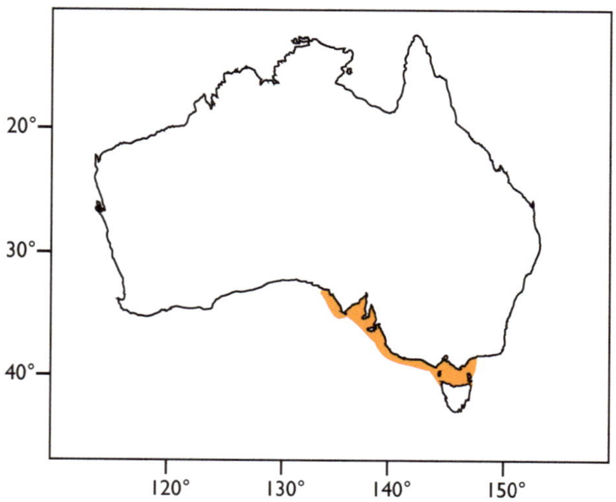


*Octopus superciliosus* distribution.

### SIZE

ML up to 26 mm, total length up to 94 mm.

### REMARKS

While currently treated under the genus *Octopus*, the generic placement of this taxon awaits a major revision of the genus (Norman and Hochberg 2005, Norman *et al.* 2014).

### DISTRIBUTION

South-eastern Australia. Known depth range 0–69 m.

### REFERENCES

Stranks (1996), Stranks (1998).

## *Octopus tetricus* Gould, 1852

*Octopus tetricus* Gould, A. A. (1852). Mollusca and Shells. United States Exploring Expedition, during the years 1838, 1839, 1840, 1841, 1842 12, 1–510 [474, fig. 588].

Type data: type not extant, repository unresolved.

Type locality: Australia: New South Wales, near Sydney.

### COMMON NAME(S)

Common Sydney Octopus, Gloomy Octopus

### DISTINGUISHING CHARACTERS

Shares many characters with *Octopus pallidus*; see 'Distinguishing characters' for *O. pallidus*.

### DIAGNOSTIC FEATURES

Large, muscular species. Funnel organ W-shaped, limbs of approximately equal length. Arms moderate to long 3.0–4.5 × ML. Lateral arms longest, dorsal arms shortest, typically 3 = 2 >4 > 1. Arm autotomy at distinct plane absent. Webs of moderate depth, deepest ~25% of arm length. Webs deepest on lateral arms, shallowest between dorsal arms. Web margins extend to arm tips. Arm suckers biserial. In large animals, ~220–260 suckers on each normal arm. Enlarged suckers present in mature males, three to five on arms 2 and 3, from about the thirteenth proximal sucker. Male hectocotylus (third right arm) shorter than opposite arm with 140–160 suckers. Ligula tiny, 1–2% of arm length; calamus small. Gills with eight or nine lamellae per demibranch. Eggs small, 2–3 mm long.

*Colour.* Active animals cream to mottled orange, or orange-red and dark brown. Transverse, narrow dark bands along arms in some colour patterns. Resting animals within lairs show grey dorsal surfaces, orange arm faces, contrasting with white suckers and eyes with a white iris. False eye-spots (ocelli) absent.

*Sculpture.* Skin sculptured in rounded rosette-shaped patches with erect central knobs separated by distinct grooves. Able to raise large papillae over dorsal surfaces, including four in diamond pattern on dorsal mantle and one above each eye. No skin ridge around lateral mantle margin.

Dorsal view (A. Reid).

### HABITAT AND BIOLOGY

Shallow coastal waters on or adjacent to rocky reefs. *Octopus tetricus* is active primarily at night although has been observed during the day, appearing alert in the mouth of its lair. Godfrey-Smith and Lawrence (2012) found that a population at Jervis Bay, New South Wales was often active during the day. Lairs are found in rock crevices or excavated under rocks on sand or mud. Crabs are the major prey item, but they will also take shellfish and finfish (at least in captivity). They can occur in reasonably high densities; the population observed over several years in an area of $12 \times 8^2$ metres at Jervis Bay ranged from two to 11 individuals with an average of 5.48 individuals. Godfrey-Smith and Lawrence (2012) speculated that a suitable den area at the centre of the site, which was otherwise fairly open and exposed, may have initially attracted an octopus or octopuses that subsequently modified the site through the addition of scallop shells that were discarded after feeding on the soft parts of the scallops. This in turn provided more suitable den sites among the shells and attracted more octopuses, or enabled juveniles to settle in the area.

Scheel *et al.* (2014) described *O. tetricus* as ecosystem engineers, modulating the availability of resources to other species and to their own species by causing physical state changes in a shell bed habitat, again in Jervis Bay, New South Wales. Here, *O. tetricus* occurs in unusually high numbers in a shell bed of its prey remains. A community of grazers and scavengers has developed on the shell bed, attracting fishes and then large predators that inhibit the activities of *O. tetricus*. Thus, positive feedbacks included the accumulation if shell debris, increasing shelter availability and aggregating fishes. Negative feedbacks included the reductions of nearby prey size and availability, aggression among octopuses and predator limitation of octopus activity.

Head-on view showing supraocular papillae (M. Dowse).

Fresh-caught specimen (K. Graham).

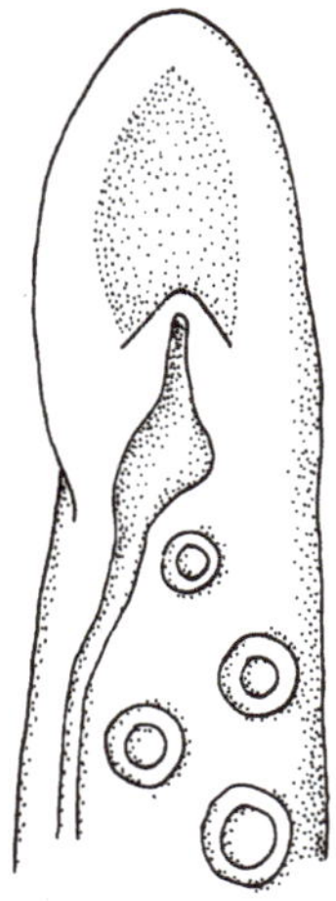

Hectocotylus (A. Reid, redrawn from Norman *et al.* 2014).

The limited observations of copulation in this species (Huffard and Godfrey-Smith 2010) suggest that males either copulate following stretching the hectocotylised arm over hard substrate to reach a female and then spermatophores are inserted into her mantle cavity, or the male lies on top of the female and envelops her with his arms and webs before copulation. Godfrey-Smith and Lawrence (2012) have observed an unusual mating behaviour in which a male stands over a female in a tent-like position on the tips of his extended arms. They also observed males 'probing' females with a non-hectocotylised arm before mating. Females lay >15 000 eggs in festoons, each egg 2–3 mm long. The species has a planktonic paralarval stage.

In a study of *O. tetricus* in Tasmanian waters (a recent range extension), Ramos *et al.* (2014) reported a lifespan of 11 months, with temperature, food availability and sex influencing growth rates. Hatching occurred throughout the year. Individuals that hatched during cooler and more productive conditions, but grew during warming conditions, had faster growth rates and reached smaller body sizes than individuals that hatched into warmer waters but grew during cooling conditions. The average total weight of individual *O. tetricus* from the study area in Tasmania was less that that recorded in warmer waters in New South Wales. This is in contrast to life-history theory expectations (if food supply is not limited) that octopuses from cooler areas grow more slowly during the exponential phase of growth and reach maturity at larger sizes compared to those found in warmer waters. This observation may be a reflection of differences in life history characteristics (body size, growth rates and lifespan) over the range of the species; reduced genetic diversity in the region of the range extension; a result of adjustment to the new physical environment, or to different community interactions. It may also be a reflection of different sampling methods (sample pots in Tasmania *v.* trawl nets used in New South Wales); see Ramos *et al.* (2014). (Perhaps food supply may, in fact, be limited in the Tasmanian population due to competition with the 'long-term' resident octopus species.)

*Interest to fisheries.* Moderate scale harvest as by-catch in prawn and finfish trawl fisheries. Frequently sold in markets in New South Wales (where it comprises ~10% of octopus landings) and Queensland. The Western Australian *O.* cf. *tetricus* sibling species, formerly a by-catch of the Western Australian rock lobster fishery, is now fished commercially on a large scale using unbaited traps with spring closures and is now widely available in fish markets around Australia. Studies are underway to develop aquaculture techniques for the species. Early results have shown that if octopuses of a similar size are kept together, territoriality is no longer a problem, aggression and cannibalism are eliminated and they can be kept at high densities without the need for hides. Without the provision of hides, cleaning and maintenance is greatly reduced (Yeoman 2015). Rearing of paralarvae to close the life cycle is still proving to be problematic.

## SIZE

Maximum ML 140 mm, total length >600 mm, weight to at least 1 kg.

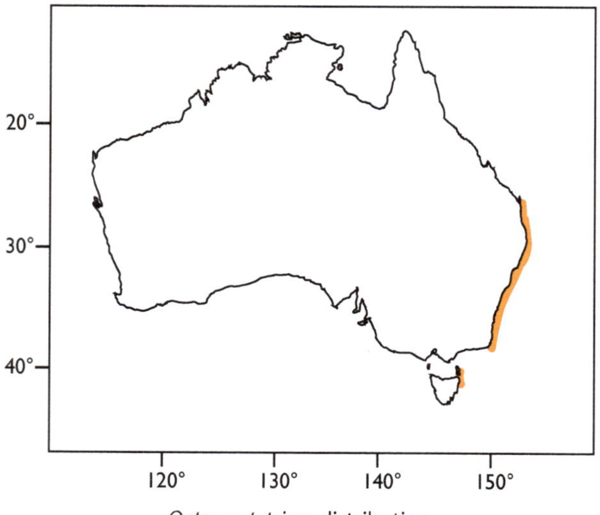


*Octopus tetricus* distribution.

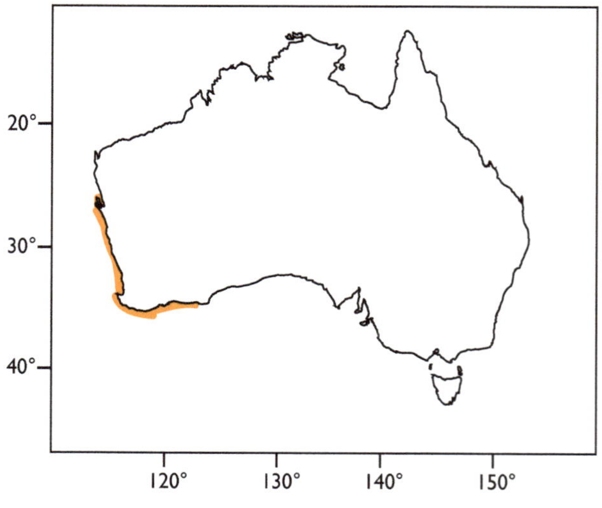


*Octopus* cf. *tetricus* distribution.

### REMARKS

The range of *Octopus tetricus*, formerly found on mainland Australia, has extended to northern Tasmania, probably is a result of anthropogenic climate change, leading to increased water temperatures and a shift further southwards of the warm East Australian Current (Pecl 2010, 2011), perhaps enabling larvae from eastern Australia to survive the distance to Tasmania.

Ramos *et al.* (2014) examined growth rates, body size and lifespan of *O. tetricus* at the leading edge of this poleward range shift in Tasmanian waters (40°S, 147°S) throughout 2011. They showed that *O. tetricus* had a relatively small body size and short lifespan of ~11 months (this was based on stylet analysis assuming a daily periodicity of growth increments as validated or assumed in studies of other octopod species). Despite cooler temperatures, this would enable a high rate of population turnover and may facilitate the population increase necessary for successful establishment in the new extended area of the range.

The identity of the purported *O. tetricus* from Tasmania was supported in a study by Amor *et al.* (2014). This study, based on a combination of morphological and molecular methods, also showed that a taxon from Western Australia, formerly known as *O. tetricus*, is now known to be a separate sister taxon to the eastern Australian form. In addition, the New Zealand form, previously known as *O. gibbsi* O'Shea, 1999, was formally recognised in Amor *et al.* (2014) as a synonym of *O. tetricus*. The Western Australian form has yet to be fully described. Amor *et al.*'s (2014) study thus identified cryptic species among what was formerly the *O. tetricus* complex.

*Octopus tetricus* and *O.* cf. *tetricus* belong to a group of taxa that includes *O. vulgaris* Cuvier, 1797, the type species of the genus *Octopus*. Thus, the generic placement of this taxon is secure.

### DISTRIBUTION

Eastern Australia from southern Queensland to eastern Victoria and north-eastern Tasmania. Also known from New Zealand (previously under the name *O. gibbsi* O'Shea, 1999). Known depth range 0–60 m.

### REFERENCES

Stranks (1998), Norman (2000), Godfrey-Smith and Lawrence (2012), Amor *et al.* (2014), Norman *et al.* (2014).

## *Octopus warringa* Stranks, 1990

*Octopus warringa* Stranks, T. N. (1990). Three new species of Octopus (Mollusca: Cephalopoda) from south-eastern Australia. *Memoirs of the Museum of Victoria* *50*(2), 457–465 [457, fig. 1].

Type data: holotype MV F.57444.

Type locality: Australia: Tasmania, Maria Island west of Darlington, 42°35'S, 148°03'E.

### COMMON NAME

Club Pygmy Octopus

### DIAGNOSTIC FEATURES

Small, mantle broadly ovoid. Eyes prominent. Funnel organ W-shaped, limbs thick; outer limbs 75% length of medium limbs. Arms 1.9–3.7 × ML; subequal, formula usually 3 > 4 > 2 > 1. Arm webs shallow. Sixth to tenth pair of arm suckers enlarged on all arms in both sexes. Male hectocotylus (third right arm) shorter than of opposite arm with 51–65 suckers. Ligula bulbus 6–10% of total

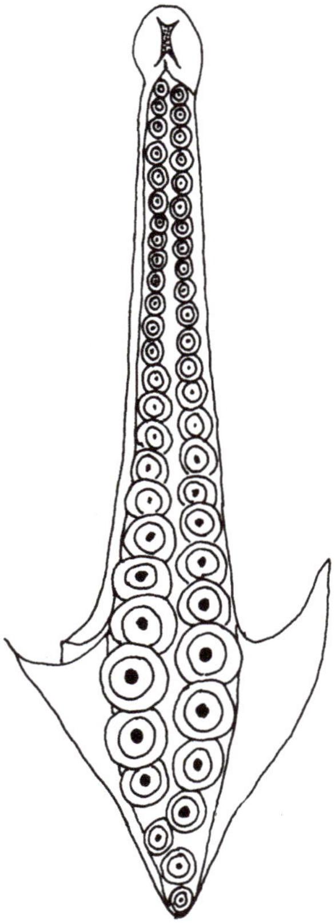

Hectocotylised arm (M. Shea, redrawn from original description).

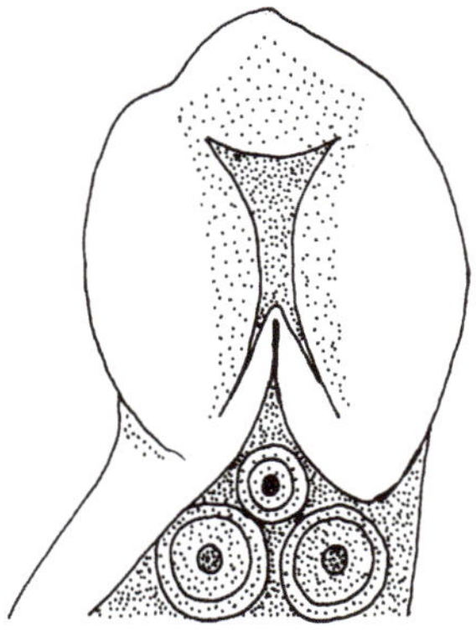

Hectocotylus, oral view (M. Shea, redrawn from original description).

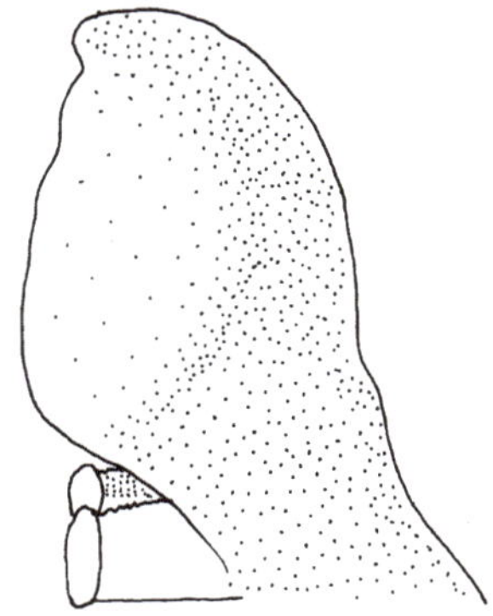

Hectocotylus, oral view (M. Shea, redrawn from original description).

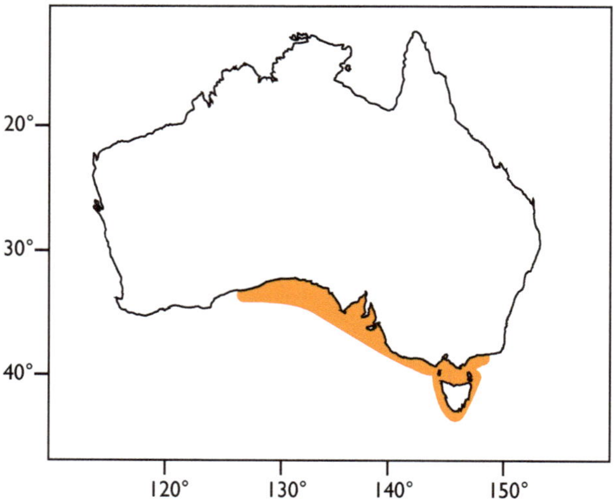


*Octopus warringa* distribution.

arm length in mature males; ligula groove long, well marked and deep, without transverse ridges; calamus short, acutely pointed. Gills with six to eight lamellae in each demibranch. Male penis very long; diverticulum coiled with three lobes. Spermatophores very long with large coiled sperm reservoir. Eggs 2–3 mm long, with long stalks and are arranged in festoons.

*Colour.* Preserved specimens light brown to purple dorsally; cream to light brown ventrally. Ocelli absent.

*Sculpture.* Dorsal and ventral integument with fine, rounded, closely set tubercles. Dorsum with branched and unbranched papillae; seven subparallel rows of three or four simple papillae along ML; singular large papilla forms posterior point on mantle; large aborescent papilla in supraocular region surrounded by three or four smaller unbranched papillae. Lateral mantle ridge absent.

### HABITAT AND BIOLOGY

Inshore species living on rocky bottoms and among sponges and polyzoans. Eggs are small in size and produced in large numbers with juveniles temporarily planktonic before setting out to the benthos. A captive animal was seen to lay ~1000 eggs and incubation time to hatching was 80 days.

### SIZE

Up to 35 mm ML; total length up to 125 mm.

### REMARKS

While currently treated under the genus *Octopus*, the generic placement of this taxon awaits a major revision of the genus (Norman and Hochberg 2005, Norman *et al.* 2014).

### DISTRIBUTION

South-eastern Australia from the Great Australian Bight to eastern Victoria, including Bass Strait and Tasmania. (Also occurs in New Zealand.) Known depth range 0–144 m.

### REFERENCES

Stranks (1990), Stranks (1996).

## *Scaeurgus* Troschel, 1857

### DIAGNOSTIC FEATURES

Mantle encircled by low keel. Funnel organ W-shaped. Arms short, 2–3 × ML. No arm autotomy. Deepest webs ~25–30% longest arms. Suckers biserial. Third left arm of males hectocotylised, slightly shorter (80%) of opposite arm. Copulatory organ with fleshy, thick-rimmed ligula. Body covered with numerous rounded papillae, denser or dorsal than ventral surfaces. Gills with 8–14 lamellae per demibranch. Ink sac and anal flaps present. Distal oviducts open approximately midway along the length of the pallial septum. Spermatophores produced in low numbers, contain a double strand of sperm cord forming a braided appearance. Rhachidian tooth of radula with one to two cusps on each side (usually two) in asymmetrical seriation. Large mineralised stylets present.

*Colour.* Four paired black spots on mantle, anterior pair on lateral mantle adjacent to mantle aperture and dorsal to lateral skin ridge; posterior pair on posterior tip of mantle, dorsal to lateral ridge, faint in some species.

*Sculpture.* Diamond of primary papillae on dorsal mantle. Single large papilla over each eye and large papilla present on posterior tip of mantle.

### Distribution

Typically found on seamounts and continental shelves at tropical and subtropical latitudes worldwide, at depths of 50–500 m (typically 200–500 m).

### REFERENCE

Norman *et al.* (2005).

### Key to *Scaeurgus* species

1a. Normal arms with 140–150 suckers; hectocotylised arm with 72–78 suckers ................................ *S. jumeau*

1b. Normal arms with 100–130 suckers; hectocotylised arm with 54–63 suckers ................................ *S. tuber*

## *Scaeurgus jumeau* Norman, Hochberg & Boucher-Rodoni, 2005

*Scaeurgus jumeau* Norman, M. D., Hochberg, F. G. and Boucher-Rodoni, R. (2005). A revision of the deep-water octopus genus *Scaeurgus* (Cephalopoda: Octopodidae) with description of three new species from the southwest Pacific Ocean. *Journal of Molluscan Studies 71*, 319–337 [325, figs 4–6, 10C, 10D, 10H].

Type data: holotype MNHN 2141.

Type locality: Norfolk Ridge, south of New Caledonia, 23°48′S, 168°17′E.

### DISTINGUISHING CHARACTERS

The arm sucker counts distinguish this species from *Scaeurgus tuber* (see key).

### DIAGNOSTIC FEATURES

Muscular species with short arms (2.2–2.7 × ML). Lateral webs slightly deeper than dorsal and ventral webs, deepest 25–32% longest arm. Normal arms with 140–150 suckers. Two or three slightly enlarged suckers on arms 2 and 3 of males (eleventh–thirteenth proximal suckers), size ~12% ML. Male hectocotylus (third right arm) with 72–78 suckers. Ligula of moderate size (7.3–9.5% arm length), peanut-shaped with thick-rimmed groove; calamus very large (50.6–68.2% length of ligula). No enlarged distal suckers on hectocotylus. Gills with 12 or 13 lamellae per demibranch. Spermatophores 18–13 mm long, 59–92% ML. Eggs 2.6–2.7 m, ~6% ML.

*Colour.* Base colour orange-brown-cream with two obvious pairs of black spots, one pair on posterior mantle, second pair positioned anteriolaterally adjacent to mantle opening. Visceral membrane with chromatophores.

*Sculpture.* Numerous rounded papillae in regular texture over all dorsal surfaces and along aboral surfaces of arms 1–3, less dense ventrally. Diamond of four small primary papillae on dorsal mantle. Single large papilla and several smaller papillae over each eye. Large papilla dorsal to each anterior black spot. Large posterior mantle papilla present. Lateral ridge present, continuous around entire lateral mantle. Short longitudinal ridges on lateral head, one under each eye.

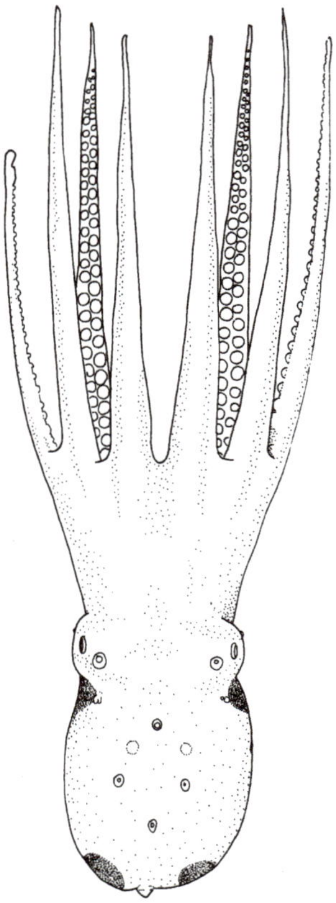

Dorsal view (A. Reid redrawn from original description).

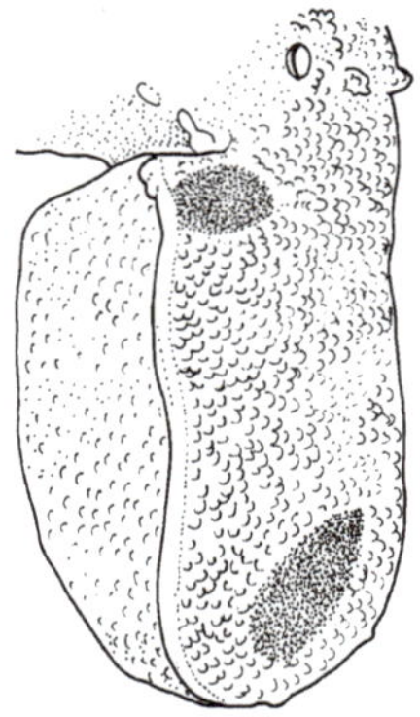

Lateral view (A. Reid redrawn from from original description).

Funnel organ (A. Reid redrawn from from original description).

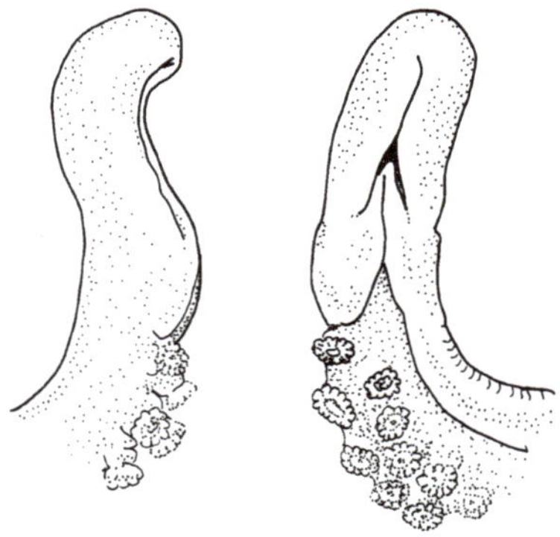

Distal tip of hectocotylus, lateral view (left) and oral view (right) (A. Reid redrawn from original description).

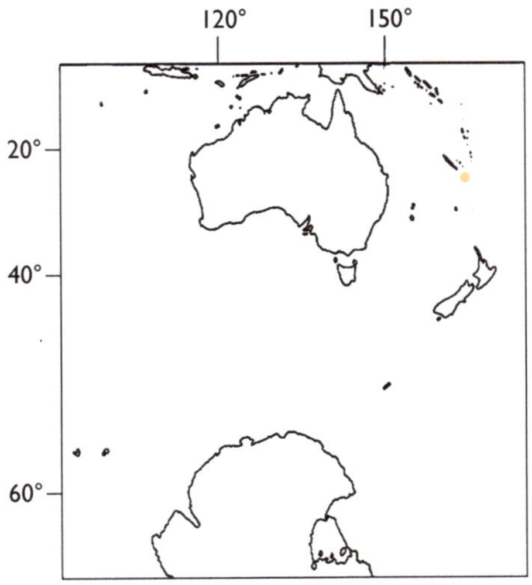



*Scaeurgus jumeau* distribution.

### HABITAT AND BIOLOGY

*Scaeurgus jumeau* is presently found on a single seamount.

### SIZE

Up to 44 mm ML.

### DISTRIBUTION

Known only from East Jumeau Bank, New Caledonia, 23°48'S, 168°17'E.

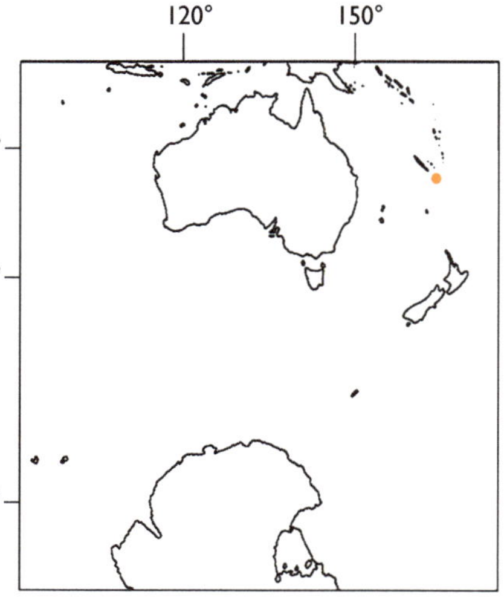

*Scaeurgus tuber* distribution.

## *Scaeurgus tuber* Norman, Hochberg & Boucher-Rodoni, 2005

*Scaeurgus tuber* Norman, M. D., Hochberg, F. G. and Boucher-Rodoni, R. (2005). A revision of the deep-water octopus genus *Scaeurgus* (Cephalopoda: Octopodidae) with description of three new species from the southwest Pacific Ocean. *Journal of Molluscan Studies 71*, 319–337 [321, figs 1-3, 10A, 10B, 10G].

Type data: holotype MNHN 2035.

Type locality: Norfolk Ridge, south of New Caledonia.

### DISTINGUISHING CHARACTERS

The arm sucker counts distinguish this species from *S. jumeau* (see key).

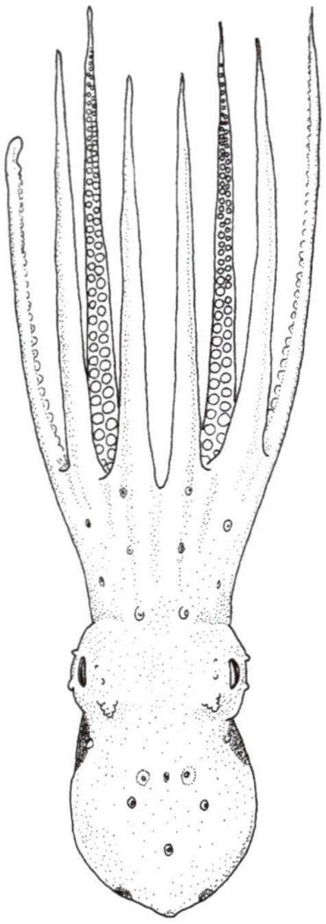
Dorsal view (A. Reid redrawn from from original description).

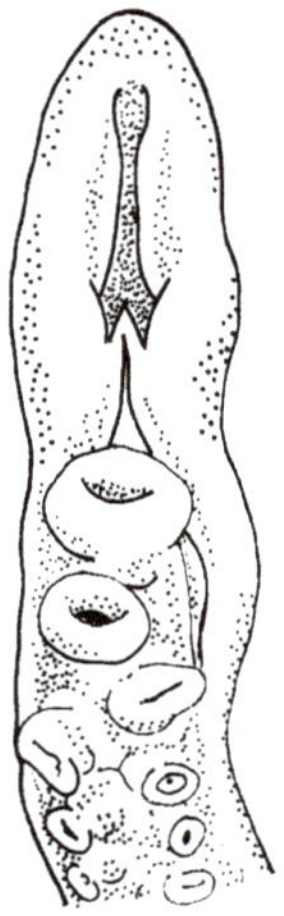
Distal tip of hectocotylus, ventral view (A. Reid redrawn from original description).

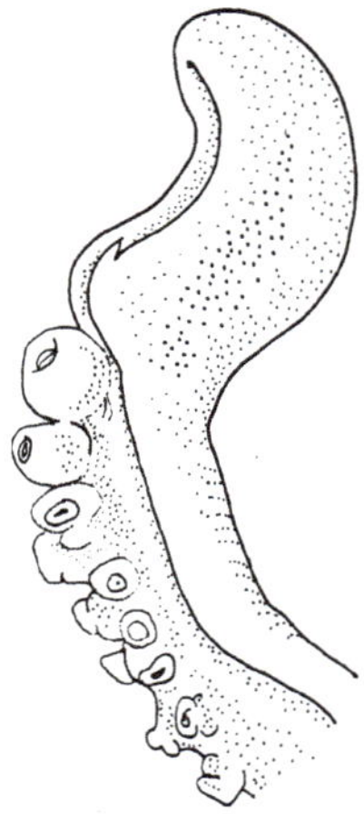
Distal tip of hectocotylus, lateral view (A. Reid redrawn from original description).

### DIAGNOSTIC FEATURES

Short arms (2.1–2.8 × ML); approximately equal in length. Lateral arm webs slightly deeper than dorsal and ventral web sectors, deepest web 24–35% longest arm. Normal arms with 100–130 suckers. Males with three or four enlarged suckers on all arms at level of eighth to twelfth proximal suckers, diameter to 16% ML. Third left arm hectocotylised with 54–63 suckers. Ligula large, 10–14% arm length with thick-rimmed groove; calamus small (20.7–43.6% length of ligula). Distal 1-3 suckers on hectocotylised arm significantly enlarged as tall, swollen cylinders. Gills with 8–10 lamellae per demibranch. Spermatophores small (9–12 mm), about two-thirds ML. Eggs 1.7–2.6 mm, ~10% ML.

*Colour.* Formalin-fixed specimens with orange-brown to cream base colour with pronounced pair of dark spots on anteriolateral mantle. Pair of paler, small black spots towards posterior end of mantle. Chromatophores on visceral mass membrane.

*Sculpture.* Skin with numerous rounded papillae in regular arrangement over dorsal surfaces and along aboral surfaces of arms 1–3. Less pronounced papillae on ventral surfaces. Diamond of four small papillae on dorsal mantle. Longitudinal row of three larger papillae down aboral midline of arms 1–3. Single large branched papilla over each eye, with several smaller papillae. Large papilla dorsal to each anterior black spot. Large papilla on posterior mantle. Lateral ridge present, well defined at margin of mantle aperture and continues as a line of broken, short ridges. Short ridges on lateral head, one under each eye.

### HABITAT AND BIOLOGY

This benthic species inhabits the upper bathyal. It has been collected from the continental slopes of New Caledonia and also three seamounts: Aztec (= Antigonia), West Jumeau and Mont A.

### SIZE

Up to 24 mm ML.

### REMARKS

This is a pygmy species, maturing at a tiny size.

### DISTRIBUTION

New Caledonia and Norfolk Ridge, 23°41′S, 168°01′E. Known depth range 230–391 m.

## *Thaumoctopus* Norman & Hochberg, 2005

*Thaumoctopus* is currently monotypic, represented by *Thaumoctopus mimicus* Norman & Hochberg, 2005.

## *Thaumoctopus mimicus* Norman & Hochberg, 2005

*Thaumoctopus mimicus* Norman, M. D., and Hochberg, F. G. (2005). The 'Mimic Octopus' (*Thaumocopus mimicus* n. gen. et sp.), a new octopus from the tropical Indo-West Pacific (Cephalopoda: Octopodidae). *Molluscan Research 25*(2), 57–70 [58, figs 1–6].

Type data: holotype MNHN 2076; paratypes MNHN 3784, ZMO D36521.

Type locality: New Caledonia, 19°49′S, 163°48.4′E.

### COMMON NAME

Mimic Octopus

### DISTINGUISHING CHARACTERS

Superficially similar to *Wunderpus photogenicus* Norman and Finn (2006), which, so far, has been found further north throughout the Indo-Malay Archipelago, but has irregular white stripes on the body and a distinctive U-shaped patch rather than oval to elongate-oval white spots. The eyes are positioned on elongate stalks in *W. photogenicus*, while in *T. mimicus* the eyes, though pronounced, are not obviously stalked. The supraocular papillae are blunt-tipped in *W. photogenicus* and there are no anterior ocular papille, while in *T. mimicus* they are narrower, with the paired (supraocular and anterior ocular papillae) appearing hornlike. In addition, *T. mimicus* has nine lamellae per demibranch, while there are five to seven in *W. photogenicus*. The hectocotylus also has a greater number of suckers; there are only 64 in *O. photogenicus*.

### DIAGNOSTIC FEATURES

Elongate, muscular. Eyes slightly pronounced. Funnel organ small, W-shaped with short lateral limbs. Arms long, thin, muscular (7–10 × ML), with well-developed margins along ventral edges, doral arm shortest, formula 4 > 3 > 2 > 1. Arm autotomy present at set level of twelfth to thirteenth proximal sucker. Webs thin and retractile, depth ~5.5% of the longest arm, extending as well-developed margins along ventral edges of all arms to arm tips, lateral webs deepest. No enlarged suckers in either sex; diameter ~11% ML; normal arms with up to 283 biserial suckers. Male hectocotylus (third right arm) ~50% length of opposite arm, with 146 suckers. Ligula small, elongate (< 2% arm length), copulatory organ of ligula with distinct groove; calamus absent, slight thickening of rim of proximal ligula groove. Gills with nine lamellae per demibranch (plus terminal lamella). Salivary glands ~2 × buccal mass length. Diverticulum short. Ink sac and anal flaps present. Spermatophores ~36% ML; width ~3% spermatophore length. Approximately 60 sperm cord whorls; ejaculatory apparatus with ~29 coils. Eggs small. Stylets present, short and non-mineralised.

*Colour.* Body dark brown with scattered white longitudinal markings with distinctive markings: base component of colour pattern includes a white teardrop ring on mid-dorsal mantle and a distinct white U-shaped patch on posteriodorsal mantle. Arms with cream to white bands against brown background colour. Arms with white band along outer margins. Small frontal white spots present.

*Sculpture.* Skin smooth with secondary papillae around lateral and posterior mantle. Paired 'horns' present over each eye, supraocular papillae larger than anterior ocular papillae. No primary papillae on dorsal mantle.

### HABITAT AND BIOLOGY

This species is famous for its ability to impersonate several marine organisms, including poisonous flatfishes and sea snakes, as a conspicuous primary defence mechanism. In tropical Indonesia they are reputed to mimic up to 13 species of other local marine organisms. Hanlon *et al.* (2008) studied and filmed this behaviour in Indonesia. Two octopuses were observed to mimic the shape, swimming action, speed, duration and sometimes colouration of swimming, drably coloured flounders that are common in the area. Before and after this mimicry they were camouflaged and motionless. When motionless they mimicked sponges, tube worms or colonial tunicates. They used flounder mimicry only when their movement would give away camouflage in this open habitat, and foraged twice daily.

Huffard *et al.* (2010) further examined mimicry in this species and questioned the possible evolution of mimicry in this octopus. The behaviour is interesting as this octopus mimics both toxic soles (Soeleidae) in a conspicuous disruptive and high contrast and/or aposematic colouration and non-toxic, drably coloured flounders (Bothidae), thus switching between acute crypsis and drawing attention with no attempt at camouflage. Their results suggest that flatfish swimming appears to have evolved concurrently with extremely long arms in a clade of sand-dwelling species. The conspicuous body colour pattern displayed by swimming *T. mimicus* may represent a secondary adaptation potentially allowing for mimicry of a toxic sole.

Live animal (D. Coker).

Live animal (D. Coker).

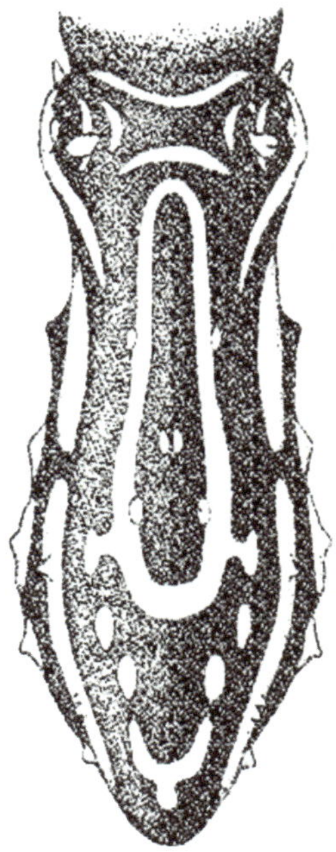

Dorsal mantle (M. Shea, redrawn from original description).

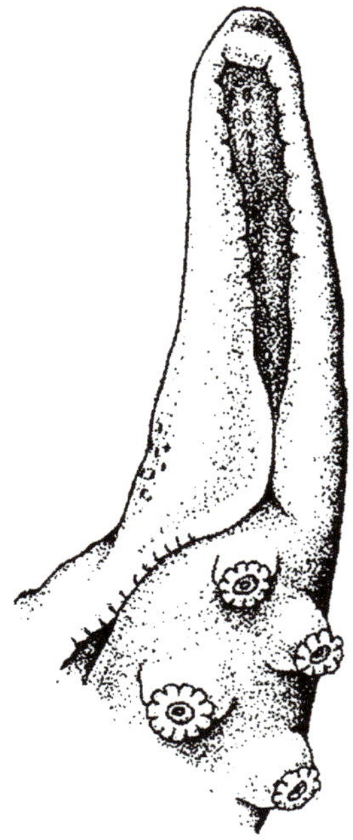

Hectocotylus, oral view (M. Shea, redrawn from original description).

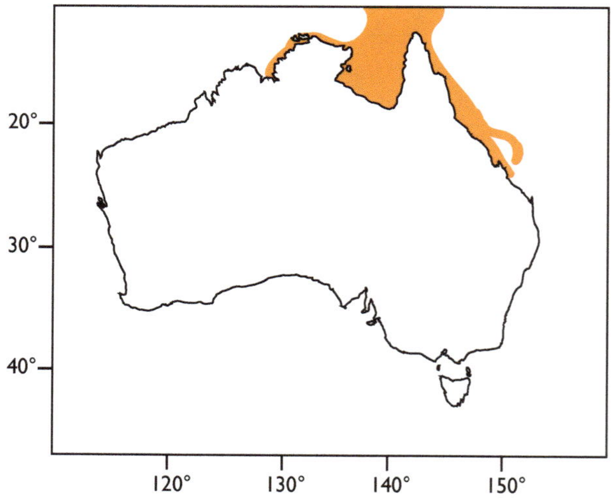


*Thaumoctopus mimicus* distribution.

It typically inhabits shallow (0.5–7.0 m) sandy and muddy habitats near coral reefs. It is a day-active species that hunts small fishes and crustaceans on the substrate or in holes on the reef flat. Three feeding strategies have been observed: probing one or more arms down burrows; gliding over the substrate with webs and arms flared to 'flush out' other animals from the substrate, and travelling along subterranean burrows. It inhabits a lair at night. Lairs appear to be vacated burrows of other animals. Homogenous low-relief habitats, such as sand flats, offer few opportunities for concealment, driving selection towards deceptive resemblance and mimicry in these octopuses.

#### SIZE

The Lizard Island animal was ~100 mm in ML (based on live animal observations). This is in contrast to the animals originally described from New Caledonia that have MLs up to 58 mm.

#### REMARKS

This species was reliably documented (with photos and video) for the first time in Australian waters in 2012 (Coker 2013). It was observed on a shallow sand flat at low tide during the afternoon at Lizard Island, Queensland, 14°41′S 145°27′E. Anecdotal observations suggest that it may occur elsewhere in northern Australia.

#### DISTRIBUTION

Indo-Malayan Archipelago from New Caledonia to Indonesia, Malaysia and the Philippines, Australia. Possibly Red Sea (photographic records). Purported geographical range includes much of the Indo-West Pacific.

#### REFERENCE

Coker (2013).

## *Order Vampyromorphida*

This order is monospecific. It is sometimes referred to as Vampyromorpha.

## Vampyroteuthidae

The family Vampyroteuthidae is monotypic, represented by *Vampyroteuthis infernalis* Chun, 1903.

### *Vampyroteuthis* Chun, 1903

As for family.

### *Vampyroteuthis infernalis* Chun, 1903

*Vampyroteuthis infernalis* Chun, C. (1903). Aus den Tiefen des Weltmeeres. Jena: Gustav Fischer, second edition. 592 pp. [88, text fig.].

Type data: holotype ZMB Moll. -63850.

Type locality: Atlantic Ocean, 1°56.7′S, 7°40.6′E.

#### COMMON NAME

Vampire Squid

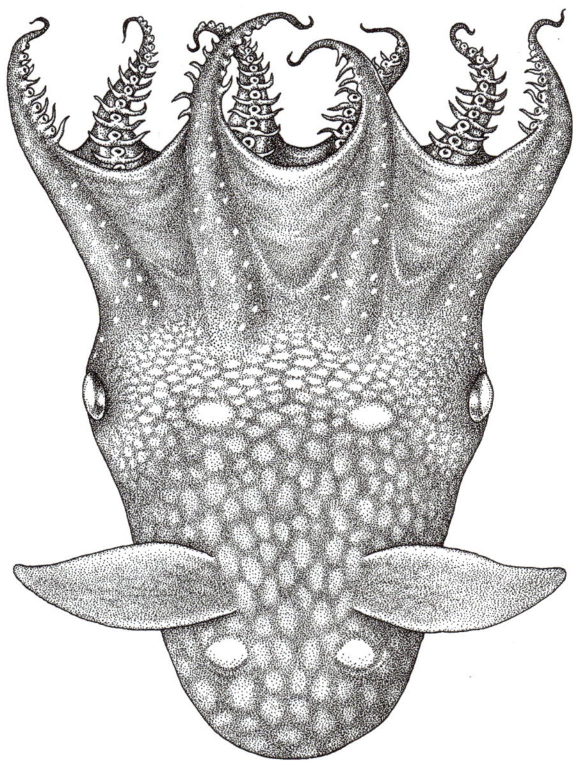

Dorsal view (K. Hollis, ABRS).

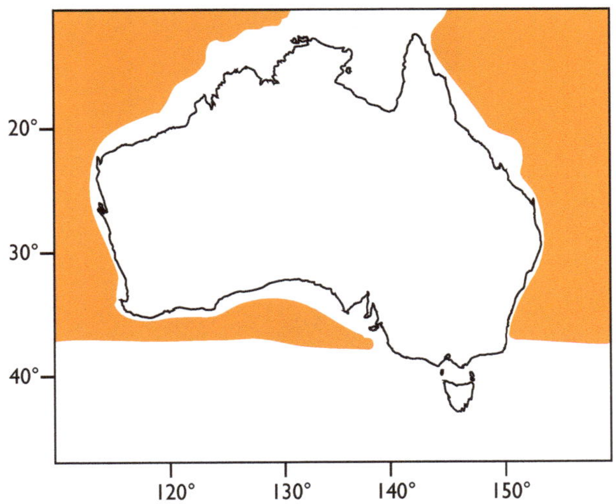


*Vampyroteuthis infernalis* distribution.

#### DIAGNOSTIC FEATURES

Body gelatinous, with jelly-like consistency. One pair of fins in adults, two pairs in juveniles. Eight arms with single series of stalked suckers starting from half way along their length and extending to arm tips. Arm webs deep, extend to near arm tips. Each arm with double series of long fingers of skin (cirri). Pair of pits in web between bases of arms 1 and 2; each pit contains a long thread-like filament. Suckers without horny sucker rings. Ink sac absent.

*Colour.* Black pigmentation.

*Photophores.* Pair of large light organs on posterior mantle at base of fins in adults. Light organs can be masked using black shutter-like lids. Smaller light organs present on arm tips and a scatter of small light organs on ventral side of head, body and arms. Arm tips capable of releasing bioluminescent particles.

## HABITAT AND BIOLOGY

*Vampyroteuthis infernalis* occupies the oxygen minimum zone of the deep sea where there are fewer predators, but typical cephalopod food is scarce.

A recent study examined *in situ* observations of *V. infernalis* obtained from deep-sea video recordings from MBARI's, remotely operated vehicles (ROVs) (Hoving and Robison 2012). This study showed that vampire squid use their retractile arm filaments to capture food, which consists of detrital matter of various sizes, including the remains of gelatinous zooplankton, discarded larvacean houses, crustacean remains, diatoms and faecal pellets, thus differing in feeding behaviour to other cephalopods.

Live animals orient with the dorsal mantle surface facing upwards and one filament extending well beyond the arms. When disturbed, they pull the arms and webs over the body to take on a shape that exposes the black skin and cirri on the oral web surfaces. They can produce light on their arm tips and squirt luminous clouds from the arm tips consisting of discrete glowing particles. Females store sperm in deep receptacles anterior to each eye.

## SIZE

Up to 130 mm ML; total length to ~300 mm.

## REMARKS

Vampire squid are considered phylogenetic relics because they have features of octopods and squids. More than one species may be present in this genus.

## DISTRIBUTION

Tropical and temperate waters worldwide, between ~35°N and 35°S. Known depth range 600–1 200 m.

## REFERENCES

Seibel *et al.* (1998), Hoving and Robison (2012), Norman and Finn (2014c).

# Appendix A Institutional acronyms

AIM Auckland Institute and Museum, Private Bag, Auckland 1, New Zealand.

ABRS Australian Biological Resources Study, Department of the Environment, John Gorton Building, King Edward Terrace, Parkes, Australian Capital Territory 2600, Australia.

AM Australian Museum, 1 William Street, Sydney South, New South Wales 2010, Australia.

ANSP Academy of Natural Sciences, 19th and The Parkway, Philadelphia, PA 19103, USA.

BMNH The Natural History Museum, Cromwell Road, London SW7 5BD, England (formerly British Museum (Natural History)).

CASIZ California Academy of Sciences, Golden Gate Park, San Francisco, CA 94118, USA.

FAO Food and Agriculture Organisation of the United Nations, Viale delle Terme di Caracalla, 00153 Rome, Italy.

FMNH Field Museum of Natural History, Roosevelt Road and Lake Shore Drive, Chicago, IL 60605, USA (formerly Chicago Natural History Museum).

IRSNB Institut Royal des Sciences Naturelles de Belgique, Rue Vautier 29, B-1040 Brussels, Belgium (formerly Musee royal d'Histoire naturelle de Belgique).

MBARI The Monterey Bay Aquarium Research Institute, 7700 Sandholdt Road, Moss Landing, CA 95039, USA.

MHNN Museum d'Histoire Naturelle (Musee Barla), 60 bis blvd Risso, 06300 Nice, France.

MLP Museo de La Plata, Paseo del Bosque, 1900 La Plata, Argentina.

MNHN Museum National d'Histoire Naturelle, Laboratoire Biologie Invertebres Marins et Malacologie, 55, rue de Buffon, 75005 Paris 05, France.

MOM Musee Oceanographique, Avenue Saint-Martin, Monaco-Ville, MC 98000, Monaco.

MSUT University Museum, University of Tokyo, Hongo 7-3-1, Bunkyoku, Tokyo 113, Japan (includes former Tokyo Imperial University (TIU) collection and University of Tokyo Department of Fisheries (FUMT) collection).

MV Museum Victoria, GPO Box 666, Melbourne, Victoria 3001, Australia (formerly National Museum of Victoria).

MZUS Musee Zoologique, Universite Louis Pasteur and de la Ville de Strasbourg, 29 Boulevard de la Victoire, F67000 Strasbourg, France.

NMGW National Museums and Galleries of Wales, Cathays Park, Cardiff CF1 3NP, Wales.

NMI National Museum of Ireland, Kildare Street, Dublin 2, Ireland.

NMNH National Museum of Natural History, 10th & Constitution Avenue, Smithsonian Institution, Washington, DC 20560, USA.

NMNZ National Museum of New Zealand *Te Papa Tongarewa*, P.O. Box 467, Wellington, New Zealand.

NRM Naturhistoriska Riksmuseet, Box 50007, S-104 05 Stockholm, Sweden.

NSMT National Science Museum, 3-23-1 Hyakunin-cho, Shinjuku-ku, Tokyo 160, Japan.

NTM Northern Territory Museum of Arts and Sciences, Box 4646, Darwin, Northern Territory 5794, Australia.

NZOI National Institute of Water and Atmospheric Research, P.O. Box 14-901, Kilbirnie, Wellington, New Zealand.

OM Otago Museum, Great King Street, Dunedin, New Zealand.

RMNH Nationaal Natuurhistorisch Museum, Raamsteeg 2, Postbus 9517, 2300 RA Leiden, The Netherlands.

SAMA South Australian Museum, North Terrace, Adelaide, South Australia 5022, Australia.

SAMC South African Museum, Government Avenue, P.O. Box 61, Cape Town 8000, South Africa.

SBMNH Santa Barbara Museum of Natural History, 2559 Puesta del Sol Road, Santa Barbara, CA 93105, USA.

TIU Tokyo Imperial University (see MSUT).

WAM Western Australian Museum, Francis Street, Perth, Western Australia 6000, Australia.

YTKC Private collection of the late Dr Iwao Taki stored at the home of Dr Yoshio Taki, Kyoto City, Japan.

ZIN Zoological Institute, Russian Academy of Sciences, Universitetskaya Nab. 1, St. Petersburg 199164, Russia.

ZMA Zoologisch Museum, Universiteit van Amsterdam, P.O. Box 4766, 1009 AT, Amsterdam, The Netherlands.

ZMB Zoologisches Museum, Museum für Naturkunde der Humboldt-Universität, Invalidenstrasse 43, D-1040 Berlin, Germany.

ZMH Zoologisches Museum, Universität Hamburg, Martin Luther King Platz 3, 2000 Hamburg 13, Germany.

ZMMGU Zoological Museum, Moscow State University, ul. Gensena 6, Moscow, Russia.

ZMUB Zoological Museum, Universitetet i Bergen, Museplass 3, 5000 Bergen, Norway.

ZMUC Københavns Universitet, Zoologisk Museum, Universitetsparken 15, DK 2100 Copenhagen, Denmark.

ZSI Zoological Survey of India, 'M' Block, New Alipore, Calcutta 700053, India.

# Appendix B Fixation and preservation methods

In order to obtain as much information as possible from cephalopods that are collected for research purposes, particularly when the identification of specimens is not known, they should be sampled and preserved in the following manner (modified after Lu and Dunning 1982). The recommendations listed below pertain to an ideal scenario and clearly may need to be modified depending on resources and equipment available at the time of collection. At all times, researchers need to be aware of animal ethics requirements, which, while not stringent at present in some Australian states, are likely to become so in the near future. Collecting permit requirements also differ among states.

Prior to collection, photograph the animal if possible and record any behavioural observations and habitat information. If possible, video it for a short period before capture. Check for bioluminescence if possible.

## Anaesthetisation

Specimens to be preserved should be as fresh and undamaged as possible. If live caught, animals must be anaesthetised. Immersion in magnesium chloride ($MgCl_2$) is currently considered the most humane method for anaesthetising cephalopods (Messenger *et al.* 1985). Use a solution of 75 g $MgCl_2$ in 1 L of seawater. Dilute 1:1 with seawater before use.

This isotonic solution is mixed 15:85 with seawater for anaesthetisation. Immersion time depends on the size and sometimes the species. A cephalopod is considered fully anaesthetised when all respiratory movements have completely ceased and it shows (1) an inability to right itself when inverted, (2) no response to a gentle pinch of the skin of the arms, fin, mantle or sensitive skin over the eye, and (3) no response when gentle pressure is applied to the eyeball.

Cephalopods immersed in magnesium chloride should be monitored for signs of distress. If the animal shows signs of distress, remove the animal from the anaesthetic solution and return it to seawater. After the signs of distress have subsided, the cephalopod can be anaesthetised in a reduced concentration of anaesthetic. Signs of distress may include (1) repetitive inking, (2) extreme colour changes, (3) defecation, (4) rapid escape response, and (5) abnormal posture.

Take basic measurements (total length, ML, head width, arm lengths, fin lengths, fin span and weight). For cuttlefish, photograph the intact animal and then carefully remove the bone before fixation and store with the specimen (it is much simpler to remove the bone without breakage before fixation).

Take kidney smears for future examination of parasites. (See below for this procedure.)

## Tissue sampling

Two tissue samples, to a maximum thickness of 0.5 cm, should be taken routinely from each specimen and placed in individual vials: (1) a small portion of mantle tissue (preferably with the skin removed) and (2) tissue from one gill only (this organ is reportedly DNA-rich). The gill in the best condition should be retained intact within the mantle cavity as gill lamellae counts are important for species identification. Tissue should be removed using a fresh scalpel blade for each specimen and sample. Reuse of blades, if necessary, should be preceded by rinsing thoroughly in seawater, and wiped using a sterile cloth. Rinsing in alcohol is not recommended because any tissue, and therefore DNA residue, may be preserved during the washing process. Fresh blades are preferred and should definitely be used if future RNA sampling is planned.

The following tissue preservation methods are suggested in decreasing order of preference in terms of existing, and possible future, molecular studies. *Note*: Advances in this field are occurring rapidly, so check recent publications to keep abreast of current developments.

- Freeze immediately at –80°C. A dry shipper can be used in the field, ensuring the shipper is kept 'charged' with liquid nitrogen for the duration of the trip to maintain temperature (check the manufacturer's specifications).

- If RNA is required (or to permit this future possibility), place in five volumes of RNA*later*® solution (i.e. one part tissue to five parts RNA*later*®), enabling future RNA and DNA work. Once in RNA*later*®, samples can be stored for up to 1 day at 37°C, 1 week at 25°C, 1 month or more at 4°C and long-term at -20°C or −80°C (www.lifetechnologies.com/au/en/home/references). Samples stored in RNA*later*® are compatible with all commonly used RNA isolation methods. It is also possible to extract genomic DNA and total protein from samples stored in this solution. (It denatures all protein.)
- Place in 10 volumes of 95% ethanol. Ensure that the ethanol is replaced after 1 day with a fresh solution: water leaching from the tissue will dilute the original solution, making it less concentrated. If possible after collection, pour ethanol off and freeze the sample at −80°C until required for analyses.
- Place the sample in a -20°C freezer for the shortest time possible before transferring to a −80°C freezer. This should be done within a week of sample collection if possible.
- Place the sample in 10 volumes of 70% ethanol (or 80% ethanol if that concentration is the only available). Replace with a fresh 70% solution after a day or so to minimise dilution from tissue water.

The rest of the specimen should be fixed in 5–10% buffered formalin (one part concentrated formalin and nine parts seawater) for at least 3 days. Enough fixative, preferably 10 × the volume of the specimen, should be used to cover the specimens completely. The fixation container should be large enough for the specimen to be laid down without folding or bending, with arms outstretched (bearing in mind the size of the long-term storage container) and deep enough to immerse the specimen in fixative.

Care should be taken to ensure that the fixative has entered the mantle cavity. This can be achieved by injection, by cutting a small slit on the ventral side of the mantle, or by gently stroking the ventral side of the mantle and applying pressure to remove any air in the mantle cavity. This is particularly important for large specimens and when working in warm climates. Rinse specimens in tap water (two or three rinses with soaking depending on the size of the specimen) and store the specimen in 70% ethanol (or 40% isopropyl alcohol). (If formalin is not available, specimens can be killed and preserved in 70–100% ethanol. However, this usually results in the tissue becoming brittle and not so useful for anatomical work.) Freezing is also possible, but results in inferior quality specimens compared to those fixed in formalin.

Retention of voucher specimens is strongly recommended to enable future comparison of morphological and molecular traits and for describing species. Vouchers should be lodged with the appropriate institution (preferably within the state of origin of the specimen, or as specified on the collection permit used).

*Always* record the following information with the specimen (without this information the value of the preserved specimen is negated, or may render it of no use for research):

- the time and date of collection;
- the locality, preferably with latitude and longitude;
- the depth of capture;
- the depth of water at the collection locality;
- the gear used (e.g. midwater trawl, dipnet, jig: note whether nets are opening-closing to determine the accuracy of collection depth data);
- environmental data such as water temperature, salinity and weather conditions (if possible);
- the names of collector(s) and the vessel name; and
- the fixation method (e.g. frozen; 10% formalin → 70% ethanol).

Ensure tissue specimens and vouchers are cross-numbered to ensure these can be matched at a later date.

## Karyology

As karyological data for cephalopods is scarce, a method for preparing chromosome squashes is provided below. Methods follow that of Vitturi *et al.* (1990), who based procedures on a method used by Colombera (1970) for tunicates. Vitturi *et al.* (1990) used fresh spermatocytes for their study. These are obtained by dissecting the testis from a fresh male specimen. If chromosome squashes are not going to be prepared immediately, store the testis tissue in a solution of 90% ethanol containing one part glacial acetic acid. To make a squash:

- Place the testis tissue in solution of 0.025 colchicine in distilled water for 20 minutes.
- Fix in 60% acetic acid orcein for a minimum of 15 minutes (longer for preserved material).
- Wash away excess orcein using 60% acetic acid.

- Place a small amount of the tissue on a microscope slide and squash the tissue between the slide and a coverslip using your thumbs. Colombera (1970) warns that squashes must not be made immediately after fixation because the tissue first hardens and then slowly softens.
- After any excess acetic acid has been absorbed, seal the edges of the coverslip (melted paraffin or nail varnish can be used). This preparation will last for a few weeks.
- Examine the slides using phase contrast microscopy.

## Cirrate octopods (reproduced from Vecchione *et al.* 2002: 80)

### Submersible observation and capture

- Video for a short period before capture. Check for bioluminescence if possible.
- Capture gently and bring to the surface. (If possible use a gelatinous zooplankton sampler.)

### On the surface

- Check for bioluminescence by placing the animal in chilled seawater in a dark room. Stimulate it mechanically to determine whether light is produced.
- Remove a tissue sample for DNA/molecular analyses and place in 80% ethanol (see the general notes above).
- Take basic measurements (total length, ML, head width, arm lengths, fin lengths, fin span, cirri length and weight).
- Photograph fresh specimens (dorsal, ventral and oral views) covered in seawater with a polarising filter.
- Collect eggs from the distal oviduct and incubate in cold seawater.
- Obtain a kidney smear.
- Fix in 10% buffered formalin for a minimum of 24 hours.
- Transfer to Steadman's solution (2.5% formalin, 1% propylene pheoxitol, 10% propylene glycol, made up in seawater) for storage.

### Method for making parasite smears from cephalopod host tissue in the field (S. Catalano, personal communication)

- Clean and sterilise all equipment in absolute ethanol.
- Remove the left and right kidneys from the anaesthetised but not fixed specimen (process one kidney at a time, keeping the samples separate and equipment sterile). Using a clean scalpel, cut a small piece off the whole kidney, hold it with fine forceps, and then brush the kidney lightly across 75% of the surface of a glass slide (leaving space on the slide for a label). Smear the issue in several horizontal lines along the length of the slide.
- Immediately put the slide in a slider mailer containing 70% ethanol to avoid dehydration and desiccation. Ensure the sample is covered with ethanol. Make four slides for each kidney. (Commercially purchased slide mailers usually contain slots for four slides, so each slide mailer will then contain four slides from a single kidney.)
- Label the slide mailer.
- Back at the laboratory, stain and mount the slides (see methods in Catalano 2013a).

# Appendix C Australian cephalopod checklist

Order Nautilida
Nautilidae Blainville, 1825
*Nautilus* Linnaeus, 1758
*Nautilus macromphalus* Sowerby, 1848
*Nautilus pompilius* Linnaeus, 1758
*Nautilus stenomphalus* Sowerby, 1849

Superorder Decapodiformes
Order Bathyteuthida
Bathyteuthidae Pfeffer, 1900
*Bathyteuthis* Hoyle, 1885
*Bathyteuthis abyssicola* Hoyle, 1885
Chtenopterygidae Grimpe, 1922
*Chtenopteryx* Appellöf, 1890
*Chtenopteryx sicula* (Verany, 1851)

Order Idiosepiida
Idiosepiidae Apellf, 1898
*Idiosepius* Steenstrup, 1881
*Idiosepius* sp. nov. Reid & Strugnell (in prep.)
*Idiosepius notoides* Berry, 1921
*Idiosepius pygmaeus* Steenstrup, 1881

Order Myopsida
Australiteuthidae Lu, 2005
*Australiteuthis* Lu, 2005
*Australiteuthis aldrichi* Lu, 2005
Loliginidae Lesueur, 1821
*Sepioteuthis* Blainville, 1823
*Sepioteuthis australis* Quoy & Gaimard, 1832
*Sepioteuthis 'lessoniana'* Lesson, 1831 (species-complex)
*Uroteuthis* Rehder, 1945
*Uroteuthis (Aestuariolus) noctiluca* (Lu, Roper & Tait, 1985)
*Uroteuthis (Photololigo) 'chinensis'* (Gray, 1849) species-complex
*Uroteuthis (Photololigo) 'duvaucelii'* (Orbigny, 1835)
*Uroteuthis (Photololigo) 'edulis'* (Hoyle, 1885) species-complex
*Uroteuthis (Photololigo) singhalensis* (Ortmann, 1891)

Order Oegopsida
Ancistrocheiridae Pfeffer, 1912
*Ancistrocheirus* Gray, 1849
*Ancistrocheirus lesueuri* (d'Orbigny, 1835)
Architeuthidae Pfeffer, 1900
*Architeuthis* Steenstrup, 1857
*Architeuthis dux* Steenstrup, 1860
Batoteuthidae Young & Roper, 1968
*Batoteuthis* Young & Roper, 1968
*Batoteuthis skolops* Young & Roper, 1968
Brachioteuthidae Pfeffer, 1908
*Brachioteuthis* Verrill, 1881
*Brachioteuthis behnii* (Steenstrup, 1882)
*Brachioteuthis picta* Chun, 1910
*Brachioteuthis 'riisei'* (Steenstrup, 1882) species-complex
*Slosarczykovia* Lipinski, 2001
*Slosarczykovia circumantarctica* Lipinski, 2001
Chiroteuthidae Gray, 1849
*Asperoteuthis* Nesis, 1980
*Asperoteuthis acanthoderma* (Lu, 1977)
*Chiroteuthis* d'Orbigny, 1839
*Chiroteuthis joubini* Voss, 1967
*Chiroteuthis picteti* Joubin, 1894
*Chiroteuthis veranyi* (Ferussac, 1834)
*Grimalditeuthis* Joubin, 1898
*Grimalditeuthis bonplandi* (Verany, 1839)
*Planctoteuthis* Pfeffer, 1912
*Planctoteuthis danae* (Joubin, 1931)
Cranchiidae Prosch, 1847
Subfamily Cranchiinae Pfeffer, 1912
*Cranchia* Leach, 1817
*Cranchia scabra* Leach, 1817
*Leachia* Lesueur, 1821
*Leachia cyclura* Lesueur, 1821

*Leachia pacificus* (Issel, 1908)
*Liocranchia* Pfeffer, 1884
*Liocranchia reinhardti* (Steenstrup, 1856)
*Liocranchia valdiviae* Chun, 1906
Subfamily Taoniinae Pfeffer, 1912
*Bathothauma* Chun, 1906
*Bathothauma lyromma* Chun, 1906
*Egea* Joubin, 1933
*Egea inermis* Joubin, 1933
*Galiteuthis* Joubin, 1898
*Galiteuthis glacialis* (Chun, 1906)
*Galiteuthis suhmi* (Hoyle, 1886)
*Helicocranchia* Massy, 1907
*Helicocranchia pfefferi* Massy, 1907
*Liguriella* Issel, 1908
*Liguriella podophthalma* Issel, 1908
*Megalocranchia* Pfeffer, 1884
*Mesonychoteuthis* Robson, 1925
*Mesonychoteuthis hamiltoni* Robson, 1925
*Sandalops* Chun, 1906
*Sandalops melancholicus* Chun, 1906
*Teuthowenia* Chun, 1910
*Teuthowenia pellucida* (Chun, 1910)
Cycloteuthidae Naef, 1923
*Cycloteuthis* Joubin, 1919
*Cycloteuthis sirventi* Joubin, 1919
*Discoteuthis* Young & Roper, 1969
*Discoteuthis discus* Young & Roper, 1969
*Discoteuthis laciniosa* Young & Roper, 1969
Enoploteuthidae Pfeffer, 1900
*Abralia* Gray, 1849
*Abralia andamanica* Goodrich, 1896
*Abralia armata* (Quoy & Gaimard, 1832)
*Abralia astrolineata* (Berry, 1914)
*Abralia astrostricta* Berry, 1909
*Abralia spaercki* Grimpe, 1931
*Abraliopsis* Joubin, 1896
*Abraliopsis* (*Micrabralia*)
*Abraliopsis (Micrabralia) gilchristi* (Robson, 1924)
*Abraliopsis (Micrabralia) lineata* (Goodrich, 1896)
*Abraliopsis* (*Abraliopsis*)
*Abraliopsis (Abraliopsis) tui* Riddell, 1985
*Enoploteuthis* d'Orbigny, 1848
*Enoploteuthis (Paraenoploteuthis) galaxias* Berry, 1918
*Enoploteuthis (Enoploteuthis) leptura* (Leach, 1817) =
*?E. (E.) magnoceani* Nesis, 1982
Gonatidae Hoyle, 1886
*Gonatus* Gray, 1849
*Gonatus antarcticus* Lönnberg, 1898
Histioteuthidae Verrill, 1881
*Histioteuthis* d'Orbigny in d'Orbigny & Ferussac, 1841
*Histioteuthis atlantica* (Hoyle, 1885)
*Histioteuthis bonnellii* (Ferrussac, 1834)
*Histioteuthis eltaninae* N. Voss, 1969
*Histioteuthis macrohista* N. Voss, 1969
*Histioteuthis meleagroteuthis* (Chun, 1910)
*Histioteuthis miranda* (Berry, 1918)
*Histioteuthis pacifica* (G. L. Voss, 1962)
*Stigmatoteuthis* Pfeffer, 1900
*Stigmatoteuthis hoylei* (Goodrich, 1896)
Joubiniteuthidae Naef, 1922
*Joubiniteuthis* Berry, 1920
*Joubiniteuthis portieri* (Joubin, 1916)
Lepidoteuthidae Pfeffer, 1912
*Lepidoteuthis* Joubin, 1895
*Lepidoteuthis grimaldii* Joubin, 1895
Lycoteuthidae Pfeffer, 1908
Lycoteuthinae Pfeffer, 1908
*Lycoteuthis* Pfeffer, 1900
*Lycoteuthis lorigera* (Steenstrup, 1875)
Mastigoteuthidae Verrill, 1881
*Idioteuthis* Sasaki, 1916
*Idioteuthis cordiformis* Chun, 1908
*Magnoteuthis* Salcedo-Vargas & Okutani, 1994
*Magnoteuthis magna* Joubin, 1913
*Mastigopsis* Grimpe, 1922
*Mastigopsis hjorti* Chun, 1913
*Mastigoteuthis* Verrill, 1881
*Mastigoteuthis psychrophila* Nesis, 1977
Neoteuthidae
*Alluroteuthis* Odhner, 1923
*Alluroteuthis antarcticus* Odhner, 1923
Octopoteuthidae Berry, 1912
*Octopoteuthis* Ruppell, 1844
*Octopoteuthis megaptera* (Verrill, 1885)
*Octopoteuthis rugosa* Clarke, 1980
*Octopoteuthis sicula* Ruppell, 1844
*Taningia* Joubin, 1931
*Taningia danae* Joubin, 1931

Ommastrephidae Steenstrup, 1857
Subfamily Ommastrephinae Posselt, 1891
*Eucleoteuthis* Berry, 1916
*Eucleoteuthis luminosa* (Sasaki, 1915)
*Hyaloteuthis* Gray, 1849
*Hyaloteuthis pelagica* (Bosc, 1802)
*Ommastrephes* d'Orbigny, 1835
*Ommastrephes bartramii* (Lesueur, 1821)
*Ornithoteuthis* Okada, 1927
*Ornithoteuthis volatilis* (Sasaki, 1915)
*Sthenoteuthis* Verrill, 1880
*Sthenoteuthis oualaniensis* (Lesson, 1830)
Subfamily Todarodinae Adam, 1960
*Martialia* Rochebrune & Mabille, 1889
*Martialia hyadesi* Rochebrune & Mabille, 1889
*Nototodarus* Pfeffer, 1912
*Nototodarus gouldi* (McCoy, 1888)
*Nototodarus hawaiiensis* (Berry, 1912)
*Todarodes* Steenstrup, 1880
*Todarodes filippovae* Adam, 1975
*Todarodes pusillus* Dunning, 1988
*Todaropsis* Girard, 1889/1890
*Todaropsis eblanae* (Ball, 1841)
Onychoteuthidae Gray, 1849
*Filippovia* Bolstad, 2010
*Filippovia knipovitchi* (Filippova, 1972)
*Kondakovia* Filippova, 1972
*Kondakovia longimana* Filippova, 1972
*Notonykia* Nesis, Roeleveld & Nikitinia, 1998
*Notonykia africanae* Nesis, Roeleveld & Nikitina, 1998
*Notonykia nesisi* Bolstad, 2007
*Onychoteuthis* Lichtenstein, 1818
*Onychoteuthis aequimanus* Gabb, 1868
*Onychoteuthis meridiopacifica* Rancurel & Okutani, 1990
*Onykia* Lesueur, 1821
*Onykia (Onykia) carriboea* Lesueur, 1821
*Onykia (Moroteuthopsis) ingens* (Smith, 1881)
*Onykia (Onykia) loennbergii* (Ishikawa & Wakiya, 1914)
*Onykia (Onykia) 'robsoni'* (Adam, 1962)
*Walvisteuthis* Nesis & Nikitina, 1986
*Walvisteuthis virilis* Nesis & Nikitina, 1986
Pholidoteuthidae Voss, 1956
*Pholidoteuthis* Adam, 1950
*Pholidoteuthis massyae* (Pfeffer, 1912)

Psychroteuthidae Thiele, 1920
*Psychroteuthis* Thiele, 1920
*Psychroteuthis glacialis* Thiele, 1920
Pyroteuthidae Pfeffer, 1912
*Pyroteuthis* Hoyle, 1904
*Pyroteuthis addolux* Young, 1972
*Pyroteuthis margaritifera* (Ruppell, 1844)
*Pterygioteuthis* Fischer, 1896
*Pterygioteuthis gemmata* Chun, 1908
*Pterygioteuthis giardi* Fischer, 1895
Thysanoteuthidae Keferstein, 1866
*Thysanoteuthis* Troschel, 1857
*Thysanoteuthis rhombus* Troschel, 1857

Order Sepiida
Sepiidae Keferstein, 1866
*Sepia* Linnaeus, 1758
*Sepia apama* Gray, 1849 species-complex
*Sepia bandensis* Adam, 1939
*Sepia baxteri* (Iredale, 1940)
*Sepia bidhaia* Reid, 2000
*Sepia braggi* Verco, 1907
*Sepia chirotrema* Berry, 1918
*Sepia cottoni* Adam, 1979
*Sepia cultrata* Hoyle, 1885
*Sepia elliptica* Hoyle, 1885
*Sepia grahami* Reid, 2001
*Sepia hedleyi* Berry, 1918
*Sepia irvingi* Meyer, 1909
*Sepia koilados* Reid, 2000
*Sepia latimanus* Quoy & Gaimard, 1832
*Sepia limata* (Iredale, 1926)
*Sepia mestus* Gray, 1849
*Sepia mira* (Cotton, 1932)
*Sepia novaehollandiae* Hoyle, 1909
*Sepia opipara* (Iredale, 1926)
*Sepia papuensis* Hoyle, 1885
*Sepia 'pharaonis'* Ehrenberg, 1831
*Sepia plana* Lu & Reid, 1997
*Sepia plangon* Gray, 1849
*Sepia reesi* Adam, 1979
*Sepia rhoda* (Iredale, 1954)
*Sepia rozella* (Iredale, 1926)
*Sepia senta* Lu & Reid, 1997
*Sepia smithi* Hoyle, 1885
*Sepia sulcata* Hoyle, 1885
*Sepia tanybracheia* Reid, 2000

*Sepia vercoi* Adam, 1979
*Sepia whitleyana* (Iredale, 1926)
*Metasepia* Hoyle, 1885
*Metasepia pfefferi* (Hoyle, 1885)
*Sepiella* Gray, 1849
*Sepiella mangkangunga* Reid & Lu, 1998
*Sepiella weberi* Adam, 1939

Order Sepiolida
Sepiadariidae Fischer, 1882 in 1880–1887
*Sepiadarium* Steenstrup, 1881
*Sepiadarium auritum* Robson, 1914
*Sepiadarium austrinum* Berry, 1921
*Sepiadarium kochii* Steenstrup, 1881
*Sepiadarium* sp.
Sepioloidea Ferussac & d'Orbigny, 1839
*Sepioloidea lineolata* (Quoy & Gaimard, 1832)
*Sepioloidea magna* Reid, 2010
Sepiolidae Leach, 1817
Heteroteuthinae Appellöf, 1898
*Heteroteuthis* Gray, 1849
*Heteroteuthis dagamensis* Robson, 1924
*Stoloteuthis* Verrill, 1881
*Stoloteuthis weberi* (Joubin, 1902)
*Iridoteuthis* Naef, 1912
*Iridoteuthis* sp.
*Sepiolina* Naef, 1912
*Sepiolina nipponensis* (Berry, 1911)
Rossiinae Appellöf, 1898
*Austrorossia* Berry, 1918
*Austrorossia australis* Berry, 1918
*Neorossia* Boletzky, 1971
*Neorossia leptodons* Reid, 1992
Sepiolinae Appellöf, 1898
*Euprymna* Steenstrup, 1887
*Euprymna hoylei* Adam, 1986
*Euprymna pardalota* Reid, 2011
*Euprymna tasmanica* (Pfeffer, 1884)
*Inioteuthis* Verrill, 1881
*Sepiola* Leach, 1817
*Sepiola* sp.

Order Spirulida
Spirulidae Owen, 1836
*Spirula* Lamarck, 1799
*Spirula spirula* (Linnaeus, 1758)

Superorder Octopodiformes
Order Octopoda

Suborder Cirrata
Cirroctopodidae
*Cirroctopus* Naef, 1923
*Cirroctopus mawsoni* (Berry, 1917)
Opisthoteuthidae
*Opisthoteuthis* Verrill, 1883
*Opisthoteuthis persephone* Berry, 1918
*Opisthoteuthis pluto* Berry, 1918

Suborder Incirrata
Superfamily Argonautoidea Naef, 1912
*Alloposidae* Verrill, 1881
*Haliphron* Steenstrup, 1861
*Haliphron atlanticus* Steenstrup, 1861
Argonautidae Cantraine, 1841
*Argonauta* Linnaeus, 1758
*Argonauta argo* Linnaeus, 1758
*Argonauta hians* [Lightfoot], 1786
*Argonauta nodosus* [Lightfoot], 1786
Ocythoidae Gray, 1849
*Ocythoe* Rafinesque, 1814
*Ocythoe tuberculata* Rafinesque, 1814
Tremoctopodidae Tryon, 1879
*Tremoctopus* Delle Chiaje, 1830
*Tremoctopus gracilis* (Eydoux & Souleyet, 1852)

Superfamily Octopodoidea d'Orbigny, 1839 in Ferrusac and d'Orbigny, 1834–1848
Amphitretidae Hoyle, 1886
Subfamily Amphitretinae Hoyle, 1886
*Amphitretus* Hoyle, 1885
*Amphitretus pelagicus* Hoyle, 1885
*Amphitretus thielei* Robson, 1930
Subfamily Bolitaeninae Chun, 1911
*Bolitaena* Verrill, 1884
*Bolitaena pygmaea* (Verrill, 1884)
*Bolitaena sheardi* (Allan, 1945)
*Japetella* Hoyle, 1885
*Japetella diaphana* Hoyle, 1885
Subfamily Vitreledonellidae Robson, 1932
*Vitreledonella* Joubin, 1918
*Vitreledonella richardi* Joubin, 1918
Eledonidae Grimpe, 1921
*Eledone* Leach, 1817
*'Eledone palari'* Lu & Stranks, 1991
Enteroctopodidae Strugnell *et al.* 2013
*Benthoctopus* Grimpe, 1921
*Benthoctopus levis* (Hoyle, 1885)
*Benthoctopus thielei* Robson, 1932

Megaleledonidae Taki, 1961
*Adelieledone* Allcock, Hochberg, Rodhouse & Thorpe, 2003
*Adelieledone adelieana* (Berry, 1917)
*Graneledone* Joubin, 1918
*Graneledone antarctica* Voss, 1976
*Megaleledone* Taki, 1961
*Megaleledone setebos* (Robson, 1932)
*Microeledone* Norman, Hochberg & Boucher-Rodoni, 2004
*Microeledone mangoldae* Norman, Hochberg & Boucher-Rodoni, 2004
*Pareledone* Robson, 1932
*Pareledone charcoti* (Joubin, 1905)
*Pareledone framensis* Lu & Stranks, 1994
*Pareledone harrissoni* (Berry, 1917)
*Pareledone prydzensis* Lu & Stranks, 1994
Octopodidae d'Orbigny, 1839
*Abdopus* Norman & Finn, 2001
*Abdopus aculeatus* (d'Orbigny, 1835)
*Abdopus capricornicus* Norman & Finn, 2001
*Abdopus tenebricus* (E. A. Smith, 1884)
*Ameloctopus* Norman, 1992
*Ameloctopus litoralis* Norman, 1992
*Amphioctopus* Fischer, 1882
*Amphioctopus exannulatus* (Norman, 1993)
*Amphioctopus* cf. *kagoshimensis* Norman & Kubodera, 2006
*Amphioctopus marginatus* (Taki, 1964)
*Amphioctopus mototi* (Norman, 1993)
*Amphioctopus polyzenia* (Gray, 1849)
*Amphioctopus rex* (Nateewathana and Norman, 1999)
*Callistoctopus* Taki, 1964
*Callistoctopus alpheus* (Norman, 1993)
*Callistoctopus aspilosomatis* (Norman, 1993)
*Callistoctopus dierythraeus* (Norman, 1993)
*Callistoctopus graptus* (Norman, 1993)
*Callistoctopus ornatus* (Gould, 1852)
*Galeoctopus* Norman, Boucher-Rodoni & Hochberg, 2004
*Galeoctopus lateralis* Norman, Boucher-Rodoni & Hochberg, 2004
*Grimpella* Robson, 1928
*Grimpella thaumastocheir* Robson, 1928
*Hapalochlaena* Robson, 1929
*Hapalochlaena fasciata* (Hoyle, 1886)
*Hapalochlaena* cf. *lunulata* (Quoy and Gaimard, 1832)
*Hapalochlaena maculosa* (Hoyle, 1883)
*Histoctopus* Norman, Boucher-Rodoni & Hochberg, 2009
*Histoctopus zipkasae* Norman, Boucher-Rodoni & Hochberg, 2009
*Octopus* s.s. Cuvier, 1797
*Octopus australis* Hoyle, 1885
*Octopus berrima* Stranks & Norman, 1993
*Octopus* cf. *bocki* Adam, 1941
*Octopus bulbus* Norman, 2001
*Octopus bunurong* (Stranks, 1990)
*Octopus cyanea* Gray, 1849
*Octopus harpedon* Norman, 2001
*Octopus kaurna* Stranks, 1990
*Octopus laqueus* Kaneko & Kubodera, 2005
*Octopus maorum* (Hutton, 1880)
*Octopus micros* Norman, 2001
*Octopus pallidus* Hoyle, 1885
*Octopus pyrum* Norman, Hochberg & Lu, 1997
*Octopus superciliosus* Quoy & Gaimard, 1832
*Octopus tetricus* Gould, 1852
*Octopus warringa* Stranks, 1990
*Scaeurgus* Troschel, 1857
*Scaeurgus jumeau* Norman, Hochberg & Boucher-Rodoni, 2005
*Scaeurgus tuber* Norman, Hochberg & Boucher-Rodoni, 2005
*Thaumoctopus* Norman & Hochberg, 2005
*Thaumoctopus mimicus* Norman & Hochberg, 2005

Order Vampyromorphida
Vampyroteuthidae
*Vampyroteuthis* Chun, 1903
*Vampyroteuthis infernalis* Chun, 1903

# References


ABRS (2009) *Australian Faunal Directory.* Australian Biological Resources Study, Canberra. http://www.environment.gov.au/biodiversity/abrs/online-resources/fauna/afd/index.html

Adam W (1979) The Sepiidae (Cephalopoda, Decapoda) in the collections of the Western Australian Museum. *Records of the Western Australian Museum* **7**(2), 111–112.

Aitken JP, O'Dor RK (2004) Respirometry and swimming dynamics of the giant Australian cuttlefish, *Sepia apama* (Mollusca, Cephalopoda). *Marine and Freshwater Behaviour and Physiology* **37**(4), 217–234. doi:10.1080/10236240400016579

Aitken JP, O'Dor RK, Jackson GD (2005) The secret life of the giant Australian cuttlefish *Sepia apama* (Cephalopoda): Behaviour and energetics in nature revealed through radio acoustic positioning and telemetry (RAPT). *Journal of Experimental Marine Biology and Ecology* **320**(1), 77–91. doi:10.1016/j.jembe.2004.12.040

Albertin CB, Bonnaud L, Titus Brown C, Crookes-Gooson WJ, da Fonseca RR, da Cristo CD, Dilkes BP, Edsinger-Gonzales E, Freeman RM, Hanlon RT, Koenig KM, Lindgren AR, Martindale MQ, Minx P, Moroz LL, Nödl M-T, Nyholm SV, Ogura A, Pungor JR, Rosenthal JJC, Schwartz EM, Shigeno S, Strugnell JM, Wollesen T, Zhang G, Ragsdale CW (2012) Cephalopod genomics: A plan of strategies and organization. *Standards in Genomic Sciences* **7**(1), 175–188. doi:10.4056/sigs.3136559

Allan J (1945) Planktonic cephalopod larvae from the eastern Australian coast. *Records of the Australian Museum* **21**(6), 317–350. doi:10.3853/j.0067-1975.21.1945.549

Allcock AL (2005) On the confusion surrounding *Pareledone charcoti* (Joubin, 1905) (Cephalopoda: Octopodidae): Endemic radiation in the Southern Ocean. *Zoological Journal of the Linnean Society* **143**(1), 75–108. doi:10.1111/j.1096-3642.2004.00146.x

Allcock AL, Barratt I, Eléaume M, Linse K, Norman M, Smith PJ, Steinke D, Stevens DW, Strugnell JM (2010) Cryptic speciation and the circumpolarity debate: A case study on endemic Southern Ocean octopuses using the COI barcode of life. *Deep-sea Research Part II, Topical Studies in Oceanography* **58**(5), 242–249. doi:10.1016/jdsr2.2010.05.016

Allcock AL, Cooke IR, Strugnell JM (2011) What can the mitochondrial genome reveal about higher-level phylogeny in the molluscan class Cephalopoda? *Zoological Journal of the Linnean Society* **161**(3), 573–586.

Allcock AL, Hochberg FG, Stranks TN (2003a) Re-evaluation of *Graneledone setebos* (Cephalopoda: Octopodidae) and allocation to the genus *Megaleledone. Journal of the Marine Biological Association of the United Kingdom* **83**(2), 319–328. doi:10.1017/S0025315403007148h

Allcock AL, Hochberg FG, Rodhouse PGK, Thorpe JP (2003b) *Adelieledone*, a new genus of octopodid from the Southern Ocean. *Antarctic Science* **15**(4), 415–424. doi:10.1017/S0954102003001512

Allcock AL, Lindgren A, Strugnell JM (2014) The contribution of molecular data to our understanding of cephalopod evolution and systematics: A review. *Journal of Natural History* **49**(21–24), 1373–1421. doi:10.1080/00222933.2013.825342

Allcock AL, Piertney SB (2002) Evolutionary relationships of Southern Ocean Octopodidae (Cephalopoda: Octopoda) and a new diagnosis of *Pareledone. Marine Biology* **140**(1), 129–135. doi:10.1007/s002270100687

Allcock AL, Strugnell JM, Prodöhl P, Piatkowski U, Vecchione M (2007) A new species of *Pareledone* (Cephalopoda: Octopodidae) from Antarctic Peninsula waters. *Polar Biology* **30**(7), 883–893. doi:10.1007/s00300-006-0248-9

Amor MD, Norman MD, Cameron HE, Strugnell JM (2014) Allopatric speciation within a cryptic species complex of Australasian octopuses. *PLoS ONE* **9**(6), e98982. doi:10.1371/journal.pone.0098982

Anderson FE (2000) Phylogeny and historical biogeography of the loliginid squids (Mollusca: Cephalopoda) based on mitochondrial DNA sequence data. *Molecular Phylogenetics and Evolution* **15**(2), 191–214. doi:10.1006/mpev.1999.0753

Anderson FE, Engelke R, Jarrett K, Valinassab T, Mohamed KS, Asokan PK, Zacharia PU, Nootmorn P, Chotiyaputta C, Dunning M (2011) Phylogeny of the *Sepia pharaonis* species complex (Cephalopoda: Sepiida) based on analyses of mitochondrial and nuclear DNA sequence data. *The Journal of Molluscan Studies* **77**(1), 65–75. doi:10.1093/mollus/eyq034

Anderson FE, Valinassab T, Ho CW, Mohamed KS, Asokan PK, Rao GS, Nootmorn P, Chotiyaputta C, Dunning M, Lu CC (2007) Phylogeny of the pharaoh cuttle *Sepia pharaonis* based on partial mitochondrial 16S sequence data. *Reviews in Fish Biology and Fisheries* **17**(2–3), 345–352. doi:10.1007/s11160-007-9042-1

Anderson TJ (1999) Morphology and biology of *Octopus maorum* Hutton, 1880 in northern New Zealand. *Bulletin of Marine Science* **65**(3), 657–676.

André J, Grist EPM, Semmens JM, Pecl GT, Segawa S (2009) Effects of temperature on energetics and the growth pattern of benthic octopuses. *Marine Ecology Progress Series* **374**, 167–179. doi:10.3354/meps07736

André J, Pecl GT, Semmens JM, Grist EPM (2008) Early life-history processes in benthic octopus: Relationships between temperature, feeding, food conversion, and growth in juvenile *Octopus pallidus*. *Journal of Marine Biology and Ecology* **354**, 81–92. doi:10.1016/j.jembe.2007.10.008

André M, Solé M, Lenoir M, Dufort M, Quero C, Mas A, Lombarte A, van der Scharr M, López-Bejar M, Morell M, Zaugg S, Houégnigan L (2011) Low-freqency sounds induce acoustic trauma in cephalopods. *Frontiers in Ecology and the Environment* **9**(9), 489–493. doi:10.1890/100124

Appellöf A (1890) Teuthologische Beitrage I: *Ctenopteryx* n.g., *Veranya sicula* Krohn, *Calliteuthis* Verrill. *Bergens Museums Aarsberetning* **1889**(3), 1–34.

Arkhipkin AI, Laptikhovsky VV (2010) Observation of penis elongation in *Onykia ingens*: Implications for spermatophore transfer in deep-water squid. *The Journal of Molluscan Studies* **76**(3), 299–300. doi:10.1093/mollus/eyq019

Arocha F (2003) A new species of *Nematolampas* (Cephalopoda: Oegopsida) from the western Central Atlantic with an overview of the family Lycoteuthidae. *Bulletin of Marine Science* **72**(3), 941–953.

Aungtonya C, Nateewathana A, Tendal OS, Nabhitabhata J (2011) New records of the bobtail squid *Euprymna hyllebergi* Nateewathana, 1997 with designation of a neotype. *Phuket Marine Biological Center Research Bulletin* **70**, 15–27.

Barord GJ, Dooley F, Dunstan A, Ilano A, Keister KN, Neumeister H, Preuss T, Schoepfer S, Ward P (2014) Comparative population assessments of *Nautilus* sp. in the Philippines, Australia, Fiji, and American Samoa using baited remote underwater video systems. *PLoS ONE* **9**(6), e100799. doi:10.1371/journal.pone.0100799

Barratt IM, Johnson MP, Allcock AL (2007) Fecundity and reproductive strategies in deep-sea incirrate octopuses (Cephalopoda: Octopoda). *Marine Biology* **150**(3), 387–398. doi:10.1007/s00227-006-0365-6

Basil JA, Lazenby GB, Nakanuku L, Hanlon RT (2002) Female *Nautilus* are attracted to male conspecific odor. *Bulletin of Marine Science* **70**(1), 217–225.

Battam H, Richardson M, Watson AWT, Buttemer WA (2010) Chemical composition and tissue energy density of the cuttlefish (*Sepia apama*) and its assimilation efficiency by *Diomedea* albatrosses. *Journal of Comparative Physiology* **180**(8), 1247–1255. doi:10.1007/s00360-010-0497-3

Beesley PL, Ross GJB, Wells A (1998) 'Mollusca: The Southern Synthesis. Fauna of Australia. Vol. 5.' (CSIRO Publishing: Melbourne.)

Bello G (2013) Description of a new sepioline species, *Sepiola bursadhaesan* sp. (Cephalopoda: Sepiolidae), from the Catalan Sea, with remarks and identification key for the *Sepiola atlantica* group. *Scientia Marina* **77**, 489–499.

Bello G, Giannuzzi-Savelli R (1993) Case 2874. *Chtenopteryx* Appellof, 1890 (Mollusca, Cephalopoda): proposed confirmation as the correct original spelling. *Bulletin of Zoological Nomenclature* **50**(4), 270–272.

Berry SS (1918) Report on the Cephalopoda obtained by the F.I.S. 'Endeavour' in the Great Australian Bight and other southern Australian localities. *Biological Results of the Fishing Experiments Carried On by the F.I.S. 'Endeavour' 1909–14* **4**(5), 201–298.

Birch H (2010) How to disappear completely. *Chemistry World* June 2010, pp. 1–5.

Boal JG, Prosser KN, Holm JB, Simmons TL, Haas RE, Nagle GT (2010) Sexually mature cuttlefish are attracted to the eggs of conspecifics. *Journal of Chemical Ecology* **36**(8), 834–836. doi:10.1007/s10886-010-9816-0

Boletzky S, Roeleveld MAC (2000) 'Ventral adhesion' to hard substrates: A thigmotactic response in sepiid cuttlefish (Mollusca, Cephalopoda). *Vie et Milieu-Life and Environment* **50**(1), 59–64.

Bolstad KS (2007) Systematics and distribution of the New Zealand onychoteuthid fauna (Cephalopoda: Oegopsida), including a new species, *Notonykia nesisi* sp. nov. *Reviews in Fish Biology and Fisheries* **17**(2–3), 305–335. doi:10.1007/s11160-007-9041-2

Bolstad KS (2008) Two new species and a review of the squid genus *Onychoteuthis* Lichtenstein, 1818 (Oegopsida: Onychoteuthidae) from the Pacific Ocean. *Bulletin of Marine Science* **83**(3), 481–529.

Bolstad KSR (2010) Systematics of the Onychoteuthidae Gray, 1847. *Zootaxa* **2696**(3), 1–186.

Bolstad KSR, Young RE (2010) *Notonykia nesisi* Bolstad 2007. Version 23 December 2010 (under construction). http://tolweb.org/Notonykia_nesisi/108189/2010.12.23 in The Tree of Life Web Project, http://tolweb.org/

Bolstad KSR, Vecchione M, Young RE, Tsuchiya K (2007) *Notonykia africanae* Nesis, Roeleveld & Nikitina 1998. Version 09 July 2007 (under construction). http://tolweb.org/Notonykia_africanae/19954/2007.07.09 in The Tree of Life Web Project, http://tolweb.org/

Bolstad KSR, Young RE, Vecchione M (2010a) *Notonykia* Nesis, Roeleveld & Nikitina, 1998. Version 23 December 2010 (under construction). http://tolweb.org/Notonykia/108188/2010.12.23 in The Tree of Life Web Project, http://tolweb.org/

Bolstad KSR, Vecchione M, Young RE (2010b) Onychoteuthidae Gray, 1847. Hooked squids. Version 23 December 2010. http://tolweb.org/Onychoteuthidae/19419/2010.12.23 in The Tree of Life Web Project, http://tolweb.org/

Bolstad KSR, Vecchione M, Young RE (2010c) *Filippovia* Bolstad, 2010. *Filippovia knipovitchi* (Filippova, 1972). Version 23 December 2010 (under construction). http://tolweb.org/Filippovia_knipovitchi/19973/2010.12.23 in The Tree of Life Web Project, http://tolweb.org/

Bolstad KSR, Vecchione M, Young RE (2010d) *Kondakovia* Filippova, 1972. Version 16 November 2010 (under construction). http://tolweb.org/Kondakovia/19953/2010.11.16 in The Tree of Life Web Project, http://tolweb.org/

Bolstad KSR, Vecchione M, Young RE, Tsuchiya K (2010e) *Onychoteuthis* Lichtenstein, 1818. Version 23 December 2010 (under construction). http://tolweb.org/Onychoteuthis/19955/2010.12.23 in The Tree of Life Web Project, http://tolweb.org/

Bolstad KSR, Vecchione M, Young RE (2011) *Onykia* Lesueur, 1821. Version 02 March 2011 (under construction). http://tolweb.org/Onykia/19956/2011.03.02 in The Tree of Life Web Project, http://tolweb.org/

Bonacum J, Landman NH, Mapes RH, White MM, White A-J, Irlam J (2011) Evolutionary radiation of present-day *Nautilus* and *Allonautilus*. *American Malacological Bulletin* **29**(1–2), 77–93. doi:10.4003/006.029.0221

Bonnaud L, Boucher Rodoni R, Monnerot M (1997) Phylogeny of cephalopods inferred from mitochondrial DNA sequences. *Molecular Phylogenetics and Evolution* **7**(1), 44–54. doi:10.1006/mpev.1996.0366

Bonnaud L, Lu CC, Boucher-Rodoni R (2006) Morphological character evolution and molecular trees in sepioids (Mollusca: Cephalopoda): Is the cuttlebone a robust phylogenetic marker? *Biological Journal of the Linnean Society. Linnean Society of London* **89**(1), 139–150. doi:10.1111/j.1095-8312.2006.00664.x

Bonnaud L, Pichon D, Boucher-Rodoni R (2005) Molecular approach of Decabrachia phylogeny: Is *Idiosepius* definitely not a sepiolid? *Phuket Marine Biological Centre Research Bulletin* **66**, 203–212.

Bonnaud L, Rodhouse PG, Boucher-Rodoni R (1998) A phylogenetic study of the squid family Onychoteuthidae (Cephalopoda: Oegopsida). *Proceedings. Biological Sciences* **265**, 1761–1770. doi:10.1098/rspb.1998.0500

Bonnaud L, Saihi A, Boucher-Rodoni R (2002) Re 28S RDNA and 18S RDNA informative for cephalopod phylogeny? *Bulletin of Marine Science* **71**(1), 197–208.

Boore JL (2006) The complete sequence of the mitochondrial genome of *Nautilus macromphalus* (Mollusca: Cephalopoda). *BMC Genomics* **7**, 182–195. doi:10.1186/1471-2164-7-182

Boucher-Rodoni R, Boucaud-Camou E, Mangold K (1987) Feeding and digestion. In 'Cephalopod Life Cycles, Vol. II: Comparative Reviews'. (Ed. PR Boyle) pp. 85–108. (Academic Press: London.)

Braid HE, McBride PD, Bolstad KSR (2014) Molecular phylogenetic analysis of the squid family Mastigoteuthidae (Mollusca, Cephalopoda) based on three mitochondrial genes. *Hydrobiologia* **725**(1), 145–164. doi:10.1007/s10750-013-1775-3

Braley M, Goldsworthy SD, Page B, Steer M, Austin JJ (2010) Assessing morphological and DNA-based diet analysis techniques in a generalist predator, the arrow squid *Nototodarus gouldi*. *Molecular Ecology Resources* **10**(3), 466–474. doi:10.1111/j.1755-0998.2009.02767.x

Brandt SB (1983) Pelagic squid associations with a warm-core eddy of the east Australian current. *Australian Journal of Marine and Freshwater Research* **34**(4), 573–585. doi:10.1071/MF9830573

Bráullio de Luna Sales J, Fernando da Silva Rodrigues-Filho L, Haimovici M, Sampaio I, Schneider H (2010) Molecular differentiation of the species of two squid families (*Loliginidae* and *Ommastrephidae*) based on a PCR study of the 5S rDNA gene. *Food Control* **16**(1), 96–98.

Brierley AS, Rodhouse PG, Thorpe JP, Clarke MR (1993) Genetic evidence of population heterogeneity and cryptic speciation in the ommastrephid squid *Martialia hyadesi* from the Patagonian Shelf and Antarctic Polar Frontal Zone. *Marine Biology* **116**(4), 593–602. doi:10.1007/BF00355478

Brown C, Garwood MP, Williamson JE (2012) It pays to cheat: Tactical deception in a cephalopod social system. *Biology Letters* **8**(5), 729–732. doi:10.1098/rsbl.2012.0435

Burford BP, Robison BH, Sherlock RE (2014) Behaviour and mimicry in the juvenile and subadult life stages of the mesopelagic squid *Chiroteuthis calyx*. *Journal of the Marine Biological Association of the United Kingdom*. **95**(6), 1221–1235. doi:10.1017/S0025315414001763

Bush SL, Robison BH, Caldwell R (2009) Behaviour in the dark: Locomotor, chromatic, postural, and bioluminescent behaviors of the deep-sea squid *Octopoteuthis deletron* Young, 1972. *The Biological Bulletin* **216**(1), 7–22.

Bustamante P, Cherel Y, Caurant F, Miramand P (1998) Cadmium, copper and zinc in octopuses from Kerguelen Islands, Southern Indian Ocean. *Polar Biology* **19**(4), 264–271. doi:10.1007/s003000050244

Butler AJ, Rees T, Beesley P, Bax NJ (2010) Marine Biodiversity in the Australian Region. *PLoS ONE* **5**(8), e11831. doi:10.1371/journal.pone.0011831

Carlini DB (1998) The phylogeny of Coleoid Cephalopods inferred from molecular evolutionary analysis of the Cytochrome C oxidase 1, muscle actin and cytoplasmic actin genes. PhD Thesis, The College of William and Mary, USA.

Carlini DB, Graves JE (1999) Phylogenetic analysis of cytochrome C oxidase I sequences to determine higher-level relationships within the coleoid cephalopods. *Bulletin of Marine Science* **64**(1), 57–76.

Carlini DB, Reece KS, Graves JE (2000) Actin gene family evolution and the phylogeny of coleoid cephalopods (Mollusca: Cephalopoda). *Molecular Biology and Evolution* **17**(9), 1353–1370. doi:10.1093/oxfordjournals.molbev.a026419

Carlini DB, Young RE, Vecchione M (2001) A molecular phylogeny of the octopoda (Mollusca: Cephalopoda) evaluated in light of morphologial evidence. *Molecular Phylogenetics and Evolution* **21**(3), 388–397. doi:10.1006/mpev.2001.1022

Cartron L, Josef N, Lerner A, McCusker SD, Darmallacq A-S, Dickel L, Shashar N (2013) Polarization vision can improve object detection in turbid waters by cuttlefish. *Journal of Experimental Marine Biology and Ecology* **447**, 80–85. doi:10.1016/j.jembe.2013.02.013

Catalano SR (2013a) First descriptions of dicyemid mesozoana (Dicyemida: Dicyemidae) from Australian octopus (Octopodidae) and cuttlefish (Sepiidae), including a new record of *Dicyemenna* in Australian waters. *Folia Parasitologica* **60**(4), 306–320. doi:10.14411/fp.2013.032

Catalano SR (2013b) Five new species of dicyemid mesozoans (Dicyemida: Dicyemidae) from two Australian cuttlefish species, with comments on dicyemid fauna composition. *Systematic Parasitology* **86**(2), 125–151. doi:10.1007/s11230-013-9443-6

Catalano SR, Furuya H (2013) Two new species of dicyemid (Dicyemida: Dicyemidae) from two Australian cephalopod species: *Sepioteuthis australis* (Mollusca: Cephalopoda: Loliginidae) and *Sepioloidea lineolata* (Mollusca: Cephalopoda: Sepiadariidae). *The Journal of Parasitology* **99**(2), 203–211. doi:10.1645/GE-3252.1

Catalano SR, Whittington ID, Donnellan SC, Gillanders BM (2014) Dicyemid fauna composition and infection patterns in relation to cephalopod host biology and ecology. *Folia Parasitologica* **61**(4), 301–310.

Ceriola L, Jackson GD (2010) Growth, hatch size and maturation in a southern population of the loliginid squid *Loliolus noctiluca*. *Journal of the Marine Biological Association of the United Kingdom* **90**(4), 755–767. doi:10.1017/S0025315409991445

Chen X, Liu B, Chen Y (2013) *Sthenoteuthis oualaniensis*, Purpleback Flying Squid. In 'Advances in Squid Biology, Ecology and Fisheries. Part II'. (Eds R Rosa, G Pierce and R O'Dor) pp. 207–222. (Nova Science Publishers, Inc.: New York.)

Cheng MW, Caldwell RL (2000) Sex identification and mating in the blue-ringed octopus *Hapalochlaena lunulata*. *Animal Behaviour* **60**(1), 27–33. doi:10.1006/anbe.2000.1447

Cheng SH, Anderson FE, Bergman A, Mahardika GN, Muchlisin ZA, Thuy DB, Calumpong HP, Mohamed KS, Sasikumar G, Venkatesan V, Barber PH (2012) Phylogenetic relationships within the *Sepioteuthis* cf. *lessoniana* complex in the Indian and Indo-Pacific Oceans using mitochondrial and nuclear DNA. In 'Cephalopod International Advisory Council Symposium'. (Ed. E Vidal). Florianópolis, Brazil, 27 October – 2 November 2012, p. 82. (Oral presentation.)

Chiao C-C, Wickiser JK, Allen JJ, Genter B, Hanlon RT (2011) Hyperspectral imaging of cuttlefish camouflage indicates good colour match in the eyes of fish predators. *Proceedings of the National Academy of Sciences of the United States of America* **108**(22), 9148–9153. doi:10.1073/pnas.1019090108

Chun C (1908) Uber Cephalopoden der Deutschen Tiefsee-Expedition. *Zoologischer Anzeiger* **33**(2), 86–89.

Chun CC (1910) Die Cephalopoden. Oegopsida. *Wissenschaftliche Ergebnisse der deutschen Tiefsee Expedition auf dem Dampfer 'Valdivia' 1898–1899* **18**(1), 1–401.

Coker D (2013) Documentation of the mimic octopus *Thaumoctopus mimicus* in the Great Barrier Reef, Australia. *Marine Biodiversity Records* **6**, e14. doi:10.1017/S175526721200125X

Collins MA, Rodhouse PGK (2006) Southern Ocean cephalopods. *Advances in Marine Biology* **50**, 191–265. doi:10.1016/S0065-2881(05)50003-8

Collins MA, Villanueva R (2006) Taxonomy, ecology and behaviour of the cirrate octopods. *Oceanography and Marine Biology – an Annual Review* **44**(6), 277–322.

Colombera D (1970) A squash method for chromosomes of ascidians. *Caryologia: International Journal of Cytology, Cytoystematics and Cytogenetics* **23**(1), 113–115. doi:10.1080/00087114.1970.10796366

Condie SA, Harris PT (2005) Interactions between physical, chemical, biological and sedimentological processes in Australia's shelf seas. In 'The Sea. Vol. 14'. (Eds AR Robinson and KH Brink) pp. 1413–1449. (Harvard College, Harvard University Press: Cambridge, Massachusetts.)

Cotton BC, Godfrey FK (1940) 'The Molluscs of South Australia. Part II Scaphopoda, Cephalopoda, Aplacophora and Crepipoda.' (Government Printer: Adelaide.)

Dawe EG, Colbourne EB, Drinkwater KF (2000) Environmental effects on recruitment of short-finned squid (*Illex illecebrosus*). *ICES Journal of Marine Science* **57**(4), 1002–1013. doi:10.1006/jmsc.2000.0585

Derby CD (2014) Cephalopod ink: production, chemistry, functions and applications. *Marine Drugs* **12**(5), 2700–2730. doi:10.3390/md12052700

Derby CD, Kicklighter CE, Johnson PM, Zhang X (2007) Chemical composition of inks of diverse marine molluscs suggests convergent chemical defenses. *Journal of Chemical Ecology* **33**(5), 1105–1113. doi:10.1007/s10886-007-9279-0

Dimmlich WF, Hoedt FE (1998) Age and growth of the myopsid squid *Loliolus noctiluca* in Western Port, Victoria, determined from statolith microstructure analysis. *Journal of the Marine Biological Association of the United Kingdom* **78**(2), 577–586. doi:10.1017/S0025315400041631

Doubleday ZA, Pecl GT, Semmens JM, Danyushevsky L (2008) Using stylet elemental signatures to determine the population structure of *Octopus maorum*. *Marine Ecology Progress Series* **360**, 125–133. doi:10.3354/meps07389

Doubleday ZA, Semmens JM (2011) Quantification of the age-pigment lipofscin in known-age octopus (*Octopus pallidus*): a potential tool for age determination. *Journal of Experimental Marine Biology and Ecology* **397**(1), 8–12. doi:10.1016/j.jembe.2010.11.010

Doubleday ZA, Semmens JM, Smolenski AJ, Shaw PW (2009) Microsatellite DNA markers and morphometrics reveal a complex structure in a merobenthic octopus species (*Octopus maorum*) in south-east Australia and New Zealand. *Marine Biology* **156**(6), 1183–1192. doi:10.1007/s00227-009-1160-y

Doubleday ZA, White J, Pecl GT, Semmens JM (2011) Age determination in merobenthic octopuses using stylet increment analysis: assessing future challenges using *Macroctopus maorum* as a model. *ICES Journal of Marine Science* **68**(10), 2059–2063. doi:10.1093/icesjms/fsr135

Duc NX (1978) Löp chän däu (Cephalopoda) vihn Bac bô. *Tuyën Täp Nghiën CúuBien* **1**, 73–93. [in Vietnamese]

Duc NX (1993) The list of squid and cuttle-fish species of Vietnam Sea. *Journal of Biology National Centre for Natural Science and Technology of Vietnam* **15**, 1–4.

Dunning MC (1993) Summer populations of *Ommatrephes bartramii* (Lesueur, 1821) and *Todarodes filippovae* Adam, 1975 (Cephalopoda; Ommastrphidae) from the Tasman Sea. In 'Recent Advances in Cephalopod Fisheries Biology'. (Eds T Okutani, RK O'Dor and T Kubodera) pp. 97–118. (Tokai University Press: Tokyo.)

Dunning MC (1998a) Loliginidae. In 'The Living Marine Resources of the Western Central Pacific. Vol. 2. Cephalopods, Crustaceans, Holothurians and Sharks. FAO Species Identification Guide for Fishery Purposes. (Eds KE Carpenter and VH Niem) pp. 788–796. (FAO: Rome.)

Dunning MC (1998b) Zoogeography of arrow squids (Cephalopods: Ommastrephidae) in the Coral and Tasman Seas, Southwest Pacific. *Smithsonian Contributions to Zoology* **586**, 385–391.

Dunning MC, Brandt SB (1985) Distribution and life history of deep-water squid of commercial interest from Australia. *Australian Journal of Marine and Freshwater Research* **36**(3), 343–359. doi:10.1071/MF9850343

Dunning MC, Förch EC (1998) A review of the systematics, distribution and biology of arrow squids of the genus *Nototodarus* Pfeffer, 1912 (Cephalopoda: Ommastrephidae). In 'Systematics and Biogeography of Cephalopods. Vol. II'. (Eds NA Voss, M Vecchione, RB Toll and MJ Sweeney) *Smithsonian Contributions to Zoology* **586**, 393–404.

Dunning MC, Lu CC (1998) Order Teuthoidea. In 'Mollusca: The Southern Synthesis. Fauna of Australia. Vol. 5'. (Eds PL Beesley, GJB Ross and A Wells) pp. 515–542. (CSIRO Publishing: Melbourne.)

Dunning M, McKinnon S, Lu CC, Yeatman J, Cameron D (1994) Demersal cephalopods of the Gulf of Carpentaria, Australia. *Australian Journal of Marine and Freshwater Research* **45**(3), 351–374. doi:10.1071/MF9940351

Dunstan A, Alanis O, Marshall J (2010) *Nautilus pompilius* fishing and population decline in the Philippines: A comparison with an unexploited Australian *Nautilus* population. *Fisheries Research* **106**(2), 239–247. doi:10.1016/j.fishres.2010.06.015

Dunstan A, Bradshaw CJA, Marshall J (2011a) *Nautilus* at risk – estimating population size and demography of *Nautilus pompilius*. *PLoS ONE* **6**(2), e16716. doi:10.1371/journal.pone.0016716

Dunstan AJ, Ward PD, Marshall NJ (2011b) Vertical distribution and migration patterns of *Nautilus pompilius*. *PLoS ONE* **6**(2), e16311. doi:10.1371/journal.pone.0016311

Dunstan AJ, Ward PD, Marshall J (2011c) *Nautilus pompilius* life history and demographics at the Osprey Reef Seamount, Coral Sea, Australia. *PLoS ONE* **6**(2), e16312. doi:10.1371/journal.pone.0016312

Espiñeira M, Vieites JM, Santaclara FJ (2010) Species authentication of octopus, cuttlefish, bobtail and bottle squids (families Octopodidae, Sepiidae and Sepiolidae) by FINS methodology in seafoods. *Food Chemistry* **121**(2), 527–532. doi:10.1016/j.foodchem.2009.12.042

Feulner PGD, Kirschbaum F, Mamonekene V, Ketmaier V, Tiedemann R (2007) Adaptive radiation in African weakly electric fish (Teleostei: Mormyridae: *Campylomormyrus*): A combined molecular and morphological approach. *Journal of Evolutionary Biology* **20**(1), 403–414. doi:10.1111/j.1420-9101.2006.01181.x

Finke E, Portner HO, Lee PG, Webber DM (1996) Squid (*Lolliguncula brevis*) life in shallow waters: Oxygen limitation of metabolism and swimming performance. *The Journal of Experimental Biology* **199**, 911–921.

Finn JK (2010) Sinking Aristotle's sailing octopus. *Australasian Science* **31**(8), 25–27.

Finn JK (2013) Taxonomy and biology of the argonauts (Cephalopoda: Argonautidae) with particular reference to Australian material. *Molluscan Research* **33**(3), 143–222. doi:10.1080/13235818.2013.824854

Finn JK (2014a) Family Alloposidae. In 'Cephalopods of the World. An Annotated and Illustrated Catalogue of Cephalopod Species Known to Date. No. 4, Vol. 3. Octopods and Vampire Squids. FAO Species Catalogue for Fishery Purposes'. (Eds P Jereb, CFE Roper, MD Norman and JK Finn) pp. 225–228. (FAO: Rome.)

Finn JK (2014b) Family Argonautidae. In 'Cephalopods of the World. An Annotated and Illustrated Catalogue of Cephalopod Species Known to Date. No. 4, Vol. 3. Octopods and Vampire Squids. FAO Species Catalogue for Fishery Purposes'. (Eds P Jereb, CFE Roper, MD Norman and JK Finn) pp. 228–237. (FAO: Rome.)

Finn JK (2014c) Family Ocythoidae. In 'Cephalopods of the World. An Annotated and Illustrated Catalogue of Cephalopod Species Known to Date. No. 4, Vol. 3. Octopods and Vampire Squids. FAO Species Catalogue for Fishery Purposes'. (Eds P Jereb, CFE Roper, MD Norman and JK Finn) pp. 237–239. (FAO: Rome.)

Finn JK (2014d) Family Tremoctopodidae. In 'Cephalopods of the World. An Annotated and Illustrated Catalogue of Cephalopod Species Known to Date. No. 4, Vol. 3. Octopods and Vampire Squids. FAO Species Catalogue for Fishery Purposes'. (Eds P Jereb, CFE Roper, MD Norman and JK Finn) pp. 240–243. (FAO: Rome.)

Finn JK, Norman MD (2010) The argonaut shell: Gas-mediated buoyancy control in a pelagic octopus. *Proceedings of the Royal Society of London. Series B, Biological Sciences* **277**, 2967–2971. doi:10.1098/rspb.2010.0155

Finn JK, Norman M (2011a) Knobbed Argonaut, *Argonauta nodosus*. In 'Taxonomic Toolkit for Marine Life of Port Phillip Bay'. Museum Victoria, accessed 30 Dec 2012, http://portphillipmarinelife.net.au/species/7832

Finn JK, Norman M (2011b) Southern Sand Octopus, *Octopus kaurna*. In 'Taxonomic Toolkit for Marine Life of Port Phillip Bay'. Museum Victoria, accessed 30 Dec 2012, http://portphillipmarinelife.net.au/species/4007

Finn JK, Norman M (2011c) Southern Keeled Octopus, *Octopus berrima*. In 'Taxonomic Toolkit for Marine Life of Port Phillip Bay'. Museum Victoria, accessed 30 Dec 2012, http://portphillipmarinelife.net.au/species/4006

Finn JK, Tregenza T, Norman MD (2009) Defensive tool use in a coconut carrying octopus. *Current Biology* **19**(23), R1069–R1070. doi:10.1016/j.cub.2009.10.052

Förch EC (1998) The marine fauna of New Zealand: Cephalopoda: Oegopsida: Architeuthidae (Giant Squid). *NIWA Biodiversity Memoir* **110**, 1–113.

Forsythe JW (1993) A working hypothesis of how seasonal temperature change may impact the field growth of young cephalopods. In 'Recent Advances in Cephalopod Fisheries Biology'. (Eds T Okutani, RK O'Dor and T Kubodera) pp. 133–143. (Tokai University Press: Kanagawa, Honshu, Japan.)

Godfrey-Smith P, Lawrence M (2012) Long-term high-density occupation of a site by *Octopus tetricus* and possible site modification due to foraging behaviour. *Marine and Freshwater Behaviour and Physiology* **45**(4), 1–8. doi:10.1080/10236244.2012.727617

Green C (2011) Population stucture of arrow squid, *Nototodarus gouldi*. *Malacological Society of Australasia Newsletter* No. 141, 4.

Gröger J, Piatkowski U, Heinemann H (2000) Beak length analysis of the Southern Ocean squid *Psychroteuthis glacialis* (Cephalopoda: Psychroteuthidae) and its use for size and biomass estimation. *Polar Biology* **23**(1), 70–74. doi:10.1007/s003000050009

Grove S, Finn J (2014) Unusual strandings of greater argonaut *Argonauta argo* in southeast Tasmania, autumn 2014. *The Malacological Society of Australasia Newsletter* No. 151, 1, 3–4.

Grubert MA, Wadley VA (2000) Sexual maturity and fecundity of *Octopus maorum* in southeast Tasmania. *Bulletin of Marine Science* **66**(1), 131–142.

Grubert MA, Wadley VA, White RWG (1999) Diet and feeding strategy of *Octopus maorum* in southeast Tasmania. *Bulletin of Marine Science* **65**(2), 441–451.

Guard M, Mgaya YD (2002) The artisanal fishery for *Octopus cyanea* Gray in Tanzania. *Ambio* **31**(7–8), 528–536. doi:10.1579/0044-7447-31.7.528

Guerra Á, Belcari P, González Á (2013) *Todaropsis eblanae*, Lesser Flying Squid. In 'Advances in Squid Biology, Ecology and Fisheries, Part II'. (Eds R Rosa and R O'Dor) pp. 149–168. (Nova Science Publishers, Inc.: New York.)

Gutowska MA, Portner HO, Melzner F (2008) Growth and calcification in the cephalopod *Sepia officinalis* under elevated seawater $pCO_2$. *Marine Ecology Progress Series* **373**, 303–309. doi:10.3354/meps07782

Guzik MT, Norman MD, Crozier RH (2005) Molecular phylogeny of the benthic shallow-water octopuses (Cephalopoda: Octopodinae). *Molecular Phylogenetics and Evolution* **37**(1), 235–248. doi:10.1016/j.ympev.2005.05.009

Hall KC, Fowler AJ, Geddes MC (2007) Evidence for multiple year classes of the giant Australian cuttlefish *Sepia apama* in northern Spencer Gulf, South Australia. *Reviews in Fish Biology and Fisheries* **17**, 367–384. doi:10.1007/s11160-007-9045-y

Hall KC, Hanlon RT (2002) Principal features of the mating system of a large spawning aggregation of the giant Australian cuttlefish *Sepia apama* (Mollusca: Cephalopoda). *Marine Biology* **140**(3), 533–545. doi:10.1007/s00227-001-0718-0

Hall NE, Hanzak J, Allcock AL, Cooke IR, Ogura A, Strugnell JM (2014) The complete mitochondrial genome of the pygmy squid *Idiosepius* (Cephalopods: Decapodiformes): the first representative from the family Idiosepiidae. *Mitochondrial DNA* Early Online, 1–2. doi:10.3109/19401736.2013.865180

Hanlon RT, Conroy LA, Forsythe JW (2008) Mimicry and foraging behaviour of two tropical sand-flat octopus species off North Sulawesi, Indonesia. *Biological Journal of the Linnean Society. Linnean Society of London* **93**(1), 23–38. doi:10.1111/j.1095-8312.2007.00948.x

Hanlon RT, Forsythe JW, Joneschild DE (1999) Crypsis, conspicuousness, mimicry and polyphenism as antipredator defences of foraging octopuses on Indo-Pacific coral reefs, with a method of quantifying crypsis from video tapes. *Biological Journal of the Linnean Society. Linnean Society of London* **66**, 1–22. doi:10.1111/j.1095-8312.1999.tb01914.x

Hanlon RT, Messenger JB (1996) 'Cephalopod Behaviour.' (Cambridge University Press: UK.)

Hanlon RT, Naud MJ, Forsythe JW, Hall K, Watson AC, McKechnie J (2007) Adaptable night camouflage by cuttlefish. *American Naturalist* **169**(4), 543–551. doi:10.1086/512106

Hanlon RT, Naud M-J, Shaw PW, Havenhand JN (2005) Transient sexual mimicry leads to fertilisation. *Nature* **433**(7023), 212. doi:10.1038/433212a

Haring E, Kruckenhauser L, Lukeneder A (2012) New DNA sequence data on the enigmatic *Spirula spirula* (Linnaeus, 1758) (Decabrachia, suborder Spirulina). *Annalen des Naturhistorischen Museums in Wien, B* **113**, 37–48.

Harvey AW, Mooi R, Goslinger TM (1999) Phylogenetic taxonomy and the status of *Allonautilus* Ward and Saunders, 1997. *Journal of Paleontology* **73**(6), 1214–1217.

Herring PJ (2007) Sex with lights on? A review of bioluminescent sexual dimorphism in the sea. *Journal of the Marine Biological Association of the United Kingdom* **87**, 829–842. doi:10.1017/S0025315407056433

Herwig JN, Depczynski M, Roberts JD, Semmens JM, Gagliano M, Heyward AJ (2012) Using age-based life history data to investigate the life cycle and vulnerability of *Octopus cyanea*. *PLoS ONE* **7**(8), e43679. doi:10.1371/journal.pone.0043679

Hibberd T, Pecl GT (2007) Effects of commercial fishing on the population structure of spawning southern calamary (*Sepioteuthis australis*). *Reviews in Fish Biology and Fisheries* **17**(2–3), 207–221. doi:10.1007/s11160-006-9028-4

Ho JD, Moltschaniwskyj NA, Carter CG (2004) The effect of variability in growth on somatic condition and reproductive status in the southern calamary *Sepioteuthis australis*. *Marine and Freshwater Research* **55**(4), 423–428. doi:10.1071/MF03149

Hochberg FG (1982) The 'kidneys' of cephalopods: A unique habitat for parasites. *Malacologia* **23**, 121–134.

Hochberg FG, Norman MD, Finn JK (2014) Family Opisthoteuthidae. In 'Cephalopods of the World. An Annotated and Illustrated Catalogue of Cephalopod Species Known to Date. No. 4, Vol. 3. Octopods and Vampire Squids. FAO Species Catalogue for Fishery Purposes'. (Eds P Jereb, CFE Roper, MD Norman and JK Finn) pp. 248–261. (FAO: Rome.)

Hoving HJ (2012) Life history strategies in oceanic squid. In 'Program and Abstracts for the 13th International Deep-Sea Biology Symposium'. Wellington, New Zealand, 3–7 December 2012. p. 77.

Hoving HJT, Laptikhovsky V, Platkowski U, Önsoy B (2008) Reproduction in *Heteroteuthis dispar* (Rüppell, 1844) (Mollusca: Cephalopoda): a sepiolid reproductive adaptation to an oceanic lifestyle. *Marine Biology* **154**(2), 219–230. doi:10.1007/s00227-008-0916-0

Hoving HJT, Lipinski MR, Roeleveld MAC, Durholtz MDY (2007) Growth and mating of southern African *Lycoteuthis lorigera* (Steenstrup, 1875) (Cephalopoda; Lycoteuthidae). *Reviews in Fish Biology and Fisheries* **17**(2–3), 259–270. doi:10.1007/s11160-006-9031-9

Hoving HJT, Robison BH (2012) Vampire squid: Detritivores in the oxygen minimum zone. *Proceedings of the Royal Society B: Biological Sciences* **279**, 4559–4567. doi:10.1098/rspb.2012.1357

Hoving HJT, Zeidberg LD, Benfield MC, Bush SL, Robison BH, Vecchione M (2013) *Grimalditeuthis bonplandi* reveal unique use of tentacles. *Proceedings of the Royal Society B: Biological Sciences* **280**(1769), 1–9. doi: 10.1098/rspb.2013.1463

Hoving H-JT, Perez JAA, Bolstad KSR, Braid HE, Evans ABE, Fuchs D, Judkins H, Kelly JT, Marian JEAR, Nakajima R, Piatkowski U, Reid A, Vecchione M, Xavier J (2014) The study of deep-sea cephalopods. *Advances in Marine Biology* **67**, 235–359. doi:10.1016/B978-0-12-800287-2.00003-2

Hoyle WE (1886) Report on the Cephalopoda. Report on the scientific results of the voyage of H.M.S. Challenger during the years 1873–7. *Zoology (Jena, Germany)* **16**(44), 1–245.

Hoyle WE (1904) Reports on the Cephalopoda. *Bulletin of the Museum of Comparative Zoology at Harvard College, in Cambridge* **43**(1), 1–72.

Huffard CL (2006) Locomotion by *Abdopus aculeatus* (Cephalopoda: Octopodidae): Walking the line between primary and secondary defences. *The Journal of Experimental Biology* **209**(19), 3697–3707. doi:10.1242/jeb.02435

Huffard CL, Godfrey-Smith P (2010) Field observations of mating in *Octopus tetricus* Gould, 1852 and *Amphioctopus marginatus* (Taki, 1964) (Cephalopoda: Octopodidae). *Molluscan Research* **30**(2), 81–86.

Huffard CL, Saarman N, Hamilton H, Simison WB (2010) The evolution of conspicuous facultative mimicry in octopuses: An example of secondary adaptation? *Biological Journal of the Linnean Society. Linnean Society of London* **101**(1), 68–77. doi:10.1111/j.1095-8312.2010.01484.x

Hunsicker ME, Essington TE, Watson R, Sumaila UR (2010) The contribution of cephalopods to global marine fisheries: Can we have our squid and eat them too? *Fish and Fisheries* **11**(4), 421–438. doi:10.1111/j.1467-2979.2010.00369.x

Hunt KV, Steer MA, Gillanders BM (2011) Validating age in southern calamary (*Sepioteuthis australis*) over seasonal and life history extremes. *Journal of the Marine Biological Association of the United Kingdom* **91**(4), 857–863. doi:10.1017/S002531541000113X

Hvorecny LM, Grudowski JL, Blakeslee CJ, Simmons TL, Roy PR, Brooks JA, Hanner RM, Beigel ME, Karson MA, Nichols RH, Holm JB, Boal JG (2007) Octopuses (*Octopus bimaculoides*) and cuttlefishes (*Sepia pharaonis, Sepia officinalis*) can conditionally discriminate. *Animal Cognition* **10**(4), 449–459. doi:10.1007/s10071-007-0085-4

Ibáñez MI, Keyl F (2010) Cannibalism in cephalopods. *Molecular Ecology Resources* **20**, 123–136.

Jackson GD (1986) A biological and ecological survey of the planktonic Cephalopoda in the waters of the central Great Barrier Reef Region with emphasis on the biology and ageing of the sepioid *Idiosepius pygmaeus*. B.Sc. (Honours) Thesis, James Cook University, Townsville, University of North Queensland, Australia.

Jackson GD (1989) Age and growth of the tropical nearshore squid *Sepioteuthis lessoniana* determined from statolith growth-ring analysis. *Fishery Bulletin* **88**, 113–118.

Jackson GD (1990) The use of tetracycline staining techniques to determine statolith growth ring periodicity in the tropical loliginid squids *Loliolus noctiluca* and *Loligo chinensis*. *The Veliger* **33**(4), 389–393.

Jackson GD (1993) Seasonal variation in reproductive investment in the tropical loliginid squid *Loligo chinensis* and the small tropical squid *Idiosepius pygmaeus*. *Fishery Bulletin* **91**, 260–270.

Jackson GD (2004) Advances in defining the life histories of myopsid squid. *Marine and Freshwater Research* **55**(4), 357–365. doi:10.1071/MF03152

Jackson GD, Choat JH (1992) Growth in tropical cephalopods: An analysis based on statolith microstructure. *Canadian Journal of Fisheries and Aquatic Sciences* **49**(2), 218–228. doi:10.1139/f92-026

Jackson GD, Domeier ML (2003) The effects of an extraordinary El niño/La niña event on the size and growth of the squid *Loligo opalescens* off Southern California. *Marine Biology* **142**(5), 925–935.

Jackson GD, Jackson CH (2004) Mating and spermatophores placement in the onychoteuthid squid *Moroteuthis ingens*. *Journal of the Marine Biological Association of the United Kingdom* **84**(4), 783–784. doi:10.1017/S0025315404009932

Jackson GD, Meekan MG, Wotherspoon S, Jackson CH (2008) Distributions of young cephalopods in the tropical waters of Western Australia over two consecutive summers. *ICES Journal of Marine Science* **65**(2), 140–147. doi:10.1093/icesjms/fsm186

Jackson GD, Moltschaniwskyj NA (2001) Temporal variation in growth rates and reproductive parameters in the small near-shore tropical squid *Loliolus noctiluca*: is cooler better? *Marine Ecology Progress Series* **218**, 167–177. doi:10.3354/meps218167

Jackson GD, Moltschaniwskyj NA (2002) Spatial and temporal variation in growth rates and maturity in the Indo-Pacific squid *Sepioteuthis lessoniana* (Cephalopoda: Loliginidae). *Marine Biology* **140**(4), 747–754. doi:10.1007/s00227-001-0746-9

Jackson GD, O'Dor RK (2001) Time, space and the ecophysiology of squid growth, life in the fast lane. *Vie et Milieu* **51**(4), 205–215.

Jackson GD, Semmens JM, Phillips KL, Jackson CH (2004) Reproduction in the deepwater squid *Moroteuthis ingens*, what does it cost? *Marine Biology* **145**(5), 905–916. doi:10.1007/s00227-004-1375-x

Jackson GD, Wotherspoon S, Jackson CH (2007) Temporal life history plasticity of the Southern Ocean squid *Todarodes filippovae* from waters off Tasmania, Australia. *Marine Biology* **150**(4), 575–584. doi:10.1007/s00227-006-0374-5

Jackson GD, Wotherspoon S, McGrath-Steer BL (2005) Temporal population dynamics in arrow squid *Nototodarus gouldi* in southern Australian waters. *Marine Biology* **146**(5), 975–983. doi:10.1007/s00227-004-1491-7

Jackson GD, Yeatman J (1996) Variation in size and age at maturity in *Photololigo* (Mollusca: Cephalopoda) from the northwest shelf of Australia. *Oceanographic Literature Review* **43**(12), 1246.

Jantzen TM, Havenhand JN (2002) Preliminary field observations of mating and spawning in the squid *Sepioteuthis australis*. *Bulletin of Marine Science* **71**(2), 1073–1080.

Jantzen TM, Havenhand JN (2003) Reproductive behaviour in the squid *Sepioteuthis australis* from South Australia:

Interactions on the spawning ground. *The Biological Bulletin* **204**, 305–317. doi:10.2307/1543601

Jennings S, Kaiser MJ (1998) The effects of fishing on marine ecosystems. *Advances in Marine Biology* **34**, 201–352. doi:10.1016/S0065-2881(08)60212-6

Jereb P (2005) Family Nautilidae. In 'Cephalopods of the World. An Annotated and Illustrated Catalogue of Cephalopod Species Known to date. No. 4, Vol. 1. Chambered Nautiluses and Sepioids (Nautilidae, Sepiidae, Sepiolidae, Sepiadariidae, Idiosepiidae and Spirulidae). FAO Species Catalogue for Fishery Purposes'. (Eds P Jereb and CFE Roper) pp. 51–55. (FAO: Rome).

Jereb P, Roper CFE (Eds) (2005) 'Cephalopods of the World. An Annotated and Illustrated Catalogue of Cephalopod Species Known to Date. No. 4, Vol. 1. Chambered Nautiluses and Sepioida (Nautilidae, Sepiidae, Sepiolidae, Sepiadariidae, Idiosepiidae and Spirulidae). FAO Species Catalogue for Fishery Purposes'. (FAO: Rome.)

Jereb P, Roper CFE (Eds) (2010) 'Cephalopods of the World. An Annotated and Illustrated Catalogue of Cephalopod Species Known to Date. No. 4, Vol. 2. Myopsid and Oegopsid Squids'. FAO Species Catalogue for Fishery Purposes'. (FAO: Rome.)

Jereb P, Roper CFE, Norman MD, Finn JK (Eds) (2014) 'Cephalopods of the World. An Annotated and Illustrated Catalogue of Cephalopod Species Known to Date. No. 4, Vol. 3. Octopods and Vampire Squids. FAO Species Catalogue for Fishery Purposes'. (FAO: Rome.)

Jereb P, Roper CFE, Vecchione M (2010a) Introduction. In 'Cephalopods of the World. An Annotated and Illustrated Catalogue of Species Known to Date. No. 4, Vol. 2. Myopsid and Oegopsid Squids. FAO Species Catalogue for Fishery Purposes'. (Eds P Jereb and CFE Roper) pp. 1–11. (FAO: Rome.)

Jereb P, Roper CFE, Vecchione M (2010b) Family Australiteuthidae. In 'Cephalopods of the World Myopsid and Oegopsid Squids. FAO Species Catalogue for Fishery Purposes'. (Eds P Jereb and CFE Roper) pp. 35–37. (FAO: Rome.)

Jereb P, Vecchione M, Roper CFE (2010c) Family Loliginidae. In 'Cephalopods of the World. An Annotated and Illustrated Catalogue of Species Known to Date. No. 4, Vol. 2. Myopsid and Oegopsid Squids. FAO Species Catalogue for Fishery Purposes'. (Eds P Jereb and CFE Roper) pp. 38–117. (FAO: Rome.)

Joubin L (1920) Céphalopodes provenant des Campagnes de la Princesse-Alice (1898–1910). *Résultata des Campagnes scientitiques accomplies sur son yacht par Albert I Prince souverain de Monaco* **54**, 1–95, 16 plates.

Kaneko N, Kubodera T (2005) A new species of shallow water octopus, *Octopus laqueus*, (Cephalopoda: Octopodidae) from Okinawa, Japan. *Bulletin of the National Science Museum, Tokyo, Series A* **31**(1), 8–15.

Kaneko N, Oshima Y, Ikeda Y (2006) Egg brooding behaviour and embryonic development of *Octopus laqueus* (Cephalopoda: Octopodidae). *Molluscan Research* **26**(3), 113–117.

Kassahn KS, Donellan SC, Fowler AJ, Hall KC, Adams M, Shaw PW (2003) Molecular and morphological analyses of the cuttlefish *Sepia apama* indicate a complex population structure. *Marine Biology* **143**(5), 947–962. doi:10.1007/s00227-003-1141-5

Kawashima Y, Nishihara H, Akasaki T, Nikaido M, Tsuchiya K, Segawa S, Okada N (2013) The complete mitochondrial genomes of deep-sea squid (*Bathyteuthis abyssicola*), bob-tail squid (*Semirossia patagonica*) and four giant cuttlefish (*Sepia apama*, *S. latimanus*, *S. lycidas* and *S. pharaonis*), and their application to the phylogenetic analysis of Decapodiformes. *Molecular Phylogenetics and Evolution* **69**, 980–993. doi:10.1016/j.ympev.2013.06.007

Kojadinovic J, Jackson CH, Cherel Y, Jackson GD, Bustamante P (2011) Multi-elemental concentrations in the tissues of the oceanic squid *Todarodes filippovae* from Tasmania and the southern Indian Ocean. *Ecotoxicology and Environmental Safety* **74**(5), 1238–1249. doi:10.1016/j.ecoenv.2011.03.015

Kubodera T (2006) *Gonatus antarcticus* Lönnberg 1898. Version 31 May 2006. http://tolweb.org/Gonatus_antarcticus/19778/2006.05.31 in The Tree of Life Web Project, http://tolweb.org/

Kubodera T (2012) How far squid can fly out of the water. In 'Cephalopod International Advisory Council Symposium'. (Ed. E Vidal). Florianópolis, Brazil, 27 October – 2 November 2012, p. 7. (Oral presentation.)

Kubodera T, Piatkowski U, Okutani T, Clarke MR (1998) Taxonomy and Zoogeography of the family Onychoteuthidae (Cephalopoda: Oegopsida). In 'Systematics and Biogeography of Cephalopods. Vol. II'. (Eds NA Voss, M Vecchione, RB Toll and MJ Sweeney) *Smithsonian Contributions to Zoology* **586**, 277–392.

Kuipers MR, Pecl GT, Moltschaniwskyj NA (2008) Batch or trickle: Understanding the multiple spawning strategy of southern calamary, *Sepioteuthis australis* (Mollusca: Cephalopoda). *Marine and Freshwater Research* **59**(11), 987–997. doi:10.1071/MF07200

Kurosaka K, Yamagimoto T, Wakabayashi T, Shigenobu Y, Ochi Y, Inada H (2012) Population genetic structure of the neon flying squid *Ommastrephes bartrami* inferred from mitochondrial DNA sequence analysis. *Nippon Suisan Gakkai Shi* **78**(2), 212–219. doi:10.2331/suisan.78.212

Lacoue-Labarthe TL, Villanueva R, Rouleau C, Oberhänsli F, Teyssié J-L, Jeffree R, Bustamante P (2011) Radioisotopes demonstrate the contrasting bioaccumulation capacities of heavy metals in embryonic stages of cephalopod species. *PLoS ONE* **6**(11), e27653. doi:10.1371/journal.pone.0027653

Lalas C (2009) Estimates of size for the large octopus *Macroctopus maorum* from measures of beaks in prey remains. *New Zealand Journal of Marine and Freshwater Research* **43**(2), 635–642. doi:10.1080/00288330909510029

Lansdell M, Young J (2007) Pelagic cephalopods from eastern Australia: Species composition, horizontal and vertical distribution determined from the diets of pelagic fishes. *Reviews in Fish Biology and Fisheries* **17**(2), 125–138. doi:10.1007/s11160-006-9024-8

Laptikovskii VV (2002) Ecology of decapod cephalopod reproduction (Cephalopoda: Teuthida, Sepiida). *Zoologihesky Zhurnal* **81**(11), 1319–1328.

Laptikhovsky V, Arkhipkin A (2003) The reproductive features of a mature female of the deep-sea planktonic squid *Galiteuthis glacialis* (Cephalopoda: Cranchiidae) from the Southern Ocean. *Polar Research* **22**(2), 395–397. doi:10.1111/j.1751-8369.2003.tb00120.x

Laptikhovsky V, Arkhipkin A, Bolstad KS (2009) A second species of the squid genus *Kondakovia* (Cephalopoda: Onychoteuthidae) from the sub-Antarctic. *Polar Biology* **32**(1), 21–26. doi:10.1007/s00300-008-0497-x

Laptikhovsky V, Collins MA, Arkhipkin A (2013) First case of possible iteroparity among coleoid cephalopods: The giant warty squid *Kondakovia longimana*. *The Journal of Molluscan Studies* **79**(3), 270–272. doi:10.1093/mollus/eyt014

Laptikovskii VV (2002) Ecology of decapod cephalopod reproduction (Cephalopoda: Teuthida, Sepiida). *Zoologihesky Zhurnal* **81**(11), 1319–1328.

Leporati SC, Pecl GT, Semmens JM (2008a) Reproductive status of *Octopus pallidus*, and its relationship to age and size. *Marine Biology* **155**(4), 375–385. doi:10.1007/s00227-008-1033-9

Leporati SC, Semmens JM, Pecl GT (2008b) Determining the age and growth of wild octopus using stylet increment analysis. *Marine Ecology Progress Series* **367**, 213–222. doi:10.3354/meps07558

Leporati SC, Ziegler PE, Semmens JM (2009) Assessing the stock status of holobenthic octopus fisheries: Is catch per unit effort sufficient? *ICES Journal of Marine Science* **66**(3), 478–487. doi:10.1093/icesjms/fsn224

Lewis AR, Choat JH (1993) Spawning mode and reproductive output of the tropical cephalopod *Idiosepius pygmaeus*. *Canadian Journal of Fisheries and Aquatic Sciences* **50**(1), 20–28. doi:10.1139/f93-003

Lindgren AR (2010a) Systematics and distribution of the squid genus *Pterygioteuthis* (Cephalopoda: Oegopsida) in the eastern tropical Pacific Ocean. *The Journal of Molluscan Studies* **76**(4), 389–398. doi:10.1093/mollus/eyq028

Lindgren AR (2010b) Molecular inference of phylogenetic relationships among Decapodiformes (Mollusca: Cephalopoda) with special focus on the squid Order Oegopsida. *Molecular Phylogenetics and Evolution* **56**(1), 77–90. doi:10.1016/j.ympev.2010.03.025

Lindgren AR, Daly M (2007) The impact of length-variable data and alignment criterion on the phylogeny of Decapodiformes (Mollusca: Cephalopoda). *Cladistics* **23**(5), 464–476. doi:10.1111/j.1096-0031.2007.00160.x

Lindgren AR, Giribet G, Nishiguchi MK (2004) A combined approach to the phylogeny of Cephalopoda (Mollusca). *Cladistics* **20**(5), 454–486. doi:10.1111/j.1096-0031.2004.00032.x

Lindgren A, Young RE, Mangold KM (1922–2003) (2011) *Pterygioteuthis* Fischer, 1896. Version 11 January 2011 (under construction). http://tolweb.org/Pterygioteuthis/19747/2001.01.11 in The Tree of Life Web Project, http://tolwb.org

Lindgren AR, Pankey MS, Hochberg FG, Oakley TH (2012) A multi-gene phylogeny of Cephalopoda supports convergent morphological evolution in association with multiple habitat shifts in the marine environment. *BMC Evolutionary Biology* **12**(1), 129. doi:10.1186/1471-2148-12-129

Lipinski MR (2001) Preliminary description of two new species of Cephalopods (Cephalopoda: Brachioteuthidae) from South Atlantic and Antarctic waters. *Bulletin of the Sea Fisheries Institute, Gdynia* **152**(1), 3–14.

Lipinski MR, Naggs FA, Roeleveld MA (2000) Catalogue of types of recent cephalopods in the collection of the Natural History Museum, London. *Annales Zoologici (Warszawa)* **50**(1), 101–120.

Lu CC, Boucher-Rodoni R Tillier A (1995) Catalogue of types of recent Cephalopoda in the Museum national d'Histoire Naturelle (France). *Bulletin du Museum national d'Histoire naturelle, Paris*, (series 4) **17**(3–4), 307–343.

Lu CC, Dunning MC (1982) Identification Guide to Australian Arrow Squid (Family Ommastrephidae). Technical Report No. 2. Victorian Institute of Marine Science.

Lu CC, Dunning MC (1998) Chapter 13. Subclass Coleoidea. pp. 499–563. In 'Mollusca: The Southern Synthesis. Fauna of Australia'. (Eds PL Beesley, GJB Ross and A Wells) pp. Part A xvi. (CSIRO Publishing: Melbourne.)

Lu CC, Phillips JU (1985) An annotated checklist of the Cephalopoda from Australian waters. *Occasional Papers from the Museum of Victoria* **2**, 21–36.

Lu CC, Reid AL (1997) Two new cuttlefishes (Cephalopoda: Sepiidae) from the North West Shelf, and a redescription of *Sepia sulcata* Hoyle, 1885. *Records of the Western Australian Museum* **18**, 277–310.

Lu CC, Roper CFE, Tait RW (1985) A revision of *Loliolus* (Cephalopoda; Loliginidae), including *L. noctiluca*, a new species of squid from Australian waters. *Proceedings of the Royal Society of Victoria* **97**(2), 59–85.

Lu CC, Stranks TN (1992) *Eledone palari*, a new species of Octopus (Cephalopoda: Octopodidae) from Australia. *Bulletin of Marine Science* **49**(1–2), 73–87.

Lu CC, Stranks TN (1994) Synopsis of *Pareledone* and *Megaleledone* species with description of two new species. *Memoirs of the Museum of Victoria* **54**, 221–242.

Lukeneder A, Harzhauser M, Müllegger S, Piller W (2008) Stable isotopes (δ18O and δ13C) in *Spirula spirula* shells from three major oceans indicate developmental changes parallelling depth distributions. *Marine Biology* **154**(1), 175–182. doi:10.1007/s00227-008-0911-5

Mangold KM, Vecchione M, Young RE (2010a) Ocythoidae Gray, 1849. *Ocythoe tuberculata* Rafinesque, 1814. Version 15 August 2010 (under construction). http://tolweb.org/Ocythoe_tuberculata/20205/2010.08 in The Tree of Life Web Project, http://tolweb.org/

Mangold KM, Vecchione M, Young RE (2010b) Tremoctopodidae Tryon, 1879. *Tremoctopus* Chiaje 1830. Blanket octopus. Version 15 August 2010. http://tolweb.org/Tremoctopus/20202/2010.08.15 in The Tree of Life Web Project, http://tolweb.org/

Mather JA, Mather DL (2004) Apparent movement in a visual display: The 'passing cloud' of *Octopus cyanea* (Mollusca: Cephalopoda). *Journal of Zoology* **263**(1), 89–94. doi:10.1017/S0952836904004911

Mäthger LM, Hanlon RT (2006) Anatomical basis for camouflaged polarized light communication in squid. *Biology Letters* **2**, 494–496. doi:10.1098/rsbl.2006.0542

Mäthger LM, Hanlon RT (2007) Malleable skin colouration in cephalopods: Selective reflectance, transmission and absorbance of light by chromatophores and iridophores. *Cell and Tissue Research* **329**(1), 179–186. doi:10.1007/s00441-007-0384-8

Mäthger LM, Hanlon RT, Håkansson J, Nilsson D-E (2013) The W-shaped pupil in cuttlefish (*Sepia officinalis*): Functions for improving horizontal vision. *Vision Research* **83**, 19–24. doi:10.1016/j.visres.2013.02.016

McArthur T, Butler EC, Jackson GD (2003) Mercury in the marine food chain in the Southern Ocean at Macquarie Island: An analysis of a top predator, Patagonian toothfish (*Dissostichus eleginoides*) and a mid-trophic species, the warty squid (*Moroteuthis ingens*). *Polar Biology* **27**(1), 1–5. doi:10.1007/s00300-003-0560-6

McCoy F (1885–90) 'Natural History of Victoria; or, Figures and Descriptions of the Living Species of All Classes of the Victorian Indigenous Animals.' (J. Ferres: Melbourne.)

McGrath BL, Jackson GD (2002) Egg production in the arrow squid *Nototodarus gouldi* (Cephalopoda: Ommastrephidae), fast and furious or slow and steady? *Marine Biology* **141**(4), 699–706. doi:10.1007/s00227-002-0864-z

Melli V, Riginella E, Nalon M, Mazzoldo C (2014) From trap to nursery. Mitigating the impact of an artisanal fishery on cuttlefish offspring. *PLoS ONE* **9**(2), e90542. doi:10.1371/journal.pone.0090542

Messenger JB, Nixon M, Ryan KP (1985) Magnesium chloride as an anaesthetic for cephalopods. *Comparative Biochemistry and Physiology Part C Toxicology & Pharmacology* **82**, 203–205. doi:10.1016/0742-8413(85)90230-0

Myers RA, Worm B (2003) Rapid worldwide depletion of predatory fish communities. *Nature* **423**, 280–283. doi:10.1038/nature01610

Milinkovitch MC, LeDuc R, Tiedman R, Dizon A (2002) Applications of molecular data in cetacean taxonomy and population genetics with special emphasis on defining species boundaries in cetaceans. In 'Marine Mammals: Biology and Conservation'. (Eds PGH Evans and JA Raga) pp. 325–359. Kluwer Academic/Plenum Press, New York.

Moltschaniwskyj NA, Doherty PJ (1994) Distribution and abundance of two juvenile tropical *Photololigo* species (Cephalopoda: Loliginidae) in the central Great Barrier Reef lagoon. *Fishery Bulletin* **92**(2), 302–312.

Moltschaniwskyj NA, Doherty PJ (1995) Cross-shelf distribution patterns of tropical juvenile cephalopods sampled with light traps. *Marine and Freshwater Research* **46**(4), 707–714. doi:10.1071/MF9950707

Moltschaniwskyj NA, Pecl GT (2003) Small-scale spatial and temporal patterns of egg production by the temperate loliginid squid *Sepioteuthis australis*. *Marine Biology* **142**(3), 509–516.

Moltschaniwskyj NA, Pecl GT (2007) Spawning aggregations of squid (*Sepioteuthis australis*) populations: a continuum of 'microcohorts'. *Reviews in Fish Biology and Fisheries* **17**(2–3), 183–195. doi:10.1007/s11160-006-9025-7

Moltschaniwskyj N, Pecl G, Lyle J (2002) An assessment of the use of short-term closures to protect spawning southern calamary aggregations from fishing pressure in Tasmania, Australia. *Bulletin of Marine Science* **71**(1), 501–504.

Mooney TA, Hanlon RT, Christensen-Dalsgaard J, Madsen PT, Katten DR, Nachtigall PE (2010) Sound detection by the longfin squid (*Loligo pealeii*) studied with auditory evoked potentials: Sensitivity to low frequency particle motion and not pressure. *The Journal of Experimental Biology* **213**, 3748–3759. doi:10.1242/jeb.048348

Nabhitabhata J (1995) Mass culture of cephalopods in Thailand. *World Aquaculture* **26**(2), 25–29.

Nabhitabhata J, Nilaphat P, Promboon P, Jaroongpattananon C, Nilaphat G, Reunreg A (2005) Performance of simple large-scale cephalopod culture system in Thailand. *Phuket Marine Biological Center Research Bulletin* **66**, 337–350.

Nabhitabhata J, Suwanamala J (2008) Reproductive behaviour and cross-mating of two closely related pygmy squids *Idiosepius biserialis* and *Idiosepius thailandicus* (Cephalopoda: Idiosepiidae). *Journal of the Marine Biological Association of the United Kingdom* **88**(5), 987–993. doi:10.1017/S0025315408001616

Naef A (1923) Die Cephalopoden. *Fauna e Flora de Golfo di Napoli*, Monograph 35, 1(1)(part 2), 149–863.

Nateewathana A (1997) The Sepiolidae (Cephalopoda) of the Andaman Sea, Thailand, with description of *Euprymna hyllebergi* sp. nov. *Phuket Marine Biological Center Special Publication* **17**(2), 465–481, 7.

Natsukari Y (1984) Taxonomical and morphological studies on the loliginid squid. V. Re-description of the type specimen of *Loligo sumatrensis* d'Orbigny, 1835. *Venus Japanese Journal of Malacology* **43**, 229–239.

Natsukari Y, Okutani T (1975) Taxonomic and morphological studies on the loliginid squids – I. Identity of Loligo chinensis Gray, 1849, re-description of the type specimen and taxonomic review. *Venus, Japanese Journal of Malacology* **34**(3,4), 85–91.

Naud MJ, Hanlon RT, Hall KC, Shaw PW, Havenhand JN (2004) Behavioural and genetic assessment of reproductive success in a spawing aggregation of the Australian giant cuttlefish, *Sepia apama*. *Animal Behaviour* **67**(6), 1043–1050. doi:10.1016/j.anbehav.2003.10.005

Naud MJ, Havenhand JN (2006) Sperm motility and longevity in the giant cuttlefish, *Sepia apama* (Mollusca: Cephalopoda). *Marine Biology* **148**(3), 559–566. doi:10.1007/s00227-005-0109-z

Naud MJ, Shaw PW, Hanlon RT, Havenhand JN (2005) Evidence for biased use of sperm sources in wild female giant cuttlefish (*Sepia apama*). *Proceedings. Biological Sciences* **272**(1567), 1047–1051. doi:10.1098/rspb.2004.3031

Neige P, Warnke K (2010) Just how many species of *Spirula* are there? A morphometric approach. In 'Cephalopods – Present and Past'. (Eds K Tanabe, Y Shigeta, T Sasaki and H Hirano) pp. 77–84. (Tokai University Press: Tokyo.)

Nesis KN (1979) A brief review of zoogeography of Australian–New Zealand pelagic realm (by Cephalopods). *Trudy Instituta Okeanologii im. P. P. Shirshova Akademii Nauk SSSR* **106**, 125–139.

Nesis KN (1982/1987) Cephalopods of the World; Squids, Cuttlefishes, Octopuses, and Allies (translated from Russian by B. S. Levitov). (T.F.H. Publications, Neptune City, NJ, USA.)

Nesis KN, Nikitina IV (1986) A new family of abyssal squids (Cephalopoda, Oegopsida) from the south-eastern Atlantic. *Zoologicheskij Zhurnal* **65**(1), 47–54.

Nilsson D-E, Warran EJ, Johnsen S, Hanlon RT, Sahashar N (2012) A unique advantage for giant eyes in giant squid. *Current Biology* **22**(8), 683–688. doi:10.1016/j.cub.2012.02.031

Norman MD (1991) *Octopus cyanea* Gray, 1849 (Mollusca: Cephalopoda) in Australian waters: Description, distribution and taxonomy. *Bulletin of Marine Science* **49**(1–2), 20–38.

Norman MD (1992) *Ameloctopus litoralis*, gen. et sp. nov. (Cephalopoda: Octopodidae), a new shallow-water octopus from tropical Australian waters. *Invertebrate Taxonomy* **6**(3), 567–582. doi:10.1071/IT9920567

Norman MD (1993a) Four new octopus species of the *Octopus macropus* group (Cephalopoda: Octopodidae) from the Great Barrier Reef, Australia. *Memoirs of the National Museum of Victoria* **53**(2), 267–308.

Norman MD (1993b) Ocellate octopuses (Cephalopoda: Octopodidae) of the Great Barrier Reef, Australia: description of two new species and redescription of *Octopus polyzenia* Gray, 1849. *Memoirs of the National Museum of Victoria* **53**(2), 309–344.

Norman MD (1998) Octopodidae: Benthic octopuses. In 'The Living Marine Resources of the Western Central Pacific. Vol. 2. Cephalopods, Crustaceans, Holothurians and Sharks'. (Eds KE Carpenter and VH Neim) pp. 800–826. (Food and Agriculture Organisation of the United Nations: Rome.)

Norman M (2000) 'Cephalopods, a World Guide.' (ConchBooks: Hackenheim, Germany.)

Norman MD (2001) New octopus from Queensland. *Memoirs of the Queensland Museum* **46**(2), 677–690.

Norman MD, Finn J (2001) Revision of the *Octopus horridus* species-group, including erection of a new subgenus and description of two member species from the Great Barrier Reef, Australia. *Invertebrate Taxonomy* **15**(1), 13–35. doi:10.1071/IT99018

Norman MD, Finn JK (2014a) Family Amphitretidae. In 'Cephalopods of the World. An Annotated and Illustrated Catalogue of Cephalopod Species Known to Date'. No. 4, Vol. 3. Octopods and Vampire Squids. FAO Species Catalogue for Fishery Purposes'. (Eds P Jereb, CFE Roper, MD Norman and JK Finn) pp. 217–219. (FAO: Rome.)

Norman MD, Finn JK (2014b) Family Bolitaenidae. In 'Cephalopods of the World. An Annotated and Illustrated Catalogue of Cephalopod Species Known to Date'. No. 4, Vol. 3. Octopods and Vampire Squids. FAO Species Catalogue for Fishery Purposes'. (Eds P Jereb, CFE Roper, MD Norman and JK Finn) pp. 221–224. (FAO: Rome.)

Norman MD, Finn JK (2014c) Family Vampyroteuthidae. In 'Cephalopods of the World. An Annotated and Illustrated Catalogue of Cephalopod Species Known to Date'. No. 2, Vol. 3. Octopods and Vampire Squids. FAO Species Catalogue for Fishery Purposes'. (Eds P Jereb, CFE Roper, MD Norman and JK Finn) pp. 268–270. (FAO: Rome.)

Norman MD, Finn JK (2014d) Family Vitreledonellidae. In 'Cephalopods of the World. An Annotated and Illustrated Catalogue of Cephalopod Species Known to Date'. No. 4, Vol. 3. Octopods and Vampire Squids. FAO Species Catalogue for Fishery Purposes'. (Eds P Jereb, CFE Roper, MD Norman and JK Finn) pp. 219–221. (FAO: Rome.)

Norman MD, Finn JK, Hochberg FG (2014) Family Octopodidae. In 'Cephalopods of the world. An annotated and illustrated catalogue of cephalopod species known to date. No. 4, Vol. 3. Octopods and Vampire Squids. FAO Species Catalogue for Fishery Purposes'. (Eds P Jereb, CFE Roper, MD Norman and JK Finn) pp. 36–215. (FAO: Rome.)

Norman MD, Hochberg FG (2005) The current state of octopus taxonomy. *Phuket Marine Biological Centre Research Bulletin* **66**, 127–154.

Norman MD, Hochberg FG, Boucher-Rodoni R (2005) A revision of the deep-water octopus genus *Scaeurgus* (Cephalopoda: Octopodidae) with description of three new species from the southwest Pacific Ocean. *The Journal of Molluscan Studies* **71**(4), 319–337. doi:10.1093/mollus/eyi033

Norman MD, Hochberg FG, Huffard C, Mangold KM (1922–2003) (2013) Octopodoidea Orbigny, 1839. Octopods, octopuses, devilfishes. Version 07 July 2013 (under construction). http://tolweb.org/Octopodoidea/20194/2013.07.07 in The Tree of Life Web Project, http://tolweb.org/

Norman MD, Kubodera T (2006) Taxonomy and biogeography of an Australian subtropical octopus with Japanese affinities (Part Two Natural History Study). In 'Proceedings of the 7th and 8th Symposia on Collection Building and Natural History Studies in Asia and the Pacific Rim'. (Eds Y Tomida *et al.*). *National Science Museum Monographs* **34**, 171–189.

Norman MD, Lu CC (1997) Redescription of the southern dumpling squid *Euprymna tasmanica* and a revision of the genus *Euprymna* (Cephalopoda: Sepiolidae). *Journal of the Marine Biological Association of the United Kingdom* **77**(4), 1109–1137. doi:10.1017/S0025315400038662

Norman M, Reid A (2000) 'A Guide to Squid, Cuttlefish and Octopuses of Australia.' (The Gould League of Australia and CSIRO Publishing: Melbourne.)

Norman MD, Sweeney MJ (1997) The shallow-water octopuses (Cephalopoda: Octopodidae) of the Philippines. *Invertebrate Taxonomy* **11**(1), 89–140. doi:10.1071/IT95026

O'Dor RK, Adamo S, Aitken JP, Anrade Y, Finn J, Hanlon RT, Jackson GD (2002) Currents as environmental constraints on the behaviour, energetics and distribution of squid and cuttlefish. *Bulletin of Marine Science* **71**(2), 601–617.

O'Dor RK, Coelho ML (1993) Big squid, big currents, and big fisheries. In 'Recent Advances in Fishery Biology'. (Eds T Okutani, RK O'Dor and T Kubodera) pp. 385–396. (Tokai University Press: Kanagawa.)

O'Dor RK, Dawe EG (1998) *Illex illecebrosus*. In 'Squid Recruitment Dynamics'. (Eds PG Rodhouse, EG Dawe and RK O'Dor) pp. 77–104. (FAO, Rome.)

O'Dor RK, Webber DM (1991) Invertebrate athletes: Tradeoffs between transport efficiency and power density in cephalopod evolution. *The Journal of Experimental Biology* **160**, 93–112.

O'Shea S (1999) 'The Marine Fauna of New Zealand: Octopoda (Mollusca: Cephalopoda).' *National Institute of Water and Atmospheric Research (NIWA) Biodiversity Memoir*, 112. (National Institute of Water and Atmospheric Research: New Zealand.)

O'Shea S, Bolstad KS (2004) First records of egg masses of *Nototodarus gouldi* McCoy, 1888 (Mollusca: Cephalopoda: Ommastrephidae), with comments on egg mass susceptibility to damage by fisheries trawl. *New Zealand Journal of Zoology* **31**(2), 161–166. doi:10.1080/03014223.2004.9518369

O'Shea S, Jackson G, Bolstad KS (2007) The nomenclatural status, ontogeny and morphology of *Pholidoteuthis massyae* (Pfeffer, 1912) new comb (Cephalopoda: Pholidoteuthidae). *Reviews in Fish Biology and Fisheries* **17**(2–3), 425–435. doi:10.1007/s11160-007-9047-9

Okutani T (1995) 'Cuttlefish and Squids of the World in Colour.' (National Cooperative Association of Squid Processors: Tokyo.)

Okutani T (2005) 'Cuttlefishes and Squids of the World.' (National Cooperative Association of Squid Processors: Tokyo.)

Payne NL, Snelling EP, Semmens JM, Gillanders BM (2013) Mechanisms of population structuring in Giant Australian Cuttlefish *Sepia apama*. *PLoS ONE* **8**(3), e58694. doi:10.1371/journal.pone.0058694

Pecl G (2010) 'Cephalopods on the Move in Tassie.' *Cephalopod International Advisory Council Newsletter.*

Pecl G (2011) Cephalopods moving southwards to avoid the heat. *Malacological Society of Australasia Newsletter* 1–2.

Pecl GT, Jackson GD (2014) The potential effects of climate change on southern calamari in Tasmanian waters. (World Wildlife Fund: USA.)

Pecl GT, Moltschaniwskyj NA (2006) Life history of the short-lived squid (*Sepioteuthis australis*): Resource allocation as a function of size, growth, maturation, and hatching season. *ICES Journal of Marine Science* **63**(6), 995–1004.

Pecl GT, Tracey SR, Semmens JM, Jackson GD (2006) Use of acoustic telemetry for spatial management of southern calamary *Sepioteuthis australis*, a highly mobile inshore squid species. *Marine Ecology Progress Series* **328**, 1–15. doi:10.3354/meps328001

Pethybridge H, Virtue P, Casper R, Yoshida T, Green CP, Jackson GD, Nichols PD (2012) Seasonal variations in diet of arrow squid (*Nototodarus gouldi*): Stomach content and signature fatty acid analysis. *Journal of the Marine Biological Association of the United Kingdom* **92**(1), 187–196.

Pfeffer G (1912) Die Cephalopoden der Plankton-Expedition. Zugleich eine monographische übersicht der Oegopsiden Cephalopoden. *Ergebnisse der Plankton-Expedition der Humboldt-stiftung* **2**, 1–815, 48 plates.

Phillips KL, Jackson GD, Nichols PD (2003a) Temporal variations in the diet of the squid *Moroteuthis ingens* at Macquarie Island: Stomach contents and fatty acid analyses. *Marine Ecology Progress Series* **256**, 135–149. doi:10.3354/meps256135

Phillips KL, Nichols D, Jackson GD (2003b) Dietary variation of the squid *Moroteuthis ingens* at four sites in the Southern Ocean: Stomach contents, lipid and fatty acid profiles. *Journal of the Marine Biological Association of the United Kingdom* **83**(3), 523–534. doi:10.1017/S0025315403007446h

Piertney SB, Hudelot C, Hochberg FG, Collins MA (2003) Phylogenetic relationships among cirrate octopods (Mollusca: Cephalopoda) resolved using mitochondrial 16S ribosomal DNA sequences. *Molecular Phylogenetics and Evolution* **27**(2), 348–353. doi:10.1016/S1055-7903(02)00420-7

Ramos JE, Pecl GT, Moltschaniwskyj NA, Strugnell JM, León RI, Semmens JM (2014) Body size, growth and life span: Implications for the polewards range shift of *Octopus tetricus* in south-eastern Australia. *PLoS ONE* **9**(8), e103480. doi:10.1371/journal.pone.0103480

Reid AL (1992) . Taxonomic review of the Australian Rossiinae (Cephalopoda: Sepiolidae), with a description of a new species, *Neorossia leptodons*, and redescription of *N. caroli* (Joubin, 1902). *Bulletin of Marine Science* **49**(3), 748–831.

Reid AL (1998) A complete description of *Sepia mira* (Cotton, 1932) (Cephalopoda: Sepiidae) from Eastern Australia. *Proceedings of the Linnean Society of New South Wales* **119**, 155–171.

Reid AL (2000) Australian cuttlefishes (Cephalopoda, Sepiidae): The 'doratosepion' species complex, with descriptions of three new species. *Invertebrate Taxonomy* **14**(1), 1–76. doi:10.1071/IT98013

Reid AL (2001) *Sepia hedleyi* Berry, 1918 (Cephalopoda: Sepiidae): A complete description and clarification of the status of *S. dannevigi* Berry, 1918 and *S. rex* (Iredale, 1926). *Records of the South Australian Museum* **34**(2), 79–97.

Reid A (2005a) Family Idiosepiidae. In 'Cephalopods of the World. An Annotated and Illustrated Catalogue of Cephalopod Species Known to date. No. 4, Vol. 1. Chambered nautiluses and sepioids (Nautilidae, Sepiidae, Sepiolidae, Sepiadariidae, Idiosepiidae and Spirulidae). FAO Species Catalogue for Fishery Purposes'. (Eds P Jereb and CFE Roper) pp. 208–210. (FAO: Rome).

Reid AL (2005b) Family Sepiadariidae. In 'Cephalopods of the World. An Annotated and Illustrated Catalogue of Cephalopod Species Known to date. No. 4, Vol. 1. Chambered nautiluses and sepioids (Nautilidae, Sepiidae, Sepiolidae, Sepiadariidae, Idiosepiidae and Spirulidae). FAO Species Catalogue for Fishery Purposes'. (Eds P Jereb and CFE Roper) pp. 204–207. (FAO: Rome).

Reid A (2005c) Family Spirulidae. In 'Cephalopods of the World. An Annotated and Illustrated Catalogue of Cephalopod Species Known to date. No. 4, Vol. 1. Chambered nautiluses and sepioids (Nautilidae, Sepiidae, Sepiolidae, Sepiadariidae, Idiosepiidae and Spirulidae). FAO Species Catalogue for Fishery Purposes'. (Eds P Jereb and CFE Roper) pp. 211–212. (FAO: Rome).

Reid AL (2009) *Sepioloidea magna* sp. nov.: A new bottletail squid (Cephalopoda: Sepiadariidae) from northern Australia. *The Beagle: Records of the Museums and Art Galleries of the Northern Territory* **25**, 103–109.

Reid A (2011) *Euprymna pardalota* sp. nov. (Cephalopoda: Sepiolidae), a new dumpling squid from northern Australia. *The Beagle: Records of the Museums and Art Galleries of the Northern Territory* **27**, 135–142.

Reid A (2016) Post-mortem drift in Australian cuttlefish sepions: its effect on the interpretation of species ranges. *Molluscan Research* doi:10.1080/13235818.2015.1064366

Reid A, Jereb P (2005) Family Sepiolidae. In 'Cephalopods of the World. An Annotated and Illustrated Catalogue of Cephalopod Species Known to date. No. 4, Vol. 1. Chambered nautiluses and sepioids (Nautilidae, Sepiidae, Sepiolidae, Sepiadariidae, Idiosepiidae and Spirulidae). FAO Species Catalogue for Fishery Purposes'. (Eds P Jereb and CFE Roper) pp. 153–203. (FAO: Rome).

Reid A, Jereb P, Roper CFE (2005) Family Sepiidae. In 'Cephalopods of the World. An Annotated and Illustrated Catalogue of Cephalopod Species Known to date. No. 4, Vol. 1. Chambered nautiluses and sepioids (Nautilidae, Sepiidae, Sepiolidae, Sepiadariidae, Idiosepiidae and Spirulidae). FAO Species Catalogue for Fishery Purposes'. (Eds P Jereb and CFE Roper). pp. 57–152. (FAO: Rome).

Reid A, Lu CC (1998) A new *Sepiella* Gray, 1849 (Cephalopoda: Sepiidae) from northern Australia, with a redescription of *Sepiella weberi* Adam, 1939. *The Beagle: Records of the Museums and Art Galleries of the Northern Territory* **14**, 71–102.

Reid AL, Lu CC (2005) A new cuttlefish, *Sepia filibrachia* n. sp., from the South China Sea with a redescription of *Sepia mestus* Gray, 1849 (Cephalopoda: Sepiidae) from eastern Australia. *Zootaxa* **911**, 1–22.

Riddell DJ (1985) The Enoploteuthidae (Cephalopoda: Oegopsida) of the New Zealand Region. *Fisheries Research Bulletin* **27**.

Robison B, Seibel B, Drazen J (2014) Deep-Sea octopus (*Graneledone boreopacifica*) conducts the longest-known egg-brooding period of any animal. *PLoS ONE* **9**(7), e103437. doi:10.1371/journal.pone.0103437

Rocha F, Guerra A, Gonzales AF (2001) A review of reproductive strategies in cephalopods. *Biological Reviews of the Cambridge Philosophical Society* **76**(3), 291–304. doi:10.1017/S1464793101005681

Rodhouse PG (1998) *Todarodes filippovae* in the Southern Ocean: an appraisal for exploitation and management. In 'Large Pelagic Squids'. (Ed. T. Okutani) pp. 207–215. (Japan Marine Fishery Research Centre, Tokyo.)

Rodhouse PG, Boyle PR (2010) Large aggregations of pelagic squid near the ocean surface at the Antarctic Polar Front, and their capture by grey-headed albatross chicks. *ICES Journal of Marine Science* **67**, 1432–1435.

Rodhouse PG, White MG (1995) Cephalopods occupy the ecological niche of epipelagic fish in the Antarctic Polar Frontal Zone. *The Biological Bulletin* **189**(2), 77–80. doi:10.2307/1542457

Rodhouse PGK (2013) Role of squid in the southern ocean pelagic ecosystem and the possible consequences of climate change. *Deep-sea Research. Part II, Topical Studies in Oceanography* **95**, 129–138. doi:10.1016/j.dsr2.2012.07.001

Roeleveld MAC (2000) Giant squid beaks: implications for systematics. *Journal of the Marine Biological Association of the United Kingdom* **80**(1), 185–187. doi:10.1017/S0025315499001769

Roeleveld MAC (2002) Tentacle morphology of the giant squid *Architeuthis* from the North Atlantic and Pacific oceans. *Bulletin of Marine Science* **71**(2), 725–737.

Roper CFE, Jereb P (2010a) Family Bathyteuthidae. In 'Cephalopods of the World. An Annotated and Illustrated Catalogue of Species Known to Date. No. 4, Vol. 2. Myopsid and Oegopsid Squids. FAO Species Catalogue for Fishery Purposes'. (Eds P Jereb and CFE Roper) pp. 124–126. (FAO: Rome.)

Roper CFE, Jereb P (2010b) Family Chtenopterigidae. In 'Cephalopods of the World. An Annotated and Illustrated Catalogue of Species Known to Date. No. 4, Vol. 2. Myopsid and Oegopsid Squids. FAO Species Catalogue for Fishery Purposes'. (Eds P Jereb and CFE Roper) pp. 146–147. (FAO: Rome.)

Roper CFE, Jereb P (2010c) Family Ancistrocheiridae. In 'Cephalopods of the World. An Annotated and Illustrated Catalogue of Species Known to Date. No. 4, Vol. 2. Myopsid and Oegopsid Squids. FAO Species Catalogue for Fishery Purposes'. (Eds P Jereb and CFE Roper) pp. 118–120. (FAO: Rome.)

Roper CFE, Jereb P (2010d) Family Architeuthidae. In 'Cephalopods of the World. An Annotated and Illustrated Catalogue of Species Known to Date. No. 4, Vol. 2. Myopsid and Oegopsid Squids. FAO Species Catalogue for Fishery Purposes'. (Eds P Jereb and CFE Roper) pp. 121–123. (FAO: Rome.)

Roper CFE, Jereb P (2010e) Family Batoteuthidae. In 'Cephalopods of the World. An Annotated and Illustrated Catalogue of Species Known to Date. No. 4, Vol. 2. Myopsid and Oegopsid Squids. FAO Species Catalogue for Fishery Purposes'. (Eds P Jereb and CFE Roper) pp. 127–128. (FAO: Rome.)

Roper CFE, Jereb P (2010f) Family Brachioteuthidae. In 'Cephalopods of the world. An annotated and illustrated catalogue of species known to date. No. 4, Vol. 2. Myopsid and Oegopsid Squids. FAO Species Catalogue for Fishery Purposes'. (Eds P Jereb and CFE Roper) pp. 129–134. (FAO: Rome.)

Roper CFE, Jereb P (2010g) Family Chiroteuthidae. In 'Cephalopods of the World. An Annotated and Illustrated Catalogue of Species Known to Date No. 4, Vol. 2. Myopsid and Oegopsid Squids. FAO Species Catalogue for Fishery Purposes'. (Eds P Jereb and CFE Roper) pp. 135–145. (FAO: Rome.)

Roper CFE, Jereb P (2010h) Family Cranchiidae. In 'Cephalopods of the World. An Annotated and Illustrated Catalogue of Species Known to Date. No. 4, Vol. 2. Myopsid and Oegopsid Squids. FAO Species Catalogue for Fishery Purposes'. (Eds P Jereb and CFE Roper) pp. 148–178. (FAO: Rome.)

Roper CFE, Jereb P (2010i) Family Cycloteuthidae. In 'Cephalopods of the World. An Annotated and Illustrated Catalogue of Species Known to Date. No. 4, Vol. 2. Myopsid and Oegopsid Squids. FAO Species Catalogue for Fishery Purposes'. (Eds P Jereb and CFE Roper) pp. 179–182. (FAO: Rome.)

Roper CFE, Jereb P (2010j) Family Enoploteuthidae. In 'Cephalopods of the World. An Annotated and Illustrated Catalogue of Species Known to Date. No. 4, Vol. 2. Myopsid and Oegopsid Squids. FAO Species Catalogue for Fishery Purposes'. (Eds P Jereb and CFE Roper) pp. 183–200. (FAO: Rome.)

Roper CFE, Jereb P (2010k) Family Histioteuthidae. In 'Cephalopods of the World. An Annotated and Illustrated Catalogue of Species Known to Date. No. 4, Vol. 2. Myopsid and Oegopsid Squids. FAO Species Catalogue for Fishery Purposes'. (Eds P Jereb and CFE Roper) pp. 223–236. (FAO: Rome.)

Roper CFE, Jereb P (2010l) Family Joubiniteuthidae. In 'Cephalopods of the World. An Annotated and Illustrated Catalogue of Species Known to Date No. 4, Vol. 2. Myopsid and Oegopsid Squids. FAO Species Catalogue for Fishery Purposes'. (Eds P Jereb and CFE Roper) pp. 237–238. (FAO: Rome.)

Roper CFE, Jereb P (2010m) Family Lepidoteuthidae. In 'Cephalopods of the World. An Annotated and Illustrated Catalogue of Species Known to Date. No. 4, Vol. 2. Myopsid and Oegopsid Squids. FAO Species Catalogue for Fishery Purposes'. (Eds P Jereb and CFE Roper) pp. 239–240. (FAO: Rome.)

Roper CFE, Jereb P (2010n) Family Lycoteuthidae. In 'Cephalopods of the World. An Annotated and Illustrated Catalogue of Species Known to Date. No. 4, Vol. 2. Myopsid and Oegopsid Squids. FAO Species Catalogue for Fishery Purposes'. (Eds P Jereb and CFE Roper) pp. 241–246. (FAO: Rome.)

Roper CFE, Jereb P (2010o) Family Mastigoteuthidae. In 'Cephalopods of the World. An Annotated and Illustrated Catalogue of Species Known to Date. No. 4, Vol. 2. Myopsid and Oegopsid Squids. FAO Species Catalogue for Fishery Purposes'. (Eds P Jereb and CFE Roper) pp. 250–256. (FAO: Rome.)

Roper CFE, Jereb P (2010p) Family Neoteuthidae. In 'Cephalopods of the World. An Annotated and Illustrated Catalogue of Species Known to Date. No. 4, Vol. 2. Myopsid and Oegopsid Squids. FAO Species Catalogue for Fishery Purposes'. (Eds P Jereb and CFE Roper) pp. 257–261. (FAO: Rome.)

Roper CFE, Jereb P (2010q) Family Octopoteuthidae. In 'Cephalopods of the World. An Annotated and Illustrated Catalogue of Species Known to Date. No. 4, Vol. 2. Myopsid and Oegopsid Squids. FAO Species Catalogue for Fishery Purposes'. (Eds P Jereb and CFE Roper) pp. 262–268. (FAO: Rome.)

Roper CFE, Jereb P (2010r) Family Onychoteuthidae. In 'Cephalopods of the World. An Annotated and Illustrated Catalogue of Species Known to Date. No. 4, Vol. 2. Myopsid and Oegopsid Squids. FAO Species Catalogue for Fishery Purposes'. (Eds P Jereb and CFE Roper) pp. 348–369. (FAO: Rome.)

Roper CFE, Jereb P (2010s) Family Pholidoteuthidae. In 'Cephalopods of the World. An Annotated and Illustrated Catalogue of Species Known to Date. No. 4, Vol. 2. Myopsid and Oegopsid Squids. FAO Species Catalogue for Fishery Purposes'. (Eds P Jereb and CFE Roper) pp. 370–373. (FAO: Rome.)

Roper CFE, Jereb P (2010t) Family Pyroteuthidae. In 'Cephalopods of the World. An Annotated and Illustrated Catalogue of Species Known to Date. No. 4, Vol. 2. Myopsid and Oegopsid Squids. FAO Species Catalogue for Fishery Purposes'. (Eds P Jereb and CFE Roper) pp. 379–383. (FAO: Rome.)

Roper CFE, Jereb P (2010u) Family Thysanoteuthidae. In 'Cephalopods of the World. An Annotated and Illustrated Catalogue of Species Known to Date. No. 4, Vol. 2. Myopsid and Oegopsid Squids. FAO Species Catalogue for Fishery Purposes'. (Eds P Jereb and CFE Roper) pp. 384–387. (FAO: Rome.)

Roper CFE, Nigmatullin C, Jereb P (2010a) Family Ommastrephidae. In 'Cephalopods of the World. An Annotated and Illustrated Catalogue of Species Known to Date. No. 4, Vol. 2. Myopsid and Oegopsid Squids. FAO Species Catalogue for Fishery Purposes'. (Eds P Jereb and CFE Roper) pp. 269–347. (FAO: Rome.)

Roper CFE, Jorgensen EM, Katugin ON, Jereb P (2010b) Family Gonatidae. In 'Cephalopods of the World. An Annotated and Illustrated Catalogue of Species Known to Date. No. 4, Vol. 2. Myopsid and Oegopsid Squids. FAO Species Catalogue for Fishery Purposes'. (Eds P Jereb and CFE Roper) pp. 200–222. (FAO: Rome.)

Roper CFE, Lu CC (1990) Comparative morphology and function of dermal structures in oceanic squids (Cephalopoda). *Smithsonian Contributions to Zoology* **493**, 1–40.

Roper CFE, Shea EK (2013) Unanswered questions about the giant squid *Architeuthis* (Architeuthidae) illustrate our incomplete knowledge of coleoid cephalopods. *American Malacological Bulletin* **31**(1), 109–122. doi:10.4003/006.031.0104

Roper CFE, Young RE (1972) First records of juvenile giant squid, *Architeuthis* (Cephalopoda: Oegopsida). *Proceedings of the Biological Society of Washington* **85**(16), 205–222.

Roper CFE, Young RE (1999) *Chiroteuthis joubini* Voss, 1967. Version 01 January 1999 (under construction). http://tolweb.org/Chiroteuthis_joubini/19475/1999.01.01 in The Tree of Life Web Project, http://tolweb.org/

Roper CFE, Young RE (2011a) *Chiroteuthis picteti* Joubin, 1894. Version 22 November 2011 (under construction). http://tolweb.org/Chiroteuthis_picteti/19477/2011.11.22 in The Tree of Life Web Project, http://tolweb.org/

Roper CFE, Young RE (2011b) *Chiroteuthis veranyi* (Ferussac, 1835). Version 22 November 2011 (under construction). http://tolweb.org/Chiroteuthis_veranyi/19479/2011.11.22 in The Tree of Life Web Project, http://tolweb.org

Roper CFE, Young RE (2013) *Chiroteuthis* Orbigny, 1841. Version 08 January 2013 (under construction) http://tolweb.org/Chiroteuthis/19462/2013.01.08 in The Tree of Life Web Project, http://tolweb.org/

Rosa R, Gonzalez L, Dierssen HM, Seibel BA (2012) Environmental determinants of latitudinal size-trends in cephalopods. *Marine Ecology Progress Series* **464**, 153–165. doi:10.3354/meps09822

Sales JBL, Shaw PW, Haimovici M, Markaida U, Cunha DB, Ready J, Figueiredo-Ready WMB, Schneider H, Sampaio I (2013) New molecular phylogeny of the squids of the family Loliginidae with emphasis on the genus *Doryteuthis* Naef, 1912: Mitochondrial and nuclear sequences indicate the presence of cryptic species in the southern Atlantic Ocean. *Molecular Phylogenetics and Evolution* **68**(2), 293–299. doi:10.1016/j.ympev.2013.03.027

Sands CJ, Jarman SN, Jackson GD (2003) Genetic differentiation in the squid *Moroteuthis ingens* inferred from RAPD analysis. *Polar Biology* **26**(3), 166–170.

Santaclara FJ, Espiñeira M, Vieites JM (2007) Genetic identification of squids (Families Ommastrephidae and Loliginidae) by PCR-RFLP and FINS methodologies. *Journal of Agricultural and Food Chemistry* **55**(24), 9913–9920. doi:10.1021/jf0707177

Sato N, Kasugai T, Munehara H (2008) Estimated life span of Japanese pygmy squid, *Idiosepius paradoxus* from statolith growth increments. *Journal of the Marine Biological Association of the United Kingdom* **88**(2), 391–394. doi:10.1017/S0025315408000581

Sato N, Yoshida M-A, Fujiwara E, Kasugai T (2013) High-speed camera observations of copulatory behaviour in *Idiosepius paradoxus*: Function of the dimorphic hectocotyli. *The Journal of Molluscan Studies* **79**(2), 183–186. doi:10.1093/mollus/eyt005

Saunders WB, Ward PD (1987) Sympatric occurrence of living *Nautilus* (*N. pompilius* and *N. stenomphalus*) on the Great Barrier Reef, Australia. *The Nautilus* **101**(4), 188–193.

Scheel D, Godfrey-Smith P, Lawrence M (2014) *Octopus tetricus* (Mollusca: Cephalopoda) as an ecosystem engineer. *Scientia Marina* **78**(4), 521–528. doi:10.3989/scimar.04073.15A

Schnell AK, Smith CL, Hanlon RT, Harcourt RT (2015) Female receptivity, mating history, and familiarity influence the mating behaviour of cuttlefish. *Behavioral Ecology and Sociobiology* **69**(2), 283–292. doi:10.1007/s00265-014-1841-5

Seibel BA, Fabry VJ (2003) Marine biotic response to elevated carbon dioxide. *Advances in Applied Biodiversity Science* **4**, 59–67.

Seibel BA, Thuesen EV, Childress JJ (1998) Flight of the vampire: Ontogenetic gait-transition in *Vampyroteuthis infernalis* (Cephalopoda: Vampyromorpha). *The Journal of Experimental Biology* **201**(16), 2413–2424.

Semmens JM, Moltschaniwskyj NA (2000) An examination of variable growth in the loliginid squid *Sepioteuthis lessoniana*: A whole animal and reductionist approach. *Marine Ecology Progress Series* **193**, 135–141. doi:10.3354/meps193135

Semmens JM, Pecl GT, Gillanders BM, Waluda CM, Shea EK, Jouffre D, Tchii T, Zumholz K, Katugin ON, Leporati SC, Shaw PW (2007) Approaches to resolving cephalopod movement and migration patterns. *Reviews in Fish Biology and Fisheries* **17**(2–3), 401–423. doi:10.1007/s11160-007-9048-8

Shaw PW (2003) Polymorphic microsatellite DNA markers for the assessment of genetic diversity and paternity testing in the giant cuttlefish, *Sepia apama*. *Conservation Genetics* **4**(4), 533–535. doi:10.1023/A:1024729100857

Sin YW, Yau C, Chu KH (2009) Morphological and genetic differentiation of two loliginid squids, *Uroteuthis* (*Photololigo*) *chinensis* and *Uroteuthis* (*Photololigo*) *edulis* (Cephalopoda: Lologinidae), in Asia. *Journal of Experimental Marine Biology and Ecology* **369**(1), 22–30. doi:10.1016/j.jembe.2008.10.029

Sinclair B, Briskey L, Aspden W, Pegg G (2007) Genetic diversity of isolated populations of *Nautilus pompilius* (Mollusca, Cephalopoda) in the Great Barrier Reef and Coral Sea. *Reviews in Fish Biology and Fisheries* **17**(2–3), 223–235. doi:10.1007/s11160-006-9030-x

Sinclair W, Newman SJ, Vianna GMS, Williams S, Aspden WJ (2011) Spatial subdivision and genetic diversity in populations on the east and west coasts of Australia: the multifaceted case of *Nautilus pompilius* (Mollusca, Cephalopoda). *Reviews in Fisheries Science* **19**(1), 52–61. doi:10.1080/10641262.2010.533794

Smith EA (1884) Mollusca. Report on the Zoological Collections made in the Indo-Pacific Ocean during the Voyage of H.M.S. 'Alert' 1881–2. London: British Museum.

Smith AM, Wolfer K, Byrne M (2012) *Argonauta* at risk: dissolution and carbonate mineralogy of egg cases. In 'Proceedings of the 12th International Coral Reef Symposium', Cairns, Australia 9–13 July 2012. www.icrs2012.com/proceedings/manuscripts/ICRS2012_8A_1.pdf

Smith SA, Wilson NG, Goetz FE, Feehery C, Andrade SCS, Rouse GW, Giribet G, Dunn CW (2011) Resolving the evolutionary relationships of molluscs with phylogenomic tools. *Nature* **480**, 364–367. doi:10.1038/nature10526

Smith T, Green C, Sherman C (2015) Patterns of connectivity and population structure of the southern calamary *Sepioteuthis australis* in southern Australia. *Marine and Freshwater Research* **66**(10), 942–947. doi:10.1071/MF14328

Soucier CP, Basil JA (2008) Chambered nautilus (*Nautilus pompilius pompilius*) responds to underwater vibrations. *American Malacological Bulletin* **24**(1), 3–11. doi:10.4003/0740-2783-24.1.3

Spady BL, Watson S-A, Chase TJ, Munday PL (2014) Projected near-future $CO_2$ levels increase activity and alter defensive behaviours in the tropical squid *Idiosepius pygmaeus*. *Biology Open* **3**(11), 1063–1070. doi:10.1242/bio.20149894

Squires ZE, Norman MD, Stuart-Fox D (2013) Mating behaviour and general spawning patterns of the southern dumpling squid *Euprymna tasmanica* (Sepiolidae): A laboratory study. *The Journal of Molluscan Studies* **79**(3), 263–269. doi:10.1093/mollus/eyt025

Sreeja V, Bijukumar A (2013) Ethological studies of the veined octopus *Amphioctopus marginatus* (Taki) (Cephalopoda: Octopodidae) in captivity, Kerala, India. *Journal of Threatened Taxa* **5**(10), 4492–4497. doi:10.11609/JoTT.o3256.4492-7

Stark KE, Jackson GD, Lyle J (2005) Tracking arrow squid movements with an automated acoustic telemetry system. *Marine Ecology Progress Series* **299**, 167–177. doi:10.3354/meps299167

Steer MA, Moltschaniwskyi NA (2007) The effects of egg position, egg mass size, substrate and biofouling on embryo mortality in the squid *Sepioteuthis australis*. *Reviews in Fish Biology and Fisheries* **17**(2–3), 173–182. doi:10.1007/s11160-006-9023-9

Steer MA, Moltschaniwskyj NA, Jordan AR (2003a) Embryonic development of southern calamary (*Sepioteuthis australis*) within the constraints of an aggregated egg mass. *Marine and Freshwater Research* **54**(3), 217–226. doi:10.1071/MF02107

Steer MA, Pecl GT, Moltschaniwskyj NA (2003b) Are bigger calamary *Sepioteuthis australis* hatchlings more likely to survive? A study based on statolith dimensions. *Marine Ecology Progress Series* **261**, 175–182. doi:10.3354/meps261175

Steer MA, Semmens JM (2003) Pulling or drilling, does size or species matter? An experimental study of prey handling by *Octopus dierythraeus* (Norman, 1992). *Journal of Marine Biology and Ecology* **290**(2), 165–178. doi:10.1016/S0022-0981(03)00076-5

Stranks TN (1988) Redescription of *Octopus pallidus* (Cephalopoda: Octopodidae) from south-eastern Australia. *Malacologia* **29**(1), 275–287.

Stranks TN (1990) Three new species of *Octopus* (Mollusca: Cephalopoda) from south-eastern Australia. *Memoirs of the National Museum of Victoria* **50**(2), 457–465.

Stranks TN (1996) Biogeography of *Octopus* species (Cephalopoda: Octopodidae) from southeastern Australia. *American Malacological Bulletin* **12**(1/2), 145–151.

Stranks TN (1998) The systematic and nomenclatural status of the Octopodinae described from Australia (Mollusca: Cephalopoda). *Smithsonian Contributions to Zoology* **586**, 529–547.

Stranks TN, Norman MD (1993) Review of the *Octopus australis* complex from Australia and New Zealand, with description of a new species (Mollusca: Cephalopoda). *Memoirs of the National Museum of Victoria* **53**(2), 345–373.

Strugnell J, Allcock AL (2010) Co-estimation of phylogeny and divergence times of Agonautoidea using relaxed phylogenetics. *Molecular Phylogenetics and Evolution* **54**(3), 701–708. doi:10.1016/j.ympev.2009.11.017

Strugnell J, Nishiguchi MK (2007) Molecular phylogeny of coleoid cephalopods (Mollusca: Cephalopoda) inferred from three mitochondrial and six nuclear loci: A comparison of alignment, implied alignment and analysis methods. *The Journal of Molluscan Studies* **73**(4), 399–410. doi:10.1093/mollus/eym038

Strugnell J, Norman M, Cooper A (2006) DNA from beach-washed shells of the Ram's Horn Squid, *Spirula spirula*. *Bulletin of Marine Science* **78**(2), 389–391.

Strugnell J, Norman M, Drummond AJ, Cooper A (2004) Neotenous origins for pelagic octopuses. *Current Biology* **14**(8), R300–R301. doi:10.1016/j.cub.2004.03.048

Strugnell J, Norman M, Jackson J, Drummond AJ, Cooper A (2005) Molecular phylogeny of coleoid cephalopods (Mollusca: Cephalopoda) using a multigene approach: The effect of data partitioning on resolving phylogenies in a Bayesian framework. *Molecular Phylogenetics and Evolution* **37**, 426–441. doi:10.1016/j.ympev.2005.03.020

Strugnell J, Norman MD, Vecchione M, Guzic M, Allcock AL (2014) The ink sac clouds octopod evolutionary history. *Hydrobiologia* **725**, 215–235. doi:10.1007/s10750-013-1517-6

Strugnell J, Voight JR, Collins PC, Allcock AL (2009) Molecular phylogenetic analysis of a known and a new hydrothermal vent octopod: Their relationship with the genus *Benthoctopus* (Cephalopoda: Octopodidae). *Zootaxa* **2096**, 442–459.

Strugnell JM, Collins MA, Allcock AL (2008a) Molecular evolutionary relationships of the octopodid genus *Thaumeledone* (Cephalopoda: Octopodidae) from the Southern Ocean. *Antarctic Science* **20**(3), 245–251. doi:10.1017/S0954102008001132

Strugnell JM, Rogers AD, Prodöhl PA, Collins MA, Allcock AL (2008b) The thermohaline expressway: the Southern Ocean as a centre of origin for deep-sea octopuses. *Cladistics* **24**(6), 853–860. doi:10.1111/j.1096-0031.2008.00234.x

Strugnell JM, Cherel Y, Cooke IR, Gleadall IG, Hochberg FG, Ibáñez CM, Jorgensen E, Laptikhovsky VV, Linse K, Norman M, Vecchione M, Voight JR, Allcock AL (2011) The Southern Ocean: Source and sink? *Deep-sea Research. Part II, Topical Studies in Oceanography* **58**(1), 196–204. doi:10.1016/j.dsr2.2010.05.015

Sugimoto C, Ikeda Y (2012) Ontogeny of schooling behaviour in the oval squid *Sepioteuthis lessoniana*. *Fisheries Science (Tokyo)* **78**(2), 287–294. doi:10.1007/s12562-011-0464-2

Sweeney MJ (n.d.) Recent cephalopod primary type specimens: A searching tool, accessed 23 January 2014, www.jiscmail.ac.uk/cgi-bin/filearea.cgi?LMGT1=FASTMOLL&a=get&f=/Type_Finder_Fastmoll-1.pdf

Sweeney MJ, Roper CFE (1998) Classification, type localities, and type repositories of recent Cephalopoda. *Smithsonian Contributions to Zoology* **586**, 561–599.

Sweeney MJ, Young RE (2002) Taxa associated with the Family Architeuthidae Pfeffer, 1900. http://tolweb.org/accessory/Architeuthidae_Taxa?acc_id=1798 in The Tree of Life Web Project, http://tolweb.org/

Tait RW (1982) A taxonomic revision of *Octopus australis* Hoyle, 1885 (Octopodidae: Cephalopoda), with a redescription of the species. *Memoirs of the National Museum of Victoria* **43**, 15–23.

Taki I (1961) On two new eledonid octopods from the Antarctic Sea. *Journal of the Faculty of Fisheries and Animal Husbandry, Hiroshima University* **3**(2), 297–316.

Takumiya M, Kobayashi M, Tsuneki K, Furuya H (2005) Phylogenetic relationships among major species of Japanese coleoid cephalopods (Mollusca, Cephalopoda) using three mitochondrial DNA sequences. *Zoological Science* **22**(2), 147–155. doi:10.2108/zsj.22.147

Talbot CM, Marshall J (2010) Polarization sensitivity and retinal topography of the striped pyjama squid (*Sepioloidea lineolata* - Quoy/Gaimard 1832). *The Journal of Experimental Biology* **213**, 3371–3377. doi:10.1242/jeb.048165

Taylor AL, McKeown NJ, Shaw PW (2012) Molecular identification of three co-occurring and easily misidentified octopus species using PCR-RFLP techniques. *Conservation Genetics Resources* **4**(4), 885–887. doi:10.1007/s12686-012-9665-y

Temple SE, Pignatelli V, Cook T, How MJ, Chiou T-H, Roberts NW, Marshall NJ (2012) High-resolution polarisation vision in a cuttlefish. *Current Biology* **22**(4), R121–R122. doi:10.1016/j.cub.2012.01.010

Thomas R, Moltschaniwskyj NA (1999) Ontogenetic changes in size and shape of statoliths: Implications for age and growth of the short-lived tropical squid *Sepioteuthis lessoniana* (Cephalopoda: Loliginidae). *Fishery Bulletin* **97**(3), 636–645.

Thomas RF (1977) Systematics, distribution and biology of cephalopods of the genus *Tremoctopus* (Octopoda: Tremotopodidae). *Bulletin of Marine Science* **27**, 353–392.

Tracey SR, Steer M, Pecl GT (2003) Life history traits of the temperate mini-maximalist *Idiosepius notoides* (Cephalopoda: Sepioidea). *Journal of the Marine Biological Association of the United Kingdom* **83**(6), 1297–1300. doi:10.1017/S0025315403008701

Triantafillos L (2004) Effects of genetic and environmental factors on growth of southern calamary, *Sepioteuthis australis*, from southern Australia and New Zealand. *Marine and Freshwater Research* **55**(4), 439–446. doi:10.1071/MF03157

Triantafillos L, Adams M (2001) Allozyme analysis reveals a complex population structure in the southern calamary, *Sepioteuthis australis* from Australia and New Zealand. *Marine Ecology Progress Series* **212**, 193–209. doi:10.3354/meps212193

Triantafillos L, Adams M (2005) Genetic evidence that the northern calamary, *Sepioteuthis lessoniana*, is a species complex in Australian waters. *ICES Journal of Marine Science* **62**(8), 1665–1670. doi:10.1016/j.icesjms.2005.06.004

Triantafillos L, Jackson GD, Adams M, Steer BLM (2004) An allozyme investigation of the stock structure of arrow squid *Nototodarus gouldi* (Cephalopoda: Ommastrephidae) from Australia. *ICES Journal of Marine Science* **61**(5), 829–835. doi:10.1016/j.icesjms.2003.12.010

Tsuchiya K (2009a) *Abralia andamanica* (Goodrich 1896). Version 26 July 2009 (under construction). http://.org/Abralia_andamanica/19655/2009.07.26 in The Tree of Life Web Project, http://tolweb.org/

Tsuchiya K (2009b) *Abralia astrolineata* Berry, 1914. Version 26 July 2009 (under construction). http://tolweb.org/Abralia_astrolineata/19651/2009.07.26 in The Tree of Life Web Project, http://tolweb.org/

Tsuchiya K (2009c) *Abralia spaercki* Grimpe, 1931. Version 26 July 2009 (under construction). http://tolweb.org/Abralia_spaercki/19650/2009.07.26 in The Tree of Life Web Project, http://tolweb.org/

Tsuchiya K (2009d) *Enoploteuthis galaxias* Berry, 1918. Version 23 July 2009 (under construction). http://tolweb.org/Enoploteuthis_galaxias/19707/2009.07.23 in The Tree of Life Web Project, http://tolweb.org/

Tsuchiya K (2013a) *Abraliopsis lineata* (Goodrich, 1896). Version 03 November 2013 (under construction). http://tolweb.org/Abraliopsis_lineata/1969/2013.11.03 in The Tree of Life Web Project http://tolweb.org.

Tsuchiya K (2013b) *Abraliopsis tui* Riddell, 1985. Version 03 November 2013 (under construction. http://tolweb.org/Abraliopsis_tui/19692/2013.11.03 in The Tree of Life Web Project, http://tolweb.org

Tsuchiya K (2013c) *Enoploteuthis* Orbigny in Rüppell 1844. Version 03 Novemer 2013 (under construction. http://tolweb.org/Enoploteuthis/19641/2013.11.03 in The Tree of Life Web Project, http://tolweb.org

Tsuchiya K, Okutani T (1988) Subgenera of *Enoploteuithis, Abralia* and *Abraliopsis* of the squid family Enoploteuthidae (Cephalopoda, Oegopsida). *Bulletin of the National Science Museum, Tokyo, Series A* **14**(3), 119–136.

Tsuchiya K, Okutani T (1992) Growth stages of *Moroteuthis robusta* (Verrill, 1881) with the re-evaluation of the genus. *Bulletin of Marine Science* **49**(1–2), 137–147.

Tsuchiya K, Piatkowski U, Okutani T (1991) Distribution and re-description of *Abraliopsis lineata* Goodrich, 1896 (Cephalopoda: Enoploteuthidae) from the Arabian Sea. *Journal of Natural History* **25**, 1121–1133. doi:10.1080/00222939100770731

Tsuchiya K, Young RE (2009) *Abralia astrostricta* Berry 1909. Version 29 December 2009 (under construction). http://tolweb.org/Abralia_astrostricta/19652/2009.12.29 in The Tree of Life Web Project, http://tolweb.org

Tsuchiya K, Young RE (2014) *Enoploteuthis magnoceani* Nesis 1982. Version 21 January 2014 (under construction). http://tolweb.org/Enoploteuthis_magnoceani/19710/2014.01.21 in The Tree of Life Web Project, http://tolweb.org/

Undheim EAB, Georgieva DN, Thoen HH, Norman JA, Morek J, Betzel C, Fry BG (2010) Venom on ice: First insights into Antarctic octopus venoms. *Toxicon* **56**(6), 897–913. doi:10.1016/j.toxicon.2010.06.013

Vafiadis P (1998) Intertidal sighting and behaviour of *Octopus maorum*. *The Victorian Naturalist* **115**(3), 100–104.

Van Camp LM, Donnellan SC, Dyer AR, Fairweather PG (2004) Multiple paternity in field- and captive-laid egg strands of *Sepioteuthis australis* (Cephalopoda: Loliginidae). *Marine and Freshwater Research* **55**(8), 819–823. doi:10.1071/MF03179

Van Camp LM, Saint KM, Donellan S, Havenhand JN, Fairweather PG (2003) Polymorphic microsatellite markers for paternity assessment in southern calamari *Sepioteuthis australis* (Cephalopoda: Loliginidae). *Molecular Ecology Notes* **3**(4), 654–655. doi:10.1046/j.1471-8286.2003.00549.x

Vecchione M, Brakoniecki TF, Natsuhari Y, Hanlon RT (1998) A provisional generic classification iof the family Loliginidae. In 'Systematics and Biogeography of Cephalopods. Vol. II'. (Eds NA Voss, M Vecchione, RB Toll and MJ Sweeney) *Smithsonian Contributions to Zoology* **586**, 215–222.

Vecchione M, Collins MA, Sweeney MJ (2002) Systematics, ecology and biology of cirrate octopods: a workshop report. *Bulletin of Marine Science* **71**(1), 79–94.

Vecchione M, Mangold K, Young RE (2008a) Opisthoteuthidae Verrill, 1896. Version 21 May 2008 (under construction). http://tolweb.org/Opisthoteuthidae/20089/2008.05.21 in The Tree of Life Web Project, http://tolweb.org/

Vecchione M, Young RE, Tsuchiya K, Bolstad KSR (2008b) *Onychoteuthis meridiopacifica* Rancurel & Okutani, 1990. Version 30 May 2008 (under construction). http://tolweb.org/Onychoteuthis_meridiopacifica/19965/2008.05.30 in The Tree of Life Web Project, http://tolweb.org/

Vecchione M, Robison BH, Roper CFE (1992) A tale of two species: Tail morphology in paralarval *Chiroteuthis* (Cephalopoda: Chiroteuthidae). *Proceedings of the Biological Society of Washington* **105**(4), 683–692.

Vecchione M, Young RE (1999) *Lycoteuthis lorigera* (Steenstrup 1875). Version 01 January 1999. http://tolweb.org/Lycoteuthis_lorigera/19738/1999.01.01 in The Tree of Life Web Project, http://tolweb.org/

Vecchione M, Young RE (2007) *Stoloteuthis weberi* (Joubin 1902). Version 26 December 2007 (under construction). http://tolweb.org/Stoloteuthis_weberi/27776/2007.12.26 in The Tree of Life Web Project, http://tolweb.org/

Vecchione M, Young RE, Roper CFE (2013) *Heteroteuthis dagamensis* Robson 1924. Version 03 November 2013 (under construction). http://tolweb.org/Heteroteuthis_dagamensis/20053/2013.11.03 in The Tree of Life Web Project, http://tolweb.org/

Vecchione M, Young RE, Tsuchiya K (2003a) *Onykia carriboea* Lesueur, 1821. Version 23 June 2003 (under construction). http://tolweb.org/Onykia_carroboea/19971/2003.06.23 in The Tree of Life Web Project, http://tolweb.org/

Vecchione M, Young RE, Tsuchiya K (2003b) *Onykia (Moroteuthopsis) ingens* (Smith, 1881). Version 23 June 2003 (under construction). http://tolweb.org/Onykia_%28Moroteuthopsis%29_ingens/19972/2003.06.23 in The Tree of Life Web Project, http://tolweb.org/

Vecchione M, Shea E, Bussarawit S, Anderson F, Alexeyev D, Lu CC, Okutani T, Roeleveld M, Chotiyaputta C, Roper CFE, Jorgensen E, Sukramongkol N (2005) Systematics of Indo-West Pacific Loliginids. *Phuket Marine Biological Center Research Bulletin* **66**, 23–26.

Vecchione M, Young RE, Tsuchiya K, Bolstad KSR (2007) *Onykia loennbergii* (Ishikawa and Wakiya 1914). Version 09 July 2007 (under construction). http://tolweb.org/Onykia_loennbergii/19974/2007.07.09 in The Tree of Life Web Project, http://tolweb.org/

Vecchione M, Young RE, Tsuchiya K (2011) *Onykia (Onykia) robsoni* (Adam, 1962). Version 20 March 2011 (under construction). http://tolweb.org/Onykia_%28Onykia%29_robsoni/19975/2011.03.20 in The Tree of Life Web Project, http://tolweb.org/

Villanueva R, Sanchez P (1993) Cephalopods of the Benguela Current off Namibia: New additions and considerations on the genus *Lycoteuthis*. *Journal of Natural History* **27**(1), 15–46. doi:10.1080/00222939300770021

Villanueva R, Young RE, Vecchione M (2008) *Opisthoteuthis* Verrill 1883. Flapjack devilfishes. Version 28 April 2008 (under construction). http://tolweb.org/Opisthoteuthis/20106/2008.04.28 in The Tree of Life Web Project, http://tolweb.org/

Vitturi R, Colombera D, Catalano E, Arnold JM (1990) Spermatocyte chromosome study of eight species of the Class Cephalopoda (Mollusca). *Journal of Cephalopod Biology* **1**(2), 101–113.

Voight JR, Grehan AJ (2000) Egg brooding by deep-sea octopuses in the North Pacific Ocean. *The Biological Bulletin* **198**(1), 94–100. doi:10.2307/1542807

von Byern J, Rudoll L, Cyran N, Klepal W (2008) Histochemical characterization of the adhesive organ of three *Idiosepius* spp. species. *Biotechnic & Histochemistry* **83**(1), 29–46. doi:10.1080/10520290801999316

von Byern J, Scott R, Griffiths C, Micossi A, Grunwald I, Norbert C (2011) Characterization of the adhesive areas in *S. tuberculata* (Mollusca, Cephalopoda). *Journal of Morphology* **272**(10), 1245–1258. doi:10.1002/jmor.10980

Voss G (1956) A review of the Cephalopods of the Gulf of Mexico. *Bulletin of Marine Science of the Gulf and Caribbean* **6**(2), 85–178.

Voss GL (1963) Cephalopods of the Philippine Islands. *United States National Museum Bulletin* **234**, 1–180. doi:10.5479/si.03629236.234.1

Voss GL (1967) Some bathypelagic cephalopods from South African waters. *Annals of the South African Museum* **50**(5), 61–88.

Voss NA (1980) A generic revision of the Cranchiidae. *Bulletin of Marine Science* **30**(2), 365–412.

Voss NA, Nesis KN, Rodhouse PG (1998) The cephalopod family Histioteuthidae (Oegopsida): Systematics, biology and biogeography. In 'Systematics and Biogeography of Cephalopods. Vol. II'. (Eds NA Voss, M Vecchione, RB Toll and MJ Sweeney) *Smithsonian Contributions to Zoology* **586**, 293–372.

Voss NA, Stephen SJ, Dong Zh (1992) Cranchiidae. In 'Larval and Juvenile Cephalopods: A Manual for their Identification'. (Eds MJ Sweeney, KM Mangold, MR Clarke and SV Boletzky) *Smithsonian Contributions to Zoology* **513**, 187–210.

Wada T, Takegaki T, Mori T, Natsukari Y (2005) Alternative male mating behaviors dependent on relative body size in captive oval squid *Sepioteuthis lessoniana* (Cephalopods, Loliginidae). *Zoological Science* **22**(6), 645–651. doi:10.2108/zsj.22.645

Wada T, Takeshi T, Mori T, Natsukari Y (2010) Sperm removal, ejaculation and their behavioural interaction in male cuttlefish in response to female mating history. *Animal Behaviour* **79**(3), 613–619. doi:10.1016/j.anbehav.2009.12.004

Wadley V, Dunning M (1998) 'Cephalopods of Commercial Importance in Australian Fisheries.' (CSIRO Marine Research, Hobart.)

Wakabayashi T, Kubodera T, Sakai M, Ichii T, Chow S (2007) Molecular evidence for synonomy of the genera *Moroteuthis* and *Onykia* and identification of their paralarvae from northern Hawaiian waters. *Journal of the Marine Biological Association of the United Kingdom* **87**(4), 959–965. doi:10.1017/S0025315407056196

Wakabayashi T, Suzuki N, Ichii M, Chow S (2012) Phylogenetic relationships among the family Ommastrephidae (Mollusca: Cephalopoda) inferred from two mitochondrial DNA sequences. *Marine Genomics* 7, 11–16. doi:10.1016/j.margen.2012.04.005

Wakabayashi T, Suzuki N, Sakai M, Ichii T, Chow S (2006) Identification of ommastrephid squid paralarvae collected in northern Hawaiian waters and phylogenetic implications for the family Ommastrephidae using mtDNA analysis. *Fisheries Science* **72**(3), 494–502. doi:10.1111/j.1444-2906.2006.01177.x

Walderon MD, Nolt KJ, Hass RE, Prosser KN, Holm JB, Nagle GT, Boal JG (2011) Distance chemoreception and the detection of conspecifics in *Octopus bimaculoides*. *The Journal of Molluscan Studies* **77**(3), 309–311. doi:10.1093/mollus/eyr009

Walsh LS, Turk PE, Forsythe JW, Lee PG (2002) Mariculture of the loliginid squid *Sepioteuthis lessoniana* through seven successive generations. *Aquaculture* **212**(1–4), 245–262. doi:10.1016/S0044-8486(02)00126-6

Waluda CM, Trathan PN, Rodhouse PG (2004) Synchronicity in southern hemisphere squid stocks and the influence of the Southern Oscillation and Trans Polar Index. *Fisheries Oceanography* **13**(4), 255–266. doi:10.1111/j.1365-2419.2004.00288.x

Ward PD, Saunders WB (1997) *Allonautilus*: A new genus of living nautiloid cephalopod and its bearing on phylogeny of the Nautilida. *Journal of Paleontology* **71**(6), 1054–1064.

Warnke KM (2012) A preliminary amplified fragment length polymorphism (AFLP) analysis of *Spirula spirula* (Cephalopoda: Spirulida). *Neues Jahrbuch für Geologie und Palaontologie. Abhandlungen* **266**(1), 3–8. doi:10.1127/0077-7749/2012/0242

Warnke KM, Meyer A, Ebner B, Lieb B (2011) Assessing divergence time of Spirulida and Sepiida (Cephalopoda) based on hemocyanin sequences. *Molecular Phylogenetics and Evolution* **58**(2), 390–394. doi:10.1016/j.ympev.2010.11.024

Williams BL, Lovenburg V, Huffard CL, Caldwell RL (2011) Chemical defence in pelagic octopus paralarvae: Tetrodotoxin alone does not protect individual paralarvae of the greater blue-ringed octopus (*Hapalochlaena lunulata*) from common reef predators. *Chemoecology* **21**(3), 131–141. doi:10.1007/s00049-011-0075-5

Winkelmann I, Campos PF, Strugnell J, Cherel Y, Smith PJ, Kubodera T, Allcock L, Kampmann M-L, Schroeder H, Guerra A, Norman M, Finn J, Ingrado D, Clarke M, Gilbert MTP (2013) Mitochondrial genome diversity and population structure of the giant squid Architeuthis: Genetics sheds new light on one of the most enigmatic marine species. *Proceedings of the Royal Society B: Biological Sciences* **28**. doi:10.1098/rspb.2013.0273

Wolfe K, Smith A, Trimby P, Byrne M (2012) Vulnerability of the paper nautilus (*Argonauta nodosa*) shell to a climate-change ocean: potential for extinction by dissolution. *The Biological Bulletin* **223**(2), 236–244.

Wolfe K, Smith A, Trimby M, Byrne M (2013) Microstructure of the paper nautilus (*Argonauta nodosa*) shell and the novel application of electron backscatter diffraction (EBSD) to address effects of ocean acidification. *Marine Biology* **160**(8), 2271–2278. doi:10.1007/s00227-012-2032-4

Wollesen T, Degnan BM, Wanninger A (2010) Expression of serotonin (5-HT) during CNS development of the cephalopod mollusk, *Idiosepius notoides*. *Cell and Tissue Research* **342**(2), 161–178. doi:10.1007/s00441-010-1051-z

Wray CG, Landman NH, Saunders WB, Bonacum J (1995) Genetic divergence and geographic diversification in *Nautilus*. *Paleobiology* **21**(2), 220–228.

Yeatman JM, Benzie JAH (1993) Cryptic speciation in *Loligo from northern Australia*. In 'Recent Advances in Fisheries Biology'. (Eds T Okutani, RK O'Dor and T Kubodera) pp. 641–652. (Tokai University Press, Tokyo.)

Yeatman JM, Benzie JAH (1994) Genetic structure and distribution of *Photololigo* spp. in Australia. *Marine Biology* **118**(1), 79–87. doi:10.1007/BF00699221

Yeoman R (2015) Brave new world of octopus farming. *FISH* **23**(1), 16–18.

Yoshida M, Tsuneki K, Furuya H (2010a) Venous branching asymmetry in the pygmy squid *Idiosepius* (Cephalopoda: Idiosepiida) with reference to its phylogenetic position and functional significance. *Journal of Natural History* **44**(33–34), 2031–2039. doi:10.1080/00222933.2010.485703

Yoshida M, Tsuneki K, Furuya H (2010b) Molecular phylogeny among East-Asian cuttlefishes using three mitochondrial genes. In 'Cephalopods Present and Past: New Insights and Fresh Perspectives'. (Eds K Tanabe, Y Shigeta, T Sasaki and H Hirano) pp. 144–155. (Springer: Dordrecht.)

Young RE (1972) The systematics and areal distribution of pelagic cephalopods from the seas off Southern California. *Smithsonian Contributions to Zoology* **97**, 1–159. doi:10.5479/si.00810282.97

Young RE (1991) Chiroteuthid and related paralarvae from Hawaiian waters. *Bulletin of Marine Science* **49**(1–2), 162–185.

Young RE (1999) *Discoteuthis* Young and Roper, 1969. Version 01 January 1999 (under construction). http://tolweb.org/Discoteuthis/19621/1999.01.01 in The Tree of Life Web Project, http://tolweb.org/

Young RE (2006) *Discoteuthis laciniosa* Young and Roper 1969. Version 26 September 2006 (under construction). http://tolweb.org/Discoteuthis_laciniosa/19626/2006.09.26 in The Tree of Life Web Project, http://tolweb.org/

Young RE (2009a) *Discoteuthis discus* Young and Roper 1969. Version 17 August 2009 (under construction). http://tolweb.org/Discoteuthis_discus/19625/2009.08.17 in The Tree of Life Web Project, http://tolweb.org/

Young RE (2009b) Joubiniteuthidae Naef, 1922. *Joubiniteuthis portieri* Joubin, 1916. Version 11 September 2009. http://tolweb.org/Joubinituthis_portieri/19450/2009.09.11 in The Tree of Life Web Project, http://tolweb.org

Young RE (2010) Nautiloidea. Nautilidae Blainville 1825. Pearly nautiluses. Version 15 August 2010 (under construction). http://tolweb.org/Nautilidae/19397/2010.08.15 in The Tree of Life Web Project, http: tolweb.org/

Young RE (2013) *Japetella diaphana* Hoyle, 1885. Version 08 January 2013 (under construction). http://tolweb.org/

Japetella_diaphana/20224/2013.01 in The Tree of Life Web Project, http://tolweb.org

Young RE, Burgess LA, Roper CFE, Sweeney MJ, Stephen SJ (1998) Classification of the Enoploteuthidae, Pyroteuthidae, and Ancistrocheiridae. In 'Systematics and Biogeography of Cephalopods. Vol. I'. (Eds NA Voss, MV Vecchione, RB Toll and MJ Sweeney) *Smithsonian Contributions to Zoology* **586**, 239–255.

Young RE, Harman RF (1998) Phylogeny of the 'Enoploteuthid' families. In 'Systematics and Biogeography of Cephalopods. Vol. I'. (Eds NA Voss, MV Vecchione, RB Toll and MJ Sweeney) *Smithsonian Contributions to Zoology* **586**, 257–270.

Young RE, Mangold KM (1922–2003) (1996) *Pyroteuthis margaritifera* (Ruppell, 1844). Version 1 January 1996 (under construction). http://tolweb.org/Pyroteuthis_margaritifera/19758/1996.01.01 in The Tree of Life Web Project, http://tolweb.org/

Young RE, Mangold KM (1922–2003) (2006a) *Galiteuthis glacialis* (Chun, 1906). Version 22 July 2006 (under construction). http://tolweb.org/Galiteuthis_glacialis/19572/2006.07.22 in The Tree of Life Web Project, http://tolweb.org/

Young RE, Mangold KM (1992–2003) (2006b) *Teuthowenia pellucida* (Chun 1910). Version 10 July 2006. http://tolweb.org/Teuthowenia_pellucida (Chun, 1910). Version 10 July 2006. http://tolweb.org/Teuthowenia_pellucida/19615/2006.07.10 in The Tree of Life Web Project, http://tolweb.org/

Young RE, Mangold KM (1922–2003) (2009) *Pyroteuthis* Hoyle, 1904. Version 29 December 2009 (under construction). http://tolweb.org/Pyroteuthis/19748/2009.12.29 in The Tree of Life Web Project, http://tolweb.org/

Young RE, Mangold KM (1922–2003) (2010) *Megalocranchia* Pfeffer, 1884. Version 15 August 2010 (under construction). http://tolweb.org/Megalocranchia/19562/2010.08.15 in The Tree of Life Web Project, http://tolweb.org/

Young RE, Mangold KM (1922–2003) (2012) *Pyroteuthis addolux* Young, 1972. Version 10 November 2012 (under construction). http://tolweb.org/Pyroteuthis_addolux/19757/2012.11.10 in The Tree of Life Web Project, http://tolweb.org/

Young RE, Mangold KM (1922–2003) (2013) *Liocranchia reinhardti* (Steenstrup 1856). Version 08 January 2013 (under construction). http://tolweb.org/Liocranchia_reinhardti/19596/2013.01.08 in The Tree of Life Web Project, http://tolweb.org/ (accessed 9 Aug. 2014).

Young RE, Roper CFE (1968) The Batoteuthidae, a new family of squid (Cephalopoda; Oegopsida) from Antarctic waters. *Antarctic Research Series* **2**, 185–202.

Young RE, Roper CFE (1999) *Chiroteuthis picteti*: Description continued. http://tolweb.org/accessory/Chiroteuthis_picteti_Description?acc_id=711 in The Tree of Life Web Project, http://tolweb.org/

Young RE, Roper CFE (2009) Batoteuthidae Young and Roper, 1968. *Batoteuthis skolops* Young and Roper, 1968. The Bush-club Squid. Version 12 August 2009 (under construction). http://tolweb.org/Batoteuthis_skolops/19452/2009.08.12 in The Tree of Life Web Project, http://tolweb.org/

Young RE, Roper CFE (2010) *Asperoteuthis acanthoderma* (Lu, 1977). Version 17 March 2010 (under construction). http://tolweb.org/Asperoteuthis_acanthoderma/19466/210.03.17 in The Tree of Life Web Project, http://tolweb.org

Young RE, Roper CFE (2011a) Chiroteuthidae Gray, 1849. Version 22 November 2011 (under construction). http://tolweb.org/Chiroteuthidae/19451/2011/11.22 in The Tree of Life Web Project, http://tolweb.org/

Young RE, Roper CFE (2011b) *Grimalditeuthis* Joubin, 1898. *Grimalditeuthis bonplandi* (Verany, 1839). Version 22 November 2011 (under construction). http://tolweb.org/Grimalditeuthis_bonplandi/19463/2011.11.22 in The Tree of Life Web Project, http://tolweb.org/

Young RE, Roper CFE, Mangold KM (1922–2003) (2008) *Heteroteuthis* Gray, 1849. Version 04 April 2008 (under construction). http://tolweb.org/Heteroteuthis/20030/2008.04.04 in The Tree of Life Web Project, http://tolweb.org/

Young RE, Roper CFE, Vecchione M (2013) Heteroteuthinae Appellof, 1898. Version 03 November 2013 (under construction). http://tolweb.org/Heteroteuthinae/20027/2013.11.03 in The Tree of Life Web Project, http://tolweb.org/

Young RE, Tsuchiya K (2009) *Abralia* Gray, 1849. Version 26 July 2009 (under construction). http://tolweb.org/Abralia/19642/2009.07.26 in The Tree of Life Web Project, http://www.tolweb.org/

Young RE, Tsuchiya K (2013) *Abraliopsis* Joubin 1896. Version 03 November 2013 (under construction). http://tolweb.org/Abraliopsis/19644/2013.11.03 in The Tree of Life Web Project, http://tolweb.org/

Young RE, Vecchione M (1996) Analysis of morphology to determine primary sister-taxon relationships within coleoid cephalopods. *American Malacological Bulletin* **12**(1–2), 91–112.

Young RE, Vecchione M (2000a) *Histioteuthis atlantica* (Hoyle 1885). Version 01 January 2000 (under construction). http://tolweb.org/Histioteuthis_atlantica/19789/2000.01.01 in The Tree of Life Web Project, http://tolweb.org/

Young RE, Vecchione M (2000b) *Histioteuthis eltaninae* (Hoyle 1885). Version 01 January 2000 (under construction). http://tolweb.org/Histioteuthis_eltaninae/19790/2000.01.01 in The Tree of Life Web Project, http://tolweb.org/

Young RE, Vecchione M (2000c) *Histioteuthis macrohista* (Voss 1960). Version 01 January 2000 (under construction). http://tolweb.org/Histioteuthis_macrohista/19787/2000.01.01 in The Tree of Life Web Project, http://tolweb.org/

Young RE, Vecchione M (2000d) *Histioteuthis pacifica* (Voss 1960). Version 01 January 2000 (under construction). http://tolweb.org/Histioteuthis_pacifica/19800/2000.01.01 in The Tree of Life Web Project, http://tolweb.org/

Young RE, Vecchione M (2003a) *Opisthoteuthis persephone* Berry, 1918. Version 13 May 2003 (under construction). http://tolweb.org/Opisthoteuthis_persephone/20164/2003.05.13 in The Tree of Life Web Project, http://tolweb.org/

Young RE, Vecchione M (2003b) *Opisthoteuthis pluto* Berry, 1918. Version 13 May 2003 (under construction). http://tolweb.org/Opisthoteuthis_pluto/20166/2003.05.13 in The Tree of Life Web Project, http://tolweb.org/

Young RE, Vecchione M (2007a) Histioteuthidae Verrill, 1881. Version 21 May 2007 (inder construction). http://tolweb.

org/Histioteuthidae/19782/2007.05.21 in The Tree of Life Web Project, http://tolweb.org

Young RE, Vecchione M (2007b) *Iridoteuthis* Naef, 1912. Version 26 December 2007 (under construction). http://tolweb.org/Iridoteuthis/20032/2007.12.26 in The Tree of Life Web Project, http://tolweb.org/

Young RE, Vecchione M (2008) *Stigmatoteuthis hoylei* (Goodrich 1896). Version 29 May 2008 (under construction). http://tolweb.org/Stigmatoteuthis_hoylei/19795/2008.05.29 in The Tree of Life Web Project, http://tolweb.org/

Young RE, Vecchione M (2013) *Octopoteuthis* Ruppell, 1844. Version 08 January 2013. http://tolweb.org/Octopoteuthis/19839/2013.01.08 in The Tree of Life Web Project, http://tolweb.org/

Young RE, Vecchione M (2014a) *Mastigoteuthis psychrophila* Nesis, 1977. Version 06 December 2014. http://tolweb.org/Mastigoteuthis_psychrophila/19522/2014.12.06 in The Tree of Life Web Project, http://tolweb.org.

Young RE, Vecchione M (2014b) *Iridoteuthis iris* (Berry, 1909). Version 21 January 2014. http://tolweb.org/Iridoteuthis_iris/111801/2014.01.21 in The Tree of Life Web Project, http://tolweb.org/

Young RE, Vecchione M, Tsuchiya K (2003) *Walvisteuthis* Nesis and Nikitina 1986. *Walvisteuthis rancureli* (Okutani 1981). Stubby hooksquid. Version 23 June 2003 (under construction). http://tolweb.org/Walvisteuthisrancureli/19957/2003.06.23 in The Tree of Life Web Project, http://tolweb.org/

Young RE, Sanchez C, Allain V (2014) *Iridoteuthis sp. A*. Version 21 January 2014 (under construction). http://tolweb.org/Iridoteuthis_sp._A/111802/2014.01.21 in The Tree of Life Web Project, http://tolweb.org/

Zeidler W (1989) The pelagic octopus *Tremoctopus violaceus* Delle Chiaje, 1830, from southern Australian waters. *The Veliger* **32**(2), 166–170.

Zylinski S, How MJ, Osorio D, Hanlon R, Marshall NJ (2011) To be seen or to hide: Visual characteristics of body patterns for camouflage and communication in the Giant Australian Cuttlefish *Sepia apama*. *The American Naturalist* **177**(5), 681–690. doi:10.1086/659626

# Index

www.ingramcontent.com/pod-product-compliance
Lightning Source LLC
LaVergne TN
LVHW071240210926
848486LV00007B/310